Compare this map of Earth's topography with Earth's population distribution shown on the map inside the back cover. Humans tend to cluster on flat areas, often near navigable waters.

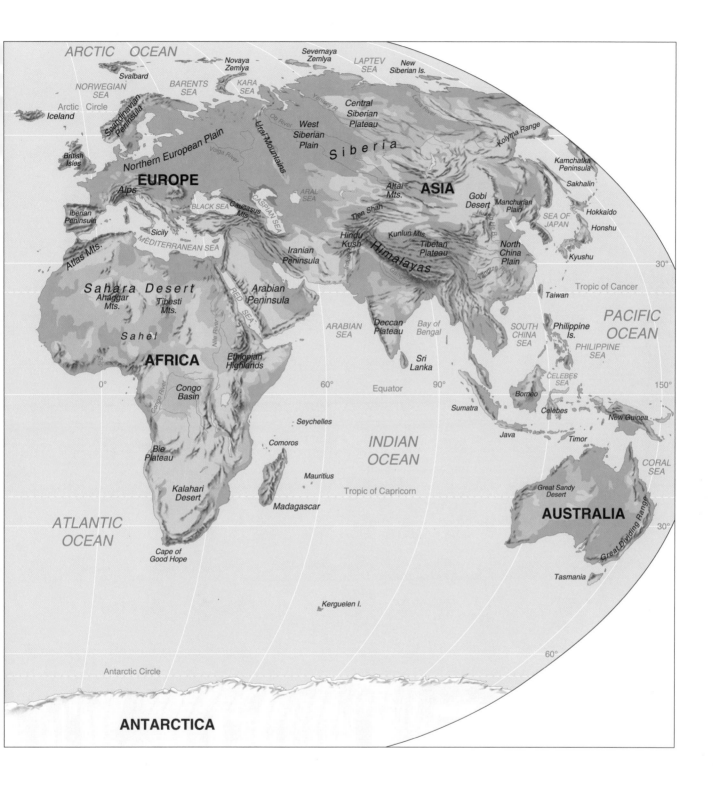

ARCTIC OCEAN

NORWEGIAN SEA

Svalbard

Novaya Zemlya

Severnaya Zemlya

LAPTEV SEA

New Siberian Is.

BARENTS SEA

KARA SEA

Arctic Circle

Iceland

Scandinavian Peninsula

Yenisey R.

Central Siberian Plateau

Lena River

Kolyma Range

British Isles

Northern European Plain

Ural Mountains

West Siberian Plain

S i b e r i a

A S I A

Kamchatka Peninsula

Volga River

Sakhalin

EUROPE

Alps

ARAL SEA

Altai Mts.

Gobi Desert

Manchurian Plain

Hokkaido

Iberian Peninsula

BLACK SEA

Caucasus Mts.

CASPIAN SEA

Tien Shan

Hindu Kush

Kunlun Mts.

Huang He R.

Honshu

SEA OF JAPAN

Sicily

MEDITERRANEAN SEA

Iranian Peninsula

Tibetan Plateau

North China Plain

Kyushu

Atlas Mts.

Himalayas

Indus R.

Yangtze R.

30°

Sahara Desert

Arabian Peninsula

Taiwan

Tropic of Cancer

Ahaggar Mts.

Tibesti Mts.

RED SEA

Ganges R.

Deccan Plateau

Bay of Bengal

SOUTH CHINA SEA

Philippine Is.

PACIFIC OCEAN

S a h e l

ARABIAN SEA

PHILIPPINE SEA

AFRICA

Nile River

Ethiopian Highlands

Sri Lanka

0°

60°

Equator

90°

150°

Congo Basin

Seychelles

Sumatra

CELEBES SEA

Celebes

New Guinea

Congo River

Comoros

INDIAN OCEAN

Java

Timor

Borneo

Bie Plateau

Mauritius

CORAL SEA

Great Sandy Desert

Kalahari Desert

Tropic of Capricorn

Madagascar

AUSTRALIA

30°

ATLANTIC OCEAN

Great Dividing Range

Cape of Good Hope

Kerguelen I.

Tasmania

60°

Antarctic Circle

ANTARCTICA

INTRODUCTION TO
GEOGRAPHY

INTRODUCTION TO
GEOGRAPHY

People, Places, and Environment

Fourth Edition

Edward F. Bergman

Professor Emeritus,
Lehman College of the City University of New York

William H. Renwick

Miami University, Oxford, Ohio

PEARSON

Prentice
Hall

Upper Saddle River, NJ 07458

Library of Congress Cataloging-in-Publication Data

Bergman, Edward F.
  Introduction to geography : people, places, and environment / Edward F. Bergman,
William H. Renwick.—4th ed.
    p.   cm.
  Includes index.
  ISBN 0-13-223899-3
  1. Geography. I. Renwick, William H. II. Title.

G128.B372 2008
910—dc22

2007001872

Publisher: Daniel E. Kaveney
Editor in Chief, Science: Nicole Folchetti
Project Manager: Tim Flem
Associate Editor: Amanda Brown
Executive Managing Editor: Kathleen Schiaparelli
Academic Assistant to Dr. Bergman: Ginger Birkeland
Marketing Manager: Amy Porubsky
Media Editor: Andrew Sobel
Creative Director: Juan Lopez
Art Director: Kenny Beck
Interior and Cover Designer: Michael J. Fruhbeis
Manufacturing Manager: Alexis Heydt-Long
Senior Managing Editor, Art Production and Management: Patricia Burns
Manager, Production Technologies: Matthew Haas
Managing Editor, Art Management: Abigail Bass
Art Production Editor: Rhonda Aversa
Illustrations: GEX
Manufacturing Buyer: Alan Fischer
Editorial Assistants: John DeSantis, Jessica Neumann
Director, Image Resource Center: Melinda Patelli
Manager, Rights and Permissions: Zina Arabia
Interior Image Specialist: Beth Brenzel
Cover Image Specialist: Karen Sanatar
Image Permission Coordinator: Joanne Dippel
Photo Researcher: Yvonne Gerin
Cover Image: China, Hong Kong city skyline, night, elevated view/Jorg Greuel/Getty Image

PEARSON
Prentice
Hall

© 2008, 2005, 2003, 2002, 1999 Pearson Education, Inc.
Pearson Prentice Hall
Pearson Education, Inc.
Upper Saddle River, New Jersey 07458

Printed in the United States of America
10 9 8 7 6 5 4 3 2

ISBN 0-13-223899-3

Pearson Education Ltd., *London*
Pearson Education Australia Pty., Limited, *Sydney*
Pearson Education Singapore, Pte. Ltd.
Pearson Education North Asia Ltd., *Hong Kong*
Pearson Education Canada, Ltd., *Toronto*
Pearson Educación de Mexico, S.A. de C.V.
Pearson Education—Japan, *Tokyo*
Pearson Education Malaysia, Pte. Ltd.

This is for all of my geography students who have given me feedback through the years.

EFB

For Debra, with thanks for her patience.

WHR

BRIEF CONTENTS

CONTENTS

8 The Human Food Supply 312

11 A World of States 438

12 National Paths to Economic Growth 484

PREFACE

Many readers of this textbook may not have studied geography since grade school, where geography may have meant simply memorizing place names. Knowing where places are, however, is not all there is to geography. Knowing place names is a tool for studying geography, just as counting is a tool for studying mathematics and reading is a tool for studying literature. In geography as in any other field of study, we begin by gathering some basic information—in the case of geography, this is the *where*.

Once we know the names and locations of environmental features, people, and activities, then geography can proceed to the challenging questions of significance: *What forces* created the physical environment that characterizes various places? *Why* are people and activities located where they are? How do the features and activities at any one place *interact* to make that place unique, and what are the *relationships* among different places? What factors or forces *cause* these distributions of human populations and activities? And how and why are these environmental features and human distributions *changing*? Exploring these questions stretches your mind and imagination. Geography helps you understand current events, and it can provide you with information that is useful in deciding where to live, in seeking or building a career, and in forming your own position on political issues.

This book introduces the principal content of geography, as well as the major tools and techniques of the field. Geography is sometimes subdivided into *physical geography*, which studies the various attributes of the physical environment such as climate, plants and animals, and landforms, and *human geography*, which studies the geography of human groups and activities. Humans and the environment, however, interact; neither can be completely understood without the other.

The contents reflect our sense of logic and our experience of the surest way to accumulate understanding. Instructors may wish to vary the order of the chapters, but many students find it easier to understand certain topics after other topics have already been covered. For example, today many rural people are migrating to cities to escape poverty. Therefore, in order to understand why the world's cities are growing fast, it helps if you have already studied changes in world agriculture and in various national farm policies, so we have placed the discussion of agriculture (Chapter 8) before that of urbanization (Chapter 10).

THREE THEMES IN THIS TEXTBOOK

This textbook emphasizes three themes throughout the study of geography. First, geography examines the interrelationships between humans and their natural environment; second, many basic principles of human geography can be studied and demonstrated in your own hometown; and third, geography is dynamic.

Geography Explores the Interrelationships Between Humankind and the Environment

The study of Earth's climates, soils, vegetation, and physical features is called physical geography, and physical geography sets the stage upon which humans act out their lives. A great deal of human effort is spent wresting a living from the environment, adjusting to it, or altering it.

The first few chapters of this book, therefore, offer an overview of Earth's physical environment. The discussion emphasizes processes operating in the landscape, such as atmospheric circulation, landform change, and vegetation growth. An understanding of these processes is necessary to comprehend the mechanisms whereby humans transform Earth's environments. The theme of human–environmental interaction then weaves through the book. People interpret and evaluate possibilities in their environment, and they alter natural conditions. Any alteration in one of Earth's natural physical and ecological systems, however, will trigger changes throughout the entire system. Because these systems are tremendously complex, many changes may be unexpected or even harmful. For example, people "tame" rivers and build on their floodplains, but that construction can increase damage during a flood. People redistribute plants and animals around Earth, but relocated species can cause unexpected ecological damage. Chapter 2 notes how cities change local climates; these changes affect human health. Pollution threatens the air we breathe, the water we drink, and the soil we till. Chapter 13 emphasizes that environmental protection is one issue that today challenges all humankind, requiring international collaboration for a positive goal.

Geography Is Not Restricted to the Study of Exotic Lands and Peoples

The basic principles of geography can be studied in your hometown and even on campus. How do local temperatures and rainfall vary throughout the year? What processes shaped the landforms, and how are these processes changing the land today? What natural hazards affect people in your area? Where have new arrivals in your community come from, and why did they move? Where are local food crops and manufactured goods sold? Are the boundaries of city council districts manipulated to give one group an unfair advantage in elections? How many religions are represented by houses of worship in your town, and are these centers of identifiable residential communities? Can you map the rents on commercial properties in your town, and how do these values reflect accessibility or, perhaps, perception of which neighborhoods are the most elegant?

Study of geography introduces a great range of careers, even though the professionals at work in many fields may not call themselves geographers. People studying or "doing" geography include planners designing new suburbs, scientists working to reduce water pollution, transportation consultants routing new highways, advertisers targeting "junk mail" to zip codes where residents have specific income levels, diplomats negotiating treaties to regulate international fishing, and still more various occupations.

Modern Geography Is Dynamic

It is important to know the current distributions of landforms, people, languages, religions, cities, and economic activities, but none of these patterns is static. Earth's surface is constantly changing, sometimes slowly as when erosion wears down a mountain chain over millions of years, and sometimes spectacularly and rapidly as when a new volcano explodes and builds on Earth's surface or a flood prompts a river to change its course. Transportation and communication ties among peoples and regions have multiplied, so social, political, and economic forces constantly redistribute human activities. Maps of these activities reveal only temporary balances between forces for change and forces for stability. What happens *at* places depends more and more on what happens *among* places, and we can understand maps of economic or cultural activity only if we understand the patterns of movement that create them. Modern geography explores the forces at work behind the maps.

Each day the news reports on events in which what happens is directly related to where it happens, and these events trigger changes in geography: A volcano erupts in Mexico; a bountiful harvest in Argentina reduces food prices and thus improves the diet available to Africans; Canadian scientists synthesize a substitute for a mineral previously imported; a new government in Africa redirects international alliances, economic links, and migration streams. American movies and music diffuse U.S. language and culture around the world, while Americans themselves adopt new foods and words, such as sushi. Developing countries join the already-developed world in sprouting new industrial sources of air pollution, poisoning their own citizens and changing the chemical composition of Earth's atmosphere sufficiently to change Earth's climate. Protestant Christianity wins converts throughout Latin America; women are accepted into the priesthood, and

governments open family planning clinics. Meanwhile, in some African and Asian countries, Islamic fundamentalists win political power and curb women's rights. These events remap world cultural, political, and economic landscapes. Today's dynamic geography doesn't just exist; it *happens*. In every topic covered in this text, it is our goal not only to *describe* distributions and locations, but to *explain* the distributions and locations.

CONTEMPORARY ISSUES IN GEOGRAPHY

What you learn in your reading and in your geography classroom can help you better understand current events and form your own opinions on important questions of the day. Each chapter of this book provides background material for understanding the news. Any number of topics in the book demonstrate this benefit, but here we might mention just two: the treatment of the topic of development, and the treatment of the issue of gender justice.

Development and Environmental Protection

Each one of Earth's billions of people aspires to a high level of material welfare, yet today many people live in conditions of deprivation. The world distribution of wealth and welfare does not coincide with the world distribution of raw material resources. If the possession of raw materials were the key to wealth, then the Republic of Congo and Mexico would count among the richest countries in the world, and Japan and Switzerland would count among the poorest. In fact, the Republic of Congo and Mexico are poor, and Japan and Switzerland are rich. An understanding of the resolution of this paradox is essential to understanding the world today.

This book goes beyond merely describing where there is wealth and where there is not. It weaves together a number of threads of understanding, making clear the *why* of the *where*. Relevant economic principles are individually introduced in the text where appropriate, including adding value to raw materials, sectoral evolution, locational determinants for manufacturing, various nations' economic policies, and the patterns of world trade. New considerations arise virtually every day, such as the discovery of new resources, new technologies, and new governments with new policies. These changes redistribute advantages. Chapter 12 suggests how and why the balance among the factors that support economic development is continuously shifting, and Chapter 13 highlights how these geographic shifts redistribute global sources of pollution. New industries in developing countries generate pollution. Furthermore, those new industries allow rising standards of living, and that in turn allows increased consumption of consumer goods and increased production of waste products.

Economic development is only one aspect of development. Human development includes adequate nutrition, education, political liberty, and the opportunity to live in a healthy, unpolluted environment. Chapter 5 analyzes the geography of health, and Chapter 8 explains the distribution of world food production and trade. Chapter 9 discusses problems and challenges in the consumption of natural resources and the treatment of wastes and disposed items. Chapter 11 maps world education and freedom. In each case, the text suggests what factors might redistribute these activities in the future.

Gender Justice

Almost everywhere, women are worse off than men. They have less power, less autonomy, less money, more work, and more responsibility. The issue of discrimination against women—gender justice—is emerging as a major question of our time. This text does not separate this issue in an isolated chapter, but examines it as it plays a role in each topic through the book.

Chapter 5, on population, notes how discrimination against women can begin even against children in the womb. Some countries record higher rates of abortion for female fetuses than male fetuses. We see that in some countries females get less health care and nutrition; the result is unexpectedly low ratios of females to males in the overall national population. Chapter 7 examines the differing attitudes toward women taught by various religions. The ramifications of those teachings reach from the issue of priesthood for women into the treatment of women in national laws. Discrimination against female farmers, discussed in Chapter 8, lowers agricultural productivity and the levels of nutrition in many countries. Chapter 11 notes the variations in the percentages of males or females in school around the world, women's role in politics, and other issues of legal rights.

This text readily admits that we do not know the answers to all questions. It invites you to learn what we know and then to join the world of scholarship by increasing the knowledge that we have today. For example, Chapter 10 notes that vast slums are growing in many cities in developing countries largely because of migration from rural areas into the cities. High percentages of people living in cities are uncounted, living in unregistered properties, and performing unrecorded economic activities. The vitality of these urban slums astounds observers and confounds economists, who struggle to understand and measure what is going on. This textbook explains what is happening as we understand it but acknowledges that we need to devise new methods of measuring and accounting for this activity. Similarly, many of Earth's physical processes are incompletely understood. The process of global warming is frequently in the news today, and yet the number of variables that might cause it and their exact interactions are beyond the range of even the greatest supercomputers. We encourage you to take up the challenge of investigating some of these issues—and, if possible, improving conditions—as possible research topics for your own careers.

MAPS, CARTOGRAMS, AND GIS

A great variety of maps illustrate this book. Some of them include relief shading. You will probably be familiar with traditional maps that illustrate distributions as mosaic patterns of color. On other maps, called flow maps, arrows represent movements of people or of goods, and on these the widths of the arrows often convey the quantity of each flow—the numbers of passengers flying major airline routes across the United States, for example (Figure 1-16).

Several cartograms have also been specially designed for this book. A cartogram is a visual device much like a map, but on a cartogram physical distance is replaced with some other measure in order to convey a visual impression of the magnitude of something. Cartograms of such things as countries' populations, for example (Figure 5-2), or of countries' economies (Figure 12-5), visually convey the relative population or wealth of different countries better than standard land area maps do.

Today maps are drawn and rapidly updated using geographic information systems (GIS) and digital cartographic tools. GIS have become essential and ubiquitous on geographers' desks, and their importance is clear in this book. An expanded discussion of GIS technology has been added in Chapter 1, and we have illustrated how GIS are used in geographic problem-solving throughout the book.

A variety of other visual devices includes tables, bar graphs, and pie graphs, with which you are probably familiar. In each case the captions have been carefully written to help you read these sophisticated images. Each contains a great deal of information.

Special care has been taken in preparing the captions for all illustrations. Each caption guides your eye through or into the figure. Many captions include questions to get you thinking about the information in the image. Many ask you to compare it with information shown on another image in the book in order to develop hypotheses of explanation.

A WORD ABOUT NUMBERS

This book contains many numbers—measurements of populations, economic conditions, production of various commodities, world trade, and more. These measures come from a variety of sources—private organizations, national governments, international organizations—and they are the best available. Such numbers, however, must always be read with two considerations in mind: reliability and date.

The compilation of measures is a tremendously difficult task. For example, the United States is the world's richest country, with many highly skilled government workers, yet the government admits that the national census is probably inaccurate by a factor of 5 to 7 percent. What degree of accuracy, then, can we expect from a poor country that does not have a sophisticated bureaucracy? It is said that the government of India does not have even an accurate count of the number of cities in India. We do not want to promote cynicism about the value or reliability of statistics, but an educated person does exercise judgment about the probable exactitude of any figure.

The second caution is that the measures themselves change. The population of the United States changes every day, yet the United States officially counts its residents only once every ten years—although periodic estimates are more frequently published. It takes a long time to gather and compile statistics, so the measures may seem out of date by the time they are published. This is especially true of international comparative statistics. For example, each year the United Nations Conference on Trade and Development (UNCTAD) publishes a handbook of statistics of world trade, but the book appears three or four years after its date, and many statistics recorded were measured years before the date of the volume. The figures used to make world maps, as in this text, are sometimes older than figures given for many individual countries because advanced countries are able to compile numbers faster than poorer countries, and then international organizations need time to compile all of the national figures. Furthermore, governments sometimes change the way they measure things. For example, for many years governments counted and published a statistic called gross national product (GNP), but today that statistic has been replaced by a slightly different measure called the gross national income (GNI). The meaning of GNI is explained in Chapter 12.

The statistics in this textbook are as up to date as could possibly be obtained from the most reliable sources as of October 2006. The text notes the direction in which many of these measures are changing, and in many cases we have dared to predict their future direction. The United States population will probably continue to rise, and the percentage of the national labor force working in manufacturing will probably continue to fall. We challenge you to go to the library or to search the World Wide Web to update those measures to today.

FEATURES TO HELP YOU STUDY

Each chapter offers a number of features to help you study:

- **A Look Ahead.** Each chapter begins with a brief outline of the main points to be made in that chapter.

- **Anecdotes.** An anecdote at the beginning of each chapter exemplifies the sort of real-world questions and problems that arise from the material discussed in that chapter.

- **Key Terms.** The key terms in each chapter are printed in **boldface** when each is first introduced. These terms are also listed at the end of each chapter along with the page on which each first appears. All key terms are defined in the Glossary at the end of the book. Some additional terms that may be unfamiliar to some readers but that are less central to the study of geography are printed in *italics*. These terms are defined carefully in the text.

- **Critical Thinking boxes.** Critical Thinking boxes discuss problems or pose questions for which there are no simple or clear answers. These require you to exercise your analytical skills to develop your own viewpoints. Each Critical Thinking box includes questions to start you thinking about the material in the box.

- **Focus On boxes.** Focus On boxes highlight individual problems or case studies of the topics in that chapter.

- **Regional Focus On boxes.** Regional Focus On boxes examine in detail particular places that exemplify points being made in the text. Many of these places are hotspots in the news.

- **Conclusion: Critical Issues for the Future.** Each chapter concludes with an examination of how contemporary issues discussed in that chapter will continue to challenge us in coming years. Some of these challenges are very real topics of current political debate; some will face humankind all through your lifetime.

- **Summary.** Each chapter closes with a summary of its main points.

- **Questions for Review and Discussion.** At the end of each chapter you will find a list of questions about the subject of that chapter. These review questions test your reading, and their answers can be found in the chapter.

- **Thinking Geographically.** Answers to the questions in this section cannot be found in the text itself. These questions encourage exploration beyond the text—into the library or out into the community. Some students may find these questions good suggestions for papers or research projects.

- **Suggestions for Further Learning.** A short list of up-to-date readings provides sources for further information.

- **Web Work.** This feature contains an annotated list of sites that you may check for updating or for additional information on the topics discussed in the text. These range from the home pages of individual newspapers to the home pages of libraries, government agencies, and of national and international organizations.

- **Three appendices.** The book also offers three appendices. The first details map projections, their creation, and their uses. The second offers substantial detail of the temperature and rainfall parameters for the various categories of climate defined in the text. The third is a table providing basic statistical information for most of the countries of the world.

THE ATTACKS ON THE UNITED STATES ON SEPTEMBER 11, 2001

On the morning of September 11, 2001, four passenger airliners taking off from U.S. East Coast airports on transcontinental flights were simultaneously hijacked. The hijackers piloted two of those planes into the twin 110-story towers of New York's World Trade Center (WTC). The conflagrations ignited by the planes' full fuel tanks, combined with the shock of the collisions, caused both towers to collapse. A third hijacked plane, also on a scheduled transcontinental flight, was smashed into the headquarters building of the U.S. Department of Defense in Arlington, Virginia (the Pentagon). Passengers on the fourth hijacked flight, hearing the fates of the three other flights from conversations with loved ones over their own mobile phones, rallied to retake control of their flight, but, during the fighting on board, that plane crashed in rural Pennsylvania. Hijackers had probably intended to plunge it into the White House or the U.S. Capitol.

The total number of people killed in these hijackings and suicidal crashes was about 3,000, including citizens of the United States, Canada, Pakistan, India, the Netherlands, the United Kingdom, Peru, and many other countries. The dead included individuals of the Christian, Jewish, Muslim, Hindu, and still other religious faiths. This was the greatest massacre on American soil since Japan's surprise attack on the U.S. naval base at Pearl Harbor on December 7, 1941—an event that brought the United States into World War II. The events of September 11 had an instant and profound effect on world politics and world economics—so profound that the phrase "post-9/11" has come to represent the cumulative impact of these changes, and that shorthand signifier has been used in this textbook.

The United States and allies launched a war against an international terrorist group named Al Qaeda ("the foundation" in Arabic), which had been responsible for earlier terrorist attacks against U.S. and allied governments and peoples. Al Qaeda's leadership was known to be operating from Afghanistan, which was ruled by a fundamentalist Islamic group called the Taliban ("student" in Arabic). Thus, the overthrow of the

Taliban was America's first objective, and that was accomplished within three months. The United States subsequently attacked and overthrew the government of Iraq, although many Americans and others doubted the necessity and wisdom of that action. Chapter 13 discusses these events, the United States's declaration of a "war against global terrorism," and how the ramifications of that "war" extend into the domestic politics of many countries and throughout international affairs. This textbook incorporates relevant explanatory material throughout in the appropriate chapters and, insofar as possible, reviews the continuing ramifications of these events.

SUPPLEMENTS

The authors and publisher have been pleased to work with a number of talented people to produce an excellent supplements package. This package includes the traditional supplements that students and professors have come to expect from authors and publishers, as well as some new kinds of supplements that involve electronic media.

For the Student

Online Study Guide: The *Introduction to Geography: People, Places, and Environment* Online Study Guide (**http://www.prenhall.com/bergman**) gives students the opportunity to further explore topics presented in the book by using the Internet. This website contains numerous review exercises (from which students get immediate feedback), exercises to expand students' understanding of geography, and resources for further exploration. In addition to being a valuable tool for student review, the site provides an excellent platform from which to start using the Internet for the study of geography. An access code for this website is bound into the front of every new copy of this textbook.

Mapping Workbook (0-13-238115-X) The Mapping Workbook contains outline maps for each chapter, based on the maps in the text, ready for student use in identification exercises.

Goode's World Atlas (0-13-612824-6) Prentice Hall and Rand McNally are pleased to announce that Prentice Hall is now distributing *Goode's World Atlas*—the leading atlas used educationally in colleges and universities worldwide. It features nearly 250 pages of maps, from definitive physical and political maps to important thematic maps that illustrate the spatial aspects of many important topics. The current 21st edition of the atlas features fully updated content and has been vetted by an academic board of professional geographers. Prentice Hall offers the atlas at a dramatically reduced price with *Introduction to Geography*. Call your local Prentice Hall representative for details.

For the Instructor

Instructor Resource Center on DVD (0-13-224324-5): This digital resource helps make you more effective, and saves you time and effort. Find all your resources in one, well-organized, easy-to-access place. Resources include all of the maps and photos from the text in electronic format, prepared PowerPoint slides, Word files of the Instructor's Manual and Test Item File, CRS questions, and more.

Prentice Hall Human Geography Videos on DVD (0-13-241656-5). This three-DVD set is designed to enhance any human geography course. It contains 14 full-length video programs covering a wide array of issues affecting people and places in the contemporary world, including international immigration, urbanization, global trade, poverty, and environmental destruction.

These DVDs are designed to function in computer-based DVD drives in addition to traditional component DVD players. The videos included on these DVDs are offered at the highest quality to allow for full-screen viewing on your computer and projection in large lecture classrooms.

Video programs include:

DVD 1

- Blue Danube?
- Staying Alive
- Cash Flow Fever

- Roma Rights
- The Barcelona Blueprint

DVD 2

- Untouchable?
- Srebrenica–Looking for Justice
- The Outsiders
- The Trade Trap
- The Coffee-Go-Round

DVD 3

- Geraldo's Brazil
- Kill or Cure?
- Slum Futures
- Warming Up in Mongolia

Average video length: 25 minutes

The Prentice Hall Human Geography videos are available either as a stand-alone or at a very substantial discount when packaged with *Introduction to Geography*. Please see your local Prentice Hall representative for details.

Transparency Set (0-13-243842-9) Images and illustrations pertaining to the text's core concepts have been selected and enhanced for optimal classroom presentations. All images within the transparency pack are also available electronically on the Instructor Resource Center on DVD (0-13-224324-5).

Test Item File (0-13-179054-4), The Test Item File provides instructors with a wide variety of test questions available in printed format, or in electronic format on the Instructor's Resource Center on DVD.

TestGen CD-ROM (0-13-243894-1), TestGen is a computerized test generator that lets you view and edit test bank questions, transfer questions to tests, and print tests in a variety of customized formats. *Available for download from the Prentice Hall online catalog,* **www.prenhall.com.**

Instructor's Manual (0-13-179053-6): The Instructor's Manual is intended as a resource for both new and experienced instructors. It includes a variety of lecture outlines, additional source materials, teaching tips, advice about how to integrate visual supplements (including the Web-based resources), and various other ideas for the classroom.

ACKNOWLEDGMENTS

Countless colleagues, librarians, and generous individuals both in government and in the private sector helped with information for this text. We wish to thank in a special way Professor James M. Rubenstein of Miami University, Ohio, with whom we have worked on earlier textbooks, for his many contributions to our thinking. Any errors in the text are the fault of the authors, but we would like to thank the following people for their special help: Emily Nerbun, NAFTA desk of the U.S. Department of Commerce; Marilyn Washington, Japan External Trade Organization; Amy Staples, National Museum of African Art; Laveta Emory, Smithsonian Institution; Priscilla Strain, National Air and Space Museum; Clyde McNair, Agency for International Development; Jim Bohn, Atlanta Regional Commmission; Don McMinn, Spriggs and Hollingsworth; Linda Carrico, U.S. Bureau of Mines; Barbara Mathe, American Museum of Natural History; Kerstin Erickson, European Union Information Service; William Usnik; Roger Wieck and William Voelkle, the Morgan Library; John Rutter, U.S. Department of Commerce; Gary Cohen; Jerry Hagstrom, *The National Journal*; Debbie Janes, U.S. Environmental Protection Administration; Melissa Starr, World Health Organization; David Sackett, Lehman Brothers; Dan Beard, National Audubon Society; Carey Pickard, the Tubman African American Museum, Macon, Georgia; Ms. Drs. Lydia Van Rietschote, Head of the Department of Sector Policy, Ministry of Education, the Netherlands; Guadalupe Morgan, the World Bank; Professor Francisco Denton, Mexico City; Patricia Dickerson, U.S. Bureau of the Census; Kurt G. Usowski, HUD; Gloria Cuaycong, U.N. Statistics Division; David Sutherland, NOAA; Louise Genereux, Statistics Canada; Randall Flynn; U.S. Board on Geographic Names; Arthur Lambert; Joseph W. Cherner; David Morgan; Gino Pepoli; plus the many friends and colleagues whose photographs are used and credited in the text. Dr. Ivor Winton of the Department of Geography, Simon Fraser University, British Columbia, Canada, offered careful reading and notes, and Michael Reid of Vancouver assisted as a researcher for Professor Bergman.

We also wish to thank our scholarly colleagues who provided thoughtful suggestions for improving the manuscript during its preparation. These include

Gillian Acheson, *Southern Illinois University-Edwardsville*

Tanya Allison, *Montgomery College*

Holly R. Barcus, *Morehead State University*

Lee Berman, *Southern Connecticut State University*

Daniel Block, *Chicago State University*

Paul L. Butt, *University of Central Arkansas*

Joseph M. Cirrincione, *University of Maryland*

Carl Dahlman, *University of South Carolina*

Bruce Davis, *Eastern Kentucky University*

Bryce Decker, *University of Hawaii at Manoa*

Stanford Demars, *Rhode Island College*

Leslie Dienes, *University of Kansas*

Gary Fowler, *University of Illinois, Chicago*

Michael Fox, *Carleton University*

Roberto Garza, *San Antonio College*

Jennifer Gebelein, *Florida International University*

Mark Guizlo, *Lakeland Community College*

Rene J. Hardy, *Shoreline College*

Erick Howenstine, *Northeastern Illinois University*

James C. Hughes, *Slippery Rock University*

Mark Jones, *University of Connecticut*

Tulasi R. Joshi, *Fairmont State College*

Walter Jung, *Central Oklahoma University*

Angelina Kendra, *Central Connecticut State University*

Rob Kent, *University of Akron*

Lori Krebs, *Salem State College*

Miriam K. Lo, *Mankato State University*

Ruben A. Mazariegos, *University of Texas-Pan American*

Ian A. McKay, *Wilfrid Laurier University*

G. L. "Jerry" Reynolds, *University of Central Arkansas*

Scott C. Robinson, *University of Nebraska Omaha*

Lallie Scott, *Northeast Oklahoma State University*

Christa Smith, *Clemson University*

James N. Snaden, *Central Connecticut State University*

David M. Solzman, *University of Illinois, Chicago*

Robert C. Stinson, *Macomb Community College*

Christopher J. Sutton, *Northwestern State University of Louisiana*

Melissa Tollinger, *East Carolina University*

James Tyner, *Kent State University*

Thomas B. Walter, *Hunter College of the CUNY*

Gerald R. Webster, *University of Alabama*

Kathy Williams, *Bronx Community College*

John Wright, *New Mexico State University*

Charles T. Ziehr, *Northeastern State University*

We owe a debt of gratitude to many people. At Prentice Hall, Dan Kaveney, Geosciences Publisher, managed the project from its beginning stages through the journey to production. Project Manager Becky Giusti brought all the pieces together into the completed volume you now hold. Amy Porubsky served as Marketing Manager, and Jessica Neumann and Amanda Brown provided invaluable editorial assistance. We have enjoyed working with all of these people, and we thank them.

Edward F. Bergman
William H. Renwick

ABOUT THE AUTHORS

Edward F. Bergman stands beside the stone wall that is the very eastern edge of the North American tectonic plate, where it is exposed in Iceland (see Figure 3-5).

Professor Bergman was born in Wisconsin and studied at the University of Wisconsin (Madison), the University of Vienna (Austria), and the University of Washington in Seattle. Today as Professor Emeritus of Lehman College of the City University of New York, he still writes and lectures widely across several continents. When not lecturing or writing, he enjoys Manhattan's cultural and social life.

William H. Renwick earned a B.A. from Rhode Island College in 1973 and a Ph.D. in geography from Clark University in 1979. He has taught at the University of California, Los Angeles, and Rutgers University, and is currently Professor and Chair of Geography at Miami University. A physical geographer with interests in geomorphology and environmental issues, his research focuses on impacts of land-use change on rivers and lakes, particularly in agricultural landscapes in the Midwest. When time permits, he studies these environments from the seat of a wooden canoe.

1

INTRODUCTION
TO GEOGRAPHY

The light swathe in the upper right of this image of the southern Sahara is a dust storm in southwestern Chad, the largest source of wind-blown dust in the world. The large dark-colored areas along the borders of Chad, Niger, Nigeria, and Cameroon was formerly Lake Chad. In the early 1960s, this lake was as large as Lake Erie, but today it is a seasonal wetland being filled with blowing sand. Lake Chad has a long history of varying water levels associated with climate change. It is also shrinking as a result of irrigation using the water of the Chari River, which flows into the lake from the southeast. Declining lake levels have aggravated problems of food insecurity in the region, which is one of many in the world where sensitive environments are changing rapidly as a result of both natural and human factors, with significant consequences for the people who inhabit them. (Image courtesy Jacques Descloitres, MODIS Rapid Response Team at NASA GSFC)

A Look AHEAD

What Is Geography?

Geography is the study of the interaction of all physical and human phenomena at individual places and of how interactions among places form patterns and organize space. Geography is one of the oldest and most diverse fields of study, so it offers an introduction to a unique breadth of possible careers.

Contemporary Approaches in Geography

Geographers employ three principal analytical methods: area analysis, spatial analysis or locational analysis, and geographic systems analysis.

Describing Earth: Communicating Geographic Information

Maps are a characteristic geographic tool, and geographic information systems are new computer tools for storing, mapping, and analyzing geographic data.

Sailors may soon successfully travel the Northwest Passage across the top of North America between the Atlantic and Pacific Oceans that explorers have sought for 500 years. Technology has made the development of strong icebreakers a reality, and an overall warming of Earth's climate, partly due to human activity, has fully opened many passages formerly closed by ice and has thinned the ice in others. The United States hopes to sail between the East Coast and Alaska's north slope, and Europeans hope for a shorter route to Asia.

Furthermore, advances in global positioning systems (GPSs) and in long-range aircraft allow transpolar flights that shorten flight times between Asia and the Americas. At any hour of the day, more people are in airplanes over the Arctic than live there on the ground.

Canada claims jurisdiction over all of the islands and water all the way to the North Pole, so the Canadian government is pressured to regulate this territory, to monitor over-flights, and to provide emergency assistance when necessary in order to strengthen its claims of sovereignty. The exercise of meaningful jurisdiction became an issue in Canadian federal elections in 2006; the party that won had promised to place more military and monitoring facilities in the Arctic regions. Canada may change its national motto from *A mari usque ad mare* ("from sea to sea") to *A mari ad mare ad mare* ("from sea to sea to sea") to emphasize its Arctic Ocean interests.

On April 20, 2006, Canada refused to allow the plane of Belarussian Prime Minister Sergei Sidorsky to land for refueling en route to Cuba. Canada's refusal did not jeopardize the Prime Minister's life, but Canada's denial of refueling rights necessitated a greatly lengthened route from Belarussia to Cuba and necessitated a refueling stop in Boston. The Canadian action was in protest of Belarussia's disputed election process of March 19, 2006, and to express Canada's strong concerns about Belarussia's commitment to democratization and human rights.

The concerns over new trans-Arctic routes exemplify the many ways in which changes in the environment, in technology, and in world politics continually rearrange the relative locations of activities and the transport routes among those locations, thus opening new possibilities and occasions for international diplomacy. ▶

WHAT IS GEOGRAPHY?

Many readers of this textbook may not have studied geography since grade school when geography may have meant simply memorizing place names. Knowing place names is a tool for studying geography, just as counting is a tool for studying mathematics and reading is a tool for studying literature. In geography, as in any other field of study, we begin by gathering some basic information. In the case of geography, this is the *where*. Knowing where places are, however, is not all there is to geography.

Once we know the names and locations of environmental features, people, and activities, we can proceed to the challenging questions of significance: *Why* are people and activities located where they are? How do the features and activities at any one place *interact* to make that place unique, and what are the *relationships* among different places? What factors or forces *cause* these distributions of things and activities? How and why are these distributions *changing*? And how, perhaps, can we *alter* those distributions for greater human convenience or profit?

Geography is the study of the interaction of all physical and human phenomena at individual places and of how interactions among places form patterns and organize space. **Physical geography** studies the characteristics of the physical environment. When geography concentrates on topics such as climate, soil, and vegetation, it is a natural science. **Human geography** studies human groups and their activities, such as language, industry, and the building of cities; human geography is a social science. **Cultural geography,** a subfield of human geography, focuses specifically on the role of human cultures. Physical and human geographers share both their approach and a great deal of information, and their analyses of the landscape always weave their understanding together. Thus, geography bridges the physical sciences and the social sciences. **Cartography** (mapmaking) and its computerized extension to **geographic information systems (GIS)** provide tools that help both physical and human geographers store, display, and analyze geographic data.

Geographers investigate the processes underlying all observed distributions and patterns. Technology has accelerated the pace of change, and knowledge of forces of change is critical to understanding and managing global problems that might arise in the future.

Geography offers a way of thinking about problems, and geographers are particularly well-equipped to understand interactions among different forces affecting a place. For example, to understand hunger in Africa, geographers examine the relationships among climate, soils, agricultural practices, population growth, food prices, environmental degradation, political unrest, and other relevant factors. To explain unrest in the Middle East, geographers study the interrelationships among the distribution of water and energy resources, the differences in religious beliefs, the historical settlement patterns, the international ties of local groups, and the conflicting strategies for achieving economic growth. The interrelationships among factors affecting places help us understand why humans behave as they do. No other scientific discipline takes this approach.

The Development of Geography

Geography in the classical Western world
Every language has a word for the study that we call "geography," but the Greek word that we use, which means "Earth description," was the title of a book by the Greek scholar Eratosthenes (c. 276–194 B.C.). He was the director of the library in Alexandria, Egypt. The library was the greatest center of learning in the Mediterranean world for hundreds of years. No copies of Eratosthenes's book have survived, but we know from other authors that Eratosthenes accepted the idea that Earth is round and that he even calculated its circumference with amazing precision. Eratosthenes also drew detailed maps of the world as it was known to him. Hipparchus (180–127 B.C.), a Greek scholar and later director of the library, was the first person to draw a grid of imaginary lines on Earth's surface to locate places precisely.

After the great age of Greek civilization gave way to the Roman era, the Roman Empire produced a number of geographical scholars and compilations of geographical learning. These culminated in the *Guide to Geography* by Ptolemy, a Greek scholar who worked at the library in Alexandria between A.D. 127 and 150. Subsequently, in the long period between the fall of the Roman Empire (A.D. 476) and the European Renaissance of the fifteenth century, Western civilization accumulated little additional geographical knowledge. Much knowledge was actually lost in Western Europe but preserved by Arab scholars.

Geography in the non-European world
Outside Europe, geography made considerable advances. The expansion of the religion of Islam (discussed in Chapter 7) fostered travel, which inspired research across the wide region from Spain to India and beyond. Islam encourages each believer to complete a pilgrimage to Mecca, in Arabia, at least once in his or her lifetime. This obligation stimulates travel. Muslim scholars, including al-Edrisi (1099?–1154), ibn-Battuta (1304?–1378), and ibn-Khaldun (1332–1406), produced impressive texts describing and analyzing both physical environments and also the customs and lifestyles of different peoples.

China also developed an extensive geographical literature. The oldest known work, *The Tribute of Yu*, dates from the fifth century B.C. and describes the physical geography and natural resources of the various provinces of the Chinese empire. It interprets world geography as a nest of concentric squares of territory. The innermost zone is China's imperial domain. The imperial domain is surrounded by the empire, and beyond the empire lies an outermost zone inhabited by "barbarians." The world's oldest map that shows distances marked with numbers was engraved on copper plate about 325 B.C. In A.D. 267, Phei Hsiu made an elaborate map of China's empire. He is often called the father of Chinese cartography. In the following centuries, Chinese Buddhists occasionally wrote accounts of travel to India to visit places sacred to the history of Buddhism, which had spread into China. Chinese maritime trade extended throughout Southeast Asia and the Indian Ocean to East Africa, but the Chinese eventually withdrew from more extensive exploration. Some scholars argue that the Chinese may even have reached North America, but the evidence for this is insubstantial since few Chinese descriptive geography texts of any Chinese explorations survive.

The Japanese and the Koreans also engaged in East Asian trade. In fact, to our knowledge, the greatest world map produced before European exploration was made in Korea in 1402 (Figure 1-1). This map, the Kangnido, was made by an unknown cartographer, but he or she drew on the combined knowledge of Korea, Japan, and China, including Islamic sources that were known to the Chinese. As a result, this map includes not only East Asia but also India, the countries of the Islamic world, Africa, and even Europe. It demonstrates a far more extensive knowledge of the world than the Europeans had at the time.

The peoples of Africa and of the Western Hemisphere may have had access to geographical accounts of the parts of the world known to them, but either these works have not survived or modern scholars have not yet fully inventoried and studied them.

The revival of European geography
Beginning in the fifteenth century, Europe's world exploration and conquest introduced so much new geographical knowledge that Europeans were challenged to devise new methods of cataloging, or organizing, it all.

One attempt at organizing the information gathered by the European voyages of discovery is the book *General Geography* (1650) by Bernhard Varen (*Varenius* in Latin; 1622–1650), a German who taught at the

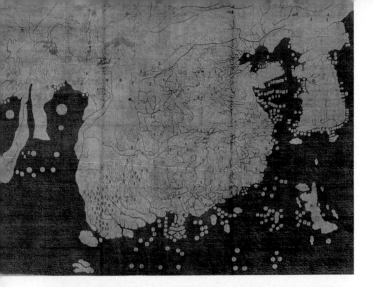

FIGURE 1-1 The Kangnido (detail). When Columbus embarked on his voyage, he knew nothing of Korea, yet the world's finest map was then hanging in the royal palace there. The Kangnido included not only China and Japan but showed India, the countries of the Middle East, and Europe itself. Spain was easy to recognize on this map, which even showed Columbus's native Genoa along the shores of the Mediterranean. The image of Europe was not exact, but no map in Europe had an image of Asia that was as good as this map's image of Europe. (Courtesy of Ryukoku University Library)

Focus ON

National Standards for the Study of Geography

In 1994 the U.S. Congress adopted Goals 2000: The Educate America Act (Public Law 103-227). This act listed geography among the fundamental subjects of a national curriculum. Geographical understanding, wrote Congress, is essential to achieve "productive and responsible citizenship in the global economy." Several academic and scholarly geographical organizations collaboratively produced an agreed-upon core of geographic material and ideas, which was published as *Geography for Life: The National Geography Standards.* These standards specify the geographical subject matter and skills that U.S. students should master.

The goals demonstrate the degree to which geographic knowledge is essential for both understanding and effectively managing environmental and human relations in the twenty-first century. They were established in the hope that all persons educated in the public school system become geographically knowledgeable. These goals were not reached by 2000, but they remain a national target. In this book, we go beyond these standards in the treatment of both subject matter and thinking skills, but we provide here the outline of the goals in order to demonstrate the great breadth of the field.

The geographically informed person knows and understands:

The World in Spatial Terms

1. How to use maps and other geographic tools and technologies to acquire, process, and report information from a spatial perspective.
2. How to use mental maps to organize information about people, places, and environments in a spatial context.
3. How to analyze the spatial organization of people, places, and environments on Earth's surface.

Places and Regions

1. The physical and human characteristics of places.
2. That people create regions to interpret Earth's complexity.
3. How culture and experience influence people's perceptions of places and regions.

Physical Systems

1. The physical processes that shape the patterns of Earth's surface.
2. The characteristics and spatial distribution of ecosystems on Earth's surface.

Human Systems

1. The characteristics, distribution, and migration of human populations on Earth's surface.
2. The characteristics, distribution, and complexity of Earth's cultural mosaics.
3. The patterns and networks of economic interdependence on Earth's surface.
4. The processes, patterns, and functions of human settlement.
5. How the forces of cooperation and conflict among people influence and control Earth's surface.

Environment and Society

1. How human actions modify the physical environment.
2. How physical systems affect human systems.
3. The changes that occur in the meaning, use, distribution, and importance of resources.

Uses of Geography

1. How to apply geography to interpret the past.
2. How to apply geography to interpret the present and plan for the future.

University of Leiden, in Holland. Varen differentiated two approaches to geography: *special geography* and *general geography*. Special geography describes and analyzes places in terms of categories such as local population, customs, politics, economy, and religion. Today we call this approach **regional geography.** Regional geographers usually begin regional studies with a description of the local physical environment—the climate, landforms, soils, and other physical attributes—and then proceed with an inventory much like Varen's, weaving these factors together as they interrelate. The Regional Focus boxes in this textbook highlight regions that illustrate geographic principles or that are in the news.

What Varen labeled general geography examines topics of universal application or occurrence. We might study, for example, the geography of climate, water, vegetation, or minerals. This notion corresponds to what we now refer to as **topical,** or, **systematic geography,** and it is the basic approach taken for the outline of this textbook. Individual geographers concentrate on topics as diverse as the geography of soils (pedology), of life forms (biogeography), of politics (political geography), of economic activities (economic geography), and of cities (urban geography). The Association of American Geographers (AAG), a professional scholarly organization, currently recognizes 55 topical specialties (Table 1-1). Regional and topical geography are complementary. A regional study covers all topics in the region under review, whereas a topical study notes how a particular topic varies across regions.

Varen's book was a standard reference for over 100 years. Even Sir Isaac Newton (1642–1727) edited two editions in Latin, one of which was studied by students at Harvard in the eighteenth century.

The human-environment tradition We cannot fully understand the physical environment without understanding humankind's role in altering it, nor can we fully understand human life without understanding the physical environment in which we live.

The German explorer and naturalist Alexander von Humboldt (1769–1859) wove together virtually all knowledge of Earth sciences and anthropology in his great multivolume book, *Cosmos.* Von Humboldt's brilliant interrelating of the phenomena of physical nature and the world of humankind had an enormous intellectual impact in the United States (Figure 1-2). This was partly because European Americans had long thought of America as the New World, a veritable Garden of Eden over which they had been given dominion.

George Perkins Marsh (1801–1882), an American scholar and diplomat, expanded on von Humboldt's theme of the interconnections between humankind and the physical environment. While Marsh served as U.S. ambassador to several Mediterranean countries, he was impressed by humankind's destruction of an environment that ancient authors had described as lush and rich. Marsh's book *Man and Nature, or Physical Geography as Modified by Human Action* (1864) was one of the earliest key works in what would become today's environmental movement.

Why study geography? Who is a geographer today? And why should anyone study geography? Today almost all of us are geographers, whether or not we call

TABLE 1-1	Topical Specialties Recognized by the Association of American Geographers	
01 Agricultural Geography	19 Gender	37 Physical Geography
02 Applied Geography	20 Geographic Information Systems	38 Planning, Environmental
03 Arid Regions	21 Geographic Theory	39 Planning, Regional
04 Biogeography	22 Geomorphology	40 Planning, Urban
05 Cartography	23 Global Change	41 Political Geography
06 Climatology	24 Hazards	42 Population Geography
07 Cultural Ecology	25 History of Cartography/Historical	43 Quantitative Methods
08 Cultural Geography	Cartography	44 Recreational Geography
09 Developmental Studies	26 Historical Geography	45 Regional Geography
10 Earth Science	27 History of Geography	46 Remote Sensing
11 Economic Development	28 Land Use and Conversion	47 Resource Geography
12 Economic Geography	29 Librarianship, Geographical	48 Rural Geography
13 Educational Geography	30 Location Theory	49 Social Geography
14 Energy	31 Marine Resources	50 Soils Geography
15 Environmental Perception	32 Marketing Geography	51 Teaching Techniques
16 Environmental Science	33 Medical Geography	52 Transportation and Communication
17 Environmental Studies	34 Military Geography	53 University/Other Administration
(Conservation)	35 Mountain Environments	54 Urban Geography
18 Field Methods	36 Oceanography	55 Water Resources

FIGURE 1-2 Alexander von Humboldt. This portrait bust of the explorer and naturalist Alexander von Humboldt was the first statue of a non-U.S. citizen to be erected in New York City's Central Park. It was installed in 1869, marking the one-hundredth anniversary of von Humboldt's birth. Today he gazes across the street at the American Museum of Natural History, where statues of the explorers Lewis and Clark gaze back. (EFB photo)

FIGURE 1-3 A geographer at work. This guide explaining the interaction of local climatic conditions, soils, vegetation, and wildlife in the Big Cypress National Preserve at Ochopee, Florida, exemplifies a "geographer in action." Demand is rising for the geographer's traditional skills of analyzing and explaining such complex environmental conditions; so, too, is demand for guides to historic and cultural attractions, as tourism is fast becoming a major global economic activity. (David Lyons/Alamy Images)

ourselves that. Geographers include people who plan the need for future schools; develop real estate; manage parklands, forests, or water supplies; plan road tours for rock groups; as well as farmers who choose which crop to plant on the basis of world markets; and journalists, diplomats, and executives of multinational corporations.

The information that any citizen needs in order to make an informed decision on an important question of the day is largely geographic. And as world events daily force rapid change upon us, the geography of our world changes too. Almost any topic of the day's news will bring about redistributions—new geographies—of people and activities. Studying geography will not always help you predict the future, but it will help you understand the forces behind these changes. Therefore, when changes come, you may be less surprised. Sometimes you will be able to see changes coming, make changes work for you, or even make the changes yourself.

Geography is one of the most diverse fields of academic study, so it offers an introduction to a unique breadth of careers. A great variety of professions involve studying or "doing" geography, even though the persons at work may not call themselves geographers (Figure 1-3). As you read through the topics in this text, which range from newspaper distribution to agricultural supplies, from urban planning to the delineation of voting districts, and from multinational marketing of consumer goods to international treaties on environmental preservation, think of the many ways you might develop a career analyzing or managing any of these activities.

Geographers' studies of *where, what, when, why,* and *why there* necessarily prompt thoughts about *where will* to predict future patterns, and of *how ought.* Geographers' expertise can inform public debates on major issues of the day—not just to help us understand the world but to help improve it.

CONTEMPORARY APPROACHES IN GEOGRAPHY

All geographers collect and analyze geographic information, but they focus on different topics and employ different analytical approaches. Most contemporary geographers employ three analytical methods:

▶ Area analysis, which integrates the geographic features of an area or a place.

▶ Spatial analysis or locational analysis, which emphasizes interactions among places.

▶ Geographic systems analysis, which emphasizes the understanding of environmental and human systems and the interactions among them.

Let us look briefly at each of these methods.

Area Analysis

Geographers have a long tradition of surveying, describing, and compiling geographic data about places. Each place in the world occupies a unique location and possesses a unique combination of human behavior and environmental processes that give it a special character. In compiling and analyzing Earth's places, geographers rely on three basic ideas: site, situation, and the concept of the region.

Site Geographers often speak of specific sites and site characteristics. **Site** describes the exact location of a place, and can be described either in terms of latitude and longitude (discussed below) or in terms of the place's characteristics. Each place has a unique combination of physical and human characteristics. Travelers do not describe the latitude and longitude of a vacation spot they have visited. Instead, they report on the interesting people and customs, different foods, and varied landscapes that made their trip memorable. They may, however, also report that they spoke the same language, used the same credit cards, and saw the same movies as at home. This mix of unique features and common features is what geographers mean when they talk about place.

Relative location and situation The location of a place relative to other places is called its **situation,** and knowledge of a place's situation helps us understand how it interacts with the rest of the world. **Relative location** influences accessibility, which is indicated by such terms as *nearer* and *farther, easier* or *more difficult to reach, between*, and *on the way* or *out of the way*. Accessibility can be a resource as valuable as mineral deposits or fertile soil.

Relative location is almost constantly changing, as changes in transportation routes and technologies as well as developments in communication continuously redistribute accessibility. The opening of a new highway, for example, stimulates the development of new housing and new roadside services, such as shopping malls. At the same time, however, the new highway can reroute traffic past older shopping malls, and thus choke them off from customers.

Great cities rise and countries often prosper if they are advantageously situated within national or international patterns of trade, but isolated areas usually lose their wealth and stagnate. One of history's greatest redefinitions of relative location occurred when Europeans learned how to sail around Africa in order to reach Asia. Much of the commerce that had previously moved by long and difficult caravan journeys by land across Asia and Africa was drawn to the seacoasts. New seacoast cities sprang up, such as Bombay (today Mumbai) in India, Rangoon in Burma (today known as Yangon in Myanmar), and Hong Kong on the coast of China. Major cities in the interior of the Eurasian and African continents were suddenly "on the back road." Some of these, such as Kashgar, a city in western China visited by the explorer Marco Polo in 1275, and Samarkand, a city thousands of years old in today's Uzbekistan, have never recovered their earlier importance (Figure 1-4).

Changes in transportation technology also redistribute accessibility. In the 1950s the Suez Canal was the principal route of oil shipments to Western Europe from the Middle East. New oil tankers, however, grew so large

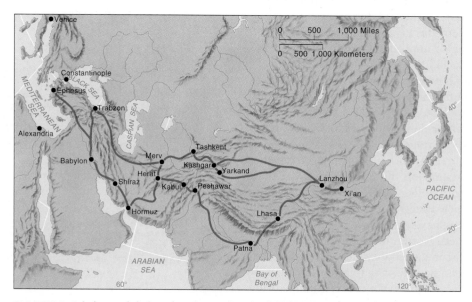

FIGURE 1-4 Internal Asian trade routes in 1400. Trabzon, Kashgar, and Merv were once great cities on major trade routes, but when Europeans opened sea routes around Africa and Asia, these and many other cities declined. Few of these cities ever recovered their former economic strength.

REGIONAL *Focus* ON

Utica, New York

This map shows how new canals, railroads, and highways have continuously rerouted the flow of national economic life in and out of Utica, New York, changing both the relative location of the city within New York State, and, at a smaller local scale of analysis, the relative location of places and land uses within the city.

Utica was founded where the Mohawk River, winding across the top of the map, could be forded. A bridge later spanned the river, and city prospered as a commercial outpost on what was then the American frontier. Utica industrialized when coal could be brought to it from Pennsylvania by the Erie and Chenango Canal (1837), and that canal's route was later paralleled by the train tracks of the New York, Ontario & Western Railroad, which can be seen entering the map from the lower left. Industry increased when the Erie Canal (cutting across the top of the map) and later the New York Central Railroad (the tracks labeled Conrail cutting across the center) were built. Notice the big buildings and the round storage buildings around Utica Harbor and along the railway downtown. In the nineteenth century, workers lived close to these factories, and downtown Utica was a busy place. Since World War II, however, local and interstate highways have provided new accessibility to the suburbs, and virtually all new industrial development has occurred there. New buildings can be identified to the north, across the river from the old downtown. Housing has also spread into the suburbs, and shopping and office malls have pulled more activities out of the downtown area. This map combines two topographic maps printed by the United States Geological Survey. Examine a topographic map for your community for signs of changing access routes and land uses.

that they were incapable of squeezing through the canal, so they had to sail around Africa. This rerouting situated Cape Town "between" the Persian Gulf and Rotterdam, Europe's chief port for importing oil. Cape Town's ship supply and repair facilities won new business.

Each day's news reports the opening of a new transportation or communication facility that will redistribute activities. In 2000, a link was completed across the Oresund Strait between Sweden and Denmark by the incorporation of a series of bridges, a man-made island, and a tunnel (Figure 1-5). The cities of Copenhagen on the Danish side and Malmö on the Swedish side are now joined into one metropolitan area of more than 3 million people. The new facility, with both a highway and railroad tracks, allows a vehicle or train to travel across Europe from northern Norway to southern Spain.

A place's relative location can also change if the territorial scale of an activity's organization changes— that is, the extent of territory within which that activity occurs. For example, when trade barriers between countries fall, activities redistribute themselves. Significant redistributions are occurring throughout North America as the economies of the United States,

FIGURE 1-5 The Oresund Bridge. Some 55,000 cyclists crossed the Oresund Bridge on June 9, 2000, as part of a three-day festival to dedicate the bridge. Today only cars and trucks are allowed. (Riehle/Laif/Aurora & Quanta Productions, Inc.)

Canada, and Mexico merge through the new trade pacts discussed in Chapter 13. Activities in Europe are readjusting from a national to the supranational scale of organization represented by the European Union.

Improvements in transportation and communication links have "shrunk" Earth, so today many activities have expanded their scale of organization to cover the whole globe, which is a process called **globalization.** An advertisement for one U.S. bank shows a satellite photograph of Earth drifting in space. The headline reads "We do business in only one place." Economic globalization has far outpaced cultural or political integration, especially since 9/11 triggered new national security concerns around the world. Tensions among the different economic, cultural, and political organizing activities generate much of today's international news.

Conversely, when a large territory is broken up into smaller individually organized territories, activities are redistributed and places' accessibility reduced. Until 1991 the U.S.S.R. organized 15 percent of Earth's land surface under one highly unified government before it broke up into 15 independent countries (see Figure 13-5). The political change required reorganizing the scale of all political, economic, and cultural activities throughout the vast territory.

The concept of a region For purposes of geographic study, we divide Earth into **regions,** which are

Critical THINKING

Tombouctou

In modern slang, Timbuktoo is a synonym for nowhere, and few Americans even know that there really is such a place. In fact, Tombouctou is a city in Mali at absolute location 16°46′ north latitude, 3° west longitude. It plays no major role in today's world, but in past centuries it was a major urban center. Tombouctou is situated on the Niger River where the river bends farthest north into the Sahara Desert. Thus, Tombouctou marks a cultural border between nomadic Arab peoples to its north and the settled black African peoples to its south. The city was settled by the Taureg people from the north in 1087. As a major contact point between north and south, it prospered as a market for slaves, gold, and salt and as a point of departure for trans-Saharan caravans to North African coastal cities. From these cities Tombouctou's fame as a center of almost mythical riches spread around the Mediterranean. It was also a religious center, and it boasted a great university.

In the fifteenth century, however, Portuguese expeditions sailing around West Africa outflanked the trade routes on which Tombouctou depended. The city declined rapidly. It was captured by the Moroccans in 1591, and by the time the French captured it in 1893, it was little more than an extensive ruin. Today Tombouctou is a small city of 10,000 people who live surrounded by remnants of the city's former greatness.

Questions

1. Examine the location of Trabzon on the Black Sea in modern-day Turkey (Figure 1-4). Among what great empires was Trabzon once a major contact and trading point? How important is it today? Do you know of any other once great cities that have declined as trade routes bypassed them? Conversely, what great port cities developed along with new trade routes in the twentieth century?

2. The geographer Ellen Churchill Semple opened her book *American History and Its Geographic Conditions* with the sentence, "The most important geographical fact in the past history of the United States has been their location on the Atlantic opposite Europe; and the most important geographical fact in lending a distinctive character to their future history will probably be their location on the Pacific opposite Asia." This was written in 1903. Have historic developments proved her correct?

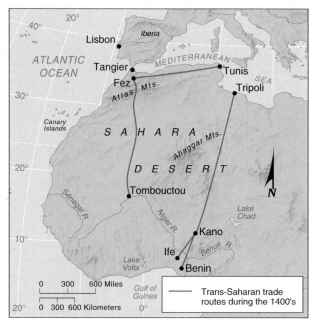

Trans-Saharan trade routes in 1400.

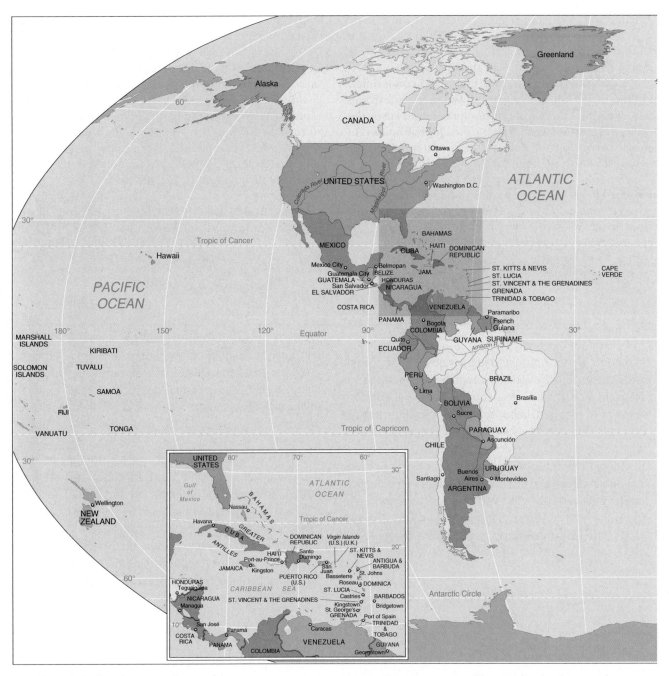

FIGURE 1-6 Political map of the world. The world includes almost 200 sovereign states. The number has increased markedly in recent decades, and many of the names of states shown here may be unfamiliar.

areas defined by one or more distinctive characteristic or feature, such as climate, soil type, language, or economic activity. Geographers define regions and describe and analyze similarities and differences among them.

Probably the most familiar of all maps of regions is the political map (Figure 1-6). People have divided Earth's land surface into almost 200 countries, ranging from Russia, occupying one-ninth of Earth's land area, to microstates such as Andorra, Monaco, and Singapore. This regionalization of Earth influences almost all other human activities. Geographers study such patterns of how people organize their societies and occupy land.

Of all the features that can be found in any landscape, a geographer chooses only a few specific criteria and then maps those criteria to define a region. The criteria chosen may be either physical or cultural.

Geographers distinguish three kinds of regions. A **formal region** is one that exhibits essential uniformity in one or more physical or cultural features, such as a country or a mountain range. A **functional region,** by contrast, is one defined by interactions among places, such as trade or communication. The city of Chicago, for example, is a formal region whose government covers the legal limits of the city's incorporation. The many commuters and shoppers who circulate daily throughout the city and

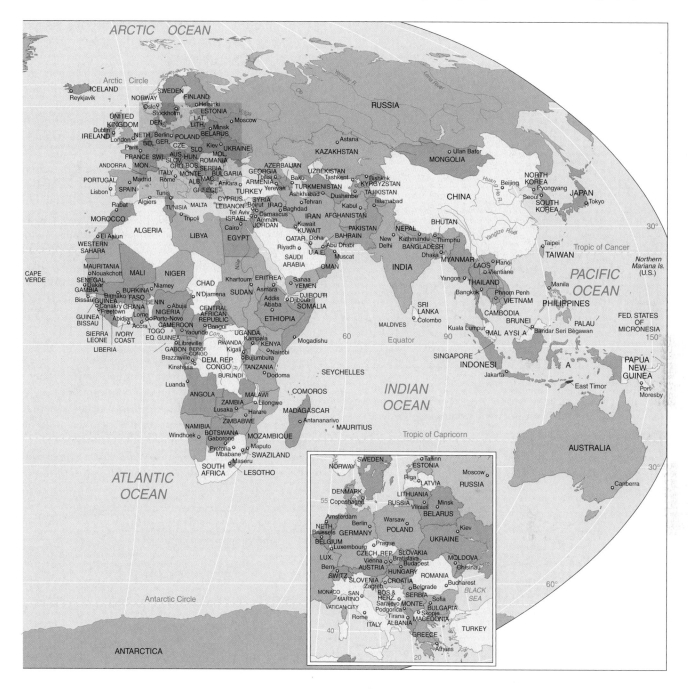

its suburbs, however, would more readily identify a region larger than Chicago; they would identify the functional region of the Chicago metropolitan area, which includes parts of Illinois, Indiana, and even Wisconsin. The desire of people in northwest Indiana to be in the same time zone as Chicago confirms that northwest Indiana is part of the Chicago functional region (Figure 1-7).

The third type of region is a **vernacular region.** Vernacular means everyday language, and vernacular regions are defined by widespread popular perception of their existence by people within or outside them (Figure 1-8).

The choice of criteria to define a region takes place in the geographer's mind. Therefore, a region is

a concept, an abstract idea. The geographer's concept of a region is similar to the historian's concept of a "period." Time, like space, is continuous, but historians divide time into units of analysis called periods to suit their purposes. Art historians define "styles," sociologists define "societies," and many other scholarly fields define their conceptual units of analysis.

Sometimes the features geographers choose to demarcate regions are clear in the landscape. For example, a region defined as a certain forest could be mapped distinctly. Individual countries are usually clearly defined on the landscape by national borders.

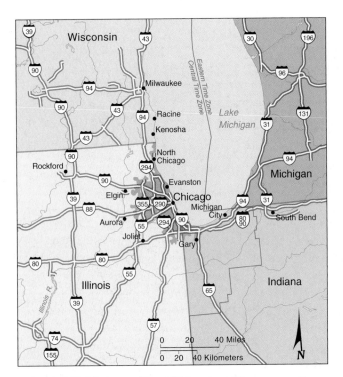

FIGURE 1-7 Metropolitan Chicago. The metropolitan region of Chicago includes parts of Wisconsin and Indiana, as well as northeastern Illinois. Commuter trains link Chicago with surrounding communities. While most of Indiana is in the Eastern Time Zone, several counties in northwestern Indiana are in the Central Time Zone to facilitate their links with Chicago, which is on Central time.

In other cases, however, the regions that geographers define are less distinct in the landscape. For example, Chapter 2 maps regions of different climates. On Earth those climatic regions are not sharply defined, but rather they merge into one another imperceptibly. On our map, however, the precise location of the line between one climatic region and its neighbor is determined by the geographer's definition and trained discrimination.

Regions defined by cultural phenomena often merge or overlap as well. For example, people who live between two cities may listen to radio stations located in both cities. If, however, a geographer wants to delineate sharply the market regions of two radio stations broadcasting from the two cities, he or she might choose to draw a line dividing the area where most households listen to the station broadcasting from City A from the area where most households listen to the station broadcasting from City B (Figure 1-9).

Sometimes people confuse regions based on natural phenomena with regions based on cultural criteria. For example, many people refer to Africa as a region. Africa is a continent, but the physical environment is not homogeneous across that vast continent, nor do its many peoples share one culture or historical experience. Historically, the Sahara Desert divided peoples more effectively than the Mediterranean Sea did. Therefore, to the ancient Romans, for example, northern Africa and southern Europe formed a single region focused on the

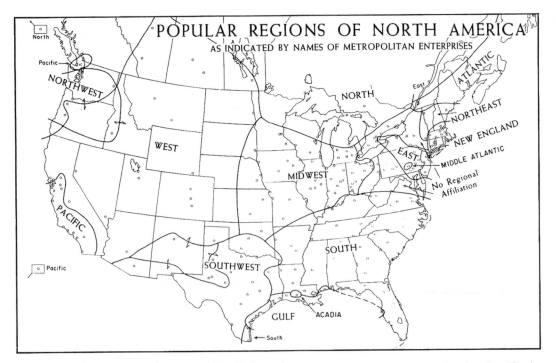

FIGURE 1-8 Vernacular regions of North America. The regions on this map reflect how local business enterprises in metropolitan areas (the small circles) define themselves in telephone directories. For example, within the region labeled "the South" on this map, businesses most probably identify themselves by that regional designation, such as the Southern Telephone Company. Thus, the regions drawn here are vernacular, defined by popular local everyday language. (After Wilbur Zelinsky, "North America's Vernacular Regions," *Annals, Association of American Geographers*, 70 (March 1980): 14. By permission of the Association of American Geographers)

Critical THINKING

Grouping Countries; Naming Groups

The countries of the world can be grouped into regions in many different ways. Grouping them by continent is fairly easy, as is grouping by membership in certain international organizations such as NATO (North Atlantic Treaty Organization) or OPEC (Organization of Petroleum Exporting Countries). At other times it is useful to categorize them by broadly similar characteristics, such as large countries, democratic countries, industrial countries, rich countries, and poor countries.

There are enormous differences in levels of wealth among the almost 200 countries of the world today, and this distinction is a particularly important one for geographers. Wealth affects virtually every aspect of a country's human environment, from food availability and disease to pollution and political power. It is useful to make generalizations about geographic patterns of wealth and power, yet by doing so we necessarily obscure important differences between countries.

Three major problems confront geographers in identifying groups of countries. One is defining a measurable characteristic of countries, such as wealth. How is wealth measured? Most often wealth is measured in monetary terms—the average amount of money earned in a country per year. In Chapter 12 we will discuss this in more detail. A second problem is the one of deciding how many groups to define. Should countries be divided into just two groups, or three, five, or twenty? One way to avoid this problem is to recognize a spectrum, such as poor at one end and rich at the other and many degrees of wealth in between. The third problem is a linguistic one: What should we call the groups, or the endpoints of the spectrum?

Although the terms *rich* and *poor* are fairly simple, descriptive words for these differences, they can be misinterpreted, because having money is not the only way to be rich. Some people prefer to use the terms *developed* and *developing*, or *more developed* and *less developed*, or *developed* and *underdeveloped*. *Core* and *periphery* are another pair of words sometimes used, as are *North* and *South*.

In this book we will use *rich* and *poor*, *developed* and *developing*. In doing so, we recognize that such generalizations are never fully accurate and should be interpreted with care.

Questions

1. What does it mean to be rich? Is monetary wealth the most important characteristic to examine in describing and grouping countries?
2. What are the hidden meanings of terms like *developed*, *underdeveloped*, or *developing*?
3. What characteristics do the terms *core* and *periphery* emphasize about rich and poor countries?

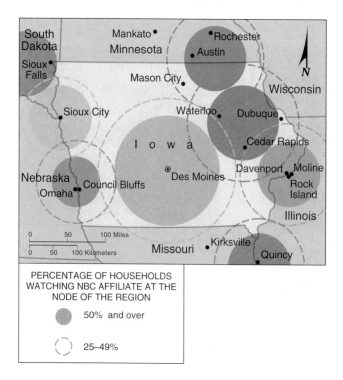

PERCENTAGE OF HOUSEHOLDS WATCHING NBC AFFILIATE AT THE NODE OF THE REGION

50% and over

25–49%

Mediterranean Sea; in fact, the word *mediterranean* means "middle of the world." Sub-Saharan Africa was a separate region, distinct and hard to reach (Figure 1-10). Whenever anyone refers to any region, we must be sure that we agree on what area is included and what the criteria of homogeneity are.

Spatial Analysis

A second approach to geographic inquiry, called spatial analysis or locational analysis, looks for patterns in the distribution of human actions and environmental processes and in movements across Earth's surface.

FIGURE 1-9 Television station viewing areas within Iowa. The colored circles on this map are known as areas of dominant influence (ADIs). Most people within any circle watch television programs from the stations in the principal city in that ADI. The construction of new housing, highways, or cable lines may shift an area on the periphery of one city's ADI to another ADI. (After James Rubenstein, *The Cultural Landscape: An Introduction to Human Geography*, updated 7th ed. Upper Saddle River, NJ: Prentice Hall, 2003)

FIGURE 1-10 Roman ruins in Morocco. The remains of ancient Roman buildings haunt the North African desert. These, in Morocco, demonstrate that to the ancient Romans, the lands surrounding the Mediterranean Sea formed one region, and that region was bounded on the south by the Sahara Desert. The African continent was not considered a region. (Courtesy of the Moroccan National Tourist Office)

Distribution The **distribution** of a phenomenon means its position, placement, or arrangement throughout space. Geographers recognize distributions and use three terms to define the distribution of any phenomenon under study: density, concentration, and pattern (Figure 1-11).

Density **Density** describes the frequency of occurrence of a phenomenon in relation to geographic area. Density is usually expressed as a number per square kilometer or square mile. Examples include road

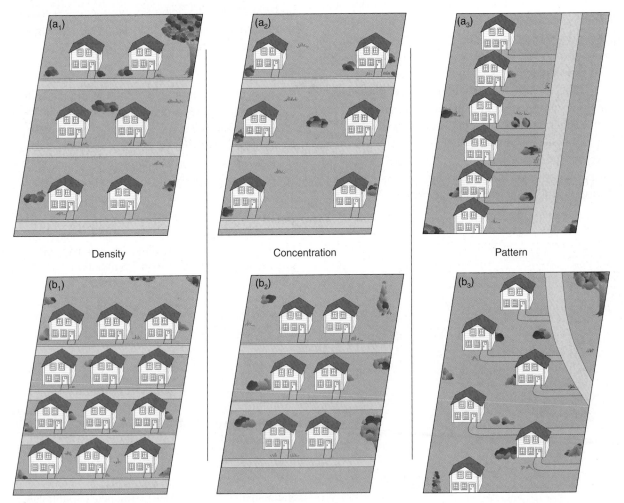

FIGURE 1-11 Density, concentration, and pattern. Assume that the area represented in all figures is one hectare. Density is the overall frequency of occurrence of a phenomenon, the number of items per unit of area. The density is six houses per hectare in a_1, whereas in b_1 it is 12 houses per hectare. Concentration refers to the extent of the spread of the feature relative to the size of the area being studied. The houses might be described as dispersed in a_2 and clustered in b_2. Pattern is the geometric arrangement of the distribution of a phenomenon. The pattern of the distribution of houses is linear in a_3 but irregular in b_3.

density (the number of kilometers of roads per square kilometer) and population density (the number of people per square kilometer).

Many statistics are reported using political unit as their measure, such as by state or country. In 2000 the population of Russia was 147 million, and the population of Belgium was 10 million, so Russia had almost 15 times as many people as Belgium. Russia, however, is much larger than Belgium, so the population density of Russia was only nine persons per square kilometer (23 per square mile), whereas that of Belgium was 333 per square kilometer (862 per square mile)—37 times that of Russia.

Concentration **Concentration** refers to the distribution of a phenomenon within a given area. If all the occurrences are found in close proximity, the distribution would be described as concentrated, but if they are scattered far from each other, the distribution would be described as dispersed. For example, in many parts of the world farmers live in villages and travel to their fields in the countryside. The population of such an agricultural landscape is concentrated in villages. In North America, on the other hand, most farmers live in isolated farmhouses located on the land they farm, so the population is dispersed.

Pattern **Pattern** refers to the geometrical arrangement of objects within an area. For example, in most modern cities, streets are arranged in a rectangular grid pattern, whereas in older cities the street layout is more irregular. In areas where rock structures exert strong controls on stream erosion, we sometimes see streams with many right-angle bends; this pattern is called a trellis pattern. Conversely, in areas without structural control, streams and their tributaries form branching or treelike patterns, called dendritic patterns.

Movement

Distance Interaction among people and places occurs across distance, but there are several ways to measure distance. **Distance** can be measured absolutely, in terms of miles or kilometers, but it can also be measured in other ways. It can, for instance, be measured in time. Most people use time as a distance measurement in their daily lives without realizing it: "The store is about 20 minutes from here." Sometimes it can be quicker to reach a place that is far away than to reach another place that is, in absolute distance, closer. This may be the result of rugged **topography** (surface relief) in one direction, direct or winding routes, or different methods of transportation (Figure 1-12).

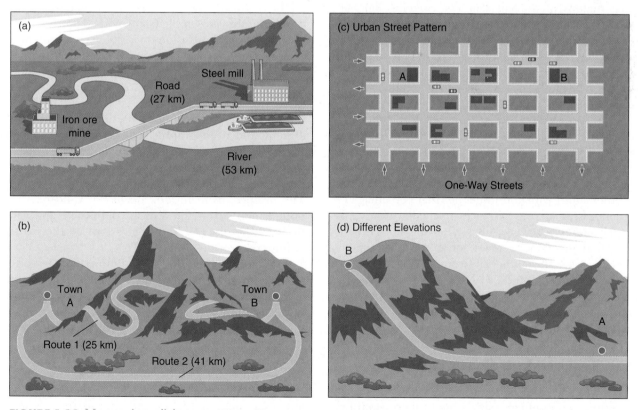

FIGURE 1-12 Measuring distance. When distance is measured by time or by cost, it may seem to be warped when compared to distance measured by absolute units "as the crow flies." For example, drawings (a) and (b) illustrate how the cheapest or fastest distance between two points may not be the shortest. Transportation by water is almost always cheaper than transportation over land, so heavy iron ore will be moved by water if possible. Traveling on a level road is often cheaper and faster than choosing a shorter route crossing rugged land. Drawings (c) and (d) illustrate how the time or cost distance from any point A to a point B may not be the same as from point B back to point A. Most drivers are at one time or another stymied by a situation such as that in drawing (c).

Geographers also frequently measure distances in terms of the cost required to overcome the distance. Transportation over land is usually more expensive than transportation over water, so two seaports that are far apart in absolute distance may be close in *cost distance.* They may even be closer to each other in cost distance than either is to cities inland from itself. That explains why many political units grew up around the shores of bodies of water. Classical Greece, for example, consisted of cities surrounding the Aegean Sea, and the Roman Empire developed around the Mediterranean Sea.

No matter how we choose to measure distance, we must make some effort to overcome distance when we want to move or transport items. We call that effort or cost the **friction of distance.** The friction of distance usually limits interactions across great distances.

Distance decay   The presence or impact of any phenomenon may diminish away from its origin, just as the volume of a sound diminishes the further it gets from its source. This phenomenon is called the **distance decay** concept. As we travel away from any city, for example, the percentage of the local population that reads that city's newspapers, tunes in to its radio and television stations, or relies on its other services shrinks. For example, we might try to define and map the functional regions of the cities of Los Angeles and San Diego. To the immediate south of Los Angeles, people probably subscribe to Los Angeles newspapers, watch Los Angeles television stations, and rely on service from the Los Angeles airport. As we move southward toward San Diego, the influence of Los Angeles diminishes and yields to that of San Diego.

Physicists have a precise mathematical formula to express the gravitational force that one object exerts on another, and geographers have tried to find analogous mathematical equations to describe patterns of human activities. For example, a geographer might try to write a mathematical equation to define what percentage of households at varying distances from downtown Los Angeles subscribes to Los Angeles newspapers. Perhaps circulation drops 10 percent for every 8 kilometers (5 miles) away from the city center. That equation could then serve as a model to compare the circulation of newspapers in metropolitan Los Angeles to the circulation of newspapers in other metropolitan areas. A **model** is an idealized, simplified representation of reality that can be used as a standard to compare individual cases in the real world. Comparing the equation model of the distribution of newspapers in metropolitan Los Angeles with patterns in New York, St. Louis, and other metropolitan regions might help us understand how shopping patterns, local professional sports loyalties, or even involvement in metropolitan politics vary from one metropolitan region to another. In some metropolitan regions, the downtown newspapers might circulate farther out into the suburbs, building a greater sense of metropolitan identity.

Diffusion   **Diffusion** is the process of an item or feature spreading through time. Any innovation—the use of a tool, a new clothing fashion, or the development of a new technology—originates at a place called a **hearth.** Tracing diffusion of innovation suggests how peoples and cultures interact and influence one another. The simplest image of diffusion suggests concentric waves spreading out from a stone dropped into a pool. Paths of cultural diffusion, however, are seldom so simple. Geographers define three basic paths of diffusion: relocation diffusion, contiguous or contagious diffusion, and hierarchical diffusion.

Relocation diffusion occurs from widely separated point to point. A nomadic tribe, for example, might wander until it finds a physical environment similar to the one that it left, and then the tribe might settle down. For example, the ancestors of today's Hungarians wandered all the way from Central Asia to the plains of today's Hungary. The ideas and practices they brought with them were spread by relocation diffusion.

Contiguous diffusion occurs from one place directly to a neighboring place. Contiguous diffusion that occurred in the past is often revealed by the dispersion of artistic styles. For example, in the fourth century B.C., Alexander the Great's conquests carried Hellenistic Greek art far into Asia. Central Asian statues of Buddha sculpted during the centuries immediately following these conquests resemble Greek gods (Figure 1-13).

Sometimes diffusion does not occur contiguously across space, but downward or upward in a hierarchy of organization. When such **hierarchical diffusion** is mapped, it shows up as a network of spots, rather than as an ink blot, spreading across a map. The Roman Catholic Church illustrates both an organizational hierarchy and a geographical hierarchy (Figure 1-14). Each parish priest is answerable to his bishop, who from his cathedral church (*cathedra* is Latin for "throne") presides over a diocese, which includes many parishes. The bishops, in turn, look up to the Vatican. An announcement from the Vatican diffuses to the cathedrals, to the dioceses and finally down to the parishes around the world.

As a general rule, more information travels up and down a hierarchy than across any level of the hierarchy. Therefore, news from a parish about an innovation in community service or in a church ceremony might reach the cathedral and even the Vatican before it reaches neighboring parishes. This principle of diffusion up and down in a hierarchy is very important, and we will return to it repeatedly. Note that the topmost point in a hierarchy, called the *apex,* may not actually generate innovations. All information, however, does clear through the apex, and most places receive information from the apex.

Advanced societies usually exhibit well-developed hierarchies of cities, so we can study how the hierarchy evolved through time. Once a hierarchy is developed, the diffusion of any specific phenomenon might be

FIGURE 1-13 Siddhartha Buddha. This richly dressed young man is the Indian prince Siddhartha before he became the Enlightened One (Buddha) and gave up his worldly wealth. The statue exhibits a mix of styles: the curly hair, stocky physique, relaxed stance, and deeply carved drapery are Greek characteristics, but the facial expression is typical of Indian mysticism. The nimbus of light around the prince's head originated with Iranian statues of sun gods. Where could all these cultural influences have come together for this extraordinary figure to have been carved? The answer is the region of Peshawar in today's Pakistan, which was at the crossroads of Eurasian travel routes, where the Khyber Pass opens from Central Asia into the plains of the Indian subcontinent. This statue dates from the late second or early third century A.D. (Indian Buddhist sculpture, "Bodhisattva," c. 2nd–3rd century. Schist, chlorite, 45 in. h, 15 in. w, 7 in. d. Eugene Fuller Memorial Collection, Seattle Art Museum. Photo by Paul Macapia)

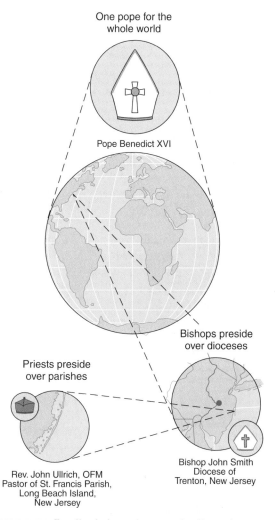

FIGURE 1-14 Territorial and organizational hierarchy. The Roman Catholic Church illustrates both a territorial and organizational hierarchy. The pope presides over the entire Church, but the Church is subdivided into dioceses, each presided over by a bishop, and the dioceses are subdivided into parishes.

traced from the few biggest cities down to the many more smaller cities and then further down to the even more numerous small towns and villages until the landscape is covered (Figure 1-15).

The map of America's most heavily traveled domestic airline routes reflects the national hierarchy of cities (Figure 1-16). To fly from one small city to another

small city, a passenger may have to fly first to a nearby big city (which is a movement up the hierarchy), then from that big city to another big city near the ultimate destination (across the hierarchy), and finally from that big city to the destination small city (down the hierarchy again).

When scientists began to study the path of the diffusion of AIDS throughout sub-Saharan Africa, they first assumed that it had diffused contiguously, as is common in poor, underdeveloped regions, where most contact is directly from one village to the next contiguous village. It was discovered, however, that AIDS was being spread across Africa in a hierarchical fashion. Epidemiologists (those who study the spread of disease) wondered how this could be occurring. They found that it was being spread by long-distance truck drivers who travel long distances between large towns and cities and often hire prostitutes in each town along the way. The knowledge of this method of

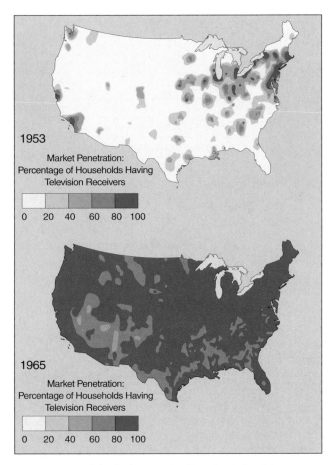

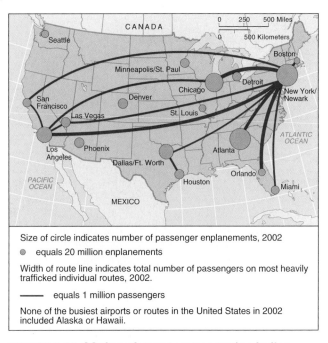

Size of circle indicates number of passenger enplanements, 2002

● equals 20 million enplanements

Width of route line indicates total number of passengers on most heavily trafficked individual routes, 2002.

——— equals 1 million passengers

None of the busiest airports or routes in the United States in 2002 included Alaska or Hawaii.

FIGURE 1-16 Major air-passenger routes in the United States. This map is a good indicator of the urban hierarchy of the United States, although it is a bit distorted by airlines' routing systems. New York is certainly the principal focus of national life. Los Angeles, Dallas, Atlanta, and Chicago are functional "capitals" of individual regions of the country. Minneapolis/St. Paul and Denver have large numbers of total enplanements, but few or no individual routes with more than 2 million passengers. These cities serve as hubs of many local regional routes with small numbers of fliers on each.

FIGURE 1-15 Market penetration by television. These maps show how the percentage of U.S. households that had television sets increased from 1953 to 1965. The spots on the first map show that only large cities had television in 1953. The second map shows that by 1965 the innovation had diffused to cover the contiguous 48 states. This occurred before cable television or satellite dishes, so viewers had to receive transmissions from metropolitan centers. How would the pattern of diffusion differ with today's technology? (Reprinted with permission from Brian J. L. Berry, "The Geography of the United States in the Year 2000," *Transactions of the Institute of British Geographers LI*, November 1970, 21–53)

diffusion in Africa caused world health officials to check truck drivers in India, another poor region suffering from soaring rates of infection, and to educate them of the necessity for safe-sex techniques. Thus, understanding of the path of diffusion of AIDS in Africa may have saved millions of lives in India and elsewhere.

Many phenomena diffuse hierarchically around the world today. The spread of a disease or of a cultural innovation such as a clothing fashion or a hit song can often be traced among the world's principal metropolises before it reaches down into the smaller cities in each country. This path of diffusion reveals the world's interconnectedness.

Barriers to diffusion   There are many barriers to cultural diffusion. Oceans and deserts, for example, increase the cost distance or time distance between places. Other topographical features, such as mountains and valleys, historically have blocked human communication and contributed to cultural isolation. In Asia the Himalayan mountains and the deeply cut valleys of the Tongtian, Mekong, Nu, and Brahmaputra rivers make overland travel between China and the Indian subcontinent extremely difficult, so these two cultures and great concentrations of population are distinctly isolated from one another.

Other barriers to cultural diffusion include political boundaries and even the boundaries between two culture realms. Hostile misunderstanding, distrust, and competition between two cultural groups can hinder any communication and exchange between them.

Physical and Human Systems

A third approach in geographic inquiry views Earth as a set of interrelated environmental and human systems. A **system** is an interdependent group of items that interact in a regular way to form a unified whole. Models of systems help geographers see how factors are interrelated—how one activity, force, or event affects others. Individual environmental systems include the climatic processes that produce precipitation, the hydrologic processes that determine what happens to the rain when it reaches Earth's surface, and the characteristics of the river valleys that receive the water runoff. These individual physical systems are themselves interrelated. For example, an increase in precipitation affects an area's soils, landforms, and vegetation. Geography also emphasizes interactions between humans and the environment. Human activities are influenced by environmental conditions, but humans in turn modify the physical environment in which they live.

In the following sections we will examine some of the ways in which geographers have studied the interactions among different aspects of the human and physical environment.

Earth's physical systems

Geographers study natural processes in terms of four systems: the atmosphere (air), the hydrosphere (water), the lithosphere (Earth's solid rocks), and the biosphere, which encompasses all of Earth's living organisms (Figure 1-17).

The **atmosphere** is a thin layer of gases surrounding Earth to an altitude of less than 480 kilometers (300 miles). Pure, dry air in the lower atmosphere contains about 78 percent nitrogen and 21 percent oxygen by volume. It also includes about 0.9 percent argon (an inert gas) and 0.38 percent carbon dioxide (a crucial percentage, as you will see in upcoming chapters). Air is a mass of gas molecules held to Earth by gravity creating pressure. Variations in air pressure from one place to another cause winds to blow, as well as create storms, and control precipitation patterns.

The **hydrosphere** is the water realm of Earth's surface. Water can exist as a vapor, liquid, or ice, such as the oceans, surface waters on land (lakes, streams, rivers), groundwater in soil and rock, water vapor in the atmosphere, and ice in glaciers. Over 97 percent of the world's water is in the oceans in liquid form. The oceans sustain a large quantity and variety of marine life in the form of both plants and animals. Seawater supplies water vapor to the atmosphere, which returns to Earth's surface as rainfall and snowfall. These are the most important sources of fresh water, which is essential for the survival of plants and animals. Water changes temperature very slowly, so oceans also moderate seasonal extremes of temperature over much of Earth's surface. Oceans also provide humans with food and a surface for transportation.

The **lithosphere** is the solid Earth, composed of rocks and sediments overlying them. Earth's core is a dense, metallic sphere about 3,500 kilometers (2,200 miles) in radius. Surrounding the core is a mantle about 2,900 kilometers (1,800 miles) thick. A thin, brittle outer shell, the crust is 8 to 40 kilometers (5 to 25 miles) thick. The lithosphere consists of Earth's crust and a portion of upper mantle directly below the crust, extending down to about 70 kilometers (45 miles). Powerful forces deep within Earth bend and break the crust to form mountain chains and shape the crust to form continents and ocean basins. The shape of Earth's crust influences climate. If the surface of Earth were completely smooth, then temperature, winds, and precipitation would form orderly bands at each latitude.

The **biosphere** consists of all living organisms on Earth. The atmosphere, lithosphere, and hydrosphere function together to create the environment of the biosphere, which extends from the depths of the oceans through the lower layers of atmosphere. On the land surface, the biosphere includes giant redwood trees, which can extend up to 110 meters (360 feet), as well as the microorganisms that live many meters down in the soil, in deep caves, or in rock fractures.

These four spheres of the natural environment interact in many ways. Plants and animals live on the surface of the lithosphere, where they obtain food and shelter. The hydrosphere provides water to drink and physical support for aquatic life. Most life forms depend on breathing air, and birds and people also rely on air for transportation. All life forms depend on inputs of solar energy.

Humans also interact with each of these four spheres. We waste away and die if we are without water. We pant if oxygen levels are reduced in the atmosphere, and we cough if the atmosphere contains pollutants. We need heat, but excessive heat or cold is dangerous. We rely on a stable lithosphere for building materials and fuel for energy. We derive our food from the rest of the biosphere.

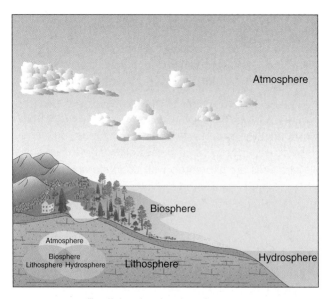

FIGURE 1-17 Earth's physical systems. Geographers regard natural processes as consisting of four open systems, including three composed of nonliving matter (the atmosphere, hydrosphere, and lithosphere), plus the biosphere, which comprises all of Earth's living organisms. The four systems are interrelated and overlap with each other.

Critical THINKING

Parks, Gardens, and Preserves

Which landscape in the photograph, the idyllic foreground or the skyscrapers in the background, is a cultural landscape? Both are. Gardens and parks are landscapes that humans have molded according to aesthetic notions. They are cultural landscapes, built on the assumption that in some ways human designs are superior to nature.

Tastes in garden design vary among cultures, and within any culture they change through time. In all cases, however, gardens and parks reflect the prevailing human notions of how best to "improve" nature. The eighteenth-century park designer Lancelot Brown earned his nickname Capability Brown because whenever he saw a natural landscape, he commented that it had great "capabilities," upon which he would improve.

In this photograph, which is artificial: the soil, drainage, topography, and selection of vegetation in the foreground, or the skyscrapers in the background? In fact, all are. New York City's 840-acre Central Park was constructed over the 20 years after 1858. The land was still beyond the fringe of urban development, but a design was approved, the land was blasted and sculpted, and vegetation was planted to achieve an ideal. (EFB photo)

In the nineteenth century Calvert Vaux and Frederick Law Olmsted "built" New York City's Central Park, including its topography and watercourses, and they chose and planted imported species of vegetation. Today naïve visitors marvel that the city has "preserved" this "natural environment" in the midst of skyscrapers, yet an army of engineers and landscape architects work full-time to maintain the park. In truth, the park is scarcely less artificial than the surrounding skyscrapers.

Minnesota's huge Mall of America encloses a seven-acre interior landscape that, according to official mall publications, "bring(s) the outdoors indoors . . . inspired by Minnesota's natural habitat—forests and meadows, river banks and marshes." Minnesota's natural conifers, however, cannot survive in a constant temperature of 21°C (70°F). Therefore, the mall's "indoor natural Minnesota forest" is populated by tropical orange jasmines, black olives, oleander, hibiscus, and other species flown in from the south. Few visitors seem to notice, and none complain. The mall is, in fact, a terrarium of exotic species.

Parks and gardens are different from reserves or preserves. These areas are closer to being natural landscapes, because they are set aside to save natural environments, usually including the natural wildlife. Most of the properties in the U.S. National Park System, such as Yellowstone National Park, the world's first national park (1872), were originally designated as reserves. These enclaves of nature attract rapidly growing numbers of visitors who demand facilities ranging from hotels to hiking trails to designated scenic viewing spots. The press of their numbers endangers the pristine character of the property. In other words, the people destroy what they come to see. In other cases, properties that were originally chosen as reserves are increasingly engineered in order to save them. For example, parts of Niagara Falls are shored up with concrete and steel to prevent them from eroding away and losing their "natural" beauty.

Questions

1. Does the fact that we call our reserves "parks" reflect confusion about what we really want?
2. Should we let our natural reserves become human-made parks?
3. How can we enjoy our reserves and yet save them?

Plants and animals interact with each other through exchange of matter, energy, and stimuli. An ecological system, or **ecosystem,** is a group of organisms and the nonliving physical and chemical environment with which they interact. The scientific study of ecosystems is called **ecology.** Ecologists study the interrelationships among life forms and their environments in particular ecosystems, as well as among various ecosystems in the biosphere.

Human-Environmental Interaction

The physical environment affects human activities in many ways. Human societies must adapt to local climatic conditions, vegetation, water resources, and other attributes of a local environment as they learn to exploit the local resources. Human societies are not, however, passive. As any human society adapts, it actively and deliberately alters its surrounding natural conditions. The vegetation that covers much of Earth, for example, is cropland or pastureland that is maintained for human benefit. Thus, interaction between the environment and humankind is reciprocal: The environment affects human life and cultures, and humans alter and transform the environment. In the world today there are no inhabited environments that have not undergone fundamental transformation by human activities, and many uninhabited places have been significantly altered as well.

Human culture and cultural landscapes

Geography is one of the social sciences that contributes to our understanding of human **culture.** The word culture is often used to mean only fine paintings or symphonic music, but to social scientists it means everything about the way people live: what sort of clothes they wear (if any); how they gather or raise food; whether they recognize marriages and, if they do, whom (and how many spouses) they think it proper to marry as well as how they celebrate marriage ceremonies; what sorts of shelters they build for themselves; which languages they speak; which religions they practice; and whether they keep any animals for pets. Human cultures vary significantly in all these aspects. Of all the ways in which the behavior and beliefs of individuals differ, a social scientist chooses a few specific criteria and then defines that bundle of attributes of shared behavior or belief as a *culture.*

People modify their local landscape in the process of making a living, housing themselves, and carrying on their lives. We say that they transform a **natural landscape,** one without evidence of human activity, into a **cultural landscape,** one that reveals the many ways people modify their local environment. Aspects of cultural landscapes include the treatment of the natural environment as well as houses and other parts of the built environment.

The evidence of human activity is so ubiquitous on Earth that we might be justified in saying that today all of Earth is a cultural landscape. Travelers quickly notice that the cultural landscape is varied. Rural China does not look like rural France or rural Nigeria. One factor in explaining these variations is that the natural environments in these places were different even before humankind set to work on them. The "raw material" that human societies transformed into cultural landscapes affected what humans would do in each.

Another factor in explaining variations in cultural landscapes, however, is that each cultural group creates a distinct cultural landscape, and the better we can learn to interpret a landscape, the more it will reveal to us about the culture that produced it. In the past, human societies developed in greater isolation from one another than today, and the extraordinary diversity of human cultures and cultural landscapes testifies to human ingenuity. Different peoples who live in very similar environments but are isolated from one another developed astonishingly different lifestyles. Conversely, some aspects of cultures that have developed in different physical environments are startlingly similar. No direct cause-and-effect relationship between a physical environment and any aspect of culture can be assumed.

DESCRIBING EARTH

Earth is nearly spherical, with a diameter of about 12,735 kilometers (8,000 miles) and a circumference of about 40,000 kilometers (25,000 miles). Its highest point is almost 8,850 meters (29,000 feet) above sea level, and the deepest spot in the oceans is about 10,920 meters (35,800 feet) below sea level. Earth's shape varies slightly from true sphericity due to the effect of centrifugal force. Any rotating body tends to bulge in the "middle" and flatten at the "ends"; thus, the equatorial diameter of Earth is slightly greater (43 kilometers, or 27 miles) than its polar diameter.

The Geographic Grid

Earth rotates continuously about an axis that penetrates Earth's surface at the North Pole and the South Pole. An imaginary plane is perpendicular to the axis of rotation and passes through Earth halfway between the poles. It is called the *plane of the equator,* and intersects Earth's surface at Earth's imaginary midline, or "waist," called the **equator.** We can use the North Pole, the South Pole, and the equator as reference points for defining positions on Earth's surface.

Latitude is the angular distance measured north and south of the equator. As shown in Figure 1-18a, we can project a line from any given point on Earth's surface to the center of Earth. The angle between this line and the equatorial plane is the latitude of that point. Latitude is expressed in degrees, minutes, and seconds. There are 360 degrees (°) in a circle, 60 minutes (′) in a degree, and 60 seconds (″) in a minute. Latitude varies from 0° at the equator to 90° at the North and South Poles.

Lines connecting all points of the same latitude are called **parallels** because they do not intersect. Parallels

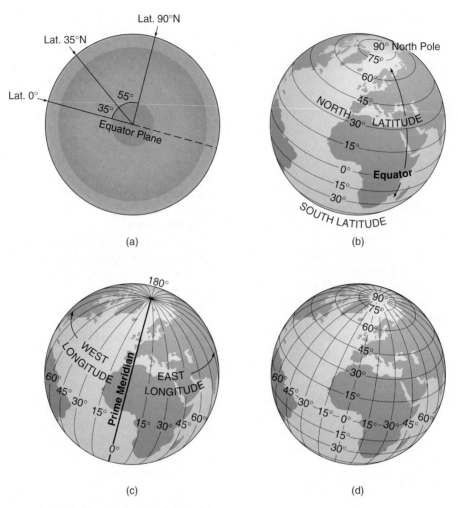

(a)

(b)

(c)

(d)

FIGURE 1-18 The geographic grid. These drawings show the development of the geographic grid: (a) the measure of latitude, (b) parallels of latitude, (c) meridians of longitude, (d) the completed grid system.

are imaginary lines. There can be an infinite number of them—one for each degree of latitude, or for every minute, or for any fraction of a second (Figure 1-18b).

Longitude is the angular distance measured east and west on Earth's surface. Longitude is also measured in degrees, minutes, and seconds. It is represented by imaginary lines, called **meridians,** extending from pole to pole and crossing all parallels at right angles (Figure 1-18c). Meridians are not parallel to one another. They are farthest apart at the equator, become increasingly close northward and southward, and converge at the poles.

The **prime meridian,** from which longitude is measured, was chosen by an international conference in 1884. It is the meridian passing through the Royal Observatory in Greenwich, England, just east of London. Longitude is measured both east and west of the prime meridian to a maximum of 180° in each direction. Thus, the location of any spot on Earth's surface can be described with great precision by reference to its latitude and longitude. For instance, if we say that the dome of the U.S. Capitol in Washington is located at 38°53′23″ north latitude and 77°00′33″ west

longitude, we have described its position within about 20 steps.

The same international conference that agreed on the prime meridian also agreed to divide the world into 24 standard time zones (Figure 1-19). When people first began keeping track of time, they defined local time in terms of when the sun rose and set, with noon defined as the time at which the sun is highest in the sky. This method of timing depends on longitude. The standardization of time from one place to another became necessary when travel between places was fast enough that differences in time from one place to another became important. In the United States, standardization became important because of the development of the transcontinental rail network.

When we standardize time across large east-west distances, the usual link between the position of the sun in the sky and time can be stretched considerably. Relative to Earth, the sun travels 360° in 24 hours, or 15° per hour, so time zones that standardize time within each 15°-wide slice of Earth's surface offer a solution. The time at the prime meridian, or 0° longitude, is designated **Greenwich Mean Time (GMT).** Earth rotates

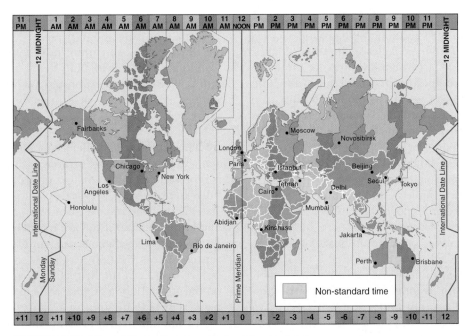

FIGURE 1-19 World Time Zones. Each of Earth's 24 time zones represents an average of 15° of longitude, or one hour of time. The number in each time zone on this map represents the number of hours by which that zone's standard time differs from Greenwich Mean Time, plus or minus. The colored stripes indicate the standard time zones; gray areas have a fractional deviation. These irregularities in the zones are dictated by political and economic factors.

eastward, so a clock advances one hour from GMT for each 15° traveled east from the prime meridian. For each 15° west of the prime meridian, a clock moves back to one hour earlier than GMT.

The meridian for 75° west longitude runs through the eastern United States, so time in the eastern United States is five hours earlier than GMT. (The 75° difference in longitude ÷ by 15° per hour = 5 hours.) When the time is 11 A.M. GMT, the time in the eastern United States is five hours earlier, or 6 A.M. The 48 conterminous U.S. states and the Canadian provinces share four standard time zones, known as Eastern, Central, Mountain, and Pacific. Most of Alaska is in the Alaska Time Zone, which is nine hours earlier than GMT. Canada has an Atlantic Time Zone, which is four hours earlier than GMT, and a Newfoundland Time Zone, which is three and a half hours earlier than GMT.

In Figure 1-19, you can see that the **International Date Line** mostly follows 180° longitude, although it deviates in several places to avoid dividing land areas. When travelers cross the International Date Line heading east (toward North America), the calendar moves back one day. When travelers cross it going west toward Asia, the calendar moves ahead one day.

Today's world time-zone map does not consist of parallel vertical stripes, but rather it bends for the convenience of individual countries to reduce the number of situations in which a given region is divided among multiple time zones. In many cases it can be much more convenient to be in one time zone than another. In the United States, for example, the Eastern Time Zone is

much wider from east to west than longitude alone would dictate: more than 20° instead of the normal 15°. This is because most people in the eastern United States would rather be on the same time as the financial and political centers on the East Coast than an hour different. For instance, most of the state of Indiana is in the Eastern Time Zone. The northwestern part of the state, however, near Chicago, is on Central time because Chicago is on Central time, and Northwest Indiana residents find it more convenient to be on the same time as Chicago.

Communicating Geographic Information: Maps

One of geographers' most characteristic tools is a map. **Maps** are two-dimensional (flat) representations of some portion of Earth's surface. They are necessarily smaller than the area portrayed, so they cannot show all the things present on the actual Earth surface. Maps show only selected information according to the particular purpose of the map. They are often used to show things that may not be visible on the landscape but are nonetheless important—political boundaries, for example. Maps are simplifications of Earth—models— designed to store and display geographic data. (See also Appendix I.)

Maps take a great variety of forms. Some are just simple drawings intended to show relative location or a path from one place to another. Perhaps the most common map with which you are familiar is a road map, showing the locations of towns and cities and the automobile routes that connect them. Airplane pilots use maps that show the elevation of the ground surface; a few things visible from airplanes such as rivers, lakes, urbanized areas, and large highways; areas of the sky in which planes are required to follow certain rules; and the locations of towers and navigational beacons, and landing fields. Photos taken from airplanes are maps, as are images collected from satellites and displayed on your TV screen as part of a weather forecast.

Cartography is both a highly technical and a somewhat artistic pursuit, combining the tools of mathematics and engineering with those of graphic design. Maps should be accurate, portraying matter as it really exists rather than as distorted, improperly located, or mislabeled information. They should be visually easy to use, prominently displaying the material a user needs without clutter from unnecessary information. This is why road maps, for example, usually do not show mountains and hills except in the simplest ways. To do so would add many extra lines to a map that is already filled with lines representing roads.

In converting geographic data from their original form on Earth's surface to a simplified form on a map, many decisions must be made about how this information is to be represented. No matter how we draw a map, we cannot possibly make it show the world exactly as it is in all its detail, nor would we want it to. Scale and projection are two fundamental properties of maps that determine how information is portrayed.

Scale Maps are smaller than the areas they represent. To know how large an area or how great a distance is represented by a given area or distance on a map, we must know its scale. The **scale** of a map is a quantitative statement of the relative sizes of an object on the map and in reality. Map scales are typically expressed in one of three ways: a written statement (1 inch = 1 mile), a representative fraction (1:63,360, meaning that one unit of length on the map corresponds to 63,360 units on the ground), or a graphic scale in which a bar on the map is labeled with ground distances.

Common map scales vary widely. The maps in this textbook showing the entire world, including the oceans, on one page, have a horizontal distance of roughly 25,000 miles (1,584,000,000 inches) on a printed

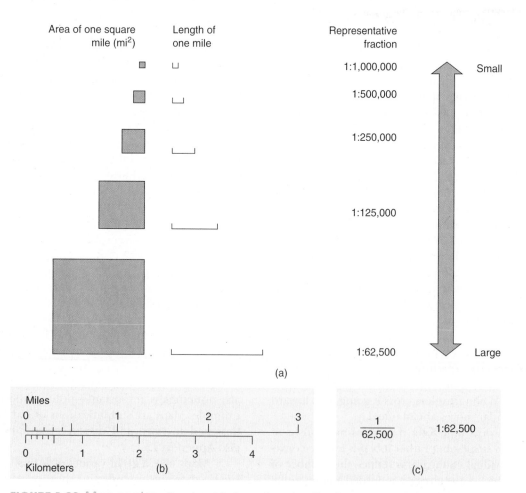

FIGURE 1-20 Map scales. Drawing (a) shows that a small-scale map can portray a large part of Earth's surface, whereas a large-scale map can show only a small part of the surface. Drawing (b) is a graphic scale, and (c) shows fractional scales.

area 6 inches wide, for a scale of 6/1,584,000,000, or about 1:264,000,000. A typical scale used for a road map of a state in the United States might be 1:1,000,000 to 1:2,000,000. A street map of a city might have a scale of 1:10,000, while a surveyor's map of a residential lot might be drawn at 1:100. Maps like the world maps in this book show the land in a very small space, so we call them **small-scale maps.** A map such as a street map that shows a given area in a large space is called a **large-scale map.** A map with a large number in the denominator of a fraction expressing scale (for example, 1/2,000,000) is a small-scale map, whereas a map with a small number in the denominator (for example, 1/100) is a large-scale map (Figure 1-20).

The amount of detailed information that can be shown about a place is a function of map scale. On a small-scale map there is not enough room to show a lot of detail. If we showed all the numbered highways of the United States on a map the size of a page in this book, the map would be virtually all highway and nothing else. On a large-scale map we have room for more detail about a place, but we cannot show the relations between that place and another that is far away. Figure 1-21 shows Seattle, Washington, viewed at a range of map scales.

Projection A second fundamental characteristic of maps that governs the way we show information is **projection,** the transferral of locations on Earth's

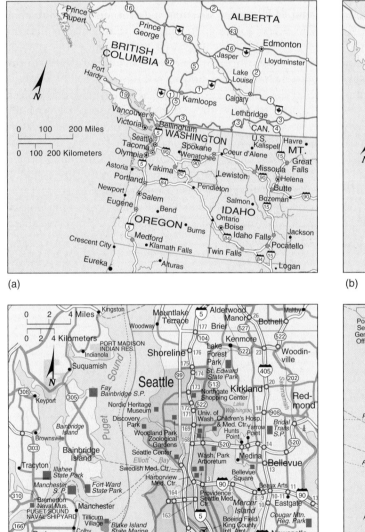

(a)

(b)

(c)

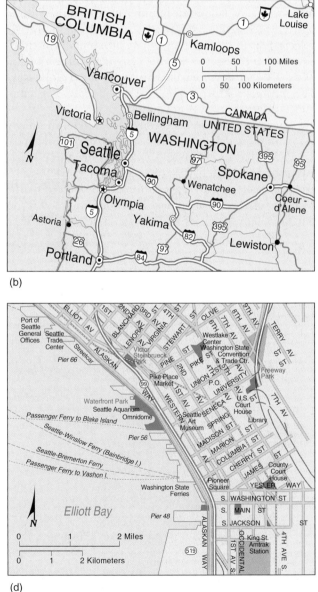

(d)

FIGURE 1-21 Seattle, Washington, at different scales. These four maps show the city of Seattle, Washington, at a variety of scales. Only major landscape features, cities, and highways are shown in Figure 1-21a. When we zoom in on this map, the amount of information displayed changes, with more detail shown at larger scales. In Figure 1-21b the names of small towns are shown, as are secondary roads. In Figure 1-21c some details of the lakefront shoreline emerge, as do the names of a few major streets. The names of all the individual streets are shown in Figure 1-21d.

surface to locations on a flat map. For practical reasons, most maps are flat. Flat maps can be printed on paper and folded, or displayed on a video monitor. However, every cartographer faces a problem when drawing our spherical Earth on a flat piece of paper. Some distortion unavoidably results. Cartographers have invented hundreds of projections, but none is completely free of distortion.

So how do we make a flat model of a spherical Earth? One way might be to draw a map on a spherical rubber balloon and then cut the balloon and stretch it in such a way that it could be glued on a page. Or we could draw our map on a transparent globe, place a light bulb inside it, and project the map onto a screen. This is, in effect, what is done, except that the conversion of the geometry of the sphere to the geometry of the screen is achieved mathematically rather than optically. Regardless of how we make a flat map, the resulting features on the map are different from their equivalents on Earth's surface. The map cannot be an exact scale model of Earth in the same way a globe can. The process of projecting the globe onto the map inevitably introduces distortion.

Several kinds of distortion occur when an image of a round Earth is fit to a flat surface, and as much as we would like to eliminate this distortion, there is no way to do so. When we are mapping relatively small areas of Earth's surface, the difference between a plane and the surface of a sphere is insignificant for most purposes. For large areas, however, and especially for world maps, the distortions are severe. All projections must distort either size (and distance), shape (and orientation), or both. Appendix I describes the methods for making different projections, but here we will examine briefly the advantages and disadvantages of some common methods.

A distortion of size and distance means that the scale in one part of a map is different from that in another part—objects in one area appear larger or smaller than they actually are in relation to objects in another part of the map. It is possible to distort size but preserve the shapes of objects, effectively blowing them up or shrinking them. Maps that distort size but preserve shape are called **conformal maps.**

A distortion of shape and orientation means that in moving from a round surface to a flat one, the shapes of objects are stretched more in some parts than in others. A change in shape inevitably means that orientation changes too, such as causing a line that on Earth runs northeast-southwest to instead appear to run east northeast–west southwest. It is possible, however, to make such changes while preserving the size of objects relative to each other. Maps that preserve size but distort shape are called **equal-area maps,** or equivalent projections.

One of the most common conformal map projections in use is the Mercator projection (Figure 1-22a). It is named for its Flemish inventor, Gerardus Mercator, who developed it in 1569 for the purpose of navigation. It serves that purpose well, because it preserves shape

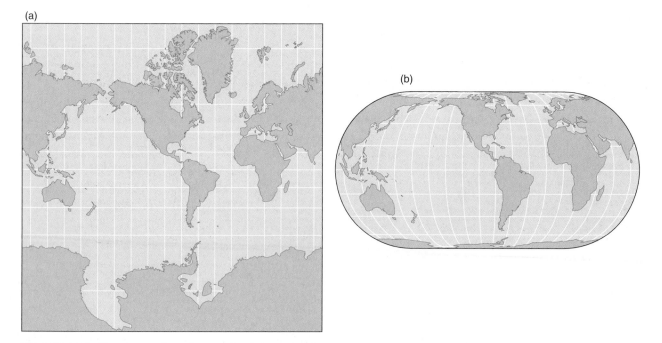

(a)

(b)

FIGURE 1-22 Conformal and equal-area projections. Flat maps necessarily distort our view of Earth. (a) The Mercator projection, which is a conformal projection, displays the correct shapes but exaggerates sizes in high-latitude areas. (b) An equal-area projection can portray accurate sizes, but shapes are distorted, especially near the edges of the map.

and, thus, orientation. On the Mercator projection, lines of latitude and longitude form a perfect rectangular grid. North is always the same direction on the map, and a straight line can be traced by following a constant compass bearing. The rectangular shape of the Mercator map adds another attraction. It fits conveniently on a printed page or on a wall. This, more than its utility in navigation, contributes to the continued widespread use of this projection.

The problem with the Mercator projection is its distortion of size. The distance around the globe at latitude 60° N is half the distance around the globe following the equator, and yet these two distances are the same on a Mercator map. If north–south distances are exaggerated in the same way, as they must be to preserve shape, then the area of a region at the latitude of the Arctic Circle is shown four times its actual size relative to a region at the equator. The distortion increases poleward. This is why on a Mercator map Greenland (which is centered a little north of 70° N) appears to be similar in size to South America (which straddles the equator), when in reality South America is nine times larger. Similarly, the Mercator projection makes North America, Europe, and northern Asia look much larger than they are relative to South America, Africa, and southern Asia. Some people object to the Mercator projection and other similar projections because they exaggerate the size and thus the suggested importance of the wealthier parts of the world relative to the poorer parts.

Size distortions are eliminated in equal-area maps, of which Figure 1-22b is an example. Equal-area maps of the world are much better than Mercator maps for most purposes, because they communicate more accurately the sizes of countries and continents; when comparing one country with another, size is usually more important than shape. Shape distortions become important, however, because they alter distances. Also, because the polar regions must be drawn much smaller than equatorial regions, there is usually considerable distortion of direction (especially north–south) near the edges of the map.

In addition to shape and size distortions, all flat maps must divide Earth's surface at some point. The surface of a globe is continuous—we can go all the way around without leaving the surface—but a flat map must have edges. If we follow the convention of placing north at the top of the map, then the usual location to split the Earth's image is in the middle of the Pacific Ocean, which causes the least disruptions of land areas. This is also convenient because the prime meridian can be placed at the center of the map with relatively little separation of land areas. Of course if you live in, say, New Zealand, you might think this layout makes it look as though islands in the central Pacific Ocean are very far away and that one would need to travel across Africa and South America to reach them. A New Zealander might prefer a map centered on 180° longitude. Maps produced in the United States, for example, often have their edges in central Asia, placing North America at the center.

Cartograms

Standard maps try to depict Earth's features in their correct locational and size relationships. Sometimes, however, it may be more important to convey a visual impression of the magnitude of something than to convey the exact spatial locations. To do this, cartographers design special images called **cartograms.** All cartograms replace physical distance or size with some other measure. The two main types of cartograms are area cartograms and linear cartograms. On an area cartogram, a region's area is drawn relative to some value other than its land surface area. In Figure 5-2 the value used is population. The greater a country's population, the bigger the country is drawn on the image. When a cartographer makes a conversion from physical space to something else, he or she tries to retain as many spatial attributes of the conventional map as possible. If recognizable shape, proximity, and continuity are preserved, it is easier for the viewer to compare the cartogram with a standard map.

Each of these attributes of space can be retained, however, only by distorting one or more of the others.

Linear cartograms draw our attention to distance. We noted that distance is sometimes measured in time or in cost, and these measures are not the same as actual mileage. A transit map, for example, is a time-distance map. The subway map of Manhattan distorts the island's true shape in order to clarify the subway lines and stops. Subway riders think in terms of stops, not distances. Therefore, on many transit maps, the distance between stops is shown as a uniform distance. Cartograms may look strange to us because the scale depends on something other than the physical units we expect. Cartograms are not, however, lies or tricks. Their purpose is to portray some other important aspect of reality. Area cartograms of such things as countries' populations (see Figure 5-2) or economic output (see Figure 12-5) visually convey the relative population or wealth of different countries better than standard land area maps do.

Focus ON

The Internet

The Internet began in 1969 as a U.S. Defense Department project to allow widely scattered scientists and engineers working on military contracts to share computers and other resources. Data are broken up into packets to be transmitted. Packets can take different routes to their destination: There is no central path, but a spiderweb of links, so if one route is destroyed or blocked, a packet simply takes another route. Scientists devised a way of sending messages, called electronic mail, or e-mail, and users invented ways for people to participate in electronic open discussions and also created document libraries accessible to all.

Meanwhile, personal computers made computer power affordable, and modems allowed them to hook up via telephone lines to commercial on-line services and bulletin boards—electronic discussion groups and software libraries—often set up by enthusiasts. By 1990 the National Science Foundation assumed responsibility for the Internet, first to connect its own supercomputer centers and later to enable links among government and academic researchers. Soon businesses were allowed to join.

The World Wide Web was introduced in 1993. It offers multimedia capability and hypertext, which is the ability to link documents to each other by way of hyperlinks. In 1995 the National Science Foundation lifted all remaining curbs on commercial use.

There are several organizations that watch over the Internet. The World Wide Web Consortium (W3C) sets standards. It was founded in 1994 by Tim Berners-Lee, the inventor of the World Wide Web, and has just over 400 members. The Internet Engineering Task Force (IETF) develops agreed-upon technical standards for the Internet. The nonprofit, privately funded corporation, the Internet Corporation for Assigned Names and Numbers (ICANN), oversees the system of domain names. These bodies are all largely self-created and self-governing. They are open in membership and in argument, and they are consensus-based in their decision making. As Web use grows, however, governance may change. Many countries have argued that the Internet should be managed internationally, and some are moving to subvert the present system of governance. For example, on March 1, 2006, China created three new internet address suffixes in the Chinese language, as national variants to .cn, .com, and .net. Chinese users now can type Chinese characters for website and e-mail addresses, thus liberating them from the Roman alphabet upon which the current addressing system is based. The new Chinese addresses still function through the ICANN-sanctioned system, and the Chinese continue to use numerical Internet protocol addresses assigned by ICANN, but the unilateral creation of new addresses threatens the Internet's basic principle that any machine should be able to communicate with any other. Furthermore, in March, 2006, ICANN, under pressure from the U.S. Commerce Department and U.S. conservative lobbies, refused to establish a dot-xxx domain for sex-related materials. Officials of the European Union and other countries protested this example of U.S. prejudices dictating Internet policies.

Today the Web is a massive multimedia encyclopedia, unedited (except by the individual contributors), and indexed only by privately created search directories (called *search engines*) and catalogs. The Web even has an on-line encyclopedia, *Wikipedia*, edited only by its contributors and readers. Search engines work by periodically visiting every address on the Internet and scanning the titles and usually a portion of the content of any Web pages that may be present. The content of the Internet ranges from the glossy advertising and

---

The choice of projection is fundamental to the way we display information and the way people who use maps receive that information. On the one hand, projections can be as dispassionate as a mathematical equation that describes the alteration of a spherical surface into a planar one. Because maps communicate information, however, the form of the map inevitably influences how and what the map says about a place. Those who use such maps therefore often have very strong opinions about what kinds of distortions are introduced in the projection process.

Insets and other conventions   Other conventions of map projections or printing can lead to misunderstanding. For example, to save space, many maps of the United States tuck a separate map of Alaska into the corner below the U.S. Southwest. This is called an *inset* map. This common practice leads many U.S. schoolchildren to think of Alaska as an island off the coast of California. If the mapmaker also prints the inset map at a smaller scale than that of the other states, as used to be common, the confusion about Alaska's location is aggravated by confusion about its size. In fact, Alaska is

information pages of multinational corporations, to pages of information about governments, to thousands of home pages created by people who have assembled some unique information or just have something they want to share. Individual on-line diaries, called Web logs, or **blogs**, first appeared in 2002 and numbered over 30 million by 2006. Thus, many Internet users are not just information consumers; they are information producers as well. At very low cost, organizations and individuals anywhere can speak out to the world. The political ramifications of this will be noted in Chapter 11. By 2000 an estimated 500 million people were using the Web.

The Internet was originally free and unfettered, immune to regulation and control by governments. Attempts by oppressive regimes to block information were futile, and it was thought that cyberspace transcended international borders. Freedom of the Internet has, however, been steadily infringed. One battle was a fight over illegal Internet trade in copyrighted material such as pop music. Legislation debated in the U.S. House of Representatives would even have allowed recording corporations to hack into home computers that trade music and video files. Threats to the Internet by terrorist organizations have prompted increased electronic surveillance by law enforcement and intelligence agencies. New legislation allows governments to collect and share e-mail traffic and Web-surfing patterns. The Internet has even been militarized. The armed forces of many countries have devoted time, money, and effort to offensive cyber-warfare capabilities, including attacks, viruses, and worms that would allow attackers to destroy the systems of their foes. Access to government information on the Web has been restricted, and postings have been limited. Formerly public documents have been removed.

The technologies for restricting access to sites and even for identifying the visitors to individual sites are increasingly successful. For example, *geolocation technology* enables websites to determine the physical locations of users by linking Internet protocol (IP) addresses of users' computers to specific countries, cities, or even postal codes. Websites can then modify themselves or be modified to address each visitor individually. Existing demographic databases describing, for example, what type of people live in an area, can target advertising. Positive aspects of geolocation are that it can help determine the appropriate language in which to present a multilingual website. Electronic commerce vendors, auction houses, pharmaceutical, and financial services industries can prevent the sale of goods that are prohibited or illegal in certain areas. On-line casinos can prohibit gamblers where gambling is illegal, and rights management policies for video or music broadcasts can be enforced.

Geolocation technology can also be used to find the Internet access point nearest to a particular location. The necessity of doing this arises from the growing popularity of 802.11b or Wi-Fi technology, which provides wireless Internet access to suitably equipped laptops within 100 meters or so of a small base station. Today jurisdictions around the world—cities, provinces or states, and even whole countries—are racing to provide complete coverage with Wi-Fi accessibility. Such Internet facility could revolutionize public safety, government, and business. An emergency room doctor could monitor the vital signs of a heart attack victim while the patient is still in the ambulance. Firefighters could access databases with electrical, sewerage, and water schematics, and police could examine criminal databanks and remotely operate mobile security cameras.

nearly one-fifth the size of the lower 48 states combined. In January 1990, in order to counter this misrepresentation and the resulting confusion in some students' minds about exactly where Alaska is, the Alaska state legislature passed a resolution "respectfully requesting that all major United States magazines, newspapers, textbook publishers, and map publishers . . . place Alaska in its correct geographical position . . . in the northwest corner."

Several world maps in this book cut out great expanses of the world's oceans so that we can map human activities on the land in greater detail. This shifts the relative positions of the continents and tucks Australia and New Zealand south of Asia, to the west of where they really are. Clear divisions on the maps mark the lines along which the map has been cut, and each portion has markings of latitude and longitude for proper positioning. The first of these maps is Figure 2-30 in Chapter 2. If you want to recheck the continents' relative positions shown more accurately, you can always refer to a globe.

It is conventional to put north at the top of a map, but top and bottom are meaningless in space. In fact, any conventional view conceals a great deal of information simply because it is conventional; we usually do not notice or consider something we have seen often before.

The inability to understand mapmakers' conventions troubled Mark Twain's fictional character Huckleberry Finn when he and Tom Sawyer were blown east in a balloon. Huck insisted that they were not over Indiana yet because the color of the land below had not changed from the green of Illinois to the pink of Indiana, as shown on his school atlas. "What's a map for?" he demanded. "Ain't it to learn you facts?" Our answer is yes, but a map works for you only if you understand its projection and the explanation of its colors and symbols as defined in its key, usually at the bottom of the map.

Geographic Information Technology

Computers have transformed many aspects of social, intellectual, and business life in the last few decades; it is only logical that their impact has been highly significant in geography. Just a few years ago, maps were being drawn with ink pens and rulers, but now they are composed on computers and printed by machine. Location and land-use information used to be collected by optical surveys and visual inspection of aerial photographs, but now it is collected using global positioning systems (GPSs) and digital satellite imagery as well as by conventional technologies. Countless hours of work with paper, pencil, and calculator were necessary to perform routine statistical analyses, but now such work is performed in seconds on a desktop computer. Geographic information systems (GIS) integrate all of these tasks by using computers to assemble automatically and rapidly collected digital data, analyze such data together with other information that may be on hand, and generate complex and specialized maps for many purposes. GIS technology has revolutionized geographic research and expanded the applications of that work to everyday life.

Automated cartography Anyone who has worked in traditional pen-and-ink technical drawing can attest that such work is tedious and, for most, not particularly exciting. A map is a highly complex technical drawing, and map production using manual techniques is very expensive. When maps had to be updated because of changes in the landscape—new roads, expanded urban areas, changed boundaries, and place names—the costs were substantial. For this reason, when *computer-assisted drawing (CAD)* technology emerged, it was quickly applied to mapmaking to develop sophisticated, specialized digital cartography systems. These systems make it possible to draw new features on maps and edit them on a video monitor before any paper map is created, in much the same way that word

processors make it possible for you to compose a term paper on screen and edit it, correcting typing mistakes, and making the needed refinements before you print a paper document. Given the complexity of mapmaking and the specialized skills it requires, this is obviously an important improvement for the cartographer.

Remote sensing satellites Another important dimension of GIS is its ability to make use of digital satellite imagery. The acquisition of data about Earth's surface from a satellite orbiting the planet or from high-flying aircraft is known as **remote sensing.** Geographic applications of remote sensing include mapping of vegetation and other surface cover; gathering data for large unpopulated areas, such as measuring the extent of the winter ice cover on oceans; and monitoring changes such as weather patterns and deforestation.

Landsat satellites, the first of which was launched by the United States in 1972, were the first satellites intended primarily for mapping characteristics of Earth's surface. Passive sensors in the Landsat satellites measure the amount of radiation emanating from Earth's surface in particular wavelengths. Sensors primarily measure the amount of radiation of the various colors of visible light, although some measure infrared (heat) energy. In a few cases, active sensors like radar send radiation to Earth and measure the radiation that is reflected back to the satellite. The most recent Landsat satellite to be launched was Landsat 7, in April 1999.

Remote sensing satellites scan Earth's surface much as a television camera scans an image in the thin lines you can see on a television screen. The sensor is moved first across the landscape in a line; it is then moved slightly to scan other lines in succession. At any moment, a sensor is recording the amount of energy from only one place, an area called a picture element, or pixel. A map created by remote sensing is essentially a grid containing many rows of pixels.

The smallest feature on Earth's surface that can be detected by a sensor is determined by the size of the pixel. This is called the resolution of the scanner. Early Landsat sensors had a pixel size of 59 meters by 59 meters (194 feet by 194 feet), compared to 30 meters by 30 meters (98 feet by 98 feet) on later Landsat versions or 10 meters by 10 meters (33 feet by 33 feet) and even 1.5 meters by 1.5 meters on the IKONOS satellite. Even smaller pixel sizes are available on military and intelligence satellites; these data are not, however, available to civilians. The importance of a smaller pixel size is that smaller objects can be detected.

Weather satellites take a broader view than Landsat satellites. They have very large pixels, covering several kilometers on a side, which enables them to rapidly map a large area such as a continent. Weather forecasters need data about a large area very quickly, because weather systems change rapidly (Figure 1-23).

Dozens of satellites orbit Earth and send back information about its atmosphere, oceans, and land

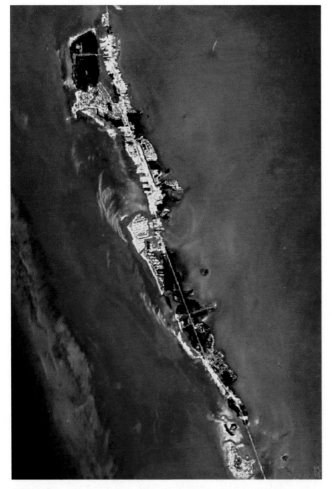

FIGURE 1-23 A satellite image of Marathon Key, Florida, showing both the above-water parts of the island and the fringing coral reefs. Images such as this are used worldwide to monitor changes in coral reefs. (Image by Scientific Visualization Studio, NASA Goddard Space Flight Center; data courtesy Landsat Project)

satellite imagery include such diverse applications as estimating nutrient levels in farmland soils, quantifying loss of wetlands in coastal areas, forecasting stream flow based on snow cover, and documenting urban growth (Figure 1-25).

Despite technological advances, the United States may strengthen agencies that specialize in traditional techniques of gathering information about foreign affairs. The United States might hire more spies, especially persons trained in foreign languages; change the hiring rules and employ double agents and others previously rejected as "undesirable"; improve the professionalism of national information services (when Robert Rubin became Secretary of the Treasury in the Clinton administration, he complained that he had been able to get better information about world affairs when he was a private banker with Goldman Sachs Corporation); and change priorities from high tech to hiring more information gatherers on the ground. For example, 90 percent of information gathered by U.S. satellites has never been analyzed; representatives on the ground, by contrast, are able to gather intelligence through overt activities such as simply reading local newspapers, listening to the local radio, speaking with local officials and people on the street, and generally moving about with attentive eyes and ears. The numbers of people performing these tasks in foreign countries—as well as the number of analysts of foreign affairs stationed in the United States—had been trimmed from the American federal budget through the 1990s, but they are being increased today. New opportunities are opening for individuals trained in geographical analysis.

Global positioning systems   Determining the location of a place on Earth's surface has become much easier with the development of **global positioning systems (GPSs).** A GPS is a navigational tool originally developed by the U.S. government for military use, but is now available for civilian purposes worldwide. The system consists of a fleet of satellites that orbit Earth, broadcasting digital codes. A portable receiver "listens" to these signals. By measuring very small differences in arrival times of the signals, it determines its own location. Inexpensive receivers are capable of determining locations to within a few meters, whereas more elaborate systems have accuracies of less than one meter (Figure 1-26). The limitations of the system are relatively few—an antenna must be outside and unobstructed by trees and the satellite signals may be withdrawn from civilian use at the discretion of the government, so that during wartime an enemy cannot use them. On May 1, 2000, however, President Clinton signed an Executive Order allowing civilian GPS devices to have the same degree of accuracy as military devices.

GPS is revolutionizing many business operations, especially those related to transportation and mapping. Surveyors, particularly those working in remote or rural

surface, so the amount of information we have available is enormous. Some early GISs were developed to make use of satellite data, and, because of the growth of powerful desktop microcomputers in the 1980s and 1990s, the use and capabilities of GISs increased exponentially.

Satellites and planes equipped with specialized sensing devices can even read history as it is recorded below the surface. Information available includes changes in plant growth or drainage patterns brought about by human activity and densely compacted soils and erosion associated with agriculture. Archaeologists have also used this kind of information to find long-lost cities.

Human activities cause new and continuing changes in the landscape, and remote sensing is critical to monitoring and evaluating these impacts. For example, accurate assessments of the extent of logging activities are critical to managing forests and the wildlife they support (Figure 1-24). Other important uses of

July 1984

May 1986

July 1988

September 1995

FIGURE 1-24 A series of Landsat images of part of the Olympic Peninsula in western Washington, showing forest harvest and regrowth. In this false-color image, green tones depict growing vegetation, while reds indicate bare ground. The brightest red patches are recent clear-cuts. Pale reds indicate young vegetation growing after the cut; over time these areas become pale greens and ultimately dark greens in this imagery. The images show that there was intensive timber harvesting between 1984 and 1986, but that by 1995 much of the harvested land was again covered with vegetation as the forest began to grow back. (National Air Survey Center Corporation)

areas, determine their locations by GPS instead of using optical devices that are useful only for distances of a mile or two at best, assuming a clear line of sight. Scientists and land managers use GPS to determine their locations when making environmental measurements. Airplanes and ships now navigate by GPS instead of relying on older, ground-based radio systems.

As GPS systems became more widely used, their price dropped rapidly, which led to a further expansion of applications. Many new automobiles now offer GPS-based mapping systems that continually track the vehicle's location and display a moving digital map on the console. Some of these devices incorporate address-locating and route information, so that the driver

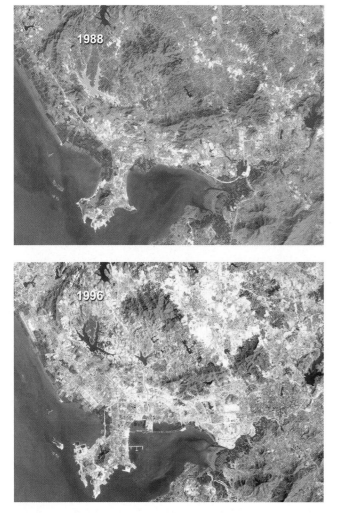

FIGURE 1-25 Two Landsat images of Shenzhen, China, taken in 1988 and 1996, showing dramatic urban development in just eight years. Although China's population growth has slowed, many people are moving from rural areas to urban ones as the industrial economy grows. Remote sensing and GIS are essential technologies for documenting and monitoring rapid changes on Earth's surface such as this. (NASA Goodard Space Flight Center, Scientific Visualization Studio)

can enter an address as a destination and the device will calculate an appropriate route from the present location to the destination, then tell the driver when and where to make turns. Hikers carry GPSs to avoid getting lost, runners carry them to monitor their speed and distance, and boaters carry them to return to their favorite fishing holes or to navigate home in a fog.

GIS: A Type of Database Software

A GIS is a special form of database software in which spatial information is an important part of the database. A simple database consists of lists of items—people, commodities, transactions, etc. Each item in the list is associated with multiple bits of information. In a

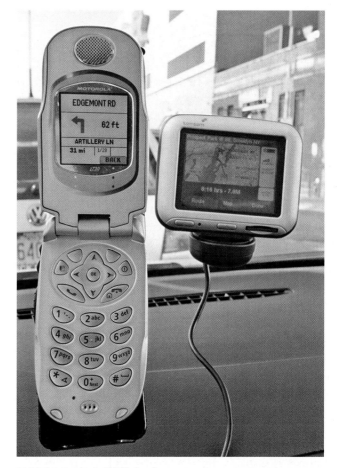

FIGURE 1-26 A GPS device. Two examples of GPS-based navigation systems. On the left is a mobile phone; on the right a windshield-mounted system. Both display maps provide turn-by-turn directions and include databases of business and other points of interest. (Richard Drew/AP Wide World Photos)

database of land parcels maintained by a city clerk's office, for example, one might expect to find information on the owner (name, address, etc.), the value of the parcel, what kinds of structures exist on the parcel, and so forth. In a GIS, one of the attributes of an item in the database is either its spatial characteristics (boundary information, in the case of a land parcel) or something else (such as an address or name of the district in which it is located) that can be used to locate the item in space.

By using such a database, we can create layers of information, each representing a different characteristic, or attribute, of a place. A single layer can be displayed by itself or combined with other layers to show relations among different kinds of information (Figure 1-27). A GIS allows the user to create rapidly many different specialized maps from a single database.

Types of geographic data in a GIS Geographic data are of many types, and each different type has certain characteristics that determine how the information can be analyzed, interpreted, and displayed. One fundamental distinction is between raster and vector data

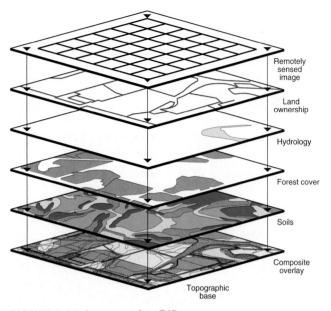

FIGURE 1-27 Layers of a GIS. A geographic information system involves storing information about a location in layers. Each layer represents a different piece of human or environmental information. The layers can be viewed individually or in combination.

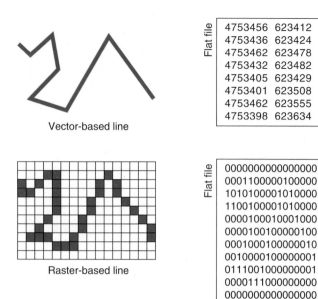

FIGURE 1-28 Raster and vector data types. Raster data is composed of grids of cells that are coded to indicate a characteristic of a given area, while vector data consists of lines defined by X and Y coordinates of points (vertices) along the line. Remotely sensed images and photographs are typically raster images, whereas maps showing regional boundaries (such as political maps) are usually stored as vector data. (After Keith Clarke, *Getting Started with Geographic Information Systems*, 4th ed. Upper Saddle River, NJ: Pearson Education, 2003)

(Figure 1-28). Raster data are arranged in a rectangular grid of cells, which are all the same size. A computer display is a raster device. For example, many computer monitors display information in a grid that is 768 rows high by 1024 columns wide, containing 786,432 grid cells. Each of these cells contains specific information—in this case, the color that is to be displayed on the monitor. The pixels in a remotely sensed image usually contain information about the amount of light or energy sensed as coming from a particular place on Earth's surface.

Vector data are based on points with X and Y coordinates that specify location. Vector data are of three basic types: points, lines, and polygons. A point is defined by a single pair of X and Y coordinates. A straight line can be defined by just two points. More complex lines, such as would represent roads or rivers, are defined by a series of points, each defining a straight-line segment of the line. Regions are represented with polygons, which are simply lines that begin and end at the same point. Only four points are needed to define the border of a rectangular region (Colorado, for example), but many more are needed to define the outline of most geographic regions.

Raster and vector data differ in the kind of information they can portray and the spatial accuracy with which they portray it. In a raster image such as a Landsat image with 30-meter by 30-meter pixels, every place within the pixel is assumed to have the same value for a given characteristic. If the pixel happens to be located in the middle of a lake, this assumption isn't very important—every place within it is water. If, however, the pixel is along the shore of that lake, then only a part

of the pixel is water and a part is land. In a satellite image, the value that will be assigned to that pixel will probably fall somewhere between the appropriate value for water and that for land, thus representing a surface cover type that does not, in fact, exist at that location. A vector line, on the other hand, could be created that would show exactly where the shoreline falls so as to accurately portray this feature. As long as pixels are small in relation to the objects being portrayed, raster formats are very useful for storing and manipulating information about complex maps such as a satellite image of a vegetated landscape, with many variations in the colors of vegetation and shapes of vegetation patches. Vectors, on the other hand, are very efficient for storing information about regions, such as political units, or lines, such as roads. Modern GIS devices are able to analyze and display both raster and vector data (including points, lines, and regions) simultaneously (Figure 1-29).

Acquiring digital geographic information

A few decades ago all maps were in paper (analog) form, but today nearly all mapping is done digitally using GIS. One of the biggest tasks in the first few decades of GIS technology has been to create the digital geographic database from which maps are made. Creation of this digital resource is done primarily in two ways. One of these is converting existing maps from

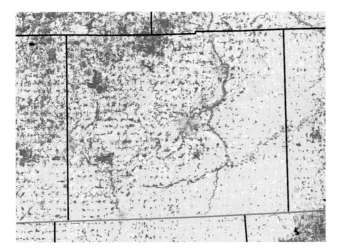

FIGURE 1-29 A land use/land cover map of Lenawee County, Michigan, showing both raster and vector data. The raster data are the U.S. Geological Survey's National Land Cover Dataset, which shows land use/land cover interpreted from 30-meter resolution Landsat imagery. Agricultural land is in yellow, urban land is in red, forest is in green, water is blue, and wetlands are in purple. The county (black) and state (red) boundaries are vector data, from the U.S. National Atlas. (WHR map)

paper to digital form. To a large extent this has been accomplished for the wealthier parts of the world, however digital data are much less available for developing countries. The other way of creating a digital database is through use of remote sensing satellites, which create digital data directly rather than going through a paper form first.

One way of digitizing paper maps involves manual tracing of lines using a specialized table, following the lines with a device somewhat like a computer mouse. As you might imagine, this is very tedious and time-consuming work. Today most conversion of maps is accomplished using scanners coupled with software that is able to "trace" features on the scanned image digitally and convert them to vectors (lines). While this technology has improved dramatically, it still requires considerable expertise as well as expensive hardware and software. This helps explain why creation of a digital database in developing counties is lagging well behind the effort in the United States.

Today a very diverse database is readily available in the United States, much of it freely available on the Internet. For example, the 1:24,000 topographic maps are available in two forms. One of these is simply a scanned image of the map called a DRG, or digital raster graphic. This version can be easily displayed on a computer using common image-viewing software. The other version, called a digital line graph or DLG, consists of most of the lines from those maps—boundaries, streams, contours, roads, etc.—in vector form. These DLGs are the form that is more often used in GIS analyses. Digital versions of older paper maps

are also available at scales of 1:100,000, 1:250,000, and 1:2,000,000.

In addition to these basic map resources, many specialized maps are now available in digital form. For example, the U.S. Geological Survey makes available on-line land use/land cover maps derived from satellite imagery, maps of hydrologic features and watersheds, and digital elevation models that are used to portray the landscape in three dimensions (Figure 1-30). The Fish & Wildlife Service distributes digital versions of its National Wetlands Inventory, and the National Oceanographic and Atmospheric Administration's marine navigational charts are available in digital form. The Census Bureau produces digital maps that show all roads and streets along with the numbering of addresses on them. These data allow you to type an address into a computer and immediately get a map showing the location of the address; such data are also essential for the emergency response systems used by police and other emergency services. Many local governments have put property maps online so that anyone can search these public records via the Internet to determine who owns a particular piece of property.

Application of GIS to spatial analysis As discussed earlier, spatial analysis includes topics such as distribution, density, concentration, pattern, and movement. GIS facilitates quantitative measurement of these geographic properties. Calculation of population density simply involves combining (or overlaying) population data with a map of regions such as counties. Using a GIS, we can determine the total number of

FIGURE 1-30 A three-dimensional perspective view of Mt. Everest. This image was made by "draping" an aerial photograph on a digital elevation model. Similar technology is used in many visualization applications, from television weather forecasts to flight simulation in movies. (National Geographic Image Collection)

people living within the boundary of the region and divide that by the area of the region to determine density. Similarly, a GIS can be used to count the total number of people living within a given distance of some specified location, as part of a measurement of concentration. Changes in the distribution of people, businesses, or forest clearances from one time period to another, as represented in two different layers of a GIS, tell us about movements of people and economic activities over time.

A recent study of the distribution of small ponds illustrates the role of GIS in modern geographic research. An inventory was made of small lakes across the conterminous (48) United States to determine how many lakes there are and their locations. The analysis began with a satellite-based land-cover map produced by the U.S. Geological Survey. Because it was derived from satellite images, it is a raster map, where individual pixels are identified as "row crops," "deciduous forest," "high-density urban," etc. Rivers, lakes, and ocean bays are identified as "water" without further distinction.

The first task was to eliminate from the map all the areas that were clearly not lakes. This was done using map layers that included rivers and the ocean shore. Anything within 1 kilometer of a river or 5 kilometers of the shore was blanked out, so these water areas would not be counted as lakes. Next, lines were drawn around each remaining group of water pixels that touched each other, creating a vector layer in which each lake a single entity. These lakes were then counted—the total is nearly 2.6 million! Figure 1-31 is a map showing their distribution.

The analysis revealed that a very large number of lakes are located in the eastern Great Plains, in areas where natural lakes are rare. In fact, Texas has more

lakes than any other state—about 10 percent of the total. Minnesota, which claims "10,000 lakes" on its license plates, in fact ranks seventh according to this study, behind Florida, Oklahoma, Kansas, Missouri, and Mississippi. The good news for Minnesotans is that they are being modest—the actual number of lakes there is more like 100,000. While the study does not distinguish between natural and artificial lakes, the great concentrations of lakes in areas where natural lakes are rare confirms that the vast majority of these features are of human origin. Most are small ponds built on farms in the latter half of the twentieth century, many of them as part of government programs. They are most numerous in the eastern Great Plains because this is an area that has enough rainfall to fill the ponds in the spring and early summer, yet also has a dry season in which this stored water is particularly important.

In addition to the basic task of counting lakes, this study demonstrates the value of GIS for monitoring environmental conditions in a fast-changing world. Each of these farm ponds is so small that, by themselves, they seem insignificant. Taken together, however, they constitute an enormous modification to the environment. In some areas of the Great Plains, for example, lakes number more than 3 per square kilometer (one for every 80 acres or so). They store water for cattle, but much lake water that would otherwise soak into the ground is lost to evaporation, principally during the summer. In some areas the amount of water that is lost to evaporation from farm ponds is enough to reduce the average annual flow in streams by 10 percent—an important amount in a water-stressed area.

This example illustrates many of the approaches of geography we have discussed in this chapter. It demonstrates that answering "where" questions remains an important part of geography. Earth's surface is constantly changing, so even though virtually all of Earth's land surface has been explored and mapped to some level, the maps must continually be redrawn. It illustrates important locational concepts such as distribution and concentration. Spatial interactions are seen in ponds' effects on downstream water availability, the transport of sediment and other substances by streams, and even the migration and distribution of wildlife. The lakes also exemplify the myriad ways in which human activity—in this case especially culture, food production, and government policy—and the physical environment interact.

Finally, it is important to note that this study was carried out using data that are freely available on the Internet, and the data were analyzed on ordinary desktop computers. Anyone with the interest, a modest amount of training in GIS, and the appropriate software could have done this study. It would not, however, have been possible with the technology existing as recently as the mid-1990s. Some of the data were not available or available only to a select few, much expensive computer

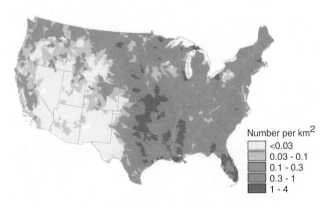

Number per km^2
- <0.03
- 0.03 - 0.1
- 0.1 - 0.3
- 0.3 - 1
- 1 - 4

FIGURE 1-31 The distribution of small lakes in the conterminous United States. The greatest number of lakes are found in the eastern Great Plains, where the vast majority are farm ponds built to store water for farming operations. Large numbers of lakes are also found in areas where they occur naturally, such as the Great Lakes States and Florida. Even in those places many of these lakes were built by humans. (WHR map)

time would have to have been devoted to the job, and some of the software needed had not yet been written. Recent advances in GIS technology have opened up many such opportunities for geographers to explore the changing face of Earth's surface.

Integration of information technologies

In the 1960s the 911 call system was developed in the United States to ease reporting of emergencies to police and fire departments and to speed their response to such reports. Similar systems were developed elsewhere. Since the 1980s these systems have made use of digital telephone-call routing technology that transmits the telephone number of the caller to the receiving telephone (caller ID). Prior to the advent of mobile telephones, telephone numbers were all associated with physical addresses, and it was a relatively simple matter to have a computer linked to the phone system so it could read the number of the incoming phone call, look it up in a list of phone numbers and addresses, and display the address on a dispatcher's computer monitor. The dispatcher could then relay the location by radio to the emergency responder without the caller having to say where he or she was. The system has evolved in many ways since then, including the display of a map on the computer monitor showing the caller's location, and the automatic transmission of the information to a computer in a police car. The growth of cellular telephone use is a problem because the phone is not associated with a fixed location, but the call is routed through a particular antenna so the approximate location can be determined automatically. Integration of GPS with mobile phones can solve this problem.

As computer technology has expanded throughout our society, information increasingly has been stored in digital rather than paper form. We still produce paper documents and communicate with each other using paper, but even the information that now is printed on paper is first entered in a computer, and remains stored there long after the paper form has gone to the landfill. How many documents that you have produced on your personal computer are still there, occupying storage space? Financial records, real estate ownership, newspaper articles, credit card transactions, medical records, automobile maintenance histories, product shipments—all are recorded in digital databases.

Sophisticated software systems called relational databases have been developed to manage this information explosion. Different categories of information may be organized in different ways, and each type of information has particular requirements regarding the way it is stored in a computer in order to maximize the efficiency of data storage and retrieval. Relational database software makes it possible to link different databases to each other. For example, an instructor in a geography course keeps records of the grades in a student's class, and these grades are submitted as a list to a central computer at the end of the course. The list has each student's name, probably a unique identification number, and the grade. The information on that list and others submitted by other instructors needs to be reorganized so that, for example, a student can get a report of all the grades earned in all the classes she/he has taken. Relational database software does that reorganizing.

Although many databases are held confidential so that information they contain cannot be easily linked with other databases, the integration of data across different aspects of society is increasing rapidly, and this integration allows powerful new tools for information users. For example, many retail businesses, such as gas stations and supermarkets, offer special discount cards for repeat customers. When a customer uses such a card the details of the transaction (time, location, what was purchased, etc.) are recorded in a database. This information is also associated with the name and address of the customer, often along with other demographic information such as age, gender, and, potentially, many other aspects of the customer's life. Credit card transaction records do this also.

A street address is a spatial location system. It identifies anything associated with that address with a specific place on Earth's surface. The combination of detailed databases with spatial information can be a powerful tool for many applications. Geographic information systems thus make it possible to map virtually anything about which information is available. Suppose, for example, a retailer that sells outdoor recreational equipment wants to open up a new store. Where should that store be located so as to maximize sales? The answer to that question could come from a GIS analysis of the distribution of the sorts of people who would buy recreational equipment. Perhaps these people tend to be young, educated, and relatively affluent. Where do these people live? Some of this information may come from Census Bureau data. Additional information may come from databases maintained by private corporations and sold expressly for this purpose. This could be combined with road maps and traffic information to give an indication of the driving time from a customer's residence to a potential store location. These sorts of analyses have become routine, and most large retail corporations and many smaller consulting firms employ geographers with GIS training to analyze data in support of business decisions.

Geographic information systems have become prominent on the Internet as well. Wherever map data have become available in digital form, these data have been integrated in on-line mapping systems that allow the user to view a map of any particular place, at any particular scale. Perhaps the best known of these systems today is the one produced by Google, which is available in both a basic on-line version and various versions with greater capabilities using both locally in-stalled software and on-line databases. These systems in-tegrate polygon, line, and point data for boundaries, roads, and cities with satellite or aerial photographic images in raster form. The software displays the data in a scale-dependent fashion, increasing the amount of detail that is shown as scale increases.

The availability of digital maps, remotely sensed images, and other data that can be integrated with GIS is expanding rapidly around the world. But large dispari-ties in information availability remain between rich and poor parts of the world, and in some countries (most no-tably China) controls on access to information reduce the utility of these technologies to ordinary citizens.

CONCLUSION: CRITICAL ISSUES FOR THE FUTURE

The twenty-first century will present new and difficult challenges to citizens, businesses, and governments. Critical issues include:

▸ Global-scale environmental change, including de-forestation, soil degradation, increased concentra-tions of atmospheric carbon dioxide, global warming, and decreased atmospheric ozone levels.

▸ Rates of population growth in some regions that will strain cultural systems, deplete scarce re-sources, and produce environmental changes.

▸ Increasing disparities in wealth and the threat of conflict between rich and poor countries and within individual countries.

▸ Systems of food production and distribution that do not meet the nutritional needs of many people in developing countries.

Confronting problems of this magnitude will re-quire a great deal of creativity and hard work from all

segments of society, including geographers. Geography offers several unique perspectives for understanding contemporary issues.

First, geography emphasizes interdisciplinary approaches to contemporary issues. Geographers are trained partly as natural scientists and partly as social scientists, so they have a wide breadth of under-standing that makes geography a unique discipline. Geographers work well in interdisciplinary teams analyzing major problems that involve combinations of human and physical factors. They can communicate with specialists in a variety of natural and social sciences, so geographers are often asked to lead group efforts.

Second, geographers have a global perspective. A century ago a famine or war in a distant part of the world had little consequence outside the immediate re-gion, but today local events have worldwide impact. New information technology and transportation sys-tems have created global markets for many commodi-ties. What happens in one place has ramifications in other areas and often throughout the world. Concern over an endangered owl in the U.S. Pacific Northwest causes a rise in lumber prices not only throughout the United States but in Canada and in East Asia as well. The 1986 accident at Chernobyl, a nuclear power plant in Ukraine (then part of the Soviet Union), spread ra-dioactive fallout around the world, heightened fears of nuclear power, and thus contributed to curtailment of nuclear power development worldwide.

Third, geography has a strong component of ap-plied research. Since the earliest days of the discipline, geographers have worked on problems of practical sig-nificance, such as finding new routes for commerce and evaluating the natural resources of sparsely inhabited lands. Geographers emphasize studying everyday phe-nomena about places. They are aware of conditions in the world around them and are accustomed to applying their knowledge to them. GIS has become central to this endeavor, and is likely to remain so.

Chapter Review

SUMMARY

Geography is the study of the interaction of all physical and human phenomena at individual places and of how inter-actions among places form patterns and organize space. Phys-ical geography studies the characteristics of the physical environment. Human geography studies human groups and activities, such as languages, industries, and cities.

Geographers divide Earth into regions, which are areas defined by one or more distinctive characteristic or feature, such as climate, soil type, language, or economic activity. Area

analysis is an approach that looks at these features of Earth's surface in a regional context.

Geographers study the distribution of objects across Earth's surface and processes by which human and environ-mental phenomena move from one place to another. Three properties of spatial distribution include density, concentra-tion, and geometric pattern. Movements of matter, people, and information result in interactions among places along networks through diffusion processes.

The place where a culture originates is called the culture hearth. Various aspects of cultures may spread out and be adopted by other peoples in a process called cultural diffusion. The impact or frequency of any cultural attribute may diminish away from its hearth area; this phenomenon is called distance decay. Some aspects of culture may also develop variations as people who carry that culture wander outward and away from one another. Tracing a course of diffusion may teach us a great deal about how peoples and cultures interact. Diffusion may be by relocation, or it may be contiguous or hierarchical. Barriers to cultural diffusion may be topographic, political, or even cultural. Diffusion does not explain the distribution of all cultural phenomena. Sometimes the same phenomenon occurs spontaneously and independently at two or more places.

Earth's physical environment results from a process of interaction among four systems: the atmosphere, hydrosphere, lithosphere, and biosphere. Cultural systems include customary beliefs, social forms, and material traits. Some geographers regard the distribution of human activities as the result of underlying regularities in cultural systems, while others emphasize the diversity and uniqueness of human behavior in explaining cultural systems. The physical environment influences human actions, although humans can adjust and choose a course of action from many alternatives.

Geographers utilize maps to depict the location of places and to interpret underlying patterns. Remote sensing from satellites and GIS devices are two recently developed tools that help geographers more fully analyze patterns on Earth's surface. GISs allow rapid integration of many different types of geographic data as well as automated map production. Rapid expansion of information technology and the integration of diverse databases with GIS technology are leading to new applications of GPS and GIS in many aspects of our lives, as we will see through this book.

KEY TERMS

atmosphere p. 19
biosphere p. 19
blog p. 29
cartogram p. 27
cartography p. 2
concentration p. 15
conformal map p. 26
contiguous diffusion p. 16
cultural geography p. 2
cultural landscape p. 21
culture p. 21
density p. 14
diffusion p. 16
distance p. 15
distance decay p. 16
distribution p. 14
ecology p. 21
ecosystem p. 21
equal-area map p. 26
equator p. 21
formal region (or uniform or homogeneous region) p. 10

friction of distance p. 16
functional region p. 10
geographic information system (GIS) p. 2
geography p. 2
global positioning system (GPS) p. 31
globalization p. 9
Greenwich Mean Time (GMT) p. 23
hearth p. 16
hierarchical diffusion p. 16
human geography p. 2
hydrosphere p. 19
International Date Line p. 23
large-scale map p. 25
latitude p. 21
lithosphere p. 19
longitude p. 22
map p. 23
meridian p. 22

model p. 16
natural landscape p. 21
parallel p. 22
pattern p. 15
physical geography p. 2
prime meridian p. 22
projection p. 25
region p. 9
regional geography p. 5
relative location p. 7
relocation diffusion p. 16
remote sensing p. 30
scale p. 24
site p. 7
situation p. 7
small-scale map p. 25
system p. 18
systematic geography p. 5
topical geography p. 5
topography p. 15
vernacular region p. 11

QUESTIONS FOR REVIEW AND DISCUSSION

1. What is geography? Describe differences among physical, human, and regional geography.
2. What contributions did ancient and medieval geographers make to the development of geographic thought?
3. What are the three ways to indicate scale on a map?
4. What are the four types of distortions that can result from map projections?
5. How do remote sensing and GIS devices contribute to geographic study?
6. What is the difference between a meridian (or longitude) and a parallel (or latitude)?
7. What is the difference between formal regions, functional regions, and vernacular regions? Give examples of each of the three types of regions.
8. What is the difference between density and concentration?
9. What do geographers mean by the concept of spatial interaction?
10. What is the difference between relocation diffusion and contiguous diffusion?
11. What are Earth's four physical systems?

THINKING GEOGRAPHICALLY

1. Cartography is not simply a technical exercise in penmanship and coloring, nor are decisions confined to scale and projection. Mapping is a politically sensitive undertaking. If you were a resident of New Zealand, what changes might you suggest to the world map projections used in this book? Look at how maps in this book distinguish the borders of India with China and Pakistan and of Israel with its neighbors. Can you identify other logical ways to draw these boundaries?

2. Imagine that a transportation device (perhaps the one in *Star Trek*) would enable all humans to travel instantaneously to any location on Earth. What impact might that invention have on spatial interaction? What if such travel were expensive, so only the rich could afford it?

3. When earthquakes, hurricanes, or other so-called "natural" disasters strike, humans tend to blame nature and see themselves as innocent victims of a harsh and cruel natural world. To what extent do environmental hazards stem from unpredictable nature, and to what extent do they originate from human activity? Should victims blame nature, other people, or themselves for the disaster? Why?

4. Geographic approaches, such as area analysis, spatial analysis, and systems analysis, are supposed to help explain contemporary issues. Find a story in your newspaper that can be explained through application of geographic concepts.

SUGGESTIONS FOR FURTHER LEARNING

Abler, Ronald F., Melvin G. Marcus, and Judy M. Olson, eds. *Geography's Inner Worlds.* New Brunswick, NJ: Rutgers University Press, 1992.

Berleant, Arnold. *Living in the Landscape: Toward an Aesthetics of Environment.* Lawrence, KS: University of Kansas Press, 1997.

Brown, Kathryn. "Mapping the future." *Science* 298 (2002): 1874–1875.

Claval, Paul. "The region as a geographical, economic and cultural concept." *International Social Science Journal* 39 (May 1987): 159–172.

Cromley, Ellen K., and Sara L. McLafferty. *GIS and Public Health.* New York: Guilford Press, 2002.

de Sherbinin, A., K. Kline, and K. Raustiala. "Remote sensing data: valuable support for environmental treaties." *Environment* 44 (2002): 1: 20–31.

Dodge, Martin, and Rob Kitchin. *Atlas of Cyberspace.* New York: Addison-Wesley, 2001.

Forman, Richard T. *Land Mosaics: The Ecology of Landscapes and Regions.* New York: Cambridge University Press, 1997.

Gaile, Gary L., and Cort J. Willmott, eds. *Geography in America.* New York: Merrill/Macmillan, 1989.

Janelle, Donald G., ed. *Geographical Snapshots of North America.* New York: Guilford, 1992.

Johnston, R. J. *Philosophy and Human Geography*, 2nd ed. London: Edward Arnold, 1986.

——, ed. *The Dictionary of Human Geography*, 3rd ed. Oxford: Basil Blackwell, 1994.

Kirby, Kathleen. *Indifferent Boundaries: Spatial Concepts of Human Subjectivity.* New York: Guildford, 1996.

Livingstone, David. *The Geographical Tradition: Episodes in the History of a Contested Enterprise.* Cambridge, MA: Blackwell, 1992.

Lock, Gary, and Zoran Stancic, eds. *Archaeology and Geographical Information Systems.* London: Taylor & Francis, 1997.

Longley, Paul A., Michael F. Goodchild, David J. Maguire, and David W. Rhind. *Geographic Information Systems and Science.* Chichester, NY: Wiley, 2001.

Mabogunjie, Akin L. "Geography as a bridge between natural and social sciences." *Nature and Resources* 20 (1984): 2–6.

MacEachren, Alan M. *How Maps Work.* New York: Guilford, 1995.

Martin, Geoffrey, Preston E. James, and Eileen W. James. *All Possible Worlds: A History of Geographical Ideas*, 3rd ed. New York: John Wiley & Sons, 1992.

Melnick, Alan L. *Introduction to Geographic Information Systems in Public Health.* Gaithersburg, MD: Aspen Publishers, 2002.

Miller, Harvey J. *Geographic Information Systems for Transportation: Principles and Applications.* New York: Oxford University Press, 2001.

Monmonier, Mark. *How to Lie with Maps*, 2nd ed. Chicago: University of Chicago Press, 1996.

National Geography Standards Project. *Geography for Life: National Geography Standards.* Washington, D.C.: National Geographic Research, 1994.

Nellis, Duane. "Geospatial information technology, rural resource development, and future geographies." *Annals of the Association of American Geographers* 95 (2005): 1–10.

Raper, Jonathan. *Multidimensional Geographic Information Science.* New York: Taylor & Francis, 2000.

Reed, Michael, ed. *Discovering Past Landscapes.* London: Croom Helm, 1984.

Smith, S. V., W. H. Renwick, J. D. Bartley, and R. W. Buddemeier. "Distribution and significance of small, artificial water bodies across the United States landscape." *Science of the Total Environment* 299 (2002): 2–36.

Strahler, Arthur N. "Systems Theory in Physical Geography." *Physical Geography* 1 (January 1980): 1–27.

United Nations Environment Programme. *One Planet Many People: Atlas of Our Changing Environment.* New York: United Nations Environment Programme, 2005.

Woodward, David, and J. B. Hartley, eds. Volumes are regularly appearing in "The History of Cartography" Project, all published by the University of Chicago Press. *Cartography in Prehistoric, Ancient and Medieval Europe and the Mediterranean* (1987); *Cartography in the Traditional Islamic and South Asian Societies* (1992);

Cartography in the Traditional East and Southeast Asian Societies (1994).

Zimmerer, Karl S. "Human geography and the 'new ecology': The prospect and promise of integration." *Annals of the Association of American Geographers* 84 (January 1994): 108–125.

WEB WORK

The World Wide Web has a tremendous variety of sites that are of interest to geographers. You might want to visit the home pages of a few geographic organizations in the United States. These include the Association of American Geographers:

http://www.aag.org

the National Geographic Society:

http://www.nationalgeographic.com/

and the National Council for Geographic Education:

http://www.ncge.org

Scholars at the University of Colorado at Boulder maintain the International Network for Learning and Teaching Geography:

http://www.colorado.edu/geography/virtdept/resources/contents.htm

Some government agencies that maintain home pages that help introduce visitors to geography include the U.S. Geological Survey:

http://www.usgs.gov

The National Geospatial-Intelligence Agency:

http://www.nga.mil/

The University of Edinburgh, Scotland, lists GIS-related sites:

http://www.geo.ed.ac.uk/home/giswww.html

The National Atlas of Canada can be found at:

http://atlas.gc.ca/

While that of the United States is at:

http://www.nationalatlas.gov/

An excellent site for explaining items of scientific interest in the news is jointly maintained by the University of Wisconsin and the National Science Foundation:

http://whyfiles.org

2

WEATHER
AND CLIMATE

A corn crop near Linton, North Dakota, in late July 2006, failing due to drought conditions. (Will Kincard/AP Wide World Photos)

A Look AHEAD

Energy and Weather

Solar energy drives the circulation of Earth's atmosphere. Energy inputs to the atmosphere vary with latitude and season, and with elevation in the atmosphere. Atmospheric circulation redistributes energy and, in the process, generates weather and storms.

Precipitation

Precipitation may be created when air rises, causing water vapor to condense in the atmosphere. Air may be forced to rise because it is heated at Earth's surface, when passing over mountains, or during storm systems.

Circulation Patterns

Variations in atmospheric pressure from one place to another cause wind. Air blows from regions of high pressure to regions of low pressure. Pressure differences help explain prevailing wind directions and seasonal differences in winds.

Climate

Climate is the summary of weather that a place experiences. Earth's climates can be grouped into 11 major types.

Climate Change

Climates have changed significantly over the last few million years, and large variations have occurred even within the last 20,000 years. Many different factors contribute to climatic change, including human activities that are raising the level of carbon dioxide in the atmosphere.

*T*he summer of 2006 was one of drought on the Great Plains of the United States. Dry conditions had prevailed for several years in a row, and in some areas the drought was the worst in several decades. The dry conditions brought crop failures, shortage of forage for cattle, and widespread fires. In Mitchell, South Dakota, the Corn Palace—a tourist attraction that celebrates the dominant crop of the region by decorating its exterior each year with corn—was unable to decorate this year because of a shortage of corn. There is growing evidence that recent dry conditions in the western United States are a symptom of global warming. If so there may be long-lasting consequences for the agricultural economy of the region. ▸

The most dramatic weather events in the news are those that cause major damage either in the short term, such as tornados, or over longer time periods, such as droughts. Floods, mostly caused by heavy rains but sometimes augmented by high sea levels during hurricanes, cause more damage than any other natural hazard. In addition to such extremes, day-to-day weather influences culture and social customs from clothing to agriculture. This chapter explores the mechanisms that cause **weather**—the day-to-day variations in temperature, precipitation, and so forth—and **climate,** the statistical summary of weather over time.

Weather conditions—such as storms, snowfall, clear skies, and warmth or cold—are caused by radiant energy received from the sun. This solar energy is redistributed from tropical latitudes to the polar regions, from lower levels to upper levels of Earth's atmosphere, and eventually back into space. The movement of energy creates air movements that carry warm air from the tropics to high latitudes, and cold, wintry air from polar regions toward warmer climates. It carries moisture from the oceans over the land and returns drier air to ocean areas.

In the process of air and water moving from one place to another, weather is created. Weather varies from day to day because atmospheric circulation is constantly changing. But varied as it is, circulation tends to follow certain patterns. A zone of rising air is usually found near the equator, causing frequent heavy rains. Over northern Africa the air usually descends, causing clear skies and aridity. In the midlatitudes the air tends to move from west to east most of the time and less often in other directions. In southern and eastern Asia the wind blows from the land to the ocean in winter and from the ocean to the land in summer, causing intense monsoon rains. All these events are linked together in a dynamic atmospheric circulation.

These patterns of weather are one of the fundamental features of Earth's surface that make places different from each other. They make agriculture possible in some places and not in others; they provide abundant water for hydroelectricity in some places and sunshine on beaches elsewhere. More than any other physical attribute of Earth, weather and climate regulate natural systems and constrain the ways in which humans use their environments. Yet even though Earth's circulation is utterly uncontrollable, it is not beyond inadvertent modification by humans. Many people believe that droughts such as the one in 2006 that led to increased wildfire intensity in the southwestern United States will become more common as a result of human effects on the atmosphere. In this chapter we will explore the processes that govern weather, the distribution of climates that results from those processes, and some of the ways in which human activities are modifying weather and climate.

ENERGY AND WEATHER

The movement of energy in the atmosphere can be likened to that in an automobile engine, which uses energy from gasoline to make a vehicle move. An engine releases energy inside cylinders through the burning of gasoline. This concentrated energy drives the vehicle and is dissipated mostly through friction.

The fuel that drives the atmosphere is **solar energy,** which heats the air, creating wind and ocean currents. These movements carry heat and moisture from one part of the planet to another before the energy is dissipated into space. In an automobile, motion takes the form of pistons moving back and forth, shafts and gears turning, and wheels spinning. In the atmosphere, air is in motion from one part of the globe to another, sometimes in steady flows and sometimes in intense spinning or vertical movements that produce storms.

The Sun is a large thermonuclear reactor, in which hydrogen atoms combine to produce helium. Most important to us is a byproduct of this reaction: the release of vast amounts of energy. The Sun radiates this energy into space in all directions, and Earth intercepts a tiny fraction of it. Yet this small, steady energy flow to Earth is sufficient to power the circulation of the atmosphere and oceans and to support all life on Earth.

As Earth revolves along its elliptical orbit around the Sun each year, it receives this energy across a void averaging about 150 million kilometers (93 million miles). Variations of about 3 percent seasonally in the distance between Earth and the Sun cause small variations in the amount of energy reaching Earth, but these are minor in comparison to the effects of latitude and the tilt of Earth's axis.

Incoming Solar Radiation

The amount of solar energy intercepted by a particular area of Earth, or **insolation,** depends on two factors: the intensity of solar radiation, or the amount arriving per unit of time, and the number of hours during the day that the solar radiation is striking.

Intensity of solar radiation

The intensity of solar radiation at a particular place on Earth depends largely on the angle at which the Sun's rays hit that place. Daily and seasonal differences in intensity are caused by variations in the **angle of incidence,** which is the angle at which solar radiation strikes a particular place at any point in time. The angle of incidence at a particular place on Earth varies during a day and among seasons.

To illustrate the angle of incidence, consider what happens when a 1-square-meter beam of sunlight strikes a spherical object such as Earth (Figure 2-1). If this beam strikes Earth's surface perpendicularly (that is, at an angle of 90°), the surface it touches also will have an area of 1 square meter. At an angle of less than 90° (that is, an oblique angle), the area illuminated by that beam will be greater than 1 square meter. The intensity of the radiation is reduced, because the beam's energy is spread over the larger area.

As the angle is reduced even further, energy is distributed over a still larger area, and the intensity of radiation becomes less and less. Compared to the level of radiation at a 90° angle, the intensity of energy is about 86 percent at an angle of 60° and 50 percent at an angle of 30°. Thus, the intensity of radiation per unit of surface area is affected greatly by the angle of the Sun's rays, that is, the position of the Sun in the sky. When the Sun is overhead, sunlight is more intense, and the level of energy received is more concentrated than when the Sun is low in the sky.

On a daily basis the Sun is most intense at noon, when the Sun is highest in the sky, and least intense at sunrise and sunset. Beachgoers are warned to take this daily variation into account: They are more likely to get sunburned around noon than early in the morning or late in the afternoon. Why? Because insolation varies at different times of day.

The angle at which the Sun's energy strikes Earth's surface varies with the season as well as within a day. Throughout the year, the area of Earth's surface where the Sun is overhead keeps shifting due to Earth's tilt and continual revolution around the Sun (Figure 2-2). As Figure 2-2 illustrates, Earth rotates once every 24 hours around its axis of rotation, which is an imaginary line passing through the North and South Poles. The axis of rotation is inclined 23.5° away from being perpendicular to the Sun's incoming rays. This axial tilt of 23.5° remains constant, regardless of Earth's point in its orbit

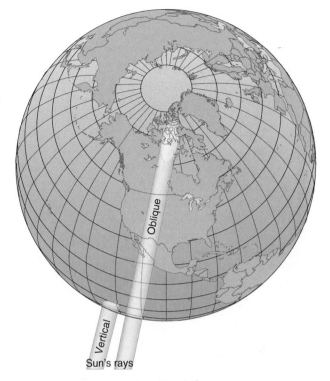

FIGURE 2-1 The angle of incidence. The angle at which the Sun's radiation strikes Earth's surface, known as the angle of incidence, varies. The angle of incidence affects the concentration of solar radiation (insolation) that a place receives. Radiation is most concentrated when the Sun's rays are perpendicular or vertical to Earth's surface, as occurs when the Sun is directly overhead. As shown, a place receiving more perpendicular insolation has this energy concentrated over a smaller area, making the area warmer. A place receiving more oblique (that is, less perpendicular) insolation has it spread over a larger area, making the area less warm.

or its daily rotation and is a key factor that determines how much insolation any point on Earth receives. Imagine our tilted Earth slowly revolving around the Sun through the four seasons of the year, and how this shifts the overhead point of the Sun. In the Northern Hemisphere summer (June and July), the Sun is higher in the sky with more intense radiation, whereas in the winter the Sun is lower in the sky with less intense radiation. In the Southern Hemisphere, the opposite is the case, and the seasons are reversed there.

The amount of insolation a place receives in a particular season depends on its latitude (Figure 2-3). At noon on the Northern Hemisphere's **vernal equinox** (March 20 or 21) and the **autumnal equinox** (September 22 or 23), the perpendicular rays of the Sun strike the equator, and the Sun is directly overhead at the equator (Figure 2-3a). At places along the **Tropic of Cancer** (23.5° north latitude) and the **Tropic of Capricorn** (23.5° south latitude), the intensity of solar radiation is reduced to 92 percent of the level at the equator. At places along latitude 45° north and south, the intensity of radiation is 71 percent of the level at the equator, and at places along latitude 60° north and south, the

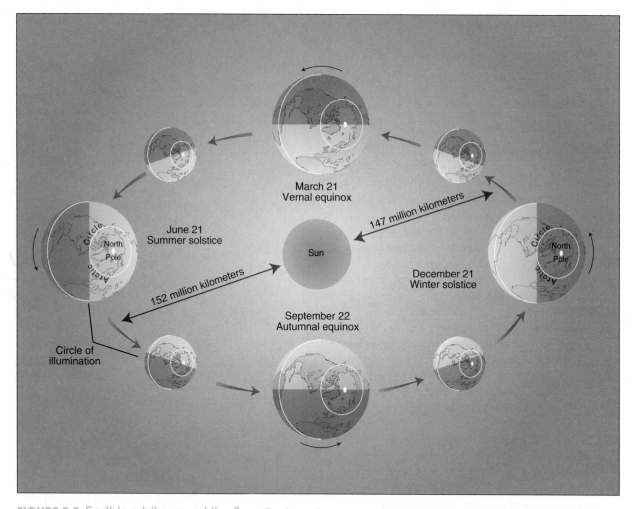

FIGURE 2-2 Earth's orbit around the Sun. Earth revolves around the Sun in an orbit that follows an imaginary plane known as the plane of the ecliptic. The plane of the ecliptic intersects all points on Earth's orbit around the Sun and passes through the Sun as well. Earth's path around the Sun is shaped like an ellipse. As it revolves around the Sun, Earth rotates around the axis of rotation, which is the line passing through the North and South Poles perpendicular to the equator. The line passing through the North and South poles is tilted at an angle of 66.5° away from the plane of the ecliptic, or 23.5° away from a line perpendicular to the plane of the ecliptic. This tilt is constant throughout the year, no matter where Earth is in its orbit. The tilt causes the Northern Hemisphere to be more exposed to the Sun's radiation from March to September (Northern Hemisphere spring and summer), while the Southern Hemisphere receives more radiation from September to March (Southern Hemisphere spring and summer).

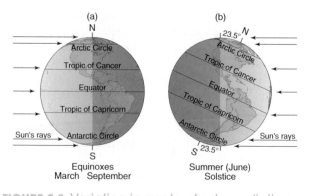

FIGURE 2-3 Variation in angle of solar radiation. The angle at which solar radiation strikes a particular latitude varies by season. These globes show how Earth receives solar radiation (insolation) on the first day of spring and fall (the equinoxes) (a) and on the first day of Northern Hemisphere summer (b). (From Christopherson, R. W., *Geosystems*, 3rd ed. Upper Saddle River, NJ: Prentice Hall, 1997)

intensity is only 50 percent of the level at the equator. At the North and South Poles on the equinoxes, the Sun is seen on the horizon and horizontal surfaces do not receive any direct radiation.

From the vernal equinox until the autumnal equinox (from late March until late September), the Sun is directly overhead at places along some latitudes in the Northern Hemisphere. The northernmost latitude receiving direct insolation is the Tropic of Cancer (23.5° north latitude), where the Sun is directly overhead at noon on the **summer solstice** (June 20 or 21). At the September equinox, the Sun is again directly overhead at noon at the equator. From then until the March equinox, places in the southern latitudes receive direct insolation, and the Sun is directly overhead of places along the Tropic of Capricorn (23.5° south latitude) at noon on the Northern Hemisphere's **winter solstice** (December 21 or 22).

Day length The total amount of heat that a particular place on Earth receives in a day is determined by the number of hours during which the Sun's energy strikes the place, as well as the intensity with which it strikes. Places on the equator always receive 12 hours of sunlight and 12 hours of night. But in higher latitudes, the amount of daylight varies considerably with the seasons. For example, Winnipeg, Manitoba, Canada, at 50° north latitude, receives about 16 3/4 hours of daylight at the summer solstice, but only about 7 1/4 hours of daylight at the winter solstice. In Winnipeg, the Sun is 63.5° above the horizon at noon on the summer solstice, compared to only 16.5° above the horizon at noon on the winter solstice. In a 24-hour day at the summer solstice, Winnipeg receives nearly six times as much solar radiation, measured at the top of the atmosphere, as it does at the winter solstice.

Variations in the length of day from place to place result from the 23.5° tilt of Earth's axis away from a perpendicular relation to the Sun. Figure 2-3b illustrates the illumination of the globe at the Northern Hemisphere summer solstice in June. Because the North Pole is located at Earth's axis of rotation, it maintains the same position with respect to the Sun throughout the 24-hour rotation of Earth at the solstice. At the summer solstice, if you stand at the North Pole, you can watch the Sun travel in a circle around you at a constant elevation of 23.5° above the horizon. On the day of the solstice, the North Pole is in full sunlight for the entire 24 hours, a phenomenon called the Midnight Sun (Figure 2-4).

Spatial and seasonal variations in radiation inputs

The amount of solar radiation reaching places at a particular latitude varies through the year because of seasonal changes in the angle of incidence, day length, and distance from the Sun. Tropical areas at low latitudes generally are warm throughout the year because they receive large amounts of insolation in every season, whereas places at high latitudes have strong seasonal contrasts in the amount of sunshine and level of temperature (Figure 2-5).

Because Earth's axis of rotation is tilted, seasonal variations in the amount of incoming solar radiation are less at places near the equator than at places in higher latitudes. Places at higher latitudes receive somewhat more solar radiation than do places at lower latitudes during the summer, but they receive much less during the winter. The tilt of Earth's axis makes life possible on a seasonal basis at relatively high latitudes; without that tilt, there would not be a warm season at mid and high latitudes, and life would probably be much more concentrated near the equator.

The effects of these variations in energy inputs are clearly visible in world temperature patterns (Figure 2-6).

In general, the highest temperatures occur in low latitudes in both maps. This is because the Sun is highest in the sky in these areas, and therefore the intensity of

FIGURE 2-4 Midnight Sun. On the day of the summer solstice, poleward of 66.5°, the Sun shines the entire 24-hour day, and the Sun appears to travel above the horizon without setting. (Brian Stablyk/Getty Images Inc.–Stone Allstock)

solar radiation is highest. High solar elevation angles occur throughout the year, and this makes the tropics (the area between the Tropic of Cancer and the Tropic of Capricorn) consistently warm.

High temperatures also occur in high latitudes such as in North America, Europe, and Asia in July. This is because at high latitudes the Sun is high in the sky in summer, and although it is not as high as in the tropics,

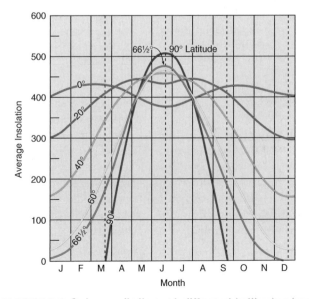

FIGURE 2-5 Solar radiation at different latitudes by season. Because Earth's axis of rotation is tilted, seasonal variations in the amount of incoming solar radiation are less at places near the equator than at places in higher latitudes. Places at higher latitudes receive somewhat more solar radiation than do places at lower latitudes during the summer, but they receive much less during the winter.

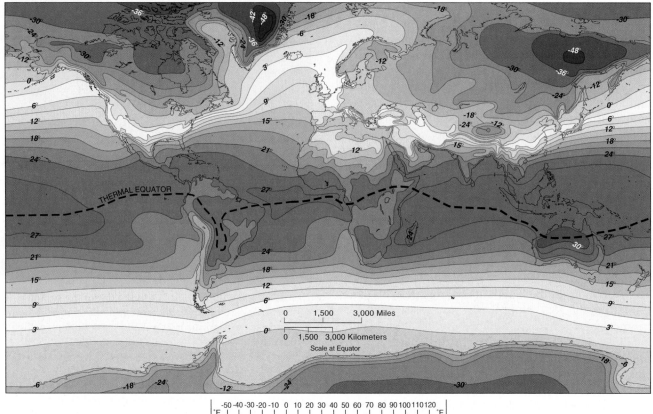

(a) January

FIGURE 2-6 Global temperatures for January (convertible to Fahrenheit by means of the scale). In the Southern Hemisphere, the warmest areas are over land, especially Australia. The coldest areas are in continental areas in the Northern Hemisphere, especially Greenland and Siberia. In the Northern Hemisphere, the oceans are warmer than land areas at similar latitudes. (Adapted from National Climatic Data Center, *Monthly Climatic Data for the World* 47, no. 1, January 1994. Prepared in cooperation with the World Meteorological Organization. Washington, D.C.: National Oceanic and Atmospheric Administration. From Christopherson, R. W., *Geosystems,* 5th ed. Upper Saddle River, NJ: Prentice Hall, 2003)

the day lengths are much greater, so the total amount of sunshine arriving at Earth's surface in a 24-hour day in midsummer is actually greater at high latitudes than in the tropics. The opposite effect is seen in winter, when high latitudes experience both low Sun angles and short days, causing temperatures to be much lower. Seasonal contrasts in temperature are greatest in high latitudes because seasonal contrasts in incoming solar radiation are greatest there.

Storage of Heat in Land and Water

When heat is absorbed by an object, its temperature rises, and when heat is released, an object cools. The ability of an object to store heat depends on what it is made of. Some materials can absorb or release a large amount of heat with only small changes in temperatures, while others heat and cool quickly with only small inputs and releases of energy. Water can absorb and

release much larger quantities of heat for a given temperature change than can land. The main reason for this is that a water body such as an ocean can be stirred by the wind and thus carry heat down below its surface to depths of 10 meters (33 feet) or more seasonally. In contrast, land surfaces warm and cool to only about 2 meters (6 feet) depth seasonally. Oceans can store a season's heat in a much larger volume of matter than can land areas, so they do not heat up as much in summer nor do they cool down as much in winter.

The effect on temperature of water's ability to store heat is clearly visible in Figure 2-6. Large landmasses in high latitudes, such as North America and Asia, have very large temperature differences between July and January. Average July temperatures in Siberia reach above 10°C (50°F), whereas average January temperatures reach −40°C (−40°F). The difference between July and January temperatures exceeds 50°C (90°F) in eastern Siberia. In contrast, the January and

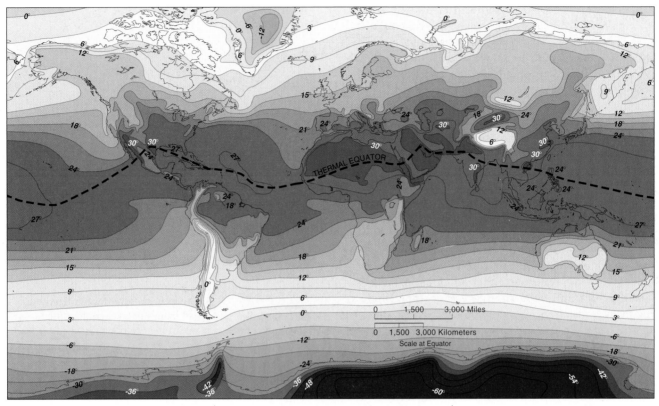

(b) July

FIGURE 2-6 Global temperatures for July. The warmest areas in the Northern Hemisphere are over land, especially central Asia and North America. The greatest temperature differences between January and July are over land areas, especially the large landmasses of the Northern Hemisphere. (Adapted from National Climatic Data Center, *Monthly Climatic Data for the World* 47, no. 7, July 1994. Prepared in cooperation with the World Meteorological Organization. Washington, D.C.: National Oceanic and Atmospheric Administration. From Christopherson, R. W., *Geosystems,* 5th ed. Upper Saddle River, NJ: Prentice Hall, 2003)

July temperatures in the Aleutian Islands of the north Pacific, at about the same latitude as Siberia, are about 15°C in July and 5°C in January, a range of only about 10°C. The moderate climates of areas near the ocean compared with the climatic extremes of midcontinent regions are a result of this heat storage in water.

Heat Transfer Between the Atmosphere and Earth

Have you ever noticed that on a winter day you feel colder sitting inside near a window or exterior walls than near the building's interior walls, even though air temperatures at both places are about the same? How does moisture on your skin help cool your body? And why does wind blowing on your skin cool it faster than still air? These effects on the temperature of your skin result from processes of heat transfer. Similar processes of heat transfer operating in the atmosphere as well as

on your skin are responsible for movement of vast amounts of energy from place to place on Earth.

Radiation The most important process of heat transfer in the environment is **radiation.** Energy transmitted by electromagnetic waves, including radio, television, light, and heat, is radiation, or radiant energy. You feel heat radiating from a fire. Heat travels at the speed of light from the fire to your skin, which senses it. Radiation can travel through the vacuum of space and through materials, although materials may restrict radiant energy flow. Radiant energy allows us to see the Sun's light energy, feel its heat energy, and listen to its radio energy using electronic machines.

Radiant energy waves have different lengths. The **wavelength** is the distance between successive waves, like waves on a pond. Wavelength affects the behavior of the energy when it strikes matter; some waves are reflected, and some are absorbed.

The Sun's energy comes to Earth as radiant energy, for this is the only way energy can travel through the vacuum of space. Two ranges of wavelengths—called **shortwave energy** and **longwave energy**—are most important for understanding how solar energy affects the atmosphere. Most insolation is shortwave, with wavelengths between 0.2 and 5 microns (a micron or micrometer is one-millionth of a meter). Wavelengths visible to the human eye account for a small portion of this shortwave energy, from about 0.4 to 0.7 microns. In contrast, most of the energy reradiated by Earth is longwave, in wavelengths between 5 and 30 microns.

It is important to note that most energy arriving from the Sun is shortwave, while most energy reradiated by Earth is longwave. As energy from the Sun passes through the atmosphere, some wavelengths are absorbed, which warms the atmosphere, while others pass through or are reflected either to be absorbed elsewhere or to travel back into space. The atmosphere is relatively transparent to incoming shortwave radiation, so this shortwave energy from the Sun easily passes through the atmosphere to reach Earth's surface. When the surface reradiates longwave energy, however, much of it is blocked and absorbed by the atmosphere. The blockage of outgoing longwave energy causes Earth's atmosphere to heat. This heating of the atmosphere is called the **greenhouse effect** because of its similarity to the way glass allows solar energy to enter a greenhouse but limits the loss of heat, causing the temperature inside to rise.

Of all the gases in the atmosphere, only a few have the property of being relatively transparent to incoming shortwave solar energy but absorbing outgoing longwave radiation. Gases with these properties are called **greenhouse gases,** and they are critical to heat exchange in the atmosphere. Among the more important ones are water vapor, **carbon dioxide (CO_2), ozone (O_3),** and methane (CH_4). Although these gases together constitute a small fraction of 1 percent of the atmosphere, they are the most important from the standpoint of atmospheric heating. Water vapor contributes the most to atmospheric heating. Human activities are increasing the amount of some greenhouse gases in the atmosphere, and this is believed to be the chief cause of **global warming.** We will return to this topic later in the chapter.

Latent heat exchange

A second important mechanism of moving energy in the environment is called **latent heat exchange.** It transfers tremendous amounts of energy from low latitudes to high ones, and it is also the most important process causing precipitation.

We can distinguish between two types of heat—sensible and latent. **Sensible heat** is detectable by your sense of touch. It is heat you can feel, from sunshine or a hot pan, and you can measure it with a thermometer. The atmosphere, oceans, rocks, and soil all have sensible heat, for you can feel their relative warmth or coldness.

FIGURE 2-7 Mist rising off the warm surface of a lake into cold air. The evaporation of water from the lake removes sensible heat from the lake, cooling the liquid water, and transferring latent heat to the air. (© Leslie & Mark Degner/CORBIS)

Latent heat is "in storage" in water and water vapor. You cannot feel latent heat, but when it is released, it has a powerful effect on its immediate environment.

Latent—which means "hidden"—is a good word to describe the heat that controls the state of water, for it is invisible stored heat. When ice melts, it must absorb heat energy from its surroundings. This is why ice melting in your hand feels so cold—it is absorbing heat from your hand. The heat becomes stored in the meltwater as latent heat.

Latent heat also is stored in water vapor (Figure 2-7). If you ever have had a finger scalded by steam, you know the startling amount of latent heat that was stored in the vapor and conducted from the condensing water into your finger! Latent heat exchange is what happens when water changes state (Figure 2-8).

Here is an analogy that describes the concept of latent heat: Consider a glass containing a mixture of ice and water on a hot summer afternoon. Because the ice water is much cooler than the air and other objects around it, heat energy is absorbed into the water. The heat begins to melt the ice, but as long as the water contains some ice, its overall temperature remains at 0°C (32°F). The water is absorbing enough heat to convert some ice to liquid, but the temperature does not change. The heat the water absorbs to melt the ice is latent heat, because it cannot be directly sensed as temperature.

When water is converted from liquid to vapor, additional heat is required—in fact, much more is needed than is required to melt ice. The latent heat of vaporization—the process that turns liquid to vapor—is about 540 calories per gram, compared to about 80 calories per gram for the latent heat of melting. A calorie is the amount of energy required to raise the temperature of 1 gram of water 1°C.

FIGURE 2-8 Dew formation. Longwave radiation from the ground to the atmosphere during the nighttime cools air near the ground. When the relative humidity reaches 100 percent, condensation occurs, and water droplets accumulate on solid surfaces such as on this dandelion seed head. (Neil Overy/Getty Images Inc.—Gallo Images)

In the ice-water example, a portion of the heat used to melt the ice came from sensible heat—heat you can sense—expressed as the temperature of the environment around the glass. The remainder came from the latent heat released by water vapor condensing to liquid on the surface of the glass. Thus, the melting of the ice water involved three transfers of heat:

▶ Latent heat was converted to sensible heat by condensation of water vapor on the glass.

▶ Sensible heat was transferred directly from the warm air to the cool glass.

▶ Sensible heat was converted to latent heat in the melting of the ice.

Now let us transfer this knowledge of latent heat to the much larger scale of the atmosphere. The amount of energy involved in latent heat transfers in the atmosphere is vast, especially for major weather systems and hurricanes. Transfer of heat from the land and water surface to the atmosphere by latent heat exchange amounts to about 40 percent of the solar energy absorbed by the

surface. Hurricanes gather strength by drawing in warm, humid air and converting the latent heat it contains into sensible heat, which drives the motion of the storm.

Insolation supplies shortwave energy to warm the seawater. As it evaporates, countless molecules of water vapor—each with its own latent heat—hover above the water, available to join the storm as it passes over. As the hurricane accumulates this water vapor, it is also gathering an enormous amount of latent heat energy. The vapor rises in the rotating storm, cools, and condenses. Latent heat is released as sensible heat, a process roughly analogous to adding gasoline to a forest fire.

The moisture in one major hurricane at its peak was estimated to weigh 27 trillion (27,000,000,000,000) metric tons (30 trillion tons). The latent heat released by this condensing water vapor in a single day roughly matched the total amount of energy consumed in the United States for six months. This provides some insight into why hurricanes are dangerous.

Heat Exchange and Atmospheric Circulation

Liquid water changes to vapor primarily at Earth's surface. This water vapor is carried aloft by rising air. In the atmosphere, vapor is converted back to liquid (clouds and rain) or solid (ice clouds and snow) through the formation of precipitation, which we will discuss later in the chapter. This movement of water back and forth between the surface and the atmosphere acts like a great conveyor belt for energy, picking up heat at the surface, carrying it aloft, and releasing it in the atmosphere. The conveyor is driven by convection.

Convection is movement in any fluid, caused when part of the fluid (whether gas or liquid) is heated. The heated portion expands and becomes less dense. Therefore, it rises up through the cooler portion. Convection causes the turbulence you see in boiling water and in turbulent clouds overhead. As air is warmed, it expands and becomes less dense. In becoming less dense—that is, weighing less per unit of volume—warm air rises above cooler, denser air, just as a hot-air balloon rises through the cooler air surrounding it.

Think of an island on a sunny day (Figure 2-9). Solar energy passes down through the atmosphere and is absorbed by the sandy surface. As the surface warms, it reradiates longwave energy, some of which is absorbed by the air just above the ground. The water surface warms less than the land does because some of the radiation is used to evaporate water, and some penetrates the water to warm lower depths of the sea. The air therefore grows much warmer over the island than over the water. This warmer air over the island expands, grows less dense, and begins to rise. As the warm air rises, cooler air descends to take its place. The cooler air is then warmed by the surface, and the convection process continues.

Convection also causes horizontal movements of air. Large horizontal transfers of air—and the latent

Critical THINKING

Climates in Urban Areas

Climates in major urban areas usually differ markedly from those of surrounding rural landscapes. Localized climate characteristics caused by topographic and land-surface characteristics are called **microclimates**. Typically urban microclimates are warmer, both in winter and in summer. The area of warm microclimates associated with an urban area is called the **urban heat island**. Several factors contribute to these differences. First, cities lack large expanses of vegetation and soil that absorb rainfall and later allow it to evaporate. Evaporation cools the land surface, and without this evaporation the surface remains warmer. Second, the walls of buildings form vertical surfaces that help to trap solar energy during the daytime. Concrete, asphalt, and similar building materials store this heat, and at night these warm surfaces shield the ground between buildings

from exposure to a large expanse of relatively cool sky and thus keep it warmer than it would otherwise be. Third, air pollutants may absorb heat and keep the air warmer. Finally, heat released from energy use, such as in heating buildings, adds to the heat available in the city (although this amount is very small in comparison to the amount of energy derived from the Sun).

Questions

1. In your experience, what parts of a city tend to be warmer, and what parts tend to be cooler? For example, in the summer are you cooler in the shade of a skyscraper or in the shade of a tree in a park?
2. What kinds of structures or land uses in cities contribute most to making urban climates different? How could cities be designed differently to reduce excess heating?

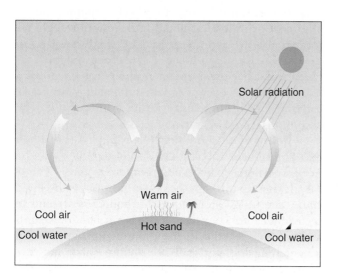

FIGURE 2-9 Convection in the atmosphere.
Convection occurs when air is warmed from beneath, expands, and rises. Cooler air descends to replace it. A common example is where land (such as an island, the Florida peninsula, or Cape Cod) is surrounded by water. Convection is strongest in the daytime when the surface is warmed by the Sun. Because of a stronger convection process, the wind tends to be gustier in the daytime than at night.

heat contained in water vapor in the air—are major components of Earth's energy system. Horizontal transfer such as this is called **advection.** Heat is advected from tropical areas toward the poles when warm winds blow poleward. Heat in ocean currents moving toward the polar regions is another form of advection.

Let us look at a small-scale example. During one day's time, warm air rising over the land allows cooler

air from the sea to advect over the land. In other words, wind in some coastal areas blows from the sea toward and over the land. On a large scale, during summertime over Eastern Europe and central Asia, the season's heating causes winds to blow onto the land from over oceans to the south and east. During winter the land area cools more than the sea, so cold air sinks over the land, warmer air rises over the ocean, and the winds tend to blow from central Asia toward the warmer Pacific Ocean. Such a circulation pattern is called a monsoon; we will discuss this later in the chapter.

PRECIPITATION

People rely on precipitation to be "normal." Normal rains and snowmelt are necessary for consistent agriculture, to feed Earth's 6.5 billion humans. People who live on lowlands along rivers rely on "normal" precipitation to keep the river within its banks and out of their basements. All plants and animals are adapted to a "normal" amount of moisture for their environment. However, "normal" does not always happen.

From 1998–2002 the southeastern U.S. suffered a severe drought. Reservoirs were drained, domestic water use was restricted, and agricultural production was reduced. Much of the U.S. Midwest suffered a severe drought in 1988 that heavily damaged crops and lowered the level of the Mississippi River so much that barges became stranded. In 1990 and 1993 the same regions of the Midwest experienced record wet years, with floods on the Mississippi, Missouri, and other rivers.

Such extremes of precipitation are rare during a typical human lifetime, but they actually occur quite frequently viewed over thousands or millions of years. And at any given time, while most places are experiencing "normal" precipitation, someplace on Earth is experiencing unusual precipitation. Because variations in the amount of rainfall sooner or later affect us all, unusually wet or dry periods make headlines.

Condensation

Precipitation is both a part of energy flows in the atmosphere and a consequence of them. Recall the "conveyor belt" of rising air that carries water vapor, with its latent heat, high enough to become cooled, condense, and cause precipitation. Precipitation is part of the flow of energy, releasing heat in the atmosphere and returning the liquid water back to Earth where it can absorb more heat. To understand this, let us look at the process of **condensation**, the conversion of water from vapor to liquid state.

Air contains water in gaseous or vapor form. Air may hold very little water vapor, as in dry desert air, or it may be filled with water vapor, as in a steamy jungle. The ability of air to hold water-vapor molecules is limited, and this limit varies according to air temperature. We measure the water-vapor content of air by the pressure that the water molecules exert. The maximum water vapor that air can hold is called the **saturation vapor pressure**. Precipitation begins when the pressure of moisture in the air exceeds the saturation amount.

Relative humidity tells us how wet air is. **Relative humidity** is the actual water content of the air expressed as a percentage of how much water the air could hold at a given temperature. For example, if a 30°C (86°F) sample of air contains half the water vapor that it could hold at that temperature, its relative humidity is 50 percent. But if cooled to 22°C (71°F), that same air sample with the same amount of water vapor would be three-fourths saturated, at 75 percent relative humidity. And if cooled to 15°C (60°F), that same air sample would be saturated, at 100 percent relative humidity. If saturated air at 0°C (32°F) is heated to 22°C (71°F), its relative humidity drops to less than 30 percent. This is why indoor air in winter is so dry. Relative humidity tends to fluctuate daily with changing air temperatures, because the amount of water vapor in the air holds fairly steady while temperature rises and falls. Relative humidity typically is lower in the warm afternoon and higher at nighttime.

When air cools, its relative humidity rises, and if the air is cooled enough, it can become saturated. If cooled still further, it becomes supersaturated and contains more water than it can hold in a vapor state. Condensation results from supersaturated air.

Condensation in the atmosphere produces clouds. Clouds form when water-vapor molecules condense around tiny particles of dust, sea salt, pollen, and other particles. The clouds contain small droplets of liquid water or particles of ice—depending on temperature—that are too light to fall. But if these water droplets or ice particles continue to grow as more moisture from the air condenses on them, they eventually become too large to be supported by air currents, and they fall to the surface as rain or snow.

Causes of Precipitation

The movement of air causes precipitation in three ways (Figure 2-10):

▸ Convection—in which air warmer than its surroundings rises, expands, and cools by this expansion.

▸ Orographic uplift—in which wind forces air up and over mountains.

▸ Frontal uplift—in which air is forced up a boundary (front) between cold and warm air masses.

Convectional precipitation On a warm, humid summer day, the sky is clear in the morning, and the Sun is bright. The Sun warms the ground quickly, and the air temperature rises. Most of the warming of the air takes place close to the ground, because the humid air is a good absorber of longwave radiation, which is being reradiated from the ground. As the air near the ground warms, it expands, becomes less dense, and rises through the surrounding cooler air above, like a hot air balloon. Convection is in progress, conveying humid air higher into the atmosphere.

Because air is a gas, it is compressed by the weight of overlying air. When it rises, air has less weight above it, and the lower pressure allows the air to expand. Compressing a gas causes an increase in temperature, whereas expanding it causes a decrease. The decrease in temperature that results from expansion of rising air is called **adiabatic cooling**; the word *adiabatic* means "without heat being involved."

As air rises, it rapidly cools adiabatically (by expansion) at a rate of about 1°C for each 100 meters (5.5°F per 1000 feet) elevation. When the air reaches several hundred meters above the ground, it has cooled to the point where it is saturated, and clouds begin to form (Figure 2-11). Rising warm air may mix with the surrounding cooler air, slowing the convection. If the convection were driven only by heating at the ground surface, it would probably not be strong enough to cause water droplets to grow big enough to fall as precipitation.

But as soon as condensation begins, latent heat—another important source of energy—is released. Latent heat further warms the rising column of air and

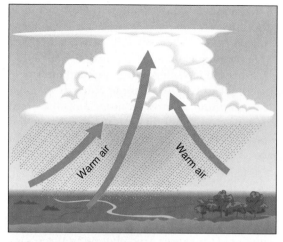

(a) Convectional Precipitation

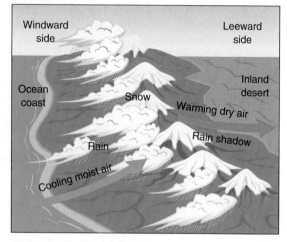

(b) Orographic Precipitation

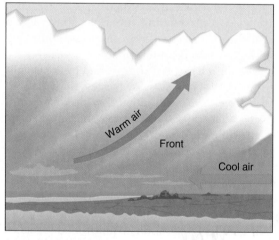

(c) Frontal Precipitation

FIGURE 2-10 Causes of precipitation. (a) Convection, (b) orographic uplift, and (c) fronts.

makes it less dense than the surrounding cool air. The process is self-reinforcing, as the cloud grows rapidly, with strong vertical motion and rapid condensation. The result is often gusty winds and intense rain: a

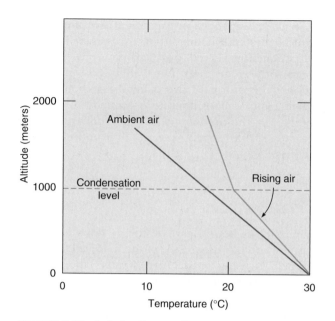

FIGURE 2-11 Adiabatic cooling. As air rises, it cools at a rate of about 1°C per 100 meters (1.8°F per 330 feet). The temperature of the rising parcel of air relative to the temperature of the air around that parcel determines its behavior. If the parcel is warmer than surrounding air, it tends to rise and therefore is unstable. If it is the same temperature or cooler, it resists movement and therefore is stable. In this example, the air parcel is warmer, and thus unstable. Condensation in the air above 1,000 meters (3,300 feet) elevation causes the air to be warmed by the release of latent (or stored) heat, warming the air. This further contributes to the instability of the air. Intense thunderstorms can form under such conditions.

thunderstorm. Thunderstorms are a common example of intense convectional storms (Figure 2-12).

Convectional storms are responsible for a large portion of the world's precipitation (Figure 2-13). In tropical climates, where strong insolation makes temperatures high, all that is needed for intense daily convectional storms is a source of humidity. In midlatitude climates, such storms occur mostly in the summer, because higher temperatures allow the air to hold more moisture, and this means more latent heat can be released, causing strong convection. They are especially common in situations where some other factor is present to favor uplift, the trigger that starts the self-reinforcing growth of the storm. Mountains and fronts, especially cold fronts, often provide the trigger.

Orographic precipitation Precipitation sometimes occurs when the wind forces air to rise over mountains. This is called **orographic precipitation**. As the air rises, it cools adiabatically (by expansion), the cooling causes condensation, and precipitation results. After air has moved up the windward side of a mountain and over the top, it then descends on the leeward side. As it does so, its relative humidity drops significantly. The leeward side of a mountain range is

FIGURE 2-12 A convectional storm over Australia. Storms such as this are most common in summer. (Paul Chesley/Getty Images Inc.—Stone Allstock)

FIGURE 2-13 Clouds revealing convection over the South Florida peninsula. The view is to the south. The land is warmer than the adjacent Atlantic Ocean and Gulf of Mexico, and this warmth stimulates convection and cloud formation. Note the absence of clouds over Lake Okeechobee, the large lake in the center of Florida, because it is cooler, like the ocean. (NASA SS/Photo Researchers, Inc.)

often much drier than the rainy windward side. A dry region on the leeward side of a mountain range is called a rain shadow. Some of the world's major deserts are arid because they are situated on the leeward side of a mountain range (Figure 2-14).

The western United States provides an excellent example of orographic effects on rainfall (Figure 2-15). Moist air from the Pacific Ocean, the region's major source of moisture, travels from west to east, producing precipitation as it rises first over coastal mountain ranges in Washington, Oregon, and California and then the Sierra Nevada and Cascade ranges. Some of the highest average U.S. rainfall totals occur in the Sierras and Cascades; but east of the mountain ranges lies the rain shadow, causing dry conditions in eastern Washington, Oregon, Nevada, and Utah. The region is arid because it is cut off from Pacific moisture by the Sierras and Cascades and from Atlantic moisture by the Rocky Mountains to the east.

Orographic precipitation also occurs in the Rocky Mountains, but the amount is less than in the Sierras because the Rockies are more isolated from moisture sources. In the eastern United States and Canada, the Gulf of Mexico and Atlantic Ocean supply ample moisture for precipitation east of the Mississippi Valley. The Appalachians receive more rain than adjacent lowlands because of orographic effects, but the eastern side of the Appalachians does not experience a rain shadow because the region has moisture sources on both sides: the Gulf of Mexico to the southwest and the Atlantic Ocean to the east.

Frontal precipitation

Frontal precipitation forms along a **front**, which is a boundary between two air masses. An air mass is a large region of air—hundreds or thousands of square kilometers—with relatively

FIGURE 2-14 The Sierra Nevada mountains in California, seen from the east. Because of orographic effects, the mean annual precipitation at the crest of the Sierras exceeds 100 centimeters (40 inches). But in the foreground, Owens Valley is in a rain shadow and has mean annual precipitation of less than 30 centimeters (12 inches). (Michele Burgess/Corbis/Stock Market)

(a)

(b)

FIGURE 2-15 Topography and precipitation in Oregon. (a) A shaded relief map showing the north-south-trending Coast Ranges and Cascades separated by the Willamette Valley in western Oregon, and plateaus of eastern Oregon. (b) Mean annual precipitation. Note that precipitation is highest in the Coast Ranges and Cascades, somewhat lower in the Willamette Valley, and quite low in eastern Oregon. The high rainfall in the mountains is due to orographic lifting of air as it moves eastward off the Pacific Ocean. By the time this air reaches the plateau east of the Cascade Range most of the moisture has been removed, and rainfall there is a fraction of that received along the coast. (*Atlas of Oregon*, 2nd ed., Copyright 2001, University of Oregon Press)

uniform characteristics of temperature and humidity. An air mass acquires these characteristics from the land or water over which it forms.

In North America, air masses that form over central Canada tend to be cool (because of Canada's relatively high latitude) and dry (because of the region's isolation from moisture sources). This is called a continental polar air mass. In contrast, air masses that form over tropical water, such as over the Gulf of Mexico, tend to be warm and moist and are called maritime tropical air masses.

When a cool air mass, such as continental polar air from Canada, meets a warm air mass, such as maritime tropical air from the Gulf of Mexico, a boundary or front may form between them. Because cool air is relatively dense, it tends to move under less dense warm air, while warm air tends to rise over cool air (Figure 2-16). Fronts are regions characterized by ascending air, cloudiness, windiness, and precipitation.

As air masses migrate across Earth's surface, the fronts move with them. When a cold air mass advances against a warmer one, the boundary is called a **cold front**. The advancing cold, denser air wedges beneath the warm air, forcing it to rise, generating clouds and usually precipitation (Figure 2-16a). A cold front can move quite fast and generate intense thunderstorms if the warm air is sufficiently moist.

The passage of a well-developed cold front in central and eastern North America includes very distinctive weather. Before the front arrives, the air is warm and moist, with the wind typically from the south or southwest. As the front arrives, intense precipitation falls, and in the summer thunderstorms usually form. As the front passes, the wind shifts to the west or northwest and the

temperature falls quickly. Then the sky clears, and cool dry air arrives, with blue skies and bright sunshine.

When, on the other hand, a warm air mass advances against cooler air, the boundary is a **warm front**. The warm air rides up over the cool air as though the cool air were a gentle ramp (Figure 2-16b). Precipitation along a warm front is less intense than along a cold front, and a warm front often passes without significant precipitation. In winter, a warm front sometimes causes freezing rain, as rain formed when the warm air aloft falls through colder air below, freezing as it reaches the ground. Air masses and the front between them may become stalled and not move, creating a stationary front. Note the cold and warm fronts depicted on the weather map (Figure 2-17).

Because a cold front normally involves rapid vertical motion of warm and relatively humid air, the clouds that form along it are typically cumulus clouds, which are tall, puffy clouds with billowy tops. In warm fronts, the warm air rises gradually over the cold air, and the stable layering of warm air over cold produces broad, flat layers of clouds known as stratus clouds.

CIRCULATION PATTERNS

Air is made of molecules, so air has mass. Earth's gravity attracts this mass of air to the surface, giving it weight. We think of air as a lightweight substance, but the thickness of air over your head has considerable weight, pressing on you and Earth's surface at an average of about 1 kilogram per square centimeter (14.7 pounds per square inch). This means that directly above each

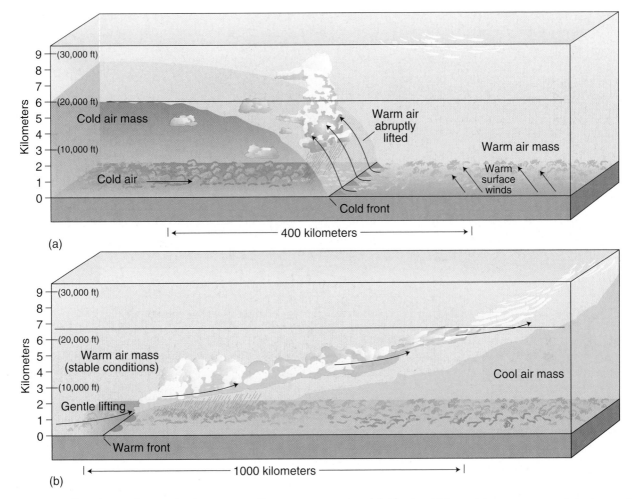

FIGURE 2-16 Fronts are boundaries separating warm and cold air. A cold front (a) generates precipitation as the cold air drives under the warm air like a wedge, lifting the warm air. This rapid vertical motion of air produces the tall, puffy cumulous clouds characteristic of a cold front. In a warm front (b), warm air rides up over cold air, causing precipitation to fall from broad, flat layers of stratus clouds. (From Christopherson, R. W., *Geosystems,* 5th ed. Upper Saddle River, NJ: Prentice Hall, 2003)

square centimeter of Earth is a column of air over 480 kilometers (300 miles) tall that weighs on average about 1 kilogram (Figure 2-18).

Atmospheric pressure varies with altitude because the higher you go, the less air exists above you. Thus, atmospheric pressure is greater at sea level than in "mile-high" Denver or atop Mount Everest (8.8 kilometers or 5.5 miles high). Because sea level is a surface that is at virtually the same height worldwide, scientists use the average atmospheric pressure at sea level as a world standard. Atmospheric pressure is measured with a *barometer.* The average atmospheric pressure at sea level read on a barometer is 1,013.2 millibars (a force that supports a column of mercury 29.92 inches tall in a mercury barometer).

Pressure and Winds

At any location, atmospheric pressure varies with conditions. For example, the air above the island in Figure 2-9 is warmer than the air around it, and therefore it is less dense. Because warm air is less dense,

it weighs less, and the atmospheric pressure over the island is lower than the pressure over the cool sea. As the warm, lighter air over the island rises, higher pressure over the cool sea forces air horizontally toward the island to replace the rising warm air. The difference in pressure between the two places is the *pressure gradient.*

Differences in pressure produce wind. In Figure 2-19, if pressure at the surface is high, air descends, and winds blow away (diverge) from the area of high pressure. But if pressure at the surface is low, air ascends vertically, and horizontal winds converge toward the area of low pressure.

Coriolis effect If Earth did not rotate, winds would simply blow in a straight line from areas of high pressure to areas of low pressure. However, on our real spinning planet, winds follow an indirect, curving path. This deflection of wind (and any other object moving above Earth's rotating surface) is called the **Coriolis effect.**

To understand Coriolis effect, imagine two people throwing a ball back and forth on the back of a flatbed

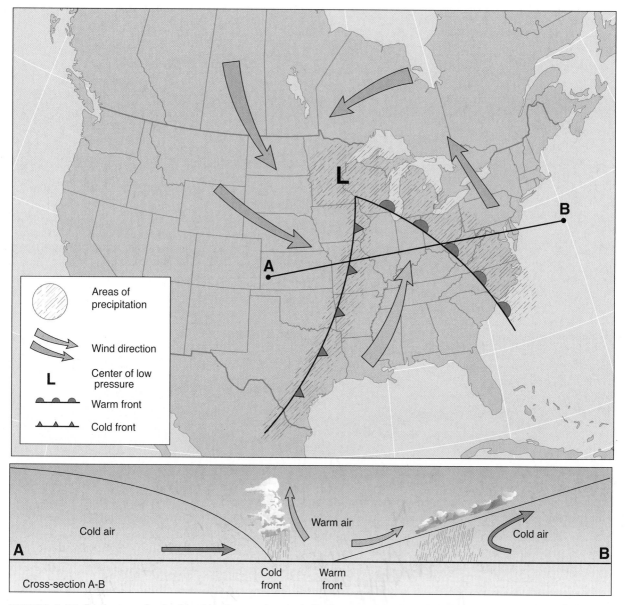

FIGURE 2-17 Patterns of winds, clouds, and precipitation around an idealized large midlatitude storm. The map view shows the position of the fronts at the ground surface. The cross section shows the locations of the fronts aloft.

truck (Figure 2-20a). Both people and the ball are moving at the same speed and in the same direction, so they do not change position relative to one another. The ball game is the same as if the truck were not moving.

On the merry-go-round (Figure 2-20b), when the ball is thrown in a straight line, the turning platform moves the other person away from where the ball arrives. As the two people throw the ball back and forth, the ball follows its expected straight-line path, but the receiver has been moved by the merry-go-round. To the players, it appears that the ball has followed a curving path. Similarly, as Earth rotates beneath moving air, the air appears to move in a curve. This deflection of motion is only apparent to those on the spinning surface. Viewed from above, the path would be straight.

Near the equator, Earth is like the flatbed truck in Figure 2-20a; the Coriolis effect is zero at the equator.

But toward the poles, rotating Earth is more like a spinning disk, and the Coriolis effect is strongest in the polar regions. This is shown in Figure 2-20c, where a plane flies from the North Pole in a straight line toward the equator but arrives far to the west of its intended destination because Earth rotated eastward during the hours that the plane was in flight.

The net result of the Coriolis effect is a deflection to the right (from the standpoint of a moving object) in the Northern Hemisphere. In the Southern Hemisphere the deflection is to the left. The effect is significant only when considerable distance is involved: A pilot must consider it in setting the flight path of a several-hour flight across the ocean, but not for a model airplane across a backyard. The Coriolis effect helps weather forecasters understand the behavior of storms.

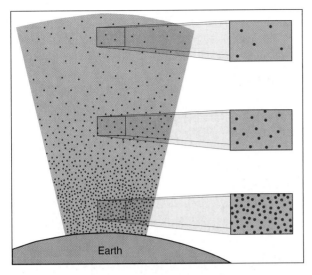

FIGURE 2-18 Density and pressure of air. Visualizing the density of air (greater near Earth's surface) and atmospheric pressure. (From Christopherson, R. W., *Geosystems,* 5th ed. Upper Saddle River, NJ: Prentice Hall, 2003)

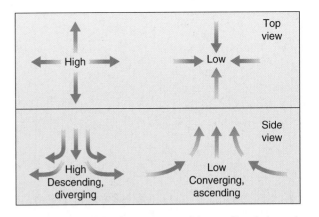

FIGURE 2-19 Wind movement is partly determined by pressure. Air descends in regions of high pressure. Air ascends in regions of low pressure. Wind blows from areas of high pressure to areas of low pressure. (From Christopherson, R. W., *Geosystems,* 5th ed. Upper Saddle River, NJ: Prentice Hall, 2003)

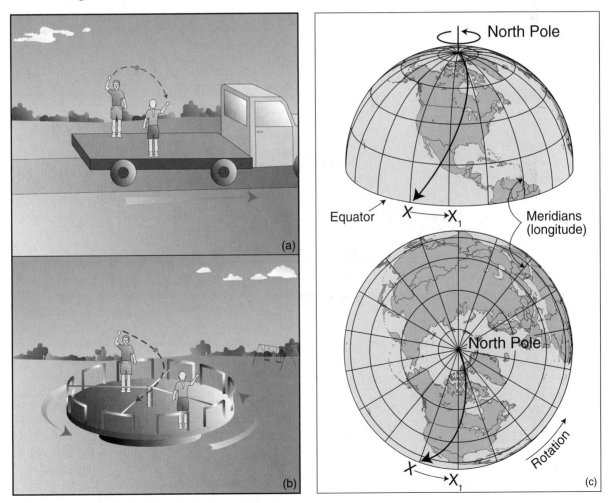

FIGURE 2-20 Coriolis effect. (a) When two people on a moving flatbed truck throw a ball to each other, the people and the ball are moving at the same speed and in the same direction, so they do not change position relative to one another. (b) When two people on a merry-go-round throw a ball to each other, the ball follows a straight-line path, but the people have moved. To the people the ball appears to be following a curve. This apparent curving of the ball demonstrates the Coriolis effect. (c) When an airplane flies in a straight line from the North Pole to the equator, it would actually land west of its intended destination because Earth would rotate eastward during the flight. (From Tarbuck, E. J., and Lutgens, F. K., *Earth Science,* 7th ed. New York: Macmillan, 1994)

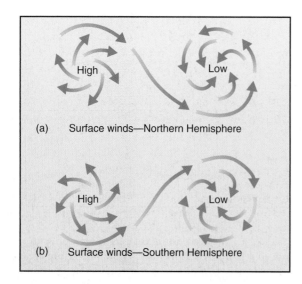

FIGURE 2-21 Wind movement with Coriolis effect added. (a) The Coriolis effect, caused by Earth's rotation, deflects wind to the right of its expected path in the Northern Hemisphere. This causes spiraling circulation around high- and low-pressure centers, in the directions shown. (b) The same pattern reversed for the Southern Hemisphere. Compare with Figure 2-19. (From Christopherson, R. W., *Geosystems,* 5th ed. Upper Saddle River, NJ: Prentice Hall, 2003)

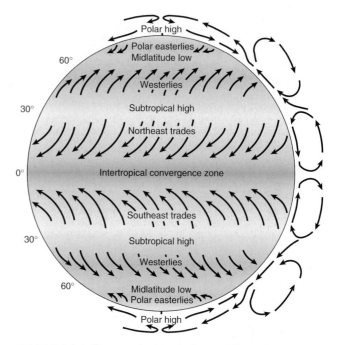

FIGURE 2-22 General circulation of the atmosphere. This simplified illustration of average atmospheric circulation patterns shows the major circulation zones by latitude: intertropical convergence zone (ITCZ), subtropical high-pressure (STH) zones, midlatitude low-pressure zones, and polar high zones.

Figure 2-21 shows the result when the Coriolis effect is added to the vertical air movements in high- and low-pressure areas. It puts a spin on the winds, adding a spiral motion to create cells that rotate clockwise and counterclockwise. In the Northern Hemisphere (Figure 2-21a), high-pressure cells are rotating clockwise, and low-pressure cells are rotating counterclockwise. The opposite patterns hold in the Southern Hemisphere (Figure 2-21b). To summarize, differences in air pressure create winds, and Earth's spin curves their flow.

Global Atmospheric Circulation

Understanding global circulation patterns is very important. These patterns control our weather, climate, and agriculture. At the global scale, atmospheric circulation operates like the convection system over the warm island we described previously. Beginning at the equator, we can identify four zones of circulation patterns:

▶ Intertropical convergence zone
▶ Subtropical high-pressure zones
▶ Midlatitude low-pressure zones
▶ Polar high-pressure zones

Intertropical convergence zone In the tropics, dependable year-round inputs of solar energy heat the air, expanding it and creating low pressure. As a result, convectional rising of air occurs daily above Earth's equator. This forms the **intertropical convergence**

zone (ITCZ), so called because it is a zone between the Tropics of Cancer and Capricorn where surface winds converge (Figure 2-22).

As air rises in the ITCZ, convectional precipitation occurs, usually as afternoon thunderstorms. Air converges toward the equator at the surface, replacing the rising air. The Coriolis effect deflects this moving air to the right in the Northern Hemisphere to form the Northeast Trade Winds and to the left in the Southern Hemisphere to form the Southeast Trade Winds. The **trade winds** are so called because they were very important to sailing ships. For example, the northeast trades in the Atlantic carried European ships to the Caribbean. Aloft, air circulates away from the ITCZ both northward and southward, toward subtropical latitudes. This air has lost most of its moisture in the daily rainfalls, and it is now warm and dry.

Subtropical high-pressure zones The warm, dry air that spreads poleward from the ITCZ descends at about 25° north and south latitudes. This creates zones of high pressure that are especially strong over the oceans. These **subtropical high-pressure (STH) zones** are areas of dry air, bright sunshine, and little precipitation.

This descending dry air associated with subtropical high-pressure cells creates an arid climate, so most of the world's major desert regions are on land in this zone, at about 25° north and south latitudes. Further, most are on the western edges of continents, including deserts in northern Mexico and southwestern United States, the Atacama in Peru and Chile, the Sahara in northern Africa, the Namib in southern Africa, and the Australian desert. The eastern sides of these continents are under the same high-pressure cells, but they tend to be more humid because the circulation brings in warm, humid tropical air.

On the poleward sides of the subtropical high-pressure cells, circulation is toward the poles. But these winds are deflected by the Coriolis effect, so winds prevail from the southwest in the Northern Hemisphere and from the northwest in the Southern Hemisphere. The prevailing westerly wind makes an eastbound transatlantic flight in the midlatitudes normally about an hour shorter than a westbound flight.

Midlatitude low-pressure zones

Poleward of the subtropical high-pressure zones are the **midlatitude low-pressure zones**. These lower-pressure areas experience convergence of warm air blowing from subtropical latitudes and cold air blowing from polar regions. The warm and cold air masses collide in swirling low-pressure cells that move along the boundary between the two air masses, which is known as the **polar front**. Winds in these regions generally blow from west to east and are therefore called westerlies. (Winds are named for the direction from which they blow.)

Polar high-pressure zones

In the polar regions, the intense cold caused by meager insolation creates dense air and high pressure. In these **polar high-pressure zones**, the air is so cold, it contains very little moisture, and convection and precipitation are limited. The snow and ice around the poles comes from snow accumulating over many thousands of years.

Seasonal Variations in Global Circulation

Global circulation patterns vary with the seasons. We can illustrate typical conditions during January and July (Figure 2-23).

Global circulation in January

During January, Earth's average atmospheric pressure and winds are arranged into broad zones according to latitude (Figure 2-23a). Conditions in the Northern and Southern Hemispheres roughly mirror each other. The ITCZ is generally at about 5° to 10° south latitude.

High-pressure cells predominate in the subtropical regions just north and south of the ITCZ, especially over the subtropical oceans. The Southern Hemisphere, where it is summer in January, displays the clearest tendency for high pressure over the oceans and low pressure over land. The Northern Hemisphere has a high-pressure region in eastern Asia, as well as in the eastern Pacific and Atlantic oceans.

In the midlatitudes, less continuous low-pressure regions in the Northern Hemisphere appear most clearly over oceans. The midlatitude low-pressure zone is much more consistent in the Southern Hemisphere, because of the absence of land between 40° and 70°. A monthly average map obscures the presence of storms, but it is useful to show the general condition of the atmosphere during Northern Hemisphere winter.

Global circulation in July

By July, January's major pressure and wind zones have moved northward. In some cases they have moved from being over land to water areas, or vice versa (Figure 2-23b). The ITCZ, which was south of the equator in January, is almost entirely in the Northern Hemisphere during July, as far north as 30° north latitude in southern Asia. The subtropical highs are still present over the oceans, although they are strengthened in the Northern Hemisphere and weakened in the Southern.

The most notable change in the subtropics and midlatitudes is the replacement of high pressure over Asia with low pressure. In fact, this region has the world's highest average pressures during January and the world's lowest average pressure during July. As a result, wind directions in the vicinity are reversed. This seasonal reversal of pressure and wind produces a **monsoon circulation**, in which winter winds from the Asian interior produce extremely dry winters in most of south and east Asia, while summer winds blowing inland from the Indian and Pacific Oceans result in wet summers. These wet summers are the well-known monsoon season of heavy rains in southern Asia.

Seasonal changes also occur in the midlatitudes. The extreme low-pressure regions over the northern oceans in January are replaced in July with generally weak low pressure and inconsistent winds, although storms with strong winds may periodically pass through the regions. In the Southern Hemisphere winter (July), the low-pressure system is especially deep and consistent around the globe because of the absence of large landmasses to break it up as in the Northern Hemisphere. The south polar high is similarly strong in July.

Remember that with the shift in seasons between January and July, solar radiation also changes. During July in the Northern Hemisphere, solar radiation inputs are highest north of the equator, while in January, insolation is greater south of the equator. Circulation patterns also shift seasonally, bringing strong seasonal contrasts in weather as wind direction and zones of frequent storms move north and south.

Ocean Circulation Patterns

When wind blows over the ocean, it exerts a frictional drag on the sea surface, creating waves and currents (Figure 2-24). Continuing wind adds energy to the waves, building them into larger ones that may travel thousands

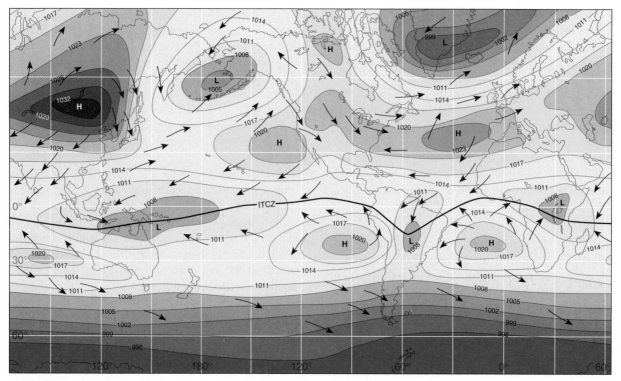

(a) January

FIGURE 2-23 Worldwide average atmospheric pressures (in millibars) and winds in a Northern Hemisphere winter in January and summer in July. (a) In January, the intertropical convergence zone (ITCZ) is generally south of the equator, especially over land areas. In the Northern Hemisphere, the subtropical high-pressure cells are weak or absent, and a large and intense high-pressure cell dominates interior Asia, with deep low-pressure areas over the North Pacific and North Atlantic.

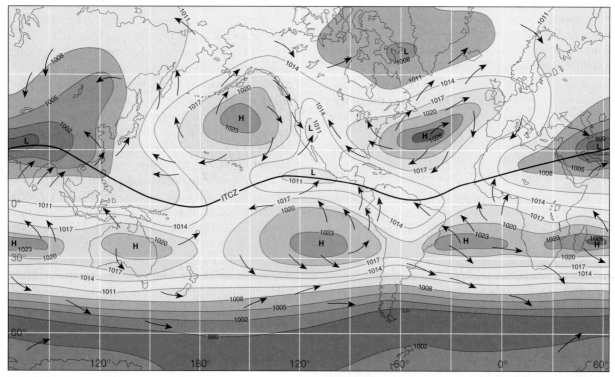

(b) July

FIGURE 2-23 (*continued*) (b) In July, the ITCZ has shifted to the north, and low pressure has replaced high pressure over interior Asia. Weak low pressure also is seen over North America, and the subtropical highs are prominent. In the Southern Hemisphere, the midlatitude low-pressure belt along latitudes 40–60° south is very strong.

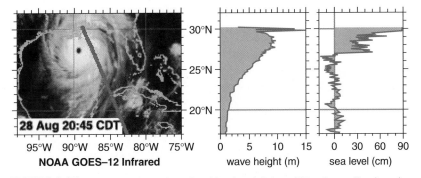

FIGURE 2-24 Wave height and sea level in the vicinity of Hurricane Katrina. As the storm approached the coast, a satellite recorded sea level and wave heights. The graph at right shows wave heights in meters, and sea level variations in centimeters. As the storm was approaching the coast sea level near the coast was already elevated by about 1 meter (3 ft), with waves 15 meters (49 feet) high. Note that the profile of sea level does not pass directly through the center of the storm; sea level would have been higher there. As the storm approached the shore the combination of elevated sea level and wind driving the water landward caused sea level at the shore to rise by 3 to 4 meters (10 to 13 feet), with wave crests rising much higher. (REDRAWN From Scharroo, Remko and Walter H. F. Smith and John L. Lillibridge, 2005. Satellite Altimetry and the Intensification of Hurricane Katrina *Eos*, Vol. 86, No. 40, 4 October 2005.)

of kilometers until they break onto a distant shore. The continuing drag of prevailing winds also causes broad currents in the ocean's surface layers. Contributing to these currents are differences in seawater temperature and salinity. These give water different densities, promoting movement of currents from areas of greater density to those of less density. This is similar to the way a high-pressure area in the atmosphere causes wind currents to blow toward a low-pressure area. Both of these currents—oceanic and atmospheric—are important in redistributing heat around Earth, a role played by the Gulf Stream, for example (Figure 2-25).

Gyres are prominent features of oceanic circulation. These wind-driven circular flows mirror the movement of prevailing winds (Figure 2-26); you can see this by comparing the oceanic circulation figure with Figure 2-23. Gyres form beneath tropical high-pressure cells. The Gulf Stream forms the western limb of the gyre in the North Atlantic.

Where ocean currents circulate warm water from low equatorial latitudes to higher latitudes, they are carrying heat poleward by advection. Such flows are balanced by cool currents traveling equatorward, most notably along the west coasts of subtropical land areas. Such cold currents cool the lower portions of the atmosphere above them, reducing convection. With less convection occurring, adjacent landmasses may be very dry. Some of the driest areas on Earth owe their aridity partly to this effect, most notably the Atacama desert of Peru and Chile. Occasional shifts in ocean circulation such as **El Niño** (see the Focus box on El Niño) can cause significant variations in weather.

The oceans also have vertical circulation driven by variations in water density associated with differences in salinity and temperature. The Gulf Stream is a part of this circulation pattern. Changing climate, particularly

FIGURE 2-25 The Gulf Stream. In this composite of satellite images of the North American East Coast from late May 1996, brown shades indicate warm water, and blue indicates cool water. The strong flow of warm water from southwest to northeast is plainly visible, including swirling eddies in the flow where the warm current moves past cooler water. (Microstock/Photri-Microstock, Inc.)

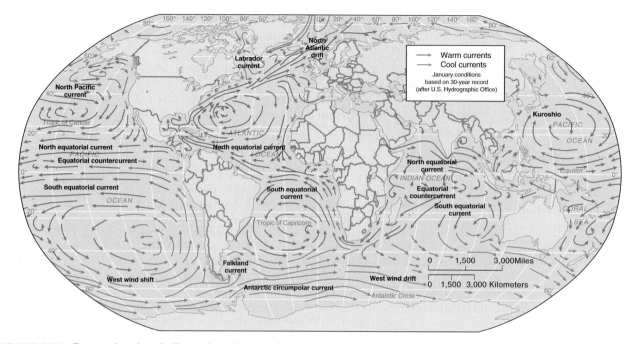

FIGURE 2-26 Oceanic circulation, showing major currents. Note how the ocean circulation is clockwise in the Northern Hemisphere and counterclockwise in the Southern Hemisphere, driven mainly by circulation around the subtropical high-pressure zones. (From Christopherson, R. W., *Geosystems,* 5th ed. Upper Saddle River, NJ: Prentice Hall, 2003)

El Niño/La Niña

Ocean currents generally are much more consistent than atmospheric winds, but oceanic flows do vary in speed and direction. One of the most important of these variations is called El Niño. The term is Spanish for "the (male) child," a reference to the Christ child, because the phenomenon occurs around Christmastime. El Niño is a circulation change in the eastern tropical Pacific Ocean that occurs every few years. In this change, the usual cool flow from South America westward is slowed and sometimes reversed, replaced by a warm-water flow from the central Pacific eastward. The counterpart of El Niño, in which especially cool waters are found in this region, is known as La Niña (the female child). The change in ocean currents is linked to a change in atmospheric circulation patterns in the Pacific Ocean known as the Southern Oscillation; the two phenomena together are referred to as El Niño-Southern Oscillation (ENSO).

Peruvian fishermen named El Niño, and they knew its effect: The reversed flow of water deprived fish populations in the eastern Pacific of their food and contributed to the collapse of the Peruvian anchovy industry in the 1970s. ENSO events are far reaching because a modification of circulation in one part of the globe may cause circulation patterns to change elsewhere in North and South America and the Pacific region. For example, El Niño events are linked to flooding in the U.S. Southwest, droughts in Australia, and reduced rainfall in India. La Niña events often bring wet weather in south and Southeast Asia, and dry conditions in the southern United States.

In 1997 a particularly strong El Niño event developed, with ocean temperatures in the eastern Pacific about 5°C (9°F) above non-El Niño levels. This led to greatly increased convection in the eastern Pacific, and rains in normally dry coastal Peru. At the same time, the western equatorial Pacific was cooler than normal, causing drought in Southeast Asia. Forest fires raged out of control in Indonesia, causing severe air-pollution problems in Kuala Lumpur, Malaysia, and intense storms pounded coastal California. By 2000 the pattern had reversed, and La Niña conditions dominated. This, along with warm water in the western Pacific, was a major contributor to the 1998–2002 drought in the southern United States.

melting of glaciers in the Arctic, is increasing the delivery of low-salinity water to the North Atlantic. There is some concern among climatologists that this may alter major ocean currents.

Storms: Regional-Scale Circulation Patterns

Variations in circulation like El Niño (also known as ENSO, for El Niño/Southern Oscillation) and the Asian monsoons are examples of recurring changes in the atmosphere. These range in scale from global phenomena like ENSO, to a continent-size land-water convection pattern such as the Asian monsoon, to regional-scale systems such as hurricanes, down to the scale of a local thunderstorm.

Storms are especially important because they are areas of concentrated convection that bring precipitation and sometimes strong winds. On a regional scale, large storms can affect areas hundreds to thousands of kilometers across. Such storms, which are called **cyclones**, are large low-pressure areas in which winds converge in a counterclockwise swirl in the Northern Hemisphere and clockwise in the Southern Hemisphere. We will look at two types of cyclones: tropical cyclones (hurricanes or typhoons) and midlatitude cyclones.

Tropical cyclones are intense, rotating convectional systems that develop over warm ocean areas in the tropics and subtropics, primarily during the warm season. When such storms have wind velocities exceeding 119 kilometers per hour (74 miles per hour), we call them **hurricanes** in North America, **typhoons** in the western Pacific, and *cyclones* in the Indian Ocean.

These storms typically develop in the eastern portion of an ocean within the trade-wind belt. They begin as areas of low pressure (rising air) and converging winds, drawing in warm, moisture-laden air. This humid tropical air contains a great deal of energy, both as sensible heat and latent heat in the water vapor, but especially latent heat. As this energy is drawn into the developing storm, condensation and the resulting release of latent heat intensifies the convection. The center of low pressure grows more intense, and as wind speed grows, the Coriolis effect causes a spiraling circulation to develop, counterclockwise in the Northern Hemisphere and clockwise in the Southern Hemisphere.

These tropical storms move with the general circulation over the subtropical Atlantic, Pacific, and Indian Oceans, from east to west. As they travel across the warm ocean surface, they often intensify. By the time they reach the western portion of the ocean, their atmospheric pressures may be as much as 10 percent lower than average sea-level pressure, with winds exceeding 150 kilometers per hour (93 miles per hour).

Hurricanes thrive on warm, moist air, so they are most intense over ocean areas during the warm season. They lose intensity over land, because they lose their source of energy. Furthermore, the smooth ocean surface favors development of high winds. In contrast, the hills and trees of land areas slow the wind by friction.

Hurricanes' greatest threat to humans is therefore in tropical and subtropical coastal areas, on the eastern margins of the continents and in southern Asia, where monsoon circulation draws air northward from the Indian Ocean. When a hurricane strikes land, the combination of intense wind and extremely low pressure causes a **storm surge,** an area of elevated sea level in the center of the storm that may be several meters high. The surge carries large waves crashing inland onto low-lying coastal areas, with devastating results. In fact, the majority of hurricane deaths and damage result from storm surges. Adding to this hazard, tornadoes commonly are spawned as a hurricane comes ashore.

Among the coastlines most vulnerable to tropical cyclones is that of the Bay of Bengal, including Bangladesh and West Bengal in eastern India. The land in this area is low (elevation less than 2 meters or 6 feet), consisting mostly of the Ganges River delta, and it is densely populated. Bangladesh has millions of people and very limited resources for coping with periodic storms. Although satellite-based warning systems help evacuate people in advance of approaching storms, transportation is difficult, and people often are unable to reach high ground. In recent decades, several severe storms struck the area, including one in 1970 that killed about 300,000 people.

In midlatitude areas another type of storm dominates the weather: the midlatitude cyclone. The midlatitude cyclone has little in common with the tropical cyclone (hurricane), for it occurs in cooler, less humid latitudes that cannot provide the immense volume of heat and water vapor needed to generate a hurricane. **Midlatitude cyclones** are centers of low pressure that develop along the polar front. They move from west to east along that front, following the general circulation in the midlatitudes. Midlatitude cyclones are usually much less intense than hurricanes but much more common.

In a midlatitude cyclone, air is drawn toward the center of low pressure from both the warm and the cold sides of the polar front. Where warm air is drawn toward cold, typically on the eastern side of the storm, a warm front develops. On the western side, the spiraling motion causes cold air to drive under the warm air, forming a cold front. As the center of low pressure moves eastward, these fronts move with it, bringing precipitation to areas over which they pass. The passage of a front at the surface usually is marked by a significant change of temperature, precipitation, and shifting

Tornadoes

Tornadoes are an extreme form of weather created when energy conditions in the atmosphere are such that extremely intense convection occurs. They generate the highest surface wind velocities known to occur on Earth, and the destruction they cause can be near total, though fortunately very localized.

Tornadoes are intense columns of rising air, usually associated with thunderstorms. As the air rises, it creates a partial vacuum (an area of very low pressure) that draws air in toward it. As this air is drawn in, it creates a swirling vortex, much like the swirling motion of water as it drains out of a sink. As the air swirls in, it gains tremendous speed, sometimes exceeding 300 kilometers per hour (185 miles per hour). This vortex usually moves horizontally, sometimes touching the ground and sometimes rising above it, leaving an erratic and usually very narrow path of destruction.

Most tornadoes occur in one of two circulation patterns: in association with intense thunderstorms embedded in hurricanes, and along especially strong cold fronts in the midlatitudes. The south-central United States has the greatest frequency of tornadoes of any place in the world. In the United States they occur mostly in spring and early summer, when contrasts between cold, dry continental air and moist, humid air from the Gulf of Mexico are greatest. Such conditions

Tornado damage in Grand View, Missouri, May 5, 2003. A series of strong storms in the first two weeks of May 2003 generated over 300 tornadoes in the Midwestern United States, killing 38 people. (© Reuters News Media Inc./CORBIS).

create cells of rapidly rising air, including very intense thunderstorms and sometimes tornadoes.

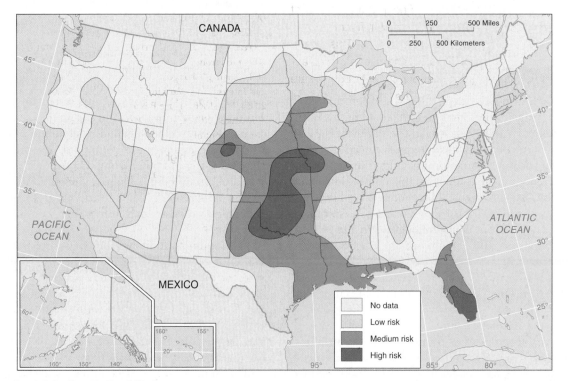

Tornado risk in the United States. Tornado risk in the United States is greatest in the southern plains, the Southeast (especially central Florida), and the Midwest. (Alaska and Hawaii not to scale.)

Global warming and hurricanes

Climate is a statistical summary of weather over time. Scientists studying the impacts of global warming on climate are reasonably certain that increased CO_2 in Earth's atmosphere is a cause of rapid warming of average temperatures that we have witnessed in the last few decades. They are also reasonably certain that this CO_2-induced warming will continue through most of the twenty-first century, and likely into the twenty-second. But if the Earth warms on average, say, 3°C that does not mean that every season of the year will be 3°C warmer, let alone every day of the year. It is possible that winters might warm a lot and summers hardly at all, or the reverse may happen. Similarly, it does not mean that every place on Earth will be 3°C warmer. Some places might get much warmer, and other places might actually get cooler. So the question of how global warming might affect the occurrence of a relatively unusual weather event such as a hurricane is very much unanswered.

However, there are some very good reasons to expect that global warming will increase either the frequency of hurricanes or their intensity or both. One of these is the link between sea surface temperatures and hurricanes. This connection is responsible for the fact that hurricanes occur primarily in the summer and autumn, when sea surface temperatures are highest. The warmer the water is the more likely a tropical depression is to grow into a hurricane. Second, we know that sea surface temperatures have risen significantly in the Atlantic and elsewhere in recent decades. This rise in sea surface temperature parallels the rise in global average temperatures during that period, although we are not certain that one is causing the other. But even if global warming is causing higher sea surface temperatures, can we then say that global warming will cause more hurricanes, or more intense hurricanes?

In the last few years several studies have focused on this question, and there does indeed appear to be a strong link between global warming and hurricanes. The evidence is generally of two types. One of these is - observational. We can document that intense hurricanes have indeed become more intense during this period of sea surface warming. As shown in figure A, the percentage of hurricanes in category 1, 2, and 3 worldwide (relatively mild hurricanes) has declined slightly since the 1970s, while the percentage of category 4 and 5 storms (the most intense storms) has increased significantly. A second type of evidence is based on models. Computerized simulation models of atmospheric circulation are routinely used to predict weather, and these models

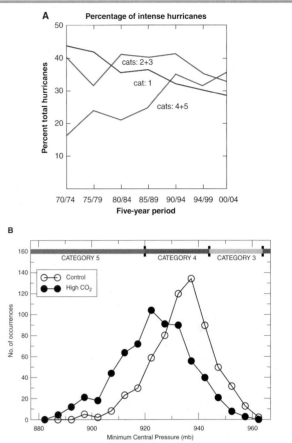

a) The increase in hurricane intensity over the last few decades is shown by an increase in the proportion of category 4 and 5 hurricanes, and a corresponding decrease in the proportion of category 1 hurricanes. b) Simulations of hurricane intensities in climate models show an increase in hurricane intensity with increasing atmospheric carbon dioxide concentrations. (Figure A is adapted from P. J. Webster, G. J. Holland, J. A. Curry, and H. R. Chang, "Changes in Tropical Cyclone Number, Duration, and Intensity in a Warming Environment," *Science 309* (September 2005): 18441846. Figure B is adapted from "Global Warming and Hurricanes, NOAA," at http://www.gfdl.noaa.gov/~tk/glob_warm_hurr.html)

have become highly sophisticated and quite effective. When such models are used to compare circulation in a CO_2-enriched atmosphere with that in a normal atmosphere (figure B), the modeled circulation in the high-CO_2 atmosphere has more intense hurricanes.

Questions

1. Hurricane intensity has increased at the same time that atmospheric and sea surface temperatures have been rising. Does this mean that global warming is *causing* the increase in hurricane intensity?
2. Models are our only way of predicting future climates. However, we cannot test predictions of the future until we observe the future. How can we judge the quality of predictions that are made by models?

winds. The repeated passage of such storms creates high-ly variable weather conditions, with alternating cold and warm air as the polar front moves back and forth across the land. Occasionally, especially in North America, strong cold fronts associated with midlatitude cyclones produce tornadoes (see the Focus box on Tornadoes).

The Weather on August 25, 1998

Global and regional circulation patterns just described are visible in satellite images of Earth, such as the one shown in Figure 2-27. These patterns may not be obvi-ous at first glance, but a trained eye can identify wind directions, fronts, and precipitation patterns even in a satellite image. Here they are shown with appropriate weather-map symbols on the image. The image shown in Figure 2-27 shows most of the Western Hemisphere

in late summer. In this view of Earth, from about 22,000 miles out in space, we see a little less than half the globe. The South and North Poles are not visible, but the view does extend to almost 80° latitude, north and south.

The intertropical convergence is visible as a bro-ken line of cloudy areas extending across the North Pacific at about 10°N, continuing east across the Caribbean Sea. The ITCZ consists mainly of convective storms, some of which are scattered and some of which are concentrated in areas of low pressure. On the northern margins of the ITCZ we see three tropical cyclones. One is in the early stages of formation in the tropical Atlantic. Another, named Hurricane Bonnie, is slamming into the North Carolina coastline. A third is visible near the left-hand edge of the image in the trop-ical North Pacific.

FIGURE 2-27 Weather pattern on Earth. Full-disk image of Earth, August 25, 1998. See the text for description of patterns shown in this image. (National Climate Data Center/National Oceanic and Atmos-pheric Administration/Seattle)

In the midlatitudes we see sweeping curves of clouds, many of which show the locations of fronts. In August it is winter in the Southern Hemisphere, and two particularly well-developed storms are visible there. A third can be seen in the North Atlantic, approaching Europe.

The atmosphere is continually in motion, and on any given day the atmosphere contains many storms, some intense and some mild. This circulation occurs in the vertical dimension as well as horizontally; areas of rising air are cloudy and may produce precipitation, while areas of sinking air are cloud-free. The locations of storms and high-pressure cells change from day to day, producing weather variability from one day to the next.

CLIMATE

The atmospheric circulation patterns we see on any given day are the result of energy distribution patterns occurring at that time, and these change rapidly from hour to hour and day to day. Geographers also study temperature, precipitation, and other atmospheric phenomena that persist for much longer periods than a thunderstorm or a heat wave.

Climate is the summary of weather conditions over several decades or more—a place's weather pattern over time. The vegetation, natural resources, and human activities that characterize a particular region of Earth are heavily influenced by its climate. At first glance, describing the climate of a place may appear to be a straightforward task of calculating annual averages of temperature and precipitation. In reality, climate is extremely difficult to describe. Air temperature and precipitation change not only day to day but year to year, and other factors such as windiness, humidity, and occurrence of storms must also be considered. We will discuss this in detail later in the chapter.

Climate changes over time. Earth has alternated between warming periods and cooling periods. Glaciers have covered large areas during cool periods, and they have melted during warm periods. The human population has survived by adapting to the alternating periods of warming and cooling. Ironically, human activity now may be causing climatic changes that will fundamentally alter the character of the planet.

People are affected by various elements of climate. The number of hours of sunshine in a day and the degree of cloudiness are important to anyone who is active out of doors—gardeners, farmers, hunters, builders, vacationers at the beach—and residents of homes heated by solar energy. Amounts of rainfall and snowfall make great differences in how we build our homes and roads from place to place. The windiness of a place is important in designing structures and harnessing wind power. Pollutants in the air can have both short-term and long-term effects on human and animal health. Counts of pollen and mold spores are important to those who suffer allergies.

Air Temperature

To understand a region's distinctive climate, the two most important measures are air temperature and precipitation. In everyday speech, we refer to "hot or cold climates" and "wet or dry climates." The most obvious differences in air temperature on Earth are those between the tropics and the poles and those between winter and summer. These variations are caused by latitudinal and seasonal variations in solar energy inputs.

Air temperature also varies with elevation. This variation occurs because the atmosphere is warmed by longwave radiation from Earth's surface and because of the adiabatic cooling of rising air described earlier. A very general rule of thumb is that temperature drops about 6.4°C per 1,000 meters (3.5°F per 1,000 feet). This is why mountain regions are cooler than adjacent lowlands (Figure 2-28). Air temperature also varies considerably within a few meters of the ground surface. In a desert the ground surface temperature can reach 70°C (160°F) during the daytime, while air that lies 2 meters above the ground rarely exceeds 55°C (130° F). At night the desert ground may be significantly cooler than the air above.

To minimize the effects of the ground surface on temperature records, climatologists usually measure air temperature at a height of about 1.4 meters (55 inches) inside a ventilated shelter that shades the thermometer and other instruments from direct sunlight. Climatologists at these weather stations usually report high and low temperatures for each day and calculate the average daily temperature as the average of the two readings.

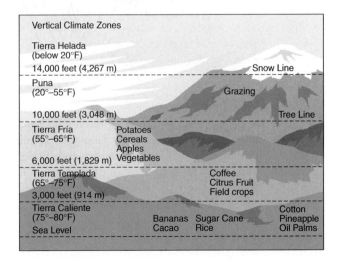

FIGURE 2-28 Vertical climate zones. Throughout much of Latin America, elevation defines these vertical climate zones. Distinct crops characterize each zone.

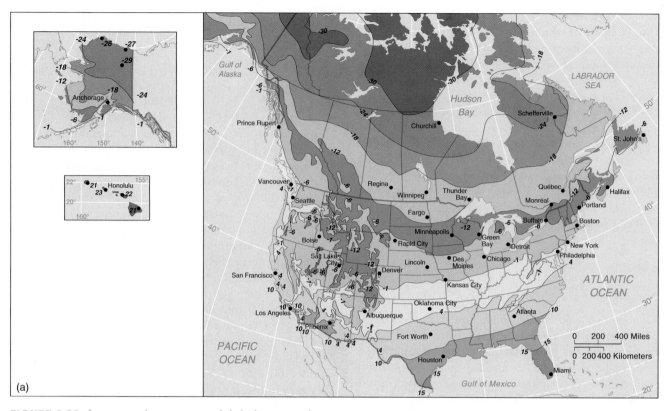

FIGURE 2-29 *Average January and July temperatures.* Average temperatures (°C) in North America during January and July, based on data collected from 1931–1960. (a) In January, the lowest temperatures occur in the interior of the continent and in high-altitude areas. Higher temperatures occur in the south and along the West Coast. (b) In July, the highest temperatures occur in the interior of the continent, especially in the desert southwest. Temperature differences between

Temperature patterns also show the effects of topography, proximity to the ocean, and moisture availability (Figure 2-29).

In both summer and winter, the higher elevations of the Rockies and Appalachians are cooler than adjacent lowlands. At very high elevations temperatures are so low that year-round snow cover is possible. Permanent snow occurs at an elevation above approximately 4,000 meters (13,100 feet) in the southern Rockies and as low as 1,000 meters (3,300 feet) in the Northern Rockies and the mountains of Alaska.

The smallest seasonal differences in temperatures in North America are found in areas near the ocean, especially along the Pacific Coast. Westerly winds bring air from ocean areas to coastal regions in the west, while on the Atlantic Coast the airflow is normally from the continent, so seasonal temperature variations are greater. The effects of moisture availability and sunshine on temperature are visible in the deserts of southwestern United States and northwestern Mexico, where generally clear skies allow ample sunshine, but water is not available to dissipate that heat through evaporation. The highest recorded temperatures in North America occur in this region.

Precipitation

The amount of precipitation is extremely variable between places and over time. A thunderstorm can unleash heavy rain in one place and a light sprinkling just a short distance away. Worldwide, precipitation generally ranges from virtually none in some desert areas to more than 300 centimeters (120 inches) in tropical areas. A few tropical mountain areas have recorded more than 10 meters (33 feet) of rainfall per year.

Precipitation usually is measured by collecting rain or snow in a cylindrical container that is marked in millimeters or hundredths of an inch. Snowfall normally is recorded as an amount of liquid water rather than the depth of snow accumulated on the ground, because the amount of water in snow varies widely with the snow's texture. Weather stations typically report daily precipitation, although the instruments are capable of recording the amount for other periods of time, such as a minute or an hour.

Figure 2-30 shows worldwide average precipitation. Note how the levels correspond to global circulation patterns: the rainy ITCZ around the equator, the midlatitude high-pressure areas with their dry

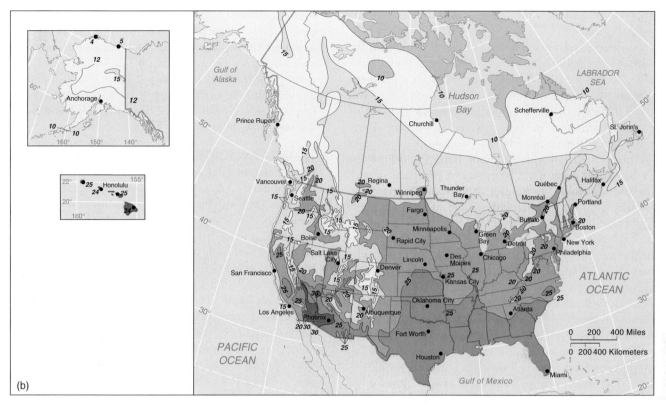

FIGURE 2-29 (*continued*)
northern and southern portions of the United States are less than in January, and storm systems traveling along the polar front are less intense as a result. Be careful when trying to interpolate between temperature lines, because large changes may occur over short distances, depending on altitude, slope, soil type, vegetation, and urban development. (Alaska and Hawaii not to scale.)

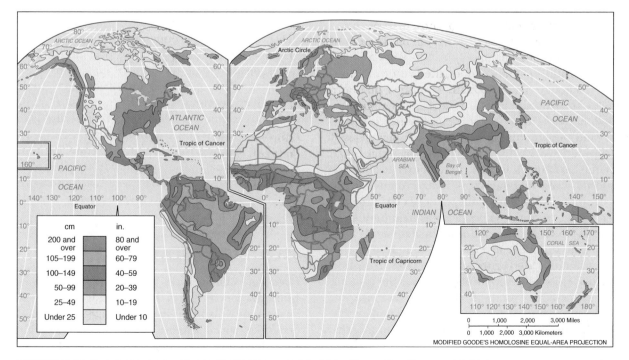

FIGURE 2-30 Average annual precipitation worldwide. Generally, climates with higher temperatures require greater rainfall to maintain vegetation. (From Christopherson, R. W., *Geosystems,* 5th ed. Upper Saddle River, NJ: Prentice Hall, 2003)

air, the mountain rain shadow areas, and the dry polar regions.

Average rainfall amounts tell us much about the climate of a region, but there is also much they do not tell. The timing and reliability of rainfall are equally important. In midlatitude areas precipitation frequently takes the form of gentle rainfall, whereas in tropical regions it may come in torrential and destructive downpours. In northern Nigeria, for example, 90 percent of all rain falls in storms of more than 25 millimeters (1 inch) per hour. That is, by way of comparison, half the average monthly rainfall of London, England. In Ghana, cloudbursts regularly occur at a rate of 200 millimeters (8 inches) per hour, four times London's monthly total. In Java, an Indonesian island, a quarter of the annual rainfall comes in downpours of 60 millimeters (2.4 inches) per hour, more than Berlin, Germany, gets in an average month.

In sudden, intense storms a greater portion of the rainfall runs off the land rather than soaking in, sometimes increasing soil erosion. Studies in the West African nation of Burkina Faso found that in one year nearly 90 percent of all erosion took place in just six hours.

The regularity of rainfall is also important. In Western Europe and eastern North America, rainfall does not vary much from month to month. New York's rainiest month receives only one and a half times as much rain as its driest month, London's rainiest month is two times greater, and Berlin's is two and a half times greater. Delhi receives about the same total annual rainfall as London, but Delhi, by contrast, receives only 10 millimeters (0.4 inch) of rain in November and more than 175 millimeters (7 inches) in both July and August. Zungeru, in central Nigeria, gets 54 times as much rain in the wettest month as in the driest.

Soil can absorb and hold only a limited amount of water for crops to use. Think of this as a bank account deposited in the wet season and drawn on in the dry season. Although many tropical countries get what seems to be adequate annual rainfall, they get it in a lump sum, like a huge win at a casino. Most is squandered, and the little that is left in the soil bank is used up within a few months. For example, in Agra, India, where the Taj Mahal is located, rainfall exceeds the current needs of vegetation only in July and August. As the rains decrease in September, the water stored in the soil is used up in just three weeks, leaving nine months of the year in which water supplies are inadequate.

Human societies can adjust to variations in rainfall if these variations are regular or predictable. Irrigation can help even out the supply of water over the growing season. In the Sahel, south of the Sahara, farmers grow fast-maturing crops such as sorghum and millet that shoot from seed to ripened ear in three or four months.

It is much more difficult, however, to cope with rains that come irregularly from year to year. In Europe and wetter portions of North America, the amount of annual rainfall varies by less than 15 percent per year on average. In many of the tropical and subtropical regions, though, it fluctuates from 15 to 20 percent, and in the semiarid and arid lands, by up to 40 percent or more. In the lean years, crops fail. If the rains are late, the growing season is cut short and yields are greatly diminished. If imported food is not available, famine can follow. That is why on the Indian subcontinent the monsoon is awaited with such hope and trepidation, and a delay of a week or two causes panic.

Water availability   Evaluating water availability involves much more than just measuring rainfall. Plants demand water and thus play a key role in what happens to the precipitation. Plants take water up through their roots and evaporate it through their leaves, releasing it into the atmosphere as water vapor, a process called **transpiration.** Therefore, most climate classification systems consider precipitation in relation to what a region's vegetation needs. Plants consume very large amounts of water, totaling about two-thirds of all precipitation that falls on Earth's land areas. A single tree may remove 100 or more liters of water from the soil in a single day. The amount of water that could evaporate from a damp soil or be transpired by plants if all of their needed moisture were available is called potential evapotranspiration (POTET; see Chapter 4). POTET indicates the moisture demand that plants make on their environment. Evapotranspiration is a fundamental part of the hydrologic cycle, which will be discussed in detail in Chapter 4.

Because heat energy must be available to evaporate water, warmer climates make possible a greater POTET. In tropical climates, where temperatures typically exceed 20°C (68°F) throughout the year, annual POTET can exceed 150 centimeters (60 inches). In contrast, in midlatitude climates, annual POTET totals 50 to 90 centimeters (20 to 35 inches), and annual POTET is only a few centimeters in polar regions.

In distinguishing between arid and humid climates, measurement of potential evapotranspiration is important. Parts of tropical East Africa and central Minnesota both receive the same average annual rainfall of about 80 centimeters (30 inches). Yet trees are sparse in East Africa and abundant in Minnesota. The reason is that East Africa is hotter and therefore has a much greater potential evapotranspiration than central Minnesota. East Africa's POTET exceeds its annual precipitation, so it cannot support as many trees.

CLASSIFYING CLIMATE

Classifying climate types is useful for many purposes. Knowing which places have certain climates allows analysis and planning by climatologists, geographers, geologists, social scientists, government, and industry. Describing and naming specific climate types gives us vocabulary to use in describing places and communicating such information to others.

Most climate classifications combine information about temperatures and precipitation to take into

TABLE 2-1	Major Climate Types and Their Köppen Equivalents

CLIMATE TYPE	CLIMATE CHARACTERISTICS
Humid tropical	
Af	Tropical, constantly warm and humid, with no dry season
Am	Tropical, constantly warm and humid, but with a short dry season
Seasonally humid tropical	
Aw	Tropical, constantly warm and humid, but with a pronounced dry low-sun season and wet high-sun season
Desert	
BWh	Hot desert climate
BWk	Cool desert climate
Semiarid	
BSh	Hot semiarid (steppe) climate
BSk	Cool semiarid (steppe) climate
Humid subtropical	
Cfa	Humid, warm subtropical climate, with hot summers and no dry season
Cw	Humid, warm subtropical climate, with hot summers and dry winters
Marine west coast	
Cfb	Marine west coast climate, with warm summers and no dry season
Cfc	Marine west coast climate, with cool summers and no dry season
Mediterranean	
Cs	Mediterranean climate, with dry, warm summers and cool, wet winters
Humid continental	
Dfa	Humid continental climate, with hot summers, cold winters, and no dry season
Dwa	Humid continental climate, with hot summers and dry, cold winters
Dfb	Humid continental climate, with warm summers, cold winters, and no dry season
Dwb	Humid continental climate, with warm summers and dry, cold winters
Subarctic	
Dfc	Moist subarctic climate, with cool summers, very cold winters, and no dry season
Dwc	Moist subarctic climate, with cool summers and very cold, dry winters
Dfd	Moist subarctic climate, with cool summers, frigid winters, and no dry season
Dwd	Moist subarctic climate, with cool summers and frigid, dry winters
Tundra	
ET	Tundra climate, with very cool, short summers and frigid winters
Icecap and ice sheets	
EF	Ice cap climate, with temperatures consistently below freezing

Source: Courtesy Institute for International Economics.

account the effects of temperature on water availability for vegetation. One example of such a classification is one devised in 1918 by German geographer Wladimir Köppen (Table 2-1 and Figure 2-31; see also Appendix II). Köppen used the distribution of plants to help him draw boundaries between climate regions. He identified five basic climate types and subdivided each to reveal important distinctions. A vegetation-based approach has the advantage of defining climates in terms that have more meaning for other environmental features rather than just selecting an arbitrary temperature or precipitation level as the boundary between two climates. Climatologists have modified Köppen's classification many times but have retained its essential elements. It remains the most widely used classification system.

Köppen's scheme uses uppercase letters A, B, C, D, and E to define the major climate types. These letters are assigned in relation to the latitudes at which climates are found, with A climates nearest the equator and E climates toward the poles. A, C, D, and E climates are distinguished primarily by temperature, with A being the warmest, but B climates are distinguished primarily by precipitation. Tropical (A) climates are warm all year. Dry (B) climates have limited moisture. Warm midlatitude climates (C) have cool winters and warm summers. Cold midlatitude climates (D) have cold winters and mild summers. Polar climates (E) are cold all year. In mountainous areas large differences in climate occur over short distances, causing detailed climatic patterns that cannot be shown on a world map. These areas are mapped as H, or highlands.

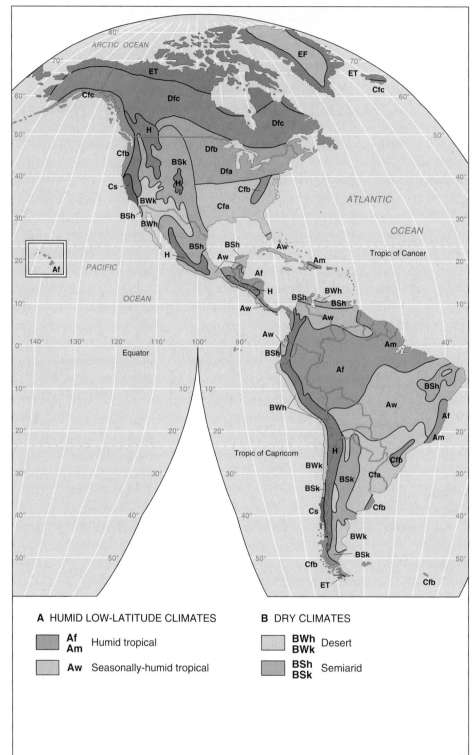

FIGURE 2-31 World climates.
This world map of climate patterns is largely based on the Köppen system of climate classification.

EARTH'S CLIMATE REGIONS

Take a moment to become acquainted with Figure 2-31, because the discussion in this entire section will refer to it. Climate zones form rough horizontal bands around Earth, following lines of latitude. Tropical (A) climates predominate in the very low latitudes, whereas polar (E) climates predominate in the higher latitudes. These horizontal bands show the powerful influence of temperature on climate. As explained earlier, higher latitudes receive less insolation than do lower latitudes. Observe the striking correspondence between this map and the patterns shown in the world maps of temperature (Figure 2-6), precipitation (Figure 2-30), and

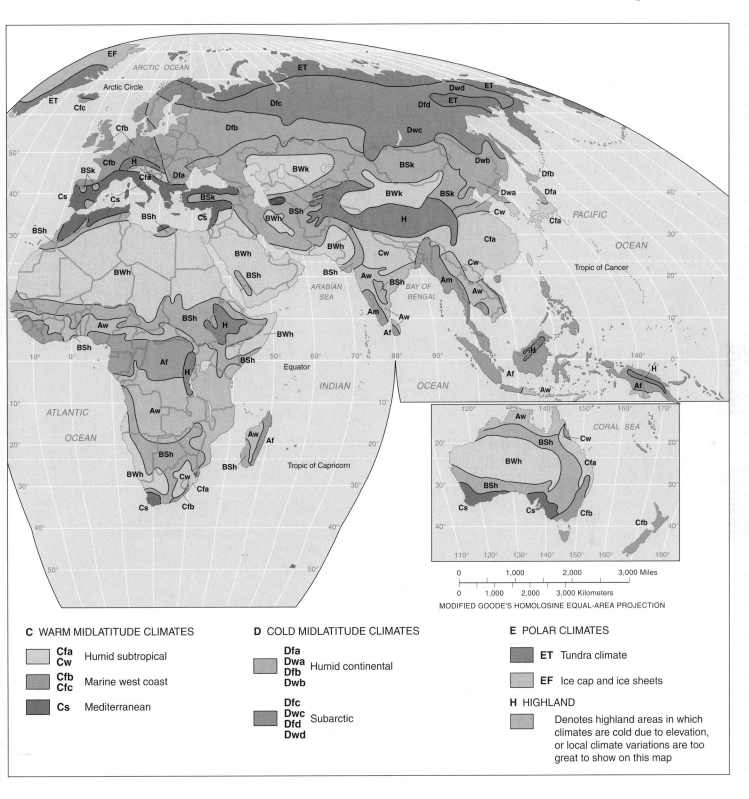

C WARM MIDLATITUDE CLIMATES

- **Cfa / Cw** Humid subtropical
- **Cfb / Cfc** Marine west coast
- **Cs** Mediterranean

D COLD MIDLATITUDE CLIMATES

- **Dfa / Dwa / Dfb / Dwb** Humid continental
- **Dfc / Dwc / Dfd / Dwd** Subarctic

E POLAR CLIMATES

- **ET** Tundra climate
- **EF** Ice cap and ice sheets

H HIGHLAND

- Denotes highland areas in which climates are cold due to elevation, or local climate variations are too great to show on this map

atmospheric circulation (Figure 2-23). In particular, clear correspondence exists between dry (B) climates and subtropical high-pressure cells. Dry climates occur on the western sides of continents in the latitudes between 20° and 35°, in both Northern and Southern Hemispheres.

You also can see the influence on climate caused by mountain ranges. The North American Rockies, the Andes of South America, and the Tibetan Plateau in Asia are mapped as H, or highland, climates. These areas are generally colder than lowland areas because of their high altitude, but they are also areas of great spatial variability in climate. East of the Rockies and Andes are rain shadow deserts. Africa's Sahara is a **desert climate,** or (B), climate because it is dominated by the subtropical high-pressure zone of the eastern Atlantic. This desert extends into interior Asia, because that region is isolated from oceanic moisture sources, both by mountains and distance from the sea.

In the following descriptions, we will focus on the core characteristics of climates, which are briefly summarized in Table 2-1 and described in detail in Appendix II. Graphs called *climographs* depict the monthly average temperature and precipitation for a typical site in each climate zone. You should recognize, however, that enormous variability occurs within each climatic region and that broad transitions, not sharp boundaries, exist between them. In high mountain regions, the high altitude causes cold conditions most of the year. Most mountain regions have much variation in elevation, with both snowcapped peaks and vegetated valleys.

Humid Low-Latitude Tropical Climates (A)

Humid tropical climates (Af, Am)

Humid tropical climates (Af and Am climates in Figure 2-31) lie mostly within 10° north and south of the equator, but can extend to 20° north and south. These areas are under the rainy ITCZ and include the world's tropical rain forests—its steamy jungles. Tropical climates share important characteristics: Throughout the year, they are warm and humid. Because tropical areas are warm throughout the year, the variation in temperature in a single day is greater than the difference between the warmest and coolest months of the year. (This is the opposite of what most midlatitude residents experience. There, temperature differences between winter and summer are far greater than between day and night.) In the tropics the air can hold a large amount of moisture because of the high temperatures, and precipitation usually is heavy.

High temperatures and ample rainfall mean that tropical climates support abundant plant growth and productive agriculture; if water supply and soil fertility are adequate, two or three harvests per year are possible. The high temperatures mean that a great deal of energy is available to evaporate water, so POTET is very high. Because precipitation is greater than evapotranspiration during most of the year, the tropics generally are humid. But many tropical areas experience distinct wet and dry seasons, as will be discussed below.

The most extensive areas of humid tropical climate are in the Amazon River basin of South America, equatorial Africa, and the islands of Southeast Asia. The small country of Singapore (Figure 2-32) typifies a humid tropical climate. Temperatures are warm throughout the year, averaging between 26°C and 28°C (79–82°F). Annual precipitation totals about

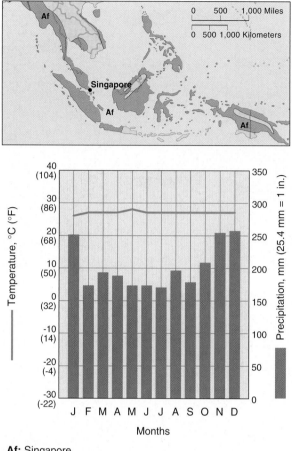

Af: Singapore
Precipitation: 241.3 cm
Temperature Range: 2°C
Lat/long: 1°10'N, 103°51'E

FIGURE 2-32 Locator map for Singapore and climograph.

240 centimeters (94 inches), and every month brings at least 150 millimeters (6 inches). Annual precipitation substantially exceeds evapotranspiration, so the area is very humid and features lush plant growth.

Humid tropical climates are influenced by the ITCZ most of the year, so rainfall in these areas primarily is convectional (Figure 2-33). Atmospheric instability results from intense daily insolation, and the high moisture content of the atmosphere favors development of severe convectional storms. Rain may fall in the afternoon or evening nearly every day, with mornings generally clear. Under these conditions, the annual totals of rainfall exceed 200 centimeters (80 inches) in many areas, and POTET is generally 120 to 170 centimeters (48 to 68 inches) per year. In oceanic areas of this climate, tropical cyclones are generated, forming hurricanes or typhoons. Southeastern Asia, the Pacific islands of Oceania, and the Caribbean are especially prone to these tropical storms. Some humid tropical climates experience a brief dry season. These are designated "Am."

FIGURE 2-33 Flooding in Kota Baru, Malasia.
Intense rainfall is characteristic of this tropical human (Af) climate, often contributing to flooding problems. (AP Wide World Photos)

Seasonally humid tropical climates (Aw)

In many areas of the humid tropics, rainfall is concentrated in part of the year, allowing for a distinct dry season. This is caused by seasonal shifts in the location of the ITCZ or by monsoonal circulation patterns (in southern and Southeastern Asia). The ITCZ moves north in the Northern Hemisphere summer (May–October) and south between November and April, and rainfall shifts with it.

Central and South America and Africa have substantial areas of seasonally wet climates caused by this shift in the ITCZ. In Managua, Nicaragua, for example (Figure 2-34), rainfall is heavily concentrated in the six-month period from May through October, with little rain falling in the other six months. Hurricanes, which occur during the high-Sun period, also contribute to high rainfall averages. Annual rainfall totals about 120 centimeters (48 inches), roughly half that of Singapore. Average monthly temperatures vary between 25° and 29°C (77–84°F). The seasonal contrasts are caused by movement of the ITCZ. The lack of moisture in the dry season contributes to strong seasonality of vegetation growth (Figure 2-35).

In southern and southeastern Asia precipitation is also seasonal, because of the monsoonal circulation pattern. In the summer the Asian landmass heats and develops an extensive low-pressure region that draws air in from all directions. The low-pressure cell draws over the land a steady flow of moist air from the Indian and Pacific Oceans, and the rainfall is intense. Monsoonal rains are critical to agriculture in Asia. The timing of this circulation change varies from year to year, and if the rain arrives too early or too late, crop yields may be severely affected.

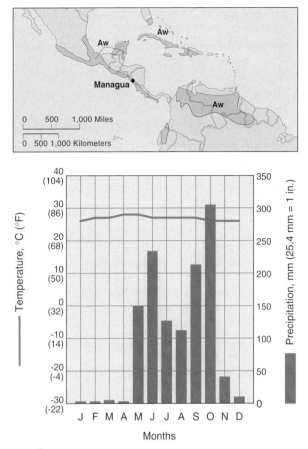

Aw: Managua
Precipitation: 120.4 cm
Temperature Range: 2°C
Lat/long: 1°0'N, 86°20'W

FIGURE 2-34 Locator map for Managua and climograph.

FIGURE 2-35 Savanna vegetation in eastern Africa. This type of vegetation is characteristic of seasonally humid tropical climates. In the background, Mt. Kilimanjaro, Tanzania, at 5,895 meters (19,340 feet) elevation, has an ice cap climate even though it is located just 3° south of the equator. Global warming has contributed to rapid shrinkage of the ice cover on Mt. Kilimanjaro; the ice may be gone by 2015. (Michele Burgess/Corbis/Stock Market)

Some of the highest average rainfall totals in the world (5–10 meters or 16–33 feet or even more) occur on the southern slopes of the Himalayas, where mountains cause orographic lifting. In winter, however, intense high pressure develops over central Asia, and the wind in southern and eastern Asia blows from the north and northwest. The air warms as it descends from the Tibetan Plateau, and extremely dry conditions result to the south of the Himalayas.

Dry Climates (BW and BS)

Dry climates cover 35 percent of Earth's land area, although fewer than 15 percent of the world's inhabitants live in such areas. Drylands are generally located in bands immediately to the north and south of the low-latitude humid climates. The most extensive region of dry lands extends from North Africa to Central Asia; the African portion is known as the Sahara (the Arabic word for "desert"). Sand dunes cover only a small portion of the Sahara and other drylands; more common is a cover of stone or gravelly soil.

Dry climates are distinguished by a formula that relates average annual temperature to average annual rainfall. Higher rainfall totals are necessary to classify a place as humid (A, C, or D) under warm conditions than under cold conditions. Dry climates can be subdivided into warm types (BWh, BSh) and cool types (BWk, BSk).

Desert climates (BWh, BWk)
The most extensive regions of warm, dry climate are found in the subtropics on the western sides of continents. It is interesting to study this in Figure 2-31. The deserts (BWh) of northwestern Mexico and the Atacama Desert of coastal Peru border the Pacific; the Sahara and Namib/Kalahari deserts of northern and southern Africa border the North and South Atlantic; and the Australian Desert borders the Indian Ocean. These BWh climate regions result from air descending in the subtropical high-pressure zones, over the eastern parts of the oceans (in other words, off the west coasts of the continents).

Cold ocean currents can also increase the aridity of these areas. Ocean currents generally circulate clockwise in the Northern Hemisphere and counterclockwise in the Southern Hemisphere. Thus, on the eastern side of the ocean, the currents flow from high latitude to low latitude, bringing cooler water toward the tropics. The cold ocean surface cools the lower layers of the atmosphere, inhibiting evaporation and convection and therefore reducing precipitation.

In warm desert climates like that of Cairo, Egypt, high temperatures exceeding 40°C (104°F) are common (Figure 2-36). The high temperatures combined with low rainfall produce the greatest differences between annual precipitation and POTET. Annual POTET exceeds 150 centimeters (60 inches) in many

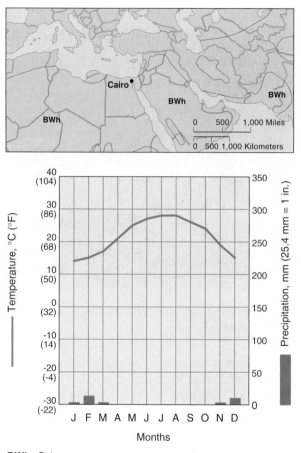

BWh: Cairo
Precipitation: 3.0 cm
Temperature Range: 14°C
Lat/long: 30°1'N, 31°14'E

FIGURE 2-36 Locator map for Cairo and climograph.

BW regions; but as precipitation totals less than 20 centimeters (8 inches), the demand greatly exceeds the supply. In these conditions only plants with specialized water-collecting and water-holding characteristics can survive (Figure 2-37).

Semiarid climates (BSh, BSk)
Not all of the world's drylands are barren of vegetation; in many areas sufficient rain falls to support extensive seasonal or perennial plant cover. A climate with enough moisture to support such vegetation is semiarid, and typically regions of **semiarid climate** lie in transitional areas between deserts and more humid regions. Semiarid climates are designated BS in the Köppen scheme. Annual precipitation is only 20–50 centimeters (8–20 inches). Most of the world's grasslands, or **steppes,** are in this climate type, which supports extensive grazing activities but only limited agriculture (Figure 2-38).

Some semiarid climates are caused by rain shadow effects that isolate the continental interiors from moisture sources. In Asia, for example, a mountain belt

FIGURE 2-37 A desert environment in Tunisia.
Potential evapotranspiration in this warm desert is much
greater than precipitation, and only vegetation adapted to
severe moisture stress can survive. (Richard Steedman/Corbis/
Stock Market)

stretches east to west from the Caucasus to China,
blocking Indian Ocean moisture from the interior. The
North American Rockies and South American Andes
similarly block Pacific moisture.

Seasonal temperature contrasts are characteristic
of midlatitude BSk climates. Long summer days and
bright sunshine warm the ground, and because of lack
of water, this heat is not dissipated through evaporation. Unlike tropical deserts, however, midlatitude dry
climates are cool part of the year, so that plants require
less moisture to survive (Figure 2-39). In some areas,
such as interior basins of the western United States or
western China, agriculture is possible with water imported from wetter mountain areas.

Warm Midlatitude Climates (C)

In the midlatitudes, seasonal variations of insolation
profoundly influence temperature. In summer these
latitudes receive more radiation per day than equatorial areas. In winter, however, insolation is cut to half or
even a third of summer levels. Therefore, the midlatitudes experience a distinct cool season, commonly
known as a winter. In subtropical locations, this winter
season may be a month or two in which frost can occur.
Winter causes life to become restricted to organisms
that can tolerate at least occasional freezing conditions.

Lower temperatures mean that annual precipitation and evapotranspiration are generally less in midlatitude climates than in the tropics. Like the tropics,
however, midlatitude climates include a wide range of
precipitation patterns, including continuous humidity
and distinctly seasonal rainfall. Precipitation in the midlatitudes is heavily influenced by the polar front—the

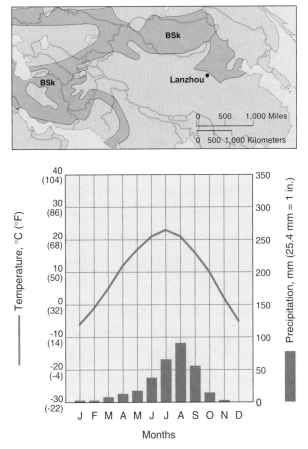

BSk: Lanzhou
Precipitation: 31.2 cm
Temperature Range: 29°C
Lat/long: 36°4'N, 103°44'E

FIGURE 2-38 Locator map for Lanzhou and
climograph.

FIGURE 2-39 Caucasus Mountains. Semiarid climates
have enough moisture for significant plant growth at least part
of the year, but moisture shortages are common. Grasses, such
as those in this area in the Caucasus Mountains of Georgia,
southwestern Asia, are particularly well adapted to semiarid
climates. (D. Thomas/Photo Researchers, Inc.)

boundary between warm tropical air and cold polar air along which midlatitude cyclones form and travel. The significance of these storm systems is that daily weather tends to be more variable in the midlatitudes than in the tropics. This climatic zone experiences rainy spells separated by dry spells, and rainfall can occur at any time of day rather than concentrated in the afternoon and evening.

In many midlatitude environments—in the central United States, for example—seasonal precipitation patterns also are influenced by temperature. Because warmer air holds more moisture, greater rainfall occurs in summer than in winter. In the southeastern United States, midlatitude cyclones are less frequent in summer than in winter. But summer rainfall is more intense because more water is present in the air, so winter and summer receive similar amounts of rainfall.

Humid subtropical climates (Cfa, Cw)

Humid subtropical climates occur in latitudes between about 25° and 40° on the eastern sides of continents and between about 35° and 50° on the western sides. These climates are relatively warm most of the year but have at least occasional freezing temperatures during the winter. Frost occurs, so many plants adapted to tropical humid climates cannot survive, and most humid subtropical climates have deciduous species that lose their leaves in autumn and become dormant in winter. Eastern China, the southeastern United States, and parts of Brazil and Argentina are the largest areas of humid subtropical climates.

The subtropical high-pressure zones that create deserts on the western margins of continents (the eastern margins of oceans) create humid conditions on the eastern sides of continents. The circulation around these high-pressure cells brings warm, moist air from low latitudes to the eastern margins of the continents (clockwise in the Northern Hemisphere and counterclockwise in the Southern Hemisphere). This moisture supplies energy to midlatitude cyclones that generate plentiful rainfall in these areas.

Precipitation may occur all year or be concentrated in summer. Summer rainfall may result from greater moisture in the warmer air, as in New Orleans (Figure 2-40), or from monsoonal effects, as in southeastern China. Cool winter temperatures bring occasional frost in low latitudes, and snowfall occurs poleward, but the ground is not frozen or snow-covered for long. The warm, humid conditions of Cfa climates support a long growing season with forest vegetation under natural conditions (Figure 2-41) and high crop yields in agricultural regions, sometimes with two harvests in a year.

In parts of southern and Southeast Asia, the monsoon circulation causes a distinct winter dry spell, producing a subtropical winter-dry climate (Cw). The reduced precipitation falls at a time of year when POTET is low, and it does not have as much of an

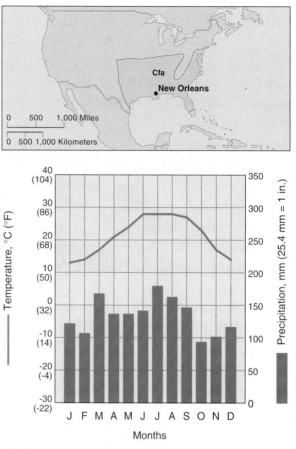

Cfa: New Orleans
Precipitation: 161.5 cm
Temperature Range: 15°C
Lat/long: 30°0'N, 90°5'W

FIGURE 2-40 Locator map for New Orleans and climograph.

FIGURE 2-41 Wetlands in Louisiana. Humid subtropical climates support lush vegetation, such as this wetland forest in Louisiana. Many of the world's most productive agricultural regions are in this climate. (Randy O'Rourke/Corbis/Stock Market)

impact on agriculture or water resources as it would if it occurred in the summer.

Marine west coast climates (Cfb, Cfc)

On continental west coasts between about 35° and 65° are mild climates with small temperature variations and plentiful moisture year-round. Major regions include western North America between northern California and coastal Alaska, southern Chile, northwestern Europe, southeastern Australia, and New Zealand. These areas are cooler than humid subtropical climates (Cfa), especially in summer. They are subdivided into cooler types (Cfc) and warmer types (Cfb).

Marine west coast climates are moderated by ocean temperatures. Typical summer temperatures are 15°C to 25°C (60–77°F), rather than the 25°C to 35°C (77–95°F) inland at these latitudes. Frost-tolerant plants stay green all year in the mild winters of −5° to +15°C (23–59°F), even if snow falls. Maritime influence on temperature is so great that winter temperatures normally associated with subtropical latitudes are found as far poleward as 55° (Figure 2-42). Palm trees survive in coastal areas of southwest England, at the same latitude

FIGURE 2-43 The coast of British Columbia. This region has mild winters despite its relatively high latitude (49° N). (Neil Rabinowitz/CORBIS—NY)

as Winnipeg, Manitoba, in south-central Canada. Kodiak, Alaska, in a maritime location at latitude 57° N, has the same January average temperature as Richmond, Virginia, which is under continental climate influence at latitude 37°N. Clearly a location on the West Coast in midlatitudes significantly raises winter temperatures relative to East Coast or interior locations.

The North Atlantic and the northeastern and southern Pacific are known for stormy weather, and these storms bring plenty of moisture to the adjacent land areas (Figure 2-43). Midlatitude cyclones are frequent, especially in winter. The low temperatures mean that air contains less moisture than at lower latitudes, so drizzle is common. Annual rainfall is about 70 to 120 centimeters (28 to 47 inches). Where mountains are near the coast, as in western North America, Chile, and Norway, orographic effects contribute to high rainfall, 2 to 3 meters (6 to 9 feet) per year. The current highest average annual rainfall in the mainland United States is in a marine climate—the west coast of the Olympic Peninsula in Washington, where annual rainfall totals 3 to 4 meters (10 to 13 feet).

Mediterranean climates (Cs)

On the western margins of continents, another distinctively seasonal climate occurs. The dry-summer **Mediterranean climate** envelops the Mediterranean region of Europe and parts of northern Africa, lending this climate type its familiar name. A similar climate occurs in extreme southern Africa, southwestern Australia, California, and central coastal Chile. Precipitation is seasonal and caused by northward and southward movement of the subtropical high-pressure zones. In summer, these zones move poleward and bring aridity; in winter they move toward the equator and are replaced by more frequent storms of the midlatitude low-pressure zone.

In Mediterranean climates, cool, rainy winters bring occasional frost or snow, with frequent midlatitude

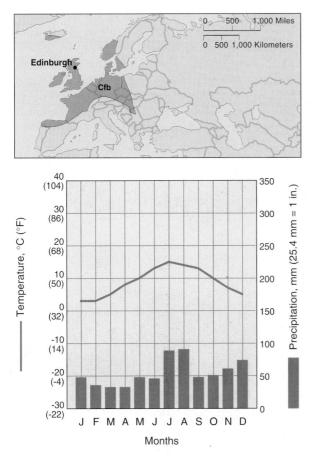

Cfb: Edinburgh
Precipitation: 65.8 cm
Temperature Range: 12°C
Lat/long: 55°57'N, 3°12'W

FIGURE 2-42 Locator map for Edinburgh and climograph.

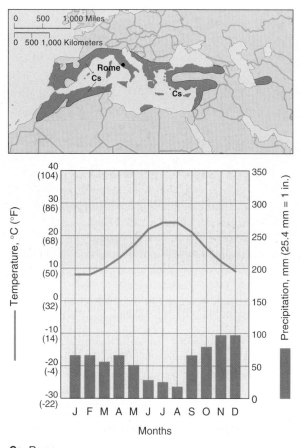

Cs: Rome
Precipitation: 71.4 cm
Temperature Range: 16°C
Lat/long: 41°54'N, 12°30'E

FIGURE 2-44 Locator map for Rome and climograph.

cyclones. Warm, dry summers have temperatures of 20°C to 35°C (68–95°F) (Figure 2-44). In Rome, Italy, for example, little or no rain may fall between May and October.

Mediterranean climates have significant water-availability problems. Precipitation occurs in winter, when human, animal, and vegetative demand for water is low. However, very little rainfall occurs in the summer, when POTET is highest. Plants with roots that can reach deeper soil moisture stay green during the summer, when shallow-rooted grasses dry out. Because water is scarce in the summer, many plants grow mostly in the winter and spring, even though there is less sunlight then (Figure 2-45). Agriculture in Mediterranean climate regions requires storage of winter precipitation in reservoirs for irrigation in summer.

Cold Midlatitude Climates (D)

Humid continental climates (Dfa, Dwa, Dfb, Dwb) Like Mediterranean climates, **humid continental climates** have strong seasonal contrasts. In the latter case, however, the contrasts are of temperature rather than moisture availability. Continental climates are so named because they are remote from the ocean and therefore deprived of the sea's input of moisture and moderating influence on temperature. These climates occur between about 35° and 60° latitude in the interior and eastern portions of Northern Hemisphere continents. Many eastern coastal areas at high latitudes also have strong seasonal contrasts because the prevailing westerly winds extend this

FIGURE 2-45 Variation in water availability. Seasonal variations in water availability change the appearance of the landscape in Mediterranean climate regions, such as coastal California. The wet season (left) is typically winter and spring; the dry season (right) is summer and autumn. (Connie Coleman/Getty Images Inc.—Stone Allstock)

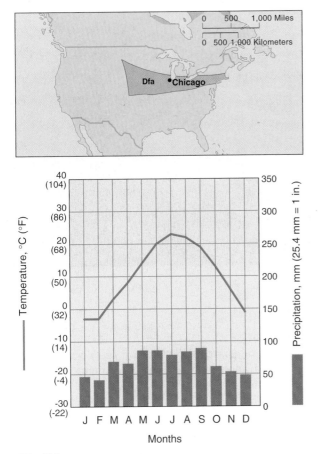

Dfa: Chicago
Precipitation: 80.8 cm
Temperature Range: 26°C
Lat/long: 41°45'N, 87°40'W

FIGURE 2-46 Locator map for Chicago and climograph.

humid continental climate to them. D climates occur exclusively in the Northern Hemisphere because very little landmass exists in the Southern Hemisphere at these latitudes.

Summer temperatures are warm because long days and high solar angles cause more radiation in a day than tropical locations receive. In winter, days are short and the Sun is low in the sky, so temperatures often fall below freezing. In more southerly humid continental areas like Chicago, Illinois (lat. 42° N), January temperatures may average below freezing, but rain still occurs in winter (Figure 2-46). In more northerly places like Moscow, Russia (lat. 56° N), winter brings many weeks of subfreezing temperatures.

In most humid continental climates, precipitation occurs throughout the year from frequent midlatitude cyclones. Typical annual rainfalls are 60 to 150 centimeters (24 to 60 inches), usually ample for plants, with an annual evapotranspiration of 50 to 90 centimeters (20 to 35 inches). In winter, plants need little water, but in summer they demand every drop that falls and sometimes more (Figure 2-47). During summer dry spells, farm fields may require supplemental irrigation.

Subarctic climates (Dfc, Dwc, Dfd, Dwd)

At the northern edge of the humid continental climates, winter temperatures are cold enough and the growing season short enough that agriculture is generally not possible (Figure 2-48). The only plants that survive in these areas are those able to weather extreme cold. No abrupt temperature or precipitation boundary exists between the humid continental climate and the **subarctic climate**; the difference literally is one of degrees. Conifers dominate; the broad expanse of spruce, fir, and larch covering much of northern Canada and Siberia is called *boreal* (meaning northern) *forest*, lending the

FIGURE 2-47 Birch woodland. Humid continental climates have strong seasonal contrasts, as seen in these two views of a birch woodland in winter and summer. (Gregory K. Scott/Photo Researchers, Inc.)

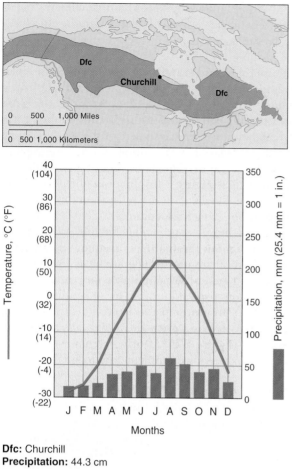

Dfc: Churchill
Precipitation: 44.3 cm
Temperature Range: 40°C
Lat/long: 58°45'N, 94°5'W

FIGURE 2-48 Locator map for Churchill and climograph.

FIGURE 2-49 Winter near Hudson Bay, Canada, a subarctic climate region. (Stephen J. Krasemann/Photo Researchers, Inc.)

name boreal forest climate to the subarctic climate (Figure 2-49). Like the humid continental climate, this climate is essentially limited to the Northern Hemisphere, because so little landmass exists at these latitudes in the Southern Hemisphere.

Extremely cold −20°C to −10°C (−4°F to +14°F) subarctic winters feature snow cover and months of temperatures well below freezing. As days lengthen in spring, temperatures rise rapidly, and 15- to 20-hour summer days bring mild temperatures approaching those of much warmer climates. The growing season is short, generally June to September, in a cool summer (10°C to 20°C, 50°F to 68°F), with warm spells above 30°C (86°F). Temperatures warm enough for plant growth occur as early as April, but frosts may come as late as June. Similarly, the first frost is often in September, even though temperatures to 20°C (68°F) occur into early October.

Annual rainfall is low, generally 20 to 50 centimeters (8 to 20 inches), but ample for plants because virtually no evapotranspiration occurs for six months of the year. Winter precipitation usually is locked up as snow and ice until the spring melt releases this water to

plants. Summer water shortages are rare, even though evapotranspiration may exceed precipitation during part of the summer.

Polar Climates (E)

High-latitude climates are characterized by two important features: low average temperatures and extreme seasonal variability. These climates occur at and around the poles and are differentiated by temperature alone, because temperature determines whether water is present in liquid form. These climates result from scant insolation, varying from zero during winter at the poles to among the highest daily totals on Earth for a brief period at midsummer.

Tundra climates (ET) Tundra temperatures are low all the year, rising above freezing only for a brief, cool summer (Figure 2-50). The annual average temperature is below freezing. Winters below −20°C (−4°F) are common, typically staying below freezing for three to five months. Midsummer daytime highs rarely exceed 10°C to 15°C (50°F to 59°F), even though the Sun can shine 20 to 24 hours a day. The soil 1.5 to 2 meters (5 to 6.5 feet) below the surface is insulated from seasonal temperature changes, so at this depth soil temperatures are close to the annual average. The condition of permanently frozen ground is known as **permafrost.**

Annual precipitation rarely exceeds 30 centimeters (12 inches). Most falls as snow, covering the ground for months except where the ground is bared by the wind. Although annual rainfall is slight, water is plentiful because evapotranspiration rates are low. Most of the year water is present as ice rather than liquid. In the brief summer when the ground thaws, the water may not be able to drain from the soil because of

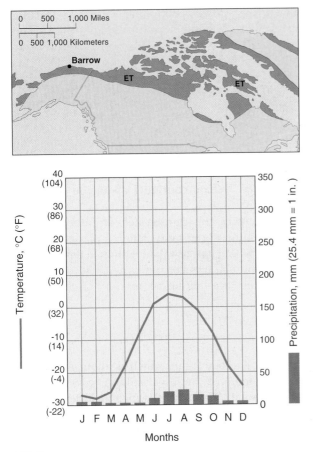

ET: Barrow
Precipitation: 11.0 cm
Temperature Range: 32°C
Lat/long: 71°28'N, 156°0'W

FIGURE 2-50 Locator map for Barrow and climograph.

FIGURE 2-51 Ellesmere Island. Flowers bloom during the brief summer growing season in Ellesmere Island, northern Canada, located in a tundra climate region. (Stephen J. Krasemann/Photo Researchers, Inc.)

permafrost below the surface; saturated soils are therefore common.

At its northern limit, the boreal forest gives way to a treeless landscape in which vegetation survives the long winter by staying close to the ground, where it is protected from winter's extremes (Figure 2-51). This low-lying hardy vegetation, called tundra, gives the **tundra climate** its name.

Ice cap climates (EF)

Near the poles, and high in some mountains at lower latitudes, are climates in which even the warmest month averages below freezing (Figure 2-52). Virtually all of Antarctica and most of the Arctic have this **ice cap climate.** Permafrost is extensive and thick beneath a year-round cover of snow or ice. The extremely low temperatures prevent the air from holding much water. Annual precipitation usually is less than 10 centimeters (4 inches), falling as snow and compacting into a thick mass of glacial ice that flows to the sea rather than melting in place (Figure 2-53). Cores of ice extracted in Antarctica indicate average annual precipitation of at most a few centimeters. Due

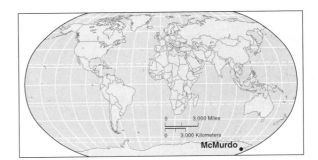

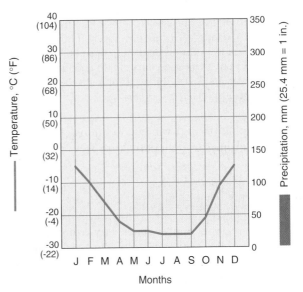

EF: McMurdo
Precipitation: 0 cm
Temperature Range: 21°C
Lat/long: 77°0'S, 170°0'E

FIGURE 2-52 Locator map for McMurdo and climograph.

FIGURE 2-53 The edge of the Antarctic ice cap. In ice cap climates, temperatures are only rarely above freezing, and glaciers form from the accumulation of snow. Virtually all of Antarctica has an ice cap climate, with little glacial melting. The glaciers flow to the sea, forming icebergs. (Robert W. Hernandez/Photo Researchers, Inc.)

to low precipitation and windy conditions, Antarctica contains desertlike dry valleys that are free from snow or ice cover.

CLIMATE CHANGE

So far we have seen that climates vary from place to place. Climates also vary over time. The world climate map is based on averages over periods of a few decades in the mid-twentieth century. However, the world climate map has not always looked like this. Considerable change has taken place through the years and is likely to continue in the future.

Humans have long been able to modify environments, sometimes radically, by building settlements, clearing forests for agriculture, domesticating animals, manufacturing goods, damming rivers, extracting resources, and spreading waste throughout the air and seas and across the land. Earth's land surface is profoundly different today from what it was only 200 years ago. For example, the forests of the eastern United States were almost completely removed between 1600 and 1900, and an entire ecological community—the tall-grass prairie of the Great Plains—has been replaced by agricultural crops since 1800. Only recently have we come to realize that humankind, in addition to modifying Earth's surface, has set in motion global-scale climatic changes. Debate continues about the nature of these changes.

To interpret the significance of future climate change, we must understand patterns of past climate changes. These can give us valuable clues in determining whether recent changes are similar to those of the past or whether they are unique because of human actions.

Climatic Change over Geologic Time

Viewed over geologic time, which is about 4.6 billion years, the climate of the last 2 million years has been quite exceptional. This period, which includes our present time, is known to geologists as the **Quaternary Period.** Earth has experienced more climatic variability during the past 2 million years than in most of the previous 200 million years. Periods of similar variability probably occurred prior to 200 million years ago, but the record of the first 4 billion years of Earth's history is not very clear. In any case, it is clear that the Quaternary glaciations are an exceptional episode in recent Earth history.

Climate during the Quaternary Period has included intervals in which global average temperature was as much as 10°C (18°F) cooler than the present and warm intervals in which the climate was as warm or warmer than today. In the last 1 million years there have been about 10 of these cold intervals, occurring fairly regularly about once per 100,000 years (Figure 2-54). During the cooler periods, great continental ice sheets, like those covering much of Antarctica and Greenland today, extended over much of North America, northern Europe, and northern Asia. Periods of glaciation were also periods of lowered sea level, because water was taken out of the sea and stored on land in the form of glacial ice. The portion of the Quaternary Period in which these glaciations occurred is called the **Pleistocene Epoch.** In everyday language we describe these times of cold climate as the Ice Ages. Even areas not covered by ice were much different in the past. Areas that today are covered with deciduous forest were frozen tundra, and lakes existed in places that now are deserts. Subtropical climates extended well into the midlatitudes.

Earth's climate shifted between warm and cool periods between 10 and 30 times during the Pleistocene

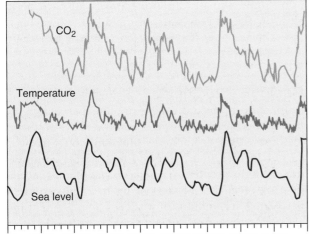

FIGURE 2-54 Atmospheric carbon dioxide, temperature, and sea level over the last 450,000 years. Carbon dioxide and temperature data are based on samples from the Vostok ice core, Antarctica. Sea levels are based on deep-sea sediment cores. The strong similarities in the patterns demonstrate that atmospheric carbon dioxide is very closely linked to climate, particularly temperature and glacial cycles. The timing of the major glaciations at an interval of about 100,000 years suggests that variations in the geometry of Earth's orbit around the Sun are key, although interactions among the biosphere, oceans, and atmosphere are also important. (Redrawn from Sigman, D. M., and Boyle, E. A., in *Nature* 407 (2000): 859)

Epoch; the exact number of cold periods is not known. The last cold period reached its maximum, as indicated by the extent of glacial ice, only about 18,000 years ago—within the period of the human archaeological record. At that time global average temperatures were about 5°C (9°F) cooler than the present. The melting of the ice back to its present extent was completed only about 9,000 years ago, or about 7000 B.C. We now appear to be within a period of relatively warm climate.

Within the past 1,000 years climate has varied, though less dramatically than in the distant past. Recent climate changes have been important in European history. For example, between A.D. 800 and 1000, seafaring pirates and adventurers from present-day Scandinavia, called the Vikings, extensively explored the North Atlantic Ocean and established settlements in Greenland and North America. Lack of sea ice during those centuries as a result of especially warm temperatures aided the Vikings' exploration.

Cooling occurred in the Middle Ages, beginning about A.D. 1200. The period from about 1500 to 1750, when temperatures were especially cool, is known as the **Little Ice Age.** Glaciers advanced in Europe, North America, and Asia. Since the early 1800s and especially after 1900, climates have warmed relatively steadily. Most of Earth's glaciers have been shrinking since the early 1800s. Short cooling intervals occurred in the 1800s, including one in 1884–1887 caused by the eruption of the

volcanic island of Krakatau in Indonesia. Krakatau spewed so much ash and sulfur dioxide into the atmosphere that it prevented some of the shortwave insolation from reaching Earth's surface. The 1930s and 1940s were relatively warm, then cooling occurred from about 1945 to 1970. Climate has warmed dramatically since 1975. It is important that these climatic fluctuations have occurred well within the period of human occupation of Earth. As noted, glacial melting from the most recent Ice Age was particularly rapid about 15,000 to 10,000 years ago. Archaeologists have found that agriculture and cities probably developed about 10,000 to 8,000 years ago. Many scholars of early human culture believe climate change was fundamental to the development of civilizations, influencing the availability of water for crops and perhaps driving major migrations.

The period of glacial melting and warming between 15,000 and 5,000 years ago coincided with expansion of human settlement in northern Europe and North America. Settlement extended gradually northward on the European mainland and became well established throughout much of the British Isles by 2000 B.C. Meltwater from the glaciers raised sea level, flooding low-lying areas and separating land areas. Water filled the English Channel, isolating the British Isles from Europe, and the land bridge that once connected North America and Asia went underwater, halting migration from Asia into North America. Rising sea level also led to rapid filling of the Black Sea about 7,500 years ago.

Modern human inventions such as heating, housing, transportation, and industry have made us less dependent on day-to-day conditions. But we remain strongly influenced by the dynamics of the atmosphere, both short term (weather) and long term (climate). Many believe that as a result of human actions, climate will change much more rapidly in the future.

Possible Causes of Climatic Variation

Why does climate change over time? Many potential causes exist, but it is difficult to say which are important and which we can ignore. Understanding the causes of climatic change is of critical importance, because if we learn that climate is affected by human activity, such as burning fossil fuels (coal, oil, and natural gas), then we have the opportunity to take action to limit such effects. On the other hand, if climatic change were entirely governed by natural processes, then there would be no need for concern about potential human impacts on the atmosphere. We will summarize three of the main possible causes of climatic change: astronomical factors, geologic processes, and human modification of Earth's surface and atmosphere.

Astronomical hypotheses
Inputs of solar energy drive atmospheric circulation and thus climate. Changes in insolation could alter climate. Humans know little

about variations in the amount of solar radiation reaching Earth, because it is difficult to observe the Sun through the filter of the atmosphere. Recent studies based on satellite observations of the Sun are beginning to suggest that variations in solar output could be responsible for some of the climatic variations observed in the last few hundred years, causing some (but not all) of the warming of the twentieth century.

In the long run—over tens of thousands of years—we know that the geometry of Earth's revolution around the Sun fluctuates (Figure 2-55). Earth spins like a top, but a slightly unbalanced one, and because of this imbalance it wobbles slightly and very slowly as it spins. The wobble causes the tilt of Earth's axis to vary between 22° and 24° instead of staying constant at around 231/2°. Other variations in the geometry of Earth's orbit around the Sun also occur, with periods of tens of thousands of years. It appears that the 100,000-year timing of major cold and warm periods over the past 1–2 million years corresponds to these variations in Earth-Sun geometry, although the mechanisms of this linkage are controversial. These variations do not explain shorter-term climatic change, however.

One short-term astronomical factor that may cause variation in the Sun's radiation output is sunspots. Sunspots are relatively cool regions on the surface of the Sun that vary in number and appear and disappear over a cycle lasting 11 years. Sunspot numbers also vary over longer periods. Sunspots affect the output of solar energy. Some climate variations appear to have periods close to the 11-year sunspot cycle. The sunspot cycle also affects concentrations of ozone in Earth's upper atmosphere.

Geologic hypotheses

Geologic factors also may cause short-term and long-term climatic change. One longtime-scale mechanism is continental drift, or plate tectonics. This remarkable phenomenon will be discussed further in Chapter 3, but briefly, continental drift is the movement of great crustal plates that make up Earth's surface at rates of up to a few centimeters per year. After millions of years continents might move great distances, opening up oceans or draining them. In addition to affecting ocean shapes and currents, the locations of continents and oceans directly affect many features of atmospheric circulation, including the ITCZ, the monsoon circulation of Asia, midlatitude cyclones, and the stormy zone around Antarctica.

Movements of continental plates have also caused the formation of major mountain ranges, such as the Himalayas, Andes, and Rocky Mountains. Continental movements are ponderously slow, so they cannot explain variations within the Quaternary Period. But they might help explain why no major glaciations occurred for more than 200 million years before then.

Volcanic eruptions can influence climate for a few years by injecting large amounts of dust and gases—especially sulfur dioxide—into the upper atmosphere. These gases reduce the amount of solar radiation filtering through the atmosphere to Earth. Past periods of more frequent volcanic eruptions may have lowered temperatures at other times.

Human causes

Processes like sunspot cycles and continental drift are beyond human control. However, humans are active participants in the changing climate. Two important ways that humans influence climate are altering the atmosphere and removing vegetation.

The carbon dioxide content of the atmosphere has increased dramatically since the start of the Industrial Revolution in the late eighteenth century (Figure 2-56). The elevated CO_2 levels cause warming, because CO_2 is a greenhouse gas. Analyses of air trapped in Antarctic ice reveal that the concentration of carbon dioxide was higher during past warm periods and lower during glacial periods (Figure 2-54). However, some past changes in CO_2 concentrations appear to be a result of changes in temperature, not a cause. Dead plant matter lying in swamps and in the soil decays during periods of warm climate, releasing more carbon dioxide to the atmosphere. During cool periods carbon accumulates on Earth's surface, reducing atmospheric CO_2. Also, the oceans may store more carbon when they are cold and less when they are warm. We can see a very close link between climate and the biosphere.

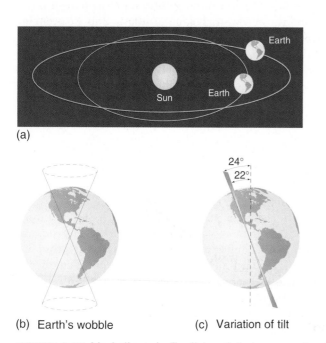

(a)

(b) Earth's wobble   (c) Variation of tilt

FIGURE 2-55 Variations in Earth's orbital geometry. Climate may vary with long-term variations in the geometry of Earth's orbit around the Sun. These include (a) the annual variation of distance between Earth and the Sun, (b) the orientation of the Earth's axis, and (c) the degree of tilt of that axis relative to our orbit around the Sun.

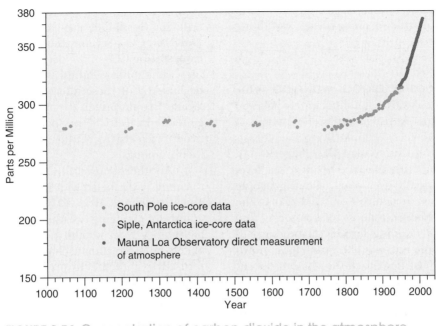

FIGURE 2-56 Concentration of carbon dioxide in the atmosphere. Carbon dioxide concentrations are now about one-third higher than two centuries ago, as a result of burning fossil fuels in factories, cars, and homes.

Global Warming

During the twentieth century, Earth's temperature increased slightly less than 1°C (1.8°F) (Figure 2-57). Throughout the 1990s the body of scientific evidence linking this temperature rise to emissions of CO_2 accumulated rapidly, so that today few scientists doubt that increased CO_2 is the principal cause of global warming. The future of global warming is always uncertain, but the consensus is that unless output of CO_2 slows dramatically, world average temperature could rise by a few degrees celsius in the next century.

Despite this scientific consensus, there are many uncertainties. One is that we do not know how rapidly atmospheric CO_2 content will increase. That will depend on whether we continue to expand our use of fossil fuels as we have in the past. Another uncertainty is that we are not sure how water in the atmosphere, which plays a major role in regulating climate, will be affected by the increased levels of greenhouse gases.

Thus, we have only limited knowledge of how global warming will affect specific regions. Will storm tracks shift? How will a place's current levels of rainfall, snowfall, and temperature change? Will the seasons

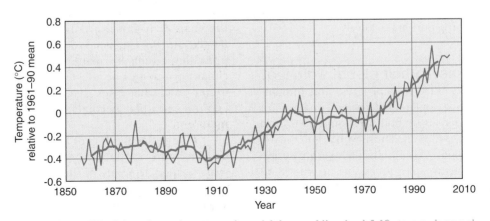

FIGURE 2-57 World surface temperature history of the last 148 years, based on records of measured temperatures. Temperatures are shown relative to the average for 1961–1990. The trend shows three distinct periods of warming, from about 1855–1880, 1910–1945, and 1975–2005. The total warming over the period of record is about 1°C (1.8°F).

change? How will plants and animals be affected? How large an effect will this have on sea level, worldwide energy use, and food supply?

The consequences of global warming

One serious effect of global warming that is widely believed to be likely is a worldwide rise in sea level of perhaps 1 to 5 meters (3 to 16 feet). People living near coasts would face danger from rising seas. The danger would not be from constant inundation, because sea level would rise very gradually over years, allowing people time to relocate, raise structures, or build dikes. The danger would be from occasional severe storms that would cause sudden flooding farther inland, such as happened in hurricane Katrina. The Dutch have shown that well-built dikes can hold back the sea, but poorer countries cannot afford such protection.

Another possibility is that climate change could reduce water supply in some regions. Consider a mid-latitude environment such as eastern Nebraska that averages 60 centimeters (24 inches) precipitation and 55 centimeters (22 inches) evapotranspiration. The 5 centimeters (2 inches) of precipitation that is not transpired by plants flow into streams and rivers. A small increase in evapotranspiration, due to a warmer climate, could sharply reduce water flow to the region's streams and rivers. Semiarid regions and densely populated subhumid areas depend on river flow for irrigation, drinking water, and waste removal. These areas might suffer severe water shortages if warming increases evapotranspiration and decreases river flow. However, if this warming brings greater precipitation, agriculture may be helped rather than harmed. Agricultural production might be especially helped in areas currently receiving little precipitation. Warming could also lengthen the growing season in high-latitude areas and make agriculture possible in areas of Canada and Siberia where today it is not. A third possibility is that warming will increase storminess (see Critical Thinking box on global warming and hurricanes).

Should we attempt to halt global warming?

Although global warming is in progress and humans are significant contributors to the problem, we have not reached global consensus on how we should try to stop it or if we should stop it at all. Those who argue for immediate action emphasize the potentially severe consequences of warming in some areas. But it is hard to convince people to spend money to prevent an event that occurs over long periods of time rather than in the short term.

Reducing carbon dioxide concentrations will be difficult because we depend on fossil fuels in our daily lives, and energy producers employ many people and earn billions of dollars each year. Significantly reducing fossil-fuel use is possible only if we consume less energy or shift to alternative energy sources. An alternative to reducing fossil-fuel use is to try to trap and store (*sequester*) CO_2. Several alternatives are being discussed, from storing CO_2 in the deep ocean or underground to increasing photosynthesis by fertilizing plants in the ocean. Any of these alternatives would be expensive. Would people prefer to cut back on their use of coal-generated electricity or spend money on new ways to produce electricity? Either alternative is expensive and inconvenient.

Another reason for not acting to curb global warming is the belief that it may be easier to adapt to climatic change than to prevent it. People already adjust to changing weather, commodity prices, and technology from year to year, so why shouldn't we be able to adjust to climate change, too? The United States refused to join the Kyoto Protocol, an international treaty that would limit CO_2 emissions (see Chapter 13) on the grounds that a strategy of adapting to global warming would be less costly and more effective than attempting to reduce it.

We must remember, though, that humans are not the only life on the planet. We are only one form of life and one part of many interacting ecosystems. Some animals and plants could not adapt to a climate change, and thus *our* human-environment interaction might be responsible for the extinction of other species. Is the possibility that we can adapt to climate change sufficient reason not to halt our contribution to global warming?

CONCLUSION: CRITICAL ISSUES FOR THE FUTURE

While we are reasonably certain that global warming is caused at least in part by human activity, there are many uncertainties. Foremost among the uncertainties is how the change in global average temperature will be reflected at the regional and local level. How will prevailing circulation patterns change? How will precipitation amounts be affected? What areas will become wetter, and what ones dryer? Will storms and other extreme weather become more frequent or more intense?

Our understanding of the dynamics of Earth's climate is improving rapidly, and within the next several years answers to many of these questions will be available. When this happens we will have to decide what actions, if any, should be taken in response to climate change. We may want to prevent further impacts, or we may decide that the costs of living with them are less than the costs of prevention. The first of these challenges primarily concerns atmospheric scientists, while the second is more of a social and political problem. Both are key geographical issues.

Chapter Review

SUMMARY

Solar radiation received by places at a particular latitude varies with season and by day length. The annual total radiation is highest and varies least over the seasons in low latitudes, while day length and angle of incidence cause greater seasonal variations at high latitudes. Energy arrives from the Sun as shortwave energy and is reradiated by Earth as longwave radiation. Insolation absorbed by Earth's surface and atmosphere is eventually returned to space. Human activities may be leading to global warming, especially through emission into the atmosphere of greenhouse gases such as carbon dioxide.

Precipitation forms when humid air rises, is cooled, and the water vapor in it condenses. Three processes create most precipitation: convection, orographic uplift over mountain ranges, and interaction between cold and warm air masses along fronts. Convectional precipitation dominates the tropics, while the main form of precipitation in the midlatitudes is frontal.

Wind blows from areas of high pressure to areas of low pressure, influenced by the Coriolis effect, which is a result of the rotational force of Earth. Global circulation patterns include bands of low pressure in the tropics, high pressure feeding trade winds in the subtropics, low pressure in the midlatitudes, and high pressure at the poles.

Climate is weather and its seasonal variations averaged over time. The most important variables in defining climate are temperature and precipitation. The Köppen scheme classifies climate into five main categories: A, B, C, D, and E. Humid tropical (A) climates occur in the tropical low-pressure zone and are dominated by the intertropical convergence. Dry climates (B) predominate in the subtropics, generally on the western sides of continents, and in continental areas isolated from moisture sources. Warm midlatitude climates (C) occur in subtropical areas on the eastern sides of continents and on west coasts at higher latitudes. Cool midlatitude climates (D) occur in continental areas, mostly in the Northern Hemisphere. Polar climates (E) occur at high latitudes. A sixth category (H) is mapped in mountain areas.

Earth's climate has varied greatly within the past 2 million years. Average global temperatures have been both higher and much lower than today. Glaciers covered much of Earth's surface during cold periods, the last of which ended about 12,000 years ago. Several possible causes for these variations have been proposed, including changes in the geometry of Earth's orbit around the Sun, geologic factors such as plate tectonics and volcanic eruptions, and changes in the composition of the atmosphere, some human-caused. Human-induced global warming, caused primarily by burning fossil fuels that release carbon dioxide into the atmosphere, is underway, demonstrating the key role humans play in Earth's physical processes.

KEY TERMS

adiabatic cooling p. 53
advection p. 52
angle of incidence p. 45
autumnal equinox p. 45
carbon dioxide (CO_2) p. 50
climate p. 44
cold front p. 56
condensation p. 53
convection p. 51
Coriolis effect p. 57
cyclone p. 65
desert climate p. 75
El Niño p. 63
front p. 55
global warming p. 50
greenhouse effect p. 50
greenhouse gases p. 50
gyre p. 63
humid continental climate p. 82
humid subtropical climate p. 80
humid tropical climate p. 76
hurricane p. 65
ice cap climate p. 85

insolation p. 45
intertropical convergence zone (ITCZ) p. 60
latent heat p. 50
latent heat exchange p. 50
Little Ice Age p. 87
longwave energy p. 50
marine west coast climate p. 81
Mediterranean climate p. 81
microclimate p. 52
midlatitude cyclone p. 66
midlatitude low-pressure zones p. 61
monsoon circulation p. 62
orographic precipitation p. 54
ozone (O_3) p. 50
permafrost p. 84
Pleistocene Epoch p. 86
polar front p. 61
polar high-pressure zones p. 61
Quaternary Period p. 86
radiation p. 49
relative humidity p. 53

saturation vapor pressure p. 53
semiarid climate p. 78
sensible heat p. 50
shortwave energy p. 50
solar energy p. 44
steppe p. 78
storm surge p. 66
subarctic climate p. 83
subtropical high-pressure (STH) zones p. 60
summer solstice p. 46
tornadoe p. 65
trade winds p. 60
transpiration p. 72
Tropic of Cancer p. 45
Tropic of Capricorn p. 45
tundra climate p. 85
typhoon p. 65
urban heat island p. 52
vernal (spring) equinox p. 45
warm front p. 56
wavelength p. 49
weather p. 44
winter solstice p. 46

QUESTIONS FOR REVIEW AND DISCUSSION

1. When do minimum and maximum daily amounts of solar radiation arriving at the top of the atmosphere occur at your latitude? How do angle of incidence and day length affect these seasonal differences?

2. What processes are responsible for the major energy exchanges between Earth's surface and the atmosphere?

3. Why, in terms of energy availability and convectional processes, do hurricanes happen (a) mostly over oceans and (b) in late summer and autumn?

4. How do climatic conditions in coastal areas differ from those in continental interiors? Why do these differences occur?

5. What are the major features of a midlatitude cyclone? What sequence of weather would one expect to observe as a midlatitude cyclone passes?

6. In what way does temperature affect the definition of a climate as humid or arid? Why?

7. For each of the 11 climate types described in this chapter, describe how the characteristics of the climate relate to the typical circulation patterns in the atmosphere.

8. Describe major variations of global average temperature that have occurred during (a) the past 1,000 years, (b) the past 10,000 years, (c) the past 100,000 years.

9. What are the major causes of global warming? What are some of its likely effects on human activities?

THINKING GEOGRAPHICALLY

1. For a two-week period, keep a daily journal of the weather, including such things as air temperature, wind direction, cloudiness, and precipitation. For the same period, clip the weather map from a daily newspaper. Then compare the two records.

2. In what ways do topography and/or land cover affect the weather where you live?

3. The following is a good group research activity. From your library or an on-line source such as the National Climatic Data Center (**http://www.ncdc.noaa.gov**), obtain a list of daily record-high and -low temperatures for your location, since record-keeping began, if possible.

Enter the data into a spreadsheet or program that will perform sorting, and sort the records by year. What percentage of record-high temperatures occur in the second half of the period of record? What percentage of record-low temperatures occur in the second half of the period of record? Can you spot any other trends?

4. How might an increase in average annual temperature of 5°C (9°F) affect your day-to-day life?

5. Compare the map of climates (Figure 2-31) with the world population map (on the rear endpaper). In what climate regions are the greatest concentration of people found? Why?

SUGGESTIONS FOR FURTHER LEARNING

Alley, R. B., and others. "Abrupt climate change." *Science* 299 (2003): 2005–2010.

Behrensmeyer, A. K. "Climate change and human evolution" *Science* 311 (27 January 2006): 476–478

Bréon, F-M. "How do aerosols affect cloudiness and climate?" *Science* 313 (4 August 2006): 623–624

Emmanuel, K. "Increasing destructiveness of tropical cyclones over the past 30 years." *Nature* 436 (2005):, 686

Epstein, P. R. "Climate and health." *Science* 285, no. 5426 (1999): 347–348.

Flannery, Tim. "The weather makers how man is changing the climate and what it means for life on earth." *Atlantic Monthly Press*, New York, 2006.

Flannery, Tim. *The Weather Makers The History and Future Impact of Climate Change.* Allen Lane, London, 2006

Grimm, Alice M., and others. "Climate variability in southern South America associated with El Niño and La Niña events." *Journal of Climate* 13, no. 1 (2000): 35–58.

Kerr, R. A. "Is Katrina a harbinger of still more powerful hurricanes?" *Science* 309 (16 September 2005): 1807.

Kerr, R. A. "A warmer Arctic means change for all." *Science* 297 (2002): 1490–1492.

Landsea, C. W., B. A. Harper, K. Hoarau, and J. A. Knaff "Can we detect trends in extreme tropical cyclones?" *Science* 313 (28 July 2006): 452–454.

Leung, L. R., and others. "Potential climate change impacts on mountain watersheds in the Pacific Northwest." *Journal of the American Water Resources Association* 35, no. 6 (1999): 1463–1472.

Linden, E., *Climate, Weather, and the Destruction of Civilizations.* New York: Simon and Schuster, 2006.

Melillo, J. M., and others. "Soil warming and carbon-cycle feedbacks to the climate system." *Science* 298 (2002): 2173– 2176.

Monon, S., J. Hansen, L. Nazarenko, and Y. Luo. "Climate effects of black carbon aerosols in China and India." *Science* 297 (2002): 2250–2253.

Oreskes, N. "The scientific consensus on climate change." *Science* 306 (3 December 2004): 1686.

Scharroo, Remko, Walter H. F. Smith, and John L. Lillibridge. Satellite Altimetry and the Intensification of Hurricane Katrina *Eos*, Vol. 86, No. 40, 4 October 2005.

Stenseth, N. C., and others. "Ecological effects of climate fluctuations." *Science* 297 (2002): 1292–1296.

Thompson, L. G., and others. "Kilimanjaro ice core records: Evidence of Holocene climate change in tropical Africa." *Science* 298 (2002): 589–593.

Webster, P. J., G. J. Holland, J. A. Curry, and H.-R. Chang. "Changes in tropical cyclone number, duration, and intensity in a warming environment." *Science* 309 (16 September 2005): 1844–1846.

Wilbanks, T. J., and others. "Possible responses to global climate change: Integrating mitigation and adaptation." *Environment* 454, no. 5 (2003): 28–38.

WEB WORK

Weather data are widely available on the World Wide Web from a great many sources. Here are just a few of the good ones. Intellicast is one of many commercial weather-information providers. Its website contains an excellent array of satellite imagery, in addition to up-to-date forecasts:

http://www.intellicast.com/

The U.S. government's atmospheric science programs are housed in the National Oceanographic and Atmospheric Administration. Its home page has links to the National Weather Service, which carries out forecasting and warning programs, and to the National Climatic Data Center, which stores and distributes historic climate data, including both statistics and satellite images:

http://www.noaa.gov/

The Space Science and Engineering Center at the University of Wisconsin provides an excellent array of recent images at:

http://www.ssec.wisc.edu/data/

Environment Canada maintains an excellent website with a wealth of Canadian environmental information, including weather and climate data:

http://www.msc.ec.gc.ca/

At its home page, the Carbon Dioxide Information Analysis Center of the Oak Ridge National Laboratory has done an excellent job of assembling data related to global warming and other global atmospheric pollution problems:

http://cdiac.esd.ornl.gov/

Two sites contain a wealth of information on global climate policy and climate change. One is Weathervane, a Digital Forum on Global Climate Policy, which is maintained by Resources for the Future:

http://www.weathervane.rff.org/

The other is the U.N. Intergovernmental Panel on Climate Change:

http://www.ipcc.ch/

3
LANDFORMS

A satellite image of Ethiopia, Eritrea, Djibouti, and Somalia and adjacent countries. The Afar region lies just west of the Bab el Mandeb, which separates the Red Sea from the Gulf of Aden. It lies at the junction of three tectonic plates that are all moving away from each other. (NASA)

A Look AHEAD

Plate Tectonics

Earth's surface is in motion, both horizontally and vertically. Earth's crust consists of several moving pieces, known as tectonic plates. The movement of these pieces causes earthquakes, volcanic eruptions, and the formation of mountain ranges at plate boundaries.

Slopes and Streams

Water, wind, ice, and gravity shape Earth's surface. Rocks are broken down into smaller pieces through exposure to air and water. The pieces then are moved downhill by gravity or are carried by streams, wind, or ice to other locations. Streams play a major role in shaping landforms. They erode material from some places, transport it further along their path, and deposit it elsewhere.

Ice, Wind, and Waves

Glaciers—rivers of ice—carve landforms in regions under continuous snow cover, such as mountaintops and the poles. In deserts, where vegetation is scarce, the wind shapes landforms. Along coastlines, waves caused by wind blowing across the ocean surface cause intensive erosion, which rapidly changes landforms in these areas.

The Dynamic Earth

Earth's surface is continually changing. Change on Earth's surface is rapid enough to affect human settlements and natural resources. In particularly dynamic environments, natural hazards like floods, earthquakes, and landslides are a major problem.

*I*n September of 2005, an enormous volcanic eruption occurred in northeast Ethiopia, on the flanks of Dabbahu, a gently sloping volcano. The eruption occurred along a 60-km-long (37-mile) fracture in Earth's crust, and widened the fracture by as much as 8 meters (25 feet). The fracture was filled by about 2.5 cubic kilometers (0.5 cubic miles) of lava— enough to cover the island of Manhattan with a layer 40 meters (131 ft) thick. Dabbahu is located in the Afar Triangle, a triple-junction of tectonic plates that are all moving away from each other. Volcanic eruptions of large volume are characteristic of this type of plate junction, but they are rarely observed on land areas because most of these junctions are under water. The Afar Triangle is thus of particular interest to geologists because we can readily observe new earth crust being created. Coincidentally, the region is also of great interest to anthropologists because it is where the oldest known fossils of early hominids have been found, buried in volcanic ash. ▶

Some changes on Earth's surface take place gradually, at rates of a few millimeters or less per year. Others, like the earthquake that shook Indonesia in May of 2006 (Figure 3-1), are sudden and dramatic. Whether gradual or catastrophic, changes in the Earth occur everywhere. In this chapter we will learn about the causes and rates of such movements and how they shape mountains, hills, and valleys. We will look at Earth's rocks, soil, and surface landforms, which comprise the lithosphere, the outermost part of one of the four "spheres." (Look back at Figure 1-17). **Geomorphology** is the study of landforms and the processes that create them.

Earthquakes and the gradual sinking of the land are newsworthy mostly because of their effects on humans. They also are notable because most of the time Earth's lithosphere appears to be fixed and unchanging relative to the other three spheres—the atmosphere, the biosphere, and the hydrosphere—which change rapidly. The atmosphere changes daily—hot to cold, wet to dry, windy to still. We can see seasonal changes in the biosphere—plants and animals appear, grow, reproduce, and die. The hydrosphere is visibly dynamic— rivers flood and later dry up. But the lithosphere seems static. Landforms like plains, hills, and valleys do not change noticeably in a lifetime; soil is soil; rock outcrops seem permanent as monuments.

Landforms, in their remarkable variety, did not just appear at some distant time as we see them today. Landforms constantly change, although often imperceptibly slowly. As happens in the atmosphere, the lithosphere is driven by continual transfers of energy and matter, and it interacts with the other three spheres. Some of the energy that modifies the lithosphere comes from deep within Earth. Heat contained in Earth's core influences movement on the planet's surface.

Occasionally, Earth's surface changes quickly. This reminds us that the lithosphere is quite dynamic— earthquakes shake the land, volcanoes spew forth hot gases and molten rock, and ocean waves pound the shore. Humans insist on getting in the way of these dramatic changes in landforms by building settlements in places where earthquakes, volcanoes, and hurricanes are likely to occur. Because of the rarity of these events, people discount their importance and thus become vulnerable when they do strike.

Our actions have profoundly changed the lithosphere, as they have changed the other spheres. Agricultural practices are depleting soil fertility. In semiarid areas, overgrazing by animals has produced desert-like conditions. In many agricultural areas, soil is eroding at rates 10 to 100 times greater than natural rates of erosion. Some estimates of human-caused earth moving suggest that more earth is moved by humans than by all natural processes combined.

Geographers studying the shape of Earth's surface—its topography—recognize that it includes many features that seem to have distinctive characteristics (Look ahead to the map on the front map insert). Elements of Earth's surface that have such identifiable form—its mountains, valleys, hills, and depressions—are

FIGURE 3-1 Earthquake damage in Indonesia. The damage shown here was caused by a severe earthquake that struck the island of Java, Indonesia, in May 2006, killing over 5,700 people and destroying hundreds of thousands of homes. This occurred only 18 months after the devasting earthquake and tsunami of December 2004 and only 6 weeks before the eruption of Mount Merapi in July 2006, also in Indonesia. (David Longstreath/AP Wide World Photos)

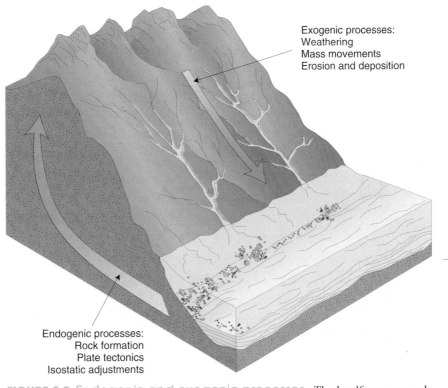

Exogenic processes:
Weathering
Mass movements
Erosion and deposition

Endogenic processes:
Rock formation
Plate tectonics
Isostatic adjustments

FIGURE 3-2 Endogenic and exogenic processes. The landforms around you are the product of interaction among endogenic and exogenic processes. Endogenic processes involve movement of Earth's crust through tectonic action. Crustal uplift may be caused by plate tectonics or by isostatic adjustments, which cause the land to rise when rock mass erodes away. Exogenic processes wear down rocks once they reach the surface.

called **landforms**. They are built through a combination of endogenic (internal) and exogenic (external) processes. **Endogenic processes** are forces that cause movements beneath or at Earth's surface, such as mountain building and earthquakes. These internal mechanisms move portions of Earth's surface horizontally and vertically, raising some parts and lowering others. Even as these internal forces are building Earth's features, these features are simultaneously attacked by **exogenic processes**, which are forces of erosion, such as running water, wind, and chemical action.

Endogenic and exogenic forces continually move and shape Earth's crust (Figure 3-2). Endogenic processes form rocks and move them to produce mountain ranges, ocean basins, and other topographic features. As these rocks become exposed, exogenic activities go to work. They erode materials, move them down hillslopes, and deposit them in lakes, oceans, and other low-lying areas. We will examine all these processes.

PLATE TECTONICS

If you were to view Earth from a great distance, as we sometimes view other planets through telescopes, the most distinctive topographic features would be the enormous mountain ranges arranged in linear patterns that extend for thousands of kilometers. Three especially prominent mountain ranges on Earth's land surface include the Rockies of North America, that continue through Central America to the Andes of South America; the Himalayas, extending across Asia; and the north–south system of mountains in eastern Africa. Large as these highly visible mountain ranges are, none rank as the world's longest. That title belongs to a mountain system beneath the oceans, the interconnecting mid-ocean ridges that are more than 64,000 kilometers (40,000 miles) long.

For millennia, people believed that Earth's continents and oceans were fixed in place for all time and that Earth was only a few thousand years old. This notion of a "fixed Earth" was challenged early in the twentieth century by German earth scientist Alfred Wegener. He argued that Earth's land areas once had been joined in a single "supercontinent," now known as Pangaea, and that over thousands of years the continents had moved apart (Figure 3-3). Because Wegener could not explain why the continents moved, his ideas were rejected at the time. But researchers in the 1960s vindicated Wegener by working out an explanation, called **plate tectonics theory**. Since then, plate tectonics theory has provided us with explanations of the origins of Earth's great mountain chains, volcanoes, and many other important phenomena.

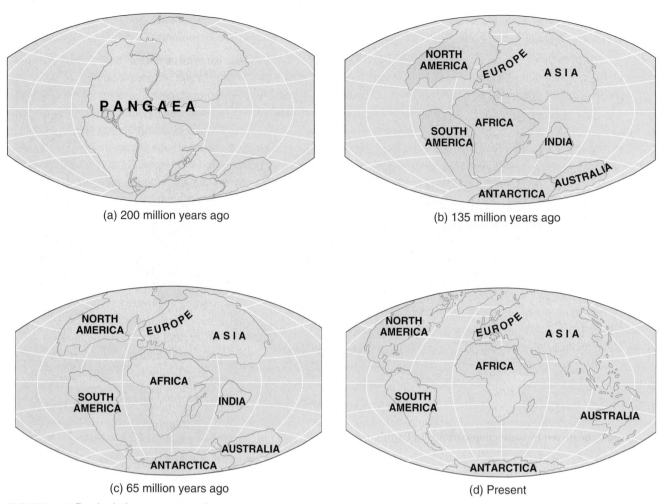

(a) 200 million years ago

(b) 135 million years ago

(c) 65 million years ago

(d) Present

FIGURE 3-3 Past plate movements. Two hundred million years ago (a) Earth's continents were all joined in one supercontinent known as Pangaea. This continent gradually broke apart, beginning with the opening of the Atlantic Ocean from south to north and the movement of India away from Antarctica toward Asia (b and c). The layout of the world of today is shown in (d).

Earth's Moving Crust

Earth resembles an egg with a cracked shell. Earth's crust is thin and rigid, averaging 45 kilometers (28 miles) in thickness. Beneath this rigid crust the rock is like a very thick fluid and is slowly deformed by movements within Earth. While far from the free-flowing substances we know as liquids, the rock just beneath the crust, known as the **mantle**, is fluid enough to move slowly along in convection currents, driven by heat within Earth's core. These currents are analogous to winds in the atmosphere, which carry heat away from Earth's surface. Geologists believe that this motion of the mantle causes pieces of Earth's rigid crust to move; those pieces are called **tectonic plates**. This is the plate tectonics theory. Movement of the plates causes earthquakes to rumble, volcanoes to erupt, and mountains to be built (Figure 3-4).

Earthquakes Thousands of **earthquakes**—sudden movements of Earth's crust—occur every day. Figure 3-5 shows major earthquakes zones around the world,

notably clustered where two plates meet. The place where Earth's crust actually moves is the **focus of an earthquake.** The focus is generally near the surface but can be as deep as 600 kilometers (372 miles) below Earth's surface. The point on the surface directly above the focus is the **epicenter** (Figure 3-6). The tremendous energy released at the focus travels worldwide in all directions and at various speeds through different layers of rock.

Most earthquakes are too small for people to feel, and they are detectable only with a **seismograph**, which is a device that records the quake's **seismic waves**, or vibrations. Earthquake intensity is measured on a 0-to-9 logarithmic scale developed by Charles F. Richter in 1935. Earthquakes with a magnitude of 3 to 4 on the Richter scale are minor; magnitude 5 to 6 quakes can break windows and topple weak buildings; and magnitude 7 to 8 quakes are devastating killers if they affect populated areas.

Several factors in addition to an earthquake's intensity determine the damage it causes. Generally, damage is greater at places closer to the epicenter and

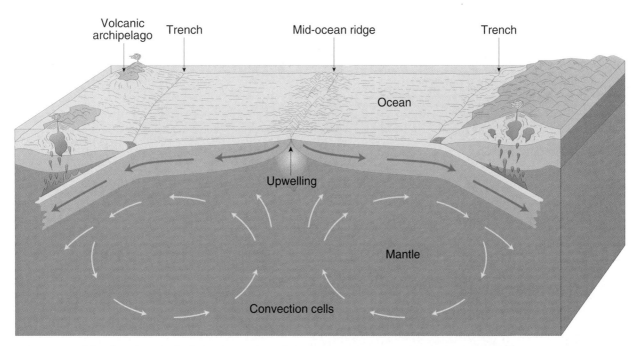

FIGURE 3-4 Mechanisms of plate movement. Plate tectonic theory explains the occurrence of moving continents, mid-ocean ridges, deep-sea trenches, and earthquakes. At the center, rock magma melted in Earth's interior rises along mid-ocean ridges, emerging to chill into new oceanic crust. Along this ridge, plates of oceanic crust are spreading apart. Where crustal plates collide (at left and right), crust is forced downward (subducted) and recycled (melted) back into the interior. This generates new magma, which migrates toward the surface, sometimes emerging as a volcano. Less dense continental crust (the continents) rides on moving plates of basaltic crust. This plate tectonic process is believed to be driven by convection currents in Earth's mantle, shown by arrows. (From E. J. Tarbuck and F. K. Lutgens, *The Earth,* 4th ed. New York: Macmillan, 1993)

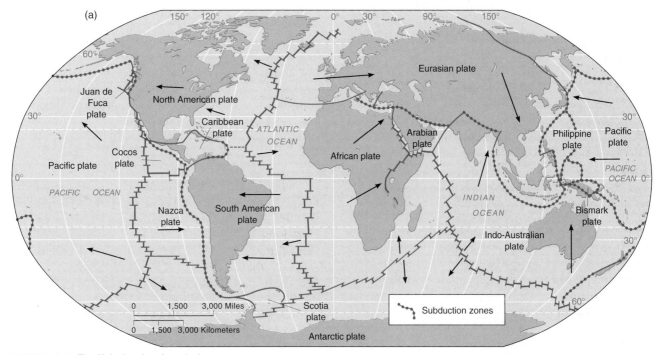

FIGURE 3-5 Earth's tectonic plates. (a) The major plates of Earth's crust move relative to one another, generally at rates of a few centimeters per year.

(b)

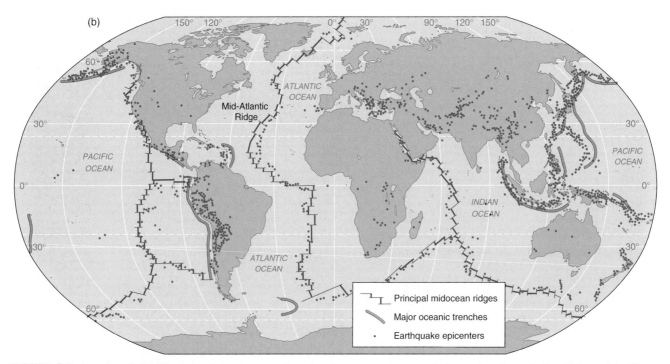

FIGURE 3-5 (*continued*) (b) These motions cause earthquakes that are concentrated along plate boundaries. Ridges with rift valleys at their centers are formed where plates are moving away from each other, generally in ocean areas. Mountain ranges are created where plates converge, sometimes with deep-ocean trenches on the seaward side of the convergence area.

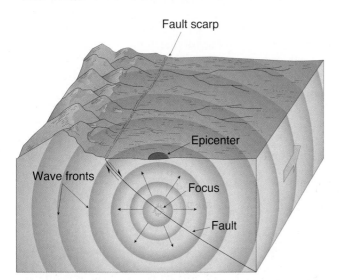

FIGURE 3-6 Earthquake focus and epicenter. The focus of an earthquake is where Earth's crust actually moves. The point on the surface directly above is the epicenter. Displacement of the surface creates a fault scarp.

at places built on ground that is subject to landsliding or collapse. Earthquake damage is also greater where buildings are not designed to absorb the shaking. To appreciate this, let us contrast two major earthquakes: one in California and one in Armenia. In 1989 an earthquake struck the San Francisco Bay area. The Bay Bridge buckled, freeways crumbled, and 67 people died. Some 100,000 buildings were damaged or destroyed, many in the Marina District, a neighborhood built on unstable fill in the San Francisco Bay. The death and destruction was limited, however, considering the quake's magnitude of 7.1 on the Richter scale. In contrast, in 1988 an earthquake in northwestern Armenia killed nearly 55,000 people, injured 15,000, and left at least 400,000 homeless. Although registering 6.9 on the Richter scale—lower than the California quake—and occurring in a city much smaller than San Francisco, this earthquake trapped many people inside collapsing buildings not designed to withstand earthquakes. People in less economically developed societies cannot finance the cost of earthquake-proofing their structures, as is routinely done with new buildings in quake-prone communities of wealthy societies like the United States.

Earthquake prediction is unreliable. Potential movement zones are closely monitored and computer models attempt to replicate conditions along plate boundaries. But predicting an earthquake is like trying to forecast the weather several months from now. It just cannot be done accurately with current technology.

Volcanoes   Like earthquakes, volcanoes are clustered along boundaries between tectonic plates (Figure 3-5). Movement within Earth and between the plates generates **magma** (molten rock). Being less dense than the surrounding rock, magma migrates toward the surface. Some reaches the surface and erupts, and is then called **lava**. A **volcano** is the surface

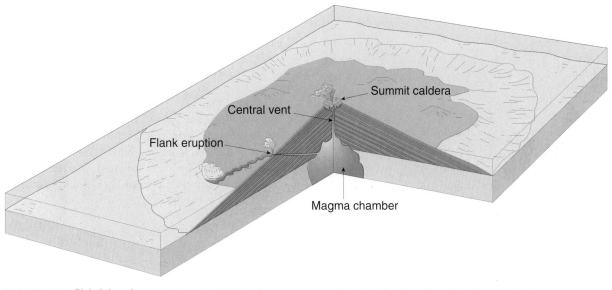

FIGURE 3-7 Shield volcano. Gentle shield volcanoes, exemplified by the Hawaiian Islands, are the largest volcanoes on Earth. Lava may erupt on the flank of the volcano or from the craterlike caldera at the top of the mountain. (From E. J. Tarbuck and F. K. Lutgens, *Earth Science*, 7th ed. New York: Macmillan, 1994)

vent where lava emerges. The magma may flow over the surface, forming a plain of volcanic rock, or it may build up to form a mountain. The chemistry of the magma/lava determines its texture and therefore the type of landform it builds.

Shield volcanoes erupt runny lava that cools to form a rock called *basalt*. They are called **shield volcanoes** because of their shape (Figure 3-7). Each of the Hawaiian Islands is a large shield volcano, although the only currently active one is Mauna Loa, on the island of Hawaii (the "Big Island"). These generally sedate volcanoes make news on the rare occasions when they grow more active, and flows of lava threaten settlements. The mid-ocean ridges are formed of similar basaltic lava.

Explosive volcanoes that cause death and destruction are more likely to be **composite cone volcanoes** (Figure 3-8). Composite cones are made up of a mixture of lava and ash. Their magma is thick and gassy, and it may erupt explosively through a vent. The eruption sends ash, glassy cinders (called *pyroclasts*), and clouds of sulfurous gas high into the atmosphere. It may also pour lethal gas clouds and dangerous mud-flows down the volcano's slopes. Repeated eruptions build a cone-shaped mountain, made up of a mixture of lava and ash layers.

Eruptions of composite cone volcanoes have killed tens of thousands of people at a time, but such disasters are much less frequent than severe earthquakes. One of the greatest volcanic explosions in recorded history was the 1883 eruption of the island of Krakatau in present-day Indonesia. Two-thirds of the island was destroyed, and the event killed about 36,000 people, most of whom died in a flood triggered by the eruption. Ash discharged into the atmosphere by Krakatau significantly

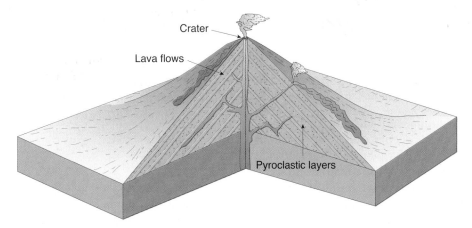

FIGURE 3-8 Composite cone volcano. Composite cone volcanoes, composed of alternating lava and ash, are relatively explosive. (From E. J. Tarbuck and F. K. Lutgens, *Earth Science*, 7th ed. New York: Macmillan, 1994)

FIGURE 3-9 Augustine volcano, Alaska. This Landsat image, taken in March of 2006, shows an eruption of the Augustine volcano in the Cook Inlet near Anchorage, Alaska. (USGS)

blocked sunlight and caused noticeable cooling of Earth's climate for a couple of years.

Thousands of volcanoes stand dormant (inactive, but with the potential to erupt) around the world. About 600 are actively spewing lava, ash, and gas—some daily—but they rarely cause damage. Others, however, have not erupted in hundreds of years, so people have settled nearby (Figure 3-9). In some areas earthquake watch centers provide warnings of volcanic eruptions. But when warnings are not available, the danger can be great. The 1985 eruption of Nevado del Ruiz, in Colombia, triggered giant mudslides that buried most of a town, killing 23,000 people. In general, predicting volcanic eruptions is more accurate than predicting earthquakes, because volcanoes give many warnings before erupting.

Types of Boundaries Between Plates

Three types of boundaries form between moving plates of Earth's crust. The type of boundary depends on whether the plates are spreading apart, pushing into each other, or grinding past each other.

Divergent plate boundaries   A boundary where plates are spreading apart is a **divergent plate boundary.** People seldom are aware of plates spreading apart, first because it happens at a rate of only a few centimeters per year (2.54 centimeters = 1 inch) and, second, because it happens mostly deep in mid-ocean. The Mid-Atlantic ridge is a well-known example. Divergent plate boundaries occur on land, too. The rift valleys of East Africa are an example, visible in Figure 3-5. These valleys are hundreds of meters deep and extend thousands of kilometers from Mozambique in the south, to the Red Sea in the north.

Where two plates are diverging on the seafloor, a phenomenon called **seafloor spreading**, lava continually

erupts. Rapidly chilled by seawater, it solidifies to form a new seafloor crust. This is how much of the seafloor crust forms, by very slowly spreading from mid-ocean ridges.

Convergent plate boundaries   A boundary where plates push together is a **convergent plate boundary.** Material from one plate—the subducting plate—is slowly forced downward by the collision, back into the mantle. Because seafloor crust is denser than continental crust, when a plate of continental crust collides with a plate of oceanic crust, the denser oceanic plate sinks beneath the lighter continental crust. The oceanic plate is carried into Earth's mantle, where some of it is remelted. This magma then migrates toward the surface, causing volcanic eruptions at sites above the plunging plate. This occurs, for example, to the south and southwest of Indonesia where the Eurasian and Indo-Australian plates converge (see Figure 3-5).

Transform plate boundaries   A boundary where the plates neither converge nor diverge, but grind past each other, is a **transform plate boundary**. California's San Andreas Fault is an example. Along this fault, the Pacific plate is moving northwest relative to the North American plate (Figure 3-5a). The boundary between these plates is not a smooth one, and ridges and mountains are built as the two plates grind against one another. The plates bind for long periods and then abruptly slip, causing the earthquakes that frequently strike California.

Vertical movements of Earth's crust   Parts of the crust move vertically as well as horizontally. As two plates collide, material may be forced downward into Earth's interior or upward to form mountains. Over millions of years, vertical movements along plate boundaries produce mountain ranges thousands of meters high.

Vertical movement of crust also occurs because the crust "floats" on the underlying mantle, much like a boat floating in water. If material is added to the crust, it sinks, and if material is removed, it rises. Deposition of sediment or accumulation of ice in glaciers can cause the crust to sink. These vertical movements caused by loading or unloading the crust are called **isostatic adjustments**. Removal of material by erosion or melting of glaciers allows the crust to adjust isostatically, or "rebound." In some places, crust that was buried under continental ice sheets has risen vertically over 100 meters (330 feet) in the last 15,000 years because of glacial melting.

Rock Formation

Although by human standards Earth's surface moves very slowly—by at most a few centimeters per year—this movement produces Earth's great diversity of rocks. As Earth's crust moves, its materials are eroded and deposited, heated and cooled, buried and exposed.

Types of rocks

Rocks can be grouped into three basic categories that reflect how they form:

Igneous rocks are formed when molten crustal material cools and solidifies. The name derives from the Greek word for fire, which is the same root as for the English word *ignite*. Examples of igneous rocks are basalt, which is common in volcanic areas, including much of the ocean floor, and granite, which is common in continental areas.

Sedimentary rocks result when rocks eroded from higher elevations (mountains, hills, plains) accumulate at lower elevations (like swamps and ocean bottoms). When subjected to high pressure and the presence of cementing materials to bind their grains together, rocks like sandstone, shale, conglomerate, and limestone are formed.

Metamorphic rocks are created when rocks are exposed to great pressure and heat, altering them into more compact, crystalline rocks. In Greek the name means "to change form." Examples include marble (which metamorphosed from limestone) and slate (which metamorphosed from shale).

Minerals

Minerals are natural substances that comprise rocks. Each type of mineral has specific chemical and crystalline properties. Earth's rocks are diverse in part because the crust contains thousands of minerals. The density of rocks depends on the kinds of materials they contain. Denser rocks are dominated by compounds of silicon, magnesium, and iron minerals; they are called **sima** (for *si*licon-*ma*gnesium). Less dense rocks are dominated by compounds of silicon and aluminum minerals; they are called **sial** (for *si*licon-*al*uminum).

Denser sima rocks make up much of the oceanic crust. Less dense sial rocks make up much of the continental crust. The lower density and greater thickness of sial rocks cause the continents to have higher surface elevations than the oceanic crust, just as a less dense dry log will float higher in water than a denser wet one.

The formation and distribution of many minerals is caused by the movements of Earth's crust. Vast areas of the continental crust, known as **shields,** have not been significantly eroded or changed for millions of years (Figure 3-10).

Shield areas often contain rich concentrations of minerals, such as metal ores and fossil fuels. Shields are located in the core of large continents such as Africa, Asia, and North America. Many of the world's mining districts exist where these continental shields are exposed at the surface.

Stress on rocks

Crustal movements along plate boundaries exert tremendous stress on rocks. Despite their rigidity, rocks bend and fold. When stressed far enough, they fracture along cracks called **faults**. The fractured pieces may then be transported to new locations. Fracturing takes place in different ways, depending on the type of boundary. Near a divergent plate boundary, rocks break apart because they are stretched; the resulting fracture is called a *normal fault* (Figure 3-11). Near a convergent plate boundary, rocks fracture because they are compressed; such fractures

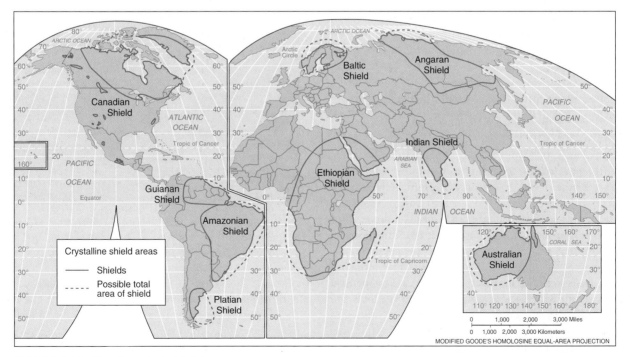

FIGURE 3-10 Continental shields. The ancient cores of the continents are exposed in several areas of Earth's surface. The rocks exposed in these areas are typically rich in mineral ores.

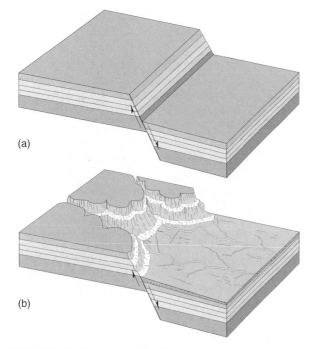

(a)

(b)

FIGURE 3-11 A normal fault. The diagrams show cutaway views of what happens beneath Earth's surface along a normal fault. (a) Rocks break apart because they are stretched. (b) Erosion alters the block that has been uplifted along a normal fault. (From E. J. Tarbuck and F. K. Lutgens, *The Earth*, 4th ed. New York: Macmillan, 1993)

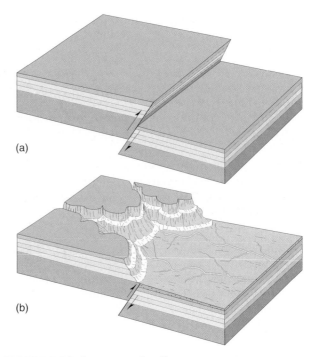

(a)

(b)

FIGURE 3-12 A reverse fault. (a) Rocks break when stressed by forces that compress them. (b) Erosion alters the surface of rocks that have been uplifted along a reverse fault. Later erosion shapes the surface, and may obscure the actual orientation of the fault. (From E. J. Tarbuck and F. K. Lutgens, *The Earth*, 4th ed. New York: Macmillan, 1993)

are called *reverse faults* (because they are the opposite of normal faults) (Figure 3-12), or *thrust faults* if there is large horizontal movement. Alternatively, the crust may rumple like a rug, creating folds. The Appalachian Mountains, the western edge of the Wasatch Range in Utah (Figure 3-13), and the Himalayas are examples of mountain ranges created by faulting and folding that happened along convergent plate boundaries. Faulting also occurs along transform boundaries. Rock movement near a transform boundary is mostly parallel to a plate boundary rather than perpendicular, as in the other two types of boundaries.

Rocks and landforms Differences in geologic structures and rock types from one place to another are a critical part of the geographic variability of Earth's surface. These geologic features influence the surface in three different ways. (1) Movement of the crust such as that along faults creates landforms like those illustrated in Figures 3-11, 3-12, and 3-13. (2) Variations in the resistance of rocks to exogenic processes cause weak rocks to be removed more rapidly while more resistant rocks remain in place. This creates **structural landforms**, in which the shape of the land reflects the underlying rock structures. The ridges and valleys of the central Appalachians are formed in this way (Figure 3-14). The ridges are formed by resistant layers, often sandstone and conglomerate, whereas the valleys are underlain by weaker materials such as shale. (3)

FIGURE 3-13 The Wasatch Fault. The Wasatch Mountains in Utah rise abruptly from the Salt Lake valley along the Wasatch Fault, an active normal fault just east of Salt Lake City. (Tom McHugh/Photo Researchers, Inc.)

Differences in the minerals contained in rocks affect the characteristics of soils that form on them. Soils overlying limestone, for example, tend to be rich in calcium, a major component of limestone. The following section describes the exogenic processes that help to expose these geologic influences on the landscape.

FIGURE 3-14 Satellite image of the central Appalachians in Pennsylvania. The winding ridges are formed by resistant layers in folded sedimentary rocks, exposed by erosion of weaker layers that now occur in valleys. (Photo Researchers, Inc.)

SLOPES AND STREAMS

Plate tectonics theory helps us to understand the causes of mountain-building motions in Earth's crust. But landforms created through endogenic processes are attacked by exogenic processes as they are being formed (Figure 3-15). The wearing down of Earth's crust through exogenic processes reshapes Earth's crust into new landforms. Exogenic processes shape Earth's surface in two principal steps: Rocks are first broken down into smaller pieces through weathering, then they are carried by gravity down the slopes of hills or are transported by water, wind, or ice from one place to another.

Weathering

Weathering is the process of breaking rocks into pieces ranging in size from boulders to pebbles, sand grains, and silt, down to microscopic clay particles and dissolved solids. Without weathering, the force of gravity and the agents of water, wind, and ice would have nothing to move. Rocks begin to break down the moment they are exposed to the weather at Earth's surface. They are attacked by water, oxygen, carbon dioxide, and temperature fluctuations. Weathering is the first step in the formation of soil, which will be discussed in more detail in Chapter 4. Weathering takes place in two ways: as chemical weathering and as mechanical weathering.

Chemical weathering Rocks may be broken down as a result of **chemical weathering**, which is a change in the elements that compose rocks when they

FIGURE 3-15 Faulted landforms. Death Valley, California, was formed by faulting that raised the land in the foreground and the distance, while it lowered the valley floor. The valley is filled with sediments eroded from the adjacent mountains. (Deborah Davis/PhotoEdit)

are exposed to air and water (Figure 3-16). Chemical weathering occurs faster in places with warm temperatures and abundant water, such as in humid tropical environments. Acids released by decaying vegetation also chemically weather rocks. Some of the dissolved products of chemical weathering are carried away by water seeping through soil and rocks, which is a process called *leaching*. The water eventually may carry these materials to rivers and thence to the sea. This is the source of the salinity (dissolved salt) of the oceans.

FIGURE 3-16 Weathering. Granite exposed along the coastline of Nova Scotia reveals the effects of weathering on rocks. Water has penetrated the rock and weakened the bonds between grains, allowing the rock to be easily broken open. Freezing and thawing along with salt spray from the sea have contributed to the weathering. (Norm R. Catto)

One example of chemical weathering is oxidation. Iron is a common element in rocks, and it combines with oxygen in the air to form iron oxide, or rust. Iron oxide has very different properties from the original iron—it is physically weaker and more easily eroded. You can see the effects of oxidation on iron or steel surfaces exposed to the weather; the rusty oxide easily flakes away.

Another example of chemical weathering is the decomposition of calcium carbonate. It is a major component of limestone and other sedimentary rocks. Calcium carbonate dissolves in water, separating into ions of calcium and carbonate that are carried by streams into the sea. In some areas of limestone bedrock, underground water may remove large quantities of rock via chemical weathering beneath the surface. Water flowing underground dissolves passageways and even carves out large caverns in the limestone. If the caverns collapse, they create depressions called sinkholes at the surface. Such underground erosion produces a distinctive form of topography called **karst**, named after a region in Croatia with this type of landscape. Karst topography is found in many parts of the world, including the Caribbean (especially Puerto Rico, Cuba, and Jamaica), several parts of the southeastern United States (especially Florida, Kentucky, and Missouri), and southeastern China.

Mechanical weathering   Rocks are also broken down by physical force. This process is called **mechanical weathering**. Rocks expand and contract with frequent changes in temperature, and this action causes them to break apart. Highway potholes in the northern United States and Canada illustrate these processes. Rainwater seeps into roadway cracks and freezes into ice crystals when the temperature turns colder. The water, which expands about 9 percent when frozen, pushes apart the pavement, and opens up the potholes in a phenomenon called frost-wedging. Plant roots growing in cracks between rocks also contribute to mechanical weathering; you probably have observed sidewalks that have been heaved by tree roots.

Mechanical and chemical weathering work together to break down rocks. Often, mechanical forces open cracks, and water seeps in to weather the rock chemically.

Moving Weathered Material

Once rocks are weathered, they may be carried from one place to another. Material most commonly moves downhill by gravity. This happens in two ways: by mass movement or by surface erosion. In **mass movement**, rocks roll, slide, or freefall downhill under the steady pull of gravity. In **surface erosion**, water—which flows downhill because of gravity—carries solid rock particles with it. Surface erosion may also result when wind or ice carries material from one place to another.

Material moves faster down steeper hills than down gentler ones. The steepness of a hill is measured through its slope, which is the difference in the elevation between two points (known as the *relief* or *rise*) divided by the horizontal distance between the two points (known as the *run*). The greater the rise and shorter the run between the two points, the faster the movement of materials down the hill. Wherever slopes occur, gravity is available to move material. Even the gentlest slope provides the potential energy necessary to move at least some material downward, either through mass movement or surface erosion. Erosion is usually much more rapid, however, on steep slopes of land than on gentle slopes.

Mass movement   The most common form of mass movement is **soil creep**. As the name suggests, creep is a very slow, gradual movement of material down the slope of a hill (Figure 3-17). A tiny movement can cause

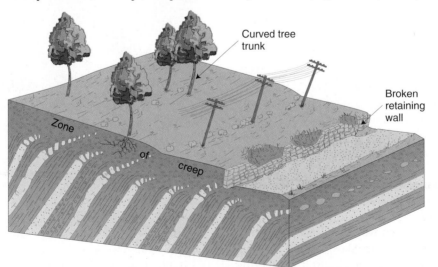

FIGURE 3-17 Soil creep. Soil creep is the most common form of mass movement. Creep, as the name suggests, is a very slow downslope movement of the uppermost part of the soil. (From E. J. Tarbuck and F. K. Lutgens, *Earth*, 7th ed. Upper Saddle River, NJ: Prentice Hall, 2002)

creep—a rodent digging, a worm burrowing, an insect pushing aside soil. Creep occurs near the surface, in the top 1 to 3 meters (3 to 10 feet) of soil.

More dangerous and dramatic mass movements, such as rock slides and mudflows, can occur on steep slopes, especially during wet conditions. Steep slopes are prone to rock slides because the force of gravity pushing down on the rocks is likely to exceed the strength of the rocks. Landslides on steep slopes can follow intense rains, because material with a high water content is heavier, weaker, and less able to resist the force of gravity. The sliding material may break down into fluid mud, which flows downhill (Figure 3-18). Houses built on very steep slopes, such as along the West Coast of North America, risk damage from landslides and mudflows.

Surface erosion
The most common form of surface erosion is caused by rainfall. Intense rain sometimes falls faster than soil can absorb it. Water that cannot infiltrate, or soak into the ground, must run off the surface. As it runs off the surface, water picks up soil particles and carries them down the slope. With enough runoff, water can carve channels into the landforms.

The smallest channels eroded by the flow of water—only a few centimeters deep—are called *rills*. Rills are so small that soil creep or a farmer's plow can obliterate them. If channels gather enough water, however, they become larger and permanent carriers of water. As these stream channels deepen, they gather water and eroded soil from adjacent slopes. When the streams gather enough water, they form ravines, valleys, and canyons.

FIGURE 3-19 Intense erosion in the central highlands of Vietnam. Land scarcity requires cultivation on steep slopes in many parts of the world. (Edward Parker/Alamy Images)

Surface erosion by water is relatively slow in most natural environments because the ground is covered by grass and trees. But on large parts of Earth's surface, humans have removed vegetation by clearing forests and plowing fields. Once the vegetative ground cover is removed, slow surface erosion can suddenly become severe erosion (Figure 3-19). Ground where vegetation has been removed can suffer more surface erosion in a few months than it experienced during the previous several thousand years. The eroded soil contributes to water pollution downstream, and the remaining soil may be less productive for agriculture.

Stream drainage
Streams collect water from two sources: groundwater and overland flow (Figure 3-20). When rain falls on the land surface, most of it soaks into the soil and accumulates as **groundwater**. Groundwater migrates slowly through the soil and underlying rocks. During dry periods, most of the water flowing into streams is supplied not by rainfall but by groundwater. If rain falls intensely, the soil may not be able to

FIGURE 3-18 Landslide. This landslide occurred on a steep slope on Santa Cruz Island, California, following heavy rain. The heavy rain weakened the rocks, triggering the landslide. Small scars near the top of the landslide reveal fractured surfaces in the rock. Once the weathered rock began to slide downhill, it was jostled and broken apart by the motion and turned to mud. By the time it reached the bottom of the slope, the landslide had become a mudflow. (WHR photo)

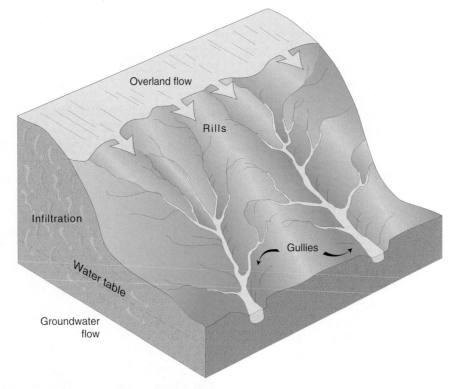

FIGURE 3-20 Sources of streamflow. Surface water (overland flow) and groundwater (water below the water table) are the two sources of water for streams. Under the pull of gravity, rainwater soaks into the soil. The water in the ground slowly migrates through soil and rock. The surface flow follows the lowest path available, eventually running into streams. Streams merge into other streams and rivers, so most water ultimately ends up in the ocean. If rainfall is intense and the soil cannot absorb it all, overland flow can become great enough to cause flooding.

absorb it as fast as it falls. Water then may run directly to streams through **overland flow**.

A stream drains groundwater and overland flow from an area called its **drainage basin**. Drainage basins may be as small as a farm field, or as large as a major portion of a continent. (Smaller basins are nested within larger basins). The greater the area of its drainage basin, the more water a stream must carry. Rivers in dry areas may be exceptions because they may lose water to evaporation. A basin with plentiful runoff from groundwater and overland flow might carve a complex network of many channels to remove the water and sediment. Small rills deliver water and material to larger streams, which join others to form still larger rivers, which flow to the sea (Figure 3-21).

The volume of water that a stream carries per unit of time is its **discharge**. Discharge ranges from a few cubic centimeters per second in rills to over 200,000 cubic meters per second at the mouth of the Amazon

FIGURE 3-21 A MODIS view of the mouth of the Yangtze River, China. The Yangtze enters from the west (left). The Yangtze carries an enormous sediment load, and construction of the Three Gorges Dam will reduce this flow of sediment to the sea, with unknown consequences for the coastal region. (Jacques Descloitres, MODIS Land Science Team/NASA Headquarters)

Critical THINKING

Wealth and Natural Hazards

When we hear about natural hazards such as wildfire, earthquake, landslides, and coastal erosion, one place that often appears in the headlines is Los Angeles, which seems to have more than the usual share of such calamities. Dramatic scenes of houses toppled by landslides are splashed across the TV screens, and often the homes are very expensive ones. Some of the most hazardous neighborhoods in Los Angeles are also among the most desired. The hills overlooking Hollywood and Malibu and the Palos Verdes Peninsula, with its spectacular ocean views, are examples. The steep slopes scattered on mountainsides around Los Angeles offer isolation from city traffic and spectacular views, on clear days at least. The steep slopes are also the places most vulnerable to important hazards such as wildfire and landsliding caused by winter rains. In contrast, the poorest neighborhoods in Los Angeles are generally flat and far removed from such hazards. South-central Los Angeles is an example. People who live in these neighborhoods have other hazards to deal with, and they are just as vulnerable to earthquakes as anyone in the area. But it does seem that in Los Angeles the rich people have chosen to live in the more hazardous environments. This may cause an increase in the property damages that occur when disasters strike, because rich people's houses cost more than poor people's apartments.

Los Angeles is just one city, and a unique one at that. But what about other cities? In some Latin American cities, the wealthier neighborhoods tend to be closer to the city center rather than on the outskirts. In the case of coastal cities like Rio de Janeiro, Brazil, the wealthy tend to live on lowlands, while many poor people live in shantytowns on steep hillsides. When landslides happen there, the poor may be more vulnerable than the rich. Although much has been said about the impacts of Hurricane Katrina on New Orleans' predominantly black population, a greater proportion of white people lost their lives in the storm than did blacks. Nonetheless poor people may be less likely than rich people to have the resources needed to rebuild their homes and lives.

Because every place is unique, it is difficult to make generalizations about the relation between wealth and vulnerability to hazards. Most people don't think about such hazards when they choose a place to live; other considerations are usually much more important. When we hear about disasters triggered by earthquakes, floods, or other natural hazards, however, the outcomes in poor countries are often very different from outcomes of similar events in rich countries.

Questions

1. Scan the news of the last few months for reports of major natural disasters. Who suffered the most? Who was affected the least? Did wealth help protect people from the danger? Did it make them more vulnerable? Or was wealth irrelevant?
2. Which environments in your community are more vulnerable to natural hazards, and which ones are less vulnerable? Who lives in the vulnerable places?

River, the world's largest. Discharge of any stream usually increases after storms and decreases during dry spells.

Drainage density is the combined length of all of the stream channels in a basin, divided by the area of the drainage basin. A basin that has soil capable of absorbing and storing most of the rain will usually have a low drainage density. Landscapes with soils that cannot absorb rainfall very rapidly are more easily eroded to form channels and tend to have higher drainage densities.

Streams shape their channels by alternately eroding and depositing material on their beds and banks. The turbulent, swirling motion of the water erodes particles from the channel. The water transports these particles, along with loose sediment in the channel and minerals dissolved in the water. This movement of material in a stream is called **sediment transport**. The amount of sediment a stream carries increases as discharge increases, and larger streams typically carry more sediment than do smaller ones. Transport also increases greatly after a heavy rain. During periods of lower flow, less sediment is transported downstream, possibly causing a channel to partially fill with sediment.

A stream is also responsible for shaping its **floodplain**, which is a nearly level surface at the bottom of the valley through which the stream is flowing (Figure 3-22). The surface of the floodplain is formed from deposits of sediment where the stream periodically floods. Flowing water tends to **meander**, or change direction from side to side, which contributes to widening the channel. The channel continually shifts from side to side as the stream erodes material from one side of the channel, where the current is swifter, and deposits it on the other side, where the slower current has less energy (Figure 3-23).

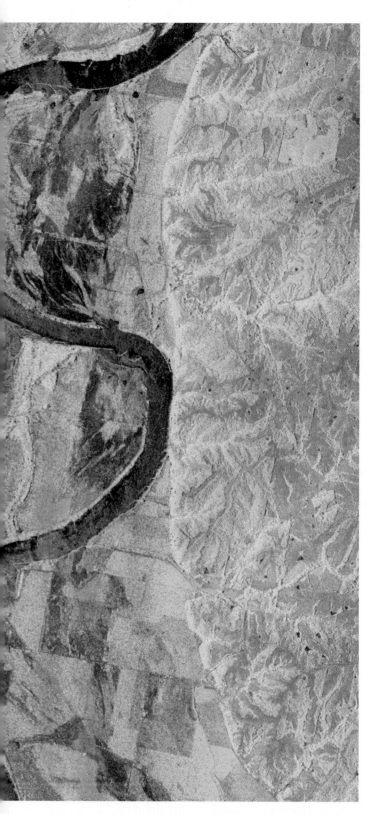

FIGURE 3-22 Floodplain of the Missouri River, Missouri. This image shows elevations in the Missouri River floodplain, as measured by satellite-borne radar. The floodplain, in shades of blue and purple, was hit hard by a flood in 1993. Upland areas are shown in shades of yellow. The yellow patches within the floodplain are areas of forest. The radar detects the height of the treetops and maps these areas as higher elevations than the surrounding fields even though the ground elevations are about the same. (NASA Headquarters)

By continually eroding and depositing material in channels and floodplains, streams tend toward a stable condition, known as its **grade**. A graded stream transports exactly as much sediment as it has collected (Figure 3-24). Streams rarely operate at a condition of grade for long, because daily changes in weather and disturbances from erosion and human activities continually upset the balance. As the stream's stable condition is upset and the transport of sediment increases or decreases, the shape of the stream channel may change. An especially heavy flow following a storm may cause the channel to shift. When increased erosion upstream generates more sediment than a stream can carry, the excess is deposited in the channel or on the floodplain (Figure 3-25).

The deposition of excess sediment slowly raises the elevation of a stream, which in turn reduces the difference in elevation between places upstream and downstream, and also reduces the stream's slope. Lowering the slope reduces the amount of sediment arriving from upstream.

Most sediment carried by a stream does not move to its final resting place in a single step. Typically, sediment becomes temporarily stored in a floodplain, then eroded, transported, and deposited a second time, then a third time, and so on for many times along its journey. Eventually, most sediment reaches the sea. Where a river enters the sea, the water velocity drops abruptly, and the sediment may form a large area of deposited sediment called a **delta**.

Rainfall also is an important erosion agent in dry areas, even though this is rare. Because deserts lack vegetation, the occasional rainfalls cause rapid erosion. Dry soils are much more prone to erosion than are vegetated ones—look at any construction site after a heavy storm. Thus, a dry landscape often is covered with gullies and dry stream channels, even though only a few centimeters of rain may fall per year (Figure 3-26).

A prominent landform in many deserts is the **alluvial fan**. Alluvial fans are broad, gently sloping deposits formed from sand and gravel where a fast-moving stream emerges from a narrow canyon onto a broad valley floor (see Figure 3-15, where large alluvial fans are shown on the far side of the valley). When the stream emerges, the water quickly spreads out, and loses velocity. Given the limited amount and frequency of rainfall, a stream in a desert is unlikely to carry sediment very far. Instead, sediment that the stream can no longer carry is deposited in an alluvial fan. Occasional floods crossing the alluvial fan can disturb the pattern and encourage wind erosion. Large depositional areas in desert valleys form important sources of wind-eroded sediment.

Increased erosion from human activity   In addition to shaping landforms over millions of years, erosional forces are important in human time scales because human modification of Earth's surface is usually

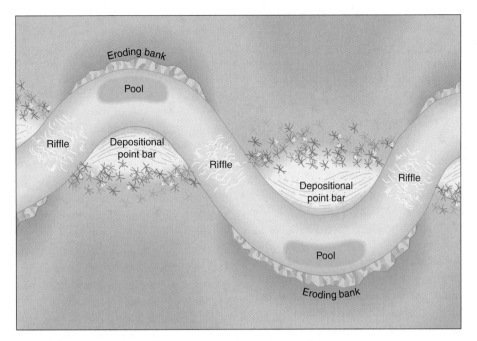

FIGURE 3-23 Features of a meandering channel. Erosion occurs primarily on the outsides of bends while deposition occurs on the insides. Depth usually alternates between deep portions or pools and shallows or riffles.

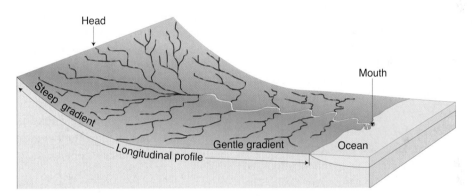

FIGURE 3-24 An idealized profile of a stream and its numerous tributaries. Note the concave-upward shape of the profile, with steeper gradient toward the heads of streams and gentler gradient downstream. As the stream flows downhill, it gathers more and more water, and its erosive power is greatly increased. The gentler gradient downstream slows the flow, balancing the erosive power of the stream with the amount of sediment it must carry. (From E. J. Tarbuck and F. K. Lutgens, *Earth Science*, 7th ed. New York: Macmillan, 1994)

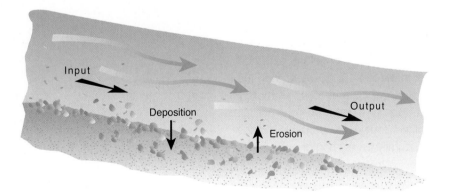

FIGURE 3-25 Stream erosion. Streams erode and deposit material in response to variations in streamflow and its ability to transport sediment. Sediment flows in and out of a portion of the channel. If more sediment enters than leaves, the excess is deposited and the channel shrinks or the floodplain accumulates sediment. On the other hand, if more sediment leaves than enters this portion of the channel, the channel erodes and enlarges.

Hurricane Katrina: Vulnerability in an Unstable Environment

On August 29, 2005, one of America's greatest cities was hit by a hurricane that broke the levees that had protected the city from the Mississippi River. As a result, over 1,800 people lost their lives, 81 billion dollars of damage was done, and 350,000 homes and 30 oil platforms were destroyed. The city population dropped by half, and may never recover. This tragedy could have been foreseen.

Despite its location in the flat, tectonically stable interior of North America, the lower Mississippi River is a remarkably dynamic environment. This dynamism, aggravated by human activity, contributes to flood hazards in that region by: (1) making the river channel unstable; (2) causing much of the land to be low in elevation; and (3) reducing the protection normally provided by coastal wetlands. New Orleans lies in the midst of this unstable environment, and because it is a major city the options for adapting to instability are limited. In 2005 we saw the consequences of that unfortunate combination of circumstances.

The Mississippi River drains nearly 3 million square kilometers (1.25 million square miles), which is about a third of the United States. Its vast floodplain, 100 kilometers (60 miles) wide in some parts, is the resting place of sediment eroded from the Rocky Mountains in the west, the Appalachians in the east, and most of the land between. Ultimately sediment is carried to the mouth of the river, where it accumulates in the Mississippi Delta. The accumulation of sediment at the mouth of the river causes this environment to be especially dynamic. Deposition of sediment each year extends the delta further out into the Gulf of Mexico, thus increasing the horizontal distance the water has to travel to reach sea level. The vertical distance to the sea remains the same, however, so the channel slope decreases through time (slope = rise/run). From time to time, usually on a time scale of a few thousand years, the river suddenly switches locations to a new, shorter route to the sea. The channel has switched locations several times in the last 10,000 years, with the mouth of the Mississippi at times reaching 300 kilometers (186 miles) to the west, near the Texas–Louisiana border. Such a switch is imminent today, and probably already would have occurred naturally were it not for engineering works that keep the channel in its present location.

At the same time, sediment in the delta settles and is compacted under its own weight. Isostatic depression of the crust also causes the delta to sink slowly over time. This means that relative to the land, sea level is rising in the delta area even faster than the world average.

In the New Orleans area the sea is rising relative to the land at an average rate of about 10 millimeters (0.4 inches) per year. The site of the city was established in 1718, and in the three centuries since then much of the city has sunken considerably. This is why so much of the city is 2 to 3 meters (7 to 10 feet) below sea level today (see figure), and why it is so vulnerable to flooding. The rate at which the land is sinking is highly variable from place to place, and in some areas the land has sunk rapidly. For example, some of the flood walls in New Orleans were originally built to a height of 4.57 meters (15 feet) above mean sea level, but at the time of the storm they had sunk to a little more than 3.66 meters (12 feet) above sea level.

The subsidence problem is compounded by the fact that over the last several decades thousands of dams have been built on streams throughout the Mississippi basin (see Figure 1-31). These dams trap sediment that would otherwise be delivered to the Mississippi River, and ultimately to the delta. So the supply of sediment that, under natural conditions, would more than adequately replace the land lost to subsidence is no longer available. Thousands of acres of coastal Louisiana wetlands are lost each year because the land is sinking and not being replaced by new deltaic sedimentation. The loss of wetlands in turn allows storm surges to penetrate much further inland than they would otherwise, which, again, contributes to the flood hazard in

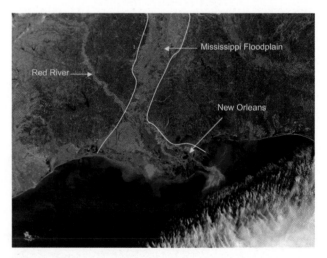

Lower Mississippi River (NASA). The lower Mississippi River enters the Gulf of Mexico at the tip of a long extension of the delta, visible in the large plume of brown sediment entering the Gulf. The Mississippi floodplain shows up as a wide swath of lighter-toned land extending northeast to southwest. The smaller floodplain of the Red River, which joins the Mississippi upstream of Baton Rouge, is also visible trending northwest to southeast. (NASA).

that region. Thus channel instability, subsidence, and coastal wetland loss combine to create a major challenge for flood managers, especially in an era of rising sea level and potentially increased tropical storms activity.

For these reasons geographers, urban planners, engineers, and others have long been concerned about New Orleans' vulnerability to a major hurricane. But despite many warnings, little was done to reduce this vulnerability. After all, what can one do? It simply isn't possible to move a city, especially one with a history as rich as that of New Orleans, out of a dangerous place. The alternative is to build defenses, which the engineers did, but until a major storm strikes it is hard to find the political will to invest billions of dollars to prevent a disaster that few can imagine.

Kartrina occurred near the middle of what was the most active hurricane season on record, with 28 named storms. The storm developed quite close to the North American mainland and moved across southern Florida as a category 1 (wind speed of 75 mph or more) storm on August 25. When it moved over water again it strengthened over a thick layer of warm water in the Gulf of Mexico. It reached category 5 (winds over 170 mph), but weakened to category 3 (winds of 111–130 mph) as it approached the coast of Louisiana and Mississippi.

Warnings were issued in advance of the storm, and residents from the central Mississippi Delta to Navarre Beach, Florida, were evacuated. Many people did leave New Orleans, but many did not. Some did not have transportation available, but many stayed because they didn't understand the magnitude of what might happen. The Superdome was opened as a temporary shelter and many of those who stayed in the city took refuge there.

The eye of the storm crossed the coast near the tip of the Mississippi Delta and continued northward, passing about 50 kilometers (30 miles) east of New Orleans. The strongest winds on the east side of the storm hit in the vicinity of Biloxi, Mississippi—not far from where a category 5 storm—Camille—had struck 37 years earlier. At Biloxi the storm surge was 8 meters (26 feet) above mean sea level (see Figure 2-24), while in the New Orleans area it was about 3 meters (10 feet). On Lake Pontchartrain to the north of New Orleans, waves were driven by winds from the north (the center of the storm was to the east, so the counterclockwise circulation would cause winds to blow from the north in this area) and wave heights there reached 2 meters (7 feet). Flood walls failed at several locations in New Orleans, allowing the sea to flow in. In most cases the walls failed because the pressure of high water on the seaward side caused the wall to slide landward, rather

than by water actually washing over the walls. In a matter of hours 80 percent of the city was flooded with water up to 6 meters (20 feet) deep, and because the flooded area was below sea level the water would not drain. Thousands of people were forced to flee to upper stories, rooftops, or attics. In the days that followed some were able make their own way to higher ground, others were rescued by boats or helicopters, and still others remained trapped in their attics. Gasoline and a wide range of everyday chemicals were released into the flood waters and dispersed, mixing with sediment to create vast amounts of contaminated mud. Initial predictions of the length of time needed to repair levees and pumping stations and to pump the water out proved pessimistic, but, still, it took about 4 weeks to accomplish the task.

The devastated area was so large and the damage so complete that fundamental questions arose concerning the future of the city. Could the flood protection system be rebuilt in such a way as to prevent this from happening again? How much would it cost, and from what source would the funds come? Given the inherent physical vulnerability of the place, was it wise to rebuild the entire devastated area, or should only certain areas be rebuilt? These are questions that could not be answered quickly. In the meantime more than 300,000 people who fled to other cities had begun to build new lives and may not return to New Orleans at all. One prominent study predicted that the city's population would ultimately drop by 70 percent, from 484,000 to 140,000.

The flood walls have been repaired, and basic services such as electricity and water have been restored to much of the city. But the citizens of New Orleans continue to wrestle with the question of whether it is wise to continue to occupy an environment that is inherently unstable and exposed, by virtue of its elevation and location, to the very real prospect of future disasters. Long-term changes in the environment—subsidence, sea-level rise associated with global warming, and loss of wetlands caused by lack of sediment delivery to the delta—will only increase New Orleans's vulnerability in the future. Following the 1993 Mississippi River flood in Iowa, Missouri, and Illinois, large areas of the floodplain were abandoned and entire towns were moved to higher ground. But New Orleans is a major city, not a small town, and its distinctive character is inseparable from the site on which it is built. To abandon the site is to lose forever one of our national treasures, yet to rebuild without enormous new investments in improvements of the flood-control system is to face increasing risk of another disaster.

FIGURE 3-26 Erosion by water in the desert. Erosion by the San Juan River has exposed folded sedimentary rocks in southeastern Utah. (Tom Bean/Corbis/Stock Market)

much more rapid than that occurring over geologic time. Human activities, such as deforestation, agriculture, and urban development, sharply increase the amount of sediment being eroded into streams.

People clear forests because they want to use trees for fuel, lumber, and paper, or they want to use the land for another purpose, especially agriculture and urban development. Erosion has increased as a result of the elimination of the vegetation cover for both reasons, but it is particularly severe where agriculture has replaced forest.

For much of Earth's agricultural land, erosion of soil into streams is a major loss. To meet the needs of a growing population, food production expands in two principal ways—by opening up new land for agriculture and by using existing farmland more intensively. But both strategies can result in erosion of the rich soil necessary for productive agriculture.

In places with rapid population growth, new lands may be opened for agriculture that may not be suited for intensive farming. For example, farmers in East Africa are clearing, tilling, and planting extremely steep mountain slopes. The rate of surface lowering by erosion in such areas can be as much as a few millimeters per year. Within a few years the rich topsoil may be completely eroded in such areas. The consequences of this erosion for food production will be discussed in Chapter 8.

Opening up new land for agriculture was a major contributor to increased erosion in the United States during the eighteenth and nineteenth centuries. More than 2 million square kilometers (800,000 square miles) of forest were cleared and replaced with plowed fields and pastures in the eastern and midwestern United States. This deforestation probably increased the rate of soil erosion by 10 to 100 times.

Erosion of agricultural land is further increasing as farmers use existing fields more intensively. The soil on about one-fifth of U.S. cropland is being lost faster than it can be replaced by natural soil formation. Also, to obtain higher yields, farmers plant profitable crops like corn, soybeans, and wheat every year, instead of periodically planting cover crops such as clover or alfalfa that restore nutrients to the soil. By failing to restore nutrients to the soil, farmers can exhaust its productivity.

Soil erosion lessened in much of the eastern United States during the twentieth century because farmland was abandoned. Its soil was depleted of nutrients, and more productive land was available in the Midwest. Much of the former farmland in the eastern United States has returned to forest, so erosion is less severe today than it was in the past. When the prairies west of the Mississippi were brought under the plow in the late 1800s, erosion became a problem there, reaching crisis proportions in the 1920s. Since the 1930s, government-sponsored soil conservation measures have significantly reduced erosion on U.S. farms, but the problem remains serious in some areas.

A heavy rainfall or rapid snowmelt may dump an overload of eroded sediment into a stream, which may be deposited on the floodplain downstream. Such sediment can bury nutrient-filled soils already in the floodplain, as can roads and buildings constructed on the floodplain. Flooding may increase if a sediment-choked stream overflows its banks.

Even after the rate of soil erosion into a stream declines or the sediment supply from upland is reduced by dams, a large quantity of sediment from the valley bottom may continue to be sent downstream. Many U.S. rivers have high sediment loads in part because they are currently excavating sediment from their floodplains that was deposited during a past time of severe agricultural erosion.

Urban development also increases erosion. Land is cleared of vegetation for building new houses, factories, and shops. Erosion rates during the construction phase often are hundreds of times greater than for undisturbed land. Evidence of this erosion is usually clear right after a storm—small channels that deepen downslope and mud washed onto adjoining areas, for example. If construction proceeds slowly, the land may be subject to high rates of erosion for several years.

Once land has been developed, surfaces once covered with vegetation are covered instead with roofs, streets, and parking lots. Because rain does not soak into these nonporous surfaces, it must run off into stream channels. Sewers collect storm water in cities and often discharge it into a stream at a rapid rate. If the sewers are unable to handle the flow during a heavy rainfall, low-lying areas in the city may flood. To handle increased discharge, streams in urban areas may enlarge their channels through erosion of their banks and damage adjacent property.

ICE, WIND, AND WAVES

In some parts of the world, running water is a less powerful erosional agent than other forces. In some cold places, for example, the land is covered with ice that flows downhill, grinding away at rocks. The wind can cause surface erosion in places with sparse vegetation, especially in the world's extensive deserts. Wind is a much less powerful erosion agent than running water, since wind can carry only smaller particles up to the size of sand, and it can carry these only short distances. In coastal areas, a combination of water and wind can cause surface erosion. Waves—which are driven by winds—pound the shore with turbulent water. Even the strongest rocks and vegetation are eventually broken down by the continual rushing forth and back of the waves.

Glaciers

Greenland, the South Pole, and many high mountain areas are currently covered with thick layers of moving ice, called **glaciers**. **Alpine glaciers** form wherever snow accumulates year after year without melting, such as on the peaks of individual mountains. **Continental glaciers** over 3 kilometers (almost 2 miles) thick cover vast areas of Greenland and Antarctica.

Glaciers are rivers of ice that flow from places where snow accumulates yearly to warmer places where the ice melts. Water enters the head of the glacier as snow (Figure 3-27). The glacier flows downhill until the ice eventually melts and leaves the glacier at its terminus as meltwater or icebergs.

Glaciers flow very slowly, usually at rates of a few meters to a few hundred meters per year. Glaciers may change size from one year to the next, depending on variations in weather and climate. They grow when they receive more snowfall at their head, and they shrink if warm temperatures increase the melting at the terminus. For the past 200 years most of the glaciers of the world have been shrinking as a result of climatic warming (Figure 3-28). Glaciers can also act erratically. For many years a glacier may move only a few meters per year, then for a few years it may suddenly surge forward at several hundred meters (hundreds to thousands of feet) per year, before slowing again.

A glacier is like a conveyor belt because it picks up sediment from areas of erosion and drops it in depositional areas. As ice accumulates and begins to flow, the glacier picks up material, and where it melts, deposits the sediment. Glacial deposits play an especially important role in shaping landforms in places with rapid melting, because large quantities of material are dropped in these places, forming **moraines** (Figure 3-29). A **terminal moraine** is a ridge of material dumped at the end of the glacier.

FIGURE 3-28 Shrinking glacier. The Qori Kalis Glacier, Peru, in 1978 (top) and the same view in 2000. Most of the world's glaciers have been shrinking rapidly in recent decades as a result of climatic warming. (Lonnie Thompson/Byrd Polar Research Center)

FIGURE 3-27 The Rhône Glacier, Switzerland. Crevasses (cracks) on the glacier's surface are caused by flow. For most of the past 200 years, this alpine glacier has been shrinking, with melt exceeding accumulation. As a result, the glacier is much smaller than it was in the 1700s, probably as a result of climatic warming in the 1800s and 1900s. (Gabe Palmer/Corbis/Stock Market)

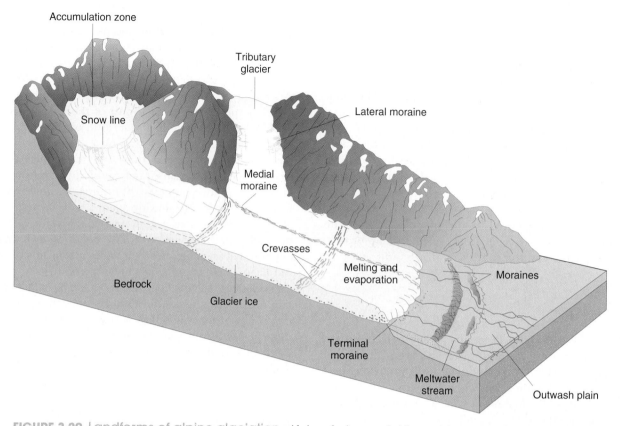

FIGURE 3-29 Landforms of alpine glaciation. Alpine glaciers are fed by snow in a zone of accumulation and flow downhill. They melt at lower elevations, leaving debris to accumulate at the end (terminal moraine), side (lateral moraine), or middle (medial moraine) of the glacier or be carried away in meltwater streams. (From R. W. Christopherson, *Geosystems*, 5th ed. Upper Saddle River, NJ: Prentice Hall, 2003)

Meltwater leaving a glacier deposits some of the debris close to the glacier in a broad, gently sloping plain, known as an **outwash plain**. The outwash plain contains a thick layer of rocks deposited close to the glacier in a layer of sand and gravel that can exceed a thickness of 100 meters (330 feet). The finer silt and clay materials are usually carried much farther and may be deposited in lakes, seas, or distant valleys.

Impact of Past Glaciations

Only 20,000 years ago—a very short time in Earth's 4.6-billion-year history—glaciers covered much of North America, Europe, and northern Asia (Figure 3-30).

These glaciers shaped landforms as they advanced. When Earth's atmosphere warmed, these rivers of ice melted back toward the poles or upslope to cold areas, and left behind debris that shapes the landforms we see today in many regions. As they advanced and retreated, alpine glaciers created distinctive landforms. In an unglaciated mountainous area, a stream may carve a V-shaped valley (Figure 3-31a).

As an alpine glacier flows through a V-shaped valley, it scours away the rock and rounds the valley bottom into a U-shape (Figure 3-31b). When the ice

melts, the U-shaped valley remains, surrounded by knife-edged ridges (Figure 3-31c).

The advance and retreat of continental glaciers have influenced the distribution of many human activities. The following are examples of how glaciation has affected three important natural resources in North America: soils, drinking-water supplies, and transport routes.

Soils The advance and retreat of glaciers in southern Ontario, Canada, and in the U.S. Midwest broke up bedrock, ground it into sand and silt, and left a great layer of debris covering the land. This rock contains important nutrients such as calcium and magnesium that are released by chemical weathering. The soils that have resulted are highly fertile and help make this region one of the most productive agricultural areas in the world. Other factors also contribute to the productivity of this region, and we will see in Chapter 8 how soil and climate interact with human input to determine regions' agricultural productivity.

Water supply Retreating glaciers left sand and gravel deposits capable of yielding a large supply of high-quality groundwater (Figure 3-32). Long Island, New York, which is more than 100 miles long and about 20 miles wide, is the top of a terminal moraine

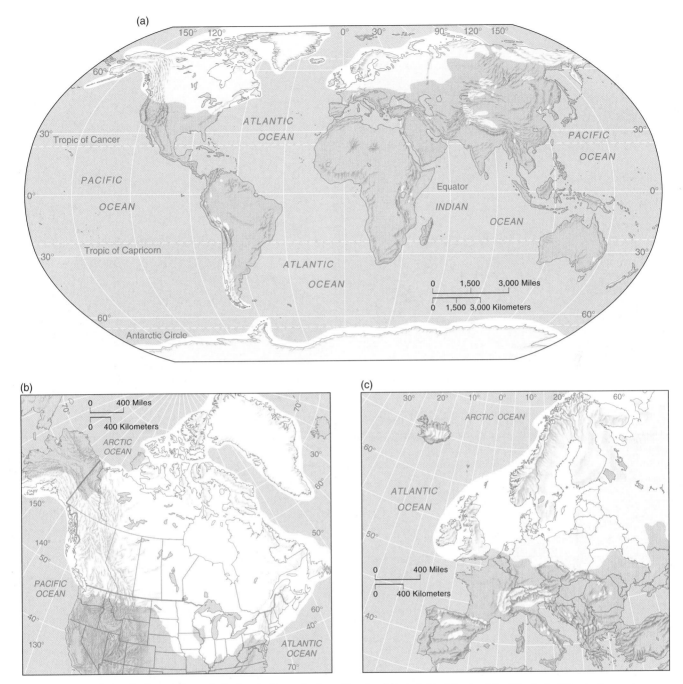

(a)

(b)

(c)

FIGURE 3-30 Extent of continental glaciation during the last Ice age. Much of the Northern Hemisphere was ice-covered 20,000 years ago. These maps show the extent of glaciation (a) worldwide, (b) North America, (c) Europe.

deposited about 20,000 years ago, marking the southern extent of the most recent glaciation. Much of Long Island consists of outwash deposits of sand and gravel that absorb precipitation readily. This groundwater is the primary source of water for 2.5 million inhabitants of Long Island.

With rapid population growth in recent years, however, demand for water has exceeded the rate at which the groundwater is being recharged. As more of Long Island is paved or built over, precipitation is blocked from entering the ground. As a result, the level of groundwater has lowered, and saltwater from the

Atlantic Ocean has seeped in to replace the depleted fresh groundwater in some places. The quality of the groundwater has also suffered from pollution by landfills, industry, and residential septic systems. Similar problems are occurring in other urbanized areas of the northeastern United States and Europe that depend on groundwater left behind by retreating glaciers.

Transportation routes Before the last continental ice sheet formed tens of thousands of years ago, the Missouri River flowed northward into Hudson Bay in Canada instead of southward to the Gulf of Mexico.

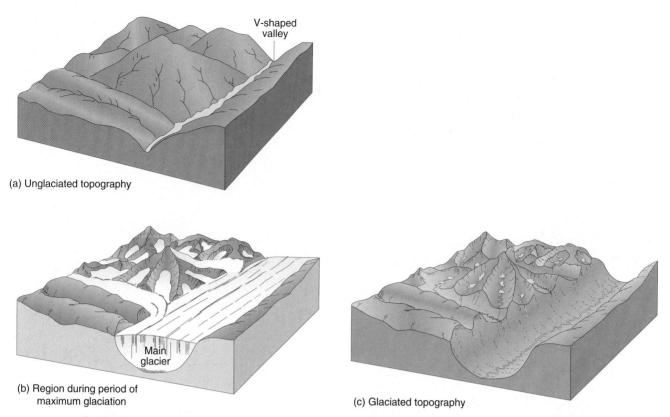

V-shaped
valley

(a) Unglaciated topography

Main
glacier

(b) Region during period of
maximum glaciation

(c) Glaciated topography

FIGURE 3-31 Typical features of glaciated landscapes. (a) Stream-eroded terrain prior to glaciation. (b) The glaciers have their way with the land, gouging, scraping, scouring, bulldozing, and plucking rock and soil. (c) After the climate warms and the glacier melts, distinctive landforms remain: U-shaped glacial valleys, new lakes and streams, and sharp-edged mountain ridges. (From E. J. Tarbuck and F. K. Lutgens, *Earth,* 4th ed. New York: Macmillan, 1993)

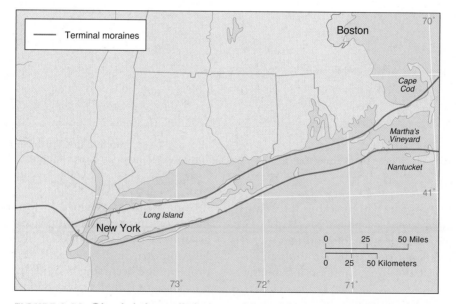

FIGURE 3-32 Glacial deposits in Long Island, New York. Long Island is covered by glacial deposits. The island is actually the top of a system of terminal moraines that also includes Martha's Vineyard, Nantucket, and Cape Cod, Massachusetts. The sandy soils of Long Island absorb water readily, and groundwater resources once were abundant. Today, however, these resources are threatened by overuse, intrusion of saltwater from the ocean, and pollution.

The Ohio River did not exist, but another river to the north carried water from the western slope of the Appalachians to the Mississippi. As the glaciers melted and began to recede, these rivers were blocked, and their water could not take the more northerly routes to the sea. Instead, these rivers, plus the meltwater from the glaciers, were forced to flow along the edge of the glacier until they emptied into the Mississippi River.

The lower Missouri and the Ohio Rivers were thus created as meltwater channels following the edge of the glaciers. When the glaciers melted, they remained in these new positions. The receding glaciers also carved the Great Lakes, which explorers eventually used to reach the interior of the continent.

European exploration of the North American continent during the seventeenth and eighteenth centuries

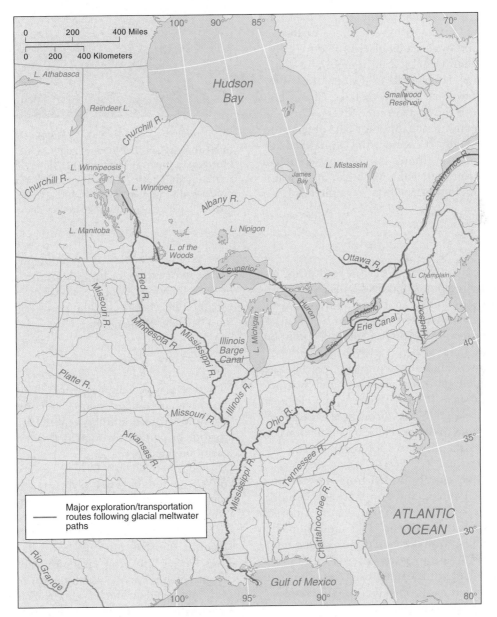

FIGURE 3-33 Glacial effects on North American drainage. Glacial meltwater channels profoundly altered the drainage of North America and created paths that were exploited by early explorers and later in the construction of canals. A series of major lakes were scoured by glaciers at the margins of the Canadian Shield, including the five Great Lakes, Lake of the Woods, Lake Winnipeg, Reindeer Lake, and others. Several of these are connected by channels formed by glacial meltwater draining from one lake to another, including the Red River–Minnesota River system, which at one time drained meltwater from Lake Winnipeg into the Mississippi River system. The Ohio and Missouri Rivers were created by meltwater flowing around the margins of the continental ice sheet (compare this map with Figure 3-30b). Some of these lakes and rivers are connected by canals, often constructed along glacial meltwater channels. Such canals include (1) the Erie Canal, and (2) the Illinois Barge Canal.

and the eventual settlement of the North American interior were heavily influenced by water transport. Exploration and settlement followed the St. Lawrence River to the Great Lakes and the major inland rivers feeding into the Mississippi, such as the Ohio, Missouri, and Illinois. Travel from the East Coast to the interior was difficult, because no convenient direct water route existed from the Atlantic to the Great Lakes and the Mississippi River system. After the United States gained its independence from Britain at the end of the eighteenth century, many canals were built in **meltwater channels** left behind by receding glaciers (Figure 3-33).

The Champlain Canal follows a meltwater channel to connect the Hudson River (which flows into the Atlantic at New York City) with Lake Champlain (which empties into the St. Lawrence River). The Erie Canal follows a meltwater channel for part of its route across New York State to connect the Mohawk River with Lake Ontario. The Illinois Barge Canal connects the Mississippi River with Lake Michigan at Chicago by following a meltwater spillway occupied by the Illinois River.

Effects of Wind on Landforms

Wind is an important shaper of Earth's landforms, especially in the dry regions. Wind erosion is also significant wherever the soil is not well covered with vegetation. This includes deserts, farmland, and coastal areas where beaches are kept free of vegetation by waves washing the shore.

Wind is not capable of moving large particles as big as gravel, but it can carry great amounts of fine-grained sediment such as sand. Where wind velocities are lower, or where topography encourages deposition, the sand accumulates in **dunes**. These accumulations of shifting sand are difficult places for vegetation to become established.

When the fine sand is eroded from the soil surface, larger rocks, pebbles, and gravel are left behind. The landforms in the desert can form a hard, armored surface, called **desert pavement**. The most popular image of a desert landform is a sand dune, but about 90 percent of Earth's desert areas are covered with desert pavement rather than sand. Wind erosion is minimized once desert pavement has formed, because the remaining rocks are too heavy to be moved by it.

Wind also can affect landforms in humid regions, because very fine-grained sediment can be carried some distance by the wind before being deposited (Figure 3-34). Thick layers of windblown silt, called **loess**, blanket many areas, including central China and the Mississippi River valley of the United States. Originally, this material was carried by meltwater from northern glaciers to nearby valley bottoms. Wind then carried the sediment to adjoining areas, forming loess, a fine agricultural soil. Running water subsequently has carved deep gullies and ravines in these loess deposits.

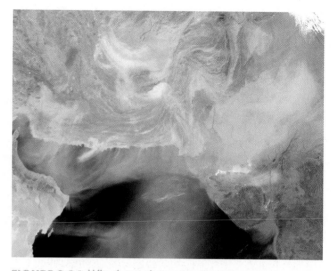

FIGURE 3-34 Wind erosion. Satellite view of a dust storm in southwest Asia. This image, taken May 2, 2003, shows dust blowing on the Arabian peninsula (left), Iran, Pakistan, and Afghanistan. Plumes of sediment laden water are also visible entering the Arabian Sea from India (right). (Jeff Schmaltz, MODIS Rapid Response Team/NASA Headquarters)

Windblown sand is common in coastal zones as well as in dry climates because of the lack of vegetation in both locations. Along coasts, however, vegetation is usually scarce because of the pounding of waves, rather than because of the climate.

Coastal Erosion

A coast is an especially active area of erosion because an enormous amount of energy is concentrated on the shorelines from pounding waves. Land may be lost to erosion, or gained through deposition, at rates up to several meters per year.

Waves   Winds blow across the sea surface, transferring their energy to the water by generating waves. Waves are a form of energy, traveling horizontally along the boundary between water and air. As the wind blows harder and longer, it transfers more energy and generates bigger waves. Waves also grow larger with an increase in the expanse of water across which the wind blows.

The speed at which a wave travels is affected by its **wavelength**. A small ripple travels very slowly, whereas an ocean swell wave may travel 10 to 50 kilometers per hour (6 to 30 miles per hour). A **tsunami**, which is an extremely long wave created by an underwater earthquake, may travel hundreds of kilometers per hour (Figure 3-35).

A wave can travel thousands of kilometers across deep ocean water relatively unchanged, but when it nears the shore, the shallow bottom restricts water motion and distorts the wave's shape. As it slows, the top part of it rushes forward and breaks (Figure 3-36). The earthquake that caused devastating tsunamis in

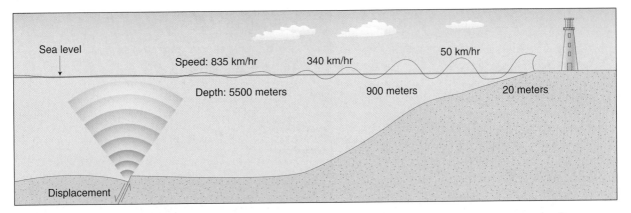

FIGURE 3-35 Tsunami. Earthquakes beneath the sea cause *tsunamis* (the Japanese word for "harbor wave"). A seafloor fault slips, violently displacing the water above, creating a very low wave at the surface that moves at a very high speed. As the wave comes ashore, it can build to heights of tens of meters. Historically, tsunamis have killed thousands in coastal settlements. After losing 150 people in a 1946 tsunami, Hawaii established the Pacific Tsunami Warning Center, which alerts nations around the Pacific when a tsunami is detected. Warning systems in the Indian Ocean were established or upgraded following the disastrous December 2004 tsunami that killed over 280,000 people. (From E. J. Tarbuck and F. K. Lutgens, *The Earth*, 4th ed. New York: Macmillan, 1993)

Southeast and southern Asia in December 2004 created waves that were detected around the globe.

The energy of a wave is released as a tremendous erosive force of rushing water on the beach. Beach pebbles and granules of sand are rolled back and forth, constantly being ground ever finer. The finest particles are carried into deep water and settle on the seafloor. The larger sand- and gravel-size particles are left behind on the shore to form a **beach**, which is a surface on which waves constantly break and move sand up and down.

A beach reflects the characteristics of the waves that form it. A coastal area pounded by larger waves will likely form a beach of very coarse material, because the waves carry the finer sand offshore and deposit it in quieter waters. The shape and size of the beach can vary if storms hit during only some but not all seasons.

Longshore current When you watch waves break on a beach, the most obvious motion of the water is perpendicular to the shoreline—the waves move up the beach and then recede. When the waves break, their energy gives a push to the water in a direction parallel to the shore, and the repeated breaking of many waves generates a **longshore current** traveling parallel to the shore (Figure 3-37). The longshore current is like a river, carrying sediment through **longshore transport** from areas where it is eroded by waves and depositing it where breaking waves lose the energy to carry

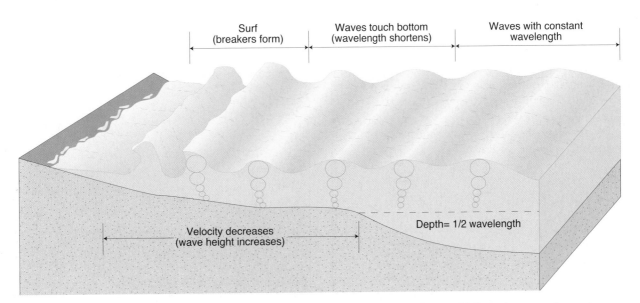

FIGURE 3-36 Breaking waves. As waves reach shallow water, they change shape, with long, low waves becoming short and tall, eventually breaking and releasing their energy on a beach. (From E. J. Tarbuck and F. K. Lutgens, *Earth*, 7th ed. Upper Saddle River, NJ: Prentice Hall, 2002)

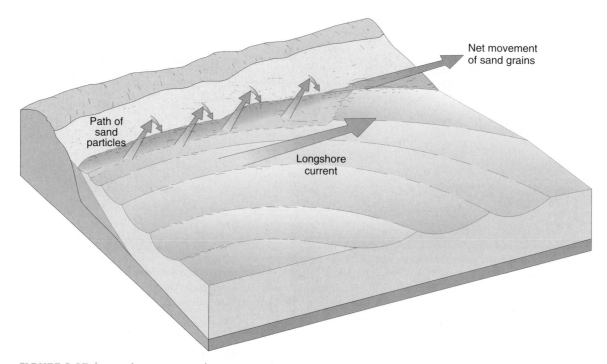

FIGURE 3-37 Longshore current. The longshore current is driven by waves breaking at an angle to the coastline. The current flows parallel to the shore. Breaking waves stir beach sand, causing it to be carried along with the current. (From E. J. Tarbuck and F. K. Lutgens, *Earth*, 7th ed. Upper Saddle River, NJ: Prentice Hall, 2002)

it—usually in deep water. Longshore currents can carry enormous amounts of sediment great distances.

Like rivers, landforms along shorelines are shaped by the balance between sediment arriving in a portion of the shore and then being removed from it. Distinctive landforms develop, such as the beaches, barrier islands, and spits, as illustrated in Figure 3-38. If more sediment is removed than arrives, the coast is eroded—which is the condition of most of the world's shorelines. But in some areas, more sediment

arrives than is removed, and the land area grows (Figure 3-39).

Sea-level change   The edge of the land—the shoreline—is defined by the elevation of the sea, or **sea level**. We think of sea level as being fixed, but on most shorelines it is not constant; it continually rises or falls relative to the adjacent land. In the short term, the sea can rise or fall several meters because of tides and storms. But two long-term factors can cause sea

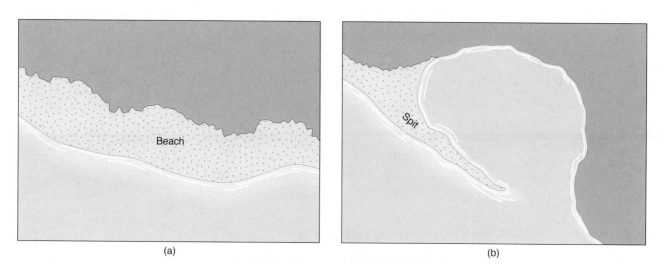

(a)   (b)

FIGURE 3-38 Coastal deposits. Deposition along coastlines creates beaches in small embayments (a), spits across the mouths of some bays (b), and

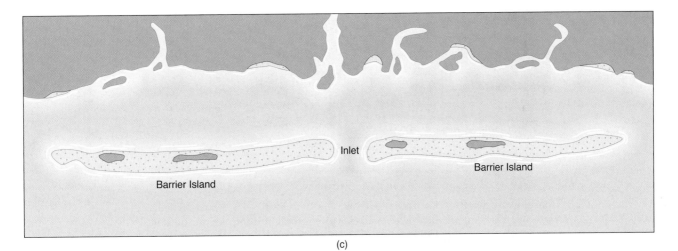

(c)

FIGURE 3-38 (*continued*) barrier islands on gently sloping coastlines (c).

level to rise or fall: climate change and movements of Earth's crust.

Over the past few hundred years, sea level as a whole has risen on Earth, at about 1 millimeter per year along coasts that are otherwise stable. Sea level has risen because the volume of seawater has increased. The increase results from the melting of glaciers, due to the overall warming trend since the eighteenth century, as well as from warming of the sea itself. This change in sea level is small compared to the sea-level rise of about 85 meters (280 feet) that occurred at the end of the most recent Ice Age.

Sea-level changes are significant at two different time scales. In the short term—a few decades—the direction of sea-level change affects how a shoreline

FIGURE 3-39 Port-Au-Prince, Haiti. This view of the coastline of Haiti's capital city shows densely populated areas close to sea level. Much of the land visible here is within 1 or 2 meters of sea level. In the center of the image is a recently built delta composed of sediment and solid waste, on which tens of thousands of people have built homes. If a major storm were to strike this coast, deaths would likely be in the thousands. (WHR Photo)

erodes. If sea level rises, the water offshore becomes deeper and waves break closer to the land, causing more erosion. If sea level falls, the shallower water causes waves to break further offshore, dissipating their energy and reducing shoreline erosion. Because many shorelines have gentle slopes, a minor increase in sea level can translate into a much larger landward migration of the shoreline (Figure 3-40).

Over thousands of years, large sea-level changes can reshape shorelines. During continental glaciation, the sea was substantially lower because more of the world's water was frozen in glacial ice on the land. With a lower sea level, rivers in coastal areas cut deep valleys as they approached the sea. When the glaciers melted, sea level rose worldwide, about 15,000 to 20,000 years ago, drowning the river valleys and creating large bays such as the lower St. Lawrence River, Chesapeake Bay, and Delaware Bay.

Sea level has fallen rather than risen relative to the adjacent land in some places. This has left inland soils, animals, and vegetation that are typical of beaches. For example, much of the U.S. West Coast has been tectonically uplifted in the last few million years as the North American and Pacific plates have ground together. As a result, the region lacks the many deep river mouths of the U.S. East Coast, but it does have former shorelines that are now above sea level, forming **marine terraces** (Figure 3-41).

Human impact on coastal processes The sea has been important to humans for thousands of years for trade, communication, and food. This long tradition has established most of the world's densest populations near the sea. The shoreline is an attractive place to build houses and recreational facilities, so coastal areas worldwide have become focal points for settlement and investment.

A drawback of coastal living is erosion from shifting shorelines. In wealthy societies like the United

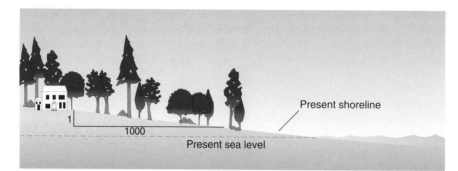

FIGURE 3-40 Sea-level and coastal erosion. Sea-level rise and shoreline position are directly related. On gently sloping shorelines, such as coastal plains, relatively low rates of sea-level rise can translate into significant rates of shoreline movement. For example, if the land rises only 1 meter for each kilometer inland (a ratio of 1:1,000), a 1-millimeter (.04-inch)-per-year rate of sea-level rise could mean horizontal shoreline displacement of 1 meter (3 feet) per year. Such rates of shoreline erosion usually result in major problems in developed coastal areas.

FIGURE 3-41 Marine terrace. The broad green area in this photo of California's Big Sur coast is a marine terrace, a former beach raised above the sea by tectonic uplift. Such terraces are common on the West Coast of the United States. (Renee Lynn/Photo Researchers, Inc.)

States, the typical response is to try to control coastal processes by modifying the coastline. These structures achieve the desired erosion control and stabilize the shoreline—but only temporarily. For example, people build groins perpendicular to the shore to slow long-shore transport of sediment (Figure 3-42). A groin constructed in one location along the shore to interrupt the movement of sand will cause less sand to be deposited on a beach farther along the shore, worsening the erosion problem there. People also build sea walls parallel to the shore, but the constant pounding of waves removes sand from around the sea walls and ultimately undermines them.

The shoreline is one of the most dynamic features of Earth's surface. On an eroding barrier island of the eastern United States, erosion over the past few decades could average between 10 centimeters and 2 meters (between 4 inches and 7 feet) per year. At that rate, someone who takes out a 30-year mortgage to buy a house that is 30 meters (100 feet) from the beach may see waves lapping at its edge by the time the last payment is made. If a hurricane hits, the house may be washed away even sooner.

Homebuilders may not know about the threat of erosion when they build new houses on an attractive beach. A coastal housing development is usually well established before its residents discover the dynamic nature of the coast. By then it is too late to abandon the substantial investment, and more money is spent to protect the structures already built. Most of the erosion and damage to these communities takes place during major storms such as hurricanes. After such a disaster, sympathy for the victims induces the government to help rebuild communities at considerable public expense. Over the long run, though, buying up the properties and relocating the families would probably be cheaper than repeatedly rebuilding shorefront communities.

THE DYNAMIC EARTH
Rates of Landform Change

We can now see how past geologic events have shaped the world we live in and how, occasionally, dramatic Earth movements occur. Most landforms change so slowly that they have become metaphors for permanence, such as "everlasting hills" and "rock of ages." Rates of horizontal movement of continents relative to one another are typically millimeters per year to centimeters per year. Vertical movements of the land are somewhat

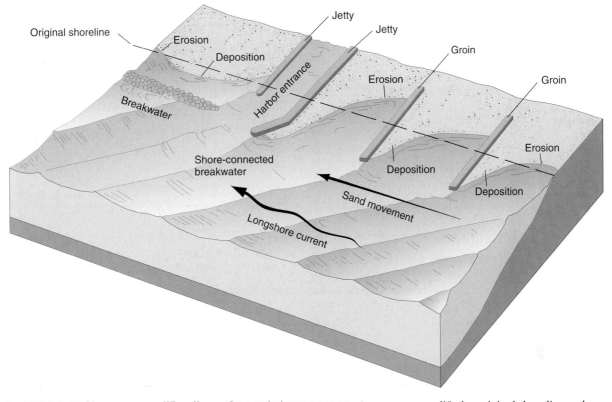

FIGURE 3-42 Human modification of coastal processes. Attempts to modify the original shoreline only change the pattern and cause problems elsewhere. (Adapted from R. W. Christopherson, *Geosystems*, 5th ed. Upper Saddle River, NJ: Prentice Hall, 2003)

slower—generally less than a few millimeters per year. Rates of landscape lowering by weathering and erosion are even slower. The surface of a mountain may erode at 5 to 50 millimeters (0.2 to 2 inches) per century, and a gentle slope at only 0.1 to 1 millimeter (.004 to .04 inches) per century.

Change may be somewhat faster where processes are strong and materials are weak. For example, massive granite usually erodes very slowly, but a steep slope of fractured shale may erode rapidly. Similarly, waves erode a shoreline of glacial sand and gravel much faster than one of solid rock. Much of the Massachusetts shoreline, which consists of glacial deposits, is eroding horizontally at a rate of 0.5 to 2 meters (2 to 7 feet) per year. But in Maine, resistant rocky coasts are eroding at less than 10 millimeters (0.4 inch) a year .

One way to assess the significance of a rate of change is to determine how long it takes to create a landform that reflects the prevailing environmental conditions. For example, how long does it take for a river to create a channel of suitable size to carry its runoff? How long does it take for waves to shape a coastline into a smooth beach? How long does it take for volcanic rock to develop into mature soil? The time required to shape the land varies greatly, depending on the processes and the materials involved (Figure 3-43).

In virtually every landscape, features that take a long time to develop can retain the marks of past events, such as climate and sea-level changes from tens of

thousands of years ago. These same landscapes also contain more dynamic features that respond to the weather or changes in land use and so they reflect the environmental conditions of the present or very recent past.

Another way to view rates of change is to see their effects on human activity in particularly sensitive areas, such as along coastlines. In the Mississippi Delta, land is sinking relative to the sea at about 10 millimeters (four-tenths of an inch) per year. The city of New Orleans was founded in 1718 at a site close to sea level. If the land sinks at 10 millimeters per year for 200 years, the total subsidence is 2 meters (7 feet), a significant amount for a low-lying city. This subsidence also complicates management of floods in the lower Mississippi Valley.

Environmental managers must understand natural rates of landform change so they can predict and manage human impacts on the land. To maintain naturally productive soils, for example, we cannot allow soil to erode severely and then expect it to recover in just a few years. On the other hand, if we pave an area and thus alter the amount of water entering a stream, we should not be surprised that the stream quickly floods and causes increased erosion.

Environmental Hazards

Some environmental processes, such as volcanic eruption, earthquakes, landslides, tornadoes, and hurricanes, occur so rapidly and violently that they cause widespread death

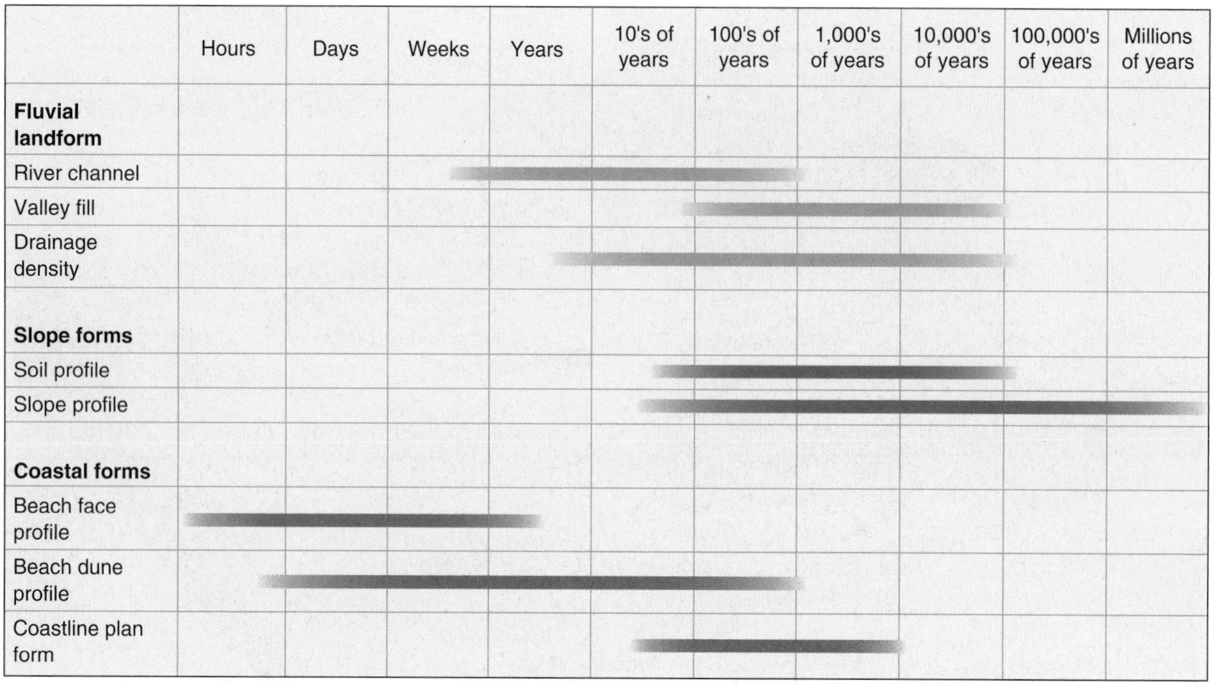

	Hours	Days	Weeks	Years	10's of years	100's of years	1,000's of years	10,000's of years	100,000's of years	Millions of years
Fluvial landform										
River channel			▬▬▬	▬▬▬	▬▬▬					
Valley fill						▬▬▬	▬▬▬	▬▬▬		
Drainage density					▬▬▬	▬▬▬	▬▬▬			
Slope forms										
Soil profile						▬▬▬	▬▬▬	▬▬▬		
Slope profile						▬▬▬	▬▬▬	▬▬▬	▬▬▬	▬▬▬
Coastal forms										
Beach face profile	▬▬▬	▬▬▬	▬▬▬	▬▬▬						
Beach dune profile		▬▬▬	▬▬▬	▬▬▬						
Coastline plan form					▬▬▬	▬▬▬	▬▬▬			

FIGURE 3-43 Length of time to form selected landscape features. The length of time needed to create a landform is short where processes are strong and materials weak. But if large amounts of material must be moved, landforms can take very long periods to be shaped by weak processes.

and destruction. From a human perspective, these changes are momentous, because at any given location they are rare in human time scales. From a geologic perspective, however, they are routine.

In some areas, rapid change is frequent enough for people to recognize the threat and avoid it. But a coastal lowland may experience a severe hurricane only once in several decades, so people feel they can safely build homes there. Because many environmental hazards are once-in-a-lifetime events, people often rebuild after a disaster rather than move to a safer area. They believe that another similar event is unlikely in the near future, making the risk too small to justify giving up a livelihood or a home. Rebuilding often is accompanied by public pressure to build new structures to protect the community, such as a levee or sea wall.

Unfortunately, structures such as sea walls on the shoreline or levees along rivers cannot protect people and property against the most extreme events in a given environment. And in many cases, these structures worsen the problem. For example, levees confine floods within a narrow channel instead of letting the water spread out across a floodplain, thus increasing the water height during a flood. Protective structures also worsen the problem by giving people a false sense of security, leading them to make greater investments, thus increasing their vulnerability. When the next flood, storm, or wildfire strikes, the damage is even greater than from the last.

As was discussed in Chapter 1, humans adapt to the natural environment, and they also modify it. Geographers studying these problems recognize that natural disasters are usually not purely natural. Rather, they are caused by a combination of natural events and human vulnerability. In a flood, a house is destroyed for a simple reason. Someone chose to build where a flood can occur. In most cases, the best approach is not to control the threat but instead to change human behavior.

In the wake of the 1993 floods on the Missouri and Mississippi rivers, some levees were neither repaired nor rebuilt. Instead, flood-prone land was bought or leased by the government so that future floods can inundate this land instead of damaging valuable properties again. Allowing land to be inundated instead of keeping out floodwaters has multiple benefits: Damage is reduced to structures in the path of the flood, and downstream communities are protected. Approaches to hazard management that focus on avoiding hazards, rather than confronting them, are usually much cheaper in the long run.

CONCLUSION: CRITICAL ISSUES FOR THE FUTURE

Earth's surface includes some features that have remained much the same for millions of years, and others that have changed dramatically in the last few hundred years. Where landforms are stable and unchanging, humans usually do not have significant impacts on the land. Human activity can destabilize landforms, however, creating danger for people living in that environment. A destabilized drainage basin might now have more frequent floods that erode riverbanks. An unstable coastline might now experience severe erosion. A deforested hillslope now might be prone to catastrophic landsliding.

As Earth's population grows and more land is developed for human use, people must learn to live with instability rather than try to control it. When Earth is moving, either in the sudden jolts of an earthquake or through gradual erosion, human structures are likely to fail. Future urban growth and land development must be sensitive to Earth's dynamic nature by avoiding hazardous areas. Floodplains and other low-lying areas should be left undeveloped or used for parks. Construction of housing on landslide-prone slopes should be discouraged, and soils that are easily eroded should be protected with conservation measures.

Geographers' training and experience in both the natural and social sciences place them in a unique position to help solve problems of human occupancy of hazardous environments. Geographers have been leaders in identifying natural hazards and developing strategies for coping with them. Such skills are increasingly important on our densely populated and dynamic Earth.

Chapter Review

SUMMARY

Major landforms of the world are created by a combination of endogenic (internal) and exogenic (external) landforming processes. Endogenic mechanisms are forces that cause movement beneath Earth's surface, raising some portions and lowering others. The most significant of these are movements associated with plate tectonics. This motion can create earthquakes, volcanoes, and mountains, depending on whether the boundaries between plates are convergent, divergent, or transform.

Exogenic processes are forces of erosion that wear down Earth's crust and reshape it into new landforms. Rocks are first broken down into smaller pieces through chemical and mechanical weathering. Then they are carried by gravity down the slopes of hills or by wind from one place to another. Streams collect water from groundwater and overland flow and transport the water down to the sea. Streams are conveyor belts for sediment from hillsides, and they play a major role in shaping landforms.

In many mountain and poleward environments, glaciers flow across the land, eroding rock and depositing it at places having higher temperatures that melt the ice. Wind plays a major role in moving material in deserts, and the absence of vegetation allows even infrequent rainfall to shape the land. Along coastlines, waves caused by wind blowing across the ocean surface cause intensive erosion and rapidly change landforms.

Most change on Earth's surface is slow in human terms, usually taking thousands of years to significantly reshape the land. But in some areas, change may be dramatic. Geographers have learned to identify areas that are subject to rapid or sudden change and to understand better how natural processes occur and how they affect human settlements. In general, we have learned that in dynamic environments it is better to live with nature and avoid hazards than to attempt to control them.

KEY TERMS

alluvial fan p. 110
alpine glacier p. 115
beach p. 121
chemical weathering p. 105
composite cone volcano p. 101
continental glacier p. 115
convergent plate boundary p. 102
delta p. 110
desert pavement p. 120
discharge p. 108
divergent plate boundary p. 102
drainage basin p. 108
drainage density p. 109
dune p. 120
earthquake p. 98
endogenic processes p. 97
epicenter p. 98
exogenic processes p. 97
fault p. 103
floodplain p. 109
focus of an earthquake p. 98
geomorphology p. 96

glacier p. 115
grade p. 110
groundwater p. 107
igneous rocks p. 103
isostatic adjustment p. 102
karst p. 106
landforms p. 97
lava p. 100
loess p. 120
longshore current p. 121
longshore transport p. 121
magma p. 100
mantle p. 98
marine terrace p. 123
mass movement p. 106
meander p. 109
mechanical weathering p. 106
meltwater channel p. 120
metamorphic rocks p. 103
moraine p. 115
outwash plain p. 116
overland flow p. 108

plate tectonics theory p. 97
sea level p. 122
seafloor spreading p. 102
sediment transport p. 109
sedimentary rocks p. 103
seismic waves p. 98
seismograph p. 98
shield p. 103
shield volcano p. 101
sial p. 103
sima p. 103
soil creep p. 106
structural landform p. 104
surface erosion p. 106
tectonic plates p. 98
terminal moraine p. 115
transform plate boundary
  p. 102
tsunami p. 120
volcano p. 100
wavelength p. 120
weathering p. 105

QUESTIONS FOR REVIEW AND DISCUSSION

1. What are endogenic and exogenic processes? Make a list of mechanisms that affect the shape of Earth's surface, and classify each as endogenic or exogenic.

2. What three types of relative motion occur at plate boundaries? What are examples of boundaries where each of these three types of motion is occurring?

3. What is isostatic adjustment, and what are its causes?

4. What environments favor rapid rates of mass movements?

5. What is a drainage basin? What measurable characteristics of drainage basins are important to understanding river processes?

6. What landforms provide evidence of sea-level rise? Sea-level fall?

7. How has soil erosion from agricultural lands modified sediment transport, erosion, and deposition in rivers?

8. What are some examples of landforms whose shape can be understood in terms of the processes acting on them? How long does it take for these landforms to develop?

THINKING GEOGRAPHICALLY

1. What were the environmental conditions where you live (or go to school) at the time of the last glacial maximum (20,000 years ago)? Were glaciers present? If not, how was the climate different? How was the vegetation different? What is the evidence that reveals these differences?

2. In your library, obtain a copy of a soil survey for your area. Soil surveys in the United States are published by the Natural Resource Conservation Service, and typically are available at most libraries. Read the sections pertaining to the development of soils in the area. Select two different sites that exemplify different kinds of landforms or deposits and visit these sites, comparing what you see with what is written in the soil survey.

3. Look at a topographic map of an area that includes a river. Can you identify the meanders of the channel?

Measure the width of the river. Now measure a distance along the channel that is long enough to include several bends of the river. Count the number of right-hand bends in the measured portion of the river, and divide that number into the distance. The result is the meander wavelength, normally 10 to 15 times the river width. Try these measurements on other maps to see how consistent the relation is.

4. Mount Saint Helens in Washington State erupted violently in 1980. The Loma Prieta earthquake in California occurred in 1989. How far apart are these two points on a map? Was there a possible connection between these two events? List reasons why there might be a connection, and why there might not.

SUGGESTIONS FOR FURTHER LEARNING

Abbott, P. L. *Natural Disasters.* McGraw-Hill, 2005.

Abrahams, Athol D., and R. A. Marston. "Drainage basin sediment budgets: An introduction." *Physical Geography* 14 (May–June 1993): 221–224.

Collins, M. J., and J. C. Knox. "Historical changes in upper Mississippi River water areas and islands." *Journal of the American Water Resources Association* 39, no. 2 (April 2003): 487–500.

Cooke, R. U., A. Warren, and A. Goudie. *Desert Geomorphology.* London: UCL Press, 1993.

Cooke, R. U., and J. C. Doornkamp. *Geomorphology in Environmental Management: A New Introduction.* Oxford: Clarendon, 1990.

Galgano, F. A., and others. "Beach erosion, tidal inlets and politics: The Fire Island story." *Shore & Beach* 67, no. 2–3 (1999): 26–32.

Goudie, A. *The Human Impact on the Natural Environment: Past, Present, and Future.* 6 ed. Oxford: Blackwell Publishing, 2005.

Hardy, R. J. "Fluvial geomorphology." *Progress in Physical Geography* Vol. 30 Issue 4 (2006): 553–567.

Hyndman, D., and D. Hyndman. *Natural Hazards and Disasters.* Belmont, CA: Brooks Cole, 2005.

Karlen, W., and others "Glacier fluctuations on Mount Kenya, East Africa, between 5700 cal. years BP and the present." *Ambio* 28, no. 5 (1999): 409–418.

Lindstrom, M. J., D. A. Lobb, and T. E. Schumacher. "Tillage erosion: An overview." *Annals of Arid Zone* 40, no. 3 (September 2001): 337–349.

McHugh, M., T. Harrod, and R. Morgan. "The extent of soil erosion in upland England and Wales." *Earth Surface Processes and Landforms* 27, no. 1 (January 2002): 99–107.

Nash, D. J. "Arid geomorphology." *Progress in Physical Geography* 27, no. 2 (June 2003): 284–303.

Parshall, T., and D. R. Foster. "Fire on the New England landscape: Regional and temporal variation, cultural and environmental controls." *Journal of Biogeography* 29, no. 10–11 (October–November 2002): 1305–1317.

Phillips, J. D. "Sources of nonlinearity and complexity in geomorphic systems." *Progress in Physical Geography* 27, no. 1 (March 2003): 1–23.

Ritchie, J. C., D. E. Walling, and J. Peters. "Application of geographic information systems and remote sensing for quantifying patterns of erosion and water quality." *Hydrological Processes* 17, no. 5 (April 15, 2003): 885–886.

Scull, P., J. Franklin, O. A. Chadwick, and D. McArthur. "Predictive soil mapping: A review." *Progress in Physical Geography* 27, no. 2 (June 2003): 171–197.

Stethem, C., B. Jamieson, P. Schaerer, D. Liverman, D. Germain, and S. Walker. "Snow avalanche hazard in Canada—a review." *Natural Hazards* 28, no. 2–3 (March 2003): 487–515.

Titov, V., A. B. Rabinovich, H. O. Mofjeld, R. E. Thomson, and F. I. González. "The Global Reach of the 26 December 2004 Sumatra Tsunami." *Science* 309 (23 September 2005): 2045–2048.

Travis, J. "Scientists' fears come true as hurricane floods New Orleans." *Science* 309 (9 September 2005): 1656–1659.

VanLooy, J. A., and C. W. Martin, "Channel and vegetation change on the Cimarron River, southwestern Kansas, 1953–2001." *Annals of the Association of American Geographers* 95, no. 4 (2005): 727–739.

Viles, H. A., and A. S. Goudie. "Interannual, decadal and multidecadal scale climatic variability and geomorphology." *Earth-Science Reviews* 61, no. 1–2 (April 2003): 105–131.

Wright, T. J., C. Ebinger, J. Biggs, A. Ayele, G. Yirgu, D. Keir and A. Stork. "Magma-maintained rift segmentation at continental rupture in the 2005 Afar dyking episode." *Nature* 442 (2006): 291–294.

WEB WORK

The U.S. Geological Survey is the U.S. government agency responsible for studying and monitoring Earth's surface, and it provides a wide range of resources concerning earthquakes, geological resources, water resources, and landforms:

http://www.usgs.gov/

The Geological Survey of Canada has a Web page that describes many of its activities and, of course, it is bilingual:

http://gsc.nrcan.gc.ca/

This site at Michigan Technical University will take you to a map of nearly all active volcanoes in the world:

http://www.geo.mtu.edu/volcanoes/

Volcano World, sponsored by the U.S. National Aeronautics and Space Administration, also has excellent imagery:

http://volcano.und.edu/

The Geography Department at the University of Edinburgh has a mapping program that will produce maps of recent earthquakes:

http://www.geo.ed.ac.uk/quakes/quakes.html

4

BIOGEOCHEMICAL
CYCLES AND
THE BIOSPHERE

A Look **AHEAD**

Wildfires near Ezaro, Spain, in 2006. Farmers have historically used fire to clear land, but in the dry conditions of 2006 many fires burned out of control and ravaged the landscape (SIPA Press).

Biogeochemical Cycles

The hydrologic cycle is a continual worldwide flow of water among the four spheres: Water in the atmosphere condenses as clouds, falls as precipitation, runs off into the ocean or is stored in the ground, and returns to the atmosphere through transpiration by plants or evaporation. Humans play major roles in all major biogeochemical cycles.

Carbon, Oxygen, and Nutrient Flows in the Biosphere

The carbon cycle is a fundamental link among the atmosphere, oceans, biosphere, and soil. Atmospheric carbon is absorbed by the biosphere during photosynthesis, stored in living organisms and in the soil, and released through respiration. Fossil-fuel combustion is a major source of carbon to the atmosphere.

Soil

Soil is the uppermost part of the solid Earth. Soil is formed through a combination of physical, chemical, and biological processes. Soil characteristics vary depending on climate; the world map of soils has patterns similar to the world climate map. In many areas, people have reduced the ability of soil to support plant life by removing nutrients and exposing soil to erosion.

Ecosystems

Earth's surface includes many ecosystems in which plants, animals, and their physical environments interact. Water, nutrients, organic matter, and energy move among living things as well as the soil, biosphere, atmosphere, and hydrosphere. Photosynthesis is the basis of food chains and supports animal life. Plants are adapted to varying availability of

light and water and have developed forms that enable them to compete and to survive environmental stress. Ecological communities may change over time through succession, especially following disturbance.

Biomes: Global Patterns in the Biosphere

Earth's ecosystems are grouped into biomes characterized by particular plant and animal types and usually named for climates or dominant vegetation types. World maps of soils, vegetation, and climate show strong similarities because climate exerts a powerful control on soil formation and plant growth. Humid tropical environments have diverse forest vegetation and nutrient-poor soils. Arid and semiarid regions have cacti, grasses, and shrubs that are adapted to low-moisture stress. Humid midlatitudes are dominated by forests: winter-deciduous forests in warmer areas and coniferous forests in colder, drier areas. High-latitude areas have either sparse vegetation or low herbaceous tundra that can survive extreme cold in winter. Throughout the populated world, vegetation and soils show extensive human impact. The biosphere is a system in which humans play a crucial role.

*T*he summer of 2006 in Europe was one of the hottest on record. Throughout Western Europe temperatures well above normal persisted for week after week in July and August. Temperatures of 30–35°C (86–95°F) were occurring in southern England, where 20–25°C (68–77°F) is the norm. The high temperatures placed severe strain on electricity supplies, both because of limited generating capacity, but also because the high temperatures contributed to droughts, making scarce the necessary water for cooling in electric generating stations. In Spain, dry conditions also contributed to Spain's worst forest fire season in recent memory. The summer of 2006, and the record-warm summer of 2003 before it, helped convince most Europeans that global warming is a reality and that weather conditions that used to be considered extreme will be the norm of the coming decades. Farmers in northwest Spain have a long tradition of deliberately setting fires to improve pasture for cattle, but that may no longer be an appropriate land management technique. Europe is not the only place where 2006 was an unusually hot and dry summer: forest fires were also a major problem in the western United States that year. ▶

Traditionally, studies of diverse topics such as vegetation, weather, human population, and industrial activity have been carried out in isolation from each other. But scientists increasingly have recognized the intimate interaction among all Earth's systems, including natural and human-dominated phenomena. **Desertification,** a process by which semiarid vegetation and soil become more desert-like as a result of human use, has occurred in semiarid lands around the world. This is an example of environmental change caused by a combination of natural and human processes operating at Earth's surface. Degradation in various forms is occurring on a significant portion of the world's agricultural land. It reduces the fertility of the soil and necessitates greater investment in agricultural development to be able to feed Earth's growing population.

In this chapter, we examine biogeochemical cycles and the biosphere, including the hydrologic, or water, cycle. Water moves among the hydrosphere, biosphere, lithosphere, and atmosphere. It plays a critical role in regulating environmental processes, determining spatial patterns of vegetation and soil on Earth's surface, and linking all of Earth's subsystems. The biosphere is the thin layer of living things on Earth's surface, of which we are an inseparable part.

The biosphere is intimately connected to Earth, ocean, and air. Plants depend on solar energy for **photosynthesis,** which is the formation of food in green plants as a result of exposure to light. Plants also depend on atmospheric circulation for temperature regulation, carbon supply, and water supply that make growth possible. At the same time, plants and animals absorb and release oxygen and carbon dioxide to the atmosphere, and in so doing, powerfully influence atmospheric composition and climate. Plants and animals help convert rocks to soil, the dynamic medium that stores water and nutrients and supports life, and they also regulate nutrient supplies in the soil. Because of these interactions among climate, soil, living things, and rocks, patterns of plant and animal life correspond closely with patterns of climate, topography, and rock types.

The flow of water among the lithosphere, biosphere, and atmosphere is an example of a biogeochemical cycle. **Biogeochemical cycles** are recycling processes that supply essential substances such as carbon, nitrogen, and other nutrients to the biosphere. Like movement of water, biogeochemical cycles also connect Earth's subsystems.

In previous chapters, we described environmental changes that occurred over the past few million years. These changes have been driven mostly by natural climatic variability. But during the last 10,000 years (since the end of the Pleistocene Epoch), and especially the last 1,000 years, the human population has expanded so rapidly that we now profoundly influence global environmental patterns as much as the most dramatic climatic variations did in the Pleistocene Epoch. Just a few thousand years ago the human species had merely a minor environmental impact. Today, however, our large numbers, our enormous consumption of natural resources, and our widespread construction projects and other environmental modifications significantly influence global biospheric processes like energy exchange, food webs, and movements of materials on the surface of Earth. As a consequence, we can understand Earth's environments today only by recognizing how humans modify and regulate them.

THE BIOGEOCHEMICAL CYCLES

Processes operating in the biosphere—the growth and decay of plants and animals and, indeed, all life processes—depend on exchanges of energy and matter.

Earth receives a constant supply of energy in the form of light from the Sun. Matter, including the essential substances water and carbon, is available in the atmosphere, hydrosphere, and lithosphere, as well as in the biosphere itself.

The law of conservation of energy states that energy cannot be created or destroyed under ordinary conditions, but it may be changed from one form to another. Similarly, the law of conservation of matter states that matter may be changed from one form to another under ordinary conditions but cannot be created or destroyed. The exception to "ordinary conditions" is nuclear reactions, in which matter is converted to energy.

Although energy and matter are not created or destroyed, they are constantly being transformed. Biogeochemical cycles are the pathways by which energy and matter are transformed and recycled in Earth systems (Figure 4-1). The dynamic earth processes that we have discussed so far (plate tectonics, mountain building, erosion) all involve the exchange of energy and matter. Processes operating on the living organisms of the biosphere also depend on exchanges of energy and matter.

This section focuses on one of the two most important biogeochemical cycles, the water, or **hydrologic, cycle.** The next section examines another important cycle, the carbon cycle. Water is essential for all life to survive on Earth. Carbon is the most important element in biological processes on Earth.

The Hydrologic Cycle

Water is central to every part of the biosphere. Some of the distinctive features of water that make it so important include the following:

▶ Earth's temperatures allow water to exist in all three states—solid, liquid, and gas (vapor)—and to change readily from one state to another. Water can permeate rock as groundwater, exist in frozen reservoirs we call glaciers, flow worldwide in the oceans, and drift worldwide through the atmosphere as vapor. Water is the only common substance that exists in all three of these states at normal environmental temperatures.

▶ Relatively large amounts of heat energy are involved in water's changes among solid, liquid, and gaseous states.

▶ Water is an excellent solvent, readily dissolving many substances, making them more mobile and available for chemical reactions with whatever water contacts.

▶ Most important, all living things are primarily water (roughly 70 percent in the case of humans).

▶ As we saw in Chapter 2, geographical variation in water availability is the key to understanding many environmental variations at the global scale.

Water cycles through the atmosphere, lithosphere, and hydrosphere by means of evaporation, condensation, precipitation, and runoff. Water falls from

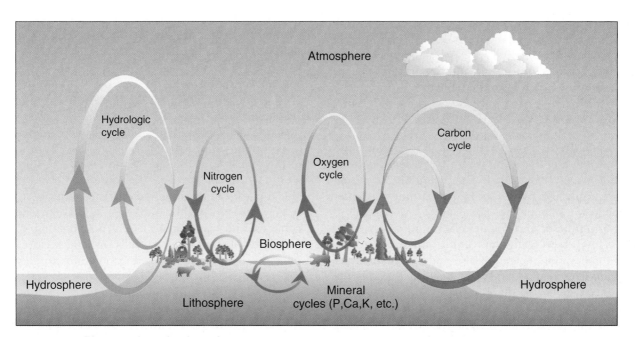

FIGURE 4-1 Biogeochemical cycles. These cycles transfer matter among the atmosphere, biosphere, hydrosphere, and lithosphere. The cycles represented here are shown in greatly simplified form. Important minerals cycling in the environment include phosphorus (P), calcium (Ca), and potassium (K), among others.

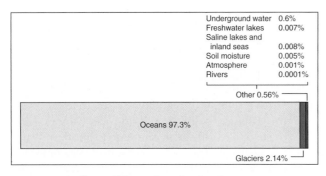

FIGURE 4-2 Quantities of water in storage. The oceans hold the most water; glaciers hold the next largest amount; and the remainder is in streams, lakes, groundwater, and the atmosphere. (Data from Jones, 1997)

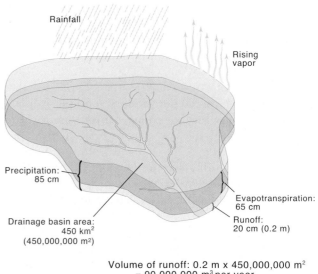

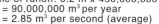

FIGURE 4-3 Calculating streamflow. Average stream-flow can be calculated from the water budget for a drainage basin, to show the close connection between water resources and climate. The average annual evapotranspiration is subtracted from the annual precipitation to determine the average annual runoff, expressed as a depth of water on the land. This value, when multiplied by the drainage basin area (in comparable units) gives the volume of runoff carried by the stream in a year.

the atmosphere to the ground and ocean through condensation and precipitation. **Runoff** carries water from the land to the sea. Evaporation converts liquid water in lakes and oceans into vapor, returning it into the atmosphere. This flow is the hydrologic cycle.

Water is stored in the atmosphere, hydrosphere, and lithosphere in solid, gaseous, and liquid forms (Figure 4-2). The salty oceans are by far the largest reservoirs of water—about 96.5 percent of all water on Earth. Nearly 2 percent is stored in glacial ice, and another 0.9 percent is saline groundwater. The remaining 0.8 percent is available as freshwater, mostly in the ground. Only about 0.014 percent of all water is in rivers and lakes. Although the atmosphere is powerful in controlling weather and climate, it contains a mere 0.001 percent of all of the world's water.

The total quantities of water present in various forms change little from year to year for Earth as a whole. But over hundreds or thousands of years, climate substantially alters the size of glaciers and, thus, the volume of water in the oceans. As explained in Chapter 3, about 20,000 years ago the amount of water stored in glacial ice was much larger, and as a result, sea level was substantially lower than today.

Water Budgets

A **water budget** is an accounting of all the inflows and outflows of water in a given system over a specified time period. Water budgets are essential tools for water-resource managers because they determine the water available for human use. For example, river water is used for domestic and commercial supplies, as cooling water for power plants and industries, hydroelectric power generation, irrigation, fish and wildlife habitats, navigation, recreation, and waste removal. Knowing the amount available for these purposes is important, yet in many areas we have only limited information on streamflows. Measuring precipitation and determining evapotranspiration allows us to predict average river flows (Figure 4-3).

Water budgets help us understand why river flows are so much greater in humid regions than arid

regions. For example, Canada's Mackenzie River has a drainage area of 1.8 million square kilometers (0.7 million square miles) and an average flow of 9,600 cubic meters per second (339,000 cubic feet per second). The Nile River in Egypt has a much larger drainage area of 3 million square kilometers (1.15 million square miles) but discharges only 950 cubic meters per second (33,500 cubic feet per second) because so much of the basin is arid. The Mackenzie's annual flow is equivalent to a depth of about 17 centimeters (6.7 inches) of water spread over the entire basin area; the Nile's annual flow is equivalent to an average depth of only 1 centimeter (0.4 inches).

Estimates of the global average rates of evaporation, condensation, precipitation, and runoff are shown in the global water budget (Figure 4-4). Each process varies geographically with the amount of water available. Excess water evaporated into the atmosphere over the oceans is carried by wind over land areas, where it condenses into clouds and falls as precipitation. About two-thirds of the water that falls on land areas evaporates there, and the remaining one-third drains into rivers, which return this excess to the sea as runoff (Figure 4-5). A significant amount of river flow (an average of about 5 percent worldwide) is returned to the atmosphere via plant-water use on irrigated lands rather than flowing to the ocean as liquid water.

The water budget shown in Figure 4-4 does not show the variability of water from place to place and season to season. Evaporation from the sea depends on

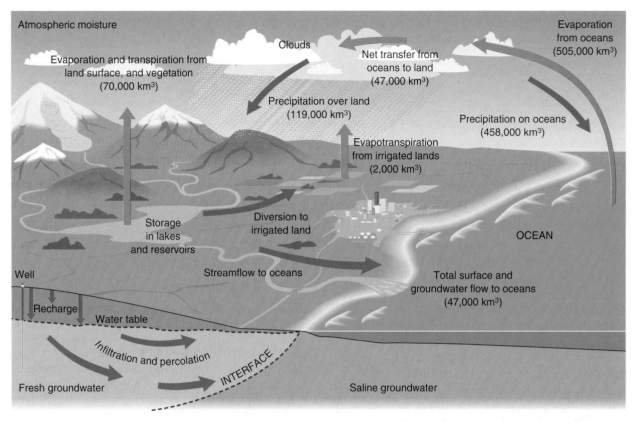

FIGURE 4-4 The hydrologic cycle: Flows of water among land, sea, and air. The quantities of water transferred in an entire year are shown. (Modified from R. W. Christopherson, *Geosystems*, 5th ed. Upper Saddle River, NJ: Prentice Hall, 2003)

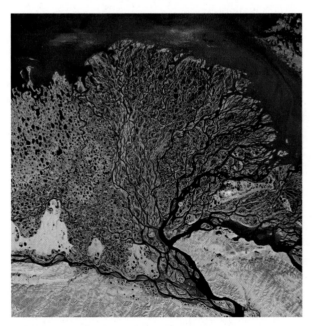

FIGURE 4-5 The Lena River Delta, Siberia. The Lena River delivers an average of about 17,000 cubic meters of water per second to the ocean, an amount similar to that carried by the Mississippi. Over the last few decades the annual flow of the Lena and other north-flowing Siberian rivers has increased substantially. This increase, along with increased melt from arctic region glaciers, may have effects on ocean circulation systems (see Chapter 2). (Image provided by the USGS EROS Data Center, Satellite Systems Branch.)

the insolation available to heat water. Precipitation and evaporation occur in both land and ocean areas, but not in equal amounts; more evaporation than precipitation occurs over the oceans.

Evapotranspiration and local water budgets

When we focus on local water budgets, we recognize the important role of plants in removing water from the soil. In well-vegetated areas, much of the conversion of water from liquid to vapor takes place in the leaves of plants. This water is replaced by water drawn from the soil through plant roots. We call this plant mechanism transpiration, and when combined with evaporation we call the process **evapotranspiration (ET).**

Evapotranspiration is low in winter and high in summer, reflecting active transpiration by plants. In the winter, when water use is low, precipitation exceeds evapotranspiration. The surplus water either runs off or becomes stored as groundwater. In springtime, plants develop leaves and water use rises rapidly, so that by early summer evapotranspiration exceeds precipitation. Initially, this excess water demand is met by water stored in the soil, but by late summer, soil moisture is exhausted and water use is limited. At this time shallow-rooted grasses may wither, while deep-rooted shrubs and trees remain green. In autumn, evapotranspiration falls with the temperature, and once again precipitation exceeds it. At this time, excess precipitation restores soil moisture to higher levels, and runoff increases again.

A local water budget compares precipitation and evapotranspiration. For a humid midlatitude site such as Chicago, Illinois, a graph of a water budget reveals that precipitation occurs fairly consistently throughout the year (Figure 4-6).

Water budgets vary tremendously from one climate to another (Figure 4-7). In some humid climates, precipitation almost always meets demand, and stored soil moisture provides the small additional amount of water needed during brief dry spells. In semiarid and arid climates, the demand for water considerably exceeds precipitation during the year, so soil moisture is never fully replenished and plants must withstand severe moisture deficits.

Soil's role in the water budget

Most of Earth's land area is covered with a layer of rock debris and organic matter known as *soil*. Soil is critical in the water budget, for it stores water and makes it available for evapotranspiration; soil is a water "bank." How effectively soil does its job depends on how fast it can absorb precipitation, how much water it can store in the root zone, and the ability of local vegetation to transpire water into the atmosphere.

Soil's **infiltration capacity** determines whether rainfall infiltrates (soaks into) the soil or runs off. Infiltration is good and most precipitation will soak in if dense vegetation exists at ground level, a layer of organic litter covers the soil, and the soil is kept porous by worms and insects. If the soil is bare or compacted, infiltration is poor, less water is stored, and runoff is greater.

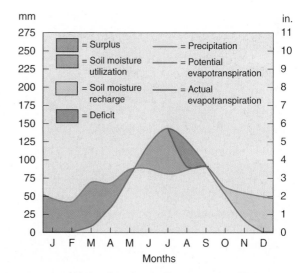

FIGURE 4-6 Water budget diagram for Chicago, Illinois, a typical humid midlatitude site.

Precipitation is shown as a blue line, and potential evapotranspiration is shown in red. In the cool months precipitation is greater than potential evapotranspiration, and the excess becomes runoff (dark blue shaded area). In the warm season potential evapotranspiration is greater than precipitation. Part of this excess demand is made up by soil-moisture use (green shading), but not all. The difference between potential evapotranspiration and actual evapotranspiration is a deficit (brown shading). The soil water that was used in early summer is recharged in autumn (light blue shading). The orange area at the bottom of the graph is water that falls as precipitation and is part of actual evapotranspiration.

Potential and Actual Evapotranspiration

Evapotranspiration (ET) occurs whenever water is available for plants to transpire and the air is not saturated with humidity. But ET rates vary tremendously over time and from place to place. Energy is necessary to evaporate water. When water vaporizes from a liquid to a gas, energy is absorbed (it becomes latent heat in the water vapor), which cools the plant leaves. Thus, the rate of evapotranspiration depends mostly on energy availability. ET occurs fastest under warm conditions and it virtually halts below freezing. Atmospheric humidity and wind speed also are significant: ET is faster on dry, windy days than on humid, calm days.

In warm weather, conditions may favor high ET, but the soil may be too dry to supply this demand. Therefore, we distinguish between potential ET and actual ET. **Potential evapotranspiration (POTET)** is the amount of water that would be evaporated if it were available. **Actual evapotranspiration (ACTET)** is the amount that actually is evaporated under existing conditions. If water is plentiful, then ACTET equals POTET. But if water is in short supply, ACTET is less than POTET.

ACTET never exceeds POTET. Comparing POTET with precipitation is a useful way to describe water availability in various climates. For example, if precipitation always is greater than POTET, plants have plenty of water, and the climate is quite humid. If precipitation is less than POTET most of the time, then plants never get as much water as they need, the natural vegetation is adapted to dryness, and the climate is arid. Farmers using irrigation to supply water to plants can calculate the difference between POTET and precipitation and thereby know how much water they must apply to crops. POTET can be calculated from mean monthly temperature data.

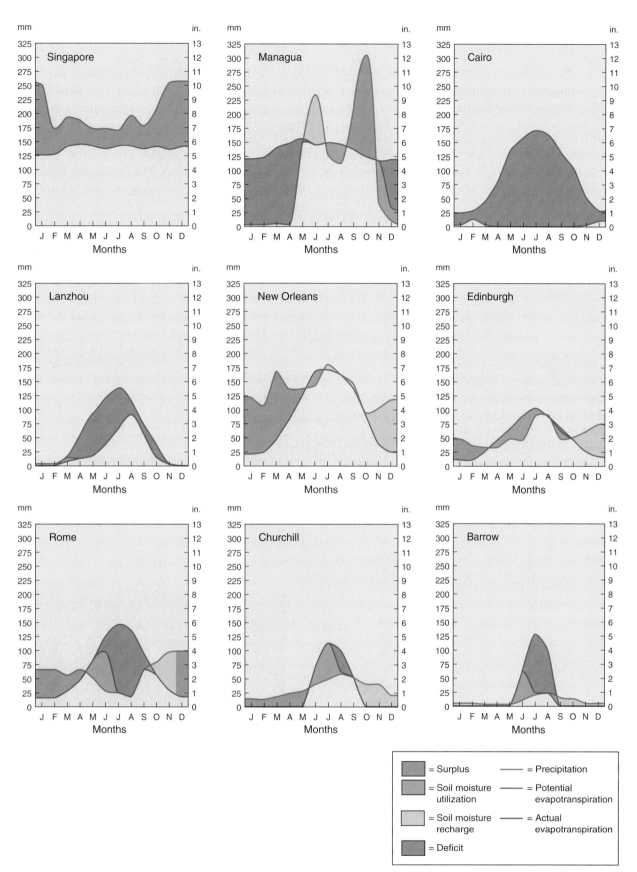

FIGURE 4-7 Water budget diagrams for selected stations. Compare with the climatic data and descriptions on pages 74 to 86 in Chapter 2.

Soil texture—the size of mineral particles in the soil—affects how much water soil can hold. Sandy soils drain quickly and hold little water (pour water on sand and watch how quickly it soaks in). Soils with high contents of finer particles—silt and clay—drain slowly (pour water on modeling clay and see how quickly it runs off). Deep-rooted plants (several meters or 10 feet or more) have a longer opportunity to absorb water as it drains through the soil but shallow-rooted plants (half a meter, or 18 inches or less) must rely on moisture contained in the upper soil. Thus, shallow-rooted grasses and herbs generally use less water (and potentially allow more runoff) than deep-rooted trees. Soil and vegetation management is critical for maintaining adequate water resources for agriculture and other human needs.

Vegetation and the Hydrologic Cycle

Forests demonstrate the close association between vegetation and the water budget. Forests occur where ample moisture is available for most of the year. Trees demand large volumes of water: A single large tree can transpire 1,000 liters per day in warm weather. Trees gather water through extensive and usually deep root systems and transpire through leaves that often are tens of meters above the ground surface. Trees use so much water that they are the dominant medium for returning rainwater to the atmosphere in large forested regions like eastern North America and the Amazon River basin.

Different types of forests have water-management implications. In the southeastern United States, areas formerly covered with slow-growing deciduous trees (that is, trees that shed their leaves in autumn) have been replanted with faster-growing evergreens to feed the timber industry. Because evergreens do not lose all their leaves in the winter, they keep growing and transpiring water year-round, whereas forests of deciduous trees do not. Evergreen needle-leaved trees also catch more precipitation on their leaves than broadleaved deciduous trees do, and this water is lost to evaporation. Studies in an experimental watershed in North Carolina showed that this measurably increased evapotranspiration and correspondingly decreased streamflow.

In contrast to trees, grasses are relatively shallow rooted, so they experience significant variations in moisture availability. When soil water is plentiful, grasses grow quickly and transpire at rates similar to trees. But during periods of limited soil moisture, grasses become dormant and transpiration virtually ceases. This helps grasses survive dry seasons, as in semiarid climates, where most trees cannot grow.

Deforestation—clear-cutting of forest—in the Amazon River basin is proceeding at a rate of about 30,000 square kilometers (11,600 square miles) per year, or about 0.6 percent of the basin's area per year. By 2000 a total of over 650,000 square kilometers (250,000 square miles) had been deforested, which is an area about the size of Texas (Figure 4-8). Deforestation may be critical to the water balance of the Amazon region. In this tropical rain forest, rainfall averages 200 to 300 centimeters (80 to 120 inches) per year, and evapotranspiration is about 110 to 120 centimeters (44 to 48 inches) per year. The excess precipitation that is not evapotranspired runs off to the Atlantic Ocean via the Amazon River, which has the world's greatest discharge. The Atlantic is the original source of water that falls as precipitation on the Amazon basin. This recycling presents a fine example of the hydrologic cycle at work.

The Amazon basin's interior, however, is more than 2,000 kilometers (1,200 miles) from the Atlantic, and most of the atmospheric water falls closer to the coast. This precipitated water infiltrates the soil, trees transpire it back into the air, and easterly winds carry it further inland. This cycle repeats, moving water step by step westward. Thus, rainfall in the Amazon basin's interior already may have been precipitated and transpired several times in its westward journey.

Forests clearly are integral to the hydrologic cycle. If we continue to cut very large areas of forests and if the grasses that replace the trees transpire much less than the trees would, precipitation in the interior could be reduced. Such precipitation changes so far are only predictions in computer models, but they help to show the important role that vegetation plays in the hydrologic cycle.

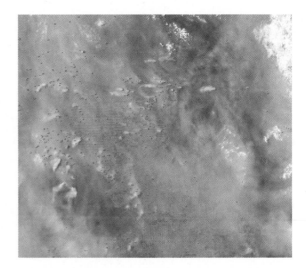

FIGURE 4-8 Fires, roads, and deforestation in the Amazon.   This satellite image was taken during the dry season in Brazil, 2003. Most of the surface is obscured by smoke, but in some areas the pattern of deforestation along parallel roads is visible. Red outlines show areas where fires were burning. (WHR photo)

CARBON, OXYGEN, AND NUTRIENT FLOWS IN THE BIOSPHERE

Carbon is not the most abundant element on Earth, but in combination with hydrogen, it is the most important for sustaining life. Compounds of carbon and hydrogen are the major component of the foods that plants produce and animals consume, and of the fossil fuels (coal, oil, natural gas) that are our most important sources of power. Living things constantly exchange carbon with the environment by photosynthesis, respiration, eating, and disposal of waste.

Exchanges of carbon among the biosphere, atmosphere, oceans, and rocks are collectively called the carbon cycle (Figure 4-9). An **oxygen cycle** is inextricably linked with the carbon cycle, and we will consider them together. We now look at these cycles.

The Carbon and Oxygen Cycles

In the **carbon cycle,** carbon in the form of atmospheric carbon dioxide (CO_2) is incorporated into carbohydrates in plant tissues through photosynthesis. Animals consume plants and use the plant carbohydrates for their own life processes. Through respiration, which occurs in almost all living organisms, carbon is returned to the atmosphere (as CO_2). Carbon dioxide is also added to the atmosphere by the combustion of fossil fuels. Thus, these three processes—photosynthesis, respiration, and combustion—cycle carbon and oxygen back and forth between living things and the environment.

Photosynthesis can be shown by the following equation:

$$\text{carbon dioxide} + \text{water} + \text{energy} \rightarrow \text{carbohydrates} + \text{oxygen}$$

For this reaction, land plants obtain carbon dioxide from the air, water from the soil, and energy from solar radiation. They store carbohydrates in tissue for later use and release oxygen to the atmosphere. Plants are the source of atmospheric oxygen, without which animals could not exist.

Respiration involves the opposite reaction to photosynthesis:

$$\text{carbohydrates} + \text{oxygen} \rightarrow \text{carbon dioxide} + \text{water} + \text{energy (heat)}$$

In respiration, carbohydrates are broken down when they combine with atmospheric oxygen to CO_2 and water. Energy is released in the process. Some of this energy is lost as heat and some is stored in chemical compounds for later use in other life processes.

The lithosphere is a major storehouse of carbon. Through geologic time, carbon enters the lithosphere slowly through rock formation, principally from oceanic sediments such as limestone but also through creation of fossil fuels such as coal.

The spatial and temporal distribution of photosynthesis is determined mostly by climate (Figure 4-10). Seasonal cycles of solar radiation are reflected directly in the CO_2 content of the atmosphere. The air we breathe has been monitored for carbon dioxide concentration since the late 1950s at the Mauna Loa observatory in Hawaii (Figure 4-11):

▶ As expected, CO_2 varies annually with the seasons by a few parts per million.

▶ The greatest CO_2 concentrations in the Northern Hemisphere are during spring, because during winter, photosynthesis is reduced (or ceases completely in cold climates) while respiration continues, returning CO_2 to the atmosphere.

▶ The lowest CO_2 concentrations are during autumn. CO_2 concentration drops during the summer because plants are actively photosynthesizing, removing CO_2 from the atmosphere and storing it in the biosphere.

▶ The Mauna Loa data also clearly reveal the steady increase in atmospheric CO_2 levels that results from fossil-fuel consumption.

The extraction and combustion of coal, oil, and natural gas since the Industrial Revolution began in the late 1700s is a relatively new process affecting the carbon cycle (Figure 4-12). Rates of carbon released into the

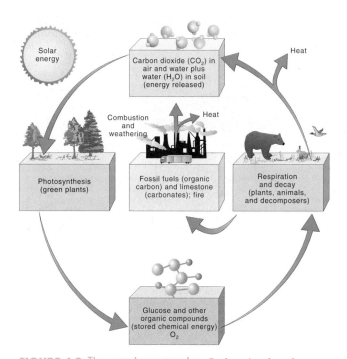

FIGURE 4-9 The carbon cycle. Carbon is taken from the atmosphere and stored in biomass through photosynthesis. It is returned to the atmosphere through respiration. Combustion of fossil fuels releases vast quantities of carbon to the atmosphere from long-term storage in rocks. (From R. W. Christopherson, *Geosystems*, 3rd ed. Upper Saddle River, NJ: Prentice Hall, 1997)

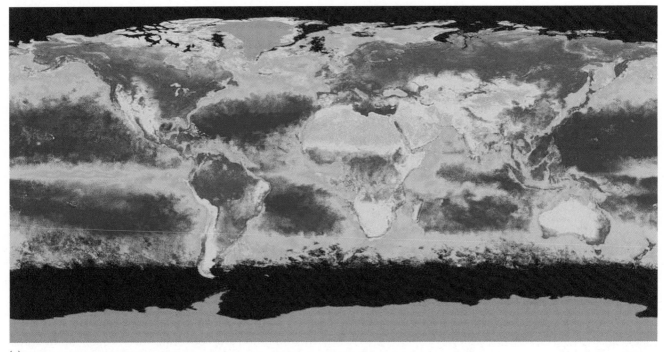

(a)

(b)

FIGURE 4-10 Seasonal variations in plant growth. A global map of plant growth in July 1987 (a) and January 1998 (b), from satellite data. On land areas, dense vegetation is shown in dark green and sparse vegetation and desert is shown in pale green and tan. Note that some high-latitude areas in the Northern Hemisphere, where it is summer, are shown in the same color as the tropical forests, while southern Africa and South America, where it is winter in July, are shown in the same colors as deserts. (NOAA)

atmosphere through this process have increased steadily over the past 200 years, and they are expected to continue to increase at least well into the twenty-first century. The long-term carbon dioxide increase that is so evident in Figures 2-54 and 2-56 illustrate this increase. Uncertainty about future fossil-fuel use makes it difficult to predict future atmospheric carbon dioxide concentrations.

Deforestation

Land-use change, especially clearing of forests, is important to these biosphere–atmosphere exchanges. In many parts of the world, most notably in tropical forests, forests are being cleared to harvest trees for lumber or fuel and to make way for agriculture. In some areas trees

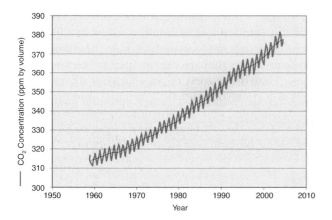

FIGURE 4-11 Atmospheric carbon dioxide concentrations measured at Mauna Loa, Hawaii. Mauna Loa, located at 19.5° north latitude in the mid-Pacific Ocean provides data that reflects general Northern Hemisphere conditions. The annual increase and decrease of carbon dioxide concentrations corresponds to seasonal patterns of plant growth and decay, with atmospheric concentrations decreasing in the Northern Hemisphere summer and increasing during the winter. (Data courtesy of Carbon Dioxide Information and Analysis Center, Oak Ridge National Laboratory)

are growing back, but at a global level, more trees are being cut than are growing back. This is important to global carbon flows because relatively large amounts of carbon can be stored in **biomass** (living and formerly living matter) in forests, and if trees are cut and burned, the carbon that they contain is released to the atmosphere. However, if the wood is used as lumber to build houses, the carbon is stored as long as the houses stand.

Focus ON

Geography, Geographic Information Systems, and the Global Carbon Budget

Concern about the accumulation of carbon dioxide in the atmosphere is mounting, and countries are moving toward implementation of an international treaty to control carbon dioxide emissions. Therefore, the need for accurate information on carbon dioxide release and uptake is increasing. One of the most important parts of the puzzle is to compile detailed knowledge of the amounts of carbon released and taken up by the biosphere. Geography and geographic information systems (GISs) are critical to providing that information.

Many different flows must be estimated in calculating the carbon dioxide budget of the atmosphere, hydrosphere, biosphere, and lithosphere. Some of these are easier to estimate than others. We know fairly closely, for example, how much carbon is emitted by fossil-fuel combustion because we know approximately how much coal, oil, and natural gas each country burns. Another large number in the budget—the amount that goes from the atmosphere to the oceans each year—can be calculated based on knowledge of the physical processes that control the exchange of carbon dioxide between water and air. On the other hand, estimating the amount of carbon released by deforestation or taken up by forest growth is much more difficult, because it is highly variable from place to place. Other key processes, too, such as the accumulation and decay of organic matter in the bottoms of reservoirs, the release of carbon dioxide from decaying peat bogs, or the buildup and loss of organic matter from agricultural soils, are very poorly known. These processes, however, are the very ones that can be managed directly by humans, so they must be accounted for in any international agreement.

Geography and GIS provide the tools for extrapolating from a few isolated measurements to estimate processes spread over large areas, so geographers are playing a big role in studies of the carbon budget. For example, we can directly measure rates of carbon uptake by photosynthesis or release through deforestation for small areas like a few square meters or a few hectares. Such measurements involve tasks like collecting, weighing, and analyzing vegetation, or carefully sampling and analyzing carbon dioxide in air samples. We could not possibly do this field research for every hectare of land in the United States, but we can do this at many individual sites representing different land-cover types—coniferous forest, deciduous forest, lakes, wetlands, farmlands, and so forth. Then, using GIS, we can make detailed maps of land cover (often based on satellite imagery) showing the spatial extent of each different kind of surface, and we can assign rates of carbon exchange for each of these types of surface based on our field measurements. The GIS then does the counting for us to estimate total carbon movements over large areas. Not only is GIS useful in that accounting process, but, in combination with satellite imagery, it can help monitor and account for changing land-cover characteristics over time (see, for example, Figure 1-25). Geography and geographers will continue to play a central role in this work.

(a)

(b)

FIGURE 4-12 Disappearing arctic lakes. A great many lakes dot arctic regions, especially low-lying coastal plains. Satellite images such as these lakes in Sibera show their disappearance (white arrows) associated with melting permafrost. Drainage of these lakes exposes sediments rich in organic carbon, and may accelerate release of this stored carbon to the atmosphere. (http://earthobservatory.nasa.gov/Newsroom/NewImages/images.php3?img-id-16986)

The most rapid deforestation, and hence the largest releases of carbon, is in tropical Central and South America, West Africa, and Southeast Asia (Figure 4-13). Deforestation rates in most midlatitudes are much lower, and in many areas, such as the eastern United States, forests are regrowing and are thus a

sink, or net storage facility, for carbon. In arctic regions large amounts of carbon are stored in soil in the form of peat deposits—accumulations of plant matter in swamps and bogs that are the first step in the formation of coal. How these areas might change if their climate warms is a point of concern. Warming would increase both respiration and photosynthesis, but we do not know whether the net change would store more carbon in the areas' biomasses or in the atmosphere.

SOIL

The interface between the lithosphere and the biosphere is the soil. **Soil** is a dynamic, porous layer of mineral and organic matter, in which plants grow. Soil is the uppermost part of the lithosphere, and at the same time is a key portion of the biosphere. Soil is a storage site for water, carbon, and plant nutrients.

Soil properties are attributable to five major factors:

1. *Climate* regulates both water movements and biological activity.
2. ***Parent material*** is the mineral matter from which soil is formed.
3. *Biological activity*—plants and animals—moves minerals and adds organic matter to the soil.
4. *Topography* affects water movement and erosion rates.
5. All these factors work over *time*, typically requiring many thousands of years to create a mature soil.

Soil Formation

The first step in soil formation is weathering, or the breakdown of rock into smaller particles and new chemical forms. Chapter 3 explained the two ways that rocks weather: mechanical weathering (such as ice expansion or tree roots growing in the cracks between rocks) and chemical weathering (such as dissolving limestone in water in the soil). The parent material from which soil is formed is important because it influences soil's chemical and physical characteristics, especially in young soils.

Water plays an important role in rock weathering and soil formation. A water budget indicates how much water moves through the soil in a place and measures the significance of water in local weathering. For example, if mean annual precipitation is 100 centimeters (39 inches) and evapotranspiration is 70 centimeters, (27 inches), then 30 centimeters (12 inches) of water would seep down through the soil in an average year. This water would be a powerful weathering agent. However, if an area has 70 centimeters of precipitation and the atmosphere is capable of evaporating 150 centimeters (59 inches) of water, virtually all water is evapotranspired by plants, so little would be left to percolate down through the soil to help weather the rocks.

In a very humid climate, much water passes through the soil and leaches out soluble minerals on its way. Because of this, soils in humid climates generally

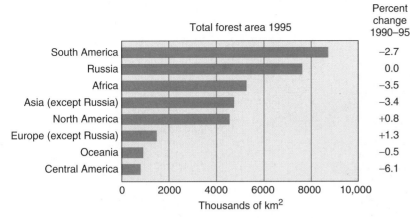

Total forest area 1995

Region	Percent change 1990–95
South America	−2.7
Russia	0.0
Africa	−3.5
Asia (except Russia)	−3.4
North America	+0.8
Europe (except Russia)	+1.3
Oceania	−0.5
Central America	−6.1

Thousands of km²

FIGURE 4-13 Forest extent and deforestation rates by region. The most extensive forests in the world are in Siberia and the Americas, but major forest areas also exist in Africa and southern Asia. Deforestation is most rapid in the tropics; forest area is increasing in Europe and North America. (Data courtesy of World Resources Institute)

have lower amounts of soluble minerals such as sodium and calcium compared to soils in dry climates. However, in semiarid areas, water enters the soil and picks up soluble minerals, which are drawn toward the surface as water is evapotranspired; soils of semiarid and arid climates often have a layer rich in calcium near the surface.

Plant and animal activities are also critical to soil formation. Plants produce organic matter that accumulates on the soil surface, and animals redistribute this organic matter through the soil. Plants and animals also play a role in weathering processes. Topography also affects the amount of water present in the soil, largely through controlling drainage and erosion. Steeply sloping areas generally have better drainage than flat or low-lying areas, and they are often more eroded. Finally, soil formation is a slow process that takes place very gradually over thousands of years. Soils that have only been forming for a few hundred or even a few thousand years have very different characteristics from those that have been modified by chemical and biological processes for tens of thousands of years.

Soil Horizons

Soil is a complex medium, containing six principal components:

1. *Rocks* and *rock particles*, which constitute the greatest portion of the soil and which may weather, releasing nutrients needed for plant growth.
2. *Humus*, which is composed of dead and decaying plant and animal matter that holds water, supports soil organisms, and supplies nutrients.
3. *Dissolved substances*, including phosphorus, potassium, calcium, and other nutrients needed for plant growth.
4. *Organisms*, including animals such as insects and worms and many microorganisms, including bacteria, and fungi.
5. *Water from rainfall*, which is necessary for plant growth and which helps to distribute other substances through the soil.
6. *Air*, which shares soil pore spaces with water and which is necessary for respiration by plant roots and soil organisms.

These substances are not uniformly distributed in soils, but found in layers called soil **horizons** (Figure 4-14). Soil horizons are formed through the vertical movement of water, minerals, and organic matter in the soil and also by variations in biological

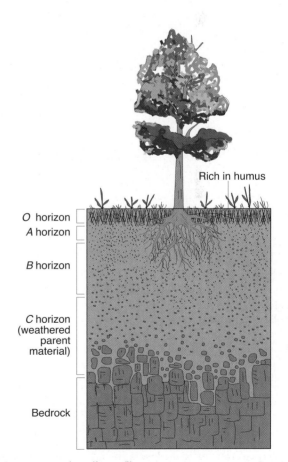

Rich in humus

O horizon
A horizon
B horizon
C horizon (weathered parent material)
Bedrock

FIGURE 4-14 A soil profile. The sequence from the bottom up shows the typical steps of soil formation. Solid rock at the bottom weathers into large pieces, which weather further into ever smaller particles. At top is the organic layer, which contains organic matter from decayed plants and animals. Designations like *A* are given to horizons in well-developed soils. Poorly developed soils lack some of these layers. (From E. J. Tarbuck and F. K. Lutgens, *Earth Science*, 7th ed. New York: Macmillan, 1994)

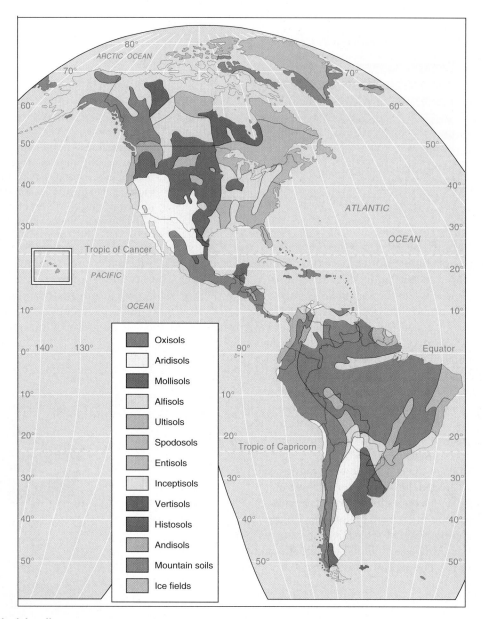

FIGURE 4-15 World soils.
Oxisol—a deeply weathered, heavily oxidized soil of the humid tropics, red in color.
Aridisol—dry desert soil with limited organic matter, limited chemical weathering, and accumulations of soluble minerals.
Mollisol—grassland soil of subhumid–semiarid land, dark with plentiful organic material. Mollisols are an important agricultural soil in the U.S. Midwest.
Alfisol—a moderately leached soil of humid subtropical and midlatitude forests, with clayish *B* horizon.
Ultisol—red- to yellow-colored, well-weathered soil with clayish *B* horizon typically found in warm, humid climates.
Spodosol—acidic soil formed under coniferous forest, with a light-colored, sandy *A* horizon and red-brown *B* horizon in which iron accumulates.

and chemical activity at different depths. In this way they are different from layers in sediments, which are deposited in sequence from the bottom up.

For example, litter—leaves, twigs, dead insects, and other organic matter—accumulates to form a horizon at the surface known as the *O* (organic) horizon. As this litter decays, organisms such as insects, worms, and bacteria consume it and carry it underground, where it helps form the *A* horizon. Waste from these burrowing animals as well as their dead carcasses add more organic

matter to the *A* horizon. In many soils the *A* horizon contains much of the nutrients that support plant life.

Water may erode materials from the soil surface, or carry some substances down from the *A* horizon to the *B* horizon. In some environments organisms play an important role in moving materials between the *A* and *B* horizons. Clay minerals formed from chemical weathering often accumulate in the *B* horizon, and in dry regions soluble minerals such as calcium accumulate in the *B* horizon. The *C* horizon, beneath the *B* horizon,

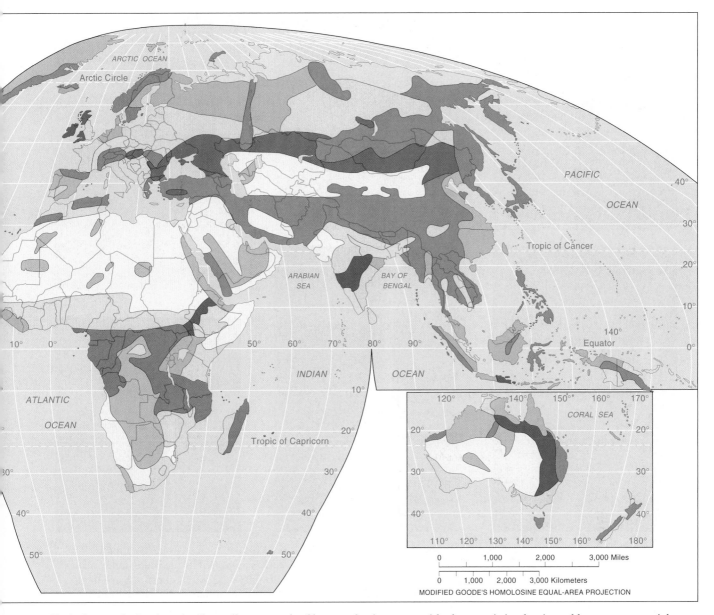

MODIFIED GOODE'S HOMOLOSINE EQUAL-AREA PROJECTION

Entisol—poorly developed soil usually as a result of low weathering rates, with characteristics dominated by parent material.
Inceptisol—weakly developed soils often found in mountainous areas where erosion limits soil development.
Vertisol—soils with high content of clay minerals that swell and shrink on wetting and drying. Deep cracks open up in the dry season.
Histosol—organic soils found in peat bogs and other areas of organic-matter accumulation.
Andisol—young volcanic soils developing on ash and lava deposits.
Mountain soils and ice fields are also mapped. Areas mapped as mountain soils have too much local variability to show individual soil types at this map scale. Ice fields have no soil.

contains weathered parent materials that have not been altered as completely by soil-forming processes as materials above it have.

Thousands of Soils

Soils are divided into 11 broad clusters, called **soil orders** (Figure 4-15 and Figure 4-16). However, these 11 orders are divided into 47 suborders, about 230 great groups, 1,200 subgroups, 6,000 families, and thousands

of soil series. In other words, Earth's dynamic lithosphere, hydrosphere, atmosphere, and biosphere have worked as a team to create an enormous diversity of soil. There is nothing simple about "ordinary dirt!"

Detailed maps developed by the U.S. Department of Agriculture's Natural Resource Conservation Service are used by farmers and land-use planners because they show the great local variability of soils. Often, two or three different great groups of soils occur within a few kilometers of each other, and these groups are subdivided into

FIGURE 4-16 Six soil orders. (a) This oxisol in Rwanda has a reddish color caused by oxidation of iron. (b) A mollisol in Nebraska has a dark, organic-rich and fertile *A* horizon. (c) This spodosol in New York is identified by the white, sandy surface horizon. (d) An inceptiosol in Massachusetts shows only weak horizon development because the soil has been subjected to erosion. (e) Areas of volcanic activity often have andisols, with distinct ash layers, like this one in Japan. (f) This entisol in Connecticut has not had much time to develop, and so does not have distinct horizons. Compare with soil descriptions on pages 144–145. (Loyal A. Quandt/National Soil Survey Center, Natural Resources Conservation Service, U.S.D.A.)

the subgroups, families, and series of more specific characteristics. Because of this great local variability, generalizations about soils must be made with caution.

Climate, Vegetation, Soil, and the Landscape

At a global scale, distinctive environmental regions derive from intimate connections among climate, vegetation, and soil:

▸ *Humid tropical* and *subtropical soils* such as oxisols and ultisols are highly weathered. They have lost soluble minerals due to heavy precipitation, which may deplete soils of essential nutrients, so fertiliza-

tion is necessary to increase crop yields. Tropical soils are usually oxidized and red in color.

▸ *Arid region soils* such as aridisols are high in soluble minerals because little leaching occurs with the scant rainfall. These soils are low in organic matter, but they can be very productive if sufficiently irrigated.

▸ *Midlatitude humid soils* such as alfisols and spodosols are moderately leached where they are dominated by deciduous forest vegetation and heavily leached where they are developed beneath the acidic litter of coniferous forests.

▸ *Midlatitude subhumid soils* such as mollisols are fertile and are the resource base for many important grain-producing regions.

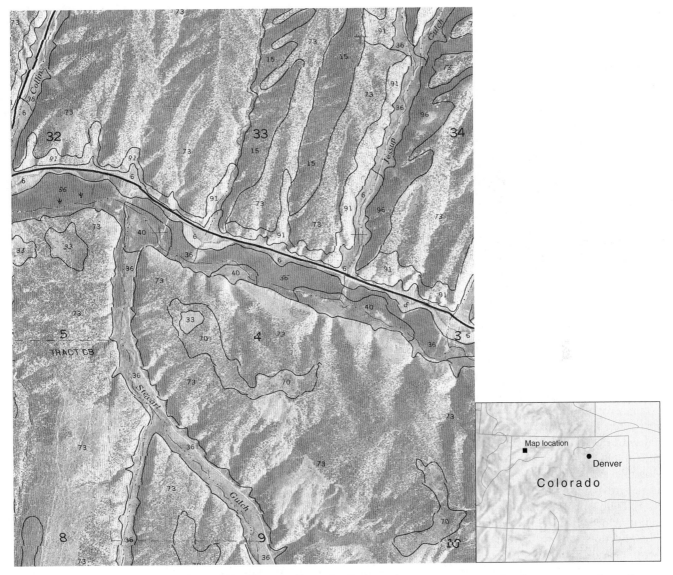

FIGURE 4-17 A soil map for part of Rio Blanco County, Colorado. This map shows the close correspondence between landforms, soils, and vegetation. Sagebrush prefers deep, well-drained soils (36); pinyon-juniper woodland prefers
thin, rocky soils (73). On gently sloping uplands, moderately thick soils have developed from sandstone bedrock; vegetation is
dense-to-open woodland of pinyon pine and juniper, with scattered sagebrush and grasses (70); small patches of drier, windblown material show a distinctive fine-textured soil with sagebrush (33, 64, 66). (Map and photo from U.S. Department of Agriculture,
Soil Conservation Service)

These climatic soil regions are interrupted by the
major mountain chains of the world. Poorly developed
soils such as inceptisols predominate because erosion
removes soil as rapidly as it is formed.

In Figure 4-17 we "zoom in" on a portion of the
world maps of Figure 4-15 to examine vegetation and soil
at a larger map scale. A close relation among climate,
vegetation, and soil also exists at the scale of local ecosystems. You can see that vegetation and soils are far more
diverse than is revealed in the large-scale map. This semiarid area of northwestern Colorado is in a sandstone and
shale upland, incised with narrow valleys. It is an open
woodland dominated by pinyon pine and Utah juniper
trees and a shrub-grass community that is mostly sagebrush (Figure 4-18). Within a few square kilometers, you

may see forest of one or more types, disturbed areas,
shrublands, and grasslands. The caption to Figure 4-17
describes the soil-vegetation relations here.

Soil Problems

In modern commercial agriculture and in many traditional systems, nutrients are added artificially to the soil.
These nutrients may be delivered in the form of animal
manure, by plowing into the soil a soil-enriching crop,
or as manufactured inorganic fertilizers. The more intensively a soil is worked, the more important such addi-
tions become. Unfortunately, increasingly intensive
agricultural activity has meant that nutrients, especially
organic matter, are not being replaced as rapidly as they

FIGURE 4-18 The landscape of Rio Blanco County, Colorado. This view is taken from the lower center of map in Figure 4-17, looking north (toward the top of the page). (WHR photo)

FIGURE 4-19 Desertification. Cattle grazing on sparse grass cover near Chaco Canyon, New Mexico. Only 300 years ago, large herds of native grazers and newly introduced cattle could be supported on extensive grasslands. But today only a few animals can be supported in this semiarid environment. (Bernard Boutrit/Woodfin Camp & Associates)

are withdrawn. Soil fertility usually declines over decades, so the problem is less apparent than if it occurred over just a few years. Variations in yield from year to year caused by weather, insects, plant diseases, and changing technology mask the effects of long-term soil degradation. Nevertheless, a long-term decline in soil quality is in progress in many parts of the world.

To this problem, add erosion accelerated by cultivation. Plowing a field rips apart the soil-holding system of plant roots and lays bare the soil to attack by rain and wind. Erosion removes the uppermost part of the soil—the *O*, or organic horizon, which is usually the most fertile part—and with it go both nutrients and the ability to store them. In many agricultural regions, topsoil loss has ranged from several centimeters (1 to 2 inches) to tens of centimeters (4 inches to a foot or more). Often, the depth of erosion caused by plowing amounts to a significant part of the entire *A* horizon. In semiarid areas the process can so degrade the soil that only plants adapted to lower soil moisture and nutrient availability, as in deserts, can survive. This is the process of desertification mentioned at the beginning of the chapter. Sometimes, desertification may be irreversible, at least over short periods of time. In such cases, grassland and other relatively productive vegetation types can be replaced with only sparse grasses—shrubs forming clumps of vegetation separated by bare ground (Figure 4-19). Soil organic matter is reduced, and when rain falls, it is more likely to run off the surface, increasing erosion and decreasing the soil's ability to support vegetation. As a result, many fewer animals can be supported by the available forage. Even though the climate today is not fundamentally altered, the land appears more desert-like than it did before excessive exploitation by humans.

Soil Fertility: Natural and Synthetic

Soil fertility refers to the ability of a soil to support plant growth through making nutrients available. Under a given set of conditions, more fertile soils produce higher crop yields than less fertile soils. Fertility, however, is a difficult thing to measure, because other factors that affect yields are also very important, such as the way the soil is plowed and the way crops are planted. Also, some soils may be relatively infertile with one set of management techniques but highly fertile when managed differently.

Plants need sunlight, carbon dioxide, oxygen, and water. But they also require nutrients for growth, including nitrogen, phosphorus, potassium, calcium, magnesium, sulfur, and others. In general, fertile soils are those that store and release sufficient quantities of nutrients and water for optimal plant growth. Nutrients are present in many different forms—dissolved in soil water, as part of organic matter, attached to the surfaces of mineral grains, and in the crystalline structure of minerals. Some forms are more available to plants than others. Organic matter, both living and dead, stores nutrients and makes them available to plants. In most soils, depletion of organic matter reduces fertility, but the addition of manure to the soil helps replace lost organic matter. Most available nutrients occur in the uppermost soil horizons: the *O*, *A*, and upper *B* horizons. Cultivation of the soil and harvesting agricultural crops accelerates the breakdown of organic matter and losses of nutrients stored in the soil. Sustaining soil fertility requires maintaining soil nutrients and maintaining the soil's ability to store and release them to plants.

Farmers long have known that measures must be taken to restore soil fertility. In traditional agricultural systems, this has meant a *fallow period*, which is a season when a field is allowed to rest and nothing is harvested from it. During this rest period the natural processes of plant growth, organic matter accumulation, mineral weathering, and biological activity can restore fertility. However, Earth's burgeoning population demands more agricultural products, so fallow periods have been reduced or eliminated. Many areas that once were fallowed are now cultivated every year, thereby reducing or eliminating the opportunity to restore natural soil fertility.

The low cost, availability, and easy application of inorganic fertilizers has led to replacing organic (natural) fertilizers with these manufactured versions. Manufactured inorganic fertilizers contain valuable nutrients, but only organic matter can help maintain the ability of the soil to store those nutrients. In the short run (a single crop season), nutrients can be added to soil to supply plants' needs, but in the long run, the ability of the soil to support plant growth is diminished.

Farmers and resource managers around the world increasingly worry about the impact of soil degradation on the world's food-producing capacity. Declining soil fertility associated with erosion is particularly acute in many tropical and subtropical areas, such as Africa. The high cost of fertilizer, and the resulting limited use of such fertilizer, is also contributing to a long-term decline in soil productivity in many areas.

The study of landscape patterns at local and regional levels, using vegetation and soil maps, almost always reveals human disturbance. As human settlement increases, environmental managers worry about our impact on ecosystems. Although global generalizations, such as reports of the amount of forest area lost in the tropics per year, are popular, impacts are best understood when at a local scale.

ECOSYSTEMS

An **ecosystem** includes all living organisms in an area and the physical environment with which they interact (Figure 4-20). An ecosystem can cover an area as small

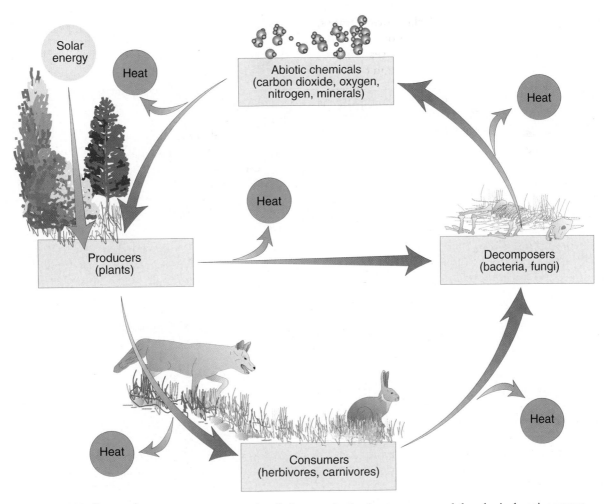

FIGURE 4-20 Ecosystems. Ecosystems consist of plants, animals, decomposers, and the physical environment with which they interact. Energy is supplied to an ecosystem from the Sun, passed through it via a food chain, and dissipated as heat.

as a field or a pond. On any scale of analysis, large or small, certain fundamental elements exist:

▶ *Producers*: green plants and other organisms that produce food for themselves and for consumers that eat them.

▶ *Consumers*: organisms that eat producers, other consumers, or both.

▶ *Decomposers*: small organisms, such as bacteria, fungi, insects, and worms, that digest and recycle dead organisms.

▶ *Materials and energy necessary for production and consumption to occur*: water, mineral nutrients, gases such as oxygen and carbon dioxide, and energy (light and heat).

Ecosystem Processes

Energy from the Sun is the starting point for understanding the operation of an ecosystem. Sunlight makes photosynthesis possible, and green plants produce food in the form of carbohydrates. The food is distributed through an ecosystem by way of a **food chain.**

Plant-eating animals, known as **herbivores,** begin the food chain by consuming plants (Figure 4-21). **Carnivores,** which are meat-eating animals, eat herbivores

and may in turn be eaten by other carnivores or by **omnivores,** such as humans, who eat both plants and animals. Most of the food that animals consume is used to keep their bodies functioning. Animals excrete some of the food, and when they die, their bodies are rich in stored-up nutrients. Decomposers attack animal excretions and dead bodies. These small organisms return chemical nutrients to the soil and the atmosphere. Some decomposers provide nutrients to new plant growth, completing the cycle.

Each step in the food chain is called a **trophic level.** At each trophic level, food is passed from one level to the next, but most of the energy is lost. As a rule of thumb, about 10 percent of the energy consumed as food at a given trophic level is converted to new biomass, and the remaining 90 percent is burned off and dissipated as heat. Because of this loss, the amount of biomass decreases as we go from the first trophic level—the green plants—to higher levels. This is why we find large numbers of herbivores such as mice and rabbits in natural systems, but very few large carnivores like wolves and lions.

Some chemicals in the environment, especially persistent pesticides such as DDT, create problems passing through a food chain. These chemicals do not break down easily in the environment, so they

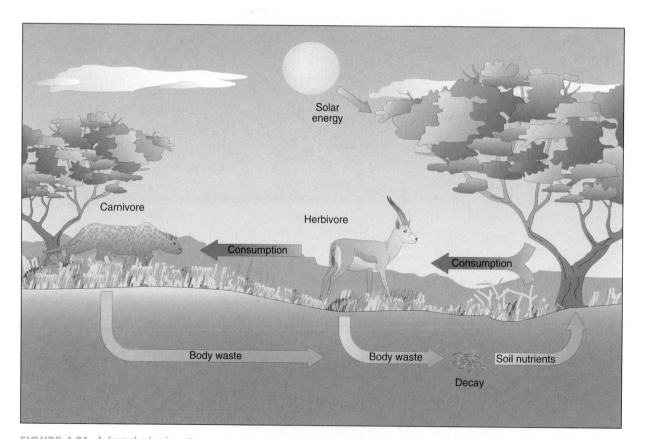

FIGURE 4-21 A food chain. Green plants are the primary producers of food, which is eaten by the primary consumers: herbivores and omnivores. Carnivores, or secondary consumers, derive their energy from other animals rather than directly from plants. At each step of the food chain, energy is lost as heat, and waste materials are broken down by decomposers. (From T. L. McKnight, *Physical Geography: A Landscape Appreciation.* 4th ed. Englewood Cliffs, NJ: Prentice Hall, 1993)

accumulate in animal tissues. If a pesticide accumulates in rather than being excreted from an animal, then at each trophic level the concentration of that pesticide increases in a process called **biomagnification.** Consider this example: If a falcon weighing 1 kilogram consumes 50 smaller birds, each weighing 1/5 of a kilogram, and if each of these smaller birds consumes in its lifetime 2 kilograms of plant-eating insects, then the falcon indirectly consumes 100 kilograms of insects. If each of these insects consumed, during its lifetime, 10 times its weight in plants, then the falcon has indirectly consumed 1,000 kilograms of plants! Therefore, even a very small concentration of pesticides at a low trophic level, such as a plant, can become a high concentration in a large carnivore. In some birds of prey such as eagles and falcons, this biomagnification causes eggs to fracture, killing chicks as they incubate. Because of these effects, DDT and similar pesticides have been banned or severely restricted in much of the world, and the pesticides used today break down relatively rapidly to reduce this problem.

Plant and animal success in ecosystems

Within any particular ecosystem, living things compete for resources such as nutrients and water. The more successful plants and animals in this competition will dominate that environment.

Success of a species means that it survives, thrives, or even comes to dominate an ecosystem for a long period. Humans are an example: surviving for the first 150,000 years of our existence as a species, then thriving over the last few thousand years, and today dominating Earth (with dramatic consequences for other species).

The success of one species over another is the result of competition. In the case of plants, this competition is for light, water, nutrients, and space. Although plants require all these factors to grow, in any ecosystem one factor usually is restricted, which forces competition and adaptation. For example, in an arid environment, plants compete for scant water but do not need to compete for the abundant sunlight. In a humid environment, water is abundant, but plants must compete for sunlight. An area with adequate water and light may have poor soils, so plants must compete for nutrients. Through evolution, the plants that have adapted their life forms, physiological characteristics, and reproductive mechanisms that allow them to succeed in particular environments have survived. The plants that compete best in an environment dominate the ground cover there.

What adaptations are effective in competing for space and nutrients? In a humid area height is effective: Plants that are able to grow tall capture the most light and shade out their competitors. An ecosystem with extreme cold, drought, fire, or short growing seasons favors organisms that have adapted to survive or even benefit from the stress. For example, some species have adapted by developing the ability to take advantage of

occasional fires. After a fire has felled other vegetation and heated the soil, long-dormant seeds respond to the flash heating and germinate quickly. They occupy the burned site, completing their life cycle and spreading new seed before the other vegetation recovers from the fire. Once the other vegetation regains dominance, the fire-resistant seeds lie dormant, awaiting the next fire opportunity.

Community succession As plants grow and multiply, they alter their environment, sometimes changing it to the point that it becomes more suitable for the growth of other kinds of plants. Thus, the plants of the community may change over time until a stable plant population emerges. This community remains until something, perhaps a fire, changes the physical environment and then different plants again begin to grow. These changes in the makeup of the plant community are called **succession.** A common example of community succession occurs when agricultural land is abandoned in eastern North America (Figure 4-22).

▶ In the first couple of growing seasons, the fields sprout fast-growing herbaceous weeds and other pioneer species. These spread seeds widely and tolerate bright sunshine.

▶ Over the next few growing seasons, the weeds and pioneer species are succeeded by slower-growing perennial shrubs and trees, such as pines, which prefer bright sunshine and gradually shade out the smaller plants beneath them.

▶ Hardwoods that tolerate shade grow beneath the pines, eventually overtaking them in height. As the pines die away, a forest of shade-tolerant species dominates the canopy. They are replaced by their own seedlings as they die, thus leading to a stable association of plants in equilibrium with local soil and climate, called a **climax community.**

The length of time necessary for succession to take place varies, depending on the growth rates and lifespans of plants, rates of seed introduction and establishment, and whether soil development proceeds along with plant growth. If the seeds of the late-successional species are present at the time succession begins and growth rates are fast, a well-developed forest may be established in a few decades. The illustration shows a more common situation, where a mature forest is achieved in 150–200 years. But if succession also involves soil development through weathering and the accumulation of organic matter, it may take thousands of years for a site to be converted from bare rock to a mature forest.

In environments subject to frequent disturbance, such as fire, windstorm, disease outbreaks, floods, and volcanic eruptions, it may take a long time to establish a climax community. Vegetation may actually change continually but never reach a climax.

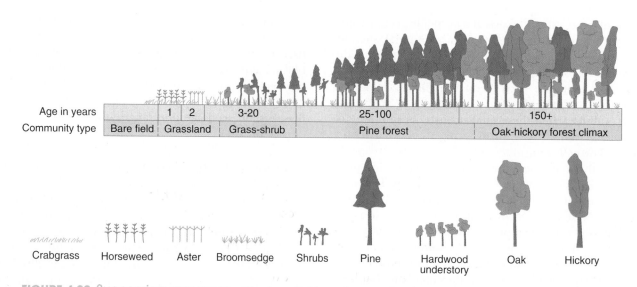

Age in years		1	2	3–20		25–100		150+	
Community type	Bare field	Grassland		Grass-shrub		Pine forest		Oak-hickory forest climax	

Crabgrass   Horseweed   Aster   Broomsedge   Shrubs   Pine   Hardwood understory   Oak   Hickory

FIGURE 4-22 Succession sequence. Clearing a field sets the stage for the succession sequence shown here, from left to right. This example is typical of conditions in the southeastern United States. In each successional stage, modification of the local microenvironment allows a new assemblage of plants and animals to become established. The sequence culminates in the climax hickory forest. (From R. W. Christopherson, *Geosystems*, 3rd ed. Upper Saddle River, NJ: Prentice Hall, 1997. Adapted from E. P. Odum, *Fundamentals of Ecology*, 3rd ed., Figure 9-4, p. 261, © 1971 Saunders College Publishing. Adapted by permission.)

Earth's recent history has been one of striking change, including dramatic climate changes and human transformation of vast areas. Even environments that have not been severely altered by humans are subject to natural disturbances such as storms, fire, or outbreaks of plant or animal diseases. These environmental changes have profoundly influenced vegetation distributions in many areas, favoring plants that are adapted to dynamic conditions. Very little of Earth's surface can be regarded as stable or undisturbed.

Biodiversity

Biodiversity is the diversity of species present in any environment. Biodiversity is important because it brings multiple food options to living things, thereby improving each one's chance for survival. Thus, biodiversity can improve the stability of an entire community.

Earth's biodiversity is vast: There are an estimated 10 million plant and animal species. Each of these has specific needs. For plants, critical needs include water, light, and nutrient availability, the absence of soil disturbance, effective seed dispersal, and good germination conditions. For animals, the most critical requirement is food supply, meaning a sufficient quantity of specific plants or animals to feed upon. In a world of rapidly changing environments, the critical needs of many species increasingly are not met.

Habitat loss is so widespread that a significant portion of the world's species may be threatened with extinction. The largest single cause of extinction is land-use change, resulting from increasing human settlement worldwide. About 39 percent of Earth's land area (excluding Antarctica) now is cropland or permanent pasture. Between 1960 and 2000, the land area of such agricultural uses increased about 10 percent, whereas the world's forest area decreased about 5 percent. Hunting, species introductions (such as new predators that outcompete indigenous species), and pollution also contribute to extinction. In Europe and North America, radical changes in the landscape have taken place over centuries of human habitation. Although a few extinctions were noted, little attention was paid to the broader, unknown impacts of those landscape changes. Rapid change now is occurring in the Amazon basin at the same time that scientists are beginning to explore tropical environments. Biologists working in the Amazon discover new species virtually every time they look, raising concern that deforestation is causing the extinction of species even before we discover them.

Biosphere reserves   To reduce the impact of land-use change on species diversity, protected areas are being established by national governments and by international agencies. One important international effort is the United Nations Biosphere Reserve Program. In South America, for example, about 7.4 percent of the entire continent (490,000 square kilometers, or 190,000 square miles) now is nationally protected, including 29 biosphere reserves (some of which overlap nationally protected areas) that cover about 508,000 square kilometers (196,000 square miles). Approximately 6.4 percent of all land area is nationally protected, and 337 biosphere reserves cover about 3.7 percent of Earth's surface. The degree of protection (restrictions on land use and disturbance) and level of enforcement, however, vary considerably.

The size of individual protected areas is important. A single animal may range over only a few square kilometers, but that small area cannot ensure species survival. A reserve must be large enough to sustain enough individuals to allow genetic diversity in the breeding population. It is difficult to decide how large a protected area should be; this disagreement can become part of the political tug-of-war among conflicting potential land users.

Today conservationists are beginning to recognize that setting aside a few patches of relatively "natural" landscape will not be sufficient to preserve Earth's remaining biodiversity. Rather, we must also consider ways of promoting biodiversity in the managed landscape—on farms, in commercial forests, and even in urban areas. One reason is the sheer size of human-dominated land areas in comparison to the small patches that have been protected. The land available for protection is insufficient to retain biodiversity across the landscape. Another reason is the belief that biodiversity in itself provides important goods for humans, perhaps through enhancing biological productivity or promoting ecological stability. If biodiversity contributes these services, then we need to maintain that biodiversity where we live and work—not just in remote preserves.

Biodiversity is a barometer of the consequences of human modification of the environment. A commitment to biodiversity means a commitment to a broad range of environmental policies—not just preserving wilderness areas, but maintaining diversity everywhere.

BIOMES: GLOBAL PATTERNS IN THE BIOSPHERE

Earth's ecosystems are grouped into **biomes** characterized by particular plant and animal types, usually named for a region's climate or dominant vegetation type. Biomes typically contain many ecosystems. A terrestrial biome's nature reflects two especially visible features: climate and vegetation. Underlying a biome's label is a diverse community of characteristic plants and animals.

The global distribution of terrestrial biomes (Figure 4-23) imitates very closely the distribution of climate regions described in Chapter 2 (Figure 2-31). This correspondence reflects a close link between climate and vegetation. To grow, plants depend on the two most important elements of a climate: energy and water.

We will now examine the vegetation and climates that distinguish major biomes: forest, savanna, woodland, scrubland, grassland, desert, and tundra. Figure 4-24 is a matrix showing some of the relationships among temperature, water availability, and vegetation characteristics. The distribution of animal species is strongly influenced by the distribution of vegetation; thus, our descriptions of biomes will focus on plants.

Forest Biomes

Forest biomes are dominated by trees: tall, woody-stemmed perennials (plants that live for at least several years) with spreading canopies. Forests occur over a wide range of humid and subhumid climates, where ample water is available most of the year. We group forests into four broad categories: *tropical rain forest, midlatitude broadleaf deciduous forest, needleleaf* or *boreal forest,* and *temperate rain forest.*

Tropical rain forest biome
Tropical forests and woodlands dominate the humid equatorial environments of Central America, the Amazon River basin, equatorial Africa, and Southeast Asia (Figure 4-25). Tall, broad-leaved trees retain their leaves all year. A **tropical rain forest** has a top layer, or canopy, and two more layers beneath (Figure 4-26). Each layer has different dominant species and associated animal communities.

The highest canopy consists of very tall trees—heights of 40 to 60 meters (130 to 200 feet)—spaced a distance apart so that sunlight reaches the next lower canopy. Underneath is the dense middle canopy. Most photosynthesis occurs in these two upper layers, where sunlight is most intense. The middle canopy blocks most light, keeping the forest floor in twilight, so plants at the lowest levels must tolerate deep shade to survive. Climbing vines are common in many tropical rain forests, as are plants that grow on the surfaces of other plants to gain height advantage without having their own long trunks or stems reaching to the ground.

Tropical rain forests are noted for biodiversity, having hundreds of tree species within a single hectare (about 2.5 acres). In contrast, midlatitude forests may have only five to 15 tree species per hectare, and high-latitude coniferous forests may have only three to five species. The tropical diversity of trees is paralleled by a great variety of animals. The diverse vertical structure adds a diversity of habitats and species. A walk through a zoo illustrates the vast number of large animal species of the world that are native to tropical climates. Non-human primates are found almost exclusively in the tropics, especially in tropical forests. The wide diversity of life in the tropical rain forest places this biome at the center of controversy over deforestation, species extinctions, and biodiversity.

In the rainforest biome, much nutrient storage is in a living biomass rather than in the soil. Organic matter decays rapidly in warm, wet conditions, and nutrients are not easily retained in the soil. When leaves fall on the forest floor and begin to decompose, the released nutrients are quickly absorbed and recycled by living plants. This is yet another reason why deforestation is a major concern. If forests are removed, their store of nutrients may be lost quickly to leaching and runoff, leaving only the impoverished soil.

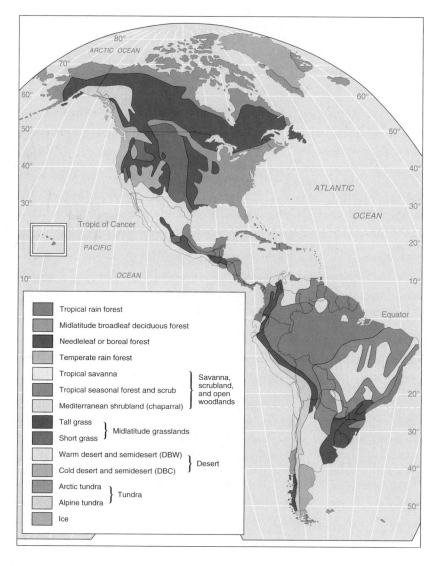

FIGURE 4-23 The major terrestrial biomes. This map shows vegetation that would exist if humans did not disturb the vegetation, as in agricultural regions. Map generalization at this scale obscures many local variations in vegetation types, especially in mountains and populated areas. (From R. W. Christopherson, *Geosystems,* 5th ed. Upper Saddle River, NJ: Prentice Hall, 2003)

Legend:
- Tropical rain forest
- Midlatitude broadleaf deciduous forest
- Needleleaf or boreal forest
- Temperate rain forest
- Tropical savanna ⎫
- Tropical seasonal forest and scrub ⎬ Savanna, scrubland, and open woodlands
- Mediterranean shrubland (chaparral) ⎭
- Tall grass ⎫ Midlatitude grasslands
- Short grass ⎭
- Warm desert and semidesert (DBW) ⎫ Desert
- Cold desert and semidesert (DBC) ⎭
- Arctic tundra ⎫ Tundra
- Alpine tundra ⎭
- Ice

Midlatitude broadleaf deciduous forest biome

The **broadleaf deciduous forest** exists in subtropical and midlatitude humid environments where seasonally cold conditions limit plant growth (Figure 4-27). The eastern United States and much of China share this biome, which explains why some Chinese ornamental species are popular on lawns in the eastern United States. On the map you can see that this biome occurs mostly in the Northern Hemisphere.

During the summer growing season, long days and a high solar angle promote rapid growth, so that annual plants may attain 60 to 75 percent of the growth of plants in the tropics, even though the growing season is only five to seven months. These plants have evolved wide, flat leaves that capture as much sunlight as possible. Broadleaf deciduous forests are much less diverse than are tropical

rain forests, but they still support many species. Because these forests are located in the midlatitudes, many animal species that make their home there either hibernate in winter or migrate to warmer locations.

Soils of this region vary with the underlying geology, but generally are leached less than are soils of the humid tropics. This is because annual precipitation

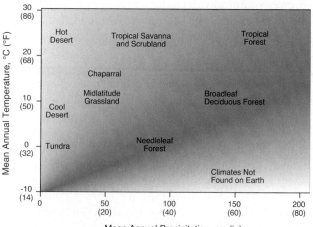

FIGURE 4-24 Generalized relation between vegetation type and climate. The amount of precipitation necessary to support a given amount of vegetation (such as forest relative to grassland) is greater in warm climates, where evapotranspiration is higher, than in cold climates. Cold climates with high rainfall amounts are not found on Earth.

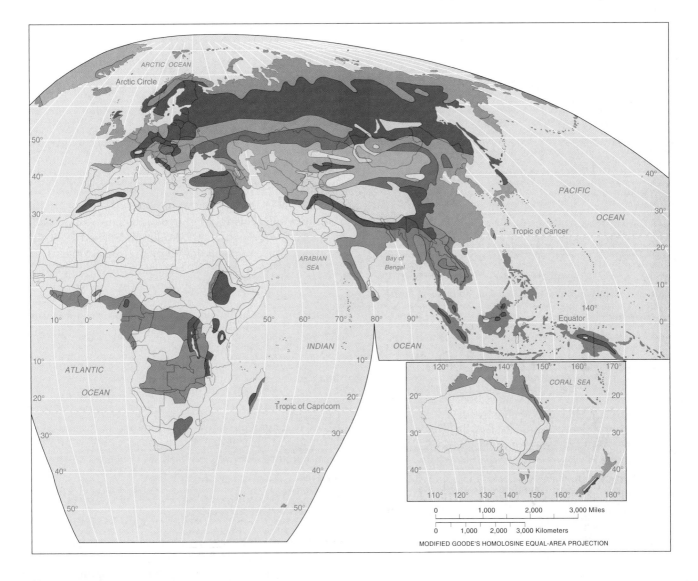

MODIFIED GOODE'S HOMOLOSINE EQUAL-AREA PROJECTION

usually is lower than in the rain forest and because lower temperatures slow chemical weathering. Lower temperatures also contribute to reduced decay rates of organic matter. Consequently, undisturbed soils typically have a substantial *O* horizon, ranging in thickness from a few centimeters (1 to 2 inches) to over 10 centimeters (4 or more inches). When these forests are cleared and the soil is brought under cultivation, the soils generally are quite fertile, but that fertility declines once the organic matter decays and is not replaced.

Needleleaf or boreal forest biome

In poleward portions of the midlatitudes, a needleleaf (coniferous) evergreen forest flourishes. The name "boreal forest" comes from the northern location of this vegetation (Latin *borealis* means "northern"). The **boreal forest** thrives in cold, continental midlatitude climates. Extensive boreal forests are restricted to the Northern Hemisphere, since there is little land in the Southern Hemisphere in

FIGURE 4-25 Tropical rain forest. Rainforest vegetation is generally dense and multistoried, as exemplified by this scene from the tropical island of Borneo. (Courtesy of Julia Nicole Paley)

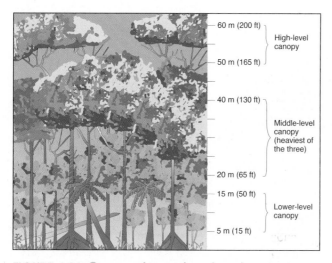

60 m (200 ft) ⎫
 ⎬ High-level
 ⎭ canopy
50 m (165 ft)

40 m (130 ft) ⎫
 ⎪ Middle-level
 ⎬ canopy
 ⎪ (heaviest of
 ⎭ the three)
20 m (65 ft)

15 m (50 ft) ⎫
 ⎬ Lower-level
 ⎭ canopy
5 m (15 ft)

FIGURE 4-26 Canopy layers in a forest. Three levels of forest canopy can be distinguished in the tropical forest. At the top is the high-level canopy, in which the tallest trees emerge above the rest. The densest canopy layer is in the middle levels. A third, lower-level canopy is composed of shade-tolerant species. (From R. W. Christopherson, *Geosystems*, 5th ed. Upper Saddle River, NJ: Prentice Hall, 2003)

latitudes 50–70°. The most extensive boreal forest regions are in central Canada and Siberia. Needleleaf forests also occur in some areas of seasonally dry soils.

During cold winters, low humidity and frozen ground cause moisture stress, but needleleaved trees survive because the leaves have low surface area in relation to their volume, and they are covered with a waxy coating to reduce water loss. This adaptation allows them to survive in colder climates that would kill broadleaved trees. Because they are evergreen, these trees photosynthesize whenever the temperature is warm enough, instead of having to wait until after the last frost to produce leaves. Most needleleaved trees are evergreen, but not all. The larch, for example, is a needleleaved deciduous tree.

FIGURE 4-27 Forest in Michigan. A forest in the transition zone between midlatitude deciduous forest and northern coniferous forest in northern Michigan. (Stan Osolinski/Corbis/Stock Market)

Coniferous forests also occur in the southeastern United States, where sandy soils drain quickly and have limited water storage. During warm summers, these soils dry out, and the moisture-retaining properties of coniferous trees allow them to compete well against broadleaved trees. The ability to sprout from roots and grow rapidly following forest fires is an added advantage of many coniferous species in areas of summer aridity.

Needleleaves accumulate beneath coniferous forests, and as they decay, they produce strong acids in water that percolate down through the soil. This acidic water leaches the upper soil of all but the most insoluble minerals, leaving a light-colored, sandy upper soil layer. Some of the iron and aluminum leached from the upper soil accumulates lower in the soil profile. This produces a distinctive soil associated with coniferous forests called a *spodosol* (see Figure 4-16).

Temperate rainforest biome On the west coasts of North and South America, lush evergreen coniferous forests grow. These temperate rain forests are found in areas of marine west coast climate, where moderate temperatures and ample rainfall allow year-round plant growth. The evergreen trees grow faster in summer, but they can photosynthesize in winter, too, despite occasional freezing temperatures. The temperate rain forest is much less diverse than the tropical rain forest and is dominated by relatively few species.

Savanna, Scrubland, and Open Woodland Biomes

Savannas and open woodlands have trees spread widely enough for sunlight to support dense grasses and shrubs beneath them. This vegetation is common in climates that have a pronounced dry season. The term **savanna** refers especially to vegetation characteristic of large seasonally dry areas in tropical Africa and South America. The rainy season brings green, lush grass, but during the dry season the grass dries out and only deeper-rooted trees continue to photosynthesize. If the dry season is pronounced, even the trees may lose their leaves, and in some areas dry-season deciduous trees are common.

Fire is common in the savanna. Although aided by the seasonal aridity, it is probably as often caused by humans as well as lightning. Fire does not destroy most grasses, which grow back quickly. Slower-growing trees might in time become dominant, shading out the grasses, but they are more susceptible to fire damage, so the savannas remain. Whether the savannas of Africa and South America are the result of human activity is often debated. Some expanses of African savanna occur in climates that support forests elsewhere, thus encouraging the hypothesis that human-caused fires have created the extensive African savannas.

Soils of tropical savannas usually are deeply weathered, as in the humid tropics, but they may be less

Critical THINKING

Human-Dominated Systems

In recent years, quantitative study of biogeochemical cycles at the global scale has determined the amounts of materials that are processed in various environments per unit of time. This research was done in part to understand the relative importance of different components of those cycles. One of the conclusions of this research is that a considerable amount of material processing—the conversion of substances from one form to another—is carried out by humans. In some cases humans do more processing than nature does.

Consider some examples. We have already seen that humans are playing such a major role in the global carbon cycle that the quantity of CO_2 in the atmosphere is steadily increasing, despite the vast amounts of carbon processed each year by the biosphere. The nitrogen cycle is another important cycle that is influenced by humans. In this cycle, nitrogen is taken from the atmosphere and converted to forms that can be used by plants and animals to build proteins and other molecules. This bioavailable nitrogen is then passed through food chains before being broken down and released back to the atmosphere. Humans *fix* nitrogen (convert it to biologically available forms; see Chapter 8) through fertilizer manufacture, and today more nitrogen is fixed by humans than by natural processes. Food production through photosynthesis is another critical biogeochemical process in which humans play a major

role. About 8 percent of global photosynthesis takes place in agricultural lands. Humans play an even larger role in consuming plant matter, either directly as food for themselves and their domestic animals and as fiber for material such as lumber, or indirectly by controlling its characteristics or fate, such as grass grown on golf courses or crop residues left in the field. Estimates of the amount of biological production that is "appropriated" by humans range from about 3 to 40 percent, depending on what human uses are included in the estimate.

These conclusions are of great importance, not only for scientific and practical purposes but in describing the physical relationship between humans and the environment. This, in turn, has implications for the way we think about nature and our role in it. In short, it goes to the very heart of the human–environment relationship.

Questions

1. How important is it that we separate human systems from natural systems in our environmental management policies? Should we be investing in wilderness preservation or the improvement of managed landscapes?
2. Are humans distinct from nature or a part of it? Should nature, as something over which humans have little or no influence, be preserved?

---

leached of nutrients as a result of reduced rainfall. High temperatures and reduced plant growth combine to lower the organic content. Many soils of these regions are quite productive if sufficient irrigation water is available.

In more arid margins of tropical savannas, a seasonal forest or thorn-scrub vegetation occurs. Drought-resistant trees and shrubs dominate, as well as cacti, and grasses are less important. Savanna-like open woodlands (also called parklands) also are found in some semiarid midlatitude areas.

Mediterranean climates promote a distinctive shrubland vegetation type called *chaparral*. This is a shrub woodland dominated by hard-leaved trees and shrubs that withstand the severe summer aridity. Fire is common because of summer aridity and flammable waxy-leaf coatings. Many plant species in the chaparral are adapted to frequent fire, and some even require fire to maintain their continued presence in the landscape (see the Critical Thinking box on page 159).

Midlatitude Grassland Biome

In Figure 4-23, extensive midlatitude grasslands are shown in deep red and brown. Major midlatitude grassland areas include interior Asia, the Great Plains of Canada and the United States, and central Argentina (Figure 4-28). The midlatitude semiarid climate of these areas features hot summers, cold winters, and moderate rainfall. In moister portions, the grass is usually taller (up to 2 meters, or 6 to 7 feet), and the biome is called **prairie.** In drier areas, such as Asia, the biome is called short-grass prairie, or **steppe.**

Grasses are well suited to this climate because they grow rapidly in the short season when temperature and moisture are favorable (generally spring and early summer). During dry or cold periods, aboveground parts of these plants die back, but the roots become dormant and survive. This trait also allows grasses to survive fire and grow back rapidly, using available moisture at the expense of trees or shrubs that might

FIGURE 4-28 Tall-grass prairie in Kansas. A small remnant of formerly extensive tall-grass prairie in the Flint Hills of Kansas. (Frank Siteman/StockBoston)

FIGURE 4-29 Former tall-grass prairie. Today most of the tall-grass prairie has been converted to row crops. (Russell Munson/Corbis/Stock Market)

invade. Many experts believe that occasional fires, either natural or deliberately set by humans, are at least partly responsible for establishing and maintaining extensive grasslands.

The soils of these midlatitude grasslands are very fertile and have a dark upper horizon with high nutrient content. Because of this fertility, many of the world's midlatitude grasslands are "breadbaskets"—they are primary production areas for wheat and other cereal grains (Figure 4-29). In North America, virtually all original *tall-grass prairies* are farmed to produce wheat, corn, soybeans, and other small grains. Much of the *short-grass prairie* present at the time of European settlement remains, although it has been heavily grazed.

Desert Biome

In **deserts,** moisture is so scarce that large areas of bare ground exist, and the sparse vegetation is entirely adapted to moisture stress. Plants with such adaptations are called **xerophytes.** Some desert plants are drought-tolerant varieties of types common in more humid areas, such as grasses. Others, such as cacti, are almost exclusive to deserts. Most desert plants are adapted to

limit evaporative losses, with thick, wax-coated leaves or with needles rather than leaves, meant to protect cacti from being consumed for the water they contain.

Another common adaptation is water-collecting root systems. These are extensive shallow roots to gather occasional rainwater or long tap roots to reach deep groundwater. Many desert plants survive long periods without moisture by becoming dormant, either as a mature plant or as a seed. When occasional rains do occur, these dormant plants spring to life and flower in a few weeks before becoming dormant again. Animal life in the desert is dispersed because of the reduced rates of plant production, but it is still diverse. Many desert animals are nocturnal, foraging and hunting at night to avoid dehydration during the hot daytime hours.

Scant moisture and the consequently slow chemical activity leave many desert soils poorly developed. Leaching of soluble nutrients is limited, and usually the horizons in desert soils are developed through vertical movements of soluble minerals, controlled by rainfall and evaporation. When water enters the soil, it may dissolve some minerals, but as evaporation draws this water back to the surface, these minerals are redeposited. As a result, high concentrations of soluble minerals near the surface (especially salts) are common. Desert soils might be fertile if irrigated, but irrigation usually introduces more minerals to the soil, possibly making it salty.

Critical THINKING

Fire and Forest Management in the Western United States

In the chaparral woodlands of California and many other forest ecosystems of the western United States, fire is an important natural part of ecosystems. Unfortunately, people have built homes and businesses in forested areas. Conflagrations like the 2006 fires near Sedona, Arizona, destroy entire communities, cause vast property damage, and occasionally take lives. Fire has also been an increasing problem in western Canada recently, and one major fire that started in the United States burned across the border into British Columbia. When a fire starts near a populated area, the natural reaction is to extinguish it at once. Paradoxically, this action to halt fire actually makes fire-prone conditions worse. In the chaparral, plant growth produces flammable biomass that does not decay as rapidly as it is produced, so it accumulates over time. The longer it accumulates, the more fuel is available for severe fires. If a fire starts in a recently burned area, little fuel is available, and the fire burns slowly at cooler temperatures, and it does not spread widely. But if a fire starts where one has not burned for a long time, a hot, dangerous fire is likely.

Consequently, many areas that routinely would burn naturally have not burned, and fuel accumulations—dry leaves and dead wood—are unusually large. When a fire starts during dry, windy conditions, it becomes very dangerous and uncontrollable. Its high temperatures make it much more destructive than fires during moist conditions.

Decades of firefighting in the western United States have not eliminated fires; they have only allowed fuel to accumulate. Normal, "natural" fires, low temperature and slow burning, have been eliminated and replaced by infrequent but catastrophic conflagrations. Fire managers now use controlled burns in some areas to reduce fuel accumulations.

Questions

1. If you lived in a fire-prone neighborhood, would you support the use of controlled burns even if there were a small risk that the fire would get out of control?
2. Can you think of other examples, perhaps where you live, of situations where poor environmental decisions in the past have locked us into protecting structures and other investments at costs that may be unreasonably large?

Tundra Biome

In cold, high-latitude environments or in high mountains above the treeline, freezing and short growing seasons severely limit plant growth. During winter, water is locked up in the soil as ice. Dehydration and abrasion by blowing snow damage exposed upper parts of plants. Therefore a **tundra** is dominated by low, tender-stemmed plants and low, woody shrubs. These survive the cold by lying dormant below the wind, often buried in snow, growing only in the short, cool summer. Animal life in this biome is very limited in winter, but in summer the tundra comes alive with insects, migratory birds, and grazers such as reindeer and caribou.

As in deserts, tundra soils often are poorly developed because of slow chemical activity. Permafrost occurs where the mean annual temperature is below 0°C (32°F). Soil horizons are weakly developed and often disrupted by frost. In poorly drained areas, organic matter often accumulates rather than decays, and peat is formed. As was mentioned previously in this chapter, peat is an accumulation of plant debris and an early stage in the formation of coal.

Because of slow plant growth in this environment, recovery from disturbances can be very slow. This,

combined with the disruption of permafrost caused by construction activities and vegetation disturbance, has led to concern about the long-term effects of oil exploration and related activities in Alaska, Siberia, and similar areas.

Natural and Human Effects on the Biosphere

Plants, soils, climate, and human activity each reflect the influence of the other. In the absence of humans, climate would have the strongest control on vegetation. Climate controls soil formation and plant growth so strongly that, with the exception of agricultural areas, world vegetation and soil patterns correspond closely to world climates. Humans, however, have had profound influences on ecosystems in most of the world's land areas, including the 37 percent of land area (excluding Antarctica) that is cropland or permanent pasture (Figure 4-30). Locally, topography and geology also exert strong influence on vegetation and soils.

Regionally and globally, climate is in turn affected by ecological processes and soil-vegetation linkages:

▸ Plants regulate evapotranspiration from land surfaces and thus moderate the hydrologic cycle. If

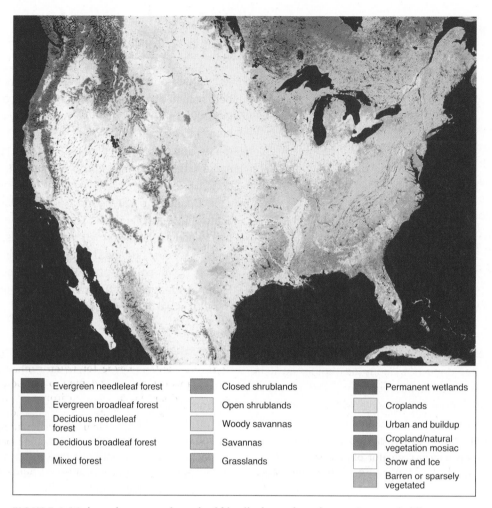

FIGURE 4-30 Land cover of part of North America, based on satellite imagery. Compare the extent of agricultural land in with the distribution of natural vegetation types, as shown in Figure 4-23. (MODIS/NASA Headquarters)

vegetation is cut back, thereby reducing the number of plants that draw water from the soil and transpire it to the atmosphere, moisture flow to the atmosphere may be reduced. The excess water runs off to the sea as river flow. In this manner, vegetation could modify the climate by affecting humidity.

▶ Soil eroded by wind from semiarid and arid areas may cloud the atmosphere with dust and modify energy exchanges there, thus modifying the climate.

▶ The biosphere stores and regulates carbon, helping to control the global carbon budget. If vegetation grows more quickly and absorbs atmospheric carbon dioxide (CO_2) faster than decaying organic matter can return carbon dioxide to the air, there is net storage of carbon in biomass. If the reverse is true, there is net release of carbon as carbon dioxide to the atmosphere. Thus, vegetation partly controls the level of this important greenhouse gas in the atmosphere.

Desertification is a good example of the impact of interactions among human and natural processes. Economic and social forces, especially population growth and demand for food, cause us to populate semiarid areas with domesticated animals or to replace native vegetation with plant crops. The reduction in vegetation cover causes increases in runoff and wind erosion that reduce the amount of water and nutrients present in the soil, further limiting vegetation growth. The loss of vegetation may also affect the atmosphere by increasing the amount of dust it contains or by reducing the amount of solar energy absorbed by the ground surface. Changes in dust or sunlight absorption are hypothesized to affect temperature and precipitation amounts, although such effects have not been confirmed by observation. These intimate relations among the soil, water, air, and living things lead some scientists to view Earth as a single interconnected dynamic system—almost a functioning organism—rather than just a set of cause-and-effect linkages. We can see all these subsystems interacting with each other in much the same way that different parts of a single living organism interact.

CONCLUSION: CRITICAL ISSUES FOR THE FUTURE

Beginning thousands of years ago, but especially within the past two centuries, humans have disrupted ecosystems worldwide, largely through spreading population and increasing demands on resources. By disturbing environments, we have unbalanced them, and they probably now are much more dynamic than they are naturally. We have increased biomass stores in some areas and reduced them elsewhere; we have altered the distribution of biomes and nearly eliminated many ecosystems; we have changed the sediment and nutrient flows of rivers and lakes.

Only a few thousand years ago, Earth was regulated by non-human processes, and people were only negligible players, no more significant than any other common species. Today we are major participants in Earth's environmental processes and are often the dominant ones. Environments in which human impacts are minor are increasingly scarce, and the future of Earth's environmental systems depends on how we manipulate biological and physical processes at the surface of Earth.

The scientific study of human interactions with the biosphere is still young, and there is much to be learned. Until recently most studies of human–environment relations have been carried out at relatively local scales. But today we recognize that human impacts are truly global, and we need to increase our understanding of Earth's systems at that scale.

Chapter Review

SUMMARY

Climate is the dominant control on local environmental processes, primarily through its influence on water availability and movement. Knowledge of the water budget is essential to understanding both physical and biological processes because the water budget is a critical regulator of ecosystem activity. This influence extends to human use of the landscape, through such issues as determining the natural vegetation cover or agricultural potential.

Carbon is the basis of life on Earth. It cycles through the atmosphere, oceans, biosphere, and soil. Photosynthesis transfers carbon from the atmosphere to the biosphere, and respiration returns it to the atmosphere. Plant growth is most prolific in warm, humid climates where water and energy are plentiful. Plant growth is less in dry or cool climates. Large amounts of carbon are exchanged between the atmosphere and the oceans, and the oceans are a major sink for carbon dioxide that is added to the atmosphere by fossil-fuel combustion.

The outermost layer of Earth's surface in land areas is soil, which is a mixture of mineral and organic matter formed by physical, chemical, and biological processes. Climate is a major regulator of soil development, through its control on water movement. Plant and animal activity in the soil produces organic matter and mixes the upper soil layers. In humid regions, water moves downward through the soil, carrying dissolved substances lower in the profile or removing them altogether. In arid regions, these substances are not so easily removed. Soil characteristics reflect these processes, and there is close correspondence between the world climate map and the world soil map.

Photosynthesis is the basis of food chains and ecological systems. Plants compete for water, sunlight, and nutrients, and plants that are best adapted to compete for limiting factors will dominate a given environment. Such adaptations help explain the world distribution of major vegetation types. Disturbances modify ecological communities, and succession following disturbance is a fundamental process of landscape change. Very large portions of the world's land surface have been modified by human activity, and humans are major players in most of the world's ecosystems.

The world vegetation map closely mirrors the world climate map. Ecologically diverse and complex forests occupy humid environments, storing most nutrients in their biomass. In arid and semiarid regions, sparse vegetation is adapted to moisture stress. Forests adapted to winter cold are in humid midlatitude climates, developing as broadleaf forests in warmer areas and coniferous forests in subarctic latitudes. In high-latitude climates, cold-tolerant short vegetation occupies areas that have a mild summer season. Vegetation is absent in ice-bound polar climates.

Overall, climate is the strongest control on *natural* vegetation, but humans have had profound influences on ecosystems in most of the world's land areas. Plants regulate evapotranspiration from land surfaces and thus moderate the hydrologic cycle. The biosphere stores and regulates carbon, helping to control the global carbon budget. Only a few thousand years ago, Earth was regulated by nonhuman processes, and people were merely negligible players. Today, however, humans are globally significant and locally dominant players in Earth's biogeochemical cycles. Environments in which human impacts are minor are increasingly scarce.

KEY TERMS

actual evapotranspiration (ACTET) p. 136
biodiversity p. 152
biogeochemical cycle p. 132

biomagnification p. 161
biomass p. 141
biome p. 153
boreal forest p. 155

broadleaf deciduous forest p. 154
carbon cycle p. 139
carnivore p. 150

QUESTIONS FOR REVIEW AND DISCUSSION

1. Diagram the hydrologic cycle, using boxes and arrows. Proportion box sizes to indicate the relative amounts of water stored at each point in the cycle. Label the arrows linking the boxes with the appropriate hydrologic process (evapotranspiration, condensation, precipitation, runoff, and so on).

2. Examine the water-budget diagram in Figure 4-7 for a climate similar to the one in which you live. Explain the seasonal patterns of precipitation, potential evapotranspiration, actual evapotranspiration, soil-moisture use and recharge, runoff, and moisture deficit.

3. What are photosynthesis and respiration? How are they related? How do they vary seasonally in relation to temperature and water availability? How do they vary in relation to major climate types?

4. What is soil texture? How does it affect the movement and storage of water in the soil? How are soil-forming processes affected by water availability?

5. For each major climate type described in Chapter 2, describe the major vegetation type generally associated with that climate.

THINKING GEOGRAPHICALLY

1. Go to the U.S. Geological Survey's surface water data website (**http://waterdata.usgs.gov/nwis/sw**) and locate monthly flow data for a river near you. Download the data and make a bar graph showing mean monthly flow. Explain the variations in river flow in terms of seasonal variations in precipitation and evapotranspiration.

2. In your library, find the soil survey for a place near you where excavation is occurring, such as for a highway or building foundation. Determine the soil characteristics in the area. Then visit the excavation site and view the soils in the field, comparing what you see with the descriptions in the soil survey.

3. Consider a typical water budget for a midlatitude continental site in which soil-moisture use is important for part of the year. How is actual evapotranspiration affected by the amount of water stored in the soil, and the rooting depth of the vegetation?

4. World maps of climate and vegetation have many similarities. If climate were to change significantly, do you think the vegetation map would also change? Why or why not?

5. How has the vegetation cover where you live changed in the past 200 years? How do you think this change may have affected the local water budget?

SUGGESTIONS FOR FURTHER LEARNING

Adams, W. M. R. Aveling, D. Brockington, B. Dickson, J. Elliott, J. Hutton, D. Roe, B. Vira, and W. Wolmer. "Biodiversity conservation and the eradication of poverty." *Science* 306 (12 November 2004): 1146–1149.

Bonnie, R., et al. "Counting the cost of deforestation." *Science* 288, no. 5472 (2000): 1763–1764.

Bradley, R. S., M. Vuille, H. F. Diaz, and W. Vergara. "Threats to water supplies in the tropical Andes." *Science* 312 (23 June 2006): 1755–1756.

Bradshaw, W. E. and C. M. Holzapfel. "Evolutionary response to rapid climate change." *Science* 312 (9 June 2006): 1477–1478.

Bryce, S. A., and others. "Ecoregions: A geographic framework to guide risk characterization and ecosystem management." *Environmental Practice* 1, no. 3 (1999): 141–155.

Cincotta, R. P., and others. "Human population in the biodiversity hotspots." *Nature* 404, no. 6781 (2000): 990–992.

Crifasi, R. R. "The political ecology of water use and development." *Water International* 27, no. 4 (December 2002): 492–503.

Dale, V. H., C. M. Crisafulli, and F. J. Swanson. "25 years of ecological change at Mount St. Helens." *Science* 308 (13 May 2005): 961–962.

Ferraro, P. J., and A. Kiss. "Direct payments to conserve biodiversity." *Science* 298 (2002): 1718–1719.

Flannagan, M. D., and C. E. Van Wagner. "Climate change and wildfire in Canada." *Canadian Journal of Forest Research* 21, no. 1 (1991): 66–72.

Foley, J. A., R. DeFries, G. P. Asner, C. Barford, G. Bonan, S. R. Carpenter, F. S. Chapin, M. T. Coe, G. C. Daily, H. K. Gibbs, J. H. Helkowski, T. Holloway, E. A. Howard, C. J. Kucharik, C. Monfreda, J. A. Patz, I. C. Prentice, N. Ramankutty, and P. K. Snyder. "Global Consequences of Land Use." *Science* 309 (22 July 2005): 570–574.

Global Envionmental Change Programmes. "Global change and the Earth system: A planet under pressure." *Science* (Stockholm: IGBP) no. 4 (2001).

Gresswell, R. E. "Fire and aquatic ecosystems in forested biomes of North America." *Transactions of the American Fisheries Society* (American Fisheries Society) 128, no. 2 (1999): 193–221.

Groombridge, B., and M. D. Jenkins. *World Atlas of Biodiversity.* Berkeley: University of California Press, 2002.

Jackson, J. B. C., et al. "Historical overfishing and the recent collapse of coastal ecosystems." *Science* 293 (2001): 629–638.

Jones, J. A. A. *Global Hydrology.* Edinburgh Gate, England: Addison Wesley Longman, 1997.

Kelly, Robert C. "The carbon conundrum: Global warming and energy policy in the third millennium." *Country Watch* (2002).

Lackner, K. S. "A guide to CO_2 sequestration." *Science* 300 (2003): 1677–1678.

Liu, J., et al. "Protecting China's biodiversity." *Science* 300 (2003): 1240–1241.

McGregor, G., G. E. Petts, A. M. Gurnell, and A. M. Milner. "Sensitivity of alpine stream ecosystems to climate change and human impacts." *Aquatic Conservation* 5, no. 3 (1995): 233–247.

Pearce, F. *When the Rivers Run Dry: Water—The Defining Crisis of the Twenty-First Century.* Boston: Beacon Press, 2006.

Post, Wilfrid M., T. -H. Peng, W. R. Emanual, A. W. King, V. H. Dale, and D. L. DeAngeles. "The global carbon cycle." *American Scientist* 78 (July–August 1990): 310–326.

Postel, Sandra L., G. C. Daily, and P. R. Ehrlich. "Human appropriation of renewable fresh water." *Science* 271 (1996): 785–778.

Running, S. W. "Is global warming causing more, larger wildfires?" *Science* 313 (18 August 2006): 927–928.

Schlesinger, William H., J. F. Reynolds, G. L. Cunningham, L. F. Huenneke, W. M. Jarrell, R. A. Virginia, and W. G. Whitford. "Biological feedbacks in global desertification." *Science* 247 (March 2, 1990): 1043–1048.

Senay, G. B. "Combining AVHRR-NDVI and land-use data to describe temporal and spatial dynamics of vegetation." *Forest Ecology and Management* 128, no. 1, 2 (2000): 83–91.

Turner, Billie Lee. *The Earth as Transformed by Human Action.* New York: Cambridge University Press, 1990.

Vitousek, Peter M., P. R. Ehrlich, A. H. Ehrlich, and P. A. Matson. "Human appropriation of the products of photosynthesis." *BioScience* 36 (June 1986): 368–373.

Williams, Michael. *Deforesting the Earth.* Chicago: University of Chicago Press, 2003.

Wilson, E. O. *The Future of Life.* New York: Knopf, 2002.

Zhang, Peichang, et al. "China's forest policy for the 21st century." American Association for the Advancement of Science, *Science* 288, no. 5474 (2000): 2135–2136.

WEB WORK

The World Wide Web is an especially rich source of information relating to the global environment. Among international governmental agencies, the United Nations Environment Program (UNEP) site provides a useful gateway to United Nations activities:

http://www.grida.no/

Data gathered by UNEP and many other governmental and nongovernmental agencies is available through CIESIN, a non-profit organization that provides information to help scientists, decision makers, and the public better understand their changing world:

http://www.ciesin.org/

For information on programs aimed at protecting the biosphere, the International Union for Conservation of Nature is a key data-collection and dissemination organization. In particular, its Biodiversity Conservation Information System (BCIS) helps coordinate conservation efforts among diverse governmental and non-governmental organizations:

http://iucn.org/

In the United States, the United States Geological Survey (USGS) is home to two key environmental research groups. The Water Resources Division collects and disseminates water resource information in the United States, and the Biological Survey is responsible for research on the nation's biological resources. You can find the USGS at:

http://www.usgs.gov/

The Natural Resources Conservation Service, formerly the Soil Conservation Service, is the key to information about soils in the United States as well as climate and water resource information of interest to the agricultural community:

http://www.nrcs.usda.gov/

Environment Canada provides a central point of access to a wealth of environmental information pertinent to Canada:

http://www.ec.gc.ca/envhome.html

POPULATION,
POPULATION INCREASE,
AND MIGRATION

The residents of Bodie, California, once home to 10,000 people, abandoned if after exhausting the gold in local mines. Today the ghost town is a California State Historic Park. Changing distributions of economic opportunities constantly trigger the abandonment of mines, farms, and whole towns, while at the same time inspiring migration and new population settlement patterns. (Alamy Images)

The Distribution and Density of Human Settlement

About 90 percent of Earth's 6.5 billion people occupy only 20 percent of Earth's surface. The distribution and density of settlement is due largely to attributes of the physical environment, to factors of history, and to varying rates of population growth.

World Population Growth

Earth's population is rising, but the rate of increase is slowing. Most population increase is occurring in the poor countries, and some people fear that Earth may be overpopulated or facing potential overpopulation.

Other Significant Demographic Patterns

Sex ratios vary greatly among national populations, and these variations may be explained as the result of both economic and cultural factors. The median age of the human population overall is rising, and this presents distinctly different challenges to rich countries and to poor countries.

Migration

All the world's modern peoples descended from a single stock that migrated away from Africa as long as 200,000 years ago. Since 1500, the great migrations that have made our world include migrations of Europeans to the Western Hemisphere, Asia, Oceania, and Africa; migrations of Africans, mostly to the Western Hemisphere; of Indians; and of Chinese.

Migration Today

Migration of people from poor countries to rich countries, of people from rural areas to cities (discussed in Chapter 10), and of people seeking greater political freedom continues today. Pressures of migration have become a political issue in many rich countries.

*I*n the summer of 2006, someone leaked test questions for the Philippine Nursing Board Examinations to an unknown number of the 42,000 applicants scheduled to take the test. The Professional Regulation Commission raced to schedule new tests to guarantee the integrity of the testing and certification process for Philippine nurses. The scandal carried extensive international ramifications. The Philippines is one of the world's leading educators of nurses, who serve in America, Europe, Japan, and other rich countries around the world. All observers agree that the Philippines could itself use more nurses. Only about 60 percent of births there are attended by any skilled health staff, and some 200 women die per 100,000 live births, compared to, for example, 17 in the United States and 6 in Canada. Philippine nurses, however, make more money in richer countries, and the money they send back home is important to the Philippine economy. Filipinos and Filipinas working abroad—a great many of whom are nurses—sent home $12 billion in 2004, a full 13 percent of the national income. Any scandal jeopardizing the international credibility of Philippine nurses is of national concern. "American hospitals, Japanese hospitals, European hospitals are watching us," said Richard Gordon, a member of the Philippine Senate.

Students in the rich countries do not seem willing to take up the career, so the United States, for example, welcomes thousands of Philippine nurses each year. The number is expected to rise as the U.S. population aged 65 or over is projected to rise from 36.3 million in 2004 to 86.7 million by 2050. Earlier in 2006, America had in fact lifted the cap on the number of foreign nurses American hospitals and clinics can hire. Almost all of the rich countries are characterized by falling numbers of births and aging populations. Although poor countries can ill afford to lose such skilled workers (a phenomenon called the *brain drain*), they need the money these workers abroad send home. Birth rates in most poor countries are also falling, but they still remain relatively high. The numbers of children born every year challenge the societies' ability to feed, clothe, educate, and eventually provide jobs for them. ▸

This chapter examines the world's **population geography**—that is, the distribution of humankind across our planet. The relative distribution of humankind is constantly changing, and this is because of two factors. First is the fact that different countries have different internal population dynamics, such as the numbers of births and deaths and the consequent rates of increase or decrease. The second factor is human **emigration** out of some places and **immigration** into others. Migration has affected the distribution of the world's peoples in the past, and continuing migration affects world affairs today.

This chapter will occasionally include ideas and information about **demography,** which is the study of individual populations in terms of specific group characteristics. These characteristics may include the distribution of ages within the group, the relative numbers of males and females, income levels, or any other characteristic. Demography means "describing people."

THE DISTRIBUTION AND DENSITY OF HUMAN SETTLEMENT

In 2006 the population of Earth was about 6.5 billion, but people are not evenly distributed across the landscape (see the map on the rear endpaper). The map shows that large areas of Earth's surface are surprisingly uninhabited. About 75 percent of the total human population lives in the Northern Hemisphere between 20° and 60° north latitude, and even great expanses of that area are sparsely populated. About 90 percent of the population is concentrated on less than 20 percent of the land area.

The map reveals three major concentrations of population: (1) East Asia, where eastern China, the Koreas, and Japan total about 1.6 billion people; (2) south Asia, where India, Pakistan, and Bangladesh together account for about 1.5 billion; and (3) Europe from the

FIGURE 5-1 Low population density. The enormous island of Greenland (2.2 million square kilometers, or 840,000 square miles—more than three times the size of Texas) is inhabited by only 56,000 people. The fact that any people live here at all is a tribute to how human culture and technology allow us to adapt to all environments. (Norman Price/Alamy Images)

Atlantic to the Ural Mountains, with almost another billion. Southeast Asia forms a secondary concentration, with about 600 million people; the eastern United States and Canada are home to almost 275 million.

These five population concentrations, plus parts of West Africa, Mexico, and areas along the eastern and western coasts of South America, are densely populated, with more than 25 people per square kilometer (60 per square mile). More than half of Earth's land area, by contrast, has fewer than 1 person per square kilometer (2.6 per square mile) (Figure 5-1). A surprising portion of Earth's land surface is virtually uninhabited: central and northern Asia; northern and western North America; and the vast interiors of South America, Africa, and Australia. The countries that occupy these spaces are enormous in area, but they contain relatively few people. The three African countries of Chad, Niger, and Mali, for example, together cover 39 times the area of South Korea, but their combined population is about half that of South Korea.

Figure 5-2 is a cartogram of world population. The size of each country reflects its population, not its land-surface area. The most populous countries are shown as the biggest, and the least populous as the smallest, no matter how big or small their land-surface areas. The single country of India, therefore, is drawn considerably larger than the entire continents of South America or Africa.

A number of factors combine to explain the distribution of the human population. In this section we will consider the impact of the physical environment and history on population patterns. In the next section we will take a close look at rates and distributions of world population increase. We will examine these factors as hypotheses of explanation for the population distribution.

Population Density

It is often thought that the population density of a place reflects the productivity of the local environment. **Arithmetic density** is the number of people per unit of area, and in this measure the countries of the world range from Mongolia's two people per square kilometer (five per square mile) to Bangladesh's 1,023 per square kilometer (2,692 per square mile) to the city-state of Singapore (Figure 5-3), with its 6,484 people per square kilometer (17,063 per square mile). The map on the rear endpaper reveals, however, that arithmetic density varies greatly even within individual countries. In Egypt, for example, the arithmetic density is nearly 2,000 people per square kilometer (5,200 per square mile) in the delta and valley of the Nile River, but it is only three persons per square kilometer (eight per square mile) in the rest of the country, which is desert.

Some scholars believe that a country's **physiological density,** which is the density of population per unit of cropland, tells us more about its ability to support its population than its arithmetic density. In this measure, the countries range from Australia's 2.89 hectares (7 acres) per capita to Singapore's total lack of any cropland to feed its 4.5 million people.

Do these statistics tell us anything about a country's standard of living? No. In the past, physiological density may have influenced the well-being of primitive societies that were entirely dependent on local agriculture. Today, however, there are several reasons why measures of population density do not reveal national welfare. We will study world agriculture in greater detail in Chapter 8 and economic development in Chapter 12, but we should note here that few countries actually farm all their potential cropland, and scholars differ about how much potential cropland each nation has. The amount of potential cropland can be increased by discovering crops that can thrive under new conditions, by treating soils in new ways, and by expanding irrigation. Chapter 4 noted, however, that the amount of potential cropland can be decreased by misusing and thus degrading soils or by expanding urban or industrial land uses.

Furthermore, the amount of land people have to sustain themselves is less important than the productivity of that land. A fertile region, for example, could presumably support a higher population density than an infertile land; in other words, the fertile area has a higher **carrying capacity.** Productivity also depends on the level of local technology and other inputs, including labor. Some countries enjoy technologically sophisticated and highly productive agriculture—even on infertile soils. Other countries suffer technological backwardness and low productivity—even on fertile soils.

FIGURE 5-2 A cartogram of world population. On this population cartogram, the size of each country reflects its population, not its actual land area. Notice that on a population cartogram the whole African continent is smaller than India, because it is less populous. South America is also relatively small.

Finally, no countries are fully dependent upon local agriculture anyway. Trade and circulation free societies from the constraints of their local environments and allow them to draw resources from around the world. Physiological density is irrelevant to the welfare of a rich city-state such as Singapore, which imports food in exchange for manufactured goods and services, or to any densely populated but industrial country, such as Japan or Belgium. Neither arithmetic density nor physiological density fully explains how the people at any place feed themselves today, or how they support

themselves. Some of the most densely populated areas on Earth are rich, yet others are poor. Conversely, some sparsely populated regions are rich, while others are poor. We can draw no clear correlation between density and welfare.

Climate The clearest relationship between population distribution and environmental factors is that population densities are low in most of the world's cold areas and dry areas. (Compare the map on the rear endpaper with Figures 2-30 and 2-31.) The major

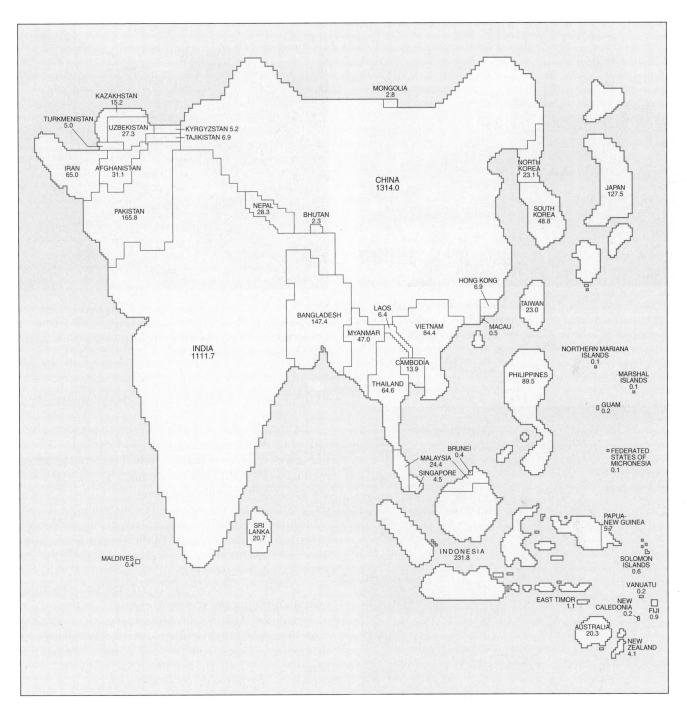

exceptions to this rule are places where rivers flowing through dry areas provide water for irrigation, such as India, Pakistan, and Egypt (Figure 5-4).

Earth's coldest areas support habitation only where mines or other special resources make it worthwhile for populations to work there, as in parts of central Asia or high in the mountains of South America. People will settle in harsh areas if it is profitable to do so.

Some warm and wet equatorial regions are also sparsely populated, such as the Amazon River basin in South America, the Congo River basin in Africa, and

the island of New Guinea. Despite the lush vegetation of these regions, conventional agriculture has not proved successful. Some soils are poor in nutrients. In other cases, agricultural crops cannot compete with other plants. A wide variety of plant and animal forms flourishes in the warm, moist climate, but many of these are hostile to humans and their agriculture. For example, insects thrive, as do the diseases they carry. In addition, these environments support countless forms of parasites, microbes, and fungi that weaken and kill humans, wilt and blight their plants, eat crops alive

FIGURE 5-3 Singapore. Modern skyscrapers encroach on historic colonial buildings in Singapore, a city that is an independent country. (© Macduff Everton/CORBIS)

FIGURE 5-5 Rice terraces. The construction and intensive cultivation of rice terraces such as these on the Indian Ocean island of Bali supports a very high human population density. (SuperStock, Inc.)

FIGURE 5-4 The Nile delta. This satellite photograph illustrates why the ancient Greek historian Herodotus called Egypt "the gift of the Nile." A thin strip of cultivable land winds northward through uninhabitable desert and then fans into a delta where it meets the Mediterranean Sea. Each year the Nile River floods and recedes, leaving a narrow plain of rich black mud, although construction of the High Dam at Aswan has significantly reduced this annual input of natural fertilizer. (Jeff Schmaltz, MODIS Rapid Response Team/NASA Headquarters)

in the fields, or quietly feast on them in granaries or storerooms.

Heat also affects humans' ability to perform physical labor. Work generates heat, and the body has to lose excess heat to work efficiently. The productivity of manual workers decreases by as much as one half when the temperature is raised to about 35° C (95° F), as is quite common in the tropics. The invention of the air conditioner has made it more comfortable for people to settle in the warm regions of the United States, however, and most of the country's population growth since 1970 has taken place in the southern Sunbelt. Cultural and technical innovation can overcome many environmental restraints, but it may be expensive to do so.

Tropical Asia is home to several great concentrations of people, but these are mostly in seasonal regions that are not wet year-round. Some of these, such as the island of Java or the Deccan Plateau of India, offer rich volcanic soils, and others are well adapted to the construction of flooded fields for growing rice, called *sawahs* (Figure 5-5).

Rice yields the highest number of calories per acre of any known crop. Sawah flooding, plus intensive care, limit the growth of competing vegetation, and algae in the water supply nitrogen as a plant nutrient. It seems almost always possible to increase the yield of a rice terrace by working it just a little bit harder, so sawahs can support increasing numbers of cultivators on a given unit of cultivated land. In addition, rice terraces create a stable ecosystem.

In conclusion, most of the world's population is concentrated in areas of seasonal environments that are not too wet, too hot, too dry, or too cold. In fact, most of Earth's inhabitants live in warm midlatitude climatic regions (designated C in the Köppen climate classification system in Figure 2-31) or in tropical areas that experience distinct seasons (categories Am and Aw).

Topography and soils

Topography often affects population distribution, although its effect is less prominent than that of climate. People tend to settle on flatlands because of the ease of cultivation, construction, and transportation. Thus, most of the densest concentrations of population in the world are found on level terrain, often near navigable waters (compare the map on the rear endpaper with the map on the front endpaper).

Flatness alone, however, is insufficient to attract a large population. Usually there is an association of other environmental attributes, such as fertile soil, available water supply, and moderate climate that helps to explain high population density. Flatlands in central Siberia, western Australia, and central Brazil are nearly unpopulated because other factors, such as the climate or soils, diminish the area's productivity.

Conversely, sloping land (hills and mountains) usually has a sparse population density, although this is not always the case. High density is found in many mountainous portions of South America, Japan, New Guinea, Southeast Asia, central Europe, and elsewhere. Specialized attractions, such as mineral resources, often help explain such concentrations. Moreover, in most mountainous areas, the settlements actually are concentrated in valley bottoms, even though they may be small and constricted. Thus the connection between topography and population density is too inconsistent to validate a direct relationship.

People usually settle areas of fertile and potentially productive soil unless there are powerful, negative factors against the choice (compare the map on the rear endpaper with Figure 4-15). For example, floodplain soils are typically fine-textured and full of nutrients, so they can be used intensively for agriculture. Thus, most river floodplains are densely populated unless they are located in extremely cold or dry areas. People do not necessarily avoid areas of poor soils, however, because some poor soils can be enriched with fertilizers.

Overall, the complexity of human history and the variety of human adaptations teach that environmental conditions alone can never explain population densities.

History

History helps explain the pattern of human settlement. The populations of China and the Indian subcontinent achieved productive agriculture and relative political stability thousands of years ago. These peoples domesticated many plants and animals early in their histories. **Domestication** is the process of taming and training animals and of sowing, caring for, and harvesting plants for human uses—largely for food. Intensive cropping and irrigation yielded generous food supplies, which supported rising populations. The Western European population multiplied when world exploration and conquest brought Europe new food crops—notably the potato. Later, during the Industrial and Agricultural Revolutions, European productivity multiplied. Migrants from Europe settled and brought European technology to more sparsely populated areas, so some of today's secondary population concentrations grew as extensions of Europe. These areas include eastern North America, coastal South America, South Africa, and Australia and New Zealand. It has been estimated that Europeans and their descendants increased from 22 percent of the world's population in 1800 to 35 percent by 1930.

Human population distribution can also be affected by the demarcation of cultural territories, particularly political territories. Two governments on opposite sides of an international border may have different environmental policies, and these can drastically alter the carrying capacity of the environment. Furthermore, a country may manipulate the distribution of its population. It might subsidize or command the settlement of harsh territories in order to occupy them effectively, or it might establish settlements along its borders for defense.

WORLD POPULATION GROWTH

Earth's population numbered 1 billion around 1800, 2 billion by 1930, 4 billion by 1975, and reached 6.5 billion in 2006. It continues to increase by about 77 million each year (Figure 5-6). The rate of increase, however, is decelerating. In the period 1965–1970, the average annual population increase was 2.06 percent, but it fell to 1.73 percent for the period 1985–1990, to 1.35 percent from 1995–2000, and 1.22 percent from 2000–2005. It has been registering 1.12 percent since 2005.

The rate of population increase represents a balance among several demographic statistics. First we will define each of these statistics, and then we will explain through the next several pages how changing

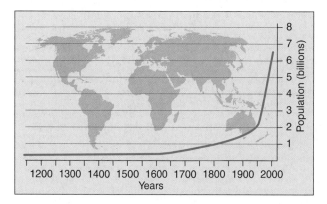

FIGURE 5-6 The increase of human population. The human population grew rapidly from about 1550 until about 1950, and then its growth accelerated to a still higher rate of increase after 1950. The total number of people is still increasing greatly today, but the rate of increase is slowing.

National Censuses

Throughout history—and in many countries today—counting a country's population has been difficult or impossible. Most countries counted their people only to record how many were on the tax rolls—or ought to have been. The Canadian census of the late-seventeenth century was probably the first census undertaken solely to count the population. The United States was the first country to require a regular census for the purpose of reapportioning seats in the representative legislature

Trying to count everyone. This census taker was one of 180,000 searching out Brazil's population in October 2000. (Larry Rohter/NYT Pictures)

(see Figure 11-27). Such a census has been taken every 10 years since 1790.

The United Nations annually publishes the *Demographic Yearbook,* but some figures in it are only estimates. Some countries lack the administrative apparatus to carry out a full census effectively. Portions of the country may be inaccessible, and the population may be moving within the country, as people migrate from the countryside into the cities. Portions of the population may avoid being counted out of superstition, distrust of the central government, or the wish to avoid taxation. In April 2000, one-third of American households failed to return their census forms, so census takers had to visit those homes. The difficulty of taking an accurate census even in the United States should warn us not to assume that all census numbers are exact. China's national census taken in 2000 is the greatest effort any nation has ever made. Over 6 million census takers were employed, and they gathered a number of social and economic statistics as well as population numbers.

In mid-2006, the world's ten most populous countries contained almost two-thirds of all humankind: China (1,300 million people), India (1,100), the United States (299), Indonesia (245), Brazil (188), Pakistan (165), Bangladesh (147), Russia (142), Nigeria (131), and Japan (127).

---

relationships among these statistics cause Earth's total population and the populations of individual countries to rise or fall.

The **crude birth rate** is the annual number of live births per 1,000 people. The **crude death rate** is the annual number of deaths per 1,000 people. The difference between the number of births and the number of deaths is the **natural increase or decrease.** For individual countries or regions, the natural increase or decrease can then be modified by subtracting the number of people who emigrated out of a particular area and adding the number of people who immigrated into that area. The result is the overall population growth or population decrease, a figure that can be expressed as a percentage of the total population. For example, Mexico currently has a birth rate of 20.7 births per 1,000 people, and a death rate of 4.7 deaths per 1,000 people, which gives Mexico a rate of natural annual increase of 1.59 percent. When the net rate of emigration of 4.3 people per 1,000 of population is taken into consideration, however, Mexico's growth rate falls to 1.16 percent.

Geographers also investigate each country's **fertility rate,** which is the number of children born per year per 1,000 females in a population. A great deal of attention is paid to each country's **total fertility rate (TFR),** which is the average number of children that would be born to each woman in a given society if, during her childbearing years (ages 15–49), she bore children at the current year's rate for women that age. This statistic carries great predictive value. A TFR of about 2.1 stabilizes a population, so this value of 2.1 children is called the **replacement rate.** If a country's TFR falls below 2.1 and the country does not experience immigration, that country's population will ultimately decrease.

Population Projections

A **population projection** is a prediction of the future, assuming that the world's current population trend remains the same or else changes in defined ways. Projecting population is an uncertain task because the smallest rise or fall in the percentage of increase today would increase or decrease the total population in the next century by hundreds of millions of people. Even if worldwide crude birth rates and TFRs continue to fall, the total number of people on Earth will continue to grow. This is because the number of young women presently reaching childbearing age is larger than ever before.

Earth's population will go on rising, but the sooner the world's TFR falls to the replacement rate, the lower the total population will be if or when the population eventually stabilizes. Many observers insist that the sooner Earth reaches a constant population—**zero population growth** overall—the better. When will this happen? At what figure?

The United Nations offers a range of projections in answer to this question. Today the world TFR is 2.6. If it drops to the replacement rate as early as 2015, then the world's population will stabilize at about 8 billion. If, however, the TFR does not fall to replacement level until 2055, then the population will not stabilize until it has reached 13 billion at the end of the twenty-first century. In 2003, the United Nations estimated that Earth's population will reach 7 billion in 2013, 8 billion in 2028, and 8.9 billion in 2050. As recently as 2000 the United Nations had estimated that Earth's population would reach 9.3 billion in 2050. The reduction of the 2050 estimate by 400 million people was attributed one-half to rising death rates in some countries (due to AIDS, discussed below) and one-half to falling fertility rates.

Projected totals are falling, but many people still fear that it might be difficult to sustain the world's economy and political order with billions more people, especially if it remains true that some of the most dense concentrations of the population are in some of the poorest areas.

Rates of Population Increase Vary

The overall population projection conceals wide variations in the rate of population increase from region to region and country to country. For example, the annual growth for the period 1990–2004 was 2.5 percent in sub-Saharan Africa, 1.6 percent in Latin America, and 1.9 percent in south Asia, but only 0.8 percent in North America. Europe reported no population growth at all. Figure 5-7 shows how the rates of population increase vary from country to country. These different growth rates suggest that there will be significant shifts in the geographical distribution of the world population. By 2050, Africa's share of Earth's population is expected to increase from 13 to 20 percent and Latin America's share from 8.5 to 9.1 percent. Asia's share, however, is projected to fall from 61 to 59 percent, North America's from 5.1 to 4.4 percent, and Europe's from 12.2 to 7 percent.

Some scholars compute the **doubling time** for each country's population—that is, the number of years it would take the country's population to double at its present rate of increase. A number compounding at 3 percent annually (the rate of population increase in Yemen) doubles in only 23 years, but a population

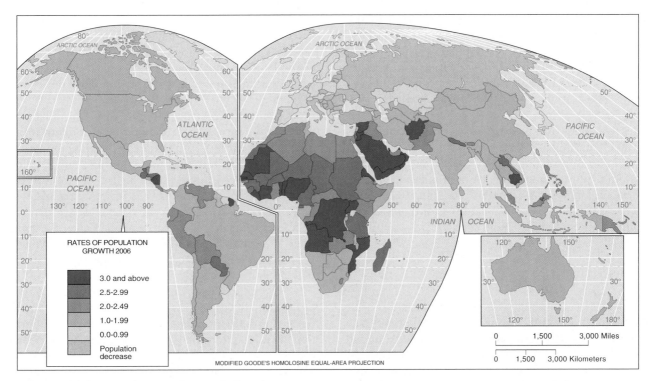

FIGURE 5-7 Rates of annual population increase. Most countries with the highest rates of population increase are in Africa and the Near East. (Data courtesy U.S. Census Bureau)

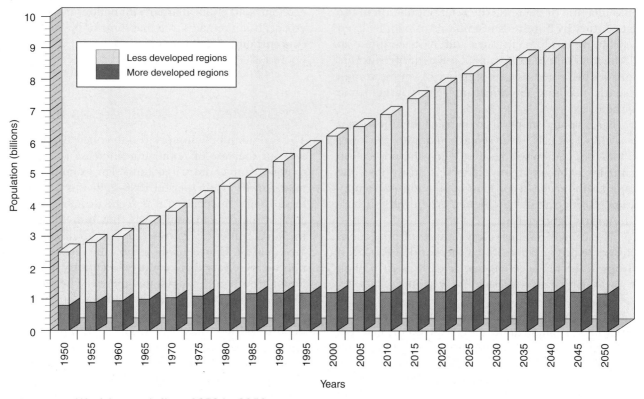

FIGURE 5-8 World population, 1950 to 2050. By one projection, the population of today's developed countries is expected to hold steady, while the populations of today's developing countries will continue to increase. (Data courtesy of the United Nations)

compounding at only 1 percent annually (as in the United States) takes 70 years to double.

Overall rates of population increase vary significantly between rich and poor countries. Ninety-five percent of the population increase projected to the year 2050 will be in countries that today are poor, those least able to support a larger population (Figure 5-8). The share of the world population living in the countries that are poor today will increase from 80 percent today to 90 percent by 2050. In these countries, economic growth is in a race with population growth. In order to achieve rising per capita (per person) incomes in these countries, the rate of economic growth must exceed the rate of population growth. If the rate of economic growth equals the rate of population growth, per capita incomes will remain static; but if economic growth rates fall below population growth rates, per capita incomes will fall.

The Age Structure of the Population

Demographic analysis reveals that in countries with high fertility rates, the populations are young. This is often represented by a graphic device called a **population pyramid** (Figure 5-9), which represents two aspects of a population: age and gender. Young people are indicated at the base of the pyramid, so a wide base indicates large numbers of young people. As the indi-

viduals of the same age (called a *cohort*) grow older together through time, the horizontal bar that represents that cohort moves toward the top of the pyramid. Numbers for males and females are set on separate sides of the pyramid to show their relative numbers in each cohort. Demographers can project the future structure of a population based on this knowledge. For example, if we know how many people are 20 years old today, we can estimate how many people will be 50 years old 30 years in the future.

Today the median age of the world's population is 27.8, but the various countries and regions demonstrate great differences in the ages of their populations (Figure 5-10). Where birth rates are high, median ages are generally lower than where birth rates are low. The median ages of the populations in many countries in Latin America, Africa, and the Near East are between 18 and 25 years. In contrast, the median ages of populations of wealthier countries are higher. In Western Europe, for example, the median age is 40.6; in the United States it is 36.5.

The **dependency ratio** of a country suggests what proportion of its people are in their most productive years. It is defined as the ratio of the combined population of children less than 15 years old and elderly people over 64 years old to the population of those between 15 and 64 years of age. The larger the percentage of dependents, the greater the financial burden on those who are working to support those too young or too old to work.

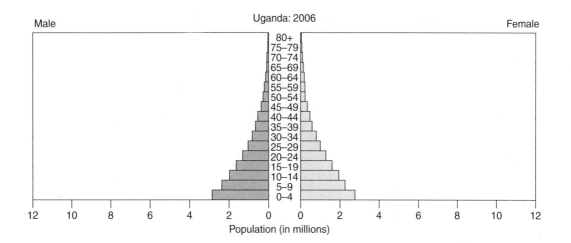

Uganda: 2006

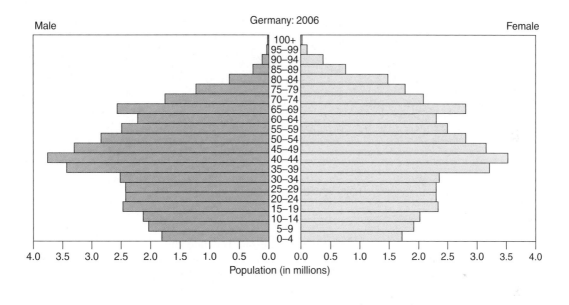

Germany: 2006

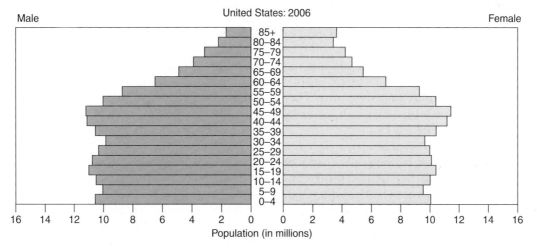

United States: 2006

FIGURE 5-9 Population pyramids. These three drawings are examples of a graphic device called a population pyramid, which shows the age and sex structure of a country's population. The pyramid for Uganda (a) shows a broad base—indicating many children—tapering to a narrow top of fewer older people. Birth rates are high, but life expectancies are limited. This pyramid is typical of those of poor countries. The "pyramid" for Germany, by contrast (b), shrinks at the base, indicating a birth rate that has been falling steadily since about 1970. Life expectancy is long. Germany's relative lack of elderly men is the continuing evidence of losses in World War II (1939–1945). The population pyramid of the United States (c) shows a bulge of people between 35 and 59 years of age—the "baby boom" of 1945–1965. Below this group the figure shrinks to reveal the "baby bust," but then, when "boomers" began to reach their childbearing years, the number of births rose slowly in a phenomenon called the "echo of the baby boom." (U.S. Census Bureau)

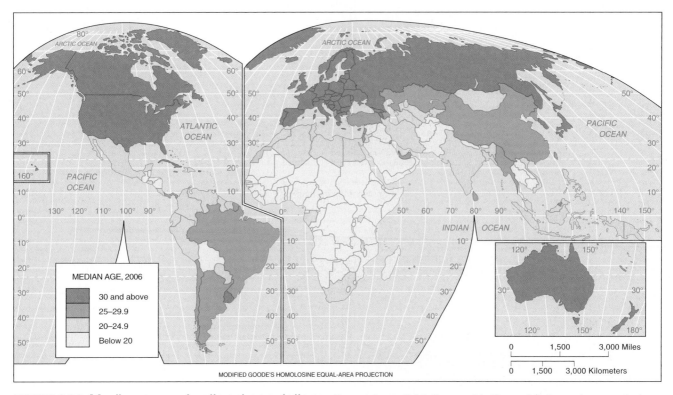

FIGURE 5-10 Median ages of national populations. Comparison of this figure with Figure 5-7 shows that populations are young where rates of population increase are high. (U.S. Census Bureau)

A dependency ratio will be high if a country has either a high percentage of young people (which is typical of the poor countries) or else a high percentage of elderly people (which is typical of the rich countries). If a country is poor, a highly dependent population compounds its financial difficulties. In Ethiopia, for example, which is a poor country, the dependency ratio is 0.87, and most of the dependents are youngsters. In the United States, by contrast, the ratio is only 0.49, and a growing share of the dependent population is made up of elderly people. Economists suggest that a country's age structure is a key to understanding its economic growth, which is usually highest when its dependency ratio is low.

In the countries with low median ages, a high percentage of the total population is drawing on national resources, but it has not yet reached its most productive working years. These countries are challenged to feed, clothe, and educate youngsters. As the youngsters reach maturity, the national economies must be able to provide jobs for them. If economic growth has not provided enough jobs, their frustration may break out in civil disorder. Economic development is urgently needed. We will see below how rising median ages challenge both rich and poor countries.

The Demographic Transition

Scholars who have studied population growth in the countries that today are rich have defined a model to describe these countries' historical experience. That model is called the **demographic transition**. The demographic transition model defines a pattern of growth that exhibits three distinct stages.

(1) In *Stage One,* both the crude birth rate and the crude death rate are high, so the population does not increase rapidly. All countries experienced this stage in the past, when the human population was a fragile number in constant danger of significant reduction, locally if not globally, by periodic epidemics.

(2) As incomes increase and medical science advances (which is usually, at first, simply an understanding of hygiene), crude death rates drop dramatically (Figure 5-11). The **infant mortality rate**, which is the number of infants per thousand who die before reaching 1 year of age, falls almost immediately (Figure 5-12). Another factor historically lowering death rates is the improvement in the quantity and quality of food that results from the Agricultural Revolution (discussed in Chapter 8). Crude birth rates, however, remain high. Thus, in *Stage Two* of the demographic transition, crude death rates are falling, but crude birth rates remain high, so the rate of natural increase is high.

Several theories have been offered to explain the persistence of high birth rates during the second stage of the demographic transition. In traditional societies, children may be economic assets. They provide more hands to help in the fields, for example. Also, people traditionally expect their children to look after them in their own old age. If infant mortality rates are high, parents have many offspring to ensure that some

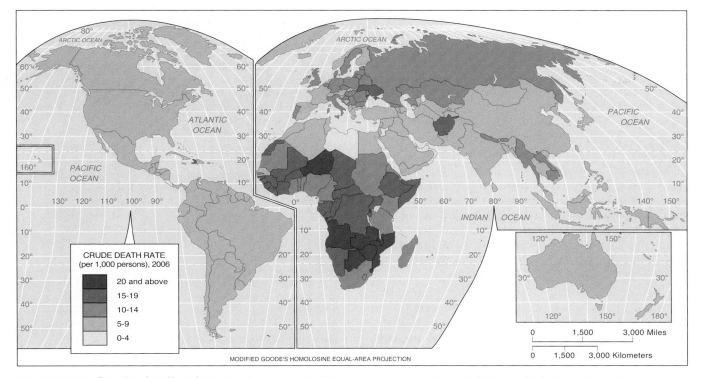

FIGURE 5-11 Crude death rates. Death rates are high in very poor countries (sub-Saharan Africa), moderately high in wealthy countries with aging populations (Eastern Europe), and low in countries with very young populations and at least modest health-care availability (Latin America). Crude death rates are falling almost everywhere, but environmental disasters, local famine, epidemics, or war can still substantially increase a country's crude death rate. (U.S. Census Bureau)

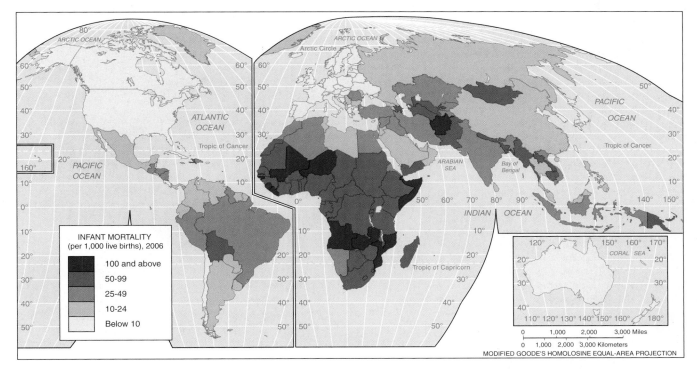

FIGURE 5-12 Infant mortality rates. This map shows the same contrast as Figure 5-11. Infant mortality rates are low in the rich countries, but they remain tragically high in the poor countries. (U.S. Census Bureau)

FIGURE 5-13 The cost of children. Traditional societies often think of children as economic assets; modern societies, however, calculate their costs. This fact was "news" in the United States eight years ago, and even people in poor countries are today coming to agree. (© 1998 *U.S. News and World Report*, L.P. Reprinted with permission. Photo © 1998 Bruce Plotkin Photography)

survive into adulthood. Not until one or two generations have passed will adults accept the reality of lower infant mortality rates and start having fewer children.

(3) Eventually, crude birth rates begin to fall. In Europe, this decline in childbearing occurred along with economic growth, urbanization, and rising standards of living and education. Children came to be seen as expenses rather than as economic assets (Figure 5-13), and many parents began to have fewer children so that they could provide a higher material standard of living for themselves and the children they did have. Thus, the rich countries of the world today demonstrate *Stage Three* of the demographic transition. Their crude death rates remain low, their crude birth rates are also low, and TFRs are at or below the replacement level.

Many rich countries' populations may be expected to decrease without substantial immigration; some are doing so already. Italy, for example, has a TFR of only 1.28 and a median age of 42.2 (the world's highest). Its population is projected to fall 12 percent between 2000

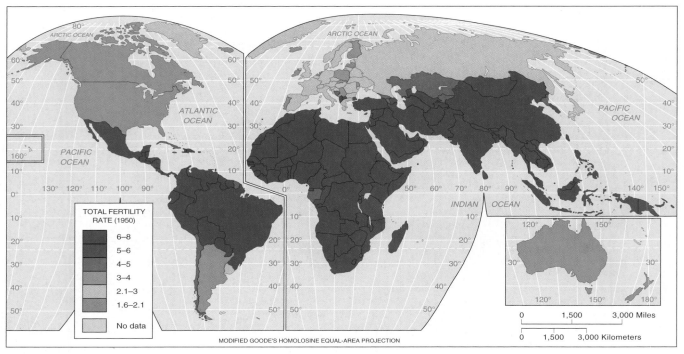

TOTAL FERTILITY RATE (1950)

- 6–8
- 5–6
- 4–5
- 3–4
- 2.1–3
- 1.6–2.1
- No data

MODIFIED GOODE'S HOMOLOSINE EQUAL-AREA PROJECTION

(a)

FIGURE 5-14 Falling total fertility rates. The "population explosion" that many observers feared in the 1950s and 1960s (a) is fizzling. Total fertility rates remain high in many poor areas, but today

and 2050, from 57 million to 50 million. Reasons cited by Italians for having fewer children typify the social changes in the rich countries: Women are pursuing careers, day-care services are limited, urban housing is short, contraceptives are increasingly available, abortion has been legalized, and people choose to enjoy material goods rather than to bear and raise children.

As this demographic transition model has been described in the rich countries, the transition has taken place over centuries. In the period from 1950 to 1980, scholars studying world population believed that the demographic transition model would eventually apply to all countries. They noted that birth rates and TFRs were still high in the poor countries, and they described these countries as being in the second stage of the demographic transition. Scholars concluded that when these countries achieved economic development, they too would enter the third stage. Therefore, the 1974 World Population Conference in Bucharest, Romania, concluded, "Economic development is the best contraceptive."

Several characteristics of the demographic transition model still do apply in today's poor countries. Crude birth rates and TFRs are falling, but they remain relatively high (Figure 5-14). The average TFR for the countries of sub-Saharan Africa during the period 2000–2005, for example, was 5.4. In many poor countries, it remains economically rational for the poor to have more children—for all the same reasons as it was a good idea in the past for the populations of today's rich countries

(Figure 5-15). It can even be argued that some African countries face labor shortages in agriculture, and because they lack capital for agricultural machinery, increases in rural labor forces could increase food output.

Furthermore, in even the world's poorest countries, infant mortality rates are dropping because antibiotics and immunization today help many babies to survive who would have died in past years (Figure 5-16). Modern medicine also keeps people alive longer, and the combined effect of lowering infant mortalities and people living longer expands **life expectancy,** which is the average number of years that a newborn baby within a given population can expect to live (Figure 5-17). Today life expectancies are lengthening fast in most of the world. The combined effect of lowering infant mortalities and lengthening life expectancies is to increase total populations.

Therefore, some circumstances in today's poor countries are comparable to the second stage of the demographic transition as it was experienced in the countries that are rich today.

Is the Demographic Transition Model Still Relevant Today?

Statistical trends recognized through the 1990s, however, triggered doubts about whether the demographic transition model really does apply to today's poor countries.

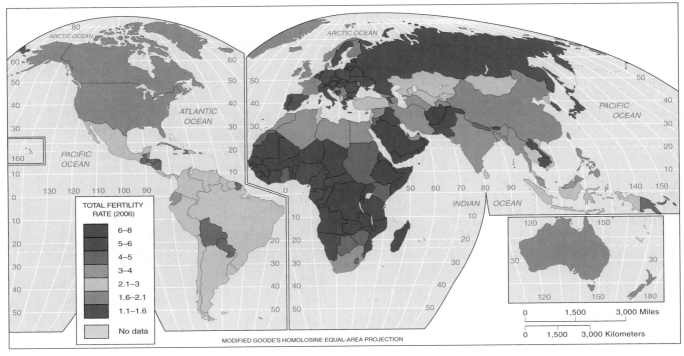

(b)

FIGURE 5-14 (*continued*) (b) women virtually everywhere are bearing fewer children. (U.S. Census Bureau)

FIGURE 5-15 Bangladeshi children at work. These children are heating and mixing rubber in a barrel at a balloon factory in Bangladesh. Children can be an economic asset if they are put to work in fields or factories. Chapter 12 will explain how international treaties are trying to stop the abuse of child labor by prohibiting products of their labor in international trade. (Pavel Rahman/AP Wide World Photos)

Already by the time of the opening of the third U.N. Conference on Population and Development, in Cairo, Egypt, in 1994, experts identified two trends occurring in the poor countries:

(1) TFRs have been dropping much faster in poor countries than they did in the countries that today are rich. From 1975 to 2006, the world TFR fell from 4.5 to 2.6. In east Asia, the decline was from 5.1 to 2.6; in Africa, the decline was from 6.6 to 5.1; and in Latin America, the decline was from 5 to 2.7.

(2) Falling TFRs are not necessarily associated with improving incomes and standards of living. Rates are falling in countries where the economy is stagnating or even declining, as in Bangladesh, Syria, and Turkey. Per capita incomes in many African economies have fallen, but TFRs have fallen, too. These observations demolish the model's assumption that economic growth is necessary to trigger falling TFRs.

These statistical trends have convinced many scholars that today's decreases in TFRs in the poor countries are fundamentally different from the fall of fertility rates that had occurred in today's rich countries. Therefore, new causes must be found to explain them. Theories of explanation focus on three new factors: the

growing role of family-planning programs, new contraceptive technologies, and the power of mass media.

(1) Family-planning programs During the demographic transition in today's rich countries, the concept of family planning was not quickly accepted. Early leaders in the population-control movement in Europe and the United States, such as Margaret Sanger (1883–1966), were even arrested for immorality or creating public disturbances. Today, by contrast, lowering the crude birth rate has become the focus of most countries' efforts to reduce overall population increases. Over 90 percent of the people of developing countries live under governments committed to reducing the rate of population growth (Figure 5-18). Today, national public health services seek out clients and have removed or lowered many of the barriers to the availability of contraception.

Empowering women is key to lowering fertility rates. The 1994 Cairo Conference, the 1995 Fourth World Conference on Women, and follow-up conferences all emphasized that key factors in successful family planning include universal access to education and health care, and relative equality in status, employment opportunities, and full political rights for women. The 1995 conference declared that "women have the right to decide freely and responsibly on matters related to their sexuality" and must be able to do so "without coercion, discrimination and violence."

(2) Contraceptive technology The demographic transition occurred in Europe without modern contraceptive methods. Today, however, technology has provided new means of birth control. The United Nations

FIGURE 5-16 Lengthening life expectancies. Chairman and founder of Microsoft Corporation, Bill Gates, center, and his wife, Melinda Gates, seated at left, receive flowers from patients at a government hospital in India. The Gates Foundation has committed huge sums to combat TB and HIV around the world. (AP Photo/Press Trust of India)

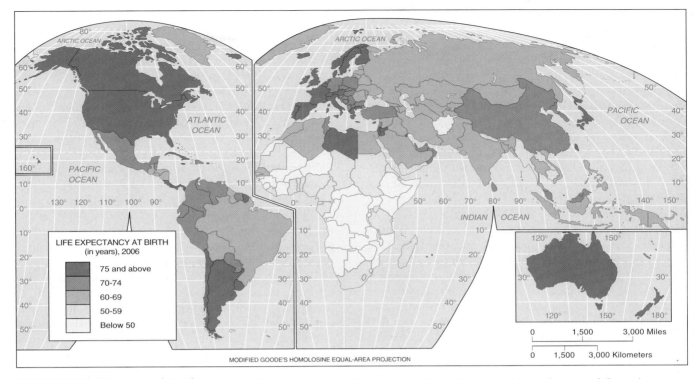

FIGURE 5-17 Life expectancies. People in North America, Europe, and the richest countries in Japan and Oceania can expect to live much longer lives than people in the poor countries. Life expectancies are lengthening everywhere, but much of the human population still cannot be expected to reach what we would call "old age." Compare this map with Figures 5-11 and 5-12. Countries with high death rates have low life expectancies. (U.S. Census Bureau)

FIGURE 5-18 Family planning. This Indonesian government billboard suggests that smaller families can be healthier, happier, and better off. It is part of the government's family-planning initiative. (Courtesy of U.S. Agency for International Development)

has replaced the slogan that it proclaimed in Bucharest in 1974 ("Economic development is the best contraceptive") with a new slogan: "A contraceptive is the best contraceptive."

Sterilization is the most widespread form of birth control in the poor countries. In Brazil, for example, an estimated 50 percent of all married women of reproductive age have been sterilized. New drugs and devices range from intrauterine devices to pills and injections, so that by 2004, 40 percent of women aged 15–49 in low-income countries, 76 percent of women in middle-income countries, and 64 percent of women in high-income countries were practicing contraception. These techniques reduce reliance on abortion, but about 60 million abortions are still performed annually, and improperly performed abortions still kill about 500,000 women each year. Efforts to develop a male contraceptive pill continue.

(3) The role of the mass media During the demographic transition in today's rich countries, information about contraceptive methods diffused slowly because educational levels were low and mass communication was limited. Today, however, communication and the mass media in developing countries have accelerated the diffusion of ideas about family planning.

Many governments, including those of Ghana, Nigeria, Gambia, Zimbabwe, and Kenya, broadcast family-planning messages on both radio and television.

Furthermore, public health officials have noted that even commercial television programs have a tremendous impact in lowering fertility rates by detaching sexuality from childbearing. Popular television programs transmit images, attitudes, and values of a modern, urban, middle-class existence in which families are small, affluent, and consumer-oriented. Popular television has been credited with lowering fertility rates in India, Pakistan, the Philippines, Kenya, Brazil, and China.

Obstacles to population control   Despite technological advances, family-planning programs face obstacles. One issue is that the manufacture and distribution of contraceptive devices and instructions for their use is expensive. Two-thirds of the users of condoms, diaphragms, and contraceptive sponges live in the industrialized world, and attitudes about family planning in the rich countries determine whether these countries offer assistance to the poor countries that may want it.

Religion can block birth-control programs. The Roman Catholic Church opposes all forms of birth control, and some other religions also preach against specific forms of birth control. Many religious groups adamantly oppose abortion.

Another obstacle to birth control is the low status of women. In numerous societies women still lack political and economic rights, have limited access to education, and exercise little control over their own lives. A 2006 ruling of the Supreme Court of Colombia gave many women hope: The Court ruled that Colombia's total ban on abortion violated international treaties ensuring rights to life and health. The Court insisted that abortions must be allowed when the mother's life is in danger, when the fetus is expected to die, or in cases of rape or incest. To ban all abortions is to value the fetus ahead of the life of the mother. Other governments may follow this legal precedent.

In areas where male children are preferred to females, birth rates remain high for several reasons. If a couple's first child is female, they will continue having children until a male child is born. In these traditional societies, a girl is often regarded as just another mouth to feed and as a temporary family member who will eventually leave to serve her husband's relatives. Also, in many societies a woman's parents must provide her with a dowry, which is a substantial financial payment at the time of her marriage. A son, on the other hand, means more muscle for the farm work and someone to care for aged parents. In several traditional Asian and African religions, only sons can burn the necessary ritual offerings to ancestors and only sons carry on the family name. As women win the right to work outside the home, as is happening in some societies, the financial preference for male children is reduced.

Throughout the world, fertility rates are lower in urban areas than in rural areas. Moreover, the larger the city, the more likely women are to use contraception. Reasons for this lower rate may be that family-planning services may be easier to reach; children cost more to raise in big cities; and space is at a premium. Today half of the world's people live in cities, and that percentage is increasing (see Chapter 10). This development will probably continue to lower TFRs.

Birth-control programs: Some examples   The most dramatic and significant change in fertility has been in China, where the annual rate of natural population increase dropped from 2.9 percent in the early 1960s to just 0.6 percent in 2006. China is home to about 20 percent of the human population, so that decline alone accounts for much of the change in world trends. In 1979 China launched a one-family/one-child policy, offering incentives to reduce childbearing, such as financial rewards and special privileges for small families, as well as penalties for exceeding the targets. Unfortunately, this policy perpetuated traditional prejudice in favor of males. Some women may have been coerced into having abortions, and there were even reports of *female infanticide* (the killing of female children). China ended the one-family/one-child policy in 2002 and instead created a policy that requires those who want more children to pay a "social compensation fee." Chinese statistics suggest that the former policy never worked, anyway: In 2000, 80 percent of Chinese children under 14 lived in families with siblings. China has managed to lower its TFR to only 1.7, but nearly a quarter of the country's people are under 15 years of age, so total births will probably remain high for many years.

India's population grew from 342 million at the country's independence in 1947 to 1.1 billion in 2006. If its growth rate continues at 1.4 percent per year, it will probably overtake China as the world's most populous country before 2050. Today the government sponsors family-planning programs, and radio and television emphasize that small families are healthier and happier. A new social security program assures people that they will be taken care of by the government in their old age. Abortions are legal, but sterilization accounts for about 90 percent of India's family-planning program. The TFR fell from 6 in 1951 to 2.73 in 2006. Programs in India's individual states also discourage large families. In Rajasthan, for example, village council members lose their seats if they have more than three children.

The Mexican population quintupled from 19.6 million in 1940 to 107 million in 2006, but today government family-planning clinics offer free contraception and sterilization, and a government slogan insists that "The small family lives better." The TFR fell from 7 in 1965 to 2.4 today, but the total population will nevertheless continue to grow by 1 million per year for at least another 20 years.

South Korea began promoting family planning in the 1960s, fearing that overpopulation would impede its economic growth. A slogan at the time warned South Koreans, who averaged six children per family, that they would become "Beggars without family planning." Today, when the TFR is 1.19, the government still insists that "Even two are a lot."

Changes in World Death Rates

A decrease in the crude birth rate is not the only event that can reduce the rate of human population increase. An increase in the crude death rate has the same effect. Natural disasters and wars claim tens of thousands of lives around the world each year. Death rates have risen in Russia since about 1990; this has been attributed to the physical and mental stresses of chaotic economic and political circumstances since the fall of the communist government and the dissolution of the U.S.S.R. (see Chapter 13). Russian income has fallen, use of tobacco and misuse of alcohol have spread, infectious diseases have multiplied, and life expectancy has fallen. At the same time, fertility rates have fallen and emigration has risen, so in the early twenty-first century, Russia's population was decreasing by about 1 million per year.

Infectious diseases The study of the incidence, distribution, and control of disease is called **epidemiology.** As countries develop, the leading causes of death shift from infectious and parasitic diseases, which are caused by disease-causing organisms called **pathogens** entering and multiplying within the body, to degenerative diseases, which result from deterioration or degeneration of the body. This change in the leading causes of death is called the **epidemiological transition.**

As recently as the 1970s, scientists tended to see infectious diseases as a series of problems to conquer, ticking off victories like notches on a gun. Smallpox was the greatest cause of death by infection in human history, but all but a few smallpox viruses, which had been kept for scientific study, were supposedly destroyed by 1995. Some countries, however, kept smallpox viruses and other pathogens for purposes of war. The countries that signed the Biological Weapons Convention of 1972 pledged to destroy their existing stocks of biological weapons, but the United States, Russia, North Korea, and possibly others held back stocks of weapons-quality pathogens.

Shortly after 9/11, spores of inhalation anthrax, a rare and particularly virulent disease bacterium, were sent through the U.S. mails. Most of the recipients and sufferers were working in offices of political or media power, so officials assumed that the United States was facing the spread of a disease deliberately introduced, that is, an attack of biological warfare. At the end of 2006, however, little progress has been made in solving these cases. Anthrax is not easy to spread, although it could have disastrous health results if a method could

be achieved of converting it into an aerosol. It is not believed that anyone has found a way to do this yet.

New drug-resistant strains of several pathogens are appearing, and scientists are challenged to develop medicines to combat them. Some pathogens evolve and mutate naturally, while others develop drug resistance when patients fail to complete their full prescriptions of antibiotic medicines. The use of antibiotics in animal feed and as sprays on fruits and vegetables during food processing also increases opportunities for organisms to evolve and develop resistance.

About 70 percent of all antibiotics produced in the United States are fed to healthy livestock to promote growth. This allows the emergence of drug-resistant bacteria, which are today widespread in commercial meats and poultry and can already be found in consumers' intestines. The drug Baytril may have given us a warning: It was once used in large-scale poultry production, but when its use in chickens reduced the effectiveness of a similar drug made for humans, Cipro, against infections, the United States in 2005 banned the use of Baytril for poultry. Only European governments, however, absolutely ban the non-therapeutic use of antibiotics in animals.

Many disease-causing pathogens are passed back and forth between humans and other animals with which we live in close proximity. Many influenza strains, for example, originate in Asia because great numbers of ducks and pigs—common animal hosts for viruses—live there in close proximity to human beings. The Great Flu Pandemic that swept the world in 1918–1919, killing as many as 40 million people, originated in China, as did the pandemics of Asian flu in 1957 and the Hong Kong flu in 1968.

In 2002 a new viral infection, called SARS (severe acute respiratory syndrome), appeared in southern China, but the Chinese government withheld information about the disease to avoid embarrassment. This action abetted the disease's global spread. Only when the Chinese government came to realize that the global panic was damaging Chinese credibility—and thus, economic development, by paralyzing business travel and tourism—did the government take emergency measures. By then, however, the disease had spread around the world, and thousands were suffering. The epidemic subsided, but the discovery of the SARS virus in other animal species suggests how the virus may have been transmitted to humans. We may never eradicate the virus if it is harbored in wild animals.

In 1996 a new lethal avian virus, H5N1, appeared in poultry in China. It spread among poultry and then to humans who lived with or handled poultry. In 1997 the virus appeared in Hong Kong, and authorities ordered the destruction of all domestic poultry there. During 2003–2004, human cases multiplied throughout Southeast Asia, and through 2005 and 2006 the flu spread across Asia into Africa and Europe, killing people and devastating poultry stocks. The first identified fatal human-to-human transmission of the virus, which may suggest a mutation, was to an Indonesian man in June 2006. Some scientists fear that the continuing close contact between humans and livestock in Asia and the global spread of poultry factory farming (see Chapter 8) could fuel a pandemic outbreak.

The single leading communicable cause of death in the world today is the human immunodeficiency virus (HIV), which causes acquired immune-deficiency syndrome (AIDS). AIDS destroys the body's ability to fight infections, and it was responsible for 2.8 million deaths in 2005, which is almost 5 percent of all deaths in that year. It is believed that the AIDS virus emerged in Africa in the 1950s; by 2006 an estimated 38.6 million people were living with HIV—the vast majority of them unaware of what they had (Figure 5-19). Social and political leaders in some countries have denied the prevalence of the disease, thus allowing it to spread. India is probably the country with the greatest number of infected people (about 5.1 million), but the disease is multiplying in east Europe, Russia, and Central Asia. Two-thirds of all people infected are in sub-Saharan Africa. In several sub-Saharan African countries, AIDS-related illnesses are the leading cause of adult death, and AIDS has shortened life expectancy by 20 or more years. AIDS causes massive human suffering and also has a pernicious economic impact. It weakens and kills adults in their prime years, thus destroying national economies. New techniques of treating the disease

offer promise, but their success is not yet proven. The high cost of these treatments has been lowered through international cooperation among pharmaceutical companies and national governments, but many countries lack the health systems necessary to distribute these drugs effectively. In 2006, U.N. Secretary General Kofi Annan noted, "The epidemic continues to outpace us. There are more new infections than ever before, more deaths than ever before, more women and girls infected than ever before."

The second leading communicable cause of death is diarrheal diseases, responsible for about 4 percent of all deaths. The most important of these is cholera. New strains of cholera appear regularly, and they are spread through contaminated water and poor sanitation. Cholera periodically surges out of Asia.

Tuberculosis is the third leading communicable cause of death, causing about 3 percent of the total. Approximately one-third of the human population carries TB bacilli in their bodies, but in the vast majority of cases the disease is latent and cannot be transmitted. New strains of drug-resistant tuberculosis, however, are multiplying; the war against this disease continues.

Malaria is responsible for about 2 percent of human deaths. The mosquitoes that carry the disease live in a broad variety of climates, so malaria occurs widely, but about 80 percent of malaria-caused deaths occur in sub-Saharan Africa. Drugs are normally used to treat malaria, but drug-resistant strains are appearing

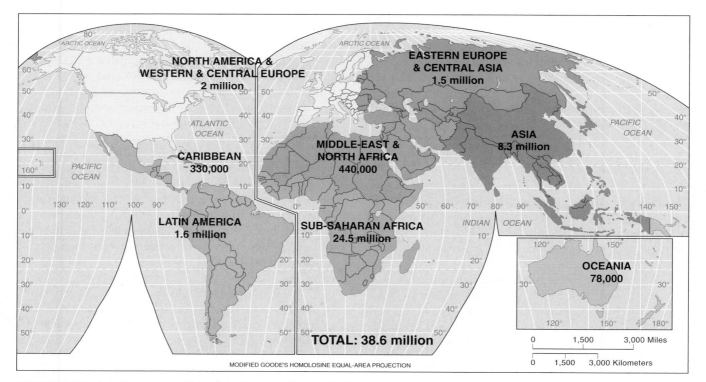

FIGURE 5-19 Adults and children estimated to be living with HIV/AIDS in mid-2006. Today scientists think that the human immunodeficiency virus (HIV) evolved in Africa and spread from other primates to humans there. It continues to spread around the world, but Africa is particularly afflicted. (The World Health Organization)

in Southeast Asia and spreading to Africa. In addition, mosquitoes are developing resistance to insecticides. Thus, health officials fear that the appearance of "supermosquitoes" armed with drug-resistant "supermalaria" could increase the number of deaths from malaria.

Threatening new viruses continue to emerge. These include not only HIV, SARS, and H5N1, but the Ebola virus (first identified in the Sudan in 1976), the Sabiá virus (identified in Brazil in 1990), the Hanta virus (identified in the U.S. Southwest in 1993), and the Nipah virus (first identified in Malaysia in 1998). These are all fatal to humans, but, luckily, most are difficult to transmit. The Ebola virus, for example, can be transmitted only through body secretions, and it kills its host so fast that there is little time for transmission to new victims.

Pathogens may spread through new medical practices such as blood transfusions and organ transplants, through sexual practices, or through needle sharing among drug addicts. Also, human alteration of the environment can create breeding grounds for new viruses and increase the number of pathways viruses can take to new populations. These include deforestation and urbanization, which bring virus-carrying rodents and people together; international trade, which unwittingly transports virus-bearing insects from country to country; and farming methods that allow viruses from birds and mammals to mingle. Global warming is increasing the portion of Earth hospitable to the mosquitoes that carry malaria; they are appearing at previously cooler higher latitudes and higher elevations. Global warming may redistribute other pathogens and diseases in ways

Critical THINKING

Americans' Health

The United States leads the world in medical research, as well as in total spending on health care, in spending per capita (about twice that of other rich countries), and in spending as a percentage of total national income (almost 15 percent). Americans' life expectancy, however, ranks below other rich countries (ranking 28th among nations in 2003), and America's infant mortality rate is higher (about the same as Malaysia, with one-ninth of America's per capita income). America's population is regularly revealed to be less healthy than those of other countries that spend substantially less on health care. How can this be explained?

Many factors combine to explain this paradox: Americans' behavioral traits (lack of exercise, obesity, smoking); the high percentage of expenditure that goes to the administration of America's mixed private-and-public health care system (31 percent of total expenditures compared to 17 percent of the cost of Canada's universal coverage); higher doctor's incomes; higher pharmeceutical company profits; great amounts of money being spent on individuals in their very last years—or even weeks—of life; plus high rates of automobile fatalities and of homicide.

Furthermore, high levels of spending reflect the country's advanced medical technology and treatment, but social inequalities, plus inequalities in health financing, limit the reach of medical advances. The American system preserves choice for those who can afford it, but the United States is the only wealthy country with no universal health care insurance system. More than half of

the population is covered through their employers, but many employers are finding the costs unsustainable, and they are dropping coverage. Almost all of the elderly enjoy Medicare, but about 45 million non-elderly Americans lack health insurance. Thirty-six percent of families living below the poverty line, 34 percent of Hispanic Americans, and 21 percent of African Americans are uninsured. Over 40 percent of the uninsured do not have a regular place to receive medical treatment, and more than one-third testify that they or a loved one went without needed medical care or medicines in the last year because of their inability to pay. Being born into an uninsured household increases the chances of infant mortality by 50 percent, and the Institute of Medicine estimates that at least 18,000 Americans die prematurely each year solely because they lack health insurance. Eliminating the gap in health care between African Americans and white Americans would save nearly 85,000 lives per year, compared to the estimated 20,000 saved each year by technological improvements in medicine.

Questions

1. Examine the numbers listed in Appendix III. How would you rate American health care among world nations?
2. How could Americans extend system coverage while maintaining choice?
3. Which large American corporations extend health care benefits to their employees and retirees? Which do not? Which of these face greatest international competition?

we cannot foresee. Global travel also triggers international outbreaks of infectious diseases. The number of transfers of pathogens from animals to humans seems to be increasing: AIDS, SARS, H5N1, Ebola, West Nile virus, Lyme disease, and still others have won headlines in recent years. Epidemiologists are not certain why transfers are increasing.

Human environmental alteration may, on the other hand, eliminate a pathogen's habitat, eradicate the pathogen, and prevent future epidemics. For example, the completion of Egypt's Aswan Dam in 1971 destroyed the floodwater habitat of the *Aedes aegypti* mosquitoes, carriers of Rift Valley fever virus. By 1980, Rift Valley fever had virtually disappeared from Egypt.

Degenerative diseases Most degenerative diseases are difficult to treat because treatment requires changes in the behavior of the sufferers, who are often adults. Degenerative diseases are leading causes of death in rich countries, but they are increasing in several developing countries, too. The World Health Organization (WHO—a United Nations Agency) estimated in 2006 that Earth's number of overweight people (1 billion) exceeded the number of undernourished people (about 800 million). This is a turning point in the history of humankind. Some 300 million people were obese, that is, in serious medical danger, and that number was rising fast. In the rich countries, causes include the switch to restaurant and take-out meals, which usually have higher calories than home-cooked meals, plus increasingly sedentary lives lacking exercise. In the developing countries, increased dietary fat—largely attributable to the spread of cheap, high-quality cooking oil—has helped to raise daily caloric intake by an average of 400 calories per capita since 1980. In these countries, too, urbanization and sedentary lives take their toll. China alone counted 65 million obese people by 2006. Overweight and obesity may trigger Type 2 diabetes, from which some 30 million people in east and south Asia alone were already suffering by 2003, plus heart diseases. From 1986 to 2006, the estimated number of diabetics on Earth swelled from 30 million to 230 million. Type 1 diabetes, which arises almost entirely from genetic factors, accounts for only 5–10 percent of cases. In many countries, including the United States, life expectancies for the next generation may actually shorten.

Tobacco products are predicted to account for an increase of 50 percent in global cases of cancer by 2025. Tobacco now accounts for about 5 million annual deaths worldwide through lung cancer and many other debilitating diseases, and more than one-half of the estimated 1.25 billion men and women who currently use tobacco are projected to die from the habit. WHO projects that tobacco will account for over 10 million deaths a year—10 percent of all deaths—by 2020 if preventative steps are not taken now. A Convention on Tobacco Control regulating crops, advertising, trade, and other aspects of marketing was signed by 168 nations and came into force in 2005, but the United States refuses to ratify it, defending the "free speech" rights of the tobacco industry.

Is Earth Overpopulated?

Many people fear that Earth may become overpopulated, or that it is already. People disagree, however, on the meaning of the word *overpopulation*. Some people complain that their neighborhood is overcrowded or that new construction and crowding are everywhere and lowering the quality of life. Studies of rats reveal that crowded rats suffer mental and physical breakdown. Does this mirror human behavior? We do not know.

Discussions of world population growth inevitably invoke the ideas of Thomas Robert Malthus (1766–1834), who asked whether humankind would always be able to feed itself. Malthus's statement of the relationship between population and food supply still demands our attention. Malthus was a professor of political economy and also a clergyman, and in 1798 he published his *Essay on Population*. His essay, stated most simply, argued that food production increases arithmetically: 1-2-3-4-5 . . . units of wheat. Population, however, increases geometrically: 2-4-8-16-32 . . . people. Therefore, the amount of food available per person must decrease as the population increases.

The human population can be kept in balance with food supplies only through checks on population increase. Malthus defined two types: positive checks and preventive checks. *Positive checks* refer to premature deaths of all types, such as those caused by war, famine, and disease. *Preventive checks*, by contrast, are human actions designed to limit births, such as a decision by a young couple to delay marriage and childbearing. According to Malthus, couples should have a sense of responsibility for the economic welfare of themselves and their children. "There are perhaps few actions that tend so directly to diminish the general happiness as to marry without the means of supporting children," he thundered. "He who commits this act, therefore, clearly offends against the will of God."

Malthus did not believe that preventive checks could control population growth. He pointed to high birth rates among the poor as evidence. Therefore, he predicted that the future of humankind would consist of endless cycles of war, pestilence, and famine. This is the pessimistic conclusion of the **Malthusian theory**. People who share Malthus's pessimism are often today called *neo-Malthusians* (new Malthusians) or *Catastrophians*.

Malthus was replying to the French philosopher Jean Antoine Condorcet (1743–1794). Condorcet had argued that increases in productivity and improvements in technical ingenuity would keep pace with the needs of Earth's rising population, so per capita material welfare could actually rise. Condorcet also suggested that, given education and prosperity, most people would

---

voluntarily limit their family size. People who share Condorcet's optimism are often referred to as *Cornucopians*; a cornucopia is a horn of plenty.

So far, Condorcet's predictions have proven more accurate than those of Malthus. The human population has sextupled since Malthus's time, yet the widespread starvation that he feared has not happened. Many people are hungry, but this deprivation does not result from insufficient food on Earth. People are hungry because the food is not priced within their reach, nor is it distributed to all. The problems are political and economic, not technological. Will these political problems ever be solved? We do not know. Chapter 8 details how humankind has managed to multiply the food supply. Could mounting numbers of people ever place demands on food and materials so great that per capita supplies will necessarily fall? That question will be answered by future developments in technology, and we cannot possibly foresee what those can be.

OTHER SIGNIFICANT DEMOGRAPHIC PATTERNS

We have seen how variations in birth rates and death rates affect both overall world population growth and the growth rates in different parts of the world. Two other demographic statistics vary so significantly around the world that they demand geographical analysis. One is the variations in the sex ratios among national populations, and the other is the aging of the human population.

Sex Ratios in National Populations

About 105 male children are naturally born for every 100 females. This ratio occurs everywhere in the world except in countries where ultrasound, amniocentesis (medical procedures that provide information on the state and sex of a fetus developing in the womb), and abortion are available, and where there is a strong preference for male children. Female fetuses are aborted at a higher rate than male fetuses, and the demographic impact is dramatic. In South Korea, for example, where fetal testing to determine sex is common, the birth ratio is 108 to 100; in China it is 112 to 100; and in Taiwan it is 109 to 100. These imbalances in sex ratios will cause problems in forming new families 20 years from today. Taiwan and some states of India have limited or banned the use of ultrasound scanners and amniocentesis.

An alternative theory that may partly explain the high ratios of male births in many poor countries is the prevalence of Hepatitis B, which increases the chance of a male birth if the disease is carried by either the mother or father. Epidemiologists are investigating this theory.

Although it is natural for more males to be born than females, women tend to live longer than men when both receive similar medical care. Therefore, a population should naturally have an approximately equal number of males and females, or perhaps slightly more females. In the world in the year 2003, there were 101.3 males for each 100 females, but Figure 5-20 shows how sex ratios vary widely.

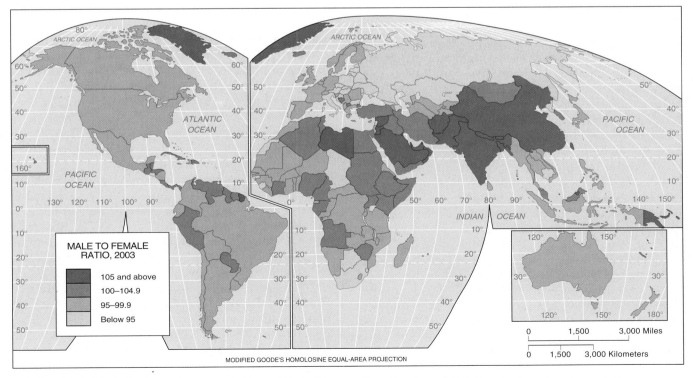

FIGURE 5-20 Male-to-female population ratios. The number of males per 100 females in national populations, as shown here, varies substantially around the world. What factors could explain these variations? (U.S. Census Bureau)

What determines national sex ratios?   The ratios of males to females are low in the rich countries of North America, in Western Europe, and in Japan. This observation suggests the hypothesis that sex ratios vary with national wealth. Wealth might be expected to reduce female mortality for two reasons: (1) Birth rates are generally lower in richer countries, thus reducing maternal mortality. (2) Improved medical care is generally available in richer countries during pregnancy and birth.

Sex ratios in the poor countries, however, do not show a consistent pattern. In some poor countries men outnumber women, but in others, in tropical Africa, for example, women outnumber men.

In fact, variations in sex ratios can be explained only by a combination of economic and cultural factors. In the countries of North America and Europe and in Japan, women may suffer many kinds of discrimination, but they are not generally discriminated against in access to medical care. Moreover, in these areas social differences tend to increase mortality among men. Men more often fall victim to violence, for example, and to workplace accidents. In most of Asia and Africa, however, women do not get the medical care, food, or social services that men get. As a result, fewer women survive than would be the case if they had equal care. The rise of death rates in the formerly communist countries of Eastern Europe afflicts mainly men, so male-to-female population ratios are falling in that region.

Great variations exist within each world region. In Asia, Japan has approximately the same male–female ratio as European countries, and in some countries of Southeast Asia women outnumber men. In south Asia, however, particularly India and Pakistan, the ratios of men to women are high. The high ratio of men to women in China may be attributable to the preference for boys.

Sharp contrasts exist even within individual countries. In the northern Indian states of Punjab and Haryana, for instance, which are among the country's richest, there are only 86 females per 100 males, whereas in the state of Kerala in southwestern India there are 103 females per 100 males, a ratio similar to that in Europe, North America, and Japan.

Women's positions are everywhere more favorable if they can earn an income outside the home. That reduces boys' advantage as potential supporters in the parents' old age, and females suffer less relative deprivation from birth. Later in life, working women can rely on their own resources.

Several of the countries with the highest ratio of males to females are Arab countries, where women's rights are not always guaranteed. Some demographers think this suggests that societies that limit women's rights do not count all women in the census. If females are actually undercounted in these countries, then the reported higher ratios of males may not be accurate. Certainly women are better off in countries that acknowledge female deprivation and seek to remedy it. Efforts to achieve this can be influenced by providing women with education and the right to vote.

The Aging Human Population

Through the twentieth century, improvements in health and hygiene triggered a rise in the median age of Earth's population for the first time in history. Humankind's median age increased from 23.5 years in 1950 to 27.8 in 2006, and it is projected to reach 37.4 by 2050. The percentage of people age 60 or older is expected to rise from the current 10 percent to 20 percent by 2050.

In the rich countries, around 20 percent of the populations are already age 60 or older; that figure may rise to nearly 33 percent by 2050. The existence of aging populations challenges the possibility of sustaining economic growth and equitably distributing wealth among generations. Higher percentages of national budgets are devoted to senior citizen centers than to elementary schools. Funds for medical research are dedicated to degenerative diseases such as cancer, heart disease, and Alzheimer's, which means that enormous sums of money are buying relatively short advances in longevity. To support elderly populations, young workers may have to increase their contributions to pension funds, or else pension benefits may have to be reduced. Alternatively, workers may have to work more years before retiring. Most rich countries are already raising the age at which people may collect pensions.

A few countries with TFRs below the replacement rate, including France, Italy, Poland, Australia, Russia, and Singapore, are encouraging higher birth rates. Japan, for example, announced in 2006 that the national TFR had fallen to 1.29 and that the national population had already begun to decrease. Furthermore, the population is aging: 21 percent of the people were over age 65, and that figure will reach 30 percent by 2025. Only 13.6 percent were 15 or younger. Therefore, Japan is encouraging women to bear children. Government-funded day care is available, and employers offer extended child-care leave. Nevertheless, social rigidities limit childbearing. Social mores insist on a traditional view of marriage and parenthood (in Japan, fewer than 2 percent of births are to unmarried mothers, compared to 40 percent in Britain, 50 percent in Scandinavian countries, and 34 percent in the United States), and Japanese law even discriminates against children born out of wedlock.

The U.S. population aged over 64 is projected to rise from 36.3 million in 2004 to 86.7 million by 2050. Because of immigration (discussed later in this chapter) and America's higher TFR, which is almost unique among rich countries, America's median age is projected to rise only from 36.5 in 2006 to 38.1 by 2050.

Historically, every country that got old, got rich first, but today this is no longer true. Many of today's poor countries still exhibit high birth rates, challenging them to care for and educate children. Even though their birth rates remain high, the numbers are falling. Therefore, their median ages will soon rise fast. About 8 percent of the populations of today's poor countries are over 60 years of age, but that percentage will rise to 20 percent by 2050. Mexico's TFR, for example, has fallen from 6.9 in 1970 to 2 today, so its median age, 25 in 2006, is expected to rise to 42 by 2050. About two-thirds of Chinese population growth between 2005 and 2025 will occur in the over-65 category, a cohort likely to double in size to about 200 million people. By then, China's median age may be higher than America's. An aging China is already facing both strains on its pension system and shortages of young labor.

Aging populations will present tremendous challenges to the poor countries. National pension systems may be inadequate. In many poor countries the elderly are today cared for by their families but this situation will shift as the ratios of old to young rise. These countries will have to build up their national incomes and devise national pension systems. They have perhaps 50 years in which to do this.

Critical THINKING

How Will You Retire?

As the American population ages and the balance between workers and retirees shifts, you might start planning for your own retirement, even if it is still many years away. The three traditional sources of income for American retirees are pensions, personal savings, and social security.

Corporate pension plans multiplied through the twentieth century. Most were *defined-benefit plans* in which a worker received a set payment for life, based on a combination of salary and years of work. Clause 401 (k) of The Tax Reform Act of 1978, however, encouraged *defined-contribution plans*, in which employers and workers themselves contribute fixed sums to savings accounts that receive favorable tax consideration. Most American corporations have switched to defined-contribution plans.

Thus, America's private pensions—like its health insurance—were linked to work. Today, however, as employers face global competitive pressures, and as the number of retirees grows relative to the number of workers, corporations have begun to reduce both their pension liabilities and their workers' health care coverage. In 1962, for example, General Motors had 464,000 U.S. employees and was paying benefits to 40,000 retirees and their spouses. By 2005, the corporation had 141,000 workers and paid benefits to 453,000 retirees. The telecommunications manufacturer Nortel announced in 2006 that it would shift its North American workers from a defined-benefit to a defined-contribution pension plan and eliminate all post-retirement health care benefits. At the same time, the corporation cut its North American workforce and hired thousands of new workers in Turkey and Mexico.

Other problems have arisen because not all workers have enjoyed satisfactory financial returns on the assets in their defined-contribution accounts, and the Pension Protection Act of 2006 only partially addressed the fact that many corporate pension accounts have been substantially underfunded. About one-half of American workers in the private sector still have no employer-sponsored retirement plan of any kind.

Local government pensions are underfunded as well, to an amount estimated to be hundreds of billions of dollars. Most state constitutions, however, guarantee these pensions, although experts wonder where the money to pay the pensioners will be found.

Many Americans find it hard to accumulate individual savings in a culture so rich in attractive consumer goods. The U.S. personal savings rate has fallen from about 8 percent of household income in the early 1990s to 1 percent in 2006. Other industrialized nations average about 7–15 percent savings rates.

Social security payments were never intended to allow retirees to live comfortably, just to prevent starvation, and social security will be strained as "boomers" reach retirement (see Figure 5-9). Suggestions to lower benefits meet strong resistance, as did the Bush Administration's 2005 attempt to allow individuals to invest their savings in investments of their own choosing, but then to eliminate the guaranteed aspects of the plan. Therefore, the United States is raising social security taxes by raising the base amount of a person's income that is taxed, although most Americans already pay more in social security taxes than in income taxes. The age at which workers may retire and receive benefits is also steadily increasing. Law current in 2006 has the retirement age for social security and Medicare slowly rising to 67 (from the current 65) by the year 2027.

Questions

1. What will be the required minimum age for retirement to collect social security when you hope to retire?
2. Try to project what the U.S. population pyramid (Figure 5-9) will look like when you retire.
3. If you have a job, do you now pay more income tax or social security tax?

One partial solution for both the rich and poor countries is the international migration of pensioners to the poor countries. Tens of thousands of Americans have already retired to Mexico and Costa Rica, where they invest and spend income, while imposing few costs on the society.

MIGRATION

Human beings do not stay put; they never have. Wanderings and migrations of people have distributed and redistributed populations throughout history and even prehistory. Significant redistributions continue.

Geographers who analyze human movements divide the causes for those movements into **push factors** and **pull factors**. Push factors drive people away from wherever they are. Push factors include starvation and political and religious persecution. Pull factors attract people to new destinations. Pull factors include economic opportunity and the promise of religious and political liberty. Physical geography can also serve as a push or pull factor. For example, we can say that migrants from northern cold regions have found the warmth of the U.S. Sunbelt attractive and that environmental disruption in the African Sahel, by contrast, has been found intolerable by emigrants from that region. The environment itself never actually pushes or pulls; people decide to migrate after considering environmental factors.

People choose a particular destination because of what they think they will find there, not because of the reality of conditions at that place. Therefore, migrants can be surprised by the reality of what they find at their new homes. Sometimes migrants find success beyond their expectations, but in other cases they are disappointed (Figure 5-21).

Migration has not always been voluntary. Some migrations have been forced. Millions of people suffered the tragedy of migrating in slavery, regardless of push or pull considerations of their own. Others have fled persecution in their homelands, and millions of others, called *internally displaced persons* (*IDPs*), have been forced to flee from one region to another within their own homeland. For example, millions of citizens of the Congo have fled the civil war that recently raged in their country; many have fled abroad, but many more are IDPs.

Many migrants complete whatever documentation is required for them formally to settle in their host countries, but today great numbers of people are crossing borders without completing legal papers. These migrants are called **undocumented, irregular, or illegal immigrants**.

Movements of people can often be explained by studying flows of trade or by mapping the most convenient transport routes away from places people want to leave. People also follow information. Successful migrants write home, and then they may provide new arrivals with employment and financial assistance. In this way, linkages

The thriving City of Eden as it appeared on Paper   *The thriving City of Eden as it appeared in Fact*

FIGURE 5-21 Disappointed migrants. Charles Dickens's comic novel *Martin Chuzzlewit* (1844) recounts these gentlemen's dismay when they migrate to the American frontier and find that it is much more primitive than land developers had suggested. Many of Dickens's American readers took offense, but in fact such misrepresentations were common. Many other migrants to the United States, by contrast, have found success beyond their original hopes. (Courtesy of the Library of Congress)

called **migration chains** are forged. Information often travels between formerly imperial powers and their former colonies, as between France and several African countries or between the United States and the Philippines. In these cases, potential migrants absorb some of the culture of their destination even before they leave home. Information flows can be astonishingly place-specific. An estimated 80 percent of the Hispanics living in metropolitan Washington, D.C., are from the tiny Central American country of El Salvador—many of them undocumented. Nobody knows exactly how word first spread throughout El Salvador to trigger this migration.

Many migrants intend to stay in their new location only until they can save enough capital to return their homeland to a higher standard of living. These **sojourners** are usually men who are either unmarried or married men who have left their families in their home country. Many sojourners eventually decide to stay, at which time they send for their wives and families. The United States has always attracted great numbers of sojourners. In various years between 1890 and 1910 the number of Italians returning to Italy was as high as 75 percent of the number of Italians arriving in the United States. Even today many Mexicans, Caribbean peoples, and Africans shuttle back and forth between their homelands and the United States.

The ease of international travel and the number of sojourners today has allowed migrants from specific villages in poor countries to form urban neighborhoods in rich countries. For example, when in 1982 a terrible fire killed 24 people in a tenement building in Los Angeles, authorities were astonished to learn that several hundred residents of the building were all neighbors from one small village, El Salitre, in the Mexican state of Zacatecas. Committees of the sojourners in the rich country not only send money home, but they decide on what expenditures will be made in their hometown—whether to build a new school or a new water-treatment plant, for example.

Some migrations have numbered in the millions, but some small migrations have been disproportionately significant because the migrants have played key roles in transforming the government, language, economics, or social customs in their new homelands. As we shall see, the impact of the international migrations of Indians and Chinese, for example, or of Caribbean blacks into the United States, are not fully realized by their numbers alone.

In several cases the impact of international migrants in their new homes has harmed the welfare of the natives, or **indigenous peoples**, of that region. The United Nations estimates that today there are 300 million indigenous people in 70 countries, defined as "descendants of the original inhabitants of a land who were subjugated by another people coming after them." These include many Native American groups, the Bushmen of South Africa, and the Lumad of the Philippines. The United Nations has voted that they should be free from discrimination and that they have the right "to consider themselves different and to be respected as such."

Prehistoric Human Migrations

Human beings are one single species, and all its members can successfully interbreed. The Human Genome Project, completed in 2000, mapped all of the 3 billion chemical bits, or nucleotides, that make up human DNA (deoxyribonucleic acid). DNA is a component of genes, which are the hereditary units that determine characteristics of organisms. All humans seem to have descended from a single stock that lived in Africa about 150,000 to 200,000 years ago (Figure 5-22). The people who are closest to the root of our common DNA are the Vasikela Kung of the northwest Kalahari Desert region. Some members of our common stock stayed "home," whereas others migrated away to Asia (probably across the Arabian Peninsula) between 56,000 and 73,000 years ago and to Europe between 39,000 and 51,000 years ago. Scientists are not certain about the timing of the first peopling of the Americas. Both archaeological and DNA evidence suggest several different occurrences, some people may have traveled on foot across a land bridge that once linked what is now Siberia and Alaska, but perhaps others may have come by boat from Asia or Europe. The discovery in 2005 of human footprints dating from about 40,000 years ago in ancient volcanic ash near Puebla, Mexico, pushed back our ideas of when there was human habitation of America by about 25,000 years.

Races DNA indicates that humans as a species are 99.9 percent genetically identical. There are greater differences between any two frogs in a pond than between any two people (but frogs have been evolving for millennia longer). Nevertheless, anthropologists traditionally used secondary biological characteristics to divide humankind into **races**. The simplest criteria for these subdivisions have been external features such as eyefolds or skin color. The crudest classification system based on these criteria divided humankind into three races: Caucasoid ("white"), Mongoloid ("yellow"), and Negroid ("black").

A second classification system was based on the analysis of blood types. Classifications of people by blood type yielded a different set of categories ("races") from classification of people by skin color or other external features. Individuals crossed over categories, and that fact cast doubt on both systems.

When the Human Genome Project discovered only 0.1 percent of genetic difference among humans, some scientists discarded the idea of race as a biological fiction. Francis Collins, however, a leader of the project, emphasized that the "well-intentioned statements" about the biological insignificance of race may have left the wrong impression. "It is not strictly true" he wrote, "that race or ethnicity has no biological connection. It must be emphasized, however, that the connection is generally quite blurry." How many races exist, which groups of people are in each, and the utility of the concept of race remains ambiguous.

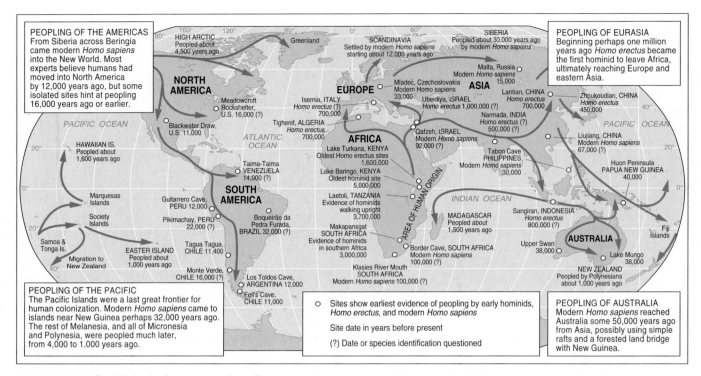

FIGURE 5-22 Prehistoric human migrations. This map illustrates the latest hypotheses about migrations of humans out of Africa to settle around Earth.

Critical THINKING

Will We Have Race-Based Medicine?

The discovery of tiny genetic differences among groups could lay the groundwork for new medical treatments and cures, help predict an individual's response to a certain drug, and allow for tailor-made medications with fewer side effects. One scientific study, for example, found variants of the PCSK9 gene (which affects cholesterol in the body) among Americans who described themselves as "Black," significantly different from the variants found in self-described "Whites."

Ignoring such differences altogether would be to the detriment of medical knowledge about the very people who might benefit. In the past, however, knowledge of such differences has often promoted racism. The eugenics movement of the late nineteenth century led to the doctrine of racial hygiene and the death camps of Nazi Germany. The discovery of differences today might lead to discrimination—perhaps from health insurance companies, requiring individuals with specific gene patterns to pay higher premiums.

In 2005, the U.S. Food and Drug Administration approved the world's first "racial" medicine, a heart-failure drug for African Americans known as BiDil. Both the National Association for the Advancement of Colored People and the Black Congressional Caucus argued for the approval of Bidil, despite the fact that the drug has never shown any health advantages for Whites. Similarly, one pharmaceutical corporation is developing a lung-cancer drug that has failed to help Caucasians but that seems to help people of Asian ancestry. More race-targeted drugs may follow.

Questions

1. How can we ensure that knowledge of minute biological differences among groups does not promote invidious discrimination?

2. Could we legislate a requirement that equal research be spent on specific diseases that afflict each "racial" group?

The concept has, however, often been used to buttress racist ideologies. **Racism** is a belief in the inherent superiority of one race over another and the linking of human ability, potential, and behavior to racial inheritance. Racism is wrong, immoral, and injurious.

The Migrations of Peoples Since 1500

For purposes of understanding contemporary geography, we will begin our study of human migrations 500 years ago, before the European voyages of exploration and conquest. In 1500, Earth's many peoples and cultures were relatively isolated from one another. The world we know is the product of their migrations and mingling. The most significant population transfers during these years have been migrations of Europeans to carve enclaves of settlement around the world, migrations of Blacks out of Africa, and migrations of other groups instigated by European expansion.

European migration to the Americas

The replacement of natives by newcomers was greatest in the Western Hemisphere (Figure 5-23). The Native American population fell victim to slavery, warfare, and most important, to European diseases, including smallpox, influenza, measles, and typhus.

We have only estimates, ranging from 8 million to 112 million, of the Native American populations at the time of the European conquests but the numbers of deaths were appalling. There may have been 25 million Native Americans in Mexico before the arrival of the Spaniards; by 1600 that number had declined to about 1.2 million. The population of the Inka Empire in South America plummeted from about 13 million in 1492 to 2 million by 1600. (You may be familiar with the spelling "Inca," which is Spanish, whereas the spelling "Inka" is Native American Quechua. For reasons discussed in Chapter 7, Inka is preferred today.) The estimated 5 million Native Americans in Brazil has shrunk to only 300,000 today. In what is today the United States and Canada, the Native American population fell from about 2 million in 1492 to a low of 530,000 in 1900.

North America received some 40 million Europeans before 1950. Today they and their descendants make up the majority of the population. The number of self-identified Native Americans in the United States tripled between 1960 and 2000 to 2.5 million, perhaps reflecting growing cultural pride, but they still constitute less than 1 percent of the national population. Canada's 1.3 million self-identified "First Nations" individuals constitute just over 4 percent of Canada's population.

Both the United States and Canada have mixed policies of both assimilating Native Americans and also segregating them (historically, sometimes against their will) on reservations, where native cultures may be preserved. In both countries Native Americans retain preferred hunting and fishing rights over large geographic areas in addition to their reservations, over which they exercise more substantial, but never complete, control. Canada has granted the Inuit people

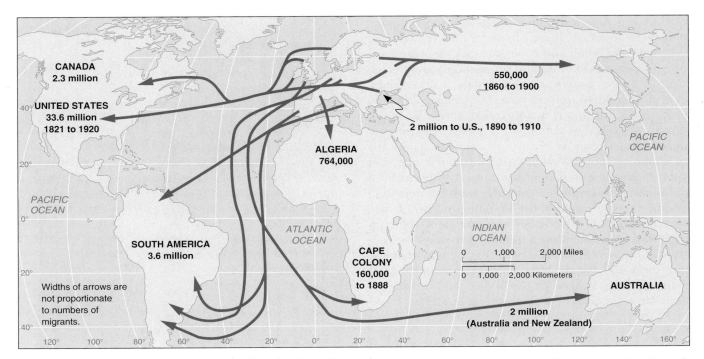

FIGURE 5-23 European migration in the nineteenth century. The nineteenth century saw the greatest migrations of Europeans. Europeans regarded the Western Hemisphere as a New World open to European settlement, and their massive migrations forever changed hemispheric demographics.

Critical THINKING

The East–West Exchange of Disease

When the Europeans first came to the Western Hemisphere, they brought diseases that were new to the native populations. The Native Americans had no immunity to these diseases, so they were decimated by them. Did the Europeans take new diseases back to Europe that decimated the European populations? No. Why not? Why was the direction of this terrible trade so one-way?

One possible reason for the virulence of Eastern Hemisphere diseases in the Western Hemisphere is that the peoples of the Eastern Hemisphere had, through history, domesticated many more animals than the peoples of the Americas had. Many disease-causing pathogens are passed back and forth between humans and other animals with which we live in close proximity,

and we and the other animal sufferers develop resistance to these pathogens. The peoples of the Americas had domesticated very few animals, so they had developed little resistance to any diseases. This helps explain why they fell victim to the new diseases arriving from the east. The only disease that, it seems, the Americans gave to the Europeans was syphilis, which a member of Columbus's crew took back to Europe.

Questions

1. Trace the origins and diffusions of some modern epidemics around the world. You might study, for example, the great Spanish influenza epidemic of 1918.
2. What factors of world trade, travel, and tourism would probably accelerate the diffusion of a new disease today, and what routes might it take?

(Eskimos) political domain over roughly one-fifth of the country, the eastern half of the former Northwest Territories, now the territory of Nunavut (see Figure 11-23). Continuing litigation of other extensive Native claims embroils national politics. In the United States, Native American lands represent 2.2 percent of the total land area.

About 4.5 million Europeans, mostly from the Mediterranean area, migrated to what came to be called Latin America, a region that developed a complex racial structure. Ibero-American (descended from inhabitants of Spain and Portugal) societies were composed as a pyramid of varying proportions, with a large base of Native Americans and Blacks; a lesser number of people of mixed race (a person of mixed white–black ancestry is usually called a *mulatto,* one of mixed black–Native American a *zambo,* and one of mixed white–Native American a *mestizo*); and a minority of Whites on top. Several Latin American colonies, and later the successor independent states, institutionalized racial status rankings, and discrimination has persisted, despite the fact noted above that all concepts of race are ambiguous. Native Americans lost their rights in their own homelands, even though they may have constituted the majority of the population. Only since the mid-1980s have some South American countries, including Peru, Venezuela, Colombia, and Brazil, set aside lands for Native Americans (Figure 5-24).

To varying degrees, racial animosities linger throughout Latin America. In Venezuela, for example, hostility lingers between the *pardos,* brown-skinned people of Amerindian or African ancestry, and the *mantuanos,* people of European appearance and pretensions. Hugo Chávez rode this distinction to political

power, winning election to the presidency in 1998 while openly referring to himself as "an Indian" against the "rotten white elites." After taking power, he suspended the Congress and Supreme Court. The *mantuano* minority launched a coup d'etat that ousted him in 2002, but he was eventually reinstated.

Native Americans still make up one-third to one half or more of the populations of Mexico, Guatemala,

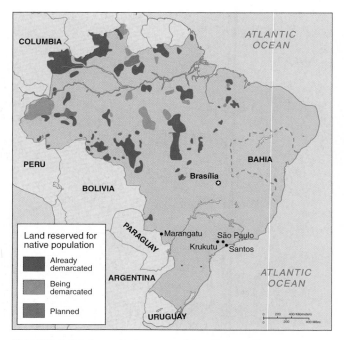

FIGURE 5-24 Lands reserved for the native population in Brazil. Brazil has only recently adopted the North American model of setting aside lands for indigenous peoples.

FIGURE 5-25 Sacsayhuaman. These Native Americans' ancestors built this fortress in Peru in the fifteenth century. Today the ruins are a tourist attraction. (Robert Frerck/Corbis/ Stock Market)

Ecuador, Bolivia, and Peru (Figure 5-25). In Mexico the Native Americans have played the most conspicuous political role, but Native Americans are waking to power elsewhere. When Alejandro Toledo won election as Peru's first Native American president in 2001, he staged an inauguration ceremony at the Inka city of Machu Picchu presided over by a native holy man. In 2005, Bolivians elected an Aymara Indian president, Evo Morales, the first indigenous chief executive in Bolivia's 180-year history.

Among the southernmost midlatitude states, Chile claims to be 95 percent mestizo, but the Native Americans were practically exterminated in Uruguay and Argentina. In Central America most people are mestizos. Costa Rica is the only country in which Whites form the majority.

The African diaspora   Black peoples have migrated out of Africa, either freely or in slavery, for centuries. These migrations are known collectively as the **African diaspora,** from the Greek word for "scattering," used first by the Jews to describe their worldwide migrations.

Modern racial slavery originated in medieval Islamic societies. Muhammad himself owned slaves, and Arabs and Persians invented the long-distance slave trade that transported millions of sub-Saharan captives out to a realm stretching from Islamic Spain across to India. The Islamic slave system lasted over a period of 12 centuries, and in total numbers it probably surpassed the African slave trade to the Western Hemisphere (Figure 5-26). The trade diminished slowly, lingering longest in the Arab countries. (It was finally banned in

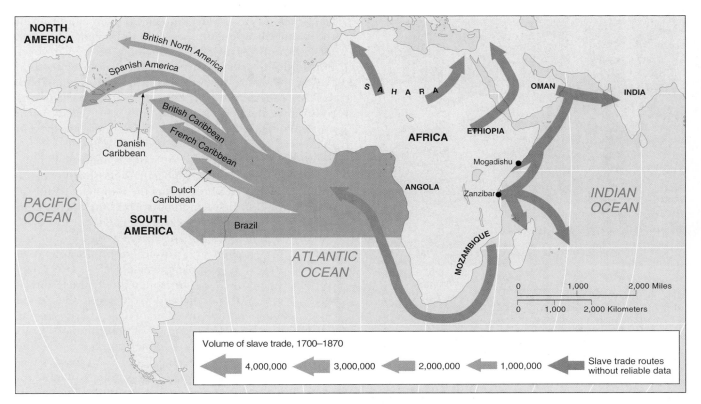

FIGURE 5-26 The movements of slaves out of Africa. Arab slave dealers first transported black slaves north and east, throughout the region from Spain to India. Transport of slaves to the Western Hemisphere started later and stopped earlier. Only about 5 percent of the slaves brought to the Western Hemisphere were brought to the area that became the United States.

Saudi Arabia in 1962 and in Mauritania in 1980.) The absence of a large black population across this enormous region today can be explained by the high mortality rate, by assimilation, and by the practice of castrating slaves. In central India, however, there are still communities of black descendants of African slaves.

International agencies insist that slavery persists in Mauritania and that Arabs from northern Sudan continue to enslave the blacks of the southern Sudan.

The transport of slaves to the Western Hemisphere
European transportation of black slaves to the Western Hemisphere had a dramatic impact on population geography. In the fifteenth century Europeans took a few hundred slaves from West Africa to Europe and the Atlantic Islands, and in the 1520s Europeans began transporting slaves to the Western Hemisphere (Figure 5-27). Black Africans served as a labor supply to replace the dying Native Americans.

During the period of the transatlantic slave trade, which ran from 1526 until 1870, about 10 to 12 million slaves reached the Western Hemisphere from Africa. The sources of the slaves reflected patterns of European colonialism in Africa. For example, the Portuguese African colonies of Angola and Mozambique provided the slaves of Portuguese-ruled Brazil, whereas West Africa supplied most of the slaves for North America. DNA samples from Western Hemisphere Blacks are

today providing evidence of their ancestors' African homelands.

The pattern of resettlement in the Western Hemisphere reflected the uses of slave labor. Slaves were brought to tropical and semitropical plantations in the U.S. Southeast, to the islands of the West Indies, and throughout Latin America, to the *tierra caliente* (hot land) at elevations of less than 914 meters (3,000 feet). In Mexico, by contrast, the slaves were put to work in the cities and mines. Mexico never had a significant rural black population, and eventually Mexico's black population died out or mixed with the other groups almost entirely. Mexico abolished slavery at independence in 1821.

Blacks in the Americas today   Descendants of slaves now constitute the majority of the population on most West Indian islands and in Belize. In South America, Blacks still occupy, for the most part, the lowland areas. After centuries of intermarriage, tropical South American countries still have significant minorities of Black, zambo, or mulatto populations.

In the United States, about 13 percent of the population is black—38 million people—giving the country one of the world's largest black national populations. The black population was originally concentrated in the Southeast, but Blacks began to migrate to the cities of the north just before the time of World War I. Most importantly for the migrants, it was a dual migration: from south to north, but also from rural to urban conditions. Between 1910 and 1970 some 6.5 million African Americans headed northward, and the percentage living in the south fell from 70 percent to 50 percent. However, the civil rights struggles of the 1960s created new opportunities for Blacks in the south, and partly for that reason but also for reasons not fully understood, African Americans have been migrating back to the south since the mid-1970s—almost exclusively to urban areas. Whereas 52 percent of African Americans lived in the south in 1980, 55 percent did in 2000. The black populations of many western states remain negligible.

West Indian and Caribbean Blacks steadily migrated into the United States during the twentieth century—almost 3 million between 1960 and 2000 alone. A high percentage of these people brought capital or job skills, and they and their descendants played a role in twentieth-century U.S. politics and culture disproportionate to their numbers. For example, the first black U.S. Secretary of State, Colin Powell, was the son of immigrants from Jamaica. The flow of migrants and sojourners to the United States, combined with the flow of U.S. tourists to the Caribbean, has caused a degree of acculturation sometimes referred to as the Americanization of the Caribbean.

Since the 1990s, immigration to the United States from Africa has sharply increased, and the number of Blacks with recent roots in sub-Saharan Africa has more than tripled. About 55,000 Africans have legally migrated to the United States each year—more than

FIGURE 5-27 A slave market. Tens of thousands of Africans passed through this old slave market on Goré Island off the coast of today's Senegal on their way to the Western Hemisphere. In 1992, Pope John Paul II stood in the doorway in the background, known as the Door of Tears, and blessed them and their descendants. Today the Senegalese government is restoring the island's historic fortifications as a museum and is building an international conference center. Furthermore, African countries are sending delegations to the Americas to perform religious ceremonies at slave burial grounds, asking for forgiveness of the slaves buried there for having sold them into slavery. (Brian Seed/Brian Seed & Associates)

were brought each year during the trans-Atlantic slave trade. The proportion of U.S. Blacks who are foreign-born rose from 4.9 percent in 1990 to 7.3 percent by 2000. Black African immigrants seem to have brought skills or to have succeeded in the United States, so statistical descriptions of America's foreign-born blacks and of their children reveal high educational attainment, high income levels, substantial representation in professional and managerial occupations, and low levels of unemployment.

Some native-born American Blacks wish to reserve the term *African American* for the descendants of slaves and to describe the newcomers by their countries of origin. U.S. Senator Barack Obama (elected from Illinois in 2004), however, born to a Kenyan father and an American mother, insists, "I'm African, I trace half of my heritage to Africa directly, and I'm American."

European migration to Asia and Oceania

The Spanish, French, Dutch, Germans, Portuguese, British, and Americans all created empires in Asia, but few people from these countries settled permanently in the Asian colonies. In no cases did any of these peoples come to make up significant proportions of mainland Asian populations. The most important European settlers in Asia and Oceania can conveniently be divided into those who migrated by land—the Russians—and those who migrated by sea—settlers from the British Isles who went to Australia and New Zealand.

Russian expansion After the Russians rose up against their Mongol rulers in the sixteenth century, they sent a steady stream of settlers to the east, over the Ural Mountains into Asia. Between 1860 and 1914 almost 1 million Russians resettled in the east, and more continued to do so within the political framework of the U.S.S.R. (look ahead to Figure 13-5). In 1990 the population of Kazakhstan, which was ostensibly a republic of the Kazak people, was 38 percent ethnic Russian. Russians migrated particularly into the cities in the non-Russian republics. In 1990 Tashkent, then the capital of the Uzbek Republic and today the capital of independent Uzbekistan, was primarily Russian.

When the U.S.S.R. dissolved in 1991, some of the newly independent republics imposed rigid language and citizenship requirements, triggering a migration of some 6 million Russians into Russia. Many of these families had not lived in Russia for generations. The position of Russians who remain in the newly independent Republics is still being defined.

Australia and New Zealand More than 2 million Europeans, mostly from the British Isles, moved to Australia and New Zealand in the nineteenth century, and another wave of 3 million Europeans (half British) arrived between 1945 and 1973. They all but exterminated the native peoples. Latest censuses count only 150,000 Aborigines in Australia (of a total population of

20 million) and 400,000 Maoris among New Zealand's 3.8 million people. Extensive litigation in both countries has recently returned lands to the native peoples and awarded them substantial payments.

In both Australia and New Zealand, commercial farming and ranching thrived from the beginning of white settlement. Wool, wheat, and later frozen meat were sent to markets within the British Empire. In Australia, white settlement in the dry areas was hazardous and only intermittent until spectacular mineral discoveries attracted people inland. In general, however, the white populations of both Australia and New Zealand have always been highly urbanized along or near the coasts (Figure 5-28).

About 88 percent of Australia's 20 million people are of British and European ancestry, but the country abandoned an explicit "white Australia" policy in 1973. Today 40 percent of Australia's 100,000 annual immigrants come from Asia, making 7 percent of the national population of Asian ancestry. A full 23 percent of Australians in 2004 were foreign-born, the highest percentage of any country.

European migration to Africa
Europeans migrated to Africa and settled in substantial numbers only where environmental conditions made possible a

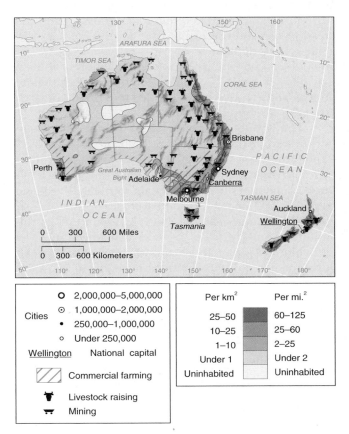

FIGURE 5-28 Population distribution in Australia and New Zealand. Vast stretches of inner Australia are virtually uninhabited.

Mediterranean or European style of life and commercial agriculture.

East Africa, Northwest Africa, and West Africa   The East African highlands attracted white settlement after the opening of the Suez Canal in 1867. Europeans claimed the fertile lands in today's Kenya, Tanzania, and Uganda, and turned much of the black population into workers on the White-owned farms or in white enterprises. Tens of thousands of Whites remain scattered throughout East Africa, but the Black-majority governments have rescinded their legal privileges.

In Northwest Africa, too, European colonial governments reserved the best lands for European settlers. During French rule of Algeria from 1830 to 1962, some 1.2 million Europeans settled there—about as many as in all 26 countries of tropical Africa combined. When Algeria, Morocco, Tunisia, and Libya gained their independence, most White-owned enterprises were nationalized or confiscated. Few Europeans remain in these countries.

West Africa never experienced substantial White settlement. Europeans were discouraged by the heat and humidity. Today Whites still come or serve as business executives, technicians, or even government advisers, but they make up insignificant percentages of the populations.

Southern Africa   Southern Africa includes large areas that Europeans found well suited to settlement and the creation of a European lifestyle. Several colonies did achieve considerable prosperity—for the Whites—but always based on repression and exploitation of the native Black majority. As these colonies won independence, White racist regimes yielded to Black-dominated governments (Figure 5-29).

The Republic of South Africa is the leading state in southern Africa. In 1671 the Dutch established Capetown colony as a staging point for voyages around Africa to Asia, and numerous Dutch farmers, called Boers, came to settle. When Britain seized that strategic colony in 1806, many Boers fled north, but the British eventually conquered all of South Africa and set up the Union of South Africa in 1910. That state was ruled by its White minority. International opposition to the government's policy of strict racial segregation, called **apartheid**, caused the government to declare itself a republic, totally free of any allegiance to Britain, in 1961.

The government divided the population into four groups: Whites (British and Boers combined, totaling 13 percent of the total population in 1996); Asians (3 percent); people of mixed race (or "Colored," 9 percent); and Blacks (75 percent). Each group held a fixed status. Most of the Asians had come to the eastern province of Natal to work on the sugar plantations, but many eventually rose into the middle class or even to substantial wealth.

Blacks were excluded from political participation. The government planned to move them into segregated

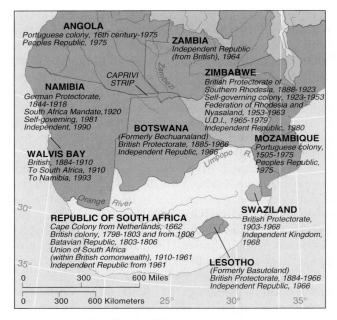

FIGURE 5-29 Southern Africa.   Substantial numbers of Europeans settled in the midlatitude areas of southern Africa when these lands were colonies, and today's independent countries are still working to achieve rising standards of living for all.

Bantustans (homelands), on the basis of ethnic roots. The Bantustans covered only 13 percent of the territory and were areas of low agricultural potential and without other resources. The Blacks obviously would never be able to support themselves in these lands, but would have to sell their labor within South Africa, where they would be immigrant workers without political rights. By the late 1980s the system of apartheid had begun to collapse under the pressures of international sanctions and internal rebellion. A new governmental system was devised, and free elections open to people of all races were held in 1994. Nelson Mandela, a black leader who had been imprisoned for 27 years during the struggle for civil rights, won the presidency—and the Nobel Peace Prize.

The government still faces tremendous pressures. Blacks understandably want the life of conspicuous affluence that many South African Whites enjoyed, but South Africa does not have the wealth to provide such a lifestyle for all its people. Furthermore, South Africa is considerably richer than its neighbors, so an estimated 2–8 million Africans from neighboring countries have illegally migrated into South Africa. South Africa's first Black Home Affairs Minister Mangosuthu Buthelezi complained that "the presence of illegal aliens impacts on housing, health services, education, crime, drugs, transmissible diseases—need I go on?"

The histories of some other southern African states echo that of the Republic of South Africa. The Portuguese colonies of Angola and Mozambique did not have many white settlers when they became independent in 1975, and most of those who did reside there left at that point. Botswana also had few white inhabitants when it gained independence in 1966; and when Namibia gained independence from South Africa in 1990, Whites made up only 5 percent of its population. Zambia won independence in 1964 with a Black-dominated government, which in 1975 nationalized all private landholdings and other enterprises. Today few Whites remain. Whites struggled to retain political power in Zimbabwe (previously known as the colony of Southern Rhodesia), as they did in South Africa, but power was transferred to Blacks in 1980, and today few Whites remain. Whites long retained their titles to much of the best agricultural land, but the lands were seized, and today land redistribution is virtually complete.

The migration of Indians

When the British ruled the Indian subcontinent, they encouraged migration to other parts of their empire. In many of the places where these Indian migrants went, they rose to positions that were socially and economically above the natives, as noted already in South Africa.

In British East Africa, Indians generally made up less than 5 percent of the populations, but when Uganda, Kenya, Malawi, and Tanzania achieved independence in the 1960s, Indians dominated the new countries' economies. Indians' high visibility made them targets of the majority Blacks' ethnic hostility, and Uganda even expelled more than 80,000 Indians. Several East African countries have recently tried to entice Indians to return, and Indians are regaining economic strength in Uganda and Kenya.

Indians settled in Myanmar (formerly known as Burma), and although several hundred thousand remain, they have faced discrimination since Myanmar achieved independence in 1948. Singapore is about 6 percent Indian. The Central American country of Belize is about 2 percent Indian. Indians still form key elements of the populations in several West Indian countries. Jamaica is only about 3 percent Indian, but Trinidad and Tobago is about 40 percent Indian, and Guyana on the South American mainland is almost 50 percent Indian. The population of the Pacific Ocean island of Fiji is almost equally split between Indians and indigenous Fijians.

Today the 20 million non-resident Indians make up one of the world's most widespread diasporas, and in many places they have enjoyed great success. Annual Indian migration to the United States is almost 50,000. Links are so strong between the U.S. computer industry and the industry in India that President Bill Clinton, while on a visit, referred to the Indian city of Bangalore (today's Bengalooru) as "Silicon Valley East," after California's concentration of high-tech industries (see Figure 12-11). First-generation Indian-Americans enjoy the highest median family income of any foreign-born group.

Several reasons have been cited for Indians' international success. One is that they have been able to take advantage of the network of commercial outposts established by the former British Empire. A second reason is their facility with English, which is increasingly the world language of business. India has more English speakers than any other country. Some Indians have themselves cited the cosmopolitan character of Indian culture as a third explanatory factor in their international success. The population of India is so diverse in terms of skin colors, languages, religions, and other cultural attributes that many Indians learn during childhood how to cooperate with people of all backgrounds. A fourth factor is that Indian migrants help one another in their new countries. If an Indian cannot borrow money from a local commercial bank to start a new business, local fellow Indians often raise the necessary capital. Hard work and frugality are additional factors often cited by Indian entrepreneurs themselves, but these characteristics also often describe groups that still do not enjoy the success that the Indians have.

Nonresident Indians constitute a valuable resource for India itself, because they have been able to transfer investment funds, technology, and marketing and financial expertise to India. The Indian government grants them privileges not available to other foreigners. For example, they may own land in India and repatriate funds elsewhere.

The Overseas Chinese

From early in the nineteenth century until about 1930, waves of South Chinese left their homeland to work in the British, French, and Dutch Asian empires (Figure 5-30). Their success mirrored that of the Indians. At the time the nations

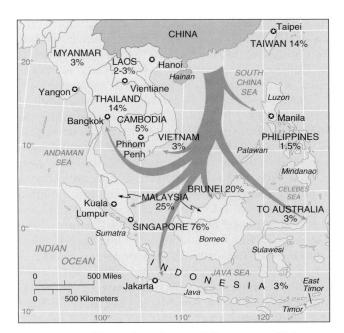

FIGURE 5-30 The migrations of Overseas Chinese. The number after each country is the percentage of its population that is Chinese. Even where the Overseas Chinese make up small minorities of the population, they exercise financial influence far beyond their numbers. Overseas Chinese control the economies of most of these countries.

of Southeast Asia gained their independence, their economies were firmly in the hands of small Chinese minorities, who were often linked across international borders by bonds of family, clan, and common home province. The Chinese commercial supremacy and their international links still survive, so that today these 50 million **Overseas Chinese** constitute a formidable economic power throughout Southeast Asia.

Just as the migrant Indians faced native ethnic hostility, so have the Chinese. For example, ethnic resentments dominate the history of Malaysia. When the British ruled Malaysia as a colony, they founded the cities of Singapore and Kuala Lumpur. The populations of both cities came to be overwhelmingly Chinese, although the majority of the population of the colony as a whole was Malay. Malaysia was granted independence in 1957, but animosity was so great between Chinese and Malays that in 1965 Singapore broke off as an independent Chinese-dominated city-state. Even in Malaysia, however, the Chinese minority still controls the economy, although the Malay-dominated government blocks Chinese and Indians from advancement in government positions and encourages Malay entrepreneurship. The population of Malaysia today is about 68 percent Malay, 25 percent Chinese, and 7 percent Indian.

In Indonesia the 7.3 million Chinese control an estimated 80 percent of the country's private industry, and the Indonesian government requires them to invest in businesses owned by other ethnic groups. A 1965 communist coup attempt, supposedly backed by China, ignited anti-Chinese sentiment, and as many as 500,000 ethnic Chinese were slaughtered. Anti-Chinese violence flared again during Indonesia's political and financial instability in the late 1990s and has lasted into the twenty-first century, making unclear the future of the Chinese in Indonesia. The loss of their skills and capital would devastate the Indonesian economy.

Overseas Chinese figure highly in their adopted countries around the world. They play an elite role in the Philippines and comprise almost the entire commercial infrastructure of Thailand. Ambitious Chinese also established themselves in New South Wales, Australia; along the west coast of the Western Hemisphere from British Columbia to Peru; and in small but significant numbers in the Caribbean. Tens of thousands of Chinese migrated to California in the mid-nineteenth century, and a new wave of Chinese migration to the United States began in 1965. Today Chinese Americans make up 1 percent of the U.S. population, up from 0.4 percent in 1980. Tens of thousands of Hong Kong Chinese emigrated before that city reverted to Chinese rule in 1997; many headed for Australia or Canada.

The Overseas Chinese have proved to be a valuable resource for their homeland, just as the nonresident Indians have. Overseas Chinese have raised investment capital, invested in homeland industries, introduced new technologies to China, and distributed Chinese products worldwide.

MIGRATION TODAY

Human migration has by no means come to an end. The United Nations estimates that in 2002 175 million people were living in countries other than the ones in which they were born. Large-scale migrations still make daily news. The United Nations' Universal Declaration of Human Rights affirms anyone's right to leave his or her homeland to seek a better life elsewhere, but it cannot guarantee that there will be any place willing to take anyone. As in the past, the major push and pull factors behind contemporary migration are economic and political. People are trying to move from the poor countries to the rich countries and from politically repressive countries to more democratic countries. In addition, millions of people are fleeing civil and international warfare. Pressures for migration are growing, and they already constitute one of the world's greatest political and economic problems.

Many migrants may be poor, uneducated, or unskilled, but usually they are also enterprising, in good health, and of working age. They seek opportunity, and it takes courage, stamina, and determination to pull up roots and head to a foreign and perhaps hostile land. In 1993, Mexico's President Carlos Salinas de Gortari said, "I don't want any more migration of Mexicans to the U.S., especially those migrants who are very courageous and take the impressive risk of going to the U.S. looking for jobs. That is precisely the kind of person I want here."

Other migrants are highly educated and skilled, and even the richest countries compete to attract them. Many were born in poor countries, and they received excellent educations either at elite institutions in their homelands or in rich countries. The poor countries want these people to return.

The first half of this chapter noted that populations are declining in many of today's rich countries, and some observers may ask why migrants from today's poor countries cannot simply migrate to fill "vacancies" in these countries. New migrants would provide the young labor force that rich countries need in order to sustain the pension benefits of aging populations. The United Nations has estimated the numbers of immigrants each rich country would have to accept each year until 2050 in order for its total population to stabilize rather than fall. These numbers range from 29,000 for France to some 350,000 for Germany. Britain would need a million immigrants per year until 2025 to maintain its 2005 dependency ratio. The barriers to such mass migration, however, are cultural, economic, and political.

Refugees

Some migrants are defined in international law as being **refugees**, and these people have special rights. A refugee, as defined by a 1951 Geneva Convention, is someone with "a well-founded fear of being persecuted in his country of origin for reasons of race, religion, nationality, membership of a particular social group or political opinion." Signatories to the Convention are obligated to accept refugees and grant them safety, called **asylum**. Through recent years, civil wars and international wars displaced hundreds of thousands of people and sent many streaming over international borders. The United Nations High Commissioner for Refugees defined almost 10 million refugees and 20.6 million people as "of concern" at the end of 2005 (Figure 5-31).

Some countries have broadened their definition of a "social group" to which they will grant asylum. For example, the United States, Canada, and several European countries have recognized persecution due to sexual orientation as a claim to refugee status. The European Parliament has accepted the legal principle that in some countries women are customarily treated so badly that they constitute a persecuted social group. Both the United States and Canada have also offered asylum to women from countries where, in the Canadian government's words, customs place women "in a more vulnerable position than men." Such countries include several Arab and African countries and Israel.

While many migrants are genuine refugees, most migrants are seeking a new home where they can work and enjoy a higher material standard of living. If they are fleeing poverty in their homelands, they are sometimes loosely called *economic refugees*. In just the five years from 1995–2000, the more-developed countries received 15 million migrants from the less-developed countries. Many "economic refugees" claim to be genuine refugees, and it is difficult for host countries to ascertain who is a genuine refugee. Claims to asylum have been abused, and many rich countries have been tightening their scrutiny of claims.

The Impact of International Migration

The impact of international migration is greater than the numbers might suggest. Immigrants are often greeted with apprehension, and immigration has become an explosive political issue in many receiving countries. This is for several reasons. (1) Official statistics substantially underestimate actual numbers, and there is evidence that undocumented migration is rising everywhere. (2) Migrants are usually in the peak years of fertility. Therefore, they are playing an increasing role in the total population growth in the rich target countries. (3) Migrant settlements are generally concentrated in a few places within a country. This usually adds to the immigrants' visibility and increases the perception of cultural differences. (4) Many migrants today seem to be assimilating to the cultures and habits of their new homelands slowly—if at all; this may be either because they are discriminated against or because they cling to their native traditions.

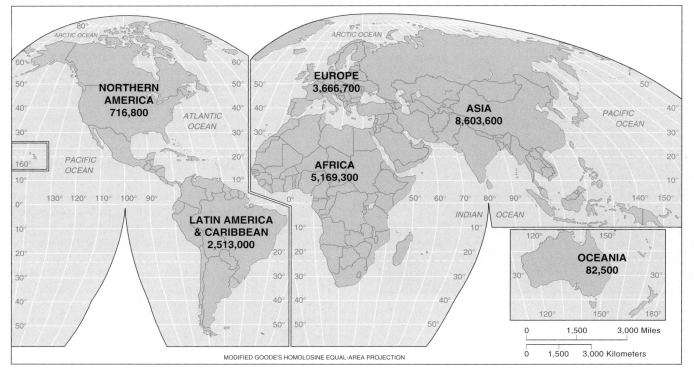

FIGURE 5-31 Persons of concern who fall under the mandate of UNHCR. This map shows the numbers of refugees, asylum-seekers, and internally displaced persons as of January 1, 2006. (Data from U.N. High Commissioner for Refugees)

The peoples of the rich and democratic countries are beginning to exhibit "compassion fatigue," and doors are closing all around the world (Figure 5-32). In 2006, only 35 percent of Germans and 49 percent of Americans agreed that immigrants were "a good influence" on their country (although 77 percent of Canadians thought so.)

Moreover, whenever unemployment rises, immigrants who were once grudgingly tolerated because they were willing to work in low-wage or dirty jobs suddenly come to be seen as competitors and become targets of wrath. The European countries are growing more selective; they welcome educated and skilled immigrants, but not poor or unskilled immigrants.

Every demographic impetus toward migration will be multiplied as populations rise in the poor countries. Either substantial flows of investment capital and technology will be directed to the poor countries and create new job opportunities in them, or else migration from the poor countries will be stimulated by the combination of large increases in population, excess labor supply, rising social and political turbulence, and persistent or worsening inequalities between rich and poor countries. Furthermore, global communications media and social networks strengthen the pull of the rich countries by spreading alluring images and information about them. Many poor people are growing desperate, and tragic stories of attempted migration can be found in the rich countries' newspapers almost every day: Would-be immigrants are drowned when flimsy boats capsize in the Mediterranean Sea or Straits of Florida, or they are found suffocated in cargo containers arriving on rich countries' highways or at airports. Altogether an estimated 200 million people are today under

FIGURE 5-32 Keeping them back. This high fence, seen here between San Diego, California, and Tijuana, Mexico, winds its way across much of the U.S.–Mexican border in the U.S. Southwest. While the fence may be an extreme measure, it is seen as the only way to control the flow of eager migrants into the United States. (Ken Cedeno/CORBIS-NY)

the sway or in the hands of traffickers in human beings who abuse would-be immigrants in sexual exploitation, economic slavery, forced labor, debt enslavement, or other forms of exploitation.

Some countries may be relieved to see people emigrate. The emigrants may be politically troublesome, or the country may hope that its emigrants will return with new skills, make investments in their homelands, or just send money home. Money sent home from workers abroad, called **workers' remittances**, is several times the total of international foreign aid, and is important support for several countries today. India receives the highest dollar amount ($22 billion), but the Philippines has the highest percentage of its national population working abroad (10 percent). Haiti receives the highest percentage of its national income (25 percent) from remittances.

Many rich countries accept foreign workers, but they do not grant them citizenship or the rights that citizenship would bring. The host countries accept them only as sojourners, and it may be unclear whether the migrant workers themselves hope to stay. The conditions under which the host countries grant citizenship, or even guarantee full civil and legal rights to aliens, have become matters of international concern. Unscrupulous employers often profit by exploiting aliens. Illegal aliens especially have faced hazards in search of work, they are seldom unionized, and they often accept lower wages and worse working conditions than legal residents or nationals will.

Migration to Europe

Some 16 million non-European legal immigrants lived in the 25 countries of the European Union in 2003 (see Chapter 13), representing about 6 percent of the total Union population: 6 percent in Germany, 3.5 percent in France, 3 percent in the Netherlands and in the United Kingdom, and 2 percent in Italy (Figure 5-33). Europe has not always easily absorbed immigrants. The European countries have not historically thought of themselves as multicultural, and even though many immigrants come from former colonies, many Europeans fear the arrival of increasing numbers of Asians and Africans. In a 1999 poll, 66 percent of EU citizens described themselves as "a little racist"; and an additional 33 percent described themselves as "very racist" or "quite racist." Compared to the Europeans around them, immigrant groups are often less educated, occupy substandard housing, and receive inferior services.

Furthermore, many immigrants from the Near East and North Africa are Muslim; and, as we shall discuss in Chapter 7, their religion sometimes clashes with traditional European Christian culture. Generalizations about Europe's immigrant Muslims, however, require many qualifications. Variations among their cultural and national backgrounds, and among the countries to which they migrate, may play a bigger part in their ability to integrate than their religion does. Britain's mainly

FIGURE 5-33 An immigrant neighborhood. This Arab butcher shop in Brussels illustrates how North African immigrants offer a variety of services and goods to their compatriots. (Courtesy of Robert Mordant)

FIGURE 5-34 Riots among immigrants to France. Burning cars lit the skies of Toulouse, Paris, and other cities throughout France as largely immigrant youths rioted in the suburbs in 2005. At least 10 percent of the children in France are Muslim, and Muslim youngsters suffer 6 times the national unemployment rate and 5 times the national incarceration rate for all young people in France. These riots were repeated in November 2006. (Lionel Bonaventure/Getty Images)

south Asian Muslims have far more in common with other Britons than they do with France's North African migrants or Germany's Turks. Overall, however, Europe's Muslims are much more likely than non-Muslims to be jobless and ill-educated, harassed by police, and discontented (Figure 5-34). If they feel equally estranged from both the Muslim lands and from their new homes (or, in the case of the second generation, from the European land of their birth), they may turn to Islam in search of an identity. European secularism sharpens that tendency. Thus, some immigrants do not want to acculturate to European cultures.

Growing radicalized communities within Europe may form ever-larger support groups and launching pads for extremism. Most Western European Muslim communities condemned the 9/11 attacks on the United States, but many hesitated to endorse any U.S. retaliation on Muslim countries. The attacks of 9/11 were actually planned in Europe among European residents of Arab background. Cells of al-Qaeda and other terrorist groups have been uncovered in Paris, London, Berlin, and other leading European cities. Some have committed terrorist acts, but the extent of others' involvement is under investigation. These discoveries have compounded many Europeans' fears of their Muslim neighbors. Western Europe may contain enclosed societies in which hate can be preached and treachery plotted while all around, non-Muslims remain utterly unaware.

America's Muslims and Muslim immigrants, by contrast, have generally assimilated so well that, even though

they may have been unnerved by law-enforcement procedures since 9/11, their lives are basically different from those of the estranged Muslim underclass of much of Europe. The United States' Muslims are in fact its first line of defense against foreign Muslim extremism.

Europeans are sharply curtailing the granting of asylum. Most European countries have decreed, for example, that people who had passed through one or more countries on their way from their homelands are no longer refugees, but are merely "country shopping" for the most favorable new site to settle. Therefore, they will not accept anyone claiming refugee status who came overland through a third country. European states have also lengthened the lists of countries from which visitors' visas are necessary, begun fining airlines for bringing people without visas, and even sending people back to the country they were last in. The reduction of legal migration to Europe has increased illegal migration, which may have reached 500,000 per year. So many desperate would-be migrants suffer while sailing toward Europe in unseaworthy vessels that in 2006, Malta's Interior Minister Tonio Borg asked, "Can we look the other way when dead bodies are coming on our shores, as if they were dead fish in polluted water?" A European fleet patrols the Mediterranean to turn ships back to Africa.

European states have also redefined the criteria for citizenship. For example, France traditionally accepted the idea that an individual's citizenship is based on his or her place of birth, a concept called *jus soli* (law of soil). The United States also accepts *jus soli*. France, however, has made it more difficult for children born in France of foreign parents to become French citizens. Some other countries, including Germany, have traditionally defined citizenship on the basis of the citizenship of one's ancestors. This concept is called *jus sanguinis* (law of blood). Germany has tightened conditions for claims to German citizenship. In Ireland, 79 percent of voters chose to stop *jus soli* after Irish hospital maternity wards reported influxes of pregnant tourists coming to win citizenship for their newborns.

Sealing borders will not solve the problem of existing immigrants and their European-born children and grandchildren. European governments are directing new attention to what the Dutch call *inburgering*, that is, the forming of citizens. Governments are requiring examinations in language, history, culture and traditions, and the rights and duties of citizenship. The Dutch require potential immigrants to watch a film about Dutch life that includes scenes of a topless woman and of men kissing. European governments are demanding that immigrant women and immigrants' daughters enjoy full rights not always guaranteed in the immigrants' homelands. Some Europeans ask whether such laws are discrimination or a last-ditch attempt at Europeans' own cultural survival.

Migrations of Asians

An estimated 5 million Asians work abroad, and in many cases it is doubtful whether many will ever return to their homelands. Meanwhile, their remittances to their homelands are an important element in the economies of their home countries.

The absence of these workers lowers the local unemployment rates in their homelands, but it also constitutes a loss of important skills needed back home. Although the home country might not have the resources to reward these workers adequately, it frequently needs their contributions. A young Filipina woman, for example, may be needed in her hometown as a schoolteacher, but by staying there she would earn $40 per month. If she is earning $150 per month as a maid in Singapore, she is better off, but the Philippines suffers. An estimated 60,000 Filipina women—many college graduates—work as maids in Singapore.

The oil wealth of some Arab states has drawn many Asian workers who seem little disposed ever to go home. In 2006, the populations of the United Arab Emirates and of Dubai were less than 20 percent native-born; that of Kuwait was only 35 percent native-born. The expatriate workers in these rich countries often come from poorer Muslim lands, and they are generally more susceptible to radical interpretations of Islam. As long as the Middle East remains an international hot spot, and as long as Arab oil supplies are important to world prosperity, any destabilizing migration into Arab oil-rich countries is of world concern.

Japan, of all the world's richest countries, limits immigration the most strictly. Between 1945 and 2000, fewer than 250,000 foreigners acquired Japanese citizenship, including those who married Japanese. The most significant minority in Japan is those people of Korean ancestry—about 0.3 percent of the total population. These people are a reminder that Japan ruled Korea from 1910 until 1945, and they suffer discrimination. However, with Japan's low TFR and high median age, Japan would need to import 600,000 workers per year to keep its working population stable between 2000 and 2050. By that time the population would be one-third foreign-born. The government has encouraged childbearing and broadened the category of jobs that foreigners in Japan (today about 1.1 million) could hold. The Minister of Justice has urged Japanese "aggressively to carry out the smooth acceptance of foreigners."

Migration to the United States and Canada

The single largest migration flow for the past 150 years has been migration to the United States. In recent years the United States has accepted 600,000 to 800,000 immigrants per year. Some 12 million illegal immigrants may also live in the United States.

The 2000 census found 10 percent of the U.S. population to be foreign-born, up from 7.9 percent in 1990. The nation's overall birth rate is low, so immigration is responsible for about one-third of U.S. population growth. If immigration continues at current levels, the foreign-born percentage of the population may account for nearly two-thirds of the population growth projected to the year 2050.

The source areas of U.S. immigrants have changed throughout history, thus changing the makeup of the national population. Immigration was totally unrestricted until the late nineteenth century, when the government enacted rules explicitly designed to keep out Chinese and other Asians. Later, in the 1920s, the government issued quotas for various nationality groups. The first quotas favored British, Irish, and Germans, who had formed, along with African Americans, the bulk of the population in 1890. By the 1920s, however, southern and Eastern Europeans had already established a strong presence. After the 1920s a maximum quota was set on immigrants from the Eastern Hemisphere (in practice, Europe), but there was in theory no ceiling on immigrants from the Western Hemisphere. In 1965 the United States changed the rules again, in a way that tended to

U.S. Census Bureau Categories

All categorizations of people in the U.S. Census are self-descriptions. People are asked "what" they are, and they choose among given categories as they wish. The choices, however, have changed several times through recent decades, reflecting changing attitudes about categorizing people.

In 1990 the U.S. Census Bureau offered five choices "White"; "Black"; "American Indian, Eskimo and Aleut"; "Asian or Pacific Islander"; and "Other." This selection, however, triggered widespread criticism. The Bureau classified Native Hawaiians, for example, as "Asian or Pacific Islanders," but many Native Hawaiians would rather be classified as "American Indians." They do share certain historical experiences; also, if Native Hawaiians were classed as "American Indians," they would enjoy access to some affirmative action programs that are closed to them as "Asians."

In 1997, therefore, the federal government began to allow people to classify themselves as being of more than one race by checking two boxes. To avoid the diminution of the minority voice in certain government programs, however, anyone who counted himself or herself as being both of a minority race and also being white, would be counted in the totals for minorities. Some observers found this step reminiscent of long-repudiated segregationist racist categorizations called "one drop" systems. In these, anyone having just "one drop" of black blood was considered black and suffered segregation.

For the 2000 census, the Census Bureau addressed Native Hawaiian concerns by separating them from Asians and categorizing them with other Pacific Islanders. Thus the six categories of "race" used in 2000 were: "White"; "Black or African American"; "American Indian and Alaska Native"; "Asian"; "Native Hawaiian and Other Pacific Islander"; and "Some Other Race."

In the 2000 census, 6.8 million people (2.4 percent of the total) chose to identify themselves as being of more than one race. People under 18 years of age were twice as likely to be identified as being multiracial than were people 18 or older. This reflects increasing ethnic mixing in the United States, but also, perhaps, that younger people, whose heroes may include proudly multiracial Mariah Carey, Derek Jeter, Hines Ward, and Tiger Woods, feel less bound by traditional ethnicities.

The U.S. government has also tried to count Spanish speakers living in the United States. The history of this question and this category illustrates again the difficulties of describing and categorizing the population. In the 1950 census the Bureau asked how many people used Spanish as their mother tongue, and in 1960 it tried asking how many people had Spanish surnames. During the 1970s, several government agencies adopted a new category: "Hispanic." The exact meaning of the term is not clear, but it would seem to include people who speak Spanish or who have ancestry in some Spanish-speaking country. Not all such people share a common "culture" (see Chapter 6) beyond sharing a language background, and some people who consider themselves Hispanic do not speak Spanish. Nevertheless, this category "Hispanic" was adopted by the Census Bureau in 1980, 1990, and 2000. People were asked to identify themselves in one of the racial categories, and then a second question asked: Are you Hispanic? A person could claim to be of any one of the six "races" in the first question, and then also claim to be Hispanic.

For some people, their identity as a Hispanic is more important to them than any racial categorization. Forcing them to identify themselves "racially" in the first question, they insist, forces them to be race-conscious in a way that Hispanic society traditionally is not, and that they do not want to be. Partly in response to this feeling, some analyses divide the U.S. population into the categories "White," "Black," "American Indian," "Asian," and "Hispanic." Whenever you see statistics reported in this way, however, you must remember that the category "Hispanic" has been created by pulling out from each of the six racial groups those people who reported themselves as "Hispanic" in the second question.

The word "Hispanic" excludes Brazilians, whose native language is Portuguese. Throughout Latin America the word *Hispanic* is reserved to refer to things Spanish, or from Spain. People throughout Latin America do use the words *Latinoamericanos* or *Latinos* to refer to themselves—as in the name of the baseball stadium in Havana, Estadio Latinamericano. Therefore, today, increasingly, U.S. "Hispanics" prefer to be called "Latinos."

An increasing share of people in the United States are growing to resent any form of racial or ethnic classification, and as cultural mixing continues, racial and ethnic categories may be meaningless within a few decades. In 2003, however, Californians voted down a Racial Privacy Initiative that would have protected people from disclosure of racial identity to government.

favor Latin Americans and Asians (Table 5-1 and Figure 5-35). In 1965 European immigrants outnumbered non-Europeans 9 to 1, but within 20 years that proportion was reversed.

The latest changes in the source areas of immigrants will again bring long-term consequences to national demography (Table 5-2). In 1990, three-quarters of all Americans were non-Hispanic whites; by 2050 that percentage may fall below half. The number and percentage of the total population that is Asian—11.9 million, or 4 percent in 2000—is projected to rise to more than 7 percent by 2040, largely through continuing immigration.

The rising percentage of the national population that is Hispanic, however, is more dramatic. The 1970 census counted about 9.6 million Latinos, a little less than 5 percent of the population. By 2000, Hispanics numbered 35.3 million, 12.5 percent of the total U.S. population. Hispanics, who exceeded non-Hispanic

Blacks as the largest minority group in the 50 states, reached 14 percent of the population by 2006 (when the total national population reached 300 million), and are projected to reach 25 percent of a total national population of 400 million by 2040.

The increase in Hispanics is due to two processes. (1) Hispanics form the largest share of all immigrants. Through recent years about 20 percent of all legal immigrants have in fact come from just one country, Mexico, and Mexicans today make up about 30 percent of all foreign-born U.S. residents. (2) Hispanic TFRs are the highest among U.S. groups (about 3). Furthermore, Hispanics have the lowest median age of any U.S. group (27), so Hispanics would make up an increasing share of the national population even if immigration had ended in 2000.

In 2000, two-thirds of all foreign-born lived in just six states (California, New York, Florida, Texas, New Jersey, and Illinois). About two-thirds of new immigrants

TABLE 5-1	Difference in Population by Race and Hispanic or Latino Origin, for the United States: 1990 to 2000				

| | 1990 CENSUS | | CENSUS 2000 | | DIFFERENCE BETWEEN 1990 AND 2000 | |
| | | | USING RACE ALONE FOR CENSUS 2000 | | USING RACE ALONE OR IN COMBINATION FOR CENSUS 2000 | |
SUBJECT	NUMBER	PERCENT OF TOTAL POPULATION	RACE ALONE	RACE ALONE OR IN COMBINATION	PERCENT DIFFERENCE (BASED ON 1990)	PERCENT DIFFERENCE (BASED ON 1990)
	(1)	(2)	(3)	(4)	(5)	(6)
Race						
Total population	**248 709 873**	**100.0**	**281 421 906**	**281 421 906**	**13.2**	**13.2**
White	199 686 070	80.3	211 460 626	216 930 975	5.9	8.6
Black or African American	29 986 060	12.1	34 658 190	36 419 434	15.6	21.5
American Indian and Alaska Native	1 959 234	0.8	2 475 956	4 119 301	26.4	110.3
Asian	6 908 638	2.8	10 242 998	11 898 828	48.3	72.2
Native Hawaiian and Other Pacific Islander	365 024	0.1	398 835	874 414	9.3	139.5
Some other race	9 804 847	3.9	15 359 073	18 521 486	56.6	88.9
Hispanic or Latino and Race Total population	**248 709 873**	**100.0**	**281 421 906**	**281 421 906**	**13.2**	**13.2**
Hispanic or Latino (of any race)	22 354 059	9.0	35 305 818	35 305 818	57.9	57.9
Not Hispanic or Latino	226 355 814	91.0	246 116 088	246 116 088	8.7	8.7
White	188 128 296	75.61	94 552 774	198 177 900	3.4	5.3
Black or African American	29 216 293	11.7	33 947 837	35 383 751	16.2	21.1
American Indian and Alaska Native	1 793 773	0.7	2 068 883	3 444 700	15.3	92.0
Asian	6 642 481	2.7	10 123 169	11 579 494	52.4	74.3
Native Hawaiian and Other Pacific Islander	325 878	0.1	353 509	748 149	8.5	129.6
Some other race	249 093	0.1	467 770	1 770 645	87.8	610.8

Source: U.S. Census Bureau.

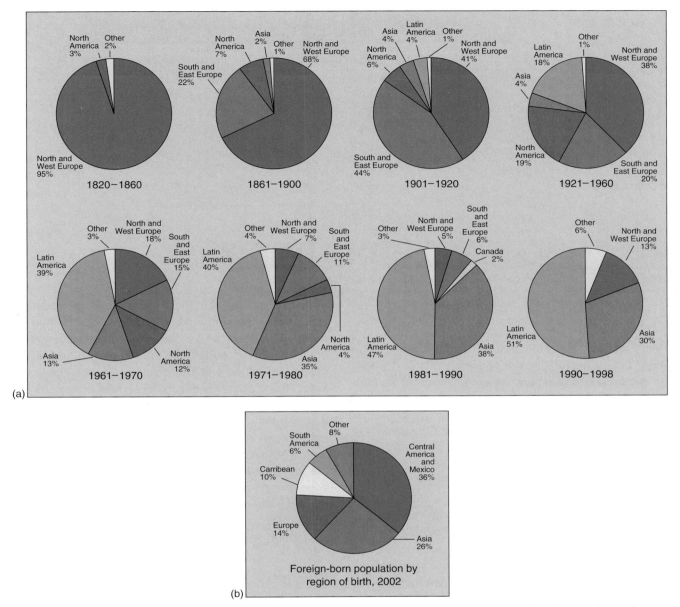

FIGURE 5-35 Sources of immigrants to the United States and foreign-born population by region of birth. The graphs in part (a) show how the source areas of immigrants to the United States have changed through history. The percentages of immigrants from Northern and Western Europe have dropped steadily, as the percentages from Latin America and Asia have risen. These trends have affected the composition of the national population. Graph (b) shows what percentage of the total foreign-born population in 2002 had come from each region. (U.S. Census Bureau)

settled in just 10 metropolitan regions (New York, Los Angeles, San Francisco, Miami, Chicago, Washington, D.C., Houston, Dallas, San Diego, or Boston). In these areas, they often replace native-born residents, so the populations of New York and Los Angeles, for example, are today 40 percent foreign-born. In 2005, Los Angeles elected Antonio Villaraigosa its first Latino mayor in 133 years.

Some plains states that are losing population are trying to lure foreign-born settlers. Iowa, for example, has a lower population than it did 20 years ago, and today it publicizes itself as an "immigration enterprise zone" in order to attract new residents.

Other features of current U.S. immigration legislation give preference to immigrants who bring either skills or capital for investment and make it easier for foreign students already in the United States to stay.

Canada has in recent years welcomed some 260,000 new immigrants each year, so foreign-born Canadians numbered 5.4 million in 2001, nearly 20 percent of the total population. Immigration now represents over half of total Canadian population growth.

TABLE 5-2	Shifting U.S. Demographics in the 21st Century U.S. Population by Race and Hispanic Origin (U.S. Census Bureau)	
	2005 POPULATION (MILLIONS)	PERCENT CHANGE FROM 2000
Non-Hispanic White	198.4	1.4
Hispanic	42.7	21.0
Non-Hispanic Black	36.3	5.8
Asian	12.4	19.2
Native American	2.2	4.8
Native Hawaiian	0.4	10.4
More than one race	4.0	17.7
Total	**296.4**	**5.3**

Critical THINKING

Is Immigration a Substitute for Education?

Many critics have long argued that the labor force provided by immigration has allowed the United States to slight the education and development of the nation's native workforce. They have suggested that the country invest in educating and training its own workers rather than rely on immigrant labor. In his 1895 speech "Let Down Your Bucket," the black educator Booker T. Washington (1856–1915) encouraged southern industrialists to seek their labor force around them among the willing Blacks who had so long "tilled your fields, cleared your forests, builded [sic] your railroads and cities," rather than "to look to the incoming of those of foreign birth and strange tongue and habits . . . to buy your surplus land, make blossom the waste places in your fields, and run your factories." This would require "helping and encouraging them . . . and education."

Today, more than 100 years later, some U.S. industrialists insist that the national labor force still has not been properly trained, so skilled immigrant labor is still needed. In May 2000, Gene Sperling, the president's economic policy adviser, said, "The best way to increase the supply of skilled workers is to improve the education and training of United States citizens and other people already in the country. However, at times, U.S. businesses need additional access to the international labor market to maintain and enhance our global competitiveness, particularly in high-growth new technology industries and particularly in tight labor markets." The U.S. government offers two special visas for skilled workers, the H-1B and the L-1. Numbers of immigrants having one or the other have risen and now total hundreds of thousands.

At the same time, as we shall see in Chapter 12, the globalization of the economy now offers U.S. firms the possibility of moving company operations abroad rather than bringing workers to the United States. U.S. corporations are increasingly choosing this option.

Questions

1. Which is cheaper, importing a trained labor force, educating native-born workers, or moving jobs abroad? Who profits? Who loses? What costs may be associated with failing to educate and train Americans?
2. Can you think of reasons why native-born citizens may not be receiving the education that equips them for job openings?

Canada has also experienced a recent shift in the source area of its immigrants. In the 1950s more than 80 percent of Canadian immigrants were drawn from Europe, but today that figure is less than 25 percent. About 60 percent of Canada's new citizens now arrive from Asia. In 1999 an ethnic Chinese from Hong Kong became Canada's governor general, the country's head of state, and she was replaced in 2005 by Haiti-born Michaëlle Jean. Canada has set a goal of increasing immigration gradually until inflows reach 1 percent of its population, up from about 0.85 percent in 2002.

In Canada, the concentration of the foreign-born population is even greater than it is in the United States. In 2001, 94 percent of all foreign-born lived in Canada's 27 metropolitan areas, with metropolitan Toronto, Vancouver, and Montreal alone accounting for three-quarters of them. A full 43 percent of Toronto's secondary school students in 2006 were born outside of Canada, and 49 percent had a first language other than English. Thus North American metropolitan areas have become increasingly "exotic" enclaves of ethnic diversity markedly different from both countries' non-metropolitan populations.

Canada selects its immigrants according to a system in which applicants are assigned points on the basis of characteristics such as their ability to speak English or French, workforce skills, family ties, and ethnic diversity. How these factors are to be weighed is a political issue in Canada, but Canada is today giving more weight to skills and less to reuniting families.

New arguments over immigration to the United States

Many Americans are beginning to question continuing immigration. These arguments have become heated as the immigrant share of the population rises, as many Americans have become concerned with securing America's borders since 9/11, and as the number of illegal aliens in the country increases. Some of the arguments have become frankly racist.

Some people argue that immigrants—legal or illegal—cost Americans money for public services. The substantial costs of immigration must indeed be borne by local governments in the areas where immigrants concentrate, exasperating local taxpayers. In California, for example, 59 percent of voters approved a 1994 ballot initiative to deny public health and education to illegal aliens. Courts, however, ruled that step unconstitutional, so today Los Angeles spends about $2 billion per year educating illegal immigrants and an additional $350 million per year for their health care.

Others counter that immigrants contribute to American economic growth. The National Academy of Sciences concluded in 1997 that immigration does produce substantial economic benefits for the United States as a whole. The poverty rate is significantly higher for immigrants than for the native-born (21.6 percent versus 12.1 percent in 2004), but intergenerational differences among immigrants and their children demonstrate upward mobility. First-generation male Mexican immigrants, for example, earn only half as much as non-Hispanic white men, but the second generation has overtaken Black men and earns three-quarters as much as Whites. Furthermore, the newcomers and their children will make up the labor force to support the growing number of pensioners.

A great many immigrants hold arduous low-paying jobs, and this may slightly reduce the wages and job opportunities of low-skilled Americans. Some scholars argue that this adverse impact on the U.S. poor is immoral.

Congress has taken up the issue of what to do about the estimated 12 million illegal immigrants already in the country, many of whom are holding jobs and paying taxes (Figure 5-36). Many are parents of children born in the United States, who are thus citizens. Is it realistic to think of identifying, arresting, and deporting millions of people? Can America be a true democracy with so many of its settled inhabitants unable to vote? On the other hand, many people are reluctant to grant amnesty to illegals. To do so would reward law breaking and discourage those around the world who are waiting and hoping to enter legally.

Immigration rules were overhauled in 1986, legalizing more than 3 million unlawful immigrants and supposedly banning the employment of illegal workers. One 2006 study, however, estimated that illegals accounted for 24 percent of the country's agricultural workers, 14 percent of construction workers, and 9 percent of manufacturing workers. Former U.S. Attorney General Edwin Meese III criticized America's "failure of political will in enforcing laws against employers."

Most arguments over immigration are often based in economics, but immigration policy has political and cultural aspects as well. For example, should the United States absorb immigrants from Mexico deliberately to preclude potential political turmoil there? Almost 10 percent of the total Mexican population lives in the United States, and their remittances sustain many Mexican communities.

And is there an "American culture," and can that culture be sustained as the racial and ethnic makeup of the population changes? The United States was dubbed the **melting pot** in a 1914 novel of that title, but various groups have maintained their cultural identities, and the traditional melting pot image cannot be sustained as demands rise for appreciation of a U.S. "multiculturalism." Perhaps the United States will redefine itself as a

FIGURE 5-36 Legislators have immigration stories. Senator Pete Dominici (NM) recalled in 2006 on the U.S. Senate floor how he had wept when his mother, an illegal alien, had been taken away by federal agents in 1943. He had kept the story to himself through his life. Several members of Congress are immigrants or children of immigrants. (Doug Mills/New York Times Agency)

cultural mosaic, which is what Canada proudly calls itself. As yet, however, many Americans seem blind to the changing composition of the national population. A study by the University of California identified as Latino only 4 percent of the characters on the four major television networks' prime-time series in 2003.

The United States must decide how widely to open its borders, and on what criteria of admission. Even the question of the definition of citizenship emerged in national politics in 1996, when the Republican Party platform called for a constitutional amendment "declaring that children born in the United States of parents who are not legally present in the United States or who are not long-term residents are not automatically citizens." Perhaps the United States will follow France and Ireland, discarding its historic *jus soli* definition of citizenship. In any event, immigration policy will continue to be a political issue.

CONCLUSION: CRITICAL ISSUES FOR THE FUTURE

Many poor countries are losing their best educated people and most skilled workers through emigration. This migration from less-developed to more-developed countries has been called the **brain drain.** In some cases a country's schools are qualifying students for professional careers faster than the economy can absorb them. In other cases the skilled or educated seek political freedom, and in still other cases the country's best students go abroad for training with the intent of returning home to help their own people, but they stay abroad and never return home. A 2005 study in the 30 rich countries that make up the Organization for Economic Cooperation and Development (OECD) revealed that from a quarter to a half of the college-educated citizens of poor countries including Ghana, Mozambique, Kenya, Uganda, and El Salvador lived abroad in an O.E.C.D. country—a fraction that rises to more than 80 percent for Haiti and Jamaica.

The United States profits most from the drain of skilled labor from other countries, and in fact, the nation's scientific establishment would suffer without them. About one-third of the U.S. Nobel Prize winners in the sciences have been foreign-born. Thousands of senior scientists, engineers, and other professionals migrate to the United States each year, and about half of the young foreigners who come for education never return home. Almost half of the scientists and engineers working in California's Silicon Valley are foreign-born.

The drain of scientists, engineers, and physicians from poor countries undermines the ability of these areas to develop. Any advanced country is reluctant to expel these skilled immigrants. That country would suffer its own brain drain, and the skilled surely would migrate to another advanced country anyway. How to keep these skilled professionals in their homelands, or to attract them back, is a problem that must be solved to avoid an ever-widening gap between the world's rich and poor countries.

Chapter **Review**

SUMMARY

The population of Earth is about 6.5 billion people, and 90 percent of the population is concentrated on less than 20 percent of the land area. Major population concentrations are in east Asia, in south Asia, and in Europe. Secondary concentrations are in Southeast Asia and eastern North America. Much of Earth's land surface is virtually uninhabited.

Factors that explain this distribution include the physical environment, technology, history, and local differences in rates of population increase. Most of the population is concentrated in areas of mild midlatitude or seasonal humid tropical climates, but people will settle in harsh areas if it is profitable. In some places local population densities may be high, but trade and circulation free members of societies from the constraints of their local environments.

The populations of China and the Indian subcontinent long ago achieved productive agricultural methods and relative political stability. Europeans multiplied when they gained material wealth and improved their food supplies as the result of world exploration and conquest, and later again during the Industrial and Agricultural Revolutions. European migrants took technology to more sparsely populated areas.

The world population is increasing, but the rate of growth is slowing. The rate of increase in each country represents a balance among the crude birth rate, the crude death rate, the fertility rate, the TFR, and migration. Most of the increase is occurring in poor, not rich, countries. The history of population growth in today's rich countries traces a demographic transition. In the first stage both the crude birth rate and the crude death rate were high. In a second stage medical science improved, so crude death rates dropped. Therefore, rates of population increase rose. In the third stage crude birth rates dropped, so today's rates of natural population increase are low. Many rich countries' populations are not even reproducing themselves.

In today's poor countries, crude birth rates are dropping fast, even without significant increases in material well-being. Theories of explanation focus on the growing role of family-planning programs, new contraceptive technologies, and the educational power of mass media. Family-planning efforts face obstacles in traditional cultures, religions, and in some cases the cost of devices.

The total population will continue to grow because the number of young women reaching their childbearing years is larger than ever before, but the sooner the world's TFR falls to the replacement rate, the lower the total population will be, at which point the projected numbers level off. The current most probable estimate is that Earth's population will reach 8.9 billion by 2050.

Natural population decreases can result either from an increase in the death rate, which could result from disease or war, or by a lowering of the birth rate. Today almost all poor countries are committed to family planning. A preference for large families is traditional in some areas, and having children may be economically rational for families in poor countries, but many people in poor countries want small families. The low status of women inhibits family planning in some societies. Urbanization usually lowers TFRs.

The Malthusian theory argues that the rate of growth in the human population will always outpace the rate of growth in food supplies, triggering cycles of famine, disease, and war.

The natural ratio between females and males in a human population should be one to one, but, worldwide, men outnumber women. However, the sex ratio varies greatly among different countries. This may be due to a combination of economic and cultural factors.

The median age of humankind as a whole is rising. In rich countries, questions arise about the possibility of sustaining economic growth and about the equitable distribution of wealth among different generations. Many poor countries must quickly build up their national incomes and devise national welfare or social security programs.

Human migrations have redistributed populations throughout history, and significant migrations continue. Consideration of push factors drives people away from wherever they are, and consideration of pull factors attracts them to new destinations, but some migrations have been forced. Sojourners intend to stay in their new location only temporarily.

The most significant population transfers during the past 500 years have been migrations of Europeans around the world, migrations of Blacks out of Africa, and migrations of other groups instigated by European expansion.

It is sometimes difficult to differentiate political refugees, who are fleeing politically repressive countries, from economic refugees, who are trying to move from the poor countries to the rich countries. The brain drain of educated people and skilled workers from poor countries undermines their ability to develop.

KEY TERMS

African diaspora p. 195
apartheid p. 198
arithmetic density p. 167
asylum p. 201
brain drain p. 210
carrying capacity p. 167
crude birth rate p. 172
crude death rate p. 172
cultural mosaic p. 210
demographic transition p. 176
demography p. 166
dependency ratio p. 174
domestication p. 171
doubling time p. 173
emigration p. 166

epidemiological transition p. 183
epidemiology p. 183
fertility rate p. 172
immigration p. 166
indigenous peoples p. 191
infant mortality rate p. 176
life expectancy p. 179
Malthusian theory p. 186
melting pot p. 209
migration chains p. 191
natural increase or decrease p. 172
Overseas Chinese p. 200
pathogens p. 183
physiological density p. 167

population geography p. 166
population projection p. 172
population pyramid p. 174
pull factors p. 190
push factors p. 190
race p. 191
racism p. 193
refugee p. 200
replacement rate p. 172
sojourner p. 191
total fertility rate (TFR) p. 172
undocumented, irregular, or illegal immigrants p. 190
workers' remittances p. 202
zero population growth p. 173

QUESTIONS FOR REVIEW AND DISCUSSION

1. Briefly describe the distribution of the human population.

2. Explain the demographic transition model, then explain the new developments in demographics that challenge the continuing validity of that model.

3. What current developments may increase the world's death rate?

4. What have been the major currents of the African diaspora?

5. Who and where are the Overseas Chinese? Why are these people of such importance in the world today?

6. Differentiate a refugee from an economic refugee.

7. What is the Malthusian theory?

THINKING GEOGRAPHICALLY

1. How has physical geography affected the distribution of people in any given continent or region?

2. Investigate any country's family-planning efforts and experience.

3. Do you know any elderly people who have retired and are now collecting pensions? Where did they spend their working years? Where are they living now?

4. Do you know how many generations of your family have moved either to this country or within this country? What push and pull factors motivated them?

5. Has your local community seen net in-migration or net out-migration in recent years? If in, where from? If out, where to? What push and pull factors did they consider?

6. Compare the cartogram in Figure 5-2, countries drawn according to their population size, with Figure 12-5, countries drawn according to their economic output, with a standard area world map. Do you think that one or the other of these criteria—population, output, or area—ought to determine the amount of time you, as a student, should devote to each country in your studies? Which ought to determine how much time on the evening news is devoted to each country? Can you suggest any other measures of each country's "importance"?

7. Are any major refugee streams reported in this week's headlines? If so, what are the push factors?

SUGGESTIONS FOR FURTHER LEARNING

Alba, Richard, and Victor Nee. *Remaking the American Mainstream: Assimilation and Contemporary Immigration.* Cambridge, MA: Harvard University Press, 2003.

Berlinski, Claire. *Menace in Europe: Why the Continent's Crisis is America's, Too.* New York: Crown Forum Books, 2006.

Brettell, Caroline, James Hollifield, and Barry Chiswick, eds. *Migration Theory: Talking Across Disciplines.* New York: Routledge, 2000.

Clarke, Kamari Maxine, and Deborah A. Thomas, eds. *Globalization and Race.* Durham, NC: Duke University Press, 2006.

Diamond, Jared. *Guns, Germs and Steel: The Fates of Human Societies.* New York: W.W. Norton & Co., 1999.

Dummett, Michael. *On Immigration and Refugees.* New York: Routledge, 2001.

Fagen, Patricia Weiss, ed. *The Uprooted: Improving Humanitarian Responses to Forced Migration.* New York: Lexington Books, 2005.

Finer, Catherine Jones, ed. *Migration, Immigration and Social Policy.* New York: Blackwell, 2006.

Garrett, Laurie. *Betrayal of Trust: The Collapse of Global Public Health.* New York: Hyperion Press, 2000.

Gilbert, Geoffrey, and Mildred Vasan, eds. *World Population: A Reference Handbook.* Contemporary World Issues. Santa Barbara: ABC-Clio, 2006.

Goldin, Liliana, ed. *Identities on the Move: Transnational Processes in North America and the Caribbean Basin.* Austin, TX: University of Texas Press, 2000.

Graham, D. *Population Geography.* New York: Routledge, 2006.

Hayter, Teresa. *Open Borders: The Case Against Immigration Controls.* New York: Pluto Press, 2001.

Hyndman, Jennifer. *Managing Displacement: Refugees and the Politics of Humanitarianism.* Minneapolis: University of Minnesota Press, 2000.

Kapur, Devesh, and John McHale. *Give Us Your Best and Brightest.* Washington, D.C.: Center for Global Development, 2005.

Kunitz, Stephen J. *The Health of Populations: General Theories and Practical Realities.* New York: Oxford University Press, 2006.

Livi-Bacci, Massimo. *A Concise History of World Population: An Introduction to Population Processes.* 4th ed. New York: Blackwell, 2006.

Nyce, Steven A., and Sylvester J. Schreiber. *The Economic Implications of Aging Societies.* Cambridge, UK: Cambridge University Press, 2006.

Ozden, Caglar, and Maurice Schiff, eds. *International Migration, Remittances, and Brain Drain.* Washington, D.C.: The World Bank, 2005.

Pagden, Anthony. *Peoples and Empires.* New York: Modern Library, 2003.

Population Reference Bureau. *Population Bulletin.* Washington, D.C., quarterly.

Poston, Dudley L., and Michael Micklin, eds. *Handbook of Population.* Handbooks of Sociology and Social Research. New York: Springer, 2006.

Preston, Samuel, Patrick Heuveline, and Michel Guillot. *Demography: Measuring and Modeling Population Processes.* Malden, MA: Blackwell, 2001.

Rumbaut, Ruben, and Alejandro Portes, eds. *Ethnicities: Children of Immigrants in America.* Berkeley: University of California Press, 2001.

U.S. Committee for Refugees. American Council for Nationalities Service. *World Refugee Survey.* Washington, D.C., annually.

Weinstein, Jay A., and Vijiyan Pillai. *Demography: The Science of Population.* Boston: Allyn & Bacon, 2000.

Williamson, Paul, William Gould, Clare Holdsworth, and Robert Woods, eds. *Population and Society.* New York: Sage Publications, 2006.

World Health Report. Geneva: World Health Organization, annually.

World Population Prospects. New York: United Nations Population Division, periodic.

Yu, Jason Xiao. *Human Population and Demography: A Guide to Reference and Information Sources. Social Sciences.* Westport, CT: Libraries Unlimited, 2006.

WEB WORK

A good place to start looking for information about population, health, and migration is at the United Nations central page for population studies:

http://www.un.org/esa/population/unpop.htm

The U.S. Census Bureau is at:

http://www.census.gov/

And it also maintains a special page for its International Programs Center:

http://www.census.gov/ftp/pub/ipc/www/

Two private organizations maintain informative home pages with links to other interesting sites. They are the Population Reference Bureau:

http://www.prb.org/

And the Population Connection:

http://www.populationconnection.org/

The World Health Organization, a U.N. agency, can be found at:

http://www.who.int/en/

World diseases and epidemics can be followed at the Web page that the U.S. Centers for Disease Control and Prevention maintains for travel warnings:

http://www.cdc.gov/travel/

The International Organization for Migration, an intergovernmental body that tracks migration figures, can be found at:

http://www.iom.int/

The Migration Information Service is at:

http://www.migrationinformation.org/

The home page for the United Nations High Commissioner for Refugees can be found at:

http://www.unhcr.org

For immigration to the United States, the Bureau of Citizenship and Immigration Services is now part of the U.S. Department of Homeland Security:

http://www.bcis.gov/graphics/index.htm

And for Canada, always try Statistics Canada at:

http://www.statcan.ca/start.html

6

CULTURAL
GEOGRAPHY

Some of the people of Leavenworth, Washington, recreate the annual celebration of "May Day" as it was in their ancestral Germany. They have even recreated a "typical" (definitely old-fashioned) German architectural setting. Americans of many backgrounds cling to historic cultural traditions even in the country known as the "melting pot." Studying the evolution and diffusion of cultures is key to understanding human geography. (Alamy Images)

A Look AHEAD

Cultural Evolution Contrasts with Cultural Diffusion

Evolutionism studies the sources of cultural change that are within individual cultures themselves, whereas diffusionism emphasizes how cultures spread out from the places they originate and are adopted by other peoples. Folk cultures preserve traditions, but popular cultures embrace changing norms.

Identity and Behavioral Geography

The behavior of any individual is partly unique to that person, but much of any person's behavior is specific to his or her cultural group. People often group themselves and interact with other groups defined by culture, race, ethnicity, or identity, even though it may be difficult for an observer to identify these groupings "objectively." Geographers study

how individuals and groups perceive and respond to their environments.

Culture Realms

The entire region throughout which a culture prevails is called a culture realm. Culture realms may be defined on the basis of religion, language, diet, customs, economic development, or still other criteria. Aspects of culture realms may be visible in the landscape.

The Global Diffusion of Western Culture

Much of today's world cultural geography is the legacy of the diffusion of European culture that accompanied European political and economic global supremacy from about 1500 until about 1950. The attacks on the United States that occurred on September 11, 2001, shocked Americans and other global citizens into reflection on America's role in the world today.

*I*magine that deep in the Brazilian rain forest there is a human community that is completely isolated. The people carry on their lives in total ignorance of the rest of us here on the planet, knowing nothing of international trade, international political affairs, or even of the government of Brazil, which exercises no effective jurisdiction over them. Certainly they are affected by the complicated affairs of the rest of us. They see our airplanes in the sky, and their local environment is changing because the rest of us are polluting the atmosphere, but they explain these things either as natural phenomena or as the work of spirits. These isolated people—let's call them the "Solitary" people—believe that they are the only humans on Earth (Figure 6-1).

The Solitary people would have developed their own way of life, including a language to communicate among themselves, a religion, and Solitary ways of organizing their families and society. They must depend entirely on their local environment for all of their needs, including food, clothing, and shelter. Thus, for example, the Solitary people would have learned over time which local plants and animals can be eaten. If the Solitary people receive no imports from the outside world, neither do they produce anything for others. Everything produced locally is consumed locally, and any surplus of food or goods is stored for emergencies or else just wastes away. Activities are probably closely related to the seasons.

As the Solitary people make use of their local environment, they can also transform it. They can clear forests or plant new ones, drain swamps or dig irrigation systems, fertilize the soil or deplete and waste it through poor agricultural practices (Figure 6-2). If they

FIGURE 6-1 A discovery of an isolated people. This haunting photograph records the first contact between these formerly isolated people and the rest of humankind. On August 4, 1938, a U.S. scientific expedition entered the supposedly uninhabited Grand Valley of the Balim River in western New Guinea. They found over 50,000 people living in Stone Age circumstances totally unaware of the existence of anyone else on Earth. We cannot say for certain whether any totally isolated groups live on Earth today, but it is less probable each year. (Neg./Transparency no.131456. Courtesy of the Department of Library Services, American Museum of Natural History)

FIGURE 6-2 A cultural landscape. This aerial photograph of the Grand Valley of the Balim was taken a few weeks before the expedition actually contacted the valley's inhabitants. It must have been a surprise to look down into a supposedly uninhabited valley and see these neat clearings and settlements outlined by irrigation ditches. This was clearly a cultural landscape, one modified by considerable human effort. (Neg./Transparency no.131457. Courtesy of the Department of Library Services, American Museum of Natural History)

burn local vegetation, they make their own contribution to world atmospheric pollution. They can even redesign local landforms, just as many societies have reduced steep mountainsides to stepped terraces for agricultural purposes. People are never entirely passive; they interact with their local environment.

CULTURAL EVOLUTION CONTRASTS WITH CULTURAL DIFFUSION

The word *culture* was introduced in Chapter 1 to describe everything about the way a people live: their clothes, diet, articles of use, and customs. Geographers study the origin, diffusion, and extent of all aspects of cultures. An object of material culture is called an *artifact,* which means literally "a thing made by skill." Culture, however, includes patterns of behavior as well as possessions. The way a group does things—its interpersonal arrangements, family structure, educational methods, and so forth—are part of its culture, and so are the facts of a group's mental or imaginative life—its poetry, music, language, and religion.

Human cultures are never static; they are constantly changing, as people learn new techniques and develop new cultural traits. The forces that cause these changes may be divided into two general categories. Some changes evolve within the society itself, but others are triggered by contact with other societies. The Solitary people, for example, were isolated, so their culture would change only through internal evolution. They might discover new uses for local plants through the years, or they might devise new religious ceremonies. **Evolutionism** is the point of view that the most important sources of cultural change are embedded in cultures, and change is internally determined. **Diffusionism,** by contrast, emphasizes how various aspects of cultures spread out from their places of origin and are adopted by other peoples. The process of spreading is called **cultural diffusion,** and the process of adopting some aspect of another culture is called **acculturation.** Cultural diffusion can be actively imposed, as when an outside power conquers a region and imposes its way of life, or it can be freely chosen, as when one group discovers and adopts some aspect of a different culture that it considers superior to its own.

This chapter will begin with an examination of two classic evolutionist theories. Both emphasize the role that technological and economic evolution play in the development of cultures and in the interaction between societies and the environment. Both theories have been used to suggest that certain developments in the past were inevitable, or even that certain events are inevitable in the future.

These imaginary Solitary people will have developed in isolation their own distinctive culture, and they will have constructed their own distinctive cultural landscape, as both of these terms were defined in Chapter 1. This chapter explores how cultures originate, develop, and spread. ▶

The second part of this chapter will then analyze diffusionistic approaches to the study of culture. Through history, the diffusion of goods, ideas, and techniques among peoples has increased, and the original individual characteristics of world cultures have come to be overlain or commingled with new shared characteristics. This mingling has created new and original cultural combinations.

The last section of this chapter analyzes the worldwide diffusion of European culture that accompanied European global conquest and economic dominance. European political dominance has retreated, but it left behind cultural, political, and economic legacies. The United States inherited Europe's leadership, yet today many people are reevaluating America's global role.

Theories of Cultural Evolution

Humankind has, for the most part, adapted itself to varying environmental conditions through cultural and technological evolution, not through biological evolution. For example, humans survive Minnesota winters by inventing furnaces and warm clothes, not by evolving furry bodies. Our cultural or technological tools allow us to survive anywhere. Scientific research stations can be maintained even in the frozen wilderness of Antarctica or in the driest desert, although supplies may have to be brought to the stations' residents. Thus, humans exist in every biome.

The theory of human stages
One theory of cultural evolution argues that all cultures evolve through certain stages of development, and it defines these stages by the way in which the culture exploits the environment. This theory was first articulated by a Roman general, Marcus Tarentius Varro (116–27 B.C.), and it remained almost unchallenged until the nineteenth century.

Varro argued that humankind originally derived its food from things that people hunted or harvested naturally. He referred to humans in this stage of evolution as **hunter-gatherers.** Varro stated that humans then domesticated animals and moved into the evolutionary stage of **pastoral nomadism.** Pastoral nomads have no fixed residences, but drive their flocks from one place to another to find grazing lands and water. If their movements are regular and seasonal—as, for example, between mountain and lowland pastures—their movements are called *transhumance*. The evolutionary stage of pastoral nomadism was in turn followed by settled agriculture. Agriculture was at first *subsistence agriculture*, which means that people raised food only for themselves, but subsistence agriculture slowly evolved into *commercial agriculture*, which means that people raised crops to sell (see Chapter 8). The final stages of social and economic evolution were urbanization and industry.

Varro's theory was generally accepted until the German naturalist and explorer Alexander von Humboldt (1769–1859), whom we met in Chapter 1, pointed out that the native peoples of South America had never experienced a stage of pastoral nomadism. Therefore, Varro's theory could not be true of all societies everywhere. The different ways of life described by Varro might be "categories" of human societies at different times and places, but they are not "stages" on a single path of social evolution.

Today there are still human groups living in each category described in this theory of cultural evolution, which we will look at in greater detail in Chapter 8. Varro's theory that all human societies advance through the same series of stages, however, carries a danger. According to this unilinear theory, societies cannot just be "different" from each other; some must be "ahead" and others "behind." Such presumptions have been used by "advanced" peoples to justify replacing "more primitive" peoples. For example, in the nineteenth century many Americans argued that it was inevitable that the Native North Americans' way of life (pastoral nomadism and subsistence agriculture) would be destroyed and replaced by American farmers. Thus, any activity causing this replacement was morally neutral. Other thinkers, however, argued that human behavior is based on conscious choices. Therefore, the results of human actions can never be scientifically inevitable and morally neutral.

Historical materialism

Historical materialism is a school of thought that tries to write a plot for human history based on the idea that human technology has increased humankind's control over the environment. Karl Marx (1818–1895) was the founder of historical materialism.

Marx said that humankind has progressively conquered the physical environment in order to improve its own material welfare. A contemporary ruler may or may not be wiser than one in the past, but a modern tractor can do more work than a horse dragging a wooden plow. Therefore, historical materialists assert, history's plot is the advance of technology. Marx insisted that all social evolution is rooted in technological evolution. This is because technology determines any society's economic system, and the economic system, in turn, determines the society's political and social life (Figure 6-3). Thus, as technology advances, technological change triggers changes in all these other aspects of society. Marx's argument that all aspects of a society are

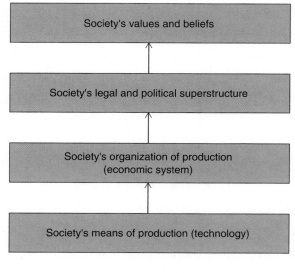

FIGURE 6-3 The theory of historical materialism. Historical materialists believe that any society's productive technology determines its economic system, and its economic system determines its political system. Ultimately, even a society's values and beliefs rest on its technological foundation. Technology is always improving, so all the other aspects of any society are constantly being dragged along and forced to change, too. Technology is the truly revolutionary force in society.

rooted in its technology and economic system has always aroused great controversy.

Marx was an optimist, and he agreed with Condorcet (see Chapter 5) that technology would eventually produce material abundance for all people. Goods would no longer be scarce, so people would no longer dispute their allocation through economic and political structures. Goods would be created and distributed "from each according to his abilities; to each according to his needs."

Some of what Marx wrote has turned out to be misguided guesses, and some of it is simply false, but much of what Marx wrote is generally accepted. Today no one doubts that science progresses almost inexorably, and that scientific progress often upsets society's economic, political, and even philosophical assumptions. The cloning of a lamb and the birth of another with human genes in 1997, for example, challenged our laws and even our ideas of life. Many of the political movements that have used Marx's name, however, disgrace his memory. The communist parties that seized power in Russia in 1917 and imposed their ideology on other peoples in the twentieth century bear no relationship to the humanitarian society that Marx envisioned material abundance would bring.

Historical materialism contrasts with Malthusianism
Recently some economists and ecologists have attacked historical materialism from a new direction. Historical materialism maintains that in the past,

technological progress offered rising standards of living to human populations: Historical materialists assume that this rate of technological advance will continue. This optimism characterizes some of our present-day leading scientists and industrialists, who would be classified among the Cornucopians discussed in Chapter 5.

Today many concerned scientists and critics, however, belong to the opposing neo-Malthusian camp. They warn that there is no guarantee that technology will continue to provide rising standards of living for the increasing human population. Confidence that technology will always solve problems of scarcity might lead humankind to disaster. Marx himself never imagined the explosive growth of the human population that has occurred since his death. During his time the human population was only about 1.5 billion, and it was growing at an annual rate of less than 1 percent. The human population today is over 6.5 billion, and it is rising at an annual rate of just over 1 percent. The neo-Malthusians ask whether the increase of the human population is, in fact, outpacing technological development or whether it might do so in the future.

These are some of the most controversial arguments of our time, and you will hear echoes of both the neo-Malthusian and the Cornucopian positions throughout this book. People of different political and economic philosophies offer different answers, and by the time you finish reading this textbook, you will have more information with which to form ideas of your own.

Cultures and Environments

Marx wrote that the key to understanding a society is the degree of control that society has over its environment. This degree of control is more important than the nature of the environment itself. Some writers, however, argue that variations in the physical environment itself explain even the variations among human cultures. The simplistic belief that human events can be explained entirely as the result of the effects of the physical environment is called **environmental determinism.** The study of the ways societies adapt to environments, by contrast, is called **cultural ecology.**

Many writers have emphasized climate as a major control on human affairs. The Greek physician Hippocrates (460?–377? B.C.) argued in his book *Airs, Waters, Places* that civilization flourishes only under certain hospitable climatic conditions. Historian Arnold Toynbee (1889–1975) turned Hippocrates' idea upside down with his own *challenge-response theory.* Toynbee argued that people need the challenge of a difficult environment to put forth their best effort and to build a civilization. A rich environment encourages only sluggishness. Today whenever anyone expresses a preference for an environment of seasonal change, as in the northern United States, over the almost tediously fine weather of southern California, that person is suggesting the challenge-response theory.

Other aspects of the natural environment have been nominated as the key factor determining a people's life. West Virginia's state motto, "Mountain people are always free," echoes a belief that rugged topography creates rugged individualists. Some political scientists have argued that societies in environments that are either particularly wet or particularly dry have necessarily developed totalitarian governments to organize and care for waterworks. Many environmentally deterministic theories are intriguing, but exceptions can always be found. For example, the Dutch have managed their waterworks democratically for hundreds of years (Figure 6-4).

The ability to find exceptions should warn us that no simple theory completely explains the relationship between the environment and human societies. Human affairs are not simple, and when we examine any world situation or historic event carefully, environmental determinism, or any other single-factor explanation, proves insufficient. Nineteenth-century French geographers proposed the theory of **possibilism** as an antidote to environmental determinism. Possibilism insists that the physical environment itself will neither suggest nor determine what people will attempt, but it may limit what people can profitably achieve. The choices and constraints involved in utilizing the natural environment are often as much cultural, economic, political, and social as they are technological. For example, Canadians have the technological capability to overcome environmental limitations and to grow bananas in greenhouses in Canada, but they choose not to do so. It is cheaper for them to buy bananas grown in Costa Rica.

Environmental determinism today   Assumptions about the physical environment's influence in human affairs still affect economic and political debates. For example, today many of Earth's tropical regions are poor, and there are three main hypotheses about why they are poor.

1. One theory emphasizes that the tropical environment handicaps human societies. Tropical heat multiplies the activity of organisms that are inimical to people and to agriculture, and it reduces human work efficiency. Tropical storms frequently are violent, and many tropical soils are poor.

2. Another school of thought, however, points out that many tropical areas were long held as colonies by countries in the temperate latitudes. This colonial experience left physical, cultural, and economic legacies that, combined with the continuing patterns and terms of world trade (discussed in Chapter 12), best explain tropical regions' current poverty.

3. A third point of view insists that the cultures of the peoples of the tropical regions retard their economic growth. Perhaps their cultures do not

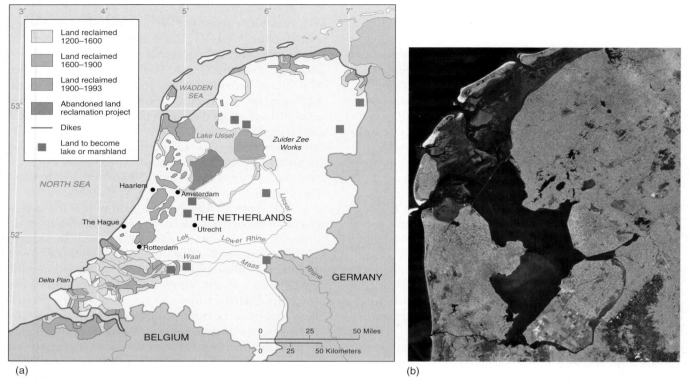

FIGURE 6-4 The Netherlands vs. the sea. More than half of the Netherlands lies below sea level, so most of the country would be under water if it were not for massive projects to modify the environment. Since the thirteenth century the Dutch have reclaimed more than 4,500 square kilometers (1,800 square miles) of *polders,* which are pieces of land created by draining water. Most polders are reserved for agriculture, although some are used for housing, and one contains Schiphol, Europe's busiest airport. The second distinctive modification of the environment is the construction of massive *dikes* to prevent the North Sea from flooding much of the country. The Dutch have built dikes in two major locations, the Zuider Zee project in the north and the Delta Plan project in the southwest. With these massive projects finished, attitudes toward the environment have changed in the Netherlands. A plan adopted in 1990 calls for returning 26 square kilometers of farms on polders to wetlands or forests. In the infrared satellite photograph showing the northern area of the map (b), the vegetation shows as red, more developed areas as green. ([a]: From W. H. Renwick and J. M. Rubenstein, *An Introduction to Geography.* Upper Saddle River, NJ: Prentice Hall, 1995; [b] Science Photo Library/Earth Satellite Corporation/Photo Researchers, Inc.)

encourage economic growth or these countries are poorly governed.

Each side in this debate may be partly correct. Which side you take will determine how you allocate responsibility for the current poverty. The first theory faults uncontrollable environmental factors; the second faults the rich, formerly colonial powers; and the third faults the peoples and governments of the tropics themselves. Furthermore, your assumptions will determine what solutions you suggest to the problem of poverty in the tropics. If the environment is to blame, nothing can be done. If the former colonial powers are to blame, perhaps they should pay reparations. If the tropical peoples themselves are to blame, they should change their habits or work harder.

All forms of environmentalist assumptions remain common in popular discussion and debate. Often they seem to be harmless figures of speech, but they are intellectually dangerous. They predispose conclusions and close the mind to alternative explanations. Watch carefully for these simplistic traps.

Cultural Diffusion

The isolation experienced by our imaginary Solitary people is very different from the way most of us live today. Through history, global communication and transportation have increased, and trade and other cultural exchanges have multiplied. Most people no longer exploit their local environments and develop their cultures in isolation. They are interconnected by transportation and communication of goods, people, ideas, and capital. More people eat imported foods and combine them in imported recipes. They wear imported clothes in imported styles, and they fashion their built environments out of imported materials in styles that originated among distant peoples. Your daily life undoubtedly exemplifies this interconnectedness. Many articles that you use and other aspects of your activities draw on materials and ideas from around the world. To describe all this movement, we use the term *circulation.*

Almost everywhere today, cultural diffusion is more important than cultural evolution. Anthropologist

Clifford Geertz wrote in 1995, "The very notion of isolation lacks, these days, much application. There are very few places—there may not be any—where the noises of the all-over present are not heard, and most anthropologists work by now in places where such noises all but drown out local harmonies." What Geertz called "the noises of the all-over present" is cultural diffusion, and what he called "local harmonies" is cultural evolution. By 1995 cultural diffusion everywhere overwhelmed cultural evolution, yet advances in transportation and communication have accelerated since then, when the World Wide Web was in its infancy. Find a picture of the most isolated people you can, and chances are high that they will be wearing or carrying something that they did not make themselves (Figure 6-5). What happens *at* places depends more and more on what happens *among*

places, and mapped patterns of economic or cultural factors can be understood only if we comprehend the patterns of movement that create them and continuously rearrange them. Geography doesn't just *exist*; it *happens*.

The anthropologist Clark Wissler (1870–1947) called the places where cultures are developed *geographical culture centers*. He stated a principle, called the *age-area principle*, that if traits diffuse outward from a single geographical culture center, the traits found farthest away from that center must be the oldest traits. This description of cultural diffusion suggests concentric waves spreading out from a stone dropped into a pool. Carl O. Sauer (1889–1975) elaborated geographical studies of cultural diffusion. Chapter 1 explained that we call the place where a distinctive culture originates the hearth area of that culture, and it outlined the paths of cultural diffusion. Geographers can investigate the distributions of any cultural attributes; they are limited only by their imagination and curiosity.

Diffusion does not explain the distribution of all phenomena. Sometimes the same phenomenon occurs spontaneously and independently at two or more places. In the history of mathematics, for example, the idea of the zero and its use as the basis of a numerical place system was conceived in two places: among the Maya, a Native Central American people, and among the ancient Hindus of India or perhaps the Babylonians before them. Diffusionists once argued that Phoenicians must have sailed across the Atlantic from their homeland in the eastern Mediterranean thousands of years ago and taught the Maya how to use the zero. There is, however, no evidence for this assumption, and it underestimates the ingenuity of the Maya. To demonstrate diffusion, we must be able to illustrate its path from one culture to another.

Folk culture   Today's rapid pace of cultural innovation and diffusion requires us to differentiate folk culture from popular culture. The term **folk culture** refers to a culture that preserves traditions. Folk groups are often bound by a distinctive religion, national background, or language, and are conservative and resistant to change. Most folk-culture groups are rural, and relative isolation helps these groups maintain their integrity. Folk-culture groups, however, also include urban neighborhoods of immigrants struggling to preserve their native cultures in their new homes. *Folk culture* suggests that any culture identified by the term is a lingering remnant of something that it is embattled by the tide of modern change.

Cultural geographers have identified culture hearths and routes of diffusion of a surprising number of folk cultures across the United States. Geographers who have investigated the architecture of houses, for example, have differentiated three types and traced the diffusion paths of these styles westward with various immigrant groups (Figure 6-6). Barns and other structures are also built in distinct architectural styles that reveal the origins of their builders. Folk geographic studies in the United States range from studies of folk songs, folk foods, folk

FIGURE 6-5 Manufactured clothes. This little boy wears a traditional robe, as is legally required in Bhutan, the world's most isolated country. Bhutan struggles to preserve its traditional culture, and it allows only a limited number of foreigners to visit each year. The collar of a modern blue shirt, however, peeks out from underneath the robe. The shirt, the doll, and other aspects of his attire and possessions are clearly Western imports. It may be that traditional Bhutanese culture has become only a façade. (Courtesy of Brett Cobb)

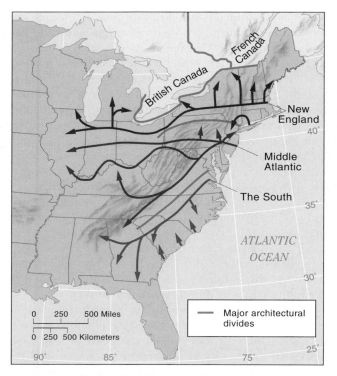

FIGURE 6-6 Paths of diffusion of house types.
Geographers have identified three main styles of houses that originated on the U.S. Eastern Seaboard and diffused westward along with pioneer settlers. (From Rooney, Zelinsky, and Louder, eds., *This Remarkable Continent*. College Station, TX: Texas A & M Press, 1982.)

FIGURE 6-7 Popular culture mingles with a folk culture.   The straw hat, suspenders, and plain black boots identify this Pennsylvania boy as Amish, but his in-line skates demonstrate how most traditional folk cultures allow adoption of some aspects of modern popular culture. (Keith Myers/New York Times Pictures)

medicine, and folklore to objects of folk material culture as diverse as locally produced pottery, clothing, tombstones, farm fencing, and even knives and guns.

In North America the Amish provide an example of a folk culture. The Amish are a sect within the Mennonite Protestant religious denomination formed in Switzerland in the sixteenth century. Early Amish immigrants to North America settled in Pennsylvania and spread across the Midwest and into Canada. The Amish are notable because they wear plain clothing and shun modern education and technology. They prosper by specializing their farm production and marketing their produce, but they severely curtail the choice of goods that they will accept in return (Figure 6-7).

Many African Americans have long felt that their traditional folk cultures were taken from them during the years they suffered in slavery, yet careful investigation has yielded a rich variety of cultural elements in some African-American communities today that are traceable to specific African origins. These range from musical styles to the meticulous sweeping of front yards (a tradition traced from villages of coastal Nigeria to rural Georgia) to the patterns of speech discussed in Chapter 7.

The United States is relatively homogeneous in culture and is modern and technologically advanced. Cultural diffusion and change churn across the country rapidly and repeatedly. In such a country the identification and study of folk cultures help us appreciate and cherish the richness of some of our remaining folk traditions. Chapter 5 noted that Canada, by contrast, more carefully nurtures and preserves its diversity as a "cultural mosaic."

Some newly independent countries are today emphasizing or even resurrecting traditional folk cultures as a way of enhancing national identity and community. When the nation of Eritrea won independence in 1993, for example, one of the new government's first acts was to open a national university mandated to research and define folk literature, folk music, and other distinct Eritrean traditions.

Popular culture   **Popular culture,** by contrast, is the culture of people who embrace innovation and conform to changing norms. Popular culture may originate anywhere, and it tends to diffuse rapidly, especially wherever people have time, money, and inclination to indulge in it.

Popular material culture usually means mass culture—that is, items such as clothing, processed food,

REGIONAL *Focus* ON

Lahic

The village of Lahic is one of the world's most extraordinary examples of an isolated community that has created and maintained a distinctive local culture. Lahic is a village of approximately 2,000 people, perched high on a ragged cliff surrounded by the towering Caucasus Mountains in the Republic of Azerbaijan. Nobody knows when this village was first settled or who the first settlers were, but the unique local language provides a clue. It is a form of Persian (Iranian) that died out in Iran hundreds of years ago. Perhaps some ancient Iranian emperor sent settlers to Lahic to defend the empire's borders against the peoples to the North.

Still today Lahic is connected to the outside world by only a rugged dirt path that winds its way around precipitous gorges and is closed by rain and snow much of the year. Wires from Baku, the capital of Azerbaijan some 128 kilometers (80 miles) away, brought electricity and telephone service to the village in the 1980s, but these services are provided only a few hours a week.

The desolate landscape of Lahic allows the people to raise livestock, but they cannot feed themselves, so long ago they developed a tradition of exchanging craftwork for food. In their isolation, the people of Lahic developed artisanal skills that have made their handcrafted products famous and prized around the world. Virtually every kind of handcrafted object has been made here and exported for hundreds of years: objects in copper and brass, carpets, handcarved and wrought iron, guns, cutlery, clothing, and leather goods. Every home is a workshop. Lahic crafts have long been treasured by collectors throughout the Middle East and Europe. Traders discovered them

many centuries ago and sold them for high prices in bazaars in Baghdad, Shiraz, and other great Middle Eastern cities. A display of copperware made in Lahic won a gold medal at the Paris World Exposition in 1878, and today Lahic crafts are on display in museums from London and Paris to Moscow and Istanbul.

A 1997 U.N. report on Azerbaijan noted that "Lahic . . . has kept almost perfectly intact a microcosm of cultural features which has preserved its communal harmony and social cohesion across the centuries. The spirit of the Middle Ages still lingers there."

These days, more young people are going off to school or to find new job opportunities in Baku and thereby abandoning their village's ancient crafts. How much longer can this time capsule survive?

Traditional artisan workmanship. This artisan in Lahic, Azerbaijan, demonstrates the skills that have earned renown for products from his isolated village and kept its economy thriving for hundreds of years. (Stanton R. Winter/ New York Times Pictures)

books, CDs, and household goods that are mass produced for mass distribution. Whereas folk culture is often produced or done by the people at-large (folk singing and dancing, cooking, costumes, woodcarving, etc.), popular culture, by contrast, is usually produced by corporations and purchased.

Mass manufacturing lowers the cost of items, but in order for a mass-produced model to succeed, consumer taste must be homogeneous. Such mass taste necessarily requires some sacrifice of individuality and cultural identity. The United States has long been the world's largest relatively homogeneous consumer market. This has lowered the cost of items and raised Americans' material standard of living. The substantial unity of the Canadian and U.S. markets has raised

Canadian standards of living, too, but some Canadians believe that it has threatened Canadian cultural identity.

The popular culture of the United States exhibits a great degree of national homogeneity, but individual consumer goods win varying degrees of acceptance (called *market penetration*) in different regions of the country. The regions are often the vernacular regions as mapped in Figure 1-8. The marketing managers of corporations are "applied cultural geographers" who specialize in this type of analysis. The soft drink Dr Pepper, for example, originated in the U.S. South, and it still enjoys its greatest acceptance there. The U.S. South is today less culturally distinct than it was decades ago, but a cultural heritage, much of which lingers from earlier folk traditions, still exists (Figure 6-8).

Focus ON

The Diffusion of Anglo-American Religious Folk Songs

The map of the diffusion of white spiritual songs from eighteenth-century New England exemplifies folk cultural geography. These songs almost seem to have slid down the Appalachian Mountains into the Upland South and then into the Lowland South, where many of them are still popular. Some individual migrants carried the songs westward, but for the most part the acceptance of these songs diffused from community to community. The songs have a simple melodic structure, which reveals their folk quality, and a first-person narrative, which reveals their Protestant Christian faith (for example, "Don't you see my Jesus coming? He'll embrace me in His arms"). These songs did not diffuse in other directions, because of cultural barriers. Further to the south, the songs' Protestant religious sentiment hindered their diffusion into largely French Roman Catholic Louisiana and Hispanic Roman Catholic southern Texas. To the north, both religious and political barriers prevented their diffusion. Quebec in Canada is also largely Roman Catholic.

At the same time as these songs were spreading southwestward, they largely disappeared from their culture hearth in the Northeast. This was probably because that region urbanized rapidly and its folk culture was replaced by a popular culture.

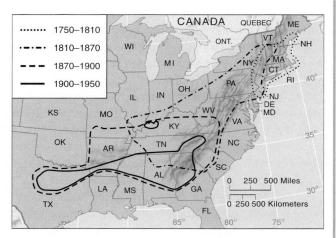

The diffusion of Protestant religious folk songs. Protestant religious folk songs diffused from New England toward the southwest until stopped by cultural barriers. (Reproduced by permission of the American Folklore Society from *Journal of American Folklore* 65:258 [Oct.–Dec. 1952]. Not for further reproduction.)

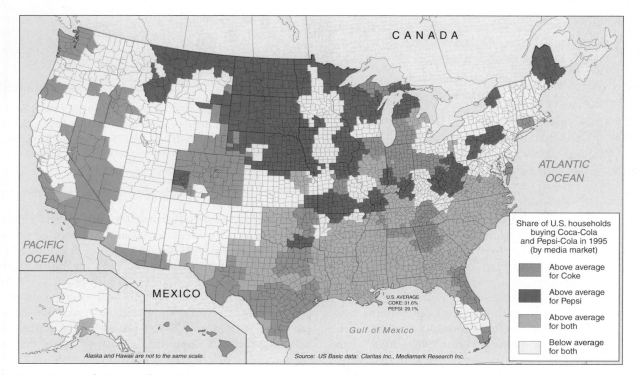

FIGURE 6-8 Coke vs. Pepsi. Both Coca-Cola and Pepsi were born in the U.S. Southeast in the nineteenth century—Coca-Cola in Atlanta, Georgia, and Pepsi-Cola in New Bern, North Carolina—and each is consumed in the region at a rate above the national average. Perhaps the warm climate explains that fact. Nationwide, Coke drinkers tend to be urban and financially better off. Pepsi drinkers are somewhat older, suburban, and rural. Market penetration information is proprietary to the corporations, and it is seldom released. (*U.S. News & World Report,* August 5, 1996)

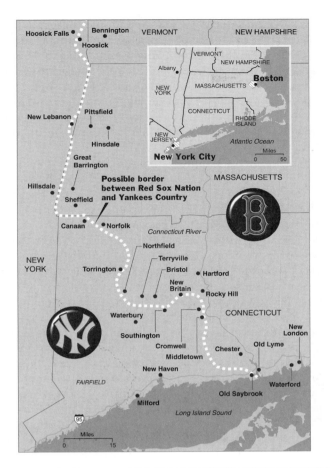

FIGURE 6-9 A "watershed" between territories of fans of two baseball teams. In 2006, researchers from Connecticut's Quinnipiac University set out to define the borderline between baseball fans loyal to the New York Yankees and fans loyal to the Boston Red Sox. Research techniques included interviews, checking sales of baseball caps and other merchandise with team logos, and even sales of locally made cookies with the different teams' logos. Rocky Hill, Connecticut, is the midpoint between the teams' homes of Yankee Stadium in the Bronx, New York, and Fenway Park in Boston, but the dividing line of team loyalties was found to lie slightly south of it. (New York Times Agency)

Geographers investigate the origins and diffusion paths of popular material culture and also of popular social culture. For example, people devise new ways of living, working, and playing, and they innovate in education and in employer–employee relations. Sport is an important part of popular culture, and popular culture regions can be defined by their sporting preferences (Figure 6-9). Regions differ in their popular entertainments—a movie that is a success in Seattle, for example, may flop in Houston. Cultural geographers want to know why. The radio stations in different regions across the United States play varying mixes of country, gospel, rock, classical, and other popular kinds of music; DJs base their decisions on market research such as that shown in Figure 6-10. Geographers

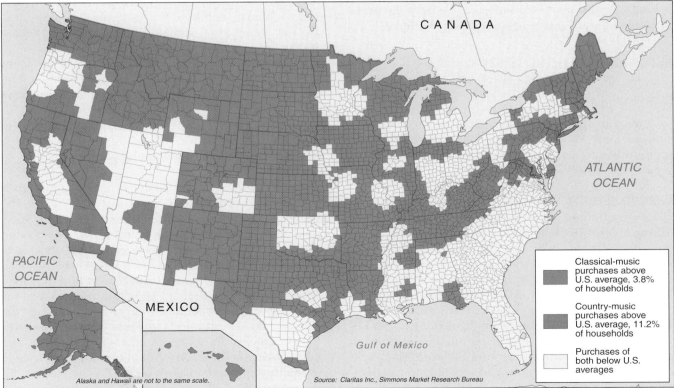

FIGURE 6-10 Tastes in music. Purchasers of classical music compact discs are generally urban, upper-income, and highly educated, whereas a much wider social spectrum and geographic distribution of people buy country music. (From Michael J. Weiss, *Latitudes & Attitudes,* Boston: Little, Brown, 1994, p. 55)

have investigated these and many more attributes of popular culture.

The marketing of popular culture can overwhelm folk culture. For example, how many of the characters and tales of traditional U.S. folk culture—or even of genuine historical heroes and heroines—have survived the onslaught of mass-produced and mass-marketed merchandise related to comic-strip characters? Who is most famous in the United States today: Paul Bunyan (a mythical folk hero), Johnny Appleseed (a real person, John Chapman, 1774–1845), or Sponge-Bob SquarePants, a cartoon product of popular culture? The answer is probably SpongeBob SquarePants.

IDENTITY AND BEHAVIORAL GEOGRAPHY

At any time a great many disputes disturb the world. Widespread international wars, civil wars, urban riots, insurrections, and street fighting persist. The press often refers to the parties of the disputes as being of different culture groups, races, ethnic groups, or other identities, but outsiders to the disputes find it difficult to understand the bases for animosity. To understand the background of many contemporary news items, it is important to understand what those categories mean and how they influence people's behavior. When geographers study identities, perception, and patterns of behavior, geographers' work borrows ideas from the study of psychology but also makes fresh contributions.

Grouping Humans by Culture, Race, Ethnicity, and Identity

When we define human groups, we must be sure that we know and agree on exactly what the criteria of inclusion are and whom the group includes. Unclear or vague references to groups may cause misunderstanding or even insults.

Culture groups
The definition of a culture may include a great number of characteristics or just a few. For example, all social scientists agree that language is an important attribute of human behavior. Two people who share a language share something very important. If, however, those two people hold different religious beliefs, feel patriotism for two different countries, and eat different diets, social scientists may insist that although those two people share one attribute of culture, they do not share one culture. The great number of English speakers who presently live around the world share few attributes other than language, so we would not say that they all share one common culture. If, however, two people do share a language, religious beliefs, political affiliation, and dietary preferences, then social scientists would agree that those two people share a culture.

A *subculture* is a smaller bundle of attributes shared among a smaller group within a larger, more generalized culture group. For example, Italian Americans, Chinese Americans, and African Americans share subsets of cultural attributes within the larger American culture. Sometimes even one single attribute—shared loyalty for a sports team, for example—can bind individuals so strongly that that single attribute is termed a subculture.

Ultimately, cultural affiliation may be a matter of the feelings or the preferences of the individuals. Two people may share so many cultural attributes that observers insist that they share a culture, yet they may hate each other. They may even kill each other because, *to them*, their differences are crucial. Each day's newspaper proves this statement. Psychologist Sigmund Freud (1856–1939) described what he called "the narcissism of the minor differences," pointing out that the most vicious and irreconcilable quarrels often arise between peoples who are to most outward appearances nearly identical. Our study of political geography in Chapter 11 cites examples of countries, which outsiders regarded as homogeneous, that have nevertheless broken out in civil war. Outsiders did not see the cultural fault lines within the countries. In other cases, people whom observers would describe as very different from each other feel strong bonds. What they share is more important to them than their differences. We must always investigate people's feelings in order to understand their loyalties and their animosities. People carry multiple identities. For example, a middle-class, Baptist, Republican, disabled, black mother of West Indian ancestry and U.S. citizenship may feel many allegiances, and at different times she may allow any one of her identities to determine her behavior. This flexibility may either facilitate or hinder interaction with others.

Many people may feel a sense of community with others of their own "race," but, as discussed in Chapter 5, the concept of race remains ambiguous.

Ethnic groups
The concept of an **ethnic group** is frequently confused with the concept of a cultural group. The word *ethnic* comes from the Greek for "people," and the definition of an ethnic group may depend upon almost any attribute of biology, culture, allegiance, or historic background. The word has historically been used in a pejorative sense: Its meanings have included "alien," "pagan," and often "primitive." Some social scientists nevertheless define ethnic groups and study the groups' characteristics or attributes. Ethnomusicology, for example, is the study of ethnic groups' music, and ethnobotany is the study of ethnic groups' knowledge of the uses of plants (Figure 6-11). The migrations discussed in Chapter 5 are sometimes described as migrations of ethnic groups.

Ethnocentrism is the term given to the tendency to judge other cultures by the standards and practices

FIGURE 6-11 An ethnobotanist. Dr. Michael Balick of the New York Botanical Garden (center) learns plant uses from these traditional healers of the Kekchi Maya people in Belize. Dr. Balick is researching potential anti-AIDS and anticancer agents for the National Cancer Institute. The healers' clothing reflects how fast these people are being acculturated and their traditional knowledge is being lost. (Photo by S. Matola, courtesy of Michael Balick and The New York Botanical Garden)

of one's own, and usually to judge them unfavorably. Practices in other cultures that may seem strange to us, however, may in fact be sensible and rational. Conversely, some aspects of our own culture may seem strange or even offensive to others. Most Americans, for example, assume that a man should have one wife and a woman one husband at a time, but any number of spouses are allowed in a series. This would shock many people. Americans, in turn, may be shocked to learn that Tibetans assume that a woman is married to all sibling brothers at once. This is called *fraternal polyandry*. Through thousands of years, fraternal polyandry has made possible sustainable population increases and prevented the fragmentation of landholdings in Tibet's poor mountain valleys. The social ramifications of fraternal polyandry confound most Americans. An American might ask how to identify the father of any given child. To a Tibetan, however, it makes no difference, and a Tibetan might consider the question prurient.

Any geography book will contain examples of ways of life that contrast with your own. None is necessarily "right" or the best for everybody. All people have to overcome the initial assumption that "different from" the way they do things themselves is "worse than" their way: people everywhere also can learn to appreciate and respect the integrity of other people's behaviors. Shrewd people even learn from others whenever they can.

Behavioral Geography

Behavioral geography is a subfield of cultural geography that studies our perception of the world around us and how our perception influences our behavior. The political commentator Walter Lippmann (1889–1974) differentiated "the world outside" from "the pictures in our heads." The world outside is the way things really are, but the pictures in our heads may be based on preconceptions, misperceptions, or incomplete understanding. Geographers call these pictures in our heads **mental maps.** Geographers borrow from psychology the theory of **cognitive behavioralism,** which argues that people react to their environment as they perceive it. In other words, people make decisions on the basis of their mental maps. Thus, we cannot understand why people make the decisions they do or act as they do by studying only the real environment in which they acted. We must discover what was in their heads when they made their decisions.

For example, in 1898 Senator George Hoar of Massachusetts cast a key vote for the annexation of Hawaii because, according to his diary, he drew a line on a map from Alaska to southern California, saw that Hawaii fell inside that line, and concluded that its possession was necessary for the country's defense. In fact, Hawaii does not lie inside such a line. Check its position on a globe. No one has been able to discover what sort of map Senator Hoar was looking at (perhaps one with an inset map of Hawaii?), but his misperception affected history. We make choices based on our perceptions of the world, but we must act in the real world. If there is a difference between what we think about the world and the way the world really is, we might try to do something impossible or fail to take advantage of some opportunity.

Where we choose to live, shop, or visit depends on our feelings about places—whether we think certain places are good or bad, beautiful or ugly, safe or dangerous (Figure 6-12). Geographers have investigated how we perceive environmental threats, either natural (earthquakes) or human-made (chemical or nuclear installations). How we perceive these threats affects our actions or reactions toward them.

Our individual notions of usefulness or of aesthetic value affect our evaluation of landscapes. Most urbanites, for example, agree that the Painted Desert in Arizona is beautiful, and it is a major tourist attraction.

(a)

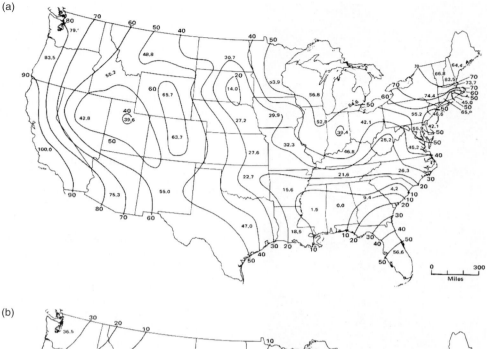

(b)

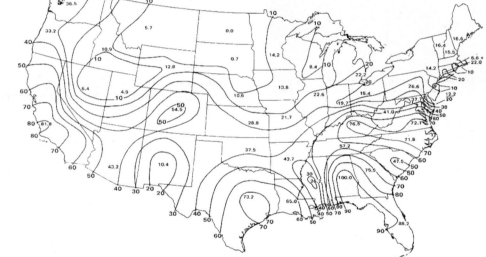

FIGURE 6-12 Residential desirability.   These are the mental maps of places' residential desirability in the heads of students at the University of California at Berkeley (a) and the University of Alabama (b). High numbers indicate a positive image in the students' minds. Students at both places seemed to like where they were. Berkeley students gave coastal California a score of 100; Alabama students rated Alabama 100. The Alabamans, however, rated the region of Berkeley (60–70) much higher than the Californians rated Alabama (0.0). Both seemed to have positive images of Colorado and low images of the Dakotas. These preferences are not necessarily based on personal experience. We all have opinions of places we have never visited, but we might be surprised if we did visit them. (From Peter R. Gould, "On Mental Maps," in Michigan Interuniversity Community of Mathematical Geographers, Discussion Paper #9, 1966. Reprinted with permission.)

A farmer, however, might be appalled at such a landscape (Figure 6-13).

Studies that have grouped men, women, children, the elderly, the disabled, and individuals categorized in many other ways have taught us that each of these groups perceives the environment differently. Therefore, these groups may almost be said to live in different environments. This knowledge is useful in the design of new environments, such as housing. Housing for the hearing impaired, for example, should have open interior sight lines. Architects and government agencies are investigating how workers are affected by their environments by monitoring heart rates and other signals of physical and mental stress levels. Results may help architects design more efficient and healthy work environments.

Differences in perception and behavior may be studied among different cultures. Human behavior is rooted in biology and physiology, but it is filtered through culture. **Proxemics** is the cross-cultural study

FIGURE 6-13 Farmland in Iowa. This black soil photographed near Martinsburg, Iowa, is so rich that a farmer might find this a beautiful view. What is your aesthetic reaction? (Courtesy of John G. Daniels)

of the use of space. People of different cultures balance the way their mind relies on input from their eyes, ears, and noses differently, even if they have the same physical sensory capacity. They do not agree, for instance, on crowding. In a room in which most Americans would feel uncomfortably crowded, most Middle Easterners would not. In designing a library, for example, it helps to know that North Americans are more distracted by sounds than by sights, whereas Japanese are more distracted by sights than by sounds. Therefore, people talk in Japanese libraries, but the libraries have study cubicles. Proxemics affects human behavior at all levels from polite interpersonal behavior to the design of buildings or even whole cities.

Some scientists believe that humans exhibit **territoriality.** Many animals lay claim to territory and defend it against members of their own species. Most aspects of territorial behavior among animals have to do with spacing, protecting against overexploitation of that part of the local environment on which the species depends for its living. Applying these studies to human behavior is complicated by the interplay between human biology and culture. People defend their standing space in a crowd (they bump back), urban gangs defend their turf, and nations defend their land; but it remains unclear if this behavior is biologically determined human territoriality.

CULTURE REALMS

The entire region throughout which a culture prevails is called a **culture realm** or **culture region.** Any aspect of culture may be used to define a culture realm. Religion is an important aspect of culture, so religion is often

chosen as a criterion. The use of the phrase "the Christian World," for instance, implies that the prevalence of Christianity across a large region unites the peoples of that region. This region might significantly be contrasted with, for instance, "the Islamic World." Each is a great culture realm, and Chapter 7 examines how the prevalence of each of those religions encourages other similarities across those realms. Other criteria that may be chosen to define culture realms include language, diet, customs, or economic development. These topics will be discussed individually in the following chapters.

Problems in Defining Great Culture Regions

Many geographers and other social scientists have drawn maps dividing the world into great culture regions. Scholars will never agree upon one single pattern of world culture regions, however, because different scholars have different purposes, so they choose different criteria. The use of different criteria yields different regions.

Our delineation of culture regions is important because the pattern we create predisposes us to stereotype the people within each region as homogeneous, rather than to recognize the differences that may exist within each region. A region, by definition, should display a certain homogeneity, but that homogeneity may exist only in the eyes of an outsider. Insiders may see the differences among themselves as greater than their similarities. For example, Westerners may label Asia as a culture region. A Korean, however, might insist that the many cultural traits that differentiate him from a Burmese are more important than anything they share. Asia does in fact exhibit greater cultural diversity within it than, for example, Europe does. Asia does not exist as a homogeneous place in the mind of a Burmese, a Japanese, a Vietnamese, or a Cambodian. It exists only in the mind of an outsider. In Chapter 12 we will see how the term *the Third World* lumps countries that are very different and thus obscures understanding of any of them.

Two questions must be asked of any pattern of culture regions. (1) Has the cartographer drawn the region correctly according to the criteria that he or she chose? For example, when mapping language regions of South America, did the cartographer remember that Brazilians speak Portuguese, not Spanish? (2) The second question to ask is, Are the chosen criteria meaningful? For example, does the fact that Brazilians speak Portuguese tell us anything important about Brazilians? Does it provoke questions about how that affects Brazil's interaction with Spanish-speaking South Americans? Any regionalization scheme is useful if it is both accurate and meaningful.

If a cartographer does not apply the same criterion consistently on a single map, the map is meaningless and confusing. For example, the World Bank has divided the world into eight "demographic regions" shown on Figure 6-14, but there are no criteria of homogeneity

Critical THINKING

Is Latin America a Region? How Did It Get Its Name?

Some criteria of homogeneity justify labeling Central and South America "Latin America." "Latin" suggests that certain important aspects of the culture spring from the cultural tradition of ancient Rome (*Latium*), and, in fact, most of Latin America was long ruled by either Spain or Portugal, both of which are Latin countries. Most Latin Americans today speak Spanish or Portuguese—both languages descended from ancient Latin—and most of the people in Latin America belong to the Roman Catholic Church. Italy itself, however, played no historic role in Latin America. Why, then, don't we call the area Luso-Hispanic America? ("Luso" meaning Portuguese, because today's Portugal was the ancient Lusitania) or Ibero-America? The more one thinks about the term *Latin America*, the more mysterious its origin becomes.

The truth is that the term *Latin America* originated as political propaganda. During the U.S. Civil War,

France's Emperor Napoleon III thought he saw an opportunity to take over Mexico. France had never had any colonial interests there, but Napoleon III invaded on the pretext that France, a Latin nation, was avenging military humiliations suffered by the Spanish and was therefore "defending Latin honor." On July 3, 1862, Napoleon published a letter in the French newspaper *Le Moniteur* introducing the term *Latin America* to justify what was really French imperialist aggression. Eventually the French were defeated and retreated, but Napoleon's sly propagandistic term has long survived him.

Questions

1. What similarities and differences are there among the countries from Mexico to Argentina? List differences and similarities among the countries in sub-Saharan Africa, in Asia, and in Europe.

2. Which do you think are more important in each case, the similarities or the differences?

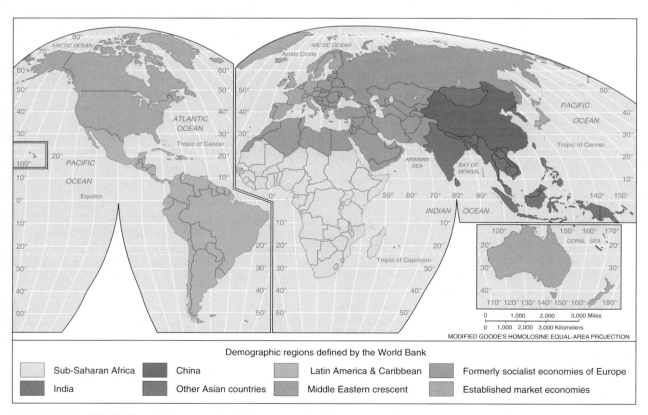

Demographic regions defined by the World Bank

Sub-Saharan Africa	China
India	Other Asian countries
	Latin America & Caribbean
	Middle Eastern crescent
	Formerly socialist economies of Europe
	Established market economies

FIGURE 6-14 World Bank "demographic regions". What internal homogeneity does each of these "demographic regions" exhibit? Virtually none. Therefore, any information provided about each of the regions on this map is not useful because the cartographer did not define clear criteria for the regionalization scheme. (From World Development Bank, *World Development Report 1993*)

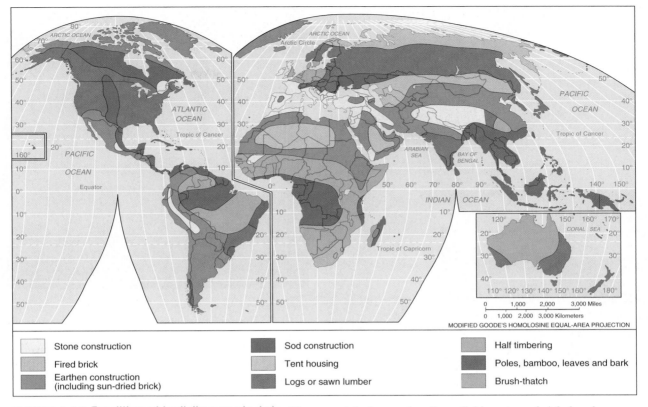

FIGURE 6-15 Traditional building materials. The materials that are locally available—stone, brick, bamboo, or wood—are used everywhere in traditional architecture, called vernacular architecture.

within any individual region, nor are there consistent criteria of heterogenity differentiating one region from another. Some regions are individual countries; other regions are defined by their national economic systems (past or present); still other regions, such as "Middle Eastern crescent," seem to be defined on the basis of religion; and the region "Other Asian Countries" is made up of countries "left over!" Each of these regions contain such great differences within it that any statistic reported to describe that region or to compare it with other regions is of little value.

Visual Clues to Culture Realms

Culture realms can reveal themselves through visual clues. These include the language of posted signs, the clothing the local people are wearing, and the goods available in local shops. In the built environment, building materials, architecture, and settlement patterns are all visible manifestations of cultures. People usually rely on local materials for building, so in one place stone may be the traditional building material, in another place brick, and in another wood (Figures 6-15 and 6-16). Innovations in the uses of these materials may diffuse across cultures.

Styles of architecture often represent adaptations to climatic conditions, so a particular style may be adopted in different regions with similar climates

FIGURE 6-16 The mosque at San, in the West African country of Mali. This building demonstrates that when people have mud and very little wood, they can still build extraordinary structures. It is in the distinctive Dyula architectural style, named for a trading people who diffused the style as they moved around West Africa. It is quite different from the Middle Eastern architectural styles that many Americans expect mosques to reflect. Compare it with Figure 7-24. (James Stanfield/National Geographic Image Collection)

FIGURE 6-17 Simla, India. The building in the foreground is a bungalow, which is a Hindi word for a low-sweeping single-story house with a roof extending out over a veranda. Such houses were first built in the mountain foothills of northern India, but the style was copied throughout the British Empire and eventually the United States. Today in English the word *bungalow* means almost any sort of small house. Simla was the summer capital of the British Indian Empire between 1864 and 1947. It was cooler up at Simla (2,200 meters/7,100 feet; notice the vegetation) than down at New Delhi in the plains about 275 kilometers (170 miles) to the south. The medieval-style cathedral in the background seems an odd presence in northern India. It is not, however, a medieval building. It was built in the nineteenth century in the medieval style that was popular for Christian churches then (and still is today). Wherever the British went, they took their culture and traditions, including architectural styles. (Courtesy of the Library of Congress)

FIGURE 6-18 A Palladian building in the United States. This house, called the Morris-Jumel Mansion, was built in 1765 in Manhattan 10 miles north of Wall Street. The choice of the Palladian architectural style is highly symbolic. Standing virtually on the frontier of Western civilization, this style boldly—almost arrogantly—announced European conquest and the coming of European civilization to the New World. Symbolism and aesthetic taste can prevail even over comfort, for although this style is appropriate for the Italian climate, in New York it would have been impossible to keep warm. Furthermore, in Italy it would have been built of stone, yet here it was built entirely of wood masquerading as stone. (Courtesy of Robert F. West/Morris-Jumel Mansion)

(Figure 6-17). Cultural preferences are sometimes so powerful, however, that an architectural style may diffuse beyond the limits of where its building materials can be found and even beyond the range where that architecture is comfortable. The style of the Italian Renaissance architect Andrea Palladio (1508–1580), for example, spread to England because of aesthetic preferences, despite the fact that Palladian buildings are uncomfortable in England's damp, cool climate. From England the preference diffused to America, even to areas that lacked both the appropriate climate and the necessary building materials. As Europe and the United States came to dominate other regions, Palladian architecture diffused throughout the world (Figure 6-18).

Public statuary and monuments may reveal local cultural values, and they may change as local values change. For example, many cities in the U.S. South erected monuments to commemorate the Confederate cause, but as African Americans gained political power there, they insisted the monuments be taken down. Russia has marked the fall of communism by erecting public statues of its pre-Communist czars (Figure 6-19). Statues also may reveal that a place is dominated by a cultural or political outsider. For example, several Eastern European countries fell under the domination of the U.S.S.R. after World War II (see Chapter 13), and statues of Russian "liberators" were often set up in their public places. Since these countries have regained their freedom, however, the local people have replaced these statues with statues of genuine local heroes. Budapest, Hungary, for example, boasts new statues of ancient Hungarian royalty and Christian saints.

Settlement patterns The designs of settlements reflect cultural differences, so a trained observer can see in the look and layout of whole towns and cities the cultural backgrounds of their builders. Rural societies can be differentiated by the way that some cluster

Focus ON

Regionalism in the United States

Interest in American regionalism has swung in and out of popularity among scholars throughout history. Today, despite a commercial landscape of Starbucks, GAP, and McDonalds that look the same from coast to coast, the study of American regions and regionalism is winning new interest. In 1976 the University of Nebraska opened its Center for Great Plains Studies, which was followed by The Center for the American West at the University of Colorado at Boulder, the Appalachian Regional Studies Center at Radford University in Virginia, the Southwest Center at the University of Arizona, the Center for the Study of Southern Culture at the University of Mississippi, the New England Studies Program at the University of Southern Maine, the Southern California Studies

Center at the University of Southern California, among others.

The University of North Carolina Press published *The Encyclopedia of Southern Culture* in 1989, and that volume was followed by similar studies for other regions. New York City and Chicago have their own encyclopedias.

Many critics doubt whether genuine folk cultural regions still exist in the United States. They ask whether we are trying to preserve something or whether we are only re-creating something because we fear homogenization. Some regional characteristics were based on characteristic livelihoods. Can we restore whaling in New England, or Western cowboys?

Each Canadian province also has its own encyclopedia, but these regions may be more culturally distinct than the U.S. states are.

FIGURE 6-19 Russia's Czar Alexander II. This new statue of Czar Alexander II (ruled 1855–1881) was unveiled in Moscow in 2005, fourteen years after the fall of Communism, which had replaced the monarchy in 1917. Today's government honored the Czar for having emancipated the serfs and achieved judicial and military reforms. (Alexander Nemenov/AFP/Getty Images)

housing settlements, whereas others isolate settlements in individual farmsteads. City planning will be discussed in Chapter 10.

In those societies that cluster housing, farmers may choose to live together in clusters ranging from a few dozen homes up to thousands. There are no dwellings in the surrounding farmland, so the farmers journey out to work in the fields each day. The farm buildings are usually concentrated together with the human settlement. There may be no economic reason for the clustering; it is a cultural choice.

Such compact villages may be found in many forms: irregular; wandering along a principal street, river, or canal; clustered about a village common; or checkerboard (Figure 6-20). Clustering may reveal family or religious bonds, communal land ownership, or the need for common security against bandits or invaders. The government may deliberately cluster the population to supervise it or provide education or health care. Clustering may also have environmental reasons—people may cluster at water sources, for example, or on dry places when the surrounding land is swampy. Clustering is more common among farmers than among livestock ranchers, except in Africa, but it generally characterizes settlement across much of Europe, Latin America, Asia, Africa, and the Middle East.

Isolated farmsteads, by contrast, characterize those areas of Anglo-America, Australia, New Zealand, and South Africa that were settled by Europeans, and also some parts of Japan and India. The conditions for isolated farmstead settlement usually include peace and security in the countryside; agricultural colonization of the region by individual pioneer families rather than by

(a) Muang Nan, Laos

(b) Ban Mae Sakud, Thailand

| 0 | 15 | 30 Meters |
| 0 | 50 | 100 Feet |

N

Houses   Sheds and farm buildings   Direction of sleeping

FIGURE 6-20 Village settlement patterns in Laos and Thailand. Arrangement of houses in two Southeast Asian communities. (a) The front gables of Laos houses, such as those in Muang Nan, Laos, face one another across a path, and the backs face each other at the rear. Ridgepoles are set perpendicular to the path and parallel to a stream if one is nearby. Inside the house, the head of a sleeping person is in the opposite direction to that of a sleeping neighbor, so that heads and feet are always together. (b) In Ban Mae Sakud, Thailand, houses are not set in a straight line, because of a local belief that evil spirits move in straight lines. Ridgepoles are set parallel to the path, and the heads of all sleeping persons face east. (From James M. Rubenstein, *The Cultural Landscape: An Introduction to Human Geography*, Updated 7th ed. Upper Saddle River, NJ: Prentice Hall, 2003.)

socially cohesive groups; agricultural private enterprise, as opposed to communalism; unit-block farms in which all of a farmer's land is in a block rather than in scattered parcels; and well-watered but well-drained land.

The history of the United States provides examples of both patterns of settlement. In the southern colonies in the seventeenth and eighteenth centuries, settlement was characterized by widely dispersed, relatively self-sufficient plantations. New England settlement, by contrast, reflects the social cohesion of the settlers' society. The people of New England were tightly bound in religious communities, so they advanced westward in tiers of adjacent, well-planned towns. The reason for the differences in the settlement patterns was not economic; it was a cultural choice.

Forces That Stabilize the Pattern of Culture Realms

Despite the force of diffusion, a number of factors tend to fix the geography of culture realms. Culture leaves its mark on the landscape. The fixed pattern of activities,

land uses, transport routes, and even individual buildings will guide, restrict, or predispose future patterns and activities. The construction of a factory, for example, represents a great investment of money, and once the factory is operating, it relies on a local workforce and develops ties to local suppliers. An industrial complex such as this cannot easily be picked up and moved. **Inertia** is the term for the force that keeps things stable. All of a people's fixed assets in place—railroads, pipelines, highways, airports, housing, and more—are called the **infrastructure.**

Historical geography is the subfield within geography that studies the geography of the past and how geographic distributions have changed. Historical geographers can sometimes read landscapes as if through time the landscape has been overlain with layer after layer of peoples using the land in different ways and organizing it for different purposes. The landscape is like an old manuscript on which a reader can discern earlier erased writings (Figure 6-21). Scholarly recreations of past land uses were defined as *sequent occupance* studies by the geographer Derwent Whittlesey in 1929. An alternative approach to historical geography

FIGURE 6-21 An abandoned city. An abandoned city prompts us to ask, "Who was here before us? What did they do? What happened to them?" This ruin is in Morocco. (Courtesy of Moroccan National Tourist Office)

focuses on how the transformation, use, and organization of the landscape is continuously changing.

Culture includes a set of values and ways of doing things, and culture groups seldom get displaced or eliminated entirely. Culture is learned behavior, and cultural norms are handed down through generations. Groups may be quick to take up new techniques or products, but tradition is a powerful force in human life, and imported ideas or lifestyles only slowly transform a people's entire cultural inheritance. This inertia is reinforced by the existence of territorially sovereign states and the enormous power that governments exert over their citizens to teach and to enforce norms of behavior. This power includes, as we have seen, the power to resurrect and promote folk traditions. Lee Kuan Yew, Senior Minister of Singapore, emphasized to an interviewer the role of culture in various Asian countries: "Culture is very deep-rooted; it's not tangible, but it's very real: the values and perceptions, attitudes, reference points, a map up here [he tapped his head], in the mind."

Today each local culture is a unique mix of what originated locally and what has been imported, and each unique culture is a local resource, just as surely as the minerals under the soil and the crops in the fields.

Most peoples value their culture, and they usually try to preserve key aspects of it. This influences the way they interact with other peoples. A people's

self-consciousness as a culture may be codified in their religion, as it is with the Jews, or in their sense of their own history. This is called **historical consciousness.** Many Americans have difficulty understanding other people's historical consciousness, because one aspect of American culture is an optimistic denial that history can shackle future opportunity. Other peoples, however, nurture their traditional cultures and their historical consciousnesses. Many people act the way they do because that is the way their ancestors acted, or to right wrongs or to account for deeds or misdeeds that took place long ago. Several peoples fighting in the Balkans in the 1990s, for example, felt that they were avenging medieval battles.

One major theme in geography is the tension between forces of change and forces for stability. Cultures evolve, cultures diffuse, and peoples can transform themselves and their behavior, but cultures and culture realms also have elements of stability. Any cultural pattern or distribution maps the current balance between those forces.

Trade and Cultural Diffusion

Cultural isolation is usually accompanied by economic self-sufficiency, but trade diminishes people's cultural isolation at the same time as it expands their economic possibilities. Trade breaks down cultural isolation and triggers cultural diffusion. Every item in trade is a product of the culture that originates it, so economic exchange is one of the most important forms of cultural diffusion. Choosing goods is an act of self-definition, of social and cultural identity (Figure 6-22). Trade, economics, and culture are intertwined.

FIGURE 6-22 New consumer goods change cultures. An Avon Lady demonstrates a product to Tembe tribesmen in Brazil's Amazon region. The tribesmen obviously have their own cultural tradition about makeup, but the introduction of new cosmetic products will change these people's culture in ways that cannot be foreseen. (John Maier, Jr./The Image Works)

The study of how various peoples make their living, how economies develop, and what peoples trade is **economic geography.** Trade releases people from dependence on their local environment. It allows them to draw resources from around the world. Fewer and fewer people anywhere rely on their local environment for all of their needs, so the link between the resources of any local environment and the well-being of the people who live there has been weakened. The Swiss, for example, live in an environment that is poor in natural resources, yet the Swiss have grown rich through trade and services.

As peoples come into contact with others and begin to trade, they usually first export only whatever they have in surplus and have no use for. They usually view imports as luxuries unnecessary to their way of life. Trade, however, triggers far-reaching cultural and economic changes.

For example, imagine three isolated communities in three different environments (Figure 6-23). Village A is located along a river plain. The people there catch fish, which they fry; they grow rice, which they eat as rice cakes; and they distill rice into sake (rice wine). Village B lies on the slope of nearby hills. The people there have domesticated grapes, which they have learned to distill into wine, and oats, which they make into oat bread. They have also domesticated goats, which clamber about the rugged hillsides. They milk the goats for dairy products, and they roast goat meat. The people in village C, up on a nearby plateau, grow corn for cornbread and to be distilled into whiskey and have domesticated cattle, which they eat as broiled steaks. Each village has developed other aspects of a unique culture, too, including its own language, religion, and customs, but let us focus on their diets as representative of their cultures.

The construction of a road and the commencement of trade among the three villages will probably trigger at least four significant changes.

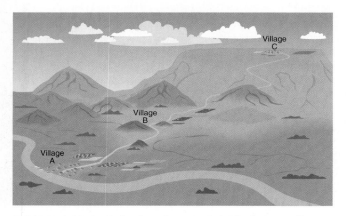

FIGURE 6-23 Isolated cultures yield to interaction. The inhabitants of three isolated villages in three different environments learn to use local resources, and they develop three distinct cultures. When a new road connects those villages, however, the lifestyles of all three villages may be transformed. Their cultural possibilities will multiply, and their economies will evolve.

1. Each village will have access to the products of each of the other villages. The people of village A, for example, will taste broiled steaks, and the people of village B will be introduced to fried fish. We might call this the simple *addition of cultural possibilities.*

2. New cultural combinations, or *cultural permutations,* will appear: fried steaks, broiled fish, roast beef, and more. A good example of cultural permutation is the appearance of pita fajitas on menus in Southern California. A pita fajita is a piece of pita bread, an item brought to southern California by immigrants from the Near East, stuffed with fajita ingredients, usually beef or chicken, which were introduced to Mexico by the Spaniards long ago and brought to Southern California by Mexican immigrants. Nobody knows who first devised this permutation. Today many U.S. McDonald's restaurants offer sweet and sour sauce, which is of Chinese derivation, with Chicken McNuggets. People often find surprising new uses for objects created by other cultures and societies, even when they do not grasp the fundamental values or technologies that created the objects in the first place.

   All aspects of the cultures of all three villages will experience these two results—cultural addition plus cultural permutation. Residents of one village may convert to the religion of another, or perhaps the religions will blend. The languages, styles of architecture, music, games, clothing, and other customs of all three villages may add up and also permutate. Those who profit from a new trading system and exchange with other cultures often challenge and overturn local politics and traditions.

3. The residents of all three villages will see how imports can raise their standard of living. In order to pay for the imports they want, they will dedicate more effort to *producing items for trade—that is, for markets.* The change from self-sufficiency to production for markets is one of the greatest transitions in history. The existence of trade and markets gives each village, for the first time, an incentive to produce surpluses of its local goods. Agronomists (economists who specialize in agriculture) say that, in general, "The market produces the surplus." In other words, if farmers have a market where they can sell their surplus for a profit, they are likely to produce a surplus. Furthermore, when people specialize in producing an item, they become more efficient, and the quantities they produce increase. Therefore, the total amount of food—of all goods—produced in all three villages will probably increase.

   Production for the market affects cultures in another subtle way: The people may produce less of those local goods that cannot be exported, but more of those local goods that win broader markets. Traditional local items may be altered in order to increase exports. For example, many Native American tribes once made traditional craft items such as baskets, jewelry, and clothing for themselves only. Then they began to produce these items for tourists, and today they concentrate on producing those items and styles that tourists favor. They even alter traditional designs if requested to do so by customers. Thus, people may slowly surrender their own traditional culture.

As trade multiplies, more of what people produce in any one place is consumed elsewhere. Conversely, a growing percentage of the things people use are produced elsewhere. Eventually, if the terms of trade are favorable, people dedicate most of their efforts to producing export products, and they rely on imports even for their necessities. They have surrendered their self-sufficiency and become dependent on trade, and they surrender their cultural isolation and experience cultural change.

4. Village B will probably develop a market larger than those in villages A or C. This is because of its situation between the two other villages. Village B is the most convenient place in the pattern or network of exchange, called the *central place*, so it will probably grow to be the largest.

World Trade and Cultural Diffusion Today

Chapter 12 examines world trade in detail, but we must note here that virtually all peoples are today experiencing this evolution from self-sufficiency and cultural isolation to trade and cultural exchange. The share of any country's territory that is devoted to export production may be small, but the shares of the national population and the national income that are involved are rising everywhere. Even within individual countries, growing cities create markets for food from surrounding rural regions. This draws the rural population out of subsistence agriculture into commercial agriculture.

Regions or peoples have not always freely chosen to enter into the system of production and exchange. Some areas were forced into specialized production by colonialism. Peoples in other regions were forced to buy goods, and they had to develop exports in order to pay for these goods. The British actually went to war to force the Chinese to buy opium (1839–1842). Opium is an addicting drug, but all kinds of less physically damaging new goods—electronic equipment, leisure activities, or styles of clothing—may trigger our desire to enjoy them. This process has been called the creation of new **felt needs**—things people begin to think they need. As people begin to depend on the availability of

desirable goods, they become more deeply enmeshed in the web of trade and circulation.

Trade leads to specialization of production and to greater production, but it does not benefit all regions equally. Some regions prosper, and others fall behind. Why international trade causes this to happen is a principal subject of research in economic geography. Many different reasons will be suggested throughout this book.

We noted earlier that items of popular culture achieve differing market penetration throughout the United States, suggesting differing regional cultures. Similarly, the international marketing divisions of large corporations study why some products have worldwide appeal, whereas others are successful only in geographically restricted markets. Salespeople strive to break down cultural differences, but the integrity of each culture realm resists complete homogenization. Therefore, cross-cultural advertising and marketing of consumer goods present fascinating cultural–geographic questions. For example, Domino's Pizza has spread around the world, but the corporation has learned that Germans prefer smaller pies than North Americans do, and Japanese like pies topped with squid and sweet mayonnaise. Wal-Mart, which is by some measures the largest and most successful corporation in world history, retreated from both Germany and South Korea in 2006, losing billions of dollars. Company executives admitted that they simply could not understand German and South Korean consumer cultures. A few consumer products, such as Coca-Cola, have achieved almost global diffusion. Is there anyplace where people refuse cola drinks (Figure 6-24)? Yes, Utah. Why there? Many of Utah's devout Mormons avoid the caffeine of the original colas and have not taken to the decaffeinated brands.

Consumer goods are only one aspect of culture. In some places people wear American blue jeans, but

FIGURE 6-24 A global product. Some consumer goods practically blanket the Earth. (Courtesy of Julia Nicole Paley)

TABLE 6-1	Transport and Communication Costs 1920–1990, Expressed in 1990 Dollars		
YEAR	**AVERAGE OCEAN FREIGHT AND PORT CHARGES PER SHORT TON OF IMPORT AND EXPORT CARGO**	**AVERAGE AIR TRANSPORT REVENUE PER PASSENGER MILE**	**COST OF A 3-MINUTE TELEPHONE CALL NEW YORK TO LONDON**
1920	$95	NA	NA
1930	60	$0.68	$244.65
1940	63	0.46	188.51
1950	34	0.30	53.20
1960	27	0.24	45.86
1970	27	0.16	31.58
1980	24	0.10	4.80
1990	29	0.11	3.32
NA—Not applicable			

Source: Courtesy Institute for International Economics.

they wear them to political demonstrations against the United States. They like blue jeans, but that does not mean that they like everything about U.S. culture. In fact, when popular culture from one country overwhelms the folk culture of another, this process may be viewed as offensive cultural aggression. We will examine below how this process feeds some people's animosity against the United States.

The Acceleration of Diffusion

In the past, travel and transportation were more difficult and expensive than they are today, whether we measure the cost in time, money, or in any other unit. The friction of distance was so high that only a few things moved far, and those things moved slowly. Over the past 200 years, however, technology has reduced the friction of distance and accelerated the diffusion of cultural elements (Table 6-1). We often hear that the world has "shrunk." As the cost and time of moving almost anything—people, food, energy, raw materials, finished goods, capital, information—have steadily fallen, things are not necessarily so fixed in place as they were in the past. Many activities have been significantly released from the constraints of any given location. If an activity can move or relocate freely, we call it a **footloose activity.**

Originally, information could move only as fast as a person could carry it, but electronics disengaged communication from transportation. When Samuel Morse demonstrated the first intercity telegraph line in 1844, a Baltimore paper wrote, "This is indeed the annihilation of space." The telephone, invented in 1879, furthered the annihilation of space by allowing a person, figuratively, to be in two places at the same time. The annihilation of space has continued with electronic mail (e-mail), facsimile machines, computer modems, and other electronic devices collectively known as the **electronic highway.** More people with personal electronic devices are plugged in everywhere with increasing regularity (Figure 6-25). Schoolchildren with

FIGURE 6-25 Where are these people? People talking on the telephone are, functionally, two places at once: where their bodies are, and the place with which they are communicating. Thus, they are literally "absent-minded," and local jurisdictions are increasingly prohibiting drivers from talking on cellphones. (Ted S. Warren/AP Wide World Photos)

personal computers can tap into networks of information that were unavailable to the world's leading scholars 25 years ago. Today the cost of global communication is virtually negligible. Between 2000 and 2006, the monthly cost of leasing a line for phone calls and data transmission from Los Angeles to Bengalooru, India (formerly Bangalore), fell from almost $60,000 to under $10,000.

The activities of communicating with other people through an electronic network or even of playing a game alone with a computer create a new mental world—a new "place" where that activity is occurring, called *virtual reality*. That extension of reality through global electronic means of communication is **cyberspace,** a word coined by William Gibson in his 1985 science fiction thriller *Neuromancer.* The term may be applied widely. Many office workers, for example, collaborate through electronic networks without sitting down together in one office. One worker may be at home tending children, a second in a car on the road, and a third in an airport waiting room. The electronic highway connects them in a virtual office in cyberspace, so office work is increasingly footloose.

The compression of space compresses time. The latest music heard on radio stations in Los Angeles and New York will be heard in Nairobi, Kenya, and Montevideo, Uruguay, before the week is over.

New means of communication grant access to news and cultural elements from all parts of the world, but we continue to be selective in what we pay attention to (Figure 6-26). Our own backgrounds; our education, perceptions, and prejudices; and the media to which we are exposed affect our understanding of other peoples and places and how we rank their relative importance. For a great variety of reasons, we are more knowledgeable about or more interested in some places than in others. These reasons are not necessarily related to the size or population of a particular region.

The fluidity of world economic and political groupings today has stimulated many theories of what new groupings might form in the future. The political scientist Samuel Huntington predicted that "the fundamental source of conflict [in coming decades] will not be primarily ideological or primarily economic. The great divisions among humankind and the dominating source of conflict will be cultural." He warned of a coming "clash of civilizations." He defines a civilization as "the highest cultural grouping of people and the broadest level of cultural identity people have short of that which distinguishes humans from other species," and his list of major contemporary civilizations includes "Western, Confucian, Japanese, Islamic, Hindu, Slavic-Orthodox, Latin American and possibly African civilization." With few exceptions, these are the major religious groupings and regions discussed in Chapter 7.

Professor Huntington suggests that these realms are the cultural equivalents of the world's geological tectonic plates, discussed in Chapter 3. If that is true, then the areas where they meet might experience virtually unending strife. Geographers have traditionally referred to such areas as *shatterbelts*. This book will explore a few such regions, including Bosnia (Muslims caught between Eastern and Western Christians), Sudan (an Arab Muslim north battling the Black Christian and animist south), and Kashmir (between Pakistan and India).

It remains unclear, however, whether the peoples within each of Professor Huntington's "civilizations" share any real affinity, or why these affinities might trigger hostilities against other groups. The Muslim extremist Osama bin Laden hoped that the attacks on the United States on 9/11 would trigger a war between Islam and the West, and thus ignite a "clash of civilizations." As the news proves each day, however, such rhetoric simplifies profound matters of identity. Few people rely on their religion as their exclusive identity; rather, they blend it with their nationality and even their common humanity. People around the world refuse to identify themselves exclusively according to typologies held either by Islamic extremists or by U.S. scholars. Some world leaders have specifically repudiated Professor Huntington's arguments (see Chapter 7). Hypotheses of global culture wars also simplify the impact of migration (see Chapter 5), and global economic and political forces.

The Challenge of Change

Maps of culture realms, of regions of economic specialization, and of political jurisdictions reveal current distributions of specific human activities. Human

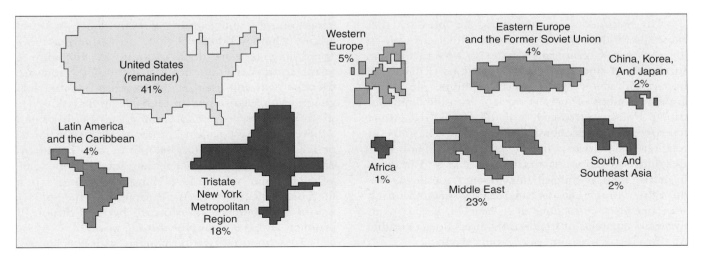

FIGURE 6-26 Relative world coverage in one newspaper. The relative sizes of world regions on this illustration correspond to the coverage devoted to each on the front page of the *New York Times* for two months in 2006. The large amount of coverage of the Middle East reflects not only the war in that region, but also that New York's large and influential Jewish population seeks news about Israel. Miami has a large and influential Latin American population, and Miami has important trade links with Latin America. Considering those facts, how do you think "the view from Miami" would differ? How about Los Angeles? How about the city in which you live? Measure your local newspaper and see.

Critical THINKING

Can Cultures Be Preserved?

In the 1930's, Brazil had scarcely begun the task of exploring its great interior expanses, let alone developing them. From 1941 to 1961 a series of expeditions, known together as the Roncador-Xingu Project, opened up 1,500 kilometers (930 miles) of trails into the forest, explored 1,000 kilometers (621 miles) of rivers, carved airstrips, and opened over 40 new towns. Along the Xingu River the Project met 14 indigenous nations that had seldom seen outsiders. At that time, native peoples were considered scarcely human and were, in fact, hunted as animals. As a result of the Project, the government created Xingu National Park, 26,000 square kilometers (10,040 square miles) where previously warring tribes learned to live together. The first director of the Park, Orlando Villas Boas (1914–2002), insisted that his task was to bring in health care workers, and also to keep other Brazilians and tourists out, to refrain from imposing white man's logic or meddling in village affairs, and to keep the people's ethnobotanical knowledge out of the hands of pharmaceutical companies.

Questions

1. Can the well-meaning goal of "protecting" other peoples become in fact a policy of denying them access to the world's ideas and goods, thus keeping them isolated in a "human zoo"?
2. Who should make the decision whether or not people should be allowed to choose for themselves whether they want the world's ideas and goods?
3. Do you agree with all of Villas Boas's goals?

---

activities, however, are not static. They are dynamic, continuously organizing and reorganizing, forming and reforming. Therefore, no pattern of the organization and use of territory is stable. Distributions and patterns are disrupted and reshaped repeatedly, just as the patterns in a kaleidoscope are. A map of human activities at any time is comparable to a weather map. We cannot understand the activities if we do not know what forces are at work. For geographers to answer the question "Why there?" we must study activities and forces.

In nature, it is variations in air pressure that cause the winds to blow from points of high pressure to low pressure. Gravity causes water to flow to the sea. In studies of human activities, variations in local job opportunities or political liberties spur people to migrate. Variations in interest rates or in profitability distribute investment capital, and the places that attract investment enjoy economic development. The force of religious belief drives people on pilgrimages to holy sites. Religious conversion triggers the flow of ideas. Variations in places' natural endowments initiate trade flows. Produce flows to markets, raw materials to manufacturing sites, tourists to attractions or to beaches. Variations in transport capability or freight rates direct flows of traffic. All these forces overcome geographic inertia.

The geography of global manufacturing, which we will examine in detail in Chapter 12, exemplifies an activity that is redistributed continuously. The factors that redistribute manufacturing include new products; new technologies; new raw materials (the replacement of metals with plastics, for example); new sources of traditional raw materials; new technologies of manufacture; new governments with new policies regarding investment, taxation, or environmental preservation; growing and shrinking labor supplies; and the opening and development of new markets. When the executives of global corporations decide where in the world to build new factories, they balance these and still more factors: Each of these factors changes every day!

This textbook will detail many redistributions: Some religions are winning new converts, expanding geographically and exerting new influence in world affairs, whereas other religions are withering away. Some languages are demonstrating the flexibility to adapt to new communication technology and are therefore gaining users at the expense of other languages. The relative rise and fall of nations' influence in world political or economic affairs wins or loses new adherents to those nations' products, cultural artifacts, or lifestyles. If the Solitary people were to be discovered today, people in other nations might adopt some aspect of Solitary culture—perhaps Solitary music or a food or medicinal crop. At the same time, the Solitary way of life would be transformed by the infusion of products and ideas from the outside world.

It is important to be familiar with the current distributions of human activities, but if you learn the reasons *why* things get distributed the way they do, then you will have learned something that will be useful for the rest of your life. Former French president François Mitterand often said, "History has accelerated." Geographic redistributions have, too.

THE GLOBAL DIFFUSION OF EUROPEAN CULTURE

Despite the rich variety of indigenous local cultures around the globe, the world is increasingly coming to look like one place. In consumer goods, architecture, industrial technology, education, and housing, the Western model is pervasive. To ethnocentric Westerners who presume the superiority of their own culture, or to local people who have accepted Western culture, this may seem natural. To them this may be "modern" life, or "progress," or "development."

A dispassionate observer, however, might expect more diversity, more styles and models of development. Why are so many people around the world imitating Western examples and adopting aspects of Western culture? In much of the world, acculturation to the Western way of life is rapidly replacing both the positive and negative features of other cultures. It is the most pervasive example of cultural diffusion in world history. It illustrates all types, paths, and processes of diffusion that we have studied, therefore an understanding of European cultural diffusion over the past 500 years is essential to understanding the world we live in today.

Europe's Voyages of Contact

Europeans came to play a central role in world history and world geography because it was they who paved the way for the modern system of global interconnectedness.

In the fifteenth century the great cultural centers of the world—the Inka and Aztec empires in the Western Hemisphere, the Mali and Songhai in Africa, the Mughal in India, Safavid Persia, the Ottoman Turks, the Chinese, and all of the lesser empires and culture realms—were still largely isolated from one another.

The European voyages of exploration and conquest connected the world (Figure 6-27). It was not inevitable that the Europeans would be the ones to do this. The Chinese were actually richer and more powerful than the Europeans, and earlier in the fifteenth century they had launched fleets of exploration that had reached the east coast of Africa. We can hardly imagine how different world history and geography would be if the Chinese had continued their initiatives and gone on to explore and conquer Africa and the Western Hemisphere. But they did not. Instead, the Chinese government focused on internal affairs, and the Europeans continued with world exploration and conquest.

The first European initiative was Prince Henry of Portugal's conquest of the city of Ceuta on the north coast of Africa in 1415. He learned about trans-Saharan caravan routes and the riches of West Africa, and thus was inspired to finance studies of naval skills. Improvements in sailing technology enabled the Portuguese to sail down the west coast of Africa to reach beyond the Sahara Desert. Prince Henry, known as "The Navigator," launched the era of European seaborne colonial empires. Soon Spain, Holland, France, and England were racing to secure colonies. In contrast to these

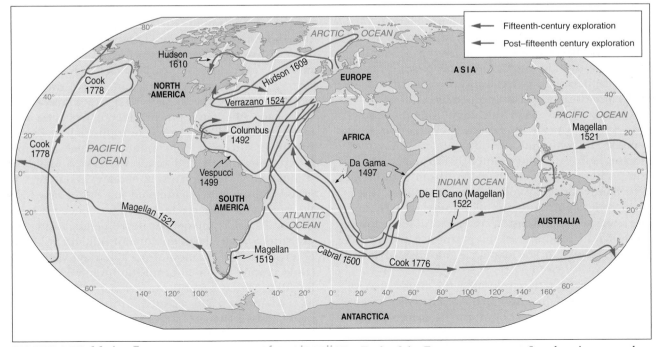

FIGURE 6-27 Major European voyages of exploration. Each of the European voyages of exploration was a daring enterprise. The Portuguese first established a string of bases around the coasts of Africa and Asia all across to Japan, where they established a trading post in 1543. Other European nations followed, and the race for discovery of new riches, of new converts to Christianity, and for help against Christianity's powerful Islamic foe was launched.

seaborne empires, Russia at the same time forged eastward overland from the Russian homeland in Europe west of the Ural Mountains. The Russian Empire pushed across Siberia to the Bering Strait, crossed over into Alaska, and eventually established colonies down the North American West Coast as far south as today's California.

This European outreach triggered the **Commercial Revolution,** between about 1650 and 1750. The development of the first oceangoing freighters that could carry heavy payloads over long distances allowed a tremendous expansion of trade. The evolution of superior ships was paralleled by the evolution of superior naval gunnery and of additional useful technologies such as clocks, which were perfected in order to determine longitude at sea. This first era of outreach ended with the death of the British explorer Captain James Cook in 1779. By then Europeans were the first people in world history who could draw a fairly accurate outline map of all the world's continents and major islands. Eventually the United States would inherit European culture and share its power.

Expansion and cultural diffusion From the time of exploration, Europe did not actually originate every "modern" idea and then impose it on the rest of the world. Europeans learned from others, too, and then transplanted around the world the ideas that they had adopted from elsewhere. "Knowledge is power," said English philosopher Sir Francis Bacon (1561–1626), and because it was the Europeans and not the Chinese, the Inka or any other group who had contacted all other civilizations, Europe became the clearinghouse of world information and products. Global diffusion was fixed hierarchically, with Europe as the apex. Europe became cosmopolitan—that is, familiar with many parts of the world. Other peoples, no matter how great their native civilizations were, remained more localized in the world Europe was creating.

Chapter 5 described the migrations of peoples triggered by European expansion. Consider, too, what Europeans did with agricultural products. They took sugarcane from Asia and planted it in the Caribbean region. They replanted bananas from Southeast Asia to South America, cocoa from Mexico to Africa, rubber from South America to Southeast Asia, and coffee from Arabia to Central America. All these foods remain major products of international trade and important factors in the export economies of many countries.

Europeans not only relocated the production of many goods around the world, they also introduced many products into world trade, both for the European home market and for other overseas markets that they pioneered. Europeans introduced many Indian goods, for instance, into China, and South American products into Africa and Asia. An Englishman introduced tea plants from China into India, leading to the development of a tea industry in India and Ceylon (today known as Sri Lanka). Europeans created world markets and profited by controlling every stage: production, transportation, and marketing.

The story of Coca-Cola, one of the world's most familiar consumer products, exemplifies cultural blending. The drink was formulated in Atlanta, Georgia, and today symbolizes westernization so powerfully that the large-scale infusion of Western products into the non-Western world is often referred to as the "Coca-colonization of the world." The two original ingredients from which Coca-Cola takes its name, however, are coca, a Native South American Quechua word for the tree whose leaves supply a stimulating drug (used today to make cocaine and crack), and cola, a word in the language of the West African Mandingo people for the nut that supplied the other original stimulating ingredient. Westerners borrowed the knowledge of both of these ingredients from their far-flung hearth areas, combined them, and marketed the drink worldwide.

Thus, modern world culture is not exclusively a Western product. It is presumptuous for Westerners to think that it is. Even non-Westerners, however, often fail to see the many contributions of non-Western peoples because the route of global diffusion was through the Western powers.

Economic Growth Increased Europe's Power

Europe pulled ahead of the rest of the world economically as it underwent the tremendous transformation of the **Industrial Revolution.** Between about 1750 and 1850, Europe evolved from an agricultural and commercial society to an industrial society relying on inanimate power and complex machinery. We cannot fully explain why Europe experienced this transformation before any other part of the world, but we can list several factors that enabled Europe to industrialize. Europe's voyages of discovery and conquest resulted in an influx of precious metals and other sources of wealth that stimulated industry and a money economy. The expansion of trade encouraged the rise of new institutions of finance and credit. In the mid-sixteenth century the joint stock company was developed, allowing many investors to share both potential profit and risk in new enterprises. The creation of stock markets where stocks could readily be bought and sold granted capital new **liquidity,** which is easy conversion from one form of asset to another. This created, in the words of English writer Daniel Defoe (1660–1731), "strange unheard-of Engines of Interest, Discounts, Transfers, Tallies, Debentures, Shares, Projects."

In 1769, James Watt designed the steam engine, which multiplied the energy available to do work. Subsequent inventions and technical innovations in

FIGURE 6-28 The first iron bridge. This is the world's first iron bridge, built over England's Severn River in 1779. It demonstrated iron's strength, and the bridge's bold design invited other uses for the new material. Its builder, John Wilkinson, launched the first iron boat in 1787, and when he died he had himself buried in an iron coffin. (Steve Vidler/Stock Photography LLC)

manufacturing, applied first to textiles and then across a broad spectrum of goods, dramatically increased productivity. Factories and industrial towns sprang up, canals and roads were built, and later the railway and the steamship expanded the capacity both to transport raw materials and to send manufactured goods to markets (Figure 6-28). New methods of manufacturing steel, chemicals, and machines played important parts in the vast changes. These innovations occurred first in Great Britain and subsequently spread to continental Europe and to North America.

Beginning in the eighteenth century, Europe also first experienced an **Agricultural Revolution.** This development, to be examined in detail in Chapter 8, both increased food production and released agricultural workers from the land, thereby creating a supply of labor for industry.

As a result of the Industrial and Agricultural revolutions, Europe and European settlements around the world drew far ahead of any other places and peoples in its productive capacity. As recently as 1800, the per capita incomes of the various regions of the world were close. If we index the Western European per capita income in the year 1800 as 100 units of wealth, then estimated per capita incomes in North America were 125, in China 107, and in the rest of the non-European world they were 94. By 1900, however, European and North American incomes were several times those of non-Western peoples.

Commercial contacts and economies

At the beginning of the age of the European voyages, European demand for foreign products such as spices, sugar, fruits, and North American furs grew rapidly. Soon the Europeans were no longer content to trade with native peoples for these goods, and the Europeans themselves established overseas estates and plantations and applied large-scale techniques to specialized production.

European commercial plantations were at first concentrated along the coasts, but in the nineteenth century the railroad allowed penetration of the continental interiors that created access to superior agricultural lands or, later, as Europe industrialized, to mineral deposits. The world's railway network expanded from 200,000 kilometers (125,000 miles) in 1870 to over 1 million kilometers (625,000 miles) by 1900. European treaty ports (ports that by treaty had to be kept open for trade) and coastal footholds became inland empires (Figure 6-29). The development of the steamship also facilitated the transport of minerals, and increased quantities of minerals supplied Europe's multiplying factories. Between 1840 and 1870 the world's merchant shipping rose from 10 million tons to 16 million tons, and then it doubled in the next 40 years.

New cities emerged in the non-European world as coordinating centers for these commercial activities and new ports sprang up along the seacoasts. The major seaport cities from the Straits of Gibraltar around Africa, across the Indian Ocean, and all throughout South Asia are the products of European contact (Figure 6-30). The same is true for most of the port cities of the Western Hemisphere. The railroads and associated commercial economies at first affected only a small percentage of the population, but over time an increasing share of the population and territory were drawn into the emerging global economy. In some countries, however, such as India, the modern commercial economy still overlies a traditional subsistence economy. There may be little exchange between the two economies unless refugees from the collapse of the traditional economy flee to the slums of the modern cities.

Political conquest

In two waves of exploration and conquest—the first extending from 1415 to 1779 and the second occurring at the end of the nineteenth century—Europe (and, in the second period, the United States) conquered most of the rest of the world. One of the original reasons for this conquest was the European nations' wish to protect their investments in foreign lands and to control these lands as markets for themselves. European countries divided up the rest of the world in order to restrain their own rivalry. Their ascendancy over the indigenous populations was guaranteed by their superior military power.

The United States and most of Latin America won independence between 1775 and 1825, but between 1875 and 1915 about one quarter of Earth's land surface was distributed or redistributed as colonies among a half-dozen imperialist states. Of all the countries in the world today, the only ones never ruled by Europeans or by the United States are Turkey, Japan (although it

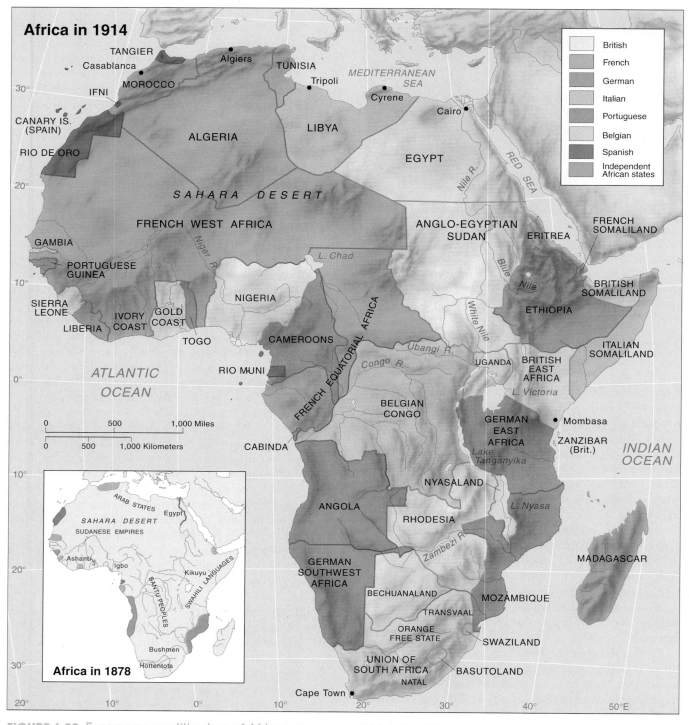

Africa in 1914

	British
	French
	German
	Italian
	Portuguese
	Belgian
	Spanish
	Independent African states

TANGIER

Casablanca

Algiers

TUNISIA

MEDITERRANEAN SEA

MOROCCO

Tripoli

IFNI

Cyrene

CANARY IS. (SPAIN)

Cairo

RIO DE ORO

ALGERIA

LIBYA

EGYPT

SAHARA DESERT

FRENCH WEST AFRICA

ANGLO-EGYPTIAN SUDAN

ERITREA

FRENCH SOMALILAND

GAMBIA

L. Chad

PORTUGUESE GUINEA

Niger R.

Blue Nile

BRITISH SOMALILAND

SIERRA LEONE

NIGERIA

ETHIOPIA

IVORY COAST

GOLD COAST

LIBERIA

TOGO

CAMEROONS

Ubangi R.

White Nile

ITALIAN SOMALILAND

ATLANTIC OCEAN

RIO MUNI

Congo R.

FRENCH EQUATORIAL AFRICA

UGANDA

BRITISH EAST AFRICA

L. Victoria

Mombasa

BELGIAN CONGO

GERMAN EAST AFRICA

ZANZIBAR (Brit.)

INDIAN OCEAN

CABINDA

Lake Tanganyika

0	500	1,000 Miles
0	500	1,000 Kilometers

NYASALAND

ANGOLA

L. Nyasa

RHODESIA

Zambezi R.

MADAGASCAR

GERMAN SOUTHWEST AFRICA

MOZAMBIQUE

BECHUANALAND

TRANSVAAL

ORANGE FREE STATE

SWAZILAND

UNION OF SOUTH AFRICA

BASUTOLAND

NATAL

Cape Town

Africa in 1878

ARAB STATES

Egypt

SAHARA DESERT

SUDANESE EMPIRES

Ashanti

Igbo

Kikuyu

BANTU PEOPLES

SWAHILI LANGUAGES

Bushmen

Hottentots

FIGURE 6-29 European partitioning of Africa. Europeans divided the African continent in order to prevent competitive war among themselves. British Prime Minister Lord Salisbury admitted, "We have been giving away mountains and rivers and lakes to each other, only hindered by the small impediment that we never knew exactly where they were." Treaty ports and trading stations along the African coast expanded into vast inland empires. Industrial Europe demanded African raw materials, and railroads allowed the Europeans to draw them out of the African interior. The native African peoples were not consulted in the political reapportioning, which was completed at a conference held in Berlin in 1884–1885.

was occupied by the United States from 1945–1952 and its constitution was imposed on it), Korea (which was ruled by Japan from 1910 to 1945 and remains split today), Thailand (left as a *buffer state* between the French and English empires), Afghanistan (a buffer between the English and Russian empires), China (which was nevertheless divided into foreign "spheres of influence"), and Mongolia (ruled by China).

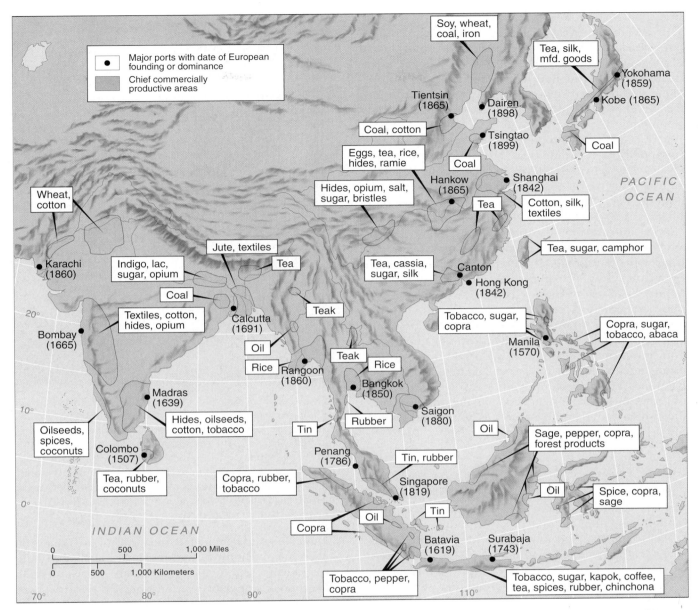

FIGURE 6-30 European intervention in Asia. Many of Asia's port cities were at first spigots that Europeans tapped into the continent to draw off Asia's wealth. As happened in Africa, however, the ports expanded into plantations for the production of goods valuable in international trade, and then the plantation regions grew into political colonies.

Therefore in most countries, European cultural attributes linger as a legacy of European rule and still predominate or overlay native pre-European traditions. European concepts of law, for example, drastically changed native societies, especially European ideas of property rights and land ownership. Before the Europeans came, land was generally considered a good that was held in common for all members of the community. Local political leaders apportioned land use and occupation by customs that brought the community together. Neither the leaders nor anyone else *owned* land. The idea that any individual could own land and single-handedly determine how to use it was largely unknown. Europeans introduced their idea of "ownership"

of an "estate" that could be bought, sold, or mortgaged by individual contract. Social cohesion was dissolved when land was no longer a common good and the regulation of its use was no longer a shared community affair.

Native American chiefs, for instance, did not by their own customs have the right to transfer land out of tribal control, and they frequently did not actually understand what Europeans meant when they "bought" Native American lands. In the history of the United States, innumerable wars were sparked when Native Americans returned to hunt or harvest unoccupied land that the Europeans insisted the Natives had sold. Throughout Africa and Asia, the Europeans often simply assigned ownership to the local political leaders. This ended traditional egalitarian systems and created new classes of rich and poor. The descendants of many of these newly enriched leaders remain great land-holders today throughout the Near East, for example. Many traditional societies crumbled under the trans-formation of land from a public asset into a private commodity.

European law tended to transform labor into a commodity, too. Traditional communities were not idyllic; they restricted individual liberties, and slavery and serfdom were not unknown. These constraints, however, were often balanced by strong webs of respon-sibilities and rights that usually kept anyone from being entirely outcast and starving. The European idea of a self-regulating market for individual labor is more ab-stract and impersonal. It cut traditional ties of both rights and responsibilities. In addition, Europeans required the use of money as a universal measure of value. Natives were forced to work for wages or to sell goods for money in order to pay taxes.

Europeans also brought their forms of administra-tion, government, centralized state authority, written arrangements, uniformity, secularization, economic planning, public accounting and treasury control, cen-tral administration, and decision making. In many cases the civil services that the Europeans left behind remain the pride of new nations, as in India.

Cultural Imperialism

European rule was marked by **cultural imperialism,** which is the substitution of one set of cultural traditions for another, either by force or by degrading those who fail to acculturate and rewarding those who do. Euro-peans seldom doubted that native cultures were inferi-or and that native peoples needed "enlightenment." Therefore, the Europeans destroyed other ways of life, including religious and political traditions, physical artifacts such as art and architecture, and even records of history and science (Figure 6-31).

One reason for this is the nature of Christianity. It is a proselytizing religion, which means that its adher-ents try to convert others to their faith. The natives in many areas accepted that there is truth in all religions, and their toleration opened them to acculturation.

Additionally, Europeans believed that their mili-tary and technological superiority presumed European superiority in all other aspects of life. Europeans did learn some things from those they conquered—farming techniques from Native Americans, for example—but they did not learn as much as they could have learned, because they often failed to appreciate the values, science, and technology of the civilizations that they conquered.

European cultural imperialism began with the sys-tematic training of local elites. The missionary schools produced converts who proselytized among their own

FIGURE 6-31 A Mayan book. This page is from a book called the *Grolier Codex*, one of only four books known to sur-vive from the considerable libraries of Central America's Mayan civilization. The rest were burned by the Spanish. This book contains astronomical records regarding the re-currence of the planet Venus. The vertical columns of mark-ings on the left are countings of days. The warrior, perhaps mythological, stands before an incense burner. (Courtesy of Justin Kerr/Barbara and Justin Kerr Studio)

people, helping to eradicate the local culture. Later the government schools turned out bureaucrats and mili-tary officers who helped govern their own people. A second channel of transmission was *reference group behavior.* People who wish to belong to, or be identified with, a dominant group often abandon their traditions in order to adopt those of the dominant group. The re-turning slaves who carried the first wave of westerniza-tion to West Africa wore black woolen suits and starched collars in the tropical heat. In India the native officer corps imitated English officers, complete with waxed mustaches.

Local elites also adopted Western ways because they were made to feel ashamed of their color and their culture. The rulers' racism and cultural imperialism meant that natives could succeed only by adopting the Whites' ways. The later autobiographies of the new na-tional leaders of every African and Asian country re-count racial humiliation (Figure 6-32). The colonial school systems implanted in children's minds an image of the power and beneficence of the "Mother Country." Children in the Congo, for instance, knew more about Belgium than about their own land and peoples. Cristo-pher Monsod, the chairman of the Philippines Election

FIGURE 6-32 Imperialism in the movies. **This is a** scene from the 1935 British movie *Saunders of the River.* In this Eurocentric and racist movie, Commissioner Saunders, a British imperialist officer, on the left, supervises and keeps peace among tribes in a British African territory. The tribes are depicted as dangerous children. In this key scene Saunders prods a submissive native in the chest with his cane while lecturing on the superiority of British civilization. The native character was portrayed by Jomo Kenyatta, who was then a young actor, but who later served as president of Kenya. What could Kenyatta have been thinking when this scene was filmed? What could he have thought years later while viewing the film in the presidential palace of independent Kenya? (Picture Desk, Inc./Kobal Collection)

Commission, remembers similarly having been taught from U.S. textbooks and says, joking, "Thanks to my American education, I know the capitals of North and South Dakota." History textbooks used in French African colonies opened with the words, "Our ancestors the Gauls. . . . " Schooling focused on each colony's ruler, so the native peoples scarcely knew that other countries existed. Residents of the Congo, for example, referred to all Whites as Belgians.

Colonial intellectual institutions usurped the power to define local tradition. For example, after Europeans founded the Asiatic Society of Bengal in 1784, Europeans' study of Indian art and literature defined Indian culture. Europeans certified what were to be considered "classics" of Indian culture. European archaeo-

logical surveys determined what monuments were fit for description and preservation as part of "the Indian heritage." Educated Indians learned about their own culture through the mediation of European ideas and scholarship.

Self-westernization   By the end of the nineteenth century, the elites of the entire non-Western world were taking the Europeans as their reference group. Even the three major nations that were never colonized—the Turks, the Chinese, and the Japanese—were all militarily humiliated, and traumatized by it. All three had had empires of their own, and their defeats forced them to reconsider all the assumptions on which their institutions and daily lives were based.

Turkey, China, and Japan all preempted or coopted Western civilization, and to some degree this saved them from Western political rule. In Turkey, Mustafa Kemal (1881–1938) seized power and forced the country to undergo self-westernization (Figure 6-33). Many historians believe that the challenges he faced, the solutions he devised, and the degree of success he achieved defined a prototype for the leaders of all nations that emerged from colonialism after World War II. In China, the Republican Revolution of 1911 attempted to modernize the country, but China's subjection to the West continued. Later another leader, Mao Zedong, applied an alternative brand of westernization—communism—although in a uniquely Chinese form.

FIGURE 6-33 Self-westernization. **In this historic** photograph Turkish president Mustafa Kemal Ataturk, whose surname means "father of the Turks," is teaching the Roman alphabet in an Istanbul public park. Ataturk recognized the technological superiority of the West, and he consciously turned his people toward westernization. One of his changes was the transformation of the writing of Turkish from the Arabic into the Roman alphabet. (Courtesy of the Turkish Tourist and Information Office)

FIGURE 6-34 A Japanese copy of Western technology. This is a Japanese woodblock cutaway view of a German battleship that visited Yokohama in 1873. This woodcut would later provide a virtual blueprint for the technologically backward Japanese Navy to copy. (Arthur M. Sackler Gallery, Smithsonian Institution, Washington, D.C.: Gift of Ambassador and Mrs. William Leonhart, S1998.65 a-c.)

The Japanese were forced by the United States to open their society to Western trade in 1854, after centuries of near-total isolation. The Meiji Restoration in 1867 launched the self-westernization of Japan: The Japanese decided to become thoroughly Western but to retain control of the process. Japan sent students to the United States and Europe and adopted Western science, technology, and even many cultural traits so rapidly and so successfully that by 1904 the country was able to defeat a Western power—Russia—in war (Figure 6-34).

Japan was later defeated in its attempt to extend its empire in World War II, but its early military successes discredited the Western powers and encouraged non-Western colonial subjects to dream of freedom. Jawaharlal Nehru (1889–1964), the first Prime Minister of independent India, wrote in his memoirs of how, in 1904, when he was 15 years old, "Japanese victories stirred up my enthusiasm. . . Nationalistic ideas filled my mind. I mused of Indian freedom . . . ." Chinese revolutionary Sun Yat-sen recorded, "We regarded the Russian defeat by Japan as the defeat of the West by the East. We regarded the Japanese victory as our own victory." Despite Japan's military defeat in 1945, it later developed into one of the world's dominant economic powers, and it continues to provide a non-Western model for economic development without complete surrender to Western culture.

The period of European relations with Africa and Asia that began 500 years ago is ending. Only two tiny Spanish holdings survive on the African continent—Ceuta and Melilla. The Western empires in Asia have been surrendered except for a few islands still held by France and by the United States. Great Britain yielded Hong Kong to China in 1997, and in 1999 Portugal returned the Chinese territory of Macau, which it had held since 1557.

The West and non-West since independence

The fixation with the West among the elite in the non-European world did not end when these countries won political independence. Those who assumed power in Africa and Asia were mostly educated in the West, and they demanded independence by quoting Western political writers whom they had read at Western universities. They often expressed anger that the West did not achieve the values it espoused and had taught them—racial equality, for example, and the equality of all people before the law. Their reference groups remained entirely European, not their own people. Nehru himself was eventually to say to U.S. Ambassador J. K. Galbraith, "You realize that I am the last Englishman to rule in India."

Few of the Western-educated elites developed indigenous models of development since they continued the diffusion of Western models. They started building in the middle of their commercial cities and went on building outward. The technical term for the favored cities is *growth poles*. The leaders hoped to convert their whole national territories to modern Western societies, but they did not realize how long it would take the majority of their people to benefit.

By the time the colonial rulers left, the traditional sources of social status in the societies, such as religion, family, and customs, had faded. The new politicians, bureaucrats, and business executives defined their status in the only way they knew, which was to exhibit Western material goods. Thus their homes, clothes, and cars mimicked Western status symbols. Even today the parliament of Kenya forbids its members to wear African clothing, but requires European-style suits and ties. This mimicry may inspire corruption in some of the world's poorest countries; their elite citizens pauperize their own countries and funnel money to banks or to buy prestigious properties in the rich countries. They flaunt their wealth in Paris and New York, and they vacation in villas along the French Riviera and in great mansions in London.

In many countries the new rulers have practiced a sort of internal colonialism on their own people. For example, the rulers who assumed power in Latin America following the independence movements of the early nineteenth century were of European stock, and their descendants still dominate this region. Peoples of Native American or African backgrounds have been treated as subject groups, forced to adopt European culture, religion, and language, and subjected to discrimination. The situation is not very different in Africa and Asia. The "colonizers" are generally the westernized elite; the "colonized" are all those who do not belong to this group, which is often the majority of the population.

Westernization Today

The diffusion of Western culture continues today. Western culture diffuses from the top of societies, from the examples and activities of the local elites. Young people also diffuse it by adopting Western dress and lifestyles as status symbols. The rich and the young are everywhere the most cosmopolitan consumers, and most consumer items in international trade are artifacts of Western popular culture.

Even the schools have become instruments of westernization. Their syllabuses emphasize modern, urban activities and values. The young sometimes emerge oblivious to their traditional culture or even despising it in favor of Western popular culture.

The media reinforce the message. Western television programs, movies, advertisements, and videos penetrate millions of homes and implant Western values. Night after night, on television screens around the world, the images of the good life are images of the life among the wealthy in the United States. This imagery has dramatically changed behavior. Chapter 5 described how popular media can affect even national birth rates. People want what they see, so new-felt needs are created.

Western media have been accused of entirely supplanting traditional culture, at least among urban middle-class adolescents. Sumner Redstone, the head of VIACOM, the parent company of MTV, has said, "Kids on the street in Tokyo have more in common with kids on the street in London than they do with their parents." Historian William McNeill agrees that the diffusion of television "is a very deep transformation of human life.

FIGURE 6-35 Television brings mixed messages. Television, particularly Western television, penetrates practically every region. How do these people interpret what they see? How does it transform their cultural inheritance? The truth is that we do not know the answers to these questions. (Courtesy of John Chiasson)

I would rank what is happening now with man's transition from a hunter and gatherer into a settled farmer. Television has replaced inherited culture" (Figure 6-35).

Tourism provides still another channel of westernization. Westerners are attracted to "different" and "unspoiled" places, but change the places they visit simply by their presence. National cultures can degenerate into commercialized spectacles and shoddy souvenirs. Many local people abandon their own material culture and adopt that of the visitors.

Global flows of professionals and of professional education are another powerful force for the diffusion of westernization. The rich Western countries export professional services, and people from around the world attend Western schools for professional education. The elite and professionals of most countries have been educated in Europe or the United States. These graduates, acculturated to Western ways, return to hold influential roles in their societies.

Western architects, civil engineers, and urban planners—or non-Western individuals educated in Western schools—are transforming built environments. Non-Western countries' cultural landscapes are increasingly "modern," and there is no prototype other than the Western. For example, the largest homebuilders in both Thailand (Anant Asavabhokhin) and in the Philippines (Manuel Villar), two of the richest men in the world, worked for Los Angeles homebuilder Kaufman & Broad, Inc., before returning to their own countries with blueprints to build homes for the rising middle classes. Some countries save "traditional" landmarks only as tourist attractions. For example, the government of Singapore bulldozed historic Bugis Street, which offered a mix of old shops and small bars, but then re-created it for tourists.

Critical THINKING

The Diffusion of "News"

Western media dominate the gathering and dissemination of news to such an extent that most people in non-Western lands learn about the affairs of other non-Western lands—and even about their own national affairs—through Western media. Nevertheless, Western countries veto efforts to achieve international information gathering and reporting as "censorship."

Alternative sources of global news are appearing. The year 2005 saw the launch of Telesur, a television network aimed to provide an alternative to U.S.-based news and analysis for Latin America. Telesur is based in Caracas, Venezuela, and financed by Venezuela, Argentina, Cuba, Brazil, and Uruguay. The U.S. government has criticized the network's news coverage as biased.

Al-Jazeera, a Qatar-based television news service, is expanding its global coverage. When founded in 2001, al-Jazeera provided only Arab-language broadcasts, but in 2006 it launched an English-language news service. Respected British journalist David Frost accepted a position with al-Jazeera, arguing, "We in the West have been broadcasting our views to the non-Western world for many years. It is only fair that these non-Western areas should have the chance to return the compliment." Then-U.S. Secretary of Defense Donald Rumsfeld criticized the network as offering an "anti-American worldview," but Secretary of State Condoleezza Rice chose to engage the network and appeared as a guest. Al-Jazeera's independence startles even many Muslim governments. It criticizes them, for example, for their lack of democracy and subjection of women. These topics are avoided in the state-owned media monopolies in most Muslim states.

America's own government-subsidized Voice of America (VOA) has won tens of millions of listeners in 53 languages. Its charter states that the service should be "a reliable and authoritative source of news" and that it should be "accurate, objective and comprehensive," but some American critics demand that it present exclusively a pro-American point of view.

The U.S. press often feels constrained to question or criticize the government during times of lofty patriotism or national challenge. Several years after the 1991 war in Iraq, however, the American people learned that the U.S. government had lied about the accuracy of U.S. bombs and other matters during that war. In October 2001 Defense Secretary Rumsfeld quoted Winston Churchill: "In wartime, truth is so precious that she should always be attended by a bodyguard of lies." It is possible to say that the American people have been warned that the American government might lie to them.

Questions

1. Do you have or could you obtain direct access to sources of international news other than U.S. sources?
2. If you have such access, have you ever been surprised by the difference between what you learned from them and the news as reported in U.S. sources?
3. Do you think that the U.S. government should be allowed to censor your access to such sources?

Countries are being transformed by world flows of capital investment, and capital is invested according to the standards of the societies that export the capital. This activity will be examined in Chapter 12.

Under this barrage of westernization, many traditional cultures and social structures are being radically transformed, and these transformations are not always in accord with the people's conscious wishes. In some cases whole cultures disappear, and their disappearance reduces cultural diversity and impoverishes all of us.

America's Role

The terrible events of 9/11 and the political realignments and wars since then have jolted Americans and many other world citizens into reflecting on the unique role of the United States. Since the collapse of Communism in the early 1990s, economic, political, and cultural dominance within the West has moved decisively to the United States. In all human history, there is no example of world dominance comparable to that of the United States today.

In the dozen years between the fall of the Berlin Wall in 1989 (see Chapter 13) and September 11, 2001, Americans tended toward self-congratulation that was grating to other peoples. Americans even stopped paying attention to the rest of the world. In the year 1989 the ABC, NBC, and CBS evening news programs gave a total of 4,032 minutes of coverage to foreign news, but in the year 2000 they gave only 1,382 minutes to foreign coverage.

On September 20, 2001, President Bush said, "Americans are asking, 'Why do they hate us?'" If some people in other countries do hate America, then Americans must investigate both "what's wrong with them," but also what America may be doing wrong, or what America could do better to communicate America's goals and hopes and to help others achieve theirs.

The Scottish poet Robert Burns asked (freely translated), "Oh! Would some power give us the gift to see ourselves as others see us! It would free us from many a blunder and foolish notion." The U.S. government does not always interpret its own actions the same way its actions are viewed by many others. For example, the U.S. government sees its "war on drugs" as a crusade against a global scourge. Many people around the world, however, and even at home, view the war on drugs first as a war against the American people themselves, imprisoning an unacceptable number of them (roughly 150 Americans per 100,000 adults, which is double the incarceration rate for *all* crimes in European countries) for what is best viewed and treated either as a sickness or as a petty crime. In addition, many see the "war" as an attempt to export an American problem to other countries. The combination of Americans' insatiable appetite for drugs and the U.S. government's outlawing of drugs has raised drugs' value to astronomical heights, thus encouraging farmers in many countries to grow drugs instead of nourishing foods, which sabotages several countries' agricultural economies. Furthermore, the U.S. government has sent troops that have intervened in civil wars and destabilized governments and societies in Panama, Colombia, Peru, Ecuador, Afghanistan, Pakistan, and other countries. U.S. Coast Guard ships patrol the rivers of Colombia even today. Cynics suggest that U.S. agencies are actually involved in the international drug trade to finance covert operations, or that the U.S. government uses the "war on drugs" simply as an excuse to occupy other countries. Whoever is correct in these descriptions of U.S. government motives and actions, the American people could at least better understand other countries' actions if they understood other people's views.

The U.S. government criticizes other countries for infringing human rights, yet some observers argue that America itself does not protect the equal rights of all its citizens. The U.S. State Department isses an annual *Global Report on Human Rights*, which criticizes other states for infringing human rights. The 2006 report chastised China, among other countries, and China responded with a report on human rights in the United States. Chinese reported that Blacks and other minorities had lower living standards, enjoyed less reliable access to health care, and faced discrimination in the workplace. Blacks received the death penalty more often than Whites convicted of the same crimes. The Chinese listed the unchecked spread of guns in private hands and secret wiretaps of U.S. citizens among other human rights violations and accused the United States of "various forms of torture" at overseas detention centers. These incidents did take place, but in several cases perpetrators were prosecuted by U.S. authorities.

Critical THINKING

Who's Listening?

The American slang word *bunk* or *bunkum* means nonsense. The story is told that U.S. Congressman Felix Walker of Buncombe County, North Carolina (served 1819–1821), made ridiculous speeches, but the speeches were never meant to be heard outside Walker's isolated community, and he defended his speeches as being "good enough for Buncombe."

Throughout history many politicians around the world have given speeches in which they said what they thought the local audience wanted to hear; isolation and lack of widespread reporting often protected the speakers from the consequences that would have occurred if wider audiences had heard those speeches. Today, however, almost any politician or general speaker anywhere may be widely quoted, out of context, and his or her remarks even used to harm his or her interests. After 9/11, for example, remarks by many U.S. politicians using the word "crusade"—a metaphor commonly heard in everyday American speech—inflamed passions throughout the Muslim world, where the word invokes the medieval Crusades during which Christian armies invaded their homelands and slaughtered their ancestors.

This textbook quotes many politicians and world leaders. In the quotations' original contexts, they may have been more temperate than the brief quotations suggest. In each case, however, it was the words exactly quoted that were widely reported, and they resulted in significant reactions by people to whom the remarks were not originally addressed.

Questions

1. Find a newspaper report of a speech by a politician. Imagine how the local, immediate audience reacted to his or her remarks. How might another audience elsewhere, possibly hostile, have interpreted those same remarks?
2. If you read or hear in a news report a brief remark that seems incendiary to you, what sources can you use to research and find the entire speech?

U.S. military power

Militarily, there is simply no other superpower than the United States. The United States currently spends more money each year on defense than the rest of the world combined, thus assuring American armed forces' technological superiority. The United States has assumed the role of global peacekeeper, and it stations troops in a growing number of foreign countries: 95 in 1950, but 148 in 2006. Many governments have invited U.S. troops to protect them against either foreign enemies or their own people. America's intentions may be peaceful, but many peoples feel intimidated by such a widespread military presence.

Military spending also imposes an enormous burden on the American people and frustrates national goals in other areas such as health care, education, and the maintenance of the national infrastructure. The president's national budget for 2007 allocated a full 52 percent of all discretionary spending to the Department of Defense, but that amount did not include tens of billions of dollars of special appropriations for the fighting in Afghanistan and Iraq. America's participation in political and military organizations and alliances will be detailed in Chapter 13.

U.S. economic power

The U.S. economy represents approximately one-quarter of the world economy, and most Americans enjoy a high standard of material comfort. Therefore, some poor and oppressed people throughout the world may inevitably resent the United States, whether or not American wealth is the cause of their own poverty. U.S. investment and U.S. products are found virtually everywhere, and conversely, the United States is the greatest market for many other countries' goods.

The diffusion of U.S. popular culture

Most of the world's most recognized brand names are American, and American corporations pursue sales abroad aggressively. Many people view the marketing of U.S. popular culture—that is, American salesmanship—as the deliberate destruction of their own traditional folk culture and its replacement by American popular culture. For example, from a distribution center in Singapore, the Disney Company floods the countries of Southeast Asia with over 16,000 Disney products. Can the children of Southeast Asia resist Donald Duck T-shirts and lunch boxes? If not, will they also remember their traditional folk fables and mythical characters (Figure 6-36)?

U.S. popular culture incorporates American cultural and political values, so it can challenge traditions and initiate cultural change even unintentionally. For example, the American women's magazine *Cosmopolitan* seeks profit, not revolution, yet its championing of women's rights and sexual freedom is revolutionary in many of the 60 countries in which it is distributed (Figure 6-37).

U.S. dominance of the world's television market has shrunk as the market has grown, but the very format of television shows—dramatic or comedic shows, for example—are an American cultural product. U.S. movies still capture about 85 percent of total global box office receipts, and U.S. studios have profited more from international audiences than from domestic audiences every year since 1993. Foreign films shown in the United States, by contrast, capture only about 1 percent of the U.S. box office. Perhaps to appease non-U.S. audiences, the 2006 film *Superman Returns* had the hero fighting for "Truth, Justice, and all that stuff," rather than the traditional "Truth, Justice, and the American Way." Many people, including many Americans, cited the winner of

FIGURE 6-36 Ronald McDonald as a substitute for Uncle Sam. These Pakistani protestors of American foreign policy wrecked a local McDonald's restaurant. Ubiquitous and conspicuous symbols of American private enterprises are often targeted during anti-American riots, even if the local business is a locally owned franchise, as it was in this case. (Mian Khursheed/Reuters Limited)

FIGURE 6-37 The diffusion of American values. In 2006, *Playboy* launched an Indonesian edition carrying pictures of a woman in lingerie; it included no nudity. In fact, the issue showed less flesh than locally produced publications did. Nevertheless, a stone-throwing mob organized by the Islamic Defenders Front, a self-appointed moral police force, attacked the offices of the magazine, and the editor and local models were arrested for indecency. (Dimas Ardian/Getty Images)

the 2006 Academy Award for best new song with a lyric lamenting "It's Hard Out Here for a Pimp" as evidence of America's immorality and decadence.

Many peoples see this onslaught of U.S. popular culture as cultural imperialism. Americans, however, generally accept the idea of *the democracy of the marketplace.* According to this theory, consumers' choice in purchasing goods is comparable to voters exercising their right to vote. People "vote" by buying products or tickets to performances. Thus, the market democratically reveals what people want. People watch Hollywood's latest movie because they want to, not because they are forced to. Americans interpret actions limiting consumer choice as comparable to restricting citizens' civil rights.

No one accepts this analogy completely—Americans debate it among themselves—but Americans do generally accept it more completely than many other peoples do. Many other peoples argue more strongly that the marketplace does not always protect values that must be preserved for the good of a society. The marketplace may not protect tradition; it may widen disparities in income or opportunity, and it may fail to protect the general welfare above economic results.

American trade representatives, accepting the theory of the democracy of the marketplace, argue that when a government limits cultural imports, its purpose is not to protect culture, but to protect markets for local producers. Other peoples say that Americans "just don't get it." Some governments insist that they have the right to censor or ban films or products in the name of their people. The Chinese government, for example, reserves the right to determine whether U.S. cultural products are "catering to the tastes of the Chinese people," and it bans many American films. To Americans, this is simply preemptive censorship.

America's insistence on consumer selection and freedom even offends many cultures' religious teachings. For example, many Americans still resent Iran's Ayatollah Khomeini having called the United States "The Great Satan" in 1979, yet few Americans understand the original interpretation of this label. In the *Koran,* the sacred text of Islam, Satan is not portrayed as the great conqueror or exploiter familiar to most Americans, but as "the insidious tempter who whispers in the hearts of men." Many fundamentalist Muslims find offensive the degree to which the U.S. economy and culture thrive on the creation of ever-new felt needs and the way in which the United States spreads that consumer culture around the world. Once this definition of Satan is understood, even American religious individuals of all faiths may admit to feeling ill-at-ease with the energy that goes into the pursuit and accumulation of material possessions in the United States. Americans themselves debate the proper role of consumerism, advertising, and even consumption-driven debt in American life.

The global dominance of U.S. popular culture has triggered a backlash. In Iran, for example, men known as *Bassijis,* "those who are mobilized," patrol the streets battling prostitution, drugs, alcohol, and atheism, as well as objects and values they interpret as imported from the West, including stereos, pop music, videos, lipstick, and indecorous dress among women. Their commander, Ali Reza Afshar, has said, "This war goes to the root of our existence. While physically there is no loss of life, our young people are being felled by cultural bullets, and this cultural corruption makes our young impotent to rebuild the nation." Many of those "cultural bullets" say "Made in the U.S.A." on them. The case of Iran, however, illustrates that those who battle American cultural influence may have an agenda they themselves wish to impose upon their people, rather than to allow true freedom.

Even other rich countries resent U.S. cultural diffusion. France requires theaters to reserve 20 weeks of screen time for French feature films. Australia demands that 55 percent of a television broadcaster's schedule be filled with domestic programs. Canada insists 60 percent of television programming be Canadian. In 2005, the United Nations Educational, Scientific, and Cultural Organization (UNESCO; see Chapter 13) approved a Convention on the Protection and Promotion of Cultural Diversity that allows each country to exclude its cultural policies, including media, from trade agreements. Governments also may use subsidies and quotas to promote their own cultures and to limit the access of other countries' cultural exports to their markets. Only the United States and Israel voted against the Convention, arguing that it could allow governments to control culture, even through censorship, and to block the free flow of ideas and information.

U.S. cultural predominance, however, may even increase if, as predicted (see Chapter 5), it is virtually the only rich country whose population will increase in coming years. America's evolving demographic mix, however, may bring surprises for the evolution of American culture itself.

U.S. cultural dominance is not just a matter of popular culture. U.S. educational institutions attract the world's leaders and future leaders. American universities educate the elite from around the world. In 2006, for example, 11 ministers in the Cabinet of the government of Taiwan held U.S. degrees. U.S. intellectual, academic, and scientific journals set world standards.

It must be reemphasized that not every American cultural export is an original U.S. cultural product. The United States is the apex of global information diffusion, so U.S. global corporations find product ideas around the globe and then introduce them into other places, as demonstrated in the example of Coca-Cola. Endless additional examples could be cited: The U.S. corporation Häagen-Dazs developed in Argentina a dulce de leche ice cream that was soon the second most popular flavor in the United States and Europe. The U.S. corporation Nike has introduced around the world a shoe designed by Kenyans: It has a separate big toe, like a mitten, to simulate running barefoot. East German coaches developed in-line skates for competitive ice skaters to practice without ice, but the skates' popularity around the world today is a product of U.S. marketing. Even MTV is less distinctively American than when it began in 1981. It is increasing the local-content percentage of its programming everywhere. By 2006 it had 44 versions around the world, being watched in almost 420 million homes with 81 percent of viewers outside the United States. MTV's chairman Tom Freston said that in opening new markets the company "starts with expatriates [from America] to transfer company culture and operating principles," but then it surrenders control of programming to local executives. "We're always trying to fight the stereotype that MTV is importing American culture," says company president Sumner Redstone. "We aren't. To do so would be cultural imperialism." MTV is "cultivating and nurturing local" artists and shows.

The U.S. political example and influence
U.S. political institutions and forms also sweep the Earth. As early as 1630, Puritan Governor John Winthrop had admonished New England settlers that "wee must Consider that wee shall be as a Citty upon a Hill, the eies of all people are upon us." Puritan New England would set an example; it would demonstrate virtue, and surely all the world would eventually follow. Later, the United States was born in revolution, and the authors of the U.S. Constitution produced the modern world's first republic—that is, a government without a king. They trusted in the people to be able to rule themselves without one, and the Founders, too, saw themselves as setting an example that would sooner or later be followed around the world. In 1836, poet Ralph Waldo Emerson was asked to commemorate the first battle of America's War of Independence, the Battle of Concord Bridge, where patriot farmers had fired upon British soldiers. Emerson wrote, "By the rude bridge that arched the flood/Their flag to April's breeze unfurled/Here once the embattled farmers stood/And fired the shot heard round the world." Emerson was already voicing the popular belief that the American War of Independence had launched the idea of republican constitutional democracy—a "shot" that would eventually diffuse around the world, challenging and eventually bringing down all other forms of government (Figure 6-38). The idea that America would lead the world by example and, correspondlingly, that its motivations in foreign affairs would always be benevolent, is often called *American exceptionalism.*

The freedom offered by U.S. political institutions, however imperfect, remains the envy of most people on Earth. The American example of democracy, the rule of law, women's rights, and other aspects of American life is still profoundly destabilizing in many places. By embodying, demonstrating, defending, and even *promulgating* these values, the United States attracts the hatred of people who feel threatened by, or who hate, these values.

In global political affairs, however, the role of the United States has been to a degree self-contradictory. On the one hand, the United States sets a revolutionary

FIGURE 6-38 The Statue of Liberty in Tiananmen Square. This papier mâché copy of the Statue of Liberty (although with her arms incorrectly placed) was erected by Chinese students demanding democracy in Tiananmen Square in front of the imperial palace in Beijing in June 1989. Thus "the shot" fired at Concord Bridge in 1775 had diffused even to the site traditionally respected by Chinese as the very center of the world. (Alan Reininger/Contract Press Images, Inc.)

example, and it often justifies its actions with regard to other countries by citing the lofty principles of democracy and the right of nations to self-determination. At the same time, however, the United States has supported and continues to support many nondemocratic governments. During the Cold War, for example, the United States felt it necessary to accept as allies many governments that suppressed the rights of their own citizens (see Chapter 13).

The United States has also repeatedly intervened militarily in what would normally be regarded as the internal affairs of other countries (such as Iran, Grenada, and Panama) when its own interests called for such actions. United States foreign policy has often switched sides on international disputes, favoring, for example, Iran or Iraq in turn when there are disputes between those two nations. Some people call such behavior by the German word *Realpolitik* (realistic politics), but others call it hypocrisy. Whatever it might be labeled, this behavior fuels resentment of U.S. power.

In the late 1980s and early 1990s the United States promoted democratization, most notably across Asia and Europe. In 1986, for example, the United States stopped supporting dictator Ferdinand Marcos in the Philippines when it became clear that tolerating his dictatorship was too high a price to pay for military bases there and for the Philippines' "stability" that was in fact toppling into chaos. The United States supported democratic revolutions (or evolution) in South Korea, Taiwan, China, and eventually, in 1989, the collapse of Communist government across Eastern Europe. In the Western Hemisphere,

Africa, and the Near East, by contrast, the United States had less forcefully promoted democracy.

Since 9/11, U.S. officials have been willing to admit these actions. In December 2003, President George W. Bush admitted that we "have in the past been willing to make a bargain: to tolerate oppression for the sake of stability. . . . Now we're pursuing a different course, a forward strategy of freedom in the Middle East. We will consistently challenge the enemies of reform. . . ." Then-U.S. Secretary of State Colin L. Powell admitted that America's role in the coup that overthrew Chile's elected President Salvador Allende in 1973 and brought military dictator General Augusto Pinochet to power for 17 years was "not a part of American history we are proud of." In the fall of 2005, Secretary of State Condoleezza Rice said, "The U.S. pursuit of stability in the Middle East at the expense of democracy has achieved neither. Now we are taking a different course. We are supporting the democratic aspirations of all people."

The revelations of abuses by American troops during the wars in Afghanistan and Iraq and at Guantánamo Bay have, for many global observers, jeopardized America's reputation as a model. America disappoints when it fails to live up to its own stated ideals. Leading Middle Eastern human-rights lawyer Azza Magour said, in reference to reports of torture of U.S.-held prisoners, "We wanted nothing more than to be with you, this rich, fair democracy. But now we ask who is giving us this lesson in freedom? If you caught your high priest in bed with a prostitute, would you still count on him getting you in the door of heaven?"

Overall, the wealth, power, and example of the United States dictate that it is the only country that is involved in the affairs of virtually every other country on Earth. People around the world consume American products and debate America's government policies, and their homelands are often host to American troops. American ubiquity in the world places unique responsibility on Americans to understand other cultures and America's impact on them.

America's Declararation of Independence calls upon "a decent respect to the opinions of mankind," and, in the twenty-first century, the United States has undertaken new initiatives in cultural diplomacy. Cultural diplomacy is different from traditional cultural exports in that it engages in the battle of ideas. The United States pioneered cultural diplomacy to combat Nazi propaganda before World War II, and later, on a larger scale, it used artistic and intellectual freedom as a weapon against Communism, both inside the Soviet bloc and across Western Europe through radio broadcasts and cultural exchanges. Intellectuals from the former Soviet bloc have often underlined the importance of these programs. With the collapse of Soviet Communism, however, the United States seems to have declared the battle of ideas won, and as a result, the budgetary support for cultural diplomacy has evaporated. By the late 1990s, the United States Information Agency was folded into the State Department, and

Congress canceled most cultural exchanges and closed American libraries and cultural centers worldwide. Then 9/11 and the Iraq war sparked the abrupt realization that the United States needed intellectual power as well as military might. The government created Radio Sawa and al-Hurra satellite television for Arab audiences and Radio Farda for Farsi speakers in Iran and Afghanistan. If the war against terrorism is a war of ideas, an America that stands for life, liberty and the pursuit of happiness should win hands down, but America must propagate its ideas to a more widespread audience. America might, for example, see that translations of *The Federalist*, arguably the finest statement of American democracy, are readily available in schools and libraries throught the world.

Business for Diplomatic Action, a group of American executives concerned about the world's growing disaffection toward America, published in 2003 a *World Citizens Guide* intended for American college-age travelers and a companion volume of the same name for business travelers in 2006. The books reflect surveys of hundreds of non-American nationals working in American offices worldwide, which asked how Americans could be better world citizens. The answers were overwhelming. Many people dislike America for four reasons: foreign policy, the negative effects of globalization, the vulgarity and violence of American popular culture, and Americans' collective personality (Americans are thought to "show no respect for others," and "Americans do not listen."). The Nations Brand Index, a ranking of global attitudes toward nations, ranks America 35th of 35 on its "cultural heritage scale." How could so many people around the world come to think that the homeland of "Duke" Ellington, Martha Graham, Frank Lloyd Wright, Georgia O'Keeffe (etc.!) has no cultural heritage? The United States cannot afford to let the world perceive America as having no culture other than raunchy rap videos, sitcoms, and blood-soaked action movies.

Global opinion polls reveal that the image of America has slipped precipitously through recent years. According to the Pew Research Center, "favourable" views of America in May 2006 were falling in all 15 countries polled and held by only 23 percent of Spaniards, 56 percent of Indians, 43 percent of Russians, 30 percent of Indonesians, and 12 percent of Turks. America's new Undersecretary of State for Public Diplomacy and Public Affairs, an office created only in 2001, faces formidable challenges.

CONCLUSION: CRITICAL ISSUES FOR THE FUTURE

Today many non-Western peoples are trying to defend or to revive their own traditional cultures and values. In societies that were colonized, however, European intervention was so profound that it is virtually impossible to reconstruct what existed before, and to do so in a present world where change occurs so fast that it seems unstoppable. Much of what is today called "traditional" is in fact the result of European codification or impressions of rules. When broadcast television was first introduced in Malawi, many people feared an influx of Western culture, but Steven Mijiga, the postmaster general, said, "If you go into any bookstore, the books are all brought here from the West. Look at our religion, our legal system. They are all brought from the West." Today European museums hold African ritual objects for which Africans have forgotten the rituals. Many non-Western people study their history as a way of reclaiming it, along with their independent identity and their self-respect (Figure 6-39). Nigerian writer Chinua Achebe quotes the African proverb, "Until the lions produce historians, the stories of the hunt will glorify the hunters." In other words, people who were colonized must produce historians and writers to tell their side of history.

All non-Western national leaders today face a problem. They can neither re-create an idealized model of what their society was like in the past, nor can they totally reject Western ideas and standards, so they must produce some synthesis of civilizations. This requires

difficult decisions and compromises. The Japanese, for example, adopted much of Western science and technology, but they have protected and maintained other aspects of traditional Japanese culture, such as an especially high regard for group effort and teamwork.

Retreat into a local cultural past can sustain local morale and cohesion in times of trouble, but if the resulting defensive attitude disregards the ideas and skills of alien peoples and cultures, the result is to be disastrously left behind by the rest of the world. Even a civilization as vast and successful as China had to face up to this hard fact beginning in the early nineteenth century, and the Chinese people have yet to recover from the shock to their self-esteem. Many people in the Islamic world, too, are rankled by the knowledge that from the ninth through the thirteenth centuries, their realm was the world's leading economic power and achieved great heights of artistic and scientific achievement. Beginning in the fifteenth century, however, residents of Islamic lands found themselves unable to keep pace with the West, and they eventually fell under its domination.

If a nation is to have long-term success, cultural continuity must somehow be combined with attention to useful new imported ideas, practices, and technologies. Many of the problems that countries face today—such as urbanization, pollution, and cultural confusion—are what Europe faced first but has been unable to solve. Some observers have spoken of the "exhaustion" of Western modernity, and they divide world history into three periods: (1) the *premodern world* of relative cultural isolation

FIGURE 6-39 Telling national history. The extraordinary 1981 international hit movie *Lion of the Desert* was financed by the government of Libya to teach Libya's history from Libya's point of view rather than Hollywood's. Anthony Quinn starred as Omar Mukhtar, who defeated Italy's attempted conquest of Libya from 1911 to 1931. Not all countries have been able to tell their own versions of history this way, either to international audiences or even to their own people. (Museum of Modern Art/Film Stills Archive)

FIGURE 6-40 Global sources of artistic inspiration. Paris's spectacular new Quai Branly Museum is devoted to the art and civilization of Africa, Asia, Oceania, and the Americas. The works of these peoples were long held in leading Western cities only as "artifacts" in museums of anthropology or even of "natural history," but newer museums and museum displays have put the artifacts of these peoples on an equal footing with European works *as art*. France's President Jacques Chirac dedicated the new museum to countering "the arrogance and ethnocentrism" of such traditional exhibits and to send "the humanist message of respect for diversity and the dialogue of cultures." (Fred Dufour/Getty Images)

before 1500; (2) the *modern world* of Western expansion and cultural diffusion from 1500–1950; and (3) the *postmodern world,* in which more world societies are no longer passive receptacles of Western influence. Instead they are active shoppers in a global cultural bazaar, picking and choosing what they want and then turning it into something of their own (Figure 6-40). As we noted previously in this chapter, new technology, such as the Internet, lowers the cost of producing and disseminating information, so virtually anyone can be an information producer.

Movies made for just a few thousand dollars win international acclaim. The biannual Panafrican Film and Television Festival in Burkina Faso attracts entries from many countries, and many of the movies shown find global distributors. If one of the world's poorest countries can become a hearth for the production of global culture, then almost anything is possible. The worldwide revival of non-Western cultures and the new cultural permutations coming into being every day allow everyone to contribute to the new world we are creating.

Chapter Review

SUMMARY

An isolated society depends entirely on its local environment for all of its needs, but the evolution of individual human cultures in isolation has, through history, yielded to the increasing interconnectedness of human societies.

Cultures diffuse, cultures change, and people can transform themselves and their behavior, but cultures and culture realms also have elements of stability. Any cultural pattern or distribution maps a current balance between the force of change and of stability. Cultures are constantly evolving through time, and these changes may result either from local initiatives and developments or else as the result of influences from other places. Each local culture is a mix between what originated locally—through cultural evolution—and what has been imported through cultural diffusion. As global communication and transportation have increased, the balance of factors that explain the local activities and culture at any place has tipped steadily away from factors of evolution and toward factors of diffusion. What happens at places depends more and more on what happens among places.

Today's rapid pace of cultural innovation and diffusion requires us to make a distinction between folk culture and popular culture. Folk culture preserves traditions. Most folk culture groups are rural, and relative isolation helps these groups maintain their integrity, but folk culture groups also include urban neighborhoods of immigrants struggling to preserve their native cultures in their new homes. Popular culture, by contrast, is the culture of people who embrace innovation and conform to changing norms. Popular culture

may originate anywhere, and it tends to diffuse rapidly, especially wherever people have time, money, and the inclination to indulge in it. The number of aspects of this culture that geographers can study is virtually limitless.

In studying economies and trade, geographers note how the people at any place make their living, where specific economic activities locate, and how economic activities and trade affect other aspects of cultural geography.

All human activities find territorial expression, and maps of culture realms, regions of economic specialization, or political jurisdictions reveal current distributions of human activities. These activities, however, are dynamic, continuously organizing and reorganizing, forming and reforming. Therefore, their distributions and patterns are disrupted and reshaped repeatedly. Geographers need to understand what forces make things move and redistribute themselves.

Despite the rich variety of indigenous local cultures around the globe, the European cultural model is widespread. Europeans came to play a central role in world history and geography because they paved the way for the modern system of global interconnectedness. Global diffusion was fixed hierarchically, with Europe as the apex. Europe conquered most of the rest of the world, and European political domination imposed European concepts of government, law, property, and other aspects of culture. The spread of Western culture continues today, and many traditional cultures and social structures are being radically transformed. In some cases whole cultures disappear.

Many non-European peoples are attempting to revive their own cultural history and values. Perhaps world cultural diffusion will be less hierarchical in the future.

KEY TERMS

acculturation p. 217
Agricultural Revolution p. 243
behavioral geography p. 227
cognitive behavioralism p. 227
Commercial Revolution p. 242
cultural diffusion p. 217
cultural ecology p. 219
cultural imperialism p. 246
culture realm p. 228
culture region p. 228
cyberspace p. 238
diffusionism p. 217

economic geography p. 236
electronic highway p. 238
environmental determinism p. 219
ethnic group p. 226
ethnocentrism p. 226
evolutionism p. 217
felt needs p. 237
folk culture p. 221
footloose activity p. 238
historical consciousness p. 235
historical geography p. 234
historical materialism p. 218

hunter-gatherers p. 217
Industrial Revolution p. 242
inertia p. 234
infrastructure p. 234
liquidity p. 242
mental maps p. 227
pastoral nomadism p. 217
popular culture p. 222
possibilism p. 219
proxemics p. 228
territoriality p. 228

QUESTIONS FOR REVIEW AND DISCUSSION

1. How did Alexander von Humboldt disprove Varro's theory of unilinear social evolution?

2. Many people believe that we have reached the end of a 500-year period of human history. What has characterized this period? What might happen next?

3. Define historical consciousness and give an example of how it can act as a force of inertia fixing cultural geography.

4. What four things usually happen when you build a road to connect three formerly isolated villages?

5. Define cultural imperialism and give specific examples of how much of the world's cultural geography is the legacy of European cultural imperialism.

6. According to the theory of historical materialism, why are political institutions almost always "behind the times"?

7. Differentiate diffusionist influences on cultural development from evolutionist influences on cultural development and give specific examples of each. What forces or factors at work in the world today favor either diffusionist influences or evolutionist influences?

8. Differentiate folk culture from popular culture. Give a few examples of each that you know about personally.

THINKING GEOGRAPHICALLY

1. What were a few distinctive products of your region 50 years ago? Religious observances? Food products? Clothing or costumes? Architectural styles? Games? Have they been exported from the region? Are they still typically produced? Attend a local street fair or celebration. What aspects of the festival are different from a similar festival 100 miles away?

2. Could there be any isolated people left on Earth? Where would they most likely be found?

3. Identify developments that are contributing to the formation of one global culture.

4. Clip a few photos from magazines and challenge your fellow students to identify the location on the basis of attributes of the cultural landscape.

5. How many cities are served by nonstop flights from your town?

6. Henry Louis Gates Jr., has written, "... for those ... who value the survival of diverse human cultures in just

the way naturalists value the diversity of flora and fauna—the march of international corporate capitalism is not without its mournful aspect. In Tibet, an intricate culture of Buddhist worship has been tragically disrupted and weakened by a brutal, deliberate state policy of cultural extirpation. In neighboring Nepal, a highly evolved practice of Hindu worship has been similarly disrupted and weakened, not by a hostile state regime but by the BBC World Service, Coca-Cola, Michael Jackson—in short, the encroachments of Western consumer culture." Do you agree that the processes he speaks of are similar or equivalent? Are they equally malevolent?

7. Explain the saying that what happens *at* places is increasingly dependent on what happens *among* places.

8. Find in today's newspaper an example of civil strife somewhere in the world. Can the strife be understood in terms of the identities or self-perceptions of the parties involved?

SUGGESTIONS FOR FURTHER LEARNING

Appiah, Kwame Anthony. *Cosmopolitanism: Ethics in a World of Strangers*. New York: W.W. Norton, 2006.

Ayers, Edward L. *All Over the Map: Rethinking American Regions*. Baltimore: Johns Hopkins University Press, 1996.

Baranowski, Shelley, and Ellen Furlough. *Being Elsewhere: Tourism, Consumer Culture, and Identity in Modern Europe and North America*. Ann Arbor: University of Michigan Press, November 2001.

Bell, Morag, Robin Butlin, and Michael Heffernan, eds. *Geography and Imperialism: 1820–1940*. Manchester, UK: Manchester University Press, 1997.

Benko, Georges, and Ulf Strohmayer, eds. *Human Geography: A History for the 21st Century*. London: Hodder Arnold, 2004.

Berry, Kate A., and others. *Geographical Identities of Ethnic America: Race, Space, and Place*. Las Vegas: University of Nevada Press, 2001.

Black, Jeremy. *Maps and History*. New Haven: Yale University Press, 1997.

Blaut, J. M. *The Colonizer's Model of the World*. New York: The Guilford Press, 1993.

Bonnemaison, Joel. *Culture and Space: Conceiving a New Geography*. London and New York: I.B. Tauris, 2005.

Braudel, Fernand. *The Structures of Everyday Life*. Vol. 1 of *Civilization and Capitalism: Fifteenth–Eighteenth Century* (English translations). 1981.

——. *The Wheels of Commerce*. Vol. 2 of *Civilization and Capitalism: Fifteenth–Eighteenth Century* (English translations). New York: Harper & Row, 1982.

——. *The Perspective of the World*. Vol. 3 of *Civilization and Capitalism: Fifteenth–Eighteenth Century* (English translations). New York: Harper & Row, 1984.

Carney, George O., ed. *The Sounds of People and Places: Readings in the Geography of American Folk and Popular Music.* Lanham, MD: University Press of America, 1993.

Carney, George O. *Baseball, Barns, and Bluegrass: A Geography of American Folklife.* Lanham, MD: Rowman & Littlefield, 1998.

Curtin, Philip. *The World and the West: The European Challenge and the Overseas Response in the Age of Empire.* New York: Cambridge University Press, 2000.

Everson, Paul, and Tom Williamson, eds. *The Archaeology of Landscape.* Manchester, UK: Manchester University Press, 1998.

Geertz, Clifford. *After the Fact: Two Countries, Four Decades, One Anthropologist.* Cambridge, MA: Harvard University Press, 1995.

Groth, Paul, and Ted Bressi, eds. *Understanding Ordinary Landscapes.* New Haven: Yale University Press, 1997.

Hodgen, Margaret T. *Early Anthropology in the Sixteenth and Seventeenth Centuries.* Philadelphia: University of Pennsylvania Press, 1964.

Low, Setha M., and Denise Lawrence-Zuniga, eds. *The Anthropology of Space and Place: Locating Culture.* New York: Blackwell, 2003.

MacDonald, Mary N. ed. *Experiences of Place.* Cambridge: Harvard University Press, 2003.

Moran, Emilio F. *People and Nature: An Introduction to Human Ecological Relations.* New York: Blackwell, 2006.

Nisbett, Richard E. *The Geography of Thought: How Asians and Westerners Think Differently . . . and Why.* New York: Free Press, 2003.

Noble, Allen G., ed. *To Build in a New Land: Ethnic Landscapes in North America.* Baltimore, MD: Johns Hopkins University Press, 1992.

Parker, Philip M. *National Cultures of the World: A Statistical Reference.* New York: Greenwood Publishing Group, 1997.

Pile, Steve, and Nigel Thrift, eds. *Mapping the Subject: Geographies of Cultural Transformation.* New York: Routledge, 1995.

Popular Culture in the Contemporary World. A series of books about individual countries with volumes on China, Germany, Russia, Latin America, and India. Santa Barbara, CA: ABC-Clio Publishing Company.

Radcliffe, Sarah A., ed. *Culture and Development in a Globalising World: Geographies, Actors and Paradigms.* New York: Routledge, 2006.

Sibley, David, and Peter Jackson, David Atkinson, and Neil Washbourne, eds. *Cultural Geography: A Critical Dictionary of Key Ideas,* International Library of Human Geography. London: I.B. Tauris, 2005.

Stea, David, and Mete Turan. *Placemaking: Production of Built Environment in Two Cultures.* Brookfield, VT: Avebury Press, 1996.

UNESCO. *The Futures of Cultures.* New York: UNESCO Publishing (Future-Oriented Studies Series), 1994.

Upton, Dell, ed. *America's Architectural Roots: Ethnic Groups that Built America.* Washington, D.C.: National Trust for Historic Preservation Press, 1986.

Weiss, Michael J. *The Clustered World: How We Live, What We Buy, and What It All Means about Who We Are.* Boston: Little Brown and Company, 2000.

The World Atlas of Architecture. London: Mitchell Beazley Publishers, 1984. Originally published as *Le Grand Atlas de l'Architecture Mondiale.* (Encyclopedia Universalis, Paris, 1981). Also published as *The World Atlas of Architecture.* (New York: Portland House, 1988.)

Yamin, Rebecca, and Karen Metheny. *Landscape Archaeology: Reading and Interpreting the American Historical Landscape.* Knoxville, TN: University of Tennessee Press, 1996.

Zelinsky, Wilbur. *A Cultural Geography of the United States: A Revised Edition.* Upper Saddle River, NJ: Prentice Hall, 1996.

WEB WORK

Many websites relate to cultural studies. You might wish to begin at:

http://www.popcultures.com/

The American Studies Web is a rich source of wide-ranging resources on the entire range of cultural topics. At this site you can find resources on many aspects of art and material culture, as well as good coverage of race and ethnicity, including African American, Asian American, Native American, and Latino and Chicano studies:

http://www.georgetown.edu/crossroads/asw/

Another interesting site is the Virtual Library and Museums of the World Homepage sponsored by the International Council of Museums:

http://www.icom.org/vlmp/

The European Research Centre on Migration and Ethnic Relations is a university-based research institute that focuses on movements of people around the world, interrelationships among ethnic groups, and problems of social integration and social exclusion. It can be found at:

http://www.ercomer.org

The Research Centre maintains the World Wide Web Virtual Library of Migration and Ethnic Relations:

http://www.ercomer.org/wwwvl/

Culture on the Edge is a web-based magazine devoted to the study of vanishing cultures:

http://www.culturesontheedge.com

Practically every large corporate producer of popular culture has a home page. Among the more interesting are:

http://www2.warnerbros.com/main/sitemap/sitemap.html

http://www.viacom.com

The home page of the Arabic news organization al-Jazeera can be found at:

http://english.aljazeera.net/HomePage

Culture is a broad topic that touches on every aspect of human existence throughout the world. A visit to the websites mentioned here will lead you to thousands of other interesting and rewarding sites.

7

THE GEOGRAPHY
OF LANGUAGES
AND RELIGIONS

This impressive building in Rome, Italy, is Europe's largest mosque (house of worship of the Muslim faith). It stands just a few miles from Vatican City, the seat of the Roman Catholic Church. Immigrants are bringing Islam to Europe, and some Europeans are converting to it, too. No Christian churches are allowed anywhere in Saudi Arabia, the seat of Islam, but Christianity allows the separation of Church and State. Religious teachings carry ramifications across many aspects of human geography. (Lars Halbauer/dpap/NewsCom)

A Look AHEAD

Defining Languages and Language Regions

The great variety of languages spoken today testifies to the relative isolation of groups in the past. The distribution of any language illustrates the pattern of dispersal of its original speakers or their cultural impact on others.

The Development and Diffusion of Languages

If groups of people who speak a common language disperse, then each breakaway group will develop its own language but retain elements of the common ancestral language. Languages that are thus related make up a language family. The geography of written language also records the origins and dispersal of peoples and cultures. Toponymy, the study of place names, records natural features or the origins or values of a place's inhabitants.

Linguistic Differentiation in the Modern World

The use of some languages is expanding because their speakers are diffusing around the world, gaining greater power

and influence in world affairs, or winning new adherents to their ideas. Other languages are disappearing.

The Teachings, Origin, and Diffusion of the World's Major Religions

Each religion originated in one place and spread out from there, but it is difficult to identify precise reasons why any religion diffuses. Some religious groups try to convert others to their religious beliefs, but others do not.

The Political and Social Impact of Religion

Religion is a major determinant of human behavior, so the geography of people holding and following a set of teachings affects politics, economics, agriculture, and diet, as well as many other aspects of human life.

*I*n 2006, Algeria passed a law prohibiting efforts to convert Muslims to any other religion. Muhammad Aissa, director of the Ministry of Religious Affairs, said that the measure was prompted by the activities of Christian evangelists in the restive ethnic Berber region of Kabylie. "We found out that . . . Christianity has been used as a tool to destabilize the country," said Mr. Aissa. Berbers, who constitute about a fifth of Algeria's 33 million people, were the original inhabitants of North Africa before the seventh-century Arab invasion. They cling to their distinct cultural identity and prefer their own language (Tamazight) or even French—the language of Algeria's former European colonial ruler—over Arabic. The central government has long felt threatened by the Berbers' distinctiveness as a destabilizing force in the country, and the Berbers' conversion to Christianity could exacerbate tensions.

In 1992, violence erupted in Algeria after the military canceled legislative elections that an Islamist group, the Islamic Salvation Front, was expected to win. The military feared that Islamists would create a theocracy, which is a form of government ruled by church officials. An estimated 200,000 people were killed during the next decade. "Never forget that the use of Islam as a political tool produced more than 10 years of terrorism," Mr. Aissa said. "We want to be immunized against the use of religion, all religions, in Algeria as political tools to destabilize the country." Christian missionaries may not have been explicitly fomenting civil war—it is doubtful that they were—in order for the central Algerian government to feel threatened by any continuing cultural disruption.

The interplay among cultural identity as expressed in language and religion, and the ramifications of those cultural attributes is often one of the most perplexing aspects of culture and politics. ▶

Language and religion are two of the most important forces that define and bond human cultures. Their influences are so pervasive that many people take them for granted and cannot objectively observe their influences in the lives of others, or even in their own lives. Nevertheless, peoples who share either of these two cultural attributes often demonstrate consistencies in other aspects of their behavior, and often they can more easily cooperate with one another in other ways, such as in international affairs. Sharing these attributes may help them understand one another in ways that are unique and profound.

Each language and each religion originated in a distinct hearth, and although the carriers of each of these cultural attributes have diffused throughout the world, each language and each religion still predominates within a definable realm. These realms of language and of religion are two of the most important of all types of culture regions.

DEFINING LANGUAGES AND LANGUAGE REGIONS

Many social scientists believe that language is the single most important cultural index. A **language** is a set of words, plus their pronunciation and methods of combining them, that is used and understood to communicate within a group of people. Each language has a unique way of dealing with facts, ideas, and concepts; variations in languages result in variations in how people think about time and space and about things and processes. Exact translation from one language to another is virtually impossible. The language an individual speaks has an influence in structuring his or her perception and logic, and the comparative study of languages is one of the richest ways of understanding human psychology.

Any group of people who communicate exclusively with one another will soon develop "their own language," whether they are nuclear physicists or a social clique at a high school. If human beings were to appear suddenly in today's highly interconnected world, one worldwide language might develop. The great variety of languages spoken today testifies to the relative isolation of groups in the past. The distribution of any language illustrates the pattern of dispersal of its original speakers or their cultural impact on others.

The number of different languages recognized varies with the accepted definition of language. The term *language* is usually reserved for major patterns of difference in communication. Minor variations within languages are called **dialects,** but scholars do not agree on the amount of distinctiveness necessary for a pattern to be considered a language. Some scholars, for instance, accept Danish, Swedish, and Norwegian as distinct languages, but a speaker of any one of them can understand the others. Therefore, other scholars insist that these are three dialects of one language. A **standard language** is the way any language is spoken and written according to formal rules of diction and grammar, although many regular speakers and writers of any language may not always follow all of the rules. A country's **official language** is the one in which official records are kept and government business is normally conducted.

A pidgin language is a system of communication that has grown up among people who do not share a common language, but who want to talk with each other. Pidgins are marginal or mixed languages, and they usually disappear after a few years or they evolve into a **creole**. A creole is a pidgin language that has survived long enough to become a mother tongue. That usually takes a generation or two. Examples of creoles include Gullah, the English-based language used among some African Americans living along the U.S. Southeast coast; Haitian French Creole; Guyanese Creole; and Seychellois, a French-based creole spoken on the Seychelles Islands.

A **lingua franca** is a second language held in common for international discourse. Today English is the world's leading lingua franca. Air controllers and pilots in international aviation, for example, all speak English. Other languages have served as lingua francas in the past. Latin long served Western civilization, and Swahili served throughout East Africa. Swahili developed among black peoples and in communication with Arab traders, so it has many Arab words. Today it is an official language in Tanzania and Kenya.

Individual languages change through time, but religious classics or classics of literature can exert a powerful force for stabilization. In English, for example, the works of William Shakespeare and the 1612 King James translation of the Bible have molded the language, yet parts of even these works may be difficult for many English speakers to read today.

Linguistic Geography

The study of different dialects across space is called *dialect geography,* or *linguistic geography*. Dialects usually diverge more in the way they are spoken than in the way they are written. This is because writing is often widely dispersed, but sounds are localized only among a group of people who speak together, called a **speech community**. Probably no student of dialects has ever developed the ability of the fictional Henry Higgins in G. B. Shaw's play *Pygmalion:* "I can place any man

within six miles. I can place him within two miles in London. Sometimes within two streets." To this day, however, there are distinct dialect regions across the United States.

Sometimes researchers survey speech and draw lines around places where speakers use a linguistic feature in the same way. These boundary lines are called **isoglosses.** Figure 7-1 reproduces a map from the early twentieth-century *Atlas linguistique de la France.* The atlas's makers identified a bundle of isoglosses running across France from east to west, dividing the country into two major dialect areas. These two areas are known as *langue d'oil* (the North) and *langue d'oc* (the South) based on the words used for "yes" in these areas during the thirteenth century, when the division was first recognized.

Isoglosses frequently parallel physical landscape features, because physical features often act as barriers to human migration and diffusion. For example, the Pyrenees mountains divide Spain from France, and the Pripet marshes separate Belarus from Ukraine. By contrast, languages often diffuse quickly across broad lowlands, and languages have historically diffused along river valleys or other routes of trade and transportation.

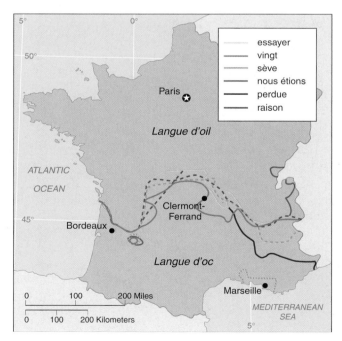

FIGURE 7-1 Dialect regions of France These six isoglosses represent the dividing lines in France between regions to the North and to the South where people pronounce six individual words differently: *essayer,* to try; *vingt,* twenty; *sève,* vigor; *nous étions,* we were; *perdue,* lost; *raison,* reason. Clearly this bundle of isoglosses represents a significant border between two spoken dialects. (Recreated from the *Atlas linguistique de la France*)

Researchers often map what is called a *geographical dialect continuum*. This is a chain of dialects or languages spoken across an area. Speakers at any point in the chain can understand people who live in adjacent areas, but they find it difficult to understand people who live farther along the chain. An example of such a chain is the North Slavic continuum, which links the Czech, Slovak, Ukrainian, Polish, and Russian languages.

The World's Major Languages

Scholars disagree on exactly how many distinct languages there are, but most suggest a count of more than 5,000, at least 30 of which are spoken by 20 million or more people as a first language (Table 7-1). Over half of all the world's people, however, speak one of the 12 major languages listed in the table. The language with the most speakers is Mandarin Chinese (Guoyo), with

FIGURE 7-2 The official languages of the countries of the world. This map reveals the worldwide distribution of some European languages. This far-reaching distribution is one legacy of European imperialism.

over 1 billion native speakers. English is the primary language of about 322 million people worldwide, and it is either the only official language or one of several official languages in approximately 50 countries (Figure 7-2). Arabic derives special transnational importance as the language of the Koran, the sacred scriptures of Islam. The Koran has been translated, but Muslims are still encouraged to study the original. Arabic is the official language in roughly 20 countries today.

In contrast to these widespread languages, some languages are extremely local. Linguists have discovered fully developed languages in New Guinea spoken by only a few hundred people living in certain valleys. These languages have developed in total isolation over long periods of time and are utterly incomprehensible to people just 20 miles away in the next valley.

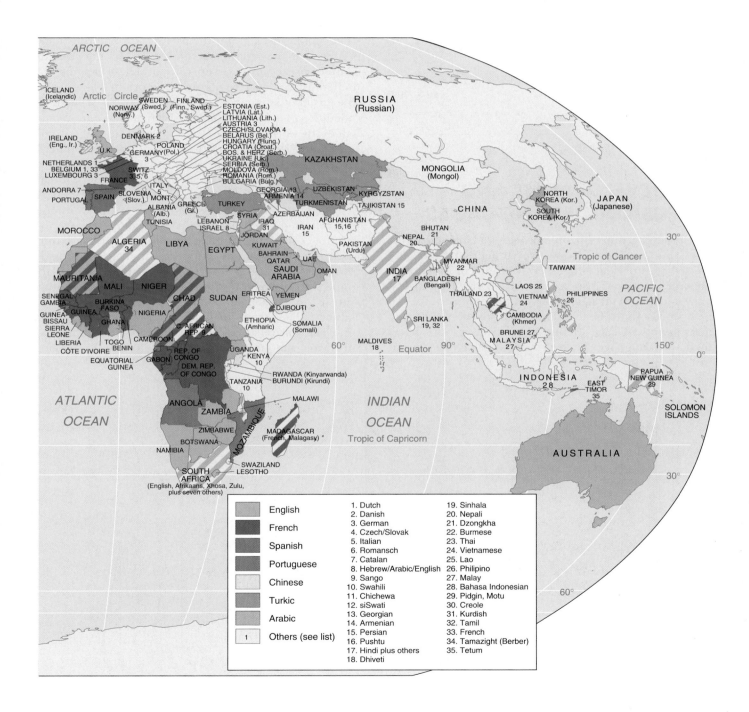

TABLE 7-1	The World's Leading Languages and the Number of Speakers of Each (in millions)		
LANGUAGE	NATIVE SPEAKERS	TOTAL SPEAKERS	
Mandarin	1,080	1,460	
Hindi	370	496	
Spanish	358	425	
English	322	514	
Malay-Indonesian	223	250	
Portuguese	210	230	
Arabic	206	254	
Bengali	171	215	
Russian	145	255	
Japanese	127	128	
French	109	239	
German	100	122	

Note: A native speaker is one for whom the language is his or her first language.

Source: U.S. Department of State

THE DEVELOPMENT AND DIFFUSION OF LANGUAGES

Any isolated group of people develops a language of its own. This language describes everything that those people see or experience together. If groups of these people break away and disperse, then each group discovers new objects and ideas, and the people have to make up new words for them. After thousands of years, the descendants of each of these breakaway groups have their own language. Each descendant language has a vocabulary of its own, but each also retains a common core of words from that earliest shared language. The ancestor that is common to any group of several of today's languages is called a **protolanguage.** The languages that are related by descent from a common protolanguage make up a **language family.**

The Indo-European Language Family

In 1786 the English philosopher Sir William Jones first pronounced his theory that a great variety of languages spoken across a tremendous expanse of Earth demonstrate similarities among themselves so numerous and precise that they cannot be attributed to chance and cannot be explained by borrowing. These languages, then, must descend from a common original language. The group of languages first identified by Sir William is called the *Indo-European family* of languages, and about half of the world's peoples today speak a language from this family (Figure 7-3). Sifting through the vocabularies of all Indo-European languages yields a common core vocabulary, which is the common ancestor of these languages, *proto-Indo-European.*

Jacob Grimm (1785–1863), one of the brothers who collected children's fairy tales, formulated rules to describe the regular shifts in sounds that occurred when the various Indo-European languages diverged from one another. There is, for example, a regular sound shift between words beginning with "p" in Latin and "f" in Germanic languages (as in "pater" and "father"). These rules are known as *Grimm's law.*

The vocabulary of proto-Indo-European tells us a surprising amount about how proto-Indo-European

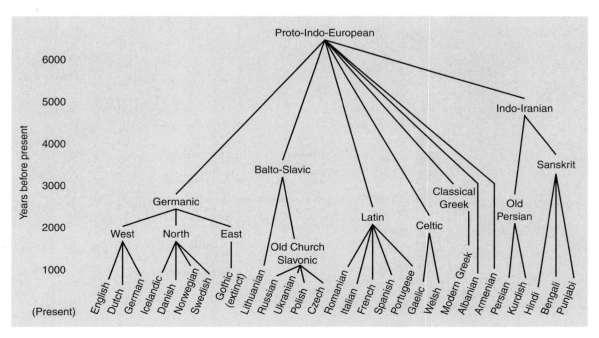

FIGURE 7-3 The Indo-European family of languages. The languages in this "family tree" are all descendants of proto-Indo-European. Common word roots can be found among them.

society was organized and how the people lived. It also hints at the language's hearth. Reconstructed proto-Indo-European has words for distinct seasons (one with snow), woody trees (including the beech and the birch), bears, wolves, beavers, mice, salmon, eels, sparrows, and wasps. These things can be found together around the Black Sea, but proto-Indo-European also includes words borrowed from the languages of the Near East. The word for wine, for instance, seems to descend from the non-Indo-European Semitic word *wanju*. Thus, the hearth area for Indo-European languages was probably in today's Turkey, some 8,000 years ago. Archaeologists disagree as to whether proto-Indo-European diffused quickly, carried by warriors, or slowly, carried by individual farmers.

Hunting common Indo-European roots of words provides a fascinating study. The proto-Indo-European root *aiw*, for example, which means "life" or "the vital force," can be found in Hindi as *ayua*, "life," but it also shows up as *aetas* in Latin, *aion* in Greek, *ewig* in German, and the words *ever* and *age* in English. Words that are related appear in surprising places. *Maharajah*, for example, a Hindi word for a great ruler, may seem exotic, yet *maha* is a distant cousin of the English words "major" and "magnitude." *Rajah* is a distant cousin of the English "reign" and "royal." In this case the cousins clearly look or sound alike, so they are called **cognates.** One must never assume any connection between two words from different languages, however, until it can be proven that they share a common root. The study of word origins and history is called **etymology.**

Proto-Indo-European provided the basic stock for all Indo-European languages, but that does not mean that one ethnic or racial group spread out to live where all these peoples are today. Sometimes a few Indo-Europeans conquered and imposed their language on a much larger group of people that previously had developed a language of their own—as occurred throughout Latin America and Africa. Sometimes peoples adopted Indo-European just to be able to communicate with Indo-Europeans. Not everyone who speaks English in the world today necessarily has an ancestor from England.

Also, cultures borrow both things and the words to name them from one another. Western culture gave Japan the word *erebata* along with the elevator itself. Japanese culture gave the West edible raw fish and a name for it: *sushi*.

Other Language Families

More people speak Indo-European languages than languages of any other family, and the Indo-European family has been studied more than any other. Researchers have benefited from written sources dating back 3,700 years—tablets in the extinct Indo-European language of Hittite. These sources give researchers considerable confidence in including individual languages in the family and in reconstructing proto-Indo-European.

The classification of all Earth's languages into families is an enormous and difficult task. Comparative linguists do not agree whether it has yet been satisfactorily completed. Figure 7-4 reflects the most generally agreed-upon state of understanding.

The same principles apply to each of these language families as to the study of Indo-European. Scholars sift vocabularies of today's individual languages to compile vocabularies of a protolanguage. The protolanguage offers clues about the origins and culture of the people who originated that protolanguage, and the sound shifts help us trace how and when various individual languages and peoples broke off linguistically and geographically from the main branch of the group. We know, for example, that the Finns, Estonians, and Hungarians are distant cousins; their languages are among the Uralic languages, and their common ancestors long ago emigrated from the region of the Ural mountains in today's Russia. Most peoples of sub-Saharan Africa speak Bantu languages, whose origins can be traced to today's Nigeria about 500 B.C. The ancestors of the Turks of today's Turkey left their homeland near Asia's Altai Mountains 1,000 years ago, while their cousins—the Uzbeks, Kazaks, Kyrgyz, Azerbaijani, and Turkmen—stayed in Central Asia. Scholars believe that early migrants to Madagascar came from the South Pacific rather than from nearby Africa, because the language of Madagascar is related to those of the South Pacific.

Some comparative linguists believe that after all languages have been classified into families, protolanguages can be constructed for each language family and that the study of those protolanguages will eventually yield superprotolanguages. They argue that the Indo-European family, for instance, is only one of six branches of a larger group, which they call *Nostratic*. Reaching even further back into prehistory, Nostratic itself is only one descendant of a protohuman language, or Mother Tongue, that was spoken in Africa 100,000 years ago and then diffused around the globe.

The first human gene that is specifically involved in language, the FOXP2 gene, was discovered in 2001. The gene had remained largely unaltered during the evolution of mammals, but it suddenly changed in humans after the hominid line split off from the chimpanzee line of descent about 100,000 years ago. These discoveries suggest that we now have both linguistic and biological evidence that all humans can be traced back to one stock that evolved in Africa. The further back into prehistory that scholars attempt to reconstruct individual languages, however, the more controversial the work becomes. Some linguists believe that the time depth of such studies is too great to reconstruct more protolanguages, let alone one prehistoric mother tongue. Geography, linguistics, anthropology, and biology each contribute to this fascinating but controversial research into prehistory.

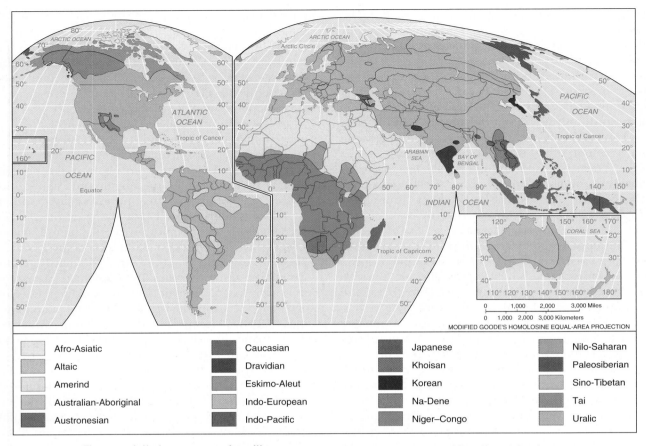

FIGURE 7-4 The world's language families. Language families are groups of languages that have common linguistic ancestors. In some areas, languages other than those shown on this map have been adopted as official languages, even though they may be spoken only by minorities.

The Geography of Writing

The geography of languages is complicated by the geography of **orthography**, which is a system of writing. There have been two indisputably independent inventions of writing—by the Sumerians in Mesopotamia before 3000 B.C. and by the Central American Olmec peoples about 650 B.C. We think that Egyptian writing of 3000 B.C. and Chinese writing (by 1300 B.C.) may also have arisen independently, but probably all other peoples who have developed writing have borrowed, adapted, or at least been inspired by existing systems (Figure 7-5).

Most languages are written in alphabets, which are systems in which letters represent sounds. There are several alphabets in use today. The alphabet in which a language is written can reflect a historic diffusion. Modern Western European languages are written in the Roman alphabet because Western Europeans were converted to Christianity from Rome. In Eastern Europe, in contrast, Russian, Belarussian, Ukrainian, Serbian, Bulgarian, and a few other languages spoken in Russia are written in the Cyrillic alphabet. This is the Greek alphabet as it was augmented and taught by the missionary Saint Cyril (d. 869), who converted many of these

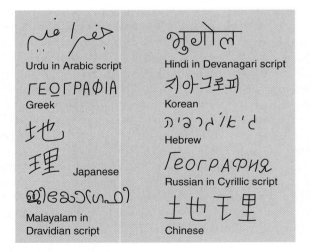

FIGURE 7-5 Examples of orthography. This figure shows the word *geography* written in several different orthographies.

peoples to Orthodox Christianity. When Bulgaria joined the European Union on January 1, 2007 (see Chapter 13), Bulgarian became the Union's first language in Cyrillic script. Scholars are now working to regularize a system of transliteration. Serbian and Croatian are one spoken

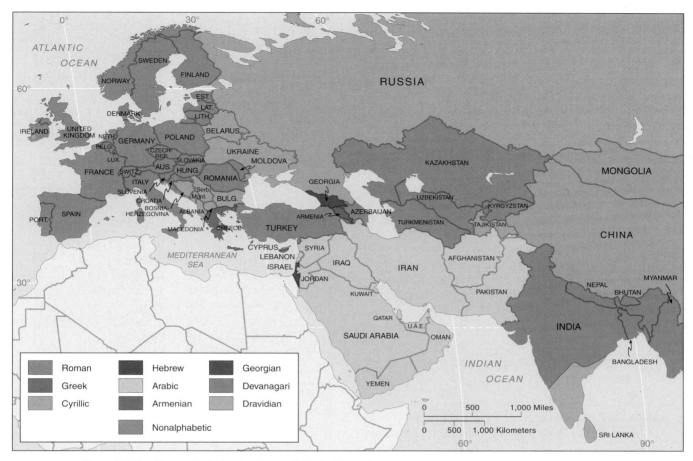

FIGURE 7-6 The distribution of alphabetic scripts in Eurasia. The distribution of scripts reveals the diffusion of religions and other cultural attributes.

language, but the Serbs write in Cyrillic, and Croats write in Roman letters. Thus, the line between Serbs and Croats demarcates a cultural borderline across Europe (Figure 7-6).

A similar orthographic–religious divide separates Pakistan from India. Spoken Urdu, the language of Pakistan, is the same as spoken Hindi, the language of northern India. The Muslim Pakistanis write in Arabic script, but the Hindus of India write in Devanagari script. Several other Indian languages are also written in Devanagari.

Sometimes a people can change its orthography as a self-conscious political act. Romania, for example, switched from the Cyrillic to the Roman alphabet early in the twentieth century. This change reflected a choice to be a "Western European" nation. In the same way, the Turkish languages of Central Asia came to be written in Arabic when these people converted to Islam, as discussed later in this chapter. Early in the twentieth century, the Turks of the country of Turkey chose to replace Arabic with the Roman alphabet to signal a deliberate Western-style modernization (refer to Figure 6-33). In 1939 Soviet dictator Joseph Stalin forced the Turkish peoples in Central Asia to replace Arabic script with Cyrillic script. His objective was to cut off these peoples

from their cultural heritage and to intensify relatively minor linguistic differences among them. In 1992, however, with the collapse of the U.S.S.R., the newly independent republics of Kazakhstan, Kyrgyzstan, Uzbekistan, Turkmenistan, and Azerbaijan all chose to abandon Cyrillic script. They did not choose to revert to Arabic, but, instead, to adopt Roman. This reflected a repudiation of their previously colonial status and a commitment to join the cultural community of the West. A Russian law of 2002 prescribes the Cyrillic alphabet as mandatory for Russian and for the languages of all ethnic republics within Russia; the restless region of Chechnya, however, defiantly uses the Roman alphabet as a symbol of cultural independence.

Not all languages are written in alphabets. The major nonalphabetic forms of writing are Chinese, in which each character represents a word or concept, and Japanese, in which each character represents a syllable. Korean writing was modeled on Chinese, and the National Academy of the Korean Language recently revised the proper English spellings of Korean words so that the sounds are more accurate when pronounced in English. Words that have long started with the letters k, t, p, and ch, for example, should now start, respectively, with g, d, b, and j. Thus the name of the Korean city of

Pusan should now be Busan, although the new spellings have been slow to find acceptance. Computer software has eased communication in ideographic languages, but the demands of contemporary international communication are nevertheless pressuring these countries to adopt the Roman alphabet. The Vietnamese changed from Chinese writing to Roman orthography in the eighteenth century when they were Christianized by missionaries.

Many traditional Native American, African, and Asian languages never had written forms until Christian or Islamic missionaries set out to translate the Bible or the Koran into these languages, using the missionaries' own orthographies. Somaliland, for example, first received a written language in 1972. These peoples may have benefited from having their languages written, but imposing any orthography on a people whose culture had been oral is an act of cultural imperialism. The transition from an oral to a literate culture is arguably the greatest transition in human cultural history.

Toponymy: Language on the Landscape

Toponymy is the study of place names. Place names record natural features as they exist in the present or as they were in the past, as well as something about the origins or values of a place's present or past inhabitants.

Some names describe environments as they are today (Oak Bay, British Columbia), whereas others describe environments as they were in the past. Many treeless towns in the North China plain, for example, are named Wang, which means oak tree. We assume that oaks grew there in the past, but that either climate change or human impact have eliminated oaks from the landscape.

Names often reveal what the people in a particular location do or believe. If cities have names like St. Paul, we are in the Christian realm; conversely, Islamabad ("the place of Islam," the capital of Pakistan) indicates the Islamic realm. City names in Russia honored Communist heroes during the period of Communist government, but today many of those places have reassumed the names they had before the Communists came to power (Figure 7-7). St. Petersburg, for example, was renamed Leningrad through the Communist period, 1917–1991. Newly independent countries often replace colonial names. In Africa, for example, Rhodesia, named for Cecil Rhodes (1853–1902), an English administrator and financier in South Africa, became Zimbabwe (the name of an ancient native city there). South Africa has recently Africanized many formerly European-language place names. The city of Bloemfontein, for example, has become Mangaung, "the place of leopards," in the Sesotho language. The government may change Pretoria,

Communist era name	Previous and now restored name
Andropov	Rybinsk
Brezhnev	Naberezhnye Chelny
Chernenko	Sharypovo
Frunze	Bishkek
Georgiu-Dezh	Lisky
Gorky	Nizhny Novgorod
Gotvald	Zmiev
Kalinin	Tver
Kuibyshev	Samara
Kirovbad	Gyanja
Leninabad	Khodjent
Leningrad	St. Petersburg
Mayakovsky	Bagdati
Ordzhonikidze	Vladikavkaz
Sverdlovsk	Yekaterinburg
Voroshilovgrad	Lugansk
Zhdanov	Mariupol

FIGURE 7-7 Politics changes place names. During the period of Communist government in the Soviet Union, all these historic cities' names were changed to honor Communist leaders. The cities have reassumed their earlier names.

originally named for Dutch settler Andreas Pretorius, to Tshwane, for a local river. Waves of nationalism in India have changed the names of the cities of Calcutta, Madras, and Bombay to Kolkata, Chennai, and Mumbai, respectively.

Toponymy can also reveal aspects of history when other evidence has been erased. Spain, for example, was long occupied by Arabic peoples, and today "guada," as in Guadalquivir or Guadaloupe, lingers as a corruption of the Arabic *wadi,* which means river. Guadalquivir is Wadi al Kabir, the great river.

The majority of place names in North America today are possessives and personal names, such as Jones Creek, but descriptive names are also common (the Red River, for example). Many Native American place names survive (Winnipeg is Cree for dirty water), but areas explored or settled by the Spanish, French, or English can be traced by trails of place names. The arc of French settlements from Quebec to New Orleans can be traced through Montreal (Mount Royal); Detroit (the "straits" between Lakes Erie and St. Clair); Fond du Lac, La Crosse, and Prairie du Chien (on routes across Wisconsin); down the Mississippi past Dubuque (named for settler Julien Dubuque); St. Louis; and Baton Rouge. Spanish-named settlements stretch across the South from St. Augustine (originally San Augustino), Florida; to San Antonio, Texas; to Sante Fe, New Mexico; to San Diego, California. Other place names recall early settler groups (Figure 7-8).

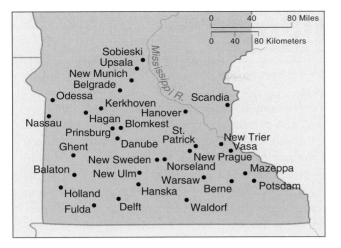

FIGURE 7-8 Minnesota place names. Some North American cities retain the names of explorers who just passed by, but these Minnesota place names supply a good clue to the origins of the immigrant settlers in the area. Each is a city in Europe or a hero to some national group. How many origins can you identify?

Commemorative names honor historic personages. Alexander von Humboldt, the geographer introduced in Chapter 1, visited North America, and he is commemorated by counties in three states; towns in six states and in Saskatchewan; and several features in the U.S. West, although he did not visit all of those places.

Today all place names in the United States are recorded by the Board on Geographic Names, created in 1890. The Board has a committee to certify foreign place names for American usage, such as all names used in this book.

LINGUISTIC DIFFERENTIATION IN THE MODERN WORLD

Diffusion and differentiation of the major languages continues. Communication within individual speech communities (usually countries) tends to individualize each dialect from the common mother language, but at the same time international communication among these groups tends to homogenize the dialects. For example, English has differentiated into many dialects. Today an English-speaking American, Nigerian, and Indian will not be able to understand one another completely without effort. Sometimes even the same words will convey different meanings when they are used by people from such diverse cultures as the United States, Nigeria, and India. In 1877 the linguist Henry Sweet (the model for Henry Higgins in *Pygmalion* and later the musical version, *My Fair Lady*) predicted that by 1977 even the English, Americans, and Australians would speak mutually incomprehensible

languages, because of their isolation from one another. These dialects have not, however, differentiated to the degree Professor Sweet predicted. This is partly because English, American, and Australian cultures have not greatly differentiated from each other in their other aspects. Also, Sir Henry never could have foreseen the great increases in world communication, nor the fact that rising percentages of national populations would receive formal education, which tends to stabilize languages.

Other major languages continue to diffuse and differentiate. Both French and Spanish are widely used. The French and the Spaniards try to "purify" their languages of linguistic borrowing but also to add new words so that their languages stay useful in the modern world. The French Academy, a scholarly institution dating back to 1635, devises new French words for new concepts and items in international trade and discourse, and these terms become official for French. Despite the efforts of the Academy, however, the French spoken in the 35 Francophone countries and Quebec are different dialects. Spanish also experiences differentiation. In 1997 Mexico, the world's most populous Spanish-speaking nation, sponsored the first International Congress on Spanish. Scholars and writers from around the world discussed the national variations of Spanish. Other international conferences bring together Germans, Austrians, and Swiss to harmonize the spelling and grammar of German.

National Languages

There is no exact correspondence between languages and the countries of the world. A few languages, such as Icelandic and Japanese, are associated almost exclusively with one country, but several languages are shared by many countries. The relationship between languages and nationalism is very complicated.

In European history, language has been interpreted as the basis of nationalism. As early as 1601 Henri IV of France seized French-speaking territories from the Duke of Savoy and declared to his new subjects, "It stands to reason that since your native tongue is French, you should be subjects of the King of France." This logic, however, never stopped the French from seizing the territory of non-French speakers whenever they could.

In 1873 the Third International Statistical Congress recommended that all censuses henceforth include a question of language. The language that a person speaks at home was thought to be the only aspect of nationality that could objectively be counted and tabulated. Therefore, a person's language was "definitive" of that person's nationality. That decision played a key role in augmenting feelings of competitive, self-conscious nationalism that eventually helped cause World War I.

Critical THINKING

Competitive Expansion and Shrinkage

The map of the geography of languages is not static. The use of some languages is expanding because the speakers of those languages are diffusing throughout the world, are gaining greater power and influence in world affairs, or are winning new adherents to their ideas. English is the world lingua franca partly because much of the world's science and business uses it. Some writers have referred to the *Anglobalization* of the world. In 2006, French President Jacques Chirac stormed out of an international meeting at which a French business executive said, "I will speak English, the language of business." In the 1970s Malaysia's then-Minister of Education Mahathir Mohammad expelled English from national schools and required the use of Malay, Chinese, or Tamil. In 2003, however, Mahathir Mohammad, who had become Prime Minister, reintroduced English for teaching all math and science classes, saying "We have to accept English whether we like it or not." Over 90 percent of secondary school students in the European Union's non-English-speaking countries study English, whereas only 33 percent study French and 13 percent German. Chile wants all school-children to be proficient in English, insisting that the use of English is an instrument of equality, and an estimated one-fifth of China's total population is studying English. Thus, by 2025 China's English-speakers may outnumber all native English speakers in the world.

Many multinational corporations have designated English as their corporate language, regardless of what the languages of their home countries might be. In computer terminology, there is virtually no language but English. U.S. popular culture spreads English through movies and music. The genuine merits of English provide other reasons for its success: It has a huge vocabulary but a simple grammar. Furthermore, it is open to change and absorption of new words: foreign words, coinages, and grammatical shifts. Nevertheless, as English grows as a second language, it is losing ground as a first language. The percentages of Earth's population speaking Hindi, Urdu, Arabic, Spanish, and Chinese as first languages are all growing faster than the percentage speaking English as its first language.

Anybody anywhere can communicate via the Internet with people who speak his or her own language, but the Internet was born in the United States, so about 80 percent of home pages (websites) are in English. The second greatest number of sites are in Japanese (only 2.5 percent), followed by German, French, and Spanish. English, however, accounts for about 97 percent of all pages linked to secure servers—that is, pages used in electronic business transactions.

Japanese websites are multiplying, but Japanese still has not become an international language. The reason is that the Japanese language developed over centuries in isolation, and it is so complicated and difficult that the Japanese actually suffer handicaps in international communication. More computer capacity is required to define nonalphabetic languages than alphabetic languages, and special computer software is required to visit Japanese-language websites. Japanese have been quick to adopt the languages of others, most notably English.

While some languages are growing and spreading, devising new ways of expressing new concepts, and winning new speakers, others are dying out. For example, among the Celtic languages of Western Europe, Manx, the original language of the Isle of Man, and Cornish, the original language of Cornwall in western Great Britain, both have been overwhelmed by English and have disappeared. Welsh is confined to a small area and population. Countless languages or dialects throughout Africa, South America, and Asia are disappearing before they are even recorded. Each language constructs perception and reality differently, so the loss of a language forecloses an opportunity to explore the workings of the human mind.

Questions

1. What government policies might preserve the national language of a small country? Investigate the policies of the Netherlands or Iceland.
2. A shrinking percentage of Americans is studying foreign languages. Can you hypothesize why? Is this good, bad, or of no significance?

Language and nation building   *Philological nationalism* is the idea that "mother tongues" have given birth to nations (Figure 7-9). This idea persists despite the fact that standard languages usually were, and still are, actually the product of self-conscious efforts at nation building by centralizing governments. Standardized languages cannot emerge before mass schooling and mass literacy or, alternatively, universal service in a national army. Usually these centralizing pressures transform the language of a small percentage

FIGURE 7-9 *Luther's German Bible.* This is the title page of the last edition of Martin Luther's German translation of the full Bible printed before his death in 1546. This book launched the Protestant Reformation and fixed the German language. It can still be read by German speakers today. The page's rich decoration includes references to Christian symbols, including the rewards of heaven and the terror of hell. (*Courtesy of the Pierpont Morgan Library, New York. PML19474/Art Resource, NY.*)

government, for example, has succeeded in codifying and establishing one common language: Bahasa Indonesian. Since the breakup of Yugoslavia into several countries, the new governments are encouraging language differentiation. Serbo-Croatian was one spoken language written in two alphabets, but today, because of political breakup, four languages are evolving: Croat and Bosnian written in Roman, and Serbian and Montenegron written in Cyrillic. Montenegro's independence from Serbia, won in 2006, may stimulate differentiation of Montenegron from the Serbian language. The governments are reviving the use of old words in official documents, and education ministries are printing checklists of forbidden words and fining people who use them. In 1996 President Tudjman of Croatia himself made a mistake. He greeted U.S. President Clinton to Zagreb using the Serbian version of happy, *srecan*, rather than the Croat word *sretan*. This mistake was broadcast live, but the government edited it out of later rebroadcasts of the event. The Czech and Slovak languages have been differentiating since the breakup of Czechoslovakia into the Czech Republic and Slovakia in 1992. Not all countries succeed in reviving or creating national languages. When Ireland achieved independence in 1922, the government tried to enforce the use of Irish Gaelic rather than English, but the people themselves did not accept it. They continued to use English. Today, however, Irish nationalism has inspired the Irish to demand that the European Union recognize Gaelic as an official language.

A minority people in a country often clings to its language as a gesture of cultural independence. For example, from 1937 until the mid-1950s, the Spanish government tried to stamp out the Basque language by forbidding its use in all public places. Through the 1960s and 1970s, however, Basques won new rights, and in 1980 the first Basque Parliament was elected, with Basque as a second official language in Spain's Basque provinces. The 1992 European Charter for Regional and Minority Languages stresses "the value of interculturalism and multilingualism" and demands that treaty participants "promote regional or minority languages," encourage their use, create political links among their speakers, guarantee access to them in criminal and civil proceedings, and encourage their presence in television and radio. Some observers, however, have criticized these efforts as treating languages as possessing inalienable rights and entitlements, as if languages merited artificial life support in inverse proportion to their importance.

Language in postcolonial societies European colonial powers imposed their languages on their colonies, regardless of how many native languages were already spoken and whether many Europeans actually settled in any given colony. The conqueror's language became the language of government, administration

of the population, the political or cultural elite, into the national language. In France in 1789, for example, 50 percent of the population did not speak French at all, and only 12 to 13 percent spoke the standard language.

Several national languages have developed as the result of deliberate political pressure. In 1919 Hebrew was actually spoken only by about 20,000 people, but the Israeli state has nurtured it for nationalistic reasons. Today Israel's Academy of Hebrew monitors the language. Some national languages were virtually invented, such as Romanian in the nineteenth century. Other new states have cultivated new national languages to unite diverse populations. The Indonesian

REGIONAL *Focus* ON

Language in the New State of East Timor

East Timor faced a language problem when it won independence in 2002 after 4 centuries of Portuguese rule, 24 years of Indonesian occupation, and 2 years of U.N. administration. Only 57 percent of its population were literate in any language. The country's new constitution designates Tetum (an indigenous lingua franca) and Portuguese as official languages, plus English and Bahasa Indonesian as "working languages." The long-term survival of Tetum is questionable: Its written form is incomplete, and there are few written materials in it, but the Roman Catholic Church adopted Tetum for its services. English and Bahasa are used in business. About 25 percent of the population speak Portuguese, and strengthening Portuguese could help promote tourism.

and law, economic development, and usually education. When the colonies gained their political independence and had to choose official languages, many kept their former ruler's language. Figure 7-2 partly reflects the map of former European empires. Some former colonial peoples find the acceptance of a major international language useful. It facilitates ties of trade, cultural exchange, travel, and even diplomacy with its former ruler and with other countries that are former colonies of that ruler. The English-speaking countries of the Caribbean, for example, find English a useful tie among themselves.

In many countries, however, the official language is not actually spoken by most of the population. Figure 7-2 shows a few international languages covering most of the globe, and yet in some of these areas the majority of the people do not actually speak that language. The difference between official languages and the languages actually spoken is particularly great in Africa. The principal official languages of Africa are those of the continent's former European rulers, but there is an extraordinary variety of native languages spoken, by some counts more than 1,000 (Figure 7-10). This large number reflects the minimal interaction of Africa's many native peoples before European conquest. Many of these languages do not have a written form, and only 40 or so have as many as 1 million speakers. Some African governments are reviving their use or study in order to rekindle appreciation of their people's pre-European heritage.

In some cases the newly independent countries retained their former ruler's language to prevent arguments about which of the native languages should become the new national language. Many African countries have retained English or French for this reason. For instance, over 250 languages are spoken in Nigeria, and advocates of a national language could not decide among Hausa, Yoruba, or Igbo, the three major native tongues. Therefore, English was chosen, although only about 20 percent of the population speaks it well and still fewer use it as their first language. Another reason for the continuing preferences for European languages is the high cost of schooling in the many indigenous languages. For example, Tanzania's choice of Swahili for primary education allowed the government to bridge linguistic divisions among the people, but it restricted the books available to Tanzanian schools.

Linguistic rivalries cause English to remain as an official language of India, although it is spoken by a small fraction of the total population. The leaders of India's independence movement intended Hindi one day to be India's official language, and Hindi is spoken by almost 400 million people. Hindi speakers, however, are concentrated in northern India, and other Indians are unwilling to accept the domination of Hindi. Whenever Hindi has been pressed on India's southern states, they have threatened to secede. India recognizes English as 1 of 16 official languages, and English remains the preeminent language of the upper classes and the upwardly mobile.

A former colony can rise to greater wealth and influence than its former ruler, in which case its dialect sets the international standard. This has happened with a few languages, including the U.S. version of English. Noah Webster published his *Spelling Book*—a self-conscious American cultural declaration of independence—in 1783, the same year that the United States won its political independence. Today American English competes with British English for precedence worldwide. Philologists in Canada, Australia, South Africa, and other English-speaking countries have published dictionaries of their national variants. The French Academy rigorously defends its privilege to guard the French language, but in 2003 the French government accepted the Quebecois term *courriel* for e-mail, rather than the French *courrier électronique*.

Portuguese boasts some 200 million native speakers, most of whom live in Brazil, a country with 17 times Portugal's population. In 1990 the governments of Brazil, Angola, Mozambique, São Tomé and Príncipe, Guinea-Bissau, and Cape Verde—all former Portuguese colonies—accepted the Brazilian version of Portuguese as their standard. Brazilian is much simpler in spelling than is the Portuguese of Portugal, and it incorporates many Native American, African, and even American words.

The fact that languages are often a cultural legacy of imperialism explains why the breadth of distribution

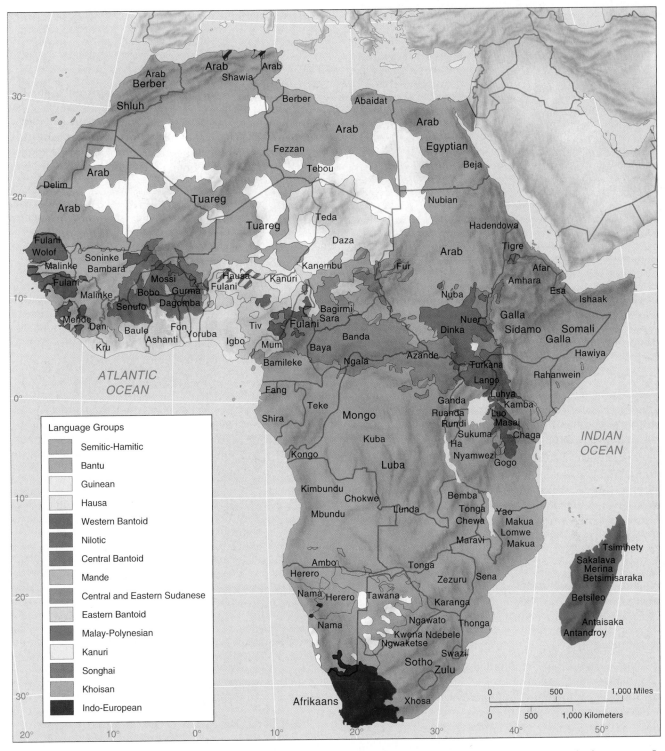

FIGURE 7-10 The languages of Africa. Figure 7-2 shows that most of the official languages in Africa are the languages of the former European rulers, but this map reveals major language groups and a sampling of the extraordinary number of native languages, many of which are actually spoken by a majority of the local population.

of a language is not always directly related to the number of people who speak it. Table 7-1 lists among the leading languages in terms of speakers several Asian languages that are geographically restricted. Hindi and Bengali are examples. These languages are not widespread across the world map, because the speakers of these languages did not create vast empires.

Polyglot states   India is not the only country that grants legal equality to two or even more languages. The several states that do so are called **polyglot states.** If various languages predominate in distinct regions of a country, the country might accept each language as official in its region. Belgium, and even its capital city of Brussels, is legally divided into French and Flemish

FIGURE 7-11 Bilingual signs in Brussels. All street signs in Brussels, the capital of polyglot Belgium, must be in both official languages: French and Flemish. The name Häagen-Dazs, which is advertised in the window, is a linguistic curiosity. The name means nothing, and was concocted by a New York businessman to sound vaguely Scandinavian to profit from associations of Scandinavia with cleanliness and health. The shop sign is in Flemish, reflecting the personal preference of its owner. (Courtesy of Robert Mordant)

zones. In the center of Brussels, all signs are in both languages (Figure 7-11). The precedence of languages is hotly contested, and several Belgian governments have fallen if they have been seen as giving preference to one language or the other. Canada is also officially polyglot, and language is a source of friction between the Anglophonic (English-speaking) majority and the Francophonic (French-speaking) minority centered in Quebec. South Africa's constitution names 11 official languages.

Polyglot states usually select one language as official for the federal center and for communications among the states, and, regardless of what the country's constitution or laws might say, one language will generally be the preferred language of the country. Those who do not use it may find their opportunity restricted or their upward mobility blocked. In Tanzania, for example, Swahili is recognized, but knowledge of English is necessary to rise in the national power structure.

Different language communities within one country may be defined not only geographically but also socially, by class. Throughout Central and South America the Native American populations and their cultures were smothered under European rule. Vestiges of this cultural imperialism remained even after independence. For example, the original constitution for Bolivia, written by Simón Bolívar, accepted as citizens only those who could speak Spanish. Today some countries are granting official recognition to indigenous languages as a way of enhancing national identity and community. Therefore, more and more countries today recognize two or more official languages—one international language plus one or more local indigenous languages. Today Bolivia, for example, recognizes three official languages: Spanish, plus the Native

American languages Aymará (spoken by 25 percent of the people) and Quechua (34 percent). Peru accepts both Spanish (spoken by 68 percent of the population) and Quechua (27 percent).

Languages in the United States

The population of the United States has always been composed of a great variety of peoples speaking a great variety of languages. English was the language of the principal colonial ruler and of the greatest number of European settlers, and it has always served as a lingua franca. A distinct American English nevertheless evolved. From the days of the earliest settlers, the American language adopted terms from Native American languages. These included the names of native animals and plants unknown in Europe or Africa, products derived from them, and also place names. Each of the many immigrant groups has in turn learned American English, but each has also contributed vocabulary, grammar, and diction to American English.

Three major dialects had developed in the 13 English colonies along the Eastern Seaboard by the time of the American Revolutionary War: northern, midland, and southern American English. Toponymy reveals the paths that settlers from each seaboard region took toward the West (Figure 7-12). Users of the northern dialect frequently named places Brook, Notch, and Corners. People moving westward from the midland areas commonly named places Gap, Cove, Hollow, Knob, and Burgh. Southern settlers favored names using Bayou, Gully, and Store.

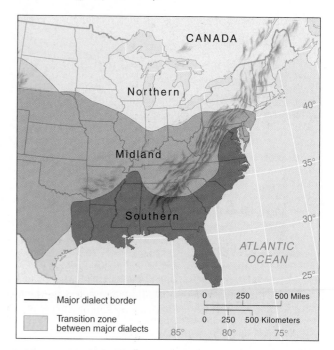

FIGURE 7-12 U.S. dialect regions. The three regional dialects of northern, midland, and southern American English developed already in the eighteenth century, and as pioneers headed westward from each dialect region, they took place names with them. (Adapted with permission from E. Bagby Atwood, *The Regional Vocabulary of Texas*. Austin: University of Texas Press, 1969.)

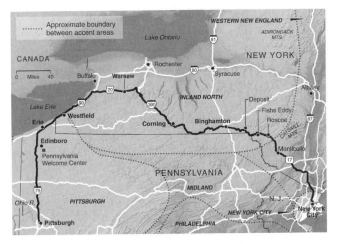

FIGURE 7-13 Pittsburgh and New York speech. Great Lakes commerce spread the Inland North dialect west to Chicago, so either in Rochester or in Chicago you might hear a person ask, "What hee-appened?" "Cot" and "caught" sound rather like "cat" and "cawt." New Englanders often drop the "r." ("Pahk the cah in Hahvahd yahd.") (From the *Atlas of North American English*. Map courtesy of the *New York Times*.)

FIGURE 7-14 Bilingual signs. Tulsa, Oklahoma, has added bilingual signs throughout its municipal buildings. Some U.S. cities, such as Miami and Los Angeles, are well known to have high proportions of Spanish speakers. The use of Spanish, however, is widespread across the United States today. (AP Photo/Tulsa/AP Wide World Photos)

Regional variations in American English persist, both in grammar and in pronunciation, despite the nationwide reach of popular media. Television homogenizes national speech patterns much less than scholars originally thought it would (Figure 7-13).

Should American English Be the Official National Language? The U.S. Constitution did not specify English as the official language, and many local governments through history and still today have found it useful to provide services and even keep official records in other languages (Figure 7-14).

In recent decades, however, a movement has grown to declare English the country's official language. In 1996 the U.S. House of Representatives approved such a law, and in 2006 the Senate did, but the House and Senate have never agreed on the same wording. Sixteen states have statutes or constitutional clauses declaring English their official language.

This movement may reflect a dedication to a national language as a bonding force of a diverse population, but some people fear that it may also reflect a resentment of the changing immigration trends discussed in Chapter 5. Immigration has increased the number of people who speak a language other than English in the United States. In 2000, 47 million people claimed to speak at home a language other than English (Table 7-2).

TABLE 7-2	The Leading Non-English Languages in Use in United States, 2000
LANGUAGE	**NUMBER OF SPEAKERS**
Spanish or Spanish Creole	28,101,052
Chinese	2,022,143
French (incl. Patois, Cajun)	1,643,838
German	1,383,442
Tagalog	1,224,241
Vietnamese	1,009,627
Italian	1,008,370
Korean	894,063
Russian	706,242
Polish	667,414
Arabic	614,582
Portuguese or Portuguese Creole	564,630
Japanese	477,997
French Creole	453,368
African languages	418,505
Hindi	317,057
Persian	312,085
Urdu	262,900
Gujarathi	235,988
Serbo-Croatian	233,865
Armenian	202,708
Mon-Khmer, Cambodian	181,889
Navajo	178,014
Miao, Hmong	168,063
Laotian	149,303
Thai	120,464
Hungarian	117,973

Source: U.S. Census Bureau

Fully 28 million of them (60 percent) spoke Spanish. Chinese ranked a distant second (2 million speakers) followed by French, German, Tagalog (the native language of the Philippines), and Vietnamese—reflecting the newest immigrant streams into the United States.

Language has become a civil rights issue in education. If, as is generally agreed, the inability to communicate in English is a handicap in the United States, then the teaching of English becomes a key route to equal opportunity for all children. In a case involving Chinese-American children in San Francisco, the U.S. Supreme Court ruled that "students who do not understand English are effectively foreclosed from any meaningful education" (*Lau v. Nichols*, 1974). This ruling triggered a national concern to identify local school districts in which English was the students' second language and to sponsor bilingual educational programs in those districts. Arguments persist, however, over whether the only purpose of bilingual programs is to ease the students' transition to English or whether they also should preserve the languages and cultures that immigrant children bring to school. Some people argue that using the education system to acculturate all children to the English language denigrates the richness of their native inheritances, while others argue that preservation of other languages threatens to lock the youngsters into second-class citizenship.

This debate intensified in 1996, when the school board in Oakland, California, declared that many of its black students spoke a distinct language, *Ebonics* (ebony phonics). Scholars have recently begun to study how some aspects of Ebonics, sometimes called Black English Vernacular (BEV), do demonstrate that the speech of parts of the African American population has preserved elements of West African languages. In Ebonics, for example, "he busy" translates into standard English as "he is busy right now," whereas "he be busy" means "he's always busy." Such grammatical structures may have diffused from West Africa to the southern United States and from there around the country. Most Americans speak at least a couple of words of Wolof, the language of Senegal: *degan*, "to understand" (as in, "Can you dig it?"), and *hippie*, "to open one's eyes" (a "hippie" is someone with his eyes open, that is, aware). Political battles continue to be fought over bilingual education in school districts across America, and referenda generally favor intensive English lessons to guarantee all students' English proficiency as fast as possible.

Equal linguistic access to the political process is another volatile issue. A 1975 law mandates bilingual ballots in voting districts where voters of selected language groups reached 5 percent or more. Today those electoral districts include vast areas of the country, including both inner-city neighborhoods and rural areas.

From the 1880s until the 1950s, federal policy tried to discourage or eliminate Native American languages. When the Europeans first arrived, more than 500 Native American languages were used throughout the territory of today's United States, but only about 200 survived in 1990, when federal legislation was passed to "encourage and support the use of Native American languages as languages of instruction." Language is such an important cultural index that many Native Americans view linguistic survival as cultural survival. Many social scientists would agree.

THE TEACHINGS, ORIGIN, AND DIFFUSION OF THE WORLD'S MAJOR RELIGIONS

A religion is a system of beliefs regarding conduct in accordance with teachings found in sacred writings or declared by authoritative teachers. Most religions involve personal commitment to worship a god or gods, but the teachings of several Asian systems of belief, including Confucianism, Taoism, and Shintoism, are basically ethical and psychological. That is, they focus on appropriate behavior (*orthopraxy*) rather than belief in a set of philosophical or theological arguments (*orthodoxy*). Several do not even address theological questions such as the nature of God or the gods, or life after death. The strictest adherence to traditional beliefs is called **fundamentalism. Secularism,** by contrast, is a lifestyle or policy that purposely ignores or excludes religious considerations.

Religion is a major determinant of human behavior, so the geography of people holding and following a set of teachings affects politics, economics, agriculture, and diet, as well as many other aspects of human life. Most religions involve joining an organized body of fellow believers, and local administrative patterns, such as Roman Catholic dioceses, often organize and focus socializing, community spirit, and, in many cases, education for their members.

Each religion originated in one place and spread out from there, but today communicants of various religions mingle around the globe, and religious affiliations cut across lines of politics, race, language, and economic status. Figure 7-15 shows the predominant faith or faiths in each region, but it cannot show minorities. At the most local scale, individual religious communities can be found grouped in identifiable neighborhoods within most of the world's large cities.

No worldwide religious census is taken, but Table 7-3 is one estimate of the adherents of each of the world's principal religions. These figures cannot indicate the depth of anyone's belief nor the degree to which a person's religion actually affects his or her behavior. Also, several systems of belief listed are not exclusive; that is, a person may adhere to more than one. The following discussion will explain additional reasons why this table must be viewed with considerable caution.

This book cannot examine the fine points of each religion's message. For that, a text in comparative theology is needed. We will, however, briefly summarize the basic

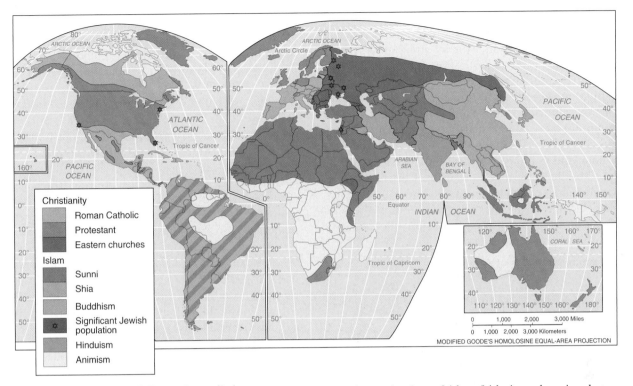

FIGURE 7-15 The world's major religions. This map shows the predominant faith or faiths in each region, but space restrictions limit it from showing minority faiths.

teachings of the world's major religions, because the distribution of people who hold those beliefs affects the distributions of other things: forms of government, dietary habits, women's rights, the organization of economies, and people's relationship with their environments.

Our coverage will identify the hearth region of each religion and review the story of its diffusion, but it is difficult for scholars to identify precise reasons why any religion diffuses. Believers in a religion will say that their religion diffused because that was God's or the gods' will. This kind of argument cannot be criticized or analyzed, for it is grounded in a faith that cannot be tested. A few religious groups **proselytize**—that is, try to convert others to their religious beliefs. Other religions do not proselytize. Some scholars differentiate *universalizing religions,* which

proselytize, from *ethnic religions,* which are identified with a particular group of people and do not proselytize.

Judaism

Judaism has only about 14 million adherents, but it was the first of the great **monotheisms**—religions that preach the existence of only one God—to emerge in history. Many scholars suggest that **polytheism,** the worship of many gods, is a primal form of religion and that among some peoples polytheism developed into monotheism.

Judaism rests on a belief in a pact between God and the Jewish people that they would follow God's law as revealed in the *Pentateuch*—the first five books of the Old Testament section of the Bible. This covenant was

RELIGION	TOTAL	AFRICA	ASIA	LATIN AMERICA	NORTHERN AMERICA	EUROPE	OCEANIA
Christians	2,019,052	368,244	317,759	486,591	261,752	559,359	25,343
Roman Catholics	1,067,053	123,467	112,086	466,226	71,391	285,554	8,327
Protestants	345,855	90,989	50,718	45,295	70,164	77,497	7,478
Orthodox	216,314	36,038	14,219	564	6,400	158,375	718
Anglicans	80,644	43,524	735	1,098	3,231	26,628	5,428
Buddhists	387,167	139	356,533	660	2,777	1,570	307
Confucians	6,313	250	6,277	–	–	11	24
Hindus	819,689	2,384	813,396	775	1,350	1,425	359
Jews	14,111	215	4,476	1,145	6,045	2,506	97
Muslims	1,207,148	323,556	845,341	1,702	4,518	31,724	307

TABLE 7-3 Adherents of the World's Major Religions (mid-2001 estimates in thousands)

Source: Adapted from the 2003 Encyclopaedia Britannica Book of the Year

granted to Abraham. Both Jews and Arabs claim descent from Abraham—Jews through his son Isaac, and Arabs through another son, Ishmael. Judaism is divided among a great variety of Orthodox (fundamentalist), Conservative, and Reform *sects,* or subdivisions.

Judaism developed historically in the Near East over many centuries. Under the Roman Empire, Jewish communities were established outside the Near East, but in A.D. 70 the Romans destroyed the temple in Jerusalem, and the Jews scattered in the Diaspora. During the Middle Ages, many European Christian rulers persecuted and expelled Jews, but Jews returned to Western Europe during the Enlightenment of the seventeenth and eighteenth centuries. On their return,

they were required to live in segregated communities called *ghettoes.* Legal emancipation came only in the nineteenth century. At that time millions of Jews came to the United States from Eastern Europe, and migration from Russia and surrounding lands continues.

An estimated 6 million Jews lived in the United States in 2006. The majority lives in New York State, other eastern states, and in California; but Jewish communities are scattered throughout the country.

Israel as a Jewish state

During the Diaspora many Jews visited Jerusalem if they could, but when the nation-state idea developed in Europe (see Chapter 11), many Jews came to accept **Zionism,** the belief that the Jews should have a homeland of their own in Palestine. When Israel came into existence in 1948, many Jews who had survived the Holocaust in Europe and many Mediterranean Jews migrated there. Israel has continued to welcome Jewish immigrants from around the world, although several Jewish sects repudiate Zionism.

Israel was intended to be a homeland for the Jewish people, but the degree to which Israel is "a Jewish state" is still being defined. In 1997, Israel declared that Arabs born in Jerusalem were aliens and thus subject to expulsion, but in 2000 the Israeli Supreme Court ruled that an Arab couple could not be barred from living in a community built solely for Jews. The Court declared: "We do not accept the conception that the values of the state of Israel as a Jewish state justify discrimination by the state between citizens on the basis of religion or nationality." In 2006, 20 percent of Israel's 6.8 million citizens were Arab (these citizens are distinct from the Palestinians in the West Bank, Gaza, and East Jerusalem). The balance of Jews and non-Jews in Israel has continuously shifted as a result of Israeli territorial expansion, immigration to Israel, and the birth rates of the various groups in Israel. In 1994 Israel surrendered some territory inhabited by Palestinians to the Palestine Liberation Organization (Figure 7-16). That action increased the Jewish percentage in Israel's

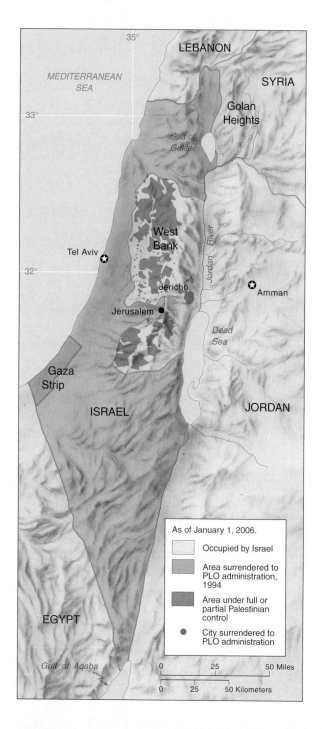

As of January 1, 2006.

☐ Occupied by Israel

▨ Area surrendered to PLO administration, 1994

■ Area under full or partial Palestinian control

● City surrendered to PLO administration

0   25   50 Miles

0   25   50 Kilometers

FIGURE 7-16 Israel and Palestinian territories. Israel was carved out of the Mideast in 1948, and it occupied larger areas during a war in 1967. Jewish settlers have occupied these territories, but their international legal status remains unclear. Israel may return the Golan Heights to Syria. In May 2000, Israeli troops pulled out of southern Lebanon, which they had occupied for 22 years. Israel has surrendered about 20 percent of the West Bank to Palestinian control and another 22 percent to joint control, but the "Palestinian" areas are criss-crossed with Israeli-held roads and checkpoints.

Zionists insist that Jerusalem is the capital of Israel, and in 2002 the U.S. Congress insisted that President George W. Bush recognize it as such. The president, however, has refused to do so, arguing that Congress cannot breach his ability to make foreign policy. No other nation recognizes Jerusalem as Israel's capital, nor does the United Nations, and the congressional action triggered a Palestinian declaration of Jerusalem as the capital of Palestine. Prime Minister Mahathir Mohamad of Malaysia called the congressional action "pouring oil on a fire."

Fundamentalism and Terrorism

The people responsible for the attacks on the United States of September 11, 2001, are often referred to as Islamic fundamentalists. Their actions were, in fact, repudiated by Muslim religious leaders throughout the world, but we must still ask what force of furious hatred this concept can carry that it can trigger slaughtering thousands of innocent people, whatever one's religious beliefs. The 9/11 attacks were defined as acts of **terrorism,** that is, violent acts for political ends that intend to frighten and to intimidate a civilian population beyond their immediate victims.

Religious fundamentalism of any kind sometimes provides meaning and direction to people who feel lost in a confusing world. Many people find comfort, or at least some kind of psychological "anchor" from strictly adhering to religious doctrines that are accepted as literal truth. These texts are followed absolutely, to the point that sometimes reason, judgment, and even conscience can be subjugated to dogma. Fundamentalism has led people to extraordinary acts of both good and evil. A "true believer" must encourage others to obey, too, or even coerce others, because other people's sins can corrupt the true believer, too. Sin must be punished and purged, even by force. When people feel that an issue is beyond matters of life and death, that the issue affects immortal souls, then they may commit acts other people would call "extremism."

The use of religion as an excuse for repression and terror is not unique to Islam. The Christian Gospels teach peace, love, and forgiveness; but throughout history, Christianity arguably has a worse record of persecution than Islam does. Christian history records centuries of persecution of Jews, mutual slaughter by Protestants and Roman Catholics (continuing in Northern Ireland, Chapter 13), hunting for religious deviation by the Inquisition and other bodies, the murder of "witches," the Crusades, forced conversions, and other similar actions.

Today Christianity is seldom associated with terrorism, but the actions taken by some Christians could be interpreted as such. The bombings of abortion clinics and murder of abortion providers by people who call themselves Christians, for example, demonstrate that some Christians continue to practice evil deeds to enforce what they see as the greater good. These Christian extremists truly believe that abortion is murder; therefore, the murder of abortion providers is logical to them. Such violence is seen as justified because it is viewed as the result of a choice between taking action and eternal damnation.

If religious texts written in primitive societies a thousand years or thousands of years ago are interpreted literally, the modern world can be terrifying. In recent decades, some Islamic fundamentalists have begun to argue that all of modern Western culture and society is sinful and evil. Believers of this fundamentalist perspective may argue that this world must be destroyed, or at least that its destruction is the consequence of its sin. Even some Christian fundamentalists hold this view. After the attacks on 9/11, the American Baptist minister Jerry Falwell argued, "God continues to lift the curtain [of His protection] and allow the enemies of America to give us probably what we deserve. I really believe that the pagans and the abortionists, and the feminists, and the gays and lesbians who are actively trying to make that an alternative lifestyle, the ACLU [American Civil Liberties Union—a civil rights organization], People for the American Way—all of them who have tried to secularize America—I point the finger in their face and say, 'You helped this happen.'" In other words, for Reverend Falwell, as for many American Christian fundamentalists, the 9/11 massacre was divine retribution for American behavior.

Some people have been confused by hearing that the hijackers lived in America for years, went bowling with their neighbors, and even drinking. Weren't they, then, "won over" to an accommodating American way of life? Obviously not. There is no room in the fundamentalist psyche for accommodation. The very dynamics that lead repressed homosexuals to murder homosexuals or that drive sexually promiscuous preachers to preach chastity are the same that lead vodka-drinking fundamentalists to destroy symbols of modern life. The act is not designed to achieve anything, but blindly to destroy the thing they feel threatened by.

remaining territory, but Israel cannot remain distinctly Jewish unless it surrenders more territory occupied by non-Jews or expels non-Jews.

After a war in 1967, Israel held on to East Jerusalem, the Gaza Strip, and the West Bank of the Jordan River (a region previously part of Jordan). Israeli governments, spurred by Zionist extremists, continued to occupy and to build new settlements in these areas in violation of international law and despite admonitions against the occupation by the United States, the United Nations, and virtually all other countries and international agencies. In 1994 Israel granted self-rule to the Palestinians in Gaza. In 2005 it began to withdraw Jewish settlers, but continuing clashes between Palestinian groups and Israeli forces have left the situation troubled.

In 2006 a quarter-million Jewish settlers still occupied the West Bank, and there Israel constructed a 225-mile wall—consisting of concrete, barbed wire, electronic fencing, motion detectors, and trenches—to separate land Israel seems to intend to retain from land that Israel may surrender. This fence does not follow Israel's pre-1967 borders, but encloses an additional 8 percent of the West Bank, so it incites Palestinian and Jordanian anger. While extending the wall, Israel has at the same time required Jewish settlers to abandon outpost settlements and return behind the wall. Thousands of settlers did return, although many struggled against their own (the Israeli) government. If the wall is a unilateral attempt to fix permanent borders, controversy will continue, because the demarcation deprives Palestinians of the best land in the area and leaves them with a jumble of unattached territories that will be difficult to coordinate into a successful Palestinian state. The Israeli Supreme Court has ruled that the wall's course should be determined by security needs alone, not the expansion of Jewish settlements.

The Palestine Liberation Organization (PLO), with both Muslims and Christians among its leaders (about 20 percent of Palestinians are Christian), has sought to establish an independent Palestinian state since its founding in 1964. The Internet Corporation for Assigned Names and Numbers has assigned the Palestinian areas a domain code (.ps). This action seems to suggest some unofficial international recognition of Palestine as a state. Thus, Palestine exists in cyberspace, but the creation of an independent Palestinian territorial entity remains the object of ongoing political confrontation and debate. The Palestinian people have fought Israeli occupation of lands they claim—often violently and even with terrorism—but the military force of Israel is so great that this contest of peoples has been an uneven one in which the "score" of dead and wounded, of destruction and persecution, enormously "favors" the Israelis.

Furthermore, anti-Israeli groups with their own militias not controlled by any national governments are based around Israel. Hamas, a group among the Palestinians, and Hezbollah, an Iranian-backed Shiite group in Lebanon, have representation in the respective official governments of the Palestinian National Authority and of Lebanon, but both also maintain militias independent of those governments. Both also deny the legality of the very existence of Israel. Both have periodically launched attacks on Israel and Israelis, and Israel has had to attack the Palestinian areas and even invaded Lebanon again in 2006 to defend itself. The process of achieving lasting peace within Israel and between Israel and its neighbors has been frustratingly slow and is marked by periodic retreats into violence, but a just settlement might encourage rising standards of living and security for everyone in the region.

Christianity

Christianity, the belief that God lived on Earth as Jesus Christ, emerged as a separate faith from Judaism, but part of the widespread appeal of Jesus' teachings rests on his emphasis on the nearness of the Kingdom of God and God's love for all. Jesus lived his entire short life in Judea; but after the death of Jesus, Saint Paul preached in many major cities in the eastern Mediterranean (Figure 7-17). Paul's letters (*epistles*) to early Christian groups make up much of the New Testament. Many Christians believe that proselytizing is a duty (Matthew 28:19).

The Bible records that the first non-Jew to convert to Christianity was an Ethiopian (Acts 8). Today many scholars believe that man was probably from Makurra, in what is today the Sudan. The Christian church of Egypt, called the Coptic Church, was founded in Alexandria in A.D. 41, and Christianity diffused south from Egypt to today's Ethiopia by the fourth century. It has survived there despite having been cut off from the Mediterranean world by the later conversion of Egypt and the Sudan to Islam (to be discussed shortly). Jesus' disciple Thomas may have traveled to India, and although that is not certain, we do know that followers of Thomas had established Christian churches in India as early as the second century. Christianity also spread to the northeast, and very early in the fourth century Armenia became the first Christian nation. The Roman emperor Constantine (288?–337) converted to Christianity and favored the religion; and by the end of the fourth century, Christianity was the Roman Empire's official religion.

Under the Roman Empire, the Christian Church adopted a geographical organization that paralleled that of the empire's secular administration. Bishops ruled over dioceses. When the western half of the Empire fell to invaders in 476, the church organization survived as one of the few stabilizing and civilizing forces in Western Europe. Bishops, including the Pope as bishop of Rome, assumed civil authority as well as ecclesiastical authority in areas where there was no other effective government. The first Western monastery was

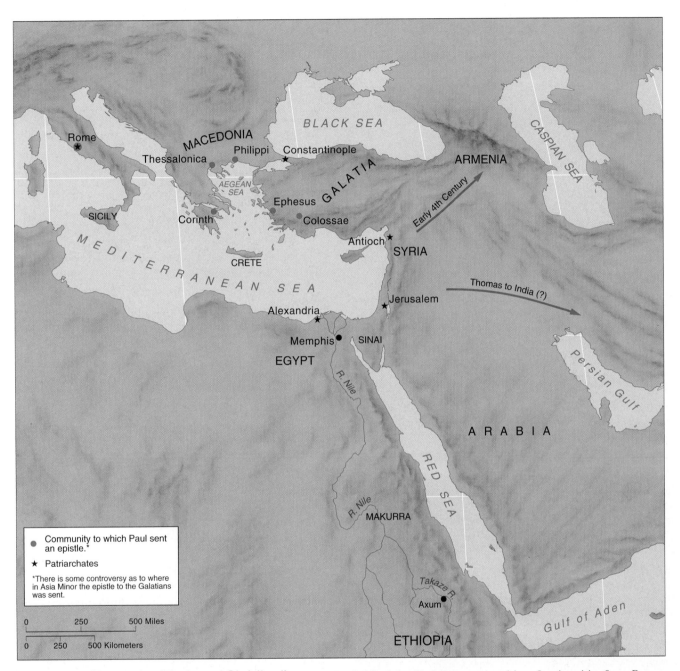

FIGURE 7-17 The early diffusion of Christianity. Saint Paul visited the Christian communities of major cities from Rome to Jerusalem. By the year 350 five patriarchs—co-equal leaders of the Church—had been recognized, with their seats at Alexandria, Jerusalem, Constantinople (today's Istanbul), Antioch (today's Antakya), and Rome. To the south, a Cathedral of Saint Mary was built at Axum in A.D. 330, where the Christian emperors of Ethiopia were crowned until 1931.

founded by Saint Benedict at Monte Cassino about 529, and as monasteries were established across Europe, the Pope brought them under his protection (Figure 7-18). Throughout the West during the Dark Ages, the Church was the focus of life, and monasteries and cathedrals rose as the monuments of the civilization.

In the eastern half of the Roman Empire, secular imperial control survived, with Constantinople (also called Byzantium, today's Istanbul) as its capital. The Patriarch of Constantinople headed the Eastern Christian Church, and missionaries from Constantinople (including Saint Cyril, mentioned earlier) converted the peoples of the Balkan Peninsula, Eastern Europe, Ukraine, Russia, and the Near East to Christianity. The Western and Eastern churches split finally when the Pope and patriarch excommunicated each other (officially declared each other to be outside the community of Christians) in 1054. Thus, Roman Catholic Western Europe came to be culturally divided from Orthodox Eastern Europe, although the mutual excommunications were revoked by the patriarch and the Pope in 1965. The Uniate Church of Ukraine bridges the gap by using the Cyrillic alphabet and following the Eastern rite, but recognizing some spiritual authority of the Pope.

Focus ON

Israel and U.S. Foreign Policy

Since 1967, the United States has stood behind U.N. Security Council Resolution 242, a formula that would guarantee peace in exchange for Israeli surrender of land. Many people around the world, however, feel that the United States, as Israel's principal ally, has not brought sufficient pressure to bear against Israel to force it to accommodate Palestinian political aspirations. The United States has consistently supported Israel with both military and economic aid. This aid has been essential to Israel's ability to win wars and maintain control of its territory. Thus, Palestinians in particular and Muslims generally associate the United States with Israeli actions, and they blame both countries.

On October 5, 2001, the terrorist Osama bin Laden condemned America's support of Israeli actions. "Anyone who lives in the United States," he said, "will not feel safe before the people of Palestine feel safe." In 2004 he reiterated that he had struck America because "things just went too far with the American–Israeli alliance's oppression and atrocities against our people." Even many of America's friends who were prompt to denounce the 9/11 attacks coupled their condemnation with condemnation of Israel. In Kuwait, for example, a country that owes its continuing independence to the United States, the Director of Kuwait University's Center for Strategic and Future Studies, Ghanim Alnajjar, said, "The story is not bin Laden. The story is the injustice to the Palestinian people."

The United States will never let Osama bin Laden or anyone else dictate national policy regarding Israel, but in 2003 President George W. Bush did announce U.S. support for the establishment of an independent Palestinian state. Even the best-intentioned diplomacy, however, struggles to reach agreement on the amount of land that Israel might surrender, explicit acceptance by Arab nations of Israel's right to exist, the guarantee of Israeli security, Israeli acceptance of or reparations to millions of Palestinians refugees, the full sovereignty of a new Palestinian state, and water rights.

FIGURE 7-18 Monte Cassino. Monte Cassino, near Naples, was the first Western Christian monastery. Founded by Saint Benedict in the sixth century, it has played an immeasurable role in the history of European civilization. Although destroyed repeatedly in wars (most recently in 1944), it since has been rebuilt. (Robert W. Madden/National Geographic Image Collection)

The fall of Constantinople to the Muslim Turks in 1453 opened Eastern Europe to Islam, but the office of the Patriarch of Constantinople survived, and all Orthodox Christians still recognize its spiritual leadership. Orthodox Christians are subdivided into national churches, such as the Serbian, Greek, and Russian, although U.S. and Canadian Orthodox Christians cling to their historic national affiliations.

The Protestant Reformation of the fifteenth and sixteenth centuries split the Christian Church in the West. The Protestant denominations (Lutherans, Anglicans or Episcopalians, Baptists, and others) were named for their protest against the Church of Rome, and a variety of Protestant denominations and national denominations sprang up in Western and Northern Europe.

When Europeans explored and conquered other lands, they often cited religious conversion as their purpose. Christian missionaries were the chief agents of the partial Europeanization—of religion, language, social mores, and the acceptance of secular authority—that made many natives more tractable to European rule. Christianity diffused rapidly and almost completely throughout the Western Hemisphere. Certainly its message of all-embracing love carried strong appeal in lands in which many of the gods had been portrayed as terrifying and cruel. Furthermore, the Christian concept of heaven after death is democratic and accessible to all classes of people. Christians also have flexibly adapted some native practices. For example, in the fourth century, Western European Christian churches timed the celebration of Christmas to replace the celebration of the winter solstice among non-Christians. Christians still accept indigenous practices in order to bring peoples to Christianity. The 1992 Conference of Latin American

Roman Catholic bishops pledged to incorporate the traditional religious symbols, rituals, and cosmologies of Latin America's Native Americans and blacks whenever "compatible with the clear sense of the faith" and "the general discipline of the church" (Figure 7-19).

In Latin America, some natives might have adopted Christianity because they were astonished and demoralized by the Spaniards' invulnerability to the diseases that were wiping out the Native Americans. Neither the Spaniards nor the Native Americans understood that those diseases had been brought by the Spaniards themselves. The Spaniards convinced the natives to acquiesce to Spanish secular authority and to seek both physical and spiritual salvation in the Spaniards' church. Similar beliefs prevailed among both the Native Americans and the Whites in North America.

Christian missions to Asia were less successful. In 1542, exactly 50 years after Christians conquered Granada, the last Arab Muslim outpost in Europe, and after Columbus's first voyage to the West, Saint Francis Xavier arrived in Goa, India, to proselytize Asia. By 1549 Saint Francis was in Japan, which nearly converted to Christianity in the early seventeenth century but then expelled or persecuted Christians when it closed its doors to the world in the mid-seventeenth century (Figure 7-20). Christians had some initial success in China, but in 1723 Christianity was banned there as well.

Conversion to Christianity, and as will be seen later, to Islam, usually weakens the continuity of a people's cultural inheritance. This is because both Christians and Muslims believe that conversion is a spiritual rebirth. It is a deliberate repudiation of much of what came before, so conversion can destroy a people's

FIGURE 7-19 Jesus as a Native American holy man. This image of Jesus as a Mescalero holy man dominates the altar at the St. Joseph Mission church on the Mescalero Apache Reservation in New Mexico. His left palm holds the symbol of the Sun; his right hand a deer-hoof rattle; at his feet an eagle feather, a grass brush, and bags of tobacco and cattail pollen used in ceremonies. He stands on Sierra Blanca, the Mescalero sacred mountain. The inscription, in Apache at the bottom and Greek at the top, reads, "Giver of life." (© 1990 Robert Lentz. Used with permission of Bridge Building Images, P.O.Box 1048, Burlington VT 05402.)

FIGURE 7-20 The Basilica of Goa. The Basilica of Bom Jesus in Goa, India, is the final resting place of the missionary Saint Francis Xavier (1506–1552). Saint Francis is today the patron saint of Roman Catholic missionaries. (Courtesy of Father Vasco do Rego/EFB)

FIGURE 7-21 Vatican City. Vatican City, made up of St. Peter's Church, Square, and the surrounding buildings and grounds, is one of the world's smallest independent states. It is entirely within the city of Rome, but it is ruled by the Pope according to a treaty signed between the papacy and Italy in 1929. The official language here is Latin, and in 2003 the Vatican produced a new Latin dictionary to keep up with the times (a motorcyle is a *birota automataria levis*). (AP Photo/Italian/AP Wide World Photos)

historical culture more completely than can political conquest.

Christian sects and their distributions

Roman Catholicism is the largest single denomination among Christians. It is headed by the Pope in Vatican City, a tiny independent city-state within Rome (Figure 7-21). Roman Catholicism has historically dominated in the Mediterranean basin, throughout Latin America, and wherever else Mediterranean peoples colonized or converted, as in former French or Portuguese African colonies and in the Philippines and Vietnam.

Protestant denominations dominate in Northern Europe and wherever Northern Europeans have settled or converted: North America, Australia, and New Zealand, and the parts of Africa that were either formerly English colonies (today usually practicing Anglicans) or German colonies (Lutherans).

Several theological points differentiate Protestantism from Roman Catholicism, but among the most important is that Roman Catholics believe that Jesus gave Saint Peter unique responsibility for founding the Christian Church, that Saint Peter became the first bishop of Rome, and that the popes retain this special responsibility. The crossed keys on the Vatican flag represent Jesus' having given Saint Peter the keys to heaven. This belief that a church or priests intercede between God and humankind is called *sacerdotalism.*

Protestants deny sacerdotalism. Evangelical Protestants (from the Latin for "bringing good news") emphasize salvation by faith through personal conver-

sion, the authority of Scripture and each individual's responsibility to read the Scriptures, and the importance of preaching, as contrasted with ritual. The evangelicals emphasize the ability of individuals to change their lives (with God's help), and this message offers to many people a new sense of personal empowerment. Thus, Evangelical Protestantism frequently brings a revolutionary force into traditionally rigid or stratified societies.

Today Evangelical Protestantism is growing and spreading throughout the world. In some places it is replacing non-Christian religions, but in many places it is replacing Roman Catholicism. The population counted as Roman Catholic in Table 7-3 is the population baptized Roman Catholic. The Church insists that once a person has been baptized as a Roman Catholic, that person is always Roman Catholic, even if the person converts to another denomination or is excommunicated. Nevertheless, Evangelical Protestantism has replaced Roman Catholicism in several countries as the most widely practiced faith, and as a rule the ardor of converts is strong. Evangelical Protestantism is also spreading in Pacific Asia (especially into South Korea and the Philippines), across Africa, and throughout Latin America. About one-quarter of the total Latin American population is Protestant today. This percentage is even higher in Brazil, which counts from baptism the world's largest Roman Catholic population, but where today full-time Protestant pastors outnumber Roman Catholic priests, and as many as 600,000 people convert to Evangelical Protestantism each year.

In 2000 more than 10,000 Protestant Evangelists from more than 200 countries and territories assembled in Amsterdam, the Netherlands. The "Amsterdam Declaration," which they adopted, urges evangelists to be sensitive to the societies in which they work and to avoid equating Christianity with any particular culture. It obliges evangelists to present the Christian gospel as authoritative, even while respecting people of other religions. A clause that encourages women to be gospel teachers could trigger change in many traditional societies that do not accept women in leadership roles and could hasten conversions from Roman Catholicism.

The spread of Evangelical Protestantism affects many other aspects of human geography. In most Latin American states the Roman Catholic Church has traditionally enjoyed special privileges and a role in education, so its teachings have been enacted into law. Many Latin American countries, for example, have prohibited divorce. Today, however, a rising share of elected officials throughout Latin America is Protestant, and elections are explicitly referred to as "Holy Wars." The diffusion of Evangelical Protestantism is also at least partly responsible for slowing population growth. Evangelical Protestants are often as opposed to abortion as Roman Catholics are, but they are not always so adamantly opposed to other forms of birth control. Roman Catholicism has lost political power throughout Latin America, Quebec, Spain, Italy, the Philippines, and elsewhere, and presidents who are Protestant or *agnostic* (doubtful or noncommital about the existence of God) have been elected, for example, in the Philippines, Guatemala, Colombia, and Chile. Governments have begun sponsoring family planning, distributing birth-control devices, legalizing divorce, and introducing sex education into school curricula. Chile, for example, legalized divorce in 2005 and in 2006 installed a new agnostic president, Michelle Bachelet, who "promised" rather than "swore" to uphold the Constitution.

Other possible results from the spread of Evangelical Protestantism have been hypothesized, and they bear watching through coming years. One is the spread of literacy. This follows from Protestants' individual responsibility to read and study Scripture. Literacy has risen in areas where Evangelical Protestantism has spread, but the exact cause-and-effect relationship is difficult to measure. If Evangelical Protestantism does raise literacy rates, this trend may carry further ramifications—greater political participation, for example. Still other possible consequences of the spread of Protestantism in the areas of women's rights and even economics will be discussed later in this chapter.

In addition to Roman Catholic, Protestant, and Orthodox Christians, many smaller sects include the Georgian and Armenian churches; the Maronite churches in the Near East; the Copts of Egypt; the Christians of India, Ethiopia, China; and a few other groups.

The future of Christianity

The geography of Christianity is undergoing significant changes. The observance of Christianity has increased throughout Eastern Europe with political liberalization since the fall of Communism. Nevertheless, within the lifetime of most living Christians, Christianity has shifted from being a religion of the Western industrialized nations to being a religion of Asia, Africa, and the Pacific. In Africa, for example, the number of Christians has grown, chiefly by conversion, from fewer than 10 million in 1900 to almost 400 million today (Figure 7-22). In Latin America, the number of Christians rose from 62 million in 1900 to almost 500 million today; in Asia from 19 million to almost 300 million. Today there are more Lutherans in South Africa than in North America, and today more than half of the world's 73 million Anglicans live in Africa, South America, or Asia. South Korea is about 30 percent Christian—80 percent of them Protestant—and it may become the second Asian nation to have a Christian majority, after the Philippines. In 2003 the World Council of Churches elected its first African Secretary General, Rev. Samuel Kobia, a Methodist from Kenya.

FIGURE 7-22 The world's largest Christian church. The Basilica of Our Lady of Peace in Yamoussoukro, the Ivory Coast, is modeled after St. Peter's in Rome (see Figure 7-21), but it is larger, capable of holding over 18,000 worshipers. The Ivory Coast's first president, Félix Houphouët-Boigny (1905?–1993), converted to Roman Catholicism as a teenager (although he always insisted on his right to have several wives), and he built the basilica as a pilgrimage center for Africa's Catholics and as a bulwark against Islam and native religions. Today the Ivory Coast is about 12 percent Christian, 60 percent Muslim, and 28 percent animist, and strife among the religious communities threatens the country's peace. (Yann Arthus-Bertrand/Getty Images, Inc.—Liaison)

Christian missionary activity is increasing: There are today about 400,000 Protestant and Roman Catholic missionaries in the world, which is six times the number in 1900. About 100,000 of these missionaries are from Protestant churches in non-Western countries. Korea, for example, has sent almost 10,000 Protestant missionaries outside its boundaries. Many missionaries witness their faith only through good works, such as building educational and charitable institutions and providing humanitarian assistance. More aggressive efforts toward conversion, however, are meeting stiffening resistance from other religions; the divisions within Christianity aggravate the tensions. The spread of evangelical Christian missionaries throughout the Muslim world, for example, has provoked anger not only among Islamic clerics, but also among the native Christian groups. Trying to convert Muslims is often interpreted as a provocation that invites violence, and conversion from Islam is in many places punishable by death. Today the estimated 40 million Christians in countries ruled by Muslims find themselves an embattled minority facing economic decline, dwindling rights, and even physical jeopardy. This oppression and decline is in contrast to the rights (if not always full equality) that the surging Muslim minority enjoys in Western societies. Already in 2003, Vatican Foreign Minister Cardinal Jean-Louis Tauran noted, "There are too many majority Muslim countries where non-Muslims are second-class citizens. Just as Muslims can build their houses of prayer anywhere in the world, the faithful of other religions should be able to do so as well." In 2006, rage exploded among Muslims around the world when Pope Benedict XVI quoted a medieval belief that Muhammad had introduced "things only evil and inhuman."

Missionary activity by Americans has become a lightning rod for anti-American political sentiments. American missionaries have been killed in the Muslim Near East and Africa, and in India by Hindus. The U.S. National Association of Evangelicals has repeatedly reaffirmed its commitment to proselytizing, but it issued a "loving rebuke" to some evangelical leaders, such as the Rev. Franklin Graham (who said Islam is "a very evil and wicked religion"), Rev. Jerry Falwell ("Muhammad was a terrorist"), and Rev. Jerry Vines (who described Muhammad as "demon-possessed") for remarks that "tarnish Christianity" and jeopardize the safety of missionaries and the indigenous Christians in some countries. In 2004, Rev. Richard Cizik, president of the association, criticized "unfortunate and particularly irresponsible" remarks that "complicate circumstances for foreign missionaries and Christian aid workers overseas who are already perceived, wrongly . . . as collaborators with U.S. intelligence agencies." Nevertheless, President G.W. Bush, speaking to the Association on March 11, 2004, said, "You are doing God's work with conviction and kindness, and, on behalf of our country, I thank you." Many listeners throughout the world interpreted this as an explicit tie between American political power and missionary Christianity.

Pope John Paul II, who served 1978–2005, doubled the number of Roman Catholic saints. Virtually all of the saints he canonized were African and Asian. In 2002, he canonized Juan Diego, a Native American who supposedly sighted the Virgin (Mary) of Guadalupe in today's Mexico in 1531. Historians insist that Juan Diego never existed, but the pope hailed Diego as "the first indigenous saint of the American continent."

The Vatican maintains that the only purpose of any interfaith dialogue is to convert others to Roman Catholicism, and this conviction of Roman Catholicism's unique truth has triggered resentment and even political and social repercussions. For example, on a trip to India John Paul II declared that "the peoples of Asia need Jesus Christ and his gospel." Prominent Hindu, Sikh, Muslim, and Buddhist priests walked out of the room. Similar statements made during visits to the Near East triggered strong objections and even street riots. Roman Catholic proselytizing in Russia—trying to convert Russians from their traditional Orthodox Church to that of Rome—has triggered political restrictions on Roman Catholicism, such as the denial of permission to build new churches. Roman Catholic Cardinal Walter Kasper's attendance as Pope Benedict

XVI's representative at the World Summit of Religious Leaders in Moscow in July, 2006, however, signaled a thaw in relations between the two churches.

Chapter 6 noted that any cultural system or artifact may adapt and permutate as it diffuses across local cultures, and the shift in Christianity's population center of gravity has inspired changes in Christian theology and practice. As Roman Catholic bishop Bonifatius Hauxiku of Namibia has said, "Our African people have accepted Christ, but this Christ walks too much among them in a European garment." Local churches are adopting local foods for the sacred rite of communion, although, historically, Christians' need for wine and bread for this rite triggered the global diffusion of grape and wheat crops. Roman Catholic bishops around the world have recommended ordaining married men, a practice forbidden by the Vatican. The bishop of Nassau, the Bahamas, emphasized that on such questions "the church should not be tied to cultural vestiges typical of the European experience."

Christian leaders struggle to express Christian belief in terms meaningful to a great diversity of cultures without abandoning essential Christian distinctiveness. In many areas Christianity is challenged by **syncretic religions,** which combine practices and beliefs from two or more religions. The African and Caribbean religions of voodoo and Santería, for example, identify African deities and spirits with Christian saints. When Mathieu Kerekou, a Christian, was sworn in as president of Benin in 1996, he left out the italicized words in the oath: "Before God, *the spirits of the ancestors,* the nation and before the people . . . I . . . swear to respect the constitution." The public of this largely voodoo nation insisted he retake the oath with the italicized words. Chapter 5 noted the Native American ceremony at which Alejandro Toledo, a Christian, was sworn in as president of Peru in 2001. Many Brazilians practice a syncretic religion called Candomblé, and in 1999 Brazil's Minister of Culture declared a Candomblé temple a national monument. Other syncretic religions include Cao-Dai from Vietnam, and Chondo-Kyo from Korea.

Islam

Through the centuries after Abraham and Ishmael, the Arabs fell away from monotheism to polytheism, but they were brought back to monotheism by Muhammad (c. 570–632). He founded the religion called Islam, which means "submission [to God's will]." "One who submits" is a Muslim. The five essential duties of a Muslim, called the *Five Pillars*, are belief in God, five daily prayers, generous giving of alms, fasting during one month (called *Ramadan*), and, if possible, a pilgrimage (*hajj*) to Mecca at least once in one's lifetime.

The Arabic word for the one God is *al-elah*, or Allah, a cognate of the Hebrew *eloh*, "God." The Muslim

Critical THINKING

Holidays

One aspect of American commercial culture that has roused objections worldwide is the commercialization of holidays, particularly the celebration of holidays that began as religious observances. Throughout Europe, for example, Christian leaders have denigrated Halloween, an increasingly popular American import, condemning it as a pagan ritual and a celebration of Satan. Their efforts, however, have been futile. Shops decked out in orange and black now offer fancy dress competitions and pumpkin decorating displays, while nightclubs sponsor parties. France's Roman Catholic Bishops' Conference derided "skeletons and witches, pumpkins and ghosts," and asked, "How long can this marketing operation called Halloween continue to distort our sense of life and death?" The celebration of St. Valentine's Day now circles the globe, with fanciful greeting cards and boxes of sweets, although Hindu protesters in India, for example, have burned down shops offering such novelties. In Muslim countries, the end of the holy month of Ramadan is now routinely celebrated with strings of lights, parties, and gifts. In 2005, Sheik Ahmed Abdelaziz Haddad, the grand mufti of Dubai, complained, "People have taken this month to be a month of shopping. . . . A Muslim who is focused on the worldly trade will miss the benefits he could get in the hereafter. [T]he commercialism of Ramadan is caused by Muslim ignorance of what is required of them to benefit their souls."

Questions

1. What do you say to the French bishops' question?
2. What holidays do people you know celebrate in ways that overlook the holidays' original religious nature?

God is the God of Abraham, Moses, and Jesus, so Judaism, Christianity, and Islam are often called "the Abrahamic religions." The essential belief of a Muslim is stated in the expression "There is no God but God," ("la Ellaha Ela Allah") with which Jews and Christians would agree. This Arabic statement, however, is often half-translated or mistranslated. For example, the U.S. Marine Corps Intelligence Service's "Iraq Culture Smart Cards," distributed to U.S. troops serving in Iraq state that Muslims believe that "Allah is the one true god" (with a small "g," rather than, for example, "Allah is the Arabic word for God" or simply "Muslims believe in God."). Such mistranslation is not only incorrect, but prejudicial. It alienates Christians and Jews from Muslims, rather than reconciling them. Some Muslims believe this insult is deliberate.

Despite Muhammad's flight from his native city of Mecca in 622 (the *Hegira*), which was a temporary setback and the year from which Muslims date their calendar, he had converted and united most Arabs by his death. Just as Christians see their religion as building on Judaism, adding the New Testament to the Old, so Muhammad envisioned his teachings as a continued evolution of monotheism. Muslims believe that Muhammad was the last of God's prophets, who also include Adam, Noah, Abraham, Moses, and Jesus. Muhammad expected Christians and Jews to be among the first to embrace Islam. He first directed Muslims to face Jerusalem in prayer, but he later changed the direction toward Mecca (Figure 7-23 and 7-24).

As Islam diffused, the Arabic of the Koran became the language of ethnically different peoples throughout

FIGURE 7-23 Mecca. This black shrouded building is the Kaaba in Mecca, where, Muslims believe, the biblical patriarch Abraham and his son Ishmael built the first sanctuary dedicated to the one God. Through millennia the temple was given over to the worship of idols, but Muhammad threw out the idols, rededicated the Kaaba, and had its walls covered with paintings of Abraham, Mary, and Jesus. (Mary, Jesus' mother, is referenced 19 times by name in the Bible, 34 in the Koran.) That building has since been replaced, but this is probably the longest continuously revered spot on Earth. To the left of the Kaaba a small stone canopy covers what believers accept as the well of Zamzam (Genesis 21: 19), where Abraham's concubine Hagar and their son Ishmael quenched their thirst. Ishmael is thought to be buried just in front of the Kaaba. (Mehmet Biben/Photo Researchers, Inc.)

FIGURE 7-24 Dome of the Rock, Jerusalem. Many sites in Jerusalem are sacred to Jews, Christians, and Muslims. All believe God prevented Abraham from sacrificing his son on this rock (Mount Moriah), thus ending human sacrifice. King David later built the Jews' Temple here. Muslims believe that Muhammad ascended to heaven from this spot. The beautiful building with the golden dome is the Dome of the Rock (begun in A.D. 643), the oldest existing monument of Muslim architecture. Its builders adopted the Byzantine style when they conquered this region. Contrast it with the Dyula-style mosque in Figure 6-16. Christians believe that Jesus ascended to heaven from the Mount of Olives in the background. In the foreground is the Western Wall, the only wall remaining of the Jewish Temple, which was destroyed by the Romans in A.D. 70, triggering the Jewish Diaspora. (A. Ramey/Woodfin Camp & Associates)

the Near East and across North Africa. They had not previously been Arabic speakers, but as they converted, all came to be known as Arabs. To the East, Persia (today's Iran) had its own ancient culture, and although the Persians converted to Islam, they retained their own language, Farsi, an Indo-European language. Therefore, today's Iranians are not Arabs, nor are any of the peoples to the east of Iran.

Within slightly more than a century after Muhammad's death, Islam stretched from the Atlantic coast of Spain across North Africa and through Southwest and Central Asia to the borders of China. An Arab army met a Chinese army in the Talas River Valley in today's Kyrgyzstan in 751, just 19 years after another Arab army

had faced a Frankish army in Tours in today's France thousands of miles to the West. Within this vast realm, dominance eventually passed from the conquering Arabs to the non-Arab majority, including the Persians and the several Turkish peoples. Islam later expanded down into South and Southeast Asia (Figure 7-25). This culture realm greatly exceeded the Christian culture realm in extent, power, and riches for hundreds of years. Learned travelers crossed its length and breadth, and their descriptive writings constitute some of the greatest works of historical geography (Figure 7-26).

Islam denigrates the earlier cultures of its converts, just as we noted that Christianity can. Everything before Islam was, in Arabic, *jahiliya*, "from the age of ignorance." This leaves little room in these people's historical consciousness for their pre-Islamic past, so they often lack interest in it. For example, despite Persia's brilliant antique history, for contemporary Iranians the glory began with the coming of Islam. Pakistan is a new Muslim state, so even though the land contains ruins of civilizations thousands of years old, contemporary Pakistanis disdain them. Many people in Muslim countries view their own ancient cultural landscapes without interest and may even discourage tourists from viewing pre-Islamic ruins. Similarly, the study of the history and art of pharaonic Egypt is the result of European historical interest. The fact that Egyptians are interested in it today demonstrates that Egyptian nationalism, a newer cultural force, competes with Islam for the primary loyalty of the Egyptian people.

As Islam filtered south across the Sahara desert from North Africa, its message reached the black peoples of the sahel and savanna. There it still competes with Christianity, which entered sub-Saharan Africa from the Europeans' coastal incursions. From Senegal to the Congo the coastal areas are Christian, and inland areas are Muslim. When several of these countries first won independence, the westernized political leaders were Christian (for example, President Léopold Senghor in Senegal and President Félix Houphouët-Boigny in the Ivory Coast), but the majority of the population was Muslim. Christian politicians continue to play prominent roles, but as democracy matures in these countries, Muslims often come to power.

Within Europe, only Albania, long ruled by Turks, is predominantly Muslim. Bosnia and Herzegovina is about 40 percent Muslim.

Islamic sects The two principal sects of Islam date back to a struggle over rule of the Islamic world that occurred shortly after Muhammad's death. Sunni Muslims accept the tradition (*Sunna*) of Muhammad as authoritative and approve the historic order of Muhammad's first four successors, or *caliphs*. About 85 percent of Muslims worldwide today are Sunni. Most of the other 15 percent are Shia Muslims, or Shiites. They believe that Muhammad's son-in-law Ali was the rightful successor to the Prophet, and they commemorate the

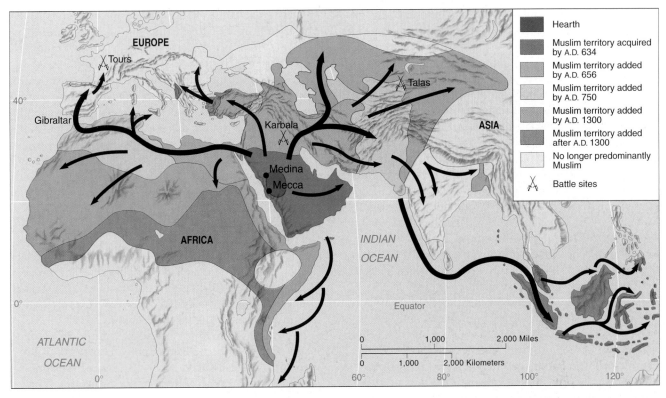

FIGURE 7-25 The diffusion of Islam.   The first nine centuries of Islam were a period of almost continual expansion, and by 1800, Arab and Indian traders had spread the faith through many of the islands of Southeast Asia.

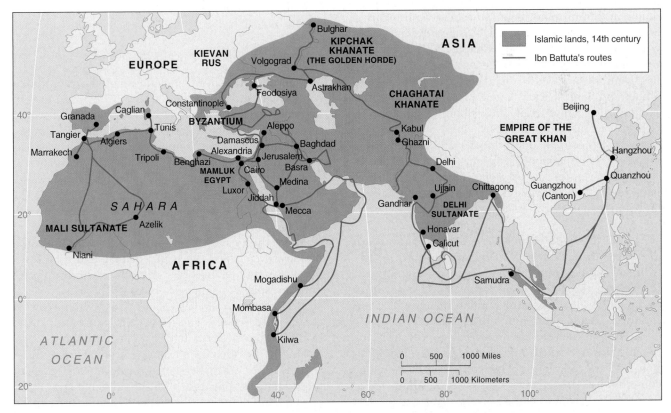

FIGURE 7-26 The travels of Ibn Battuta.   In the fourteenth century the Muslim scholar Ibn Battuta traveled from Morocco to China, where he was happily received by rulers. Several even appointed him as a judge thousands of miles from his own home. His geographic writings are a superlative description of life across the vast Islamic realm of his day.

martyrdom of Muhammad's grandsons in a Muslim civil war battle at Karbala in today's southern Iraq in 680. Through the centuries, differences in ceremony and in law have further differentiated the Sunnis from the Shiites.

The ruling family of Saudi Arabia is of the Wahhabi sect, a puritanical movement that is the only modern separatist Sunni sect. Muhammad Ibn Abd al-Wahhab (1703–1792) sought to purify Islam from what he saw as the corruptions of mysticism, rationalism, and Shiite theology. Western collaboration with Wahhabites during World War I to defeat the Ottoman Empire gave birth to Saudi Arabia, a Wahhabite state whose immense oil wealth since World War II has allowed the aggressive diffusion of Wahhabism through the construction of schools and mosques throughout the Muslim realm. The diffusion of this fundamentalism has had political repercussions, as we shall see below.

Shiites form the majority in Iran, Azerbaijan, and in Iraq and are important minorities in Kuwait, Lebanon, Bahrain, Syria, Saudi Arabia, and Pakistan. Animosity between Sunni and Shia can be fierce. Countries that contain significant shares of both sects are split by the enactment into national laws of either Sunni or Shiite interpretations of Muslim religious teachings, and several, including Lebanon and Pakistan, have suffered civil disturbances between the two groups. Ever since the overthrow of the government of Iraq in 2003, fighting between the historically dominant Sunni minority and the newly enfranchised Shiite majority has threatened U.S. peacekeeping and nation-building efforts (see Chapter 13). Some observers suggest that, in the long run, the United States has been only a pawn in a 650-year-long clash between Shiites and Sunnis: Today a newly resurgent Shiite Iraq stands with Shiite Iran and Shiite Hezbollah in Lebanon against Sunni Pakistan (where Shiites are regularly referred to as "mosquitoes') and Saudi Arabia (where school textbooks refer to Shia Islam as "heresy"). Even the sacred *hajj* to Mecca has regularly been disrupted by violent clashes between Sunnis and Shiites.

The Muslim world today

Today Muslims are distributed from Morocco to Indonesia, north to the frontiers of Siberia and south to Zanzibar, with outposts throughout the world. Even Rome boasts a splendid mosque. The leading Muslim states, in numbers, are Indonesia, Bangladesh, India, and Pakistan.

Muslims believe in proselytizing, as Christians do, and this necessarily injects an element of competition into the relationship between the two religious communities. Islam may evolve as Muslims move outside the Islamic realm, just as Christianity has evolved as it has diffused. For example, speaking at the dedication of a new mosque in Lyons, France, the mosque's imam (Islamic prayer leader), Algerian-born Chellali Benchellali, proclaimed that he stood

"between two cultures. . . . I think this is the beginning of a French Islam." At a gathering of European imams held in Vienna in 2006, the Bosnian representatives insisted that "Muslims who live in Europe have the right—no, the duty—to develop their own European culture of Islam." Sayed Ghaemmagami, the Grand Mufti of the Shiiites in Germany, argued that "the existence of an Islamic diaspora . . . requires new thinking about relations between Muslims and others, so Muslims should engage with their new countries and not set up parallel structures." Institutes for Islamic studies are proliferating across Europe. If European Muslims create an Islam that is open to democracy, sexual equality, and modernity, it will affect Islam around the world.

In other non-Arab regions, local Islamic practices differ. Many Indonesians, for example, retain ancient local gods and goddesses and folk practices in a form of syncretism. These are declining, however, as a result of urbanization and Wahhabite missionary work.

Although many Americans tend to equate Islam with Arabs, Arabs actually constitute only a fraction of Muslims. There are more Muslims in Indonesia and Pakistan than in all the Arab countries combined. Americans' association of Islam with Arabs is a political preconception, caused by the distinctive role of Islam in Arab countries and by U.S. news media focus on Arab lands as sources of oil and antagonists of Israel.

Hinduism and Sikhism

Hinduism is the most ancient religious tradition in Asia. The oldest Hindu sacred texts (the *Vedas*) date to about 1800 B.C., but the religion originated somewhere in Central Asia long before that. It entered the Indian subcontinent with the coming of Central Asian peoples about the time of the writing of the *Vedas*. Today it is confined almost exclusively to India and Nepal, where it is the official state religion (Figure 7-27).

Hindus believe in one Supreme Consciousness, Brahman, whose aspects are realized in three deities: Brahma, the creator; Vishnu, the preserver; and Siva, the destroyer. These three are coequal, and their functions are interchangeable. All other Hindu "gods," saints, or spirits are emanations of Brahman.

Hinduism classifies people in a hierarchy of classes called **castes.** The four main castes are (1) the Brahman, or priestly caste; (2) the Kshatriya, or warrior caste; (3) the Vaisya, or tradesman and farmer caste; and (4) the Sudra, or servant and laborer caste. Each of these castes is split into hundreds of subcastes, many of which are defined by occupation. People are expected to mix socially, marry, and stay in the caste into which they were born. A group of people called **untouchables** (*Dalit*) is considered so low that their status is below the formal structure of the four castes.

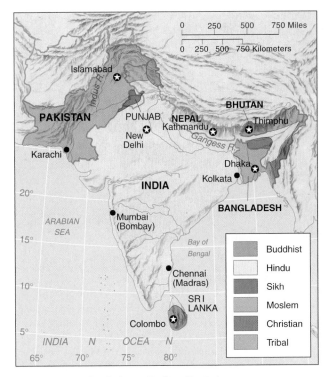

FIGURE 7-27 Religions in South Asia. Hinduism is today restricted almost exclusively to India, Nepal, and to places where Indians have migrated, including Jamaica, Trinidad, Guyana, and Fiji in the Pacific Ocean. Sikhs form a majority in the Indian state of Punjab, but smaller Sikh communities can be found throughout India. The regions labeled Christian converted from animism in the nineteenth century.

FIGURE 7-28 The Golden Temple in Amritsar. The Sikhs' 400-year-old Golden Temple in Amritsar, India, stands in the middle of a sacred lake (in Sanskrit *amrita saras*, "pool of immortality"). The building was occupied by armed Sikh extremists in 1983; the following year Indian troops stormed it and drove out the occupants in a bloody encounter. Sikh fanatics retaliated by assassinating India's Prime Minister Indira Gandhi. (*Jenhangir Gazdar/Woodfin Camp & Associates*)

To some degree caste discrimination still structures Indian life, but it was legally abolished in the Indian Constitution of 1950, and an untouchable, K. R. Narayanan, was even sworn in as president of India in 1997. (India's president is elected by parliament, not by popular vote.)

Hindus believe in *reincarnation*—that is, individual rebirth after death. The caste into which you are born is not haphazard, but depends upon your behavior in an earlier life. This teaching, called *karma*, discourages ambition, because only by docilely keeping to your place in this life can you hope to enjoy a better position in your next life. The goal of Hindus is liberation from the cycle of death and rebirth.

Sikhism is an offshoot of Hinduism based on the teachings of Guru (teacher) Nanak (c. 1469–1539). Nanak tried to reconcile Hinduism and Islam, teaching monotheism and the realization of God through religious exercises and meditation. Nanak opposed the maintenance of a priesthood and the caste system. Under a series of gurus, the Sikhs had their own state in northern India, but they were eventually conquered by the British. Many Sikhs dream of the restoration of an independent Sikh state. The Sikh's holy temple is in the city of Amritsar in the Indian state of Punjab (Figure 7-28). This state is largely Sikh, and it might provide a territorial base for independence.

Buddhism

Siddhartha Gautama (c. 563–483 B.C.) was a Hindu prince born in what is today's Nepal who, through meditation, achieved the status and title of Buddha, or Enlightened One. He taught Four Noble Truths: (1) life involves suffering, (2) the cause of suffering is desire, (3) elimination of desire ends suffering, and (4) desire can be eliminated by right thinking and behavior. This cessation of suffering is called *nirvana*, or total transcendence.

As Buddhism diffused out of India, sects and schools arose. The Theravada school ("doctrine of the elders") diffused to the South (Figure 7-29). This school centers on the idea of a monk striving for his own deliverance. Mahayana Buddhism (the "great vehicle" because it carries more people to nirvana) diffused

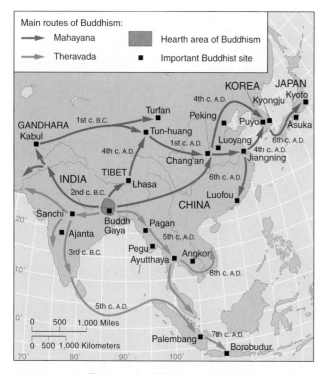

FIGURE 7-29 The early diffusion of Buddhism.
Buddhism was localized near its hearth for hundreds of years, but it spread widely due to the support of the third-century Indian emperor Asoka. It has largely been abandoned in India.

FIGURE 7-30 The Dalai Lama.   The Dalai Lama, the Buddhist spiritual leader of Tibet, was forced into exile by the government of China. He has won worldwide acclaim—including the Nobel Peace Prize in 1989—for his efforts to promote religious freedom. Here he is shown greeting Congresswoman Nancy Pelosi (Dem. CA), who became the Speaker of the U.S. House of Representatives in January, 2007, the first woman ever to win that position. (Carrie Devorah/WEN/NewsCom)

northward. Mahayana idealizes the concept of the *bodhisattva,* someone who merits nirvana but postpones it until all others have achieved enlightenment. In Bhutan, Tibet, and Mongolia, Buddhism evolved a special form called *Lamaism,* which is known for its elaborate rituals and complex priestly hierarchy (Figure 7-30). Chinese Buddhists produced a new theory of spontaneous enlightenment, called *Ch'an.* This diffused into Japan as *zen.*

Buddhism has several hundred million followers, but its adherents are hard to count because it is not an exclusive system of belief. Its practice has declined in India, but today it is the state religion in Thailand and Sri Lanka, and it may achieve that status in Mongolia. Buddhist philosophy has also won considerable influence in the modern Western world. Much popular "New Age" philosophy derives from Buddhism.

Other Eastern Religions

Confucianism is a philosophical system based on the teachings of K'ung Fu-tzu (c. 551–479 B.C.). He taught a system of "right living" preserved in a collection of sayings, *The Analects,* which governed much of China's political and moral culture for 2,000 years.

Confucianists believe that people may attain heavenly harmony by cultivating knowledge, patience, sincerity, obedience, and the fulfillment of obligations between parents and children, subject and ruler. These moral precepts permeate life in many Eastern societies, and today the word *Confucian* is often popularly used as a synonym for these qualities. Confucianism influenced Western philosophy and political theory at the time of the Enlightenment, when it appeared as the realization of Plato's utopian dream of a state ruled by philosophers.

Taoism is derived from the book *Tao-te Ching* (third century B.C.) attributed to Lao-tze. It advocates a contemplative life in accord with nature (Figure 7-31).

Shinto is the ancient religion native to Japan. Its rituals and customs involve reverence for ancestors, celebration of festivals, and pilgrimage to shrines, but there is no dogma or formulated code of morals. Shinto is nationalistic, and it traditionally recognized the emperor as divine. The Emperor Hirohito renounced this divinity in 1946, following Japan's defeat in World War II. Nevertheless, signs at the Yakasuni Shinto Shrine still proclaim, "Some 1,068 people, who were wrongly accused as war criminals by the Allied court were enshrined here. . . . War . . . was necessary in order for us to protect the independence of Japan and to prosper together with Asian neighbors." Periodic visits to Yakasuni by Japan's prime ministers roil antagonism between Japan and Japan's World War II enemies, especially China and the Koreas. In 2005 Chinese Foreign Minister Li Zhapxing asked, "The leader of a certain country is still worshipping war criminals. . . . What sort of behavior is this?" In 2006 Japan's new Prime Minister Shinzo Abe promised to end official visits to Yakasuni, although he has often gone privately in the past.

FIGURE 7-31 Lao-tze and Confucius with the infant Buddha. This fourteenth-century Chinese painting shows Lao-tze (on the left) and Confucius protecting the infant Sakayumi, the future Buddha. Lao-tze's appearance reflects his legendary life spent wandering and communing with nature, in contrast to the courtly scholar Confucius. Images such as this, emphasizing the compatibility of Confucianism, Taoism, and Buddhism, became popular as Buddhism spread across East Asia. (© The British Museum)

Animism and Shamanism

Animism is a belief in the ubiquity of spirits or spiritual forces (Figure 7-32). Animistic religions may be basically monotheistic, but they recognize hierarchies of divinities who assist God and personify natural forces. Millions of Africans believe in animism.

Animism is frequently accompanied by **shamanism.** A shaman is a medium who characteristically goes into auto-hypnotic trances, during which he or she is thought to be in mystical communion with the spirit world. Shamanism exists among the peoples of Siberia, the Inuit, some Native American tribes, in Southeast Asia, and in Oceania. Almost everywhere both animism and shamanism are yielding to Muslim and Christian proselytizing. Meanwhile, the world's pharmaceutical companies race to learn shamans' ethnobotanical knowledge.

FIGURE 7-32 An animist altar. This animist altar stands in a Dogon village in the West African country of Mali. It is covered with a patina of millet gruel laid on as an offering to the spirits of the harvest. (Sanga region, Mali. Photograph by Eliot Elisofon, 1972. Eliot Elisofon Photographic Archives, National Museum of African Art, Smithsonian Institution.)

THE POLITICAL AND SOCIAL IMPACT OF THE GEOGRAPHY OF RELIGION

The distribution of religions affects the distributions of many other facets of life and culture. Religion influences what people eat; their tolerance for others; when and how they work; when and how they celebrate; which types of behavior they encourage and discourage; and many other aspects of life. The influence of religion is so pervasive that it is difficult to isolate and identify precisely. Clearly, however, many people choose to obey their religion's teachings. The following pages catalog just a few of the many ways that the geography of religion influences the geography of other aspects of human lives and societies.

Religion and Politics

Almost all countries guarantee freedom of religion, and most governments observe a form of secularism. Many governments nevertheless favor one religion over others, either implicitly or explicitly. Therefore, the world political map partly fixes or stabilizes the map of world religions, just as it fixes the map of world languages.

FIGURE 7-33 King Muhammad VI of Morocco.
King Mohammed VI alternates between wearing modern
westernized clothes and the robes traditional to his role as
hereditary religious leader. (Benito-Buu-Benyatouille
Hires/Getty Images, Inc.—Liaison)

FIGURE 7-34 The citadel of the Church of Jesus
Christ of Latter-Day Saints (often called the
Mormon church) in Salt Lake City. In 1847 this site
was surrounded by hundreds of miles of wilderness, but Mormon leader Brigham Young recognized it as an oasis that
Mormon industriousness could make bloom and where Mormons might escape the persecution they had suffered in the
East. The building at left in the form of an overturned hull
of a ship is the Tabernacle, and the neo-Gothic building in
the center is the Temple. Mormonism is one of the fastest-
growing religions in the world. In 1950 there were 1 million
Mormons, nearly half of whom lived in Utah. Today the religion counts 12 million members around the world, fewer
than 15 percent of whom live in Utah. (© The Church of Jesus
Christ of Latter-Day Saints. Used by permission.)

Theocracy is a form of government where
a church rules directly. The Vatican is a theocracy.
Morocco may be a modified theocracy, because the
king's legitimacy derives partly from his descent from
Muhammad (Figure 7-33). Many theocracies have existed in history: Tibet, for example, and Massachusetts
Bay Colony in British North America. Utah was once a
Mormon theocracy, and although it is today increasingly cosmopolitan, no Utah politician can ignore Mormon Church leaders (Figure 7-34).

In Israel Israel has steadily enacted more religious
strictures into law since its founding in 1948. Recent laws,
for example, ban the production or sale of pork and prohibit activities on the Sabbath. Also, the government has
become more Orthodox, recognizing, for example, conversions to Judaism performed only by Orthodox rabbis.
This has distressed some U.S. Jews, 85 percent of whom
practice Conservative or Reform Judaism. Many of the
Russian Jews who have recently migrated to Israel are not
Orthodox, and they press for secularization, including
civil marriage, eliminating the "nationality" designation

on identity cards, and the operation of public transit on
the Sabbath—all policies opposed by the Orthodox.

In Christianity The Christian Bible records Jesus'
words, "Render therefore unto Caesar the things which
be Caesar's, and unto God the things which be God's"
(Luke 20:25). For 2,000 years, governments among
Christian peoples have argued about exactly where to
divide responsibility between religion and government,
but they accept the idea that church and state, religion
and politics, must be separated.

The separation between religion and the political
order allowed Europe to develop secular government,
secular knowledge, and a secular culture. Nevertheless,
several countries are today explicitly Christian or at least
support various Christian sects (called *established* churches) with public funds. These include Argentina, Peru, Ireland, Norway, Denmark, Iceland, the United Kingdom,
Finland, and Germany (although in 1998 Gerhard
Schroeder became the first-ever German chancellor to
refuse to end his oath of office with the words "so help me
God."). At the same time, other countries are disestab-
lishing official churches. For example, Italy disestablished

Roman Catholicism in 1984, as did Spain in 1988, and Colombia in 1991. Sweden disestablished Lutheranism in 2000, and the government of Greece stopped listing religious affiliation on state identity cards in 2000. Many countries in Eastern Europe have revived their national churches as a means of defining and emphasizing their national identities since the fall of Communist governments. The Uniate Church of Ukraine, for example, survived persecution by the Communists and revived in the late 1980s.

Liberal democracy, as it is known in the West, was shaped by a thousand years of European history and, beyond that, by Europe's double heritage of Judeo–Christian religion and ethics and Greco–Roman statecraft and law. No such system ever originated in any other cultural tradition, and it remains to be seen whether such a system can survive when transplanted and adapted in another culture.

In Islam

Theoretically, no distinction can be made between church and state in Islam, which teaches that the only purpose of government is to ensure that each person can lead a good Muslim life. Church and state should be the same, thus making Islam inherently political. (The centrality of religion in Islamic cultures is illustrated even in city planning, see Chapter 10) Therefore, many countries with largely Muslim populations are today officially Islamic, including Mauritania, Afghanistan, Libya, Saudi Arabia, Yemen, Oman, Qatar, Bahrain, Iraq, Iran, Comoros, Maldives, Pakistan, Bangladesh, and Malaysia. These countries may be considered modified theocracies. Islamic states enact Islamic teachings into law, called **Sharia** law, and establish Sharia courts to rule whether their secular laws conform to Islamic teaching. If the Sharia court rules against a secular law, that law is usually repealed. All legislation in Iran, for example, must be approved by a body of 12 religious judges, who enforce fundamentalist beliefs. Indonesia is officially a secular country, but it has recognized Sharia rulings on family law. Under Sharia law, and in many Muslim countries, *blasphemy* (an irreverent or impious act or utterance) is punishable, as it was in most western Christian countries until the concept of "free speech" evolved. Furthermore, *apostasy* (abandonment of one's faith, or conversion) may even be punishable by death.

Fundamentalist Islamic political parties are contesting elections in several countries that are not now officially Islamic. Many observers fear that these parties see democracy as a means to an end—the creation of an Islamic state—rather than as a system to be valued for itself. Turkey is officially secular, but the government pays for mosques, supervises the (compulsory) religious education, specifies qualifications for imams, appoints all imams to mosques, and pays their salaries. The Turkish Constitution declares the army to be the official guardian of the country's secular status, and Turkey's army has repeatedly intervened in democratic

processes in order to frustrate the rise of fundamentalists. The anecdote opening this chapter recounts the continuing tension in Algeria. Secular government probably survives in Egypt only through brutal repression of Muslim fundamentalists.

The role of education   Some observers insist that the rise of fundamentalist Islamic parties is due to the failure of the secular governments in many Muslim countries to provide basic services, such as health care and education. Many see education as secularists' strongest weapon against fundamentalism. Mohammed Charfi, who served as education minister in Tunisia from 1989 to 1994, presided over hundreds of Arabists, historians, philosophers, and other specialists to remove aspects of political Islam from Tunisian schoolbooks. "I left not a single schoolbook untouched," he wrote. "The reform is not against Islam, but rather designed as a modern version of Islam consonant with democracy. School is the best place to fight fundamentalism, and school is where it is born and dies. No one could be a fundamentalist if they read Spinoza, Voltaire, and Freud." The school curriculum in Jordan underwent a similar revision in 2005. In schools in Saudi Arabia, by contrast, religion occupies from one-quarter to one-third of class time, and textbooks still in use in 2006 teach that "every religion other than Islam is false," and even that "it is forbidden for a Muslim to be a loyal friend to someone who does not believe in God and His prophet." Fundamentalists of other faiths in other countries may hold such intolerant beliefs, but such beliefs are seldom taught in other nations' public schools.

In many Muslim lands, the only education available (and available only to males) is in Islamic schools, called *madrassas,* where modern mathematics and science, as well as humanistic concerns, are repudiated. Students spend hours poring over the Koran, and the "A" students memorize the most passages. Fundamentalist governments, individuals, and Islamic foundations have financed madrassas in many countries in Africa, Southeast Asia, and other places where secular public education is weak, thus arousing fears of networks of anti-Western indoctrination. Pakistan's President Pervez Musharraf has acknowledged that much of the backwardness of the Muslim world is rooted in the education systems. "Do we believe," he has asked, "that religious education alone is enough for governance, or do we want Pakistan to emerge as a dynamic Islamic state? The verdict of the masses is in favor of a progressive Islamic state." President Musharraf has promised educational reform.

The Muslim Umma   To a Muslim, the international community of all Muslims is called the *Umma.* To a fundamentalist, the division of this community into many political states is blasphemy. In 1998, the terrorist Osama bin Laden issued a *fatwa* (holy directive) saying, "The call to wage war against America was called

Are We Seeing an Islamic Reformation?

The early twenty-first century is witnessing a war of ideas within the Islamic world comparable to Western Europe's Protestant Reformation. The question in Islam, as it was in Christendom, is whether one class of interpreters holds a monopoly on scripture or God's will, or whether each individual is, in Martin Luther's words, "a priest." Libya's President Qaddafi has taken the most "Protestant Muslim" position: He has repressed Islamic fundamentalist movements (for his own political ends) by insisting that all Muslims speak directly to Allah, so clergymen are unnecessary intermediaries.

Hashem Aghajari is a former Islamic revolutionary and college professor in Iran. He is not the only significant theologian in the Islamic world, nor are his ideas necessarily the best, nor will they necessarily triumph, but his ideas exemplify the birth of a new conversation among Muslims. In 2002, Professor Aghajari gave a speech noting that "the Protestant movement wanted to rescue Christianity from the clergy and the church hierarchy," so Muslims must do the same. "Just as people at the dawn of Islam conversed with the Prophet, we have the right to do this today. . . . For years, young people were afraid to open a Koran. They said, 'We must go ask the mullahs [religious leaders] what the Koran says.' The religious leaders taught that if you understand the Koran on your own, you have committed a crime [but] we need a religion that respects the rights of all—a progressive religion, rather than a traditional religion that tramples the people."

Professor Aghajari was in effect accusing the mullahs of introducing sacerdotalism into Islam. He was arrested and sentenced to death for apostasy, but, after widespread protests, the Iranian Supreme Court dropped those charges against him that carried the death penalty. In 2005 he was sentenced to three years in jail, two years on probation, and five years' suspension of his social rights.

---

because America has spearheaded the crusade against the Islamic community, sending tens of thousands of troops to the land of the two holy mosques . . . and . . . meddling in its affairs and its politics. . . ." Bin Laden referred to an "Islamic community," that spans many ethnicities and states, and to the Aqsa Mosque in Jerusalem and the Haram Mosque in Mecca as being in the same "land." Bin Laden thus saw all the governments across the entire Muslim world (including Israel) as illegitimate. This belief made bin Laden, ultimately, a threat to all individual Muslim rulers and states (compare this to Figure 7-26). By his definition, the stationing of U.S. troops in the Near East is blasphemous, even if American troops were first based on Saudi territory in 1991 to protect Saudi and Kuwaiti independence against Iraq. Bahrain and Oman have also provided bases for U.S. troops, but Americans' 2003 conquest and occupation of Iraq has fired many Muslims' indignation. Even though U.S. troops left Saudi Arabia in 2003, Saudi Arabia, Jordan, and other Muslim lands have continued to suffer disruption by al-Qaeda-linked bands.

Bangladesh is officially an Islamic country, but Islam there is generally characterized as moderate. Nevertheless, in August 2005, Bangladesh suffered an extraordinarily sophisticated, coordinated bombing attack. Approximately 450 bombs went off nearly simultaneously at military bases, airports, government and police offices, universities, markets, and 63 of 64 district courthouses. The bombers' intentions were clearly to warn people, rather than to harm them (only 2 people were killed and 100 injured), but a fundamentalist group claimed responsibility in a pamphlet arguing that rigid Sharia law must be instituted. The situation in Bangladesh has become so dangerous for Americans that all Peace Corps volunteers were pulled from the country in 2006.

Osama bin Laden never represented the Islamic community in general. The Organization of the Islamic Conference, a coalition of 56 Islamic nations representing 1.2 billion Muslims, declared in October 2001 that the "terrorist acts [of 9/11] contradict the teaching of all religions and human and moral values." In 2003, Grand Sheik of Al-Azhar Mohammed Sayed Tantawi, considered by many to be the Sunni Muslim world's highest religious authority, said Muslims should "wholeheartedly open our arms to the people who want peace with us. . . . I do not subscribe to the idea of a clash of civilizations," he insisted, revealing his familiarity with popular Western thought (see Chapter 6).

The strife within the Islamic Umma between nationalism and internationalism, over questions of sacerdotalism, and over efforts to define the role of religion in modern life, will occupy headlines for the foreseeable future.

The geography of religion in the United States

Several of the American colonies were founded as theocracies, and most retained established churches

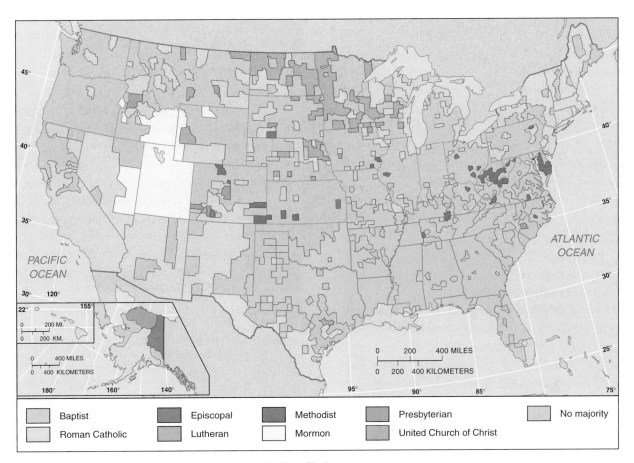

Baptist	Episcopal	Methodist	Presbyterian	No majority
Roman Catholic	Lutheran	Mormon	United Church of Christ	

FIGURE 7-35 Church membership in the United States. This map of Americans' church affiliations reveals the predominance of Baptist membership in the Southeast and of Mormons in Utah and adjacent regions. The predominance of Lutherans in the northern midwest reflects the Scandinavian and north German backgrounds of that region's pioneer settlers, and the distribution of Roman Catholics reflects, among other factors, the migration to the United States of Hispanics and southern Europeans. (Modified from Martin B. Bradley, Norman M. Green, Jr., Dale E. Jones, Mac Lynn, and Lou McNeil, *Churches and Church Membership in the United States 1990.* Nashville, TN: Glenmarry Research Center, 1992.)

until the War for Independence. In forging the new United States, however, James Madison included in the Bill of Rights a prohibition against the federal government's establishing any religion—the world's first such clause: "Congress shall make no law respecting an establishment of religion, or prohibiting the free exercise thereof. . . . " The 1797 treaty between the United States and Tripoli—ratified by Congress—declares that "The government of the United States is not, in any sense, based on the Christian religion." Congress added the words "under God" to the pledge of allegiance in 1954, but ongoing litigation disputes the constitutionality of that action. American coins do insist "In God We Trust."

People of every faith can worship in the United States as they wish, and they can even proselytize. The consequence of the United States' separation of church and state has not been the death of religion, but, rather, the birth of what is arguably the most religious nation in the world. In a 2001 Gallup Poll, 82 percent of Americans identified themselves as Christian; 10 percent of other faiths, and only 8 percent as agnostic or atheist. This is a unique degree of religious commitment and religious homogeneity. Surveys in different

countries do not always ask the same questions, but for comparison about 2 percent of people in England regularly attend church, and fewer than one French person in 10 goes to church as often as once a year. In 2002, 49 percent of Danes, 52 percent of Norwegians, 55 percent of Swedes, and 64 percent of Czechs said God "does not matter" to them, whereas 82 percent of Americans said God was "very important." The U.S. Census Bureau does not study religions, but Figure 7-35 reproduces one map of the variations of church membership across the country. Americans' religiosity, despite the constitutional separation of church and state, is, perhaps, particularly baffling to Islamic fundamentalists, for whom the unity of church and state is important.

Immigration is changing the religious composition of the U.S. population. In one survey of immigrants in the late 1990s, 41 percent described themselves as Roman Catholic (compared to about 25 percent of the current national population), and 9 percent were Protestant (compared to about 50 percent of the current population). Nearly 30 percent describe themselves as something other than Protestant, Catholic, or

Jewish; 7 percent claimed to be Muslim. In 1995, Paterson, New Jersey, with a total city population approximately 15 percent Muslim and a school-age population almost 25 percent Muslim, became the first U.S. city ever to close its public schools for a Muslim holiday. In November 2006, the people of Minnesota's Fifth Congressional District elected to send the first U.S. Muslim Representative to Congress: Keith Ellison. There is no evidence that his personal religion played a role in his election.

Churches in the United States, particularly in central-city, African-American neighborhoods, often provide an array of social services, including counseling, meals, education, and even health care. Thus, churches are potential building blocks of government community-assistance programs: Bills funding "faith-based" programs first passed Congress in 1996. Such programs have increased, and today the federal government even pays to maintain houses of worship if social services are provided in them. The constitutionality of these programs is under legal review.

Indirect Religious Influences on Government

In all countries the prevailing religion can dictate national ethics or morals. Religion so deeply affects what people assume to be natural or desirable human behavior that religious prejudices are taken for granted and unconsciously translated into laws. Religion should never be underestimated as an institutional, economic, and political force. Any organized religion has its own bureaucracy, sources of income, and, in the pulpit, its own channel of communication that often reaches more of the population than does government information (Figure 7-36).

In several countries churches operate a school system parallel to that of the government. In the United States, religious schools are accredited, but not funded, by the government. The provinces of Canada vary on this subject, but in Ontario, for example, the provincial government finances both Roman Catholic schools and secular public schools. In virtually all countries but the most fundamentalist Muslim, the governments insist that religious schools cover such modern subjects as science and math along with religious studies.

Relations between the government and the leading religious organizations constitute a major political issue in most countries. A national church may sustain a suppressed nationalism, as in Ukraine or in Ireland. Religious leaders may lend legitimacy to rulers or, conversely, provide a rallying point for political opposition. The Polish Roman Catholic Church stood against the Communist party and state as an alternative repository of Polish identity between 1945 and 1989. Attending mass was an act of political defiance. From 1959 until 1992 the constitution of Communist Cuba insisted that Cuba was officially atheist, yet the Roman Catholic

FIGURE 7-36 The main square of Mexico City. Here in Mexico City, as in many other Latin American capitals, the cathedral (to the left) and the presidential palace (to the right) confront one another across the main square. (Courtesy of the Mexico Government Tourism Office)

Church has remained the only force outside the state-party machine with any significant structure or loyalty. After 1992 the government granted more freedom to churches, and Pope John Paul II visited Cuba in 1998. In Hong Kong, Roman Catholic Bishop Joseph Zen fought the diminution of civil liberties; thereafter Hong Kong merged with China in 1997. The Vatican honored him by elevating him to the rank of cardinal in 2006. In 2003, then-U.S. Secretary of State Colin Powell singled out Archbishop Pius Ncube of Zimbabwe for his bravery in the face of government oppression there. Even in the United States, Roman Catholic Cardinal Roger Mahony of Los Angeles stood firm against a storm of criticism in 2006 when he announced that priests would defy legislation then under consideration in Congress that would have required churches to check the legal status of anyone to whom they gave assistance and that would have criminalized giving aid to illegal immigrants.

In many countries a church or religion may form a political party, as the Christian Democrats have in several European countries, or as various Islamic groups have done. India's Hindu nationalist party Bharatiya Janata has headed coalition governments.

In many countries a religion can become a big landowner or financier, but religious organizations do not always maximize the productive capacity of their property. Real property that is restricted in ownership or in the purposes for which it can be used is called *mortmain,* literally "dead hand." Church property may also enjoy preferential tax rates or escape taxation entirely. Throughout Latin America the Catholic Church is a principal landowner. In the United States,

Critical THINKING

Liberation Theology

A philosophical dispute within the Roman Catholic Church has brought about revolutionary changes in politics throughout Latin America. **Liberation theology**—named after a 1968 book by the Peruvian Father Gustavo Gutiérrez—puts the problems of overcoming poverty at the heart of Christian theology. It recommends political activism, using the church institutional framework to organize the population in the struggle for social justice and equality. Following this idea, many Latin American bishops have proposed national redistributions of land and income, provided legal services for the poor, and attacked elite power structures from their pulpits. Cuban leader Fidel Castro has praised liberation theology and supported liberation theologians throughout Latin America.

Liberation theology provoked political repercussions throughout Latin America. In some countries it triggered militant opposition among conservatives threatened by changes on either religious or political grounds. For example, in El Salvador during the 1980s, nuns, priests, and even Archbishop Oscar Romero were assassinated by right-wing elements of the military regime. The most conservative Latin American governments have favored Evangelicals as a counterforce to liberation theology. In Guatemala, for example, government death squads murdered politically active priests during the 1980s, and one Roman Catholic bishop was forced into exile. Meanwhile, Protestant sects, with government encouragement, have converted more than 30 percent of the population.

Popes John Paul II and Benedict XVI have tried to arrest liberation theology. John Paul II forbade Father Gutiérrez to write or teach for several years and appointed conservative bishops who closed church legal-services offices, land-rights offices, and liberal seminaries.

The arguments over liberation theology are ostensibly religious arguments restricted to church affairs, but in fact this theological dispute may activate politics in Latin America and other areas for years to come.

Questions

1. Investigate the official position papers of the U.S. Council of Roman Catholic Bishops. Do you think they tend toward liberation theology?
2. How has the Catholic Church interacted with Cuban leader Fidel Castro? Has it supported him?

---

church-owned properties are usually concentrated in the cities. Because they are free from taxation, their concentration reduces cities' ability to raise property-tax income. In turn, however, the churches themselves may provide a wide range of services to a public beyond their own communicants.

In some countries a community of citizens might be alienated from their national government if co-religionists in an adjacent country enjoy more complete religious freedom. For example, the Islamic vigor of Iran provided religious inspiration to the Muslim populations of the Central Asian Republics when they were under the rule of the Soviet Union.

France has long enforced a rigid separation of church and state, but the growth of Islam in France (today having at least 5 million adherents) has forced the government to rethink its role. In 2003 a government-sponsored elected Consultative Council of Islam was welcomed to the presidential palace, and, although the law forbids the government from building houses of worship, foreign subsidies to mosques and madrassas have been so great (mostly from Saudi Arabia), that local authorities are donating land for mosques.

States split by two or more religions When two religious communities compete to write a nation's laws, the country must devise compromises or else suffer internal conflict. Attempts to enforce Sharia in the Sudan, for example, have intensified conflict between the Arab Muslim-dominated north and the black Christian and animist south. Similarly, in Nigeria, several northern states have adopted Sharia, triggering rioting between Christians and Muslims. Christians have migrated out of Sharia states, and Muslims into them. Religion-based civil war broke out in the Ivory Coast in 2002. Muslims are the majority, but Christians have monopolized political power since independence in 1960. In the Philippines, tensions between Christians and Muslims have fostered revolts on several islands against the central government.

Sometimes states will enforce two separate systems of law, depending on the religion of the people involved in a case. Senegal and several other African states will allow Muslim men—but only Muslim men—more than one wife. In India, Muslims are allowed to follow Muslim law in many matters of education, marriage, divorce, and property. The government of Malaysia enforces mandatory tithing (giving 10 percent

Religious Tensions on the Indian Subcontinent

The Indian subcontinent illustrates how religious divisiveness can cause problems both within and among countries. The British ruled the subcontinent (today's India, Pakistan, and Bangladesh) as one colony, "India," and they intended to grant it independence as one secular state. Many of the Muslims within this immense territory, however, demanded that the colony be divided so that they could create a Muslim state. When independence came in 1947, the Muslim population established the country of Pakistan. It originally included a western territory (today's Pakistan) and a separate eastern territory. In the months before the partition was realized, millions of Muslims fled toward areas that were designated to become Pakistan, and millions of Hindus to regions scheduled to become parts of India. Tens of thousands died in terrible acts of violence. Muslim–Hindu tensions were exacerbated by the fact that for hundreds of years in many areas Muslim minorities had ruled Hindu majorities.

In 1971 eastern Pakistan broke off from Pakistan to form a separate Islamic country, today's Bangladesh, but religio–political turmoil continues in what is left of Pakistan. Sunnis (about 70 percent) scorn Shiites (about 20 percent), and tribal loyalties and identifications subdivide the population still further. Non-Muslims (about 10 percent of the population) are guaranteed 10 of the 307 seats in Parliament, but they do not represent geographical districts, so they have to campaign across the entire country.

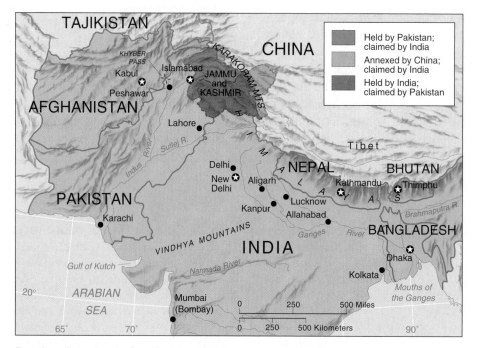

Border disputes in South Asia. Antagonism and border disputes persist between India and Pakistan and between India and China. These three nuclear powers contain about 40 percent of the human population.

of one's income to charity) among Muslims, about half of the country's total population.

In some countries, struggles that are ostensibly religious actually conceal other social divisions. Tension in Northern Ireland is, on the surface, a conflict between a Protestant majority and a Roman Catholic minority. That split, however, involves much more than theology. It is a competition for jobs and opportunity, and it is complicated by deep historic political and economic differences and the fact that both sides draw on outside support.

China recognizes five "approved" religions: Protestant Christianity, Catholic Christianity, Buddhism, Islam, and Taoism. The government oversees religious activity carefully, and the Chinese Catholic Church does not recognize the authority of the Vatican.

Religion and women's rights Religions differ in the attitudes that their teachings advocate toward women, and national laws often reflect these religious teachings. In many cases, however, the actual practice

India itself emerged as a secular state with a majority Hindu population. Between 10 and 13 percent of India's population, however, is Muslim, which is a greater number of Muslims than in Pakistan. India's Muslims were persecuted during India's wars with Pakistan in 1965 and again in 1971 and Muslims and Hindus still regularly battle on the streets of India's cities. The rise of the Hindu nationalist party Bharatiya Janata has coincided with a growing climate of intolerance of minorities and the introduction of new Hindu-oriented syllabi in schools. Avul J. P. Abdul Kalam, a Muslim nuclear scientist, was inaugurated as India's president in 2002 (his family converted to Islam, but Hindus still consider him a Hindu with, in fact, Brahmin status), but growing hostility between Muslims and Hindus in India threatens domestic peace. Furthermore, the conversion of many untouchables to Islam or Buddhism further upsets India's internal religious balance.

Conflict continues on the Indian–Pakistani border. In 1947 India and Pakistan fought over the border state of Jammu and Kashmir. Its population was mostly Muslim, but the state was eventually split, with India taking about two-thirds. The population of Indian Kashmir today is about 60 percent Muslim, but India refuses to allow the people of Kashmir to vote on whether they want to be a part of India or Pakistan. It is important to India's image as a secular country to have a state with a Muslim majority. Fighting in Kashmir has claimed an estimated 70,000 lives from 1989 to 2006 and has been blamed for occasional terrorist attacks throughout India.

Water is another source of contention between India and Pakistan. A 1960 treaty regulates sharing of the water of the Indus River and its eastern tributaries that rise in China or in the Indian-controlled part of Kashmir. India, however, has threatened to revoke the treaty and cut off the supply of water to Pakistan. Throughout much of the Indus watershed region, local aquifers are being depleted, water tables are falling, waterways are polluted, and soils are becoming saline from overuse of underground water supplies.

The Taj Mahal. A Muslim ruler built the Taj Mahal, a mausoleum in Agra, India, for his beloved wife, in 1648. India's Muslim rulers were so rich and powerful that their name, Mughals, has come down to us in English as a word for a business executive: a mogul. (Courtesy of U.S. Agency for International Development)

of any religion varies from place to place. These alternate practices reflect variations in other aspects of the cultures among places. Therefore, it is difficult to determine exactly how much difference in women's rights can be attributed to the predominant local religion and, in turn, how much of that difference has been translated into local law.

For example, the Catholic Church is centralized, and it imposes a worldwide ban on the ordination of women. The Vatican has excommunicated women ordained by bishops in Argentina and Austria. The worldwide Anglican Church, however, is less centralized. Its official head, the Archbishop of Canterbury, has approved the ordination of women, but local communities of Anglicans have different attitudes, and these reflect local variations in general social attitudes toward women's rights. In U.S. culture, for example, women's rights are protected in many areas, and the American Anglican Church (the Episcopal Church) ordains women. Reverend Katharine Jefferts Schori was even elected Presiding Bishop of the U.S. Episcopal Church in 2006. Anglican communicants from some other world regions, however—most notably Africa, where women's rights are not always assured—have refused to ordain women or to recognize ordained women.

As the rich and more democratic and free countries have embraced women's rights and gay rights as well, some of the Protestant sects have splintered. In 2005 the Anglican Church of Nigeria, for example, deleted from its constitution all references to "the see of Canterbury." Protestant churches in Nigeria and other African countries have even spurned the financial aid that supported local schools and hospitals.

Turmoil in Pakistan

Pakistan cooperated with the United States in 2001 when the United States overthrew the Taliban in Afghanistan and attempted to root out Al Qaeda, but Pakistan is itself a troubled state. Some observers doubt the amount of control that the government of Pakistan has over its territory, its people, or even its own army, and Pakistani–Indian animosity periodically threatens to explode in full-scale war.

During the long anti-Soviet war in Afghanistan (see Chapter 13), refugees from Afghanistan overloaded Pakistan's ability to absorb and help them. The Afghan fighters, in an effort to raise funds, turned Afghanistan into a virtual heroin farm. Pakistan became the transshipment point, creating an estimated 600,000 heroin addicts in the Pakistani city of Karachi alone. The violence also engendered what is called a "Kalashnikov culture," from the name of a popular automatic weapon that has been left behind in the hands of thousands of poor, unemployed, hardened young men. Still today Pakistan seems incapable of controlling or sealing its

borders with Afghanistan, Iran, or India, so drug runners, guerilla fighters, and religious extremist agitators circulate. Within Pakistan, the traditional animosities between Sunnis and Shiites also continue to disturb the country, and anti-American fundamentalist groups have consistently won local and regional elections.

Pakistan's government supported radical madrassas in both Pakistan and Afghanistan in order to create a cadre of devout Muslims who would strengthen Pakistan's hand against India. This plan succeeded, perhaps, too well. Extremist graduates of madrassas have embroiled Pakistan not only in repeated attacks on India, but, it is believed, in the Near East. Graduates of madrassas in Pakistan's own armed forces may secretly back the attacks on India. An Islamic fundamentalist coup d'etat in Pakistan would place nuclear weapons in the hands of extremists. Dr. Abdul Qadeer Khan, an engineer who fathered Pakistan's nuclear weapons, was involved in an illegal international network that spread nuclear materials to Libya, Iran, and North Korea.

Variations in practice can also be found across the Islamic realm. Islam originated in an Arabic culture that granted women few rights, but the list of women's rights in the Koran was liberal for the time and place. Islam outlawed female infanticide, made the education of girls a sacred duty, and established a woman's right to own and inherit property. At the same time, Islam also fixed certain discriminatory practices. It teaches that a woman's testimony in court is half that of a man's and that men are entitled to four spouses, whereas women may have only one. In practice, women's rights vary throughout the countries where Islam predominates. Different countries interpret Sharia differently: In Dubai, for example, Sharia courts have ruled that a man may divorce a wife by sending her a text message over a mobile phone; Sharia courts in Singapore have ruled that he may not. Women are frequently secluded, or veiled in public, in Islamic countries, but this is compulsory only in Saudi Arabia and Iran. Also, only in the most fundamentalist Arab societies are women generally forbidden to work outside the home.

Future struggle is nevertheless almost inevitable between the expansion of women's rights in Islamic societies and the Islamic fundamentalist backlash that demands the repeal of laws that had banned polygyny (having more than one wife), permitted birth control, and given women the right to divorce in some countries.

Almost everywhere, women struggle for equal access to education.

The Old Testament of the Bible contains restrictions on women's rights that are comparable to those found in the Koran, but few Jews or Christians today observe them. In Israel, however, the law denies women equality with men. For example, Israel has no civil marriage or divorce, and in Jewish law only a husband can initiate a divorce.

It has been hypothesized that religious teachings about women's rights may affect women's role in politics, but this hypothesis is difficult to prove. Women have served as heads of government in Islamic countries, as well as in countries that are officially Christian (Catholic and Protestant), Jewish, and Buddhist. What sort of evidence for such a hypothesis could be sought? Reading different religions' scriptures alone does not prove whether those teachings are obeyed. Could we examine the percentage of national legislatures that are female? Geographers have noted that within the United States, the populations of Utah and of the states of the Southeast are, by some measures, the most committed to organized religion. The legislatures of these states have the smallest percentages of women legislators. Perhaps there is a cause-and-effect relationship between these facts, but clearly more research is needed on this topic.

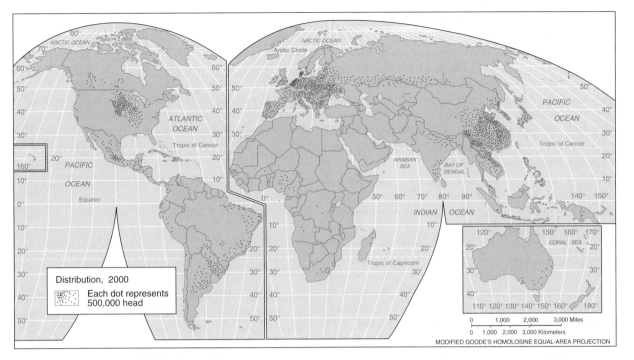

FIGURE 7-37 World distribution of hogs. The virtual absence of hogs across North Africa, the Near East, and all the way to the borders of China is not caused by any attribute of the physical environment, but by the predominance of Islam and Judaism across this region. Both religions prohibit the consumption of pork. An increase in protein supply would be beneficial to the local populations, but the people demand alternative meats and protein-rich plants, such as lentils and peas. (Data from U.N. Food and Agriculture Organization)

Religion and Dietary Habits

Different religions teach different beliefs concerning the sacredness of plant and animal life, and this affects the farming and dietary practices of their adherents. Buddhists are generally vegetarians. Hindus believe that cattle are sacred; therefore, they will not eat beef, and they use cattle only as draft animals and as a source of milk. McDonald's restaurants in India feature lamburgers and vegetarian sandwiches. Outside observers have tried to measure in economic terms the advantages and disadvantages of preserving India's cattle. Aging and unproductive cattle may overgraze the land or may compete with the human population for limited food supplies. Sacred cows roam the streets of Indian cities unmolested, befouling them and congesting them. On the other hand, the cattle provide dung as a fuel and manure, and those that die of natural causes are eaten by non-Hindus anyway. Whatever the balance of these economic arguments might be, to a religious Hindu this is a religious matter not open to economic debate.

Other religions prohibit certain foods. Jews and Muslims both refuse pork, so there are few hogs in Israel or anywhere across the Islamic realm (Figure 7-37). Muslims are prohibited from drinking alcohol. Altogether, a significant share of humankind restricts its diet because of religious beliefs.

These restrictions affect even world trade in food. Several of the greatest importers of plants and seeds that are rich in protein (peas, beans, and lentils, for example) are countries that limit their consumption of meat for religious reasons. Annual per capita consumption of meat in India, for example, is only 2 kilograms (4.4 pounds), about one-tenth the U.S. figure, and India is a major importer of protein-rich vegetable products.

Religion and Economics

The religion that predominates in any region may have a significant impact on that society's economy. Elaborate burial practices, for example, can drain a society of capital: Many peoples have buried their dead along with their possessions. Some societies lavish expenditure on houses of worship and other religious institutions. In some Buddhist societies high percentages of the male labor force spend several years, or even their whole lives, as monks.

Furthermore, religious teachings may affect the way people view the accumulation of money. For example, the scriptures of most religions bar the charging of interest on a loan as taking unfair advantage of another person. Christian teachings, however, have evolved to differentiate between two reasons for borrowing money: for needs and for investment. It is considered sinful to charge interest on money borrowed to buy food, but not sinful to charge interest if the borrower invests the money for his or her own profit—as when the borrower buys a truck to increase the profit in his or her own business. Religious leaders in some societies do not accept such a distinction. Islamic financial systems forbid interest, but instead encourage risk sharing. Islamic banking systems have sprung up throughout North America and Europe that do not, for example, offer mortgages, but buy homes and devise rent-to-buy contracts for Muslim customers. In practice, financial structures and compromises have to be defined in each Islamic country.

Weber and the Protestant ethic German sociologist Max Weber linked Protestantism and capitalism in *The Protestant Ethic and the Spirit of Capitalism* (1904). According to Weber, Protestantism encourages individualism, and with the rise of Protestantism in Western Europe, acquisitiveness, as the result of exercising individual ability in the marketplace, became a recognized virtue. Today this ethic seems to characterize adherents of almost any religion, so it is usually referred to as the *work ethic*.

If, however, there is any true advantage in the Protestant Christian attitude toward moneymaking, then we could hypothesize that the worldwide expansion of Protestantism might encourage individualistic capitalism. This explains why the spread of Protestantism is widely reported and happily received in America's business press, which clearly accepts the hypothesis of a link. *The Wall Street Journal* and other publications all devote extensive coverage to the spread of Protestantism. They report it as economic news, not just religious news.

The Catholic Church and capitalism In 1991 Pope John Paul II wrote, "On the level of individual nations and of international relations, the free market is the most efficient instrument for utilizing resources and effectively responding to needs." The Catholic Church's 1992 *Universal Catechism* notes, however, that markets do not always meet human needs, and it insists that governments regulate markets "according to a just hierarchy of values." It also demands that every worker receive a "just salary." The Church's attitudes toward markets, speculation, and worker–management relations may influence the behavior of individual Roman Catholics and of governments in Roman Catholic lands.

Religion and economics in Asia The Confucian tradition in East Asian societies may exemplify another religious influence on economic development. Confucianism recommends societal leadership by an intelligent elite with the moral obligation to guide the people. Several East Asian leaders have praised their own societies' traditional communitarianism and decried Western Christian individualism. Confucianism enhances the status of jobs in government bureaucracies, so ambitious young graduates compete for positions with the Japanese Ministry of Finance or the Korean Economic Planning Board the way ambitious young Americans compete for jobs at investment banks. The ability to attract talent gives these Asian governments more legitimacy and competence in dealing with businesses than the U.S. government can bring to bear, which may have helped to build their successful economies.

Also, diligence, obedience, and high savings rates characterize the peoples of neo-Confucian Japan, South Korea, Taiwan, and Singapore, and of every Chinese settlement in the world. It is risky to hypothesize a cause-and-effect relationship, but the coincidence must be noted.

Religions, Science, and the Environment

Different religions teach different attitudes toward environmental issues, and cultural landscapes often reflect the religions of their creators. Most religions address the questions of how the world came into existence, what power created it, and what humankind's relationship should be to it and to other living things. Many major faiths not normally classified as animist nevertheless designate specific features of the environment as holy. Some religions revere rivers (Hindus revere the Ganges, Christians the Jordan); some revere high places (the Shinto revere Mount Fuji, Jews and Christians Mount Ararat, Hindus and Buddhists Mount Kailash) (Figure 7-38).

FIGURE 7-38 Mount Fuji. Volcanic Mount Fuji, Japan's highest peak at 3,776 meters (12,389 feet), is sacred to many Japanese. It has been celebrated for centuries in Japanese paintings and verse. (Courtesy of Japan National Tourist Organization)

Early religious beliefs often formed as attempts to explain natural phenomena, such as the changes in the seasons or floods. Many religions celebrate natural events, time their annual ceremonies in accordance with astronomical events, or celebrate human affairs connected to the environment. Most religions celebrate harvest festivals of thanksgiving, for example.

Religions recognize different holidays and pace human activities through the year differently, which is often reflected in national laws across religious realms. Most Christian countries use the solar calendar. Jews use a calendar of 12 months, with an additional month added seven times in 19 years. Muslims use the lunar calendar. Muslims celebrate the Sabbath on Friday, Jews celebrate on Saturday, and most Christians on Sunday. Many governments prohibit work or other activities on that day. Muslims recognize an entire month as holy and accordingly curtail their activities.

Most religious traditions award humankind a special role in the world, but, it has been argued, some religions suggest adaptive approaches to the physical world, while others suggest more exploitative approaches. Today, however, environmental degradation afflicts areas of all religious persuasions. In 2006, 86 U.S. Evangelical Christian leaders launched a political initiative to fight global warming, saying "millions of people could die in this century because of climate change, most of them our poorest global neighbors." Signers of the statement included the presidents of 39 Evangelical colleges, leaders of aid groups and churches, the Salvation Army, and pastors of megachurches, including Rick Warren, author of the bestseller *The Purpose-Driven Life.* The leaders called for federal legislation that would require reductions in carbon dioxide emissions. The statement was only the first stage of an "Evangelical Climate Initiative" including television and radio announcements in states with influential legislators, informational campaigns in churches, and educational events at Christian colleges.

Some scholars have suggested that continuing advances in science, particularly in the life sciences, may be influenced by cultures' religious beliefs. Asian scientists, it has been argued, face less cultural resistance to their work than their colleagues in the West. In Roman Catholic areas and some fundamentalist Christian regions, for example, religious authorities argue against the destruction of microscopic embryos because in their teaching a fertilized egg is already a person. In 2006, Vatican spokesman Cardinal Alfonso Lopez Trujillo stated that "Excommunication will be applied to the women, doctors and researchers who eliminate embryos [and to the] politicians that approve the law." The independence of mind of Western scientists, however, was demonstrated by the Italian Dr. Cesare Galli, the first scientist to clone a horse, who said, "I can bear excommunication. I was raised as a Catholic . . . but I am able to make my own judgment on some issues

and I do not need to be told by the Church what to do or to think."

In Confucian and Buddhist societies, by contrast, the defining moment of life is birth, not conception, and Buddhists and Hindus view life not as beginning with conception, but as a cycle of reincarnations. One leading South Korean bioscientist said in 2004, "Cloning is a different way of thinking about the recycling of life. It's a Buddhist way of thinking."

Religious landscapes Most landscapes reveal the predominant local religion. The architectural styles of houses of worship, monasteries, and similar institutions are often revealing—Muslim mosques, for example, usually have free-standing towers, called minarets.

Any religion's places of worship may become goals of pilgrimages, which attract visitors and play a major role in cultural diffusion (Figure 7-39). Fairs and markets grow up at these sites and along the pilgrimages' prescribed path. Muslims' hajj pilgrimages bring together people from around the world. At the local level, religious administrative patterns may organize and focus community spirit. More Roman Catholic New Yorkers can name the parish they live in than can name their city council district, and it is through the parish that

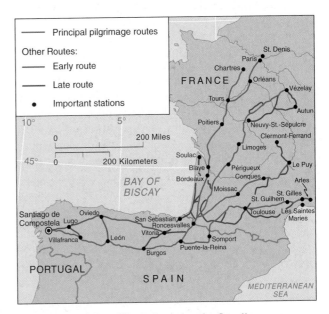

FIGURE 7-39 Traditional routes to Santiago.
The Cathedral of Santiago, in the northwest corner of Spain, is believed to hold the relics of Saint James. Pilgrims from throughout Europe and the world have sought it out. In medieval times these traditional routes and staging points played important roles in trade and in the diffusion of ideas. The routes are marked by a chain of historic churches, inns, universities, and trade fairs. Still today a Pilgrim's Certificate is given to anyone who walks at least 100 kilometers (62.5 miles) along the medieval pilgrim's trail, but the certificate must be stamped by parishes along the way.

FIGURE 7-40 Great stupa in Nepal. The eyes on this great stupa in Nepal emphasize that God sees everything. The symbolism is virtually universal. An eye can even be found on the back of the U.S. dollar bill. (Courtesy of Julia Nicole Paley)

they make friends and meet potential spouses. As noted earlier, the diffusion of religions can affect the diffusion of languages. Muslims everywhere are encouraged to read the Koran in its original Arabic, and the translation of the Bible into German by Martin Luther defined the German language and, arguably, the German nation.

Different religions' burial practices create different mortuary landscapes. Hindus and Buddhists cremate their dead, whereas Christians and Muslims usually bury their dead and erect monuments to them. Buddhist landscapes are marked by *stupas* (called pagodas east of India). These serve as temples today, but the form was originally devised as a funeral mound to contain the remains of a revered bodhisattva (Figure 7-40).

Religion means "rebinding" of people with their gods and of people with one another. The geography of such binding will always be a key consideration of human geography.

CONCLUSION: CRITICAL ISSUES FOR THE FUTURE

This chapter opened by noting that peoples who share either language or religion can often more easily cooperate with one another in other ways, such as in international affairs. Sharing these attributes may help them understand one another. Much of the detail in this chapter, however, cataloged not cooperation and sharing but divisiveness. The same strong ties that bond groups of people together often separate them from other groups and cause conflict.

Martin Luther, who triggered the Protestant Reformation that split the Christian Church, noted the divisiveness of language: "The Frenchman," he wrote, "has only hatred and scorn for the German. The Italians have only hatred and scorn for all others. So we can see that this division of languages has led to division of habits, ways of thought and priorities that have put barriers between the very essence of peoples. It can justly be called the source of all misfortunes."

Religious contention also disturbs the world's peace. Groups' efforts to impose the teachings of their religion on other people often provoke violence. The Sudan, it was noted earlier, is rent by war between the Islamic government that wants to impose Sharia on the entire country, and the Christians and animists of the South. Pope John Paul II, speaking of Sharia law while visiting the Sudanese capital Khartoum in February 1993, said, "You absolutely cannot impose this law on those of other faiths. Islamic law can be applied only to the Muslim faithful." The Church of Rome nevertheless assumes that its own teachings should be written into law wherever it can accomplish that purpose.

What of the future? We can only hope that people can live together in peace.

Chapter Review

SUMMARY

Language and religion define and bond human cultures. Peoples who share either of these two cultural attributes often demonstrate consistencies in other aspects of behavior and often cooperate with one another in other ways.

The distribution of any language illustrates the pattern of dispersal of its original speakers or their cultural impact on others. If human beings were to appear suddenly in today's

highly interconnected world, one worldwide language might develop. The great variety of languages spoken today testifies to the relative isolation of groups in the past. There are more than 5,000 distinct languages, but about 60 percent of all people speak 1 of just 12 languages. The language with the most speakers is Mandarin Chinese, but English is the world's lingua franca. In contrast to these widespread languages, some languages are extremely local.

A few languages are associated almost exclusively with one country, but several languages are shared by many countries, and some polyglot countries officially recognize several languages. The idea that languages gave birth to nations persists despite the fact that standard languages usually were the product of self-conscious efforts at nation building by centralizing governments.

Today's world distribution of European languages as official languages partly reflects patterns of colonization. Europeans forced their languages on their colonies, and when the colonies gained their independence, many kept their former ruler's language. Across much of the world, a country's official language is not actually spoken by most of the population.

Orthography can also reflect a historic diffusion process. Toponymy records information about the environment or the beliefs of a region's inhabitants.

Languages can be categorized into families on the basis of common ancestry, and they continue to diffuse and differentiate today. The use of some languages is expanding because their speakers are diffusing throughout the world, gaining greater power and influence in world affairs, or winning new adherents to their ideas. Other languages are dying out.

The population of the United States has always been composed of a great variety of peoples speaking a great variety of languages. Regional differences persist, but some people think English should be the official national language.

Religion is a vital force in human affairs, and understanding the messages of different religions contributes substantially to understanding many other aspects of human life.

Judaism, Christianity, and Islam form one ancient religious tradition, originating in the Near East and diffusing from there. Christianity came to predominate in Europe, and it spread worldwide with European influence. Today most Christians are outside Europe and North America. Christianity continues to win new converts, but its practices and doctrines that derive from European culture are under pressure for change. New syncretic religions are thriving. Islam is newly influencing the politics of the countries in which it predominates.

East Asia is dominated by Hinduism and its offshoots Buddhism and Sikhism. Other Eastern religions include Confucianism, Taoism, and in Japan, Shinto. The teachings of several of these systems of belief are exclusively ethical and psychological, and they do not address theological questions. Many people are adherents to more than one of these. Animism and shamanism have many followers in the less-developed parts of the world.

Religious teachings bear great influence on political organization. Theocracy is a form of government where the church rules directly, but religious teachings will often be translated into law in societies that claim to be secular. Religions' teachings differ regarding politics, women's rights, government, diet, economics, and environmental attitudes, so the geography of religious communities influences the geography of these other aspects of human culture. Christianity recognizes a distinction between church and state, but Islam teaches that the only purpose of government is to allow people to be good Muslims, and fundamentalist Islamic societies enforce Sharia law. The actual practice of any religion varies from place to place, often reflecting variations in other aspects of the cultures among places. Each society defines a balance among the teachings of its religion, other aspects of its cultural tradition, and the demands of modern life, but that balance may shift through time.

KEY TERMS

caste p. 292
cognate p. 267
creole p. 263
dialect p. 262
etymology p. 267
fundamentalism p. 278
isogloss p. 262
language p. 262
language family p. 266
liberation theology p. 301

lingua franca p. 262
monotheism p. 279
official language p. 262
orthography p. 268
polyglot state p. 275
polytheism p. 279
proselytize p. 279
protolanguage p. 266
secularism p. 278
shamanism p. 295

Sharia p. 297
speech community p. 262
standard language p. 262
syncretic religions p. 288
terrorism p. 281
theocracy p. 296
toponymy p. 270
untouchable p. 292
Zionism p. 280

QUESTIONS FOR REVIEW AND DISCUSSION

1. Why is it that several former colonies retained their former rulers' language? What advantages and disadvantages did this offer?

2. Why are more countries today recognizing more official languages? What does this tell us about the internal politics of those countries?

3. Name a language that is expanding in use today and one that is shrinking. In each case, explain why.

4. What does it mean to say that Judaism, Christianity, and Islam are three religions, but form one religious tradition?

5. What are the observed and hypothesized consequences of the conversion of much of Latin America from Roman Catholicism to Protestant Christianity?

6. How do religious teachings affect economic behavior? How do they affect the systems of national laws?

7. What is a lingua franca? What languages have served as lingua francas?

8. What is the caste system?

9. What is the Indo-European family of languages? Who first identified it? What is the geographic distribution of these languages? What is the history of the family?

10. Describe the diffusion of Islam from A.D. 622 to 751.

11. From what religious tradition did the Sikh religion spring? Where? When?

12. Where did Christianity diffuse before it diffused throughout Western Europe?

THINKING GEOGRAPHICALLY

1. How many languages are spoken in your community? Does the local school system have programs that are bilingual or to teach English as a second language? Where do your community's non-English speakers come from?

2. The Dutch have considered making English the official language for college-level education in the Netherlands. Can you consider why?

3. If you speak two or more languages, can you think of specific ideas or images that may be difficult to translate from one to another? Try to explain this difficulty to your fellow students.

4. How many religions have temples, churches, or synagogues in your community? What, roughly, is the statistical breakdown of religious affiliations of the people of your community? Does this reflect the source areas of settlers to your region? Does each place of worship anchor a residential neighborhood of that religion's communicants? How many religiously oriented parochial schools exist in your community? With what religions are they associated?

5. Almost the whole world uses Arabic numerals, but the use of Arabic script is largely restricted to Islamic lands. Why?

6. Investigate the competition between Christianity and Islam for converts in an African country. Consider Nigeria, Senegal, or the Ivory Coast.

7. The fourteenth-century Islamic scholar Ibn Battuta could travel from Morocco to China unmolested—even being welcome across most of the region. What political, religious, and other criteria of territorial organization have fragmented that region today so that a contemporary traveler might have trouble retracing Ibn Battuta's travels?

8. What do the place names in your community reveal about its environment or history?

SUGGESTIONS FOR FURTHER LEARNING

Abate, Frank. *American Places Dictionary.* 4 vols. Detroit: Omnigraphics, 1994.

Asher, Ron. *Atlas of World Languages.* New York: Routledge, rev. ed. 2006.

Comrie, Bernard, Stephen Matthews, and Maria Polinsky, eds. *The Atlas of Languages: The Origin and Development of Languages Throughout the World.* New York: Facts on File, 2003.

Dictionary of American Regional English. vol. 1, A–C. Edited by Frederic G. Cassidy and Joan Houston Hall, 2003; vol. 2, D–H. Edited by Frederic G. Cassidy and Joan Houston Hall, 1999; vol. 3, I–O. Edited by Frederic G. Cassidy and Joan Houston Hall, 1996; vol. 4, P–Sk. Edited by Joan Houston Hall, 2002. Cambridge, MA: Belknap Press (Harvard University).

Fardon, Richard, and Graham Furniss, eds. *African Languages: Development and the State.* New York: Routledge, 1994.

Gaustad, Edwin, and Philip Barlow. *New Historical Atlas of Religion in America.* New York: Oxford University Press, 2001.

Haspelmath, Martin, Matthew S. Dryer, David Gil, and Bernard Comrie, eds. *The World Atlas of Language Structures.* New York: Oxford University Press, 2005.

Haynes, Jeff. *Religion, Globalization and Political Culture in the Third World.* New York: St. Martins Press, 1999.

Holloway, Joseph, and Winifred K. Vass. *The African Heritage of American English.* Bloomington, IN: Indiana University Press, 1993.

Juergensmeyer, Mark. *Terror in the Mind of God: The Global Rise of Religious Violence.* Berkeley: University of California Press, rev. ed. 2003.

Labov, William, Sharon Ash, and Charles Boberg. *Atlas of North American English.* New York: Mouton de Gruyter, 2005.

Monmonier, Mark. *From Squaw Tit to Whorehouse Meadow: How Maps Name, Claim, and Inflame.* Chicago: University of Chicago Press, 2006.

Nettle, Daniel, and Suzanne Romaine. *Vanishing Voices: The Extinction of the World's Languages.* New York: Oxford University Press, 2000.

Ostler, Nicholas. *Empires of the Word: A Language History of the World.* New York: HarperCollins, 2005.

Partridge, Christopher H., ed. *Introduction to World Religions.* Minneapolis, MN: Augsburg Fortress, 2005.

Regenstein, Lewis. *Replenish the Earth: A History of Organized Religion's Treatment of Animals and Nature.* New York: Crossroad, 1991.

Smith, Huston. *The World's Religions: Our Great Wisdom Traditions.* San Francisco: Harper SanFrancisco, rev. ed. 1991.

Thomas, Scott. *The Global Resurgence of Religion and the Transformation of International Relations: The Struggle for the Soul of the Twenty-First Century.* London: Palgrave Macmillan, 2005.

Van Voorst, Robert E. *Anthology of World Scriptures.* Belmont, CA: Wadsworth, 5th ed. 2005.

Wolfram, Walt and Ben Ward, eds. *American Voices: How Dialects Differ From Coast to Coast.* New York: Blackwell, 2005.

World Wide Fund for Nature. *World Religions and Ecology.* Multivolumes. New York: Cassell, 1992.

Wurm, Stephen A. *Atlas of the World's Languages in Danger of Disappearing.* New York: UNESCO, rev. ed. 2001.

WEB WORK

The World Wide Web is rich in resources for the study of material about languages and religions. The home page of Ethnologue, a searchable database of language resources, can be found at:

http://www.ethnologue.com

The latest redesign of The Human-Languages Page and the Languages catalog of the WWW Virtual Library is called:

http://www.ilovelanguages.com

The Modern Language Association has an online language map, indicating population of foreign-language speakers throughout the United States at:

http://www.mla.org/census_main.

For research on religion, you might start at the Academic Info: Religion, Main page at:

http://www.academicinfo.net/religindex.html

The British Broadcasting Corporation maintains an informative page on religions at:

http://www.bbc.co.uk/religion/religions/

The award-winning Facets of Religion, a virtual library, can be found at:

http://www.edunet.ie/resources/religioninfo/religion.html

Ekklesia is a British think tank and religious news service:

http://www.ekklesia.co.uk/

The Virtual Religion Index of Rutgers University is at:

http://virtualreligion.net/vri/

A guide to Jewish teachings and traditions, with links to other related sites, may be found at:

http://www.torah.org/

The Vatican's home page offers information in several languages—but not Latin:

http://www.vatican.va/

A good source of information on Hinduism is the home page of Hinduism Today:

http://www.himalayanacademy.com/

Islami-City is a website presented by the Islamic Information Network. It contains a wide variety of news and information and has an internal search engine for researching specific topics:

http://www.islamicity.org/

One attempt to count followers of world religions is at:

http://www.adherents.com

8

THE HUMAN
FOOD SUPPLY

Strips of wheat, vegetable, oil seed, and fallow fields in northern China define a modern art patchwork. Chinese agriculture has enjoyed tremendous strides as the Chinese have experimented with new systsems of landholding and marketing and with new crops in China's many environments. Modern scientific methods of farming—including the promise of biotechnology—have probably secured our ability to feed Earth's population, but raising food in environmentally sustainable ways and assuring all humans a healthy diet remain challenges. (Georg Gerster/Photo Researchers Inc.)

A Look AHEAD

Food Supplies Over the Past 200 Years

Humans have managed to avoid the global starvation predicted by Thomas Malthus 200 years ago by opening new lands to agriculture, redistributing and improving crops, and applying technology to agriculture.

Agriculture Today

Ten principal types of agriculture may be identified today. Four of these (nomadic herding, low-technology subsistence farming, intensive rice farming, and Asian mixed cereal and pulse farming) are basically subsistence farming—that is, just to feed the farmers. Five other types (mixed farming with livestock, prairie cereals, ranching, Mediterranean, and

plantation agriculture) are predominantly commercial. Irrigated lands may be either subsistence or commercial.

Livestock Around the World

Humans consume most grain directly, but much grain is also consumed indirectly when people eat grain-fed livestock. Livestock present many environmental problems. Dairying is a special kind of livestock raising, and value can be added to the raw milk by processing it into milk products.

Food Supplies in the Future

We can be cautiously optimistic about food supplies for the future, but some people fear the newest biotechnologies. Increasingly intensive agriculture also carries environmental

hazards. Challenges remain in spreading technology throughout the world and designing political and economic systems that will allocate food to all.

The World Distribution of Food Supplies and Production

As many as 840 million people are undernourished today, and sub-Saharan Africa is the most troubled region. Problems deterring an increase in food production include inefficient application of fertilizers, lack of incentive for many farmers, and problems of land ownership.

The Harvest of Fish

We may have reached the limit of fish we can draw from the sea. The possibility triggers angry confrontations among fishing nations, but aquaculture offers new possibilities for fish supplies.

*I*n the summer of 2006, a strain of rice that had been produced by genetic modification (called GM food) and that had never been approved for human consumption was found in small amounts in commercial supplies throughout southern rice-growing region of the United States. The unapproved strain was found in supplies destined for human consumption. GM rice was not grown commercially in the United States, so it was not clear how the GM rice had come to contaminate supplies, although it could have happened through the use of a common means of transportation.

Science is achieving unprecedented strides in its ability to understand and manipulate life. When this new knowledge is applied to food production it is creating astonishing possibilities, including the increase of productivity, plant and animal resistance to diseases and pests, and even food's inherent nutritional content. Some people, however, fear the scientific manipulation of something once thought so "natural" as food.

Furthermore, the political and economic context for the development of these possibilities will determine when they might be applied, where, and for whose benefit. Even today, producing enough food for everyone on Earth is not as much of a problem as is the distribution of food. There are great disparities in the map of food production and supplies. Some countries produce substantial surpluses of food, while others lack supplies sufficient to feed their entire populations a nutritious diet. Even in many of the countries with surpluses, some people go hungry. Problems of food production and distribution can be understood only as the result of interaction among environmental, technological, political, and economic factors.

FOOD SUPPLIES OVER THE PAST 200 YEARS

Chapter 5 examined a pessimistic prediction that Thomas Malthus made 200 years ago. Malthus argued that the population would always tend to increase faster than the food supply, so cycles of mass starvation would limit the population increase.

Since Malthus published his theory, the human population has increased from one billion to over six billion. The mass starvation he predicted, however, has not occurred. Chapter 5 noted that today there are more overweight people than undernourished people. Humankind has avoided starvation not by lowering the rate of population increase, but by increasing the food

GM crops have been biologically altered to build in desirable characteristics ranging from insect resistance to faster growth to greater vitamin content. Despite scientific studies that have found GM food to be safe, many consumers fear their possible long-term effects on human health and on the natural products of fields and farms. Thus many countries will not accept GM crops that they have not approved themselves, and the discovery of the rogue strain in U.S. fields jeopardized U.S. rice exports. Japan and the European Union immediately suspended imports of U.S. rice. ▶

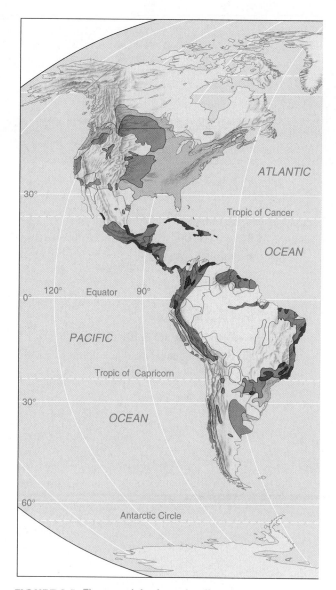

FIGURE 8-1 The world of agriculture. These are regions of different types of agricultural land use. The success of agriculture depends largely upon the possibilities of the physical environment, so we should not be surprised that the

supply faster than the population, so per capita supplies have actually increased. How has this been done?

New Crops and Cropland

Extensive areas that were scarcely utilized during Malthus's lifetime have been opened to productive agriculture (Figure 8-1). The vast prairies of both North and South America, as well as food surplus regions in Australia and South Africa, have been developed since Malthus published his theory. Since the middle of the last century alone, over 900 million hectares (2.2 billion acres) of Earth's surface have been converted to croplands. Most of these lands were opened by irrigation.

A second factor is that many food crops have been transplanted to new areas where they have thrived, in some cases better than in their areas of origin (Figure 8-2). The world's first farmers raised barley and wheat in the Middle East about 10,000 years ago. Later, about 9,000 years ago, Mexicans began cultivating corn and beans, Chinese began cultivating rice roughly 8,000 years ago, and about 7,000 years ago South Americans began growing potatoes. Natives in what is today the eastern United States raised sunflowers 5,000 years ago, and peoples in the highlands of New Guinea cultivated taro and banana between 7,000 and 10,000 years ago. It is unknown whether the first sub-Saharan African farmers began to cultivate millet, sorghum, and rice independently about

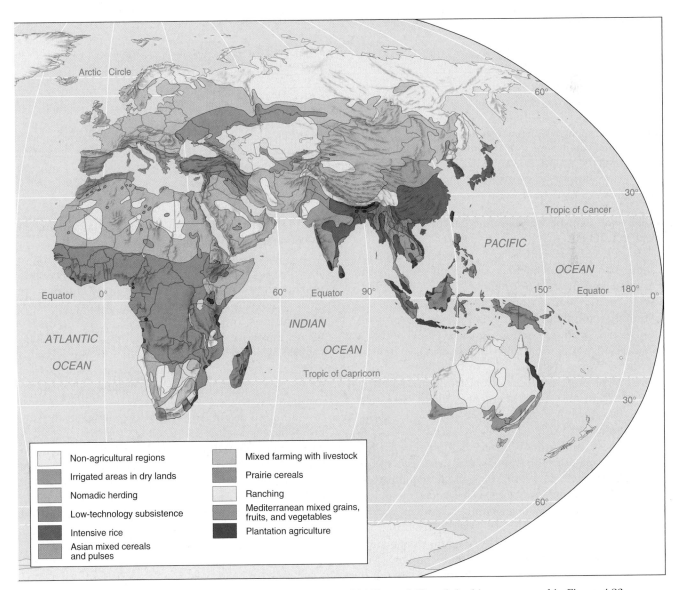

Legend:
- Non-agricultural regions
- Irrigated areas in dry lands
- Nomadic herding
- Low-technology subsistence
- Intensive rice
- Asian mixed cereals and pulses
- Mixed farming with livestock
- Prairie cereals
- Ranching
- Mediterranean mixed grains, fruits, and vegetables
- Plantation agriculture

categories mapped here largely coincide with the climates mapped in Figure 2-31 and the biomes mapped in Figure 4-23. "Nomadic herding" and "low-technology subsistence farming" cover impressive areas on this map, but their productivity is low. These areas are generally sparsely populated, as you can see by comparing this figure with the map on the rear endpaper.

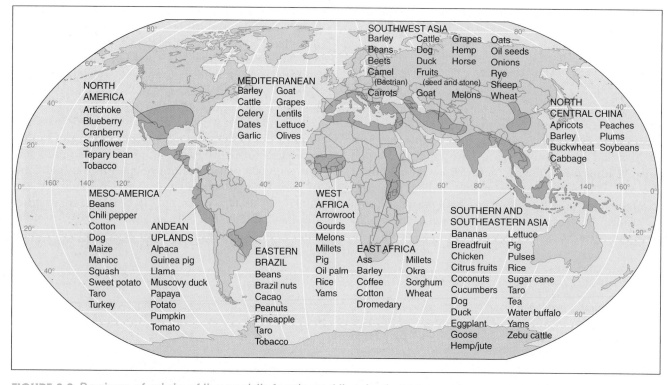

FIGURE 8-2 Regions of origin of the world's foods and livestock. This map identifies ten regions where significant numbers of plants or animals were domesticated. A few examples, such as barley, were domesticated in two places. The fact that the people in so many different regions and biomes learned to exploit the different plants and animals they found locally is a tribute to human ingenuity. Many of these plants and animals, however, have been widely redistributed. Therefore, comparing this map with the following maps of crop and livestock distribution today reveals that an area that produces the most of any crop or animal today is not necessarily a place where that plant or animal was originally domesticated.

4,000 years ago, or whether they learned from Near Eastern teachers.

Even before Malthus wrote, the Western Hemisphere had contributed important food crops to the Eastern Hemisphere. For example, the potato is native to the Andes region of South America. It yields the second highest number of calories per acre of any crop, although its protein content is only 6 percent. By Malthus's day it had already become a major food in Northern Europe (Figure 8-3). Today it is the world's fourth most important food crop, when measured by total tonnage harvested, after corn, wheat, and rice. China has recently recognized the potato's versatility and is today the world's largest producer. Russia is in second place, and India is not far behind. The crop is becoming a mainstay throughout Africa and Asia. Potato yields have increased due to **genetic engineering**, which is the manipulation of a species' genetic material. Two main techniques of genetic engineering are selective breeding, which has been carried on for millennia, and recombinant DNA, a new form of genetic modification discussed later in this chapter. New genetically modified potatoes are being introduced to the tropics, where they are ready to harvest only 40 to 90 days after planting. Potato planting continues to spread, and the value of the world's potato crop increases each year. The true treasure of the Andes was

not the gold that the Spanish conquerors sought, but the potatoes they trampled.

Maize is another Western Hemisphere native. "Corn" is the name English-speaking people have traditionally given to the major cereal crop of any given region. Therefore, in England corn means wheat; in Scotland and Ireland, oats; but in North America it means maize, or Indian corn. Maize was cultivated in Central and South America long before the arrival of Europeans, who introduced it into northeastern North America, where almost one-half of the world's maize is grown today (Figure 8-4). Maize is susceptible to frost, so the length of the frost-free period limits its distribution. Maize is a staple for people in South America and Africa, but about 70 percent of the U.S. crop is fed to livestock. Maize's average protein content is 10 percent.

Still another transplant is manioc, or cassava. Portuguese traders introduced manioc into Africa in the sixteenth century. It has a very low protein content and should be supplemented with other high-protein foods, but it remains a staple across lowland Africa.

Transportation and Storage

Improvements in worldwide transportation have allowed regional specialization in food production, and specialization can multiply productivity. Railroads, trucks,

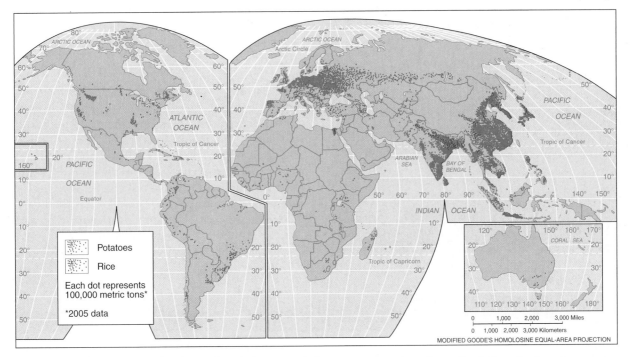

FIGURE 8-3 Current production of potatoes and rice. The potato is native to South America, but the country with the greatest production today is China. Rice remains the primary staple food of Asia, where it was first domesticated. (Data from United Nations Food and Agriculture Organization)

and cargo ships—many of them refrigerated— move quantities of food with a speed and efficiency that could not have been imagined in Malthus's day (Figure 8-5).

Transportation also allows the shifting of food from surplus to deficit areas, so fewer people need die from local famines. In the past it was not uncommon for surplus food to rot in the sun 100 miles from a starving population because the food could not be transported. Improved transportation also allows us to maintain smaller inventories of food at any specific time and place, yet remain secure.

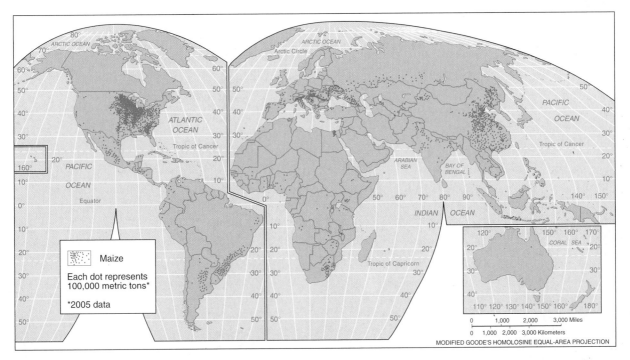

FIGURE 8-4 World maize production. Maize was first domesticated in Central America, but world production today is concentrated in North America. (Data from FAO)

Modern techniques of food transportation and storage allow Alaskans to enjoy fresh produce even when outside temperatures read −40°C (−40°F). (Courtesy of Carr Gottstein Foods Company, Anchorage, Alaska)

An increasing share of the world's food production is entering world trade—$674 billion worth in 2004. Table 8-1 lists the largest importers and exporters of some of the major foods traded. Some of the major exporters of certain crops are not among that crop's major producers. This indicates that in those countries, food production for export is a highly commercialized business. For example, the United States produced only about 1.6 percent of the total world rice crop in 2005, but the United States accounted for 10.5 percent of all world rice exports because most rice harvested in Asia is

TABLE 8-1 Top Ten Importers and Exporters of Wheat, Rice, and Corn, 2004 (in thousands of metric tons; data from FAO)

LEADING EXPORTERS	EXPORTS	LEADING IMPORTERS	IMPORTS
WHEAT			
United States	28,464	EU-25	7,393
Canada	15,142	Egypt	8,150
Australia	15,826	Brazil	5,309
European Union-25 (EU-25)	14,367	Japan	5,744
Russia	7,951	Algeria	5,398
Argentina	13,502	Indonesia	4,661
Ukraine	4,351	Korea	3,591
Kazakhstan	2,700	Nigeria	3,014
Turkey	2,217	Iraq	3,010
China	1,171	Mexico	3,717
RICE			
Thailand	7,274	Philippines	1,900
Vietnam	5,174	Nigeria	1,800
United States	3,950	Iraq	800
India	4,500	Indonesia	500
Pakistan	2,650	Saudi Arabia	1,250
Egypt	1,100	EU-25	1,050
Uruguay	762	Iran	950
China	656	Senegal	1,200
Australia	100	South Africa	850
Argentina	350	Bangladesh	800
CORN			
United States	45,223	Japan	16,485
Argentina	13,752	Korea	8,638
China	7,589	Mexico	5,921
Ukraine	2,334	Egypt	5,398
Brazil	1,431	Taiwan	4,500
South Africa	1,517	EU-25	2,951
Serbia and Montenegro	500	Malaysia	2,400
Romania	587	Colombia	2,256
European Union 25 (EU-25)	164	Iran	2,558
Paraguay	386	Algeria	2,046

FIGURE 8-6 New food-storage facilities. This officer of the U.S. Agency for International Development (AID) is overseeing the construction of a silo in Peru. Many countries produce enough food to nourish their people, yet much of the crop is wasted or lost to pests or spoilage because of inadequate storage and distribution facilities. (Courtesy of U.S. Agency for International Development)

consumed where it is produced. As much as one-sixth of world wheat production enters into world trade.

The technology of food storage has also continually improved, decreasing both spoilage and loss to pests (Figure 8-6). Improvements have been continuous from the introduction of the silo in the nineteenth century to antiseptic packaging and freeze drying. (Actually, the Inka freeze dried food 500 years ago, but the rest of the world learned how only recently.)

The Green Revolution

Starting about 1950, an intensive effort was launched to develop new grain varieties and associated agronomic systems and to establish them in developing countries. It focused on certain crops (rice, wheat) and certain techniques (breeding for response to fertilizer inputs). It was driven largely by private foundations such as the Ford and Rockefeller Foundations. This focused effort was known as the **green revolution.** Many projects continue. For example, specialists at Texas A&M University and at the International Rice Research Institute in the Philippines have produced higher-yielding strains of rice that are more resistant to disease and pests and that mature in only 110 days. This allows for two or even three crops per year. New potato strains are developed at the International Potato Center in Lima, Peru. Wheat yields have multiplied from new hybrids. India, where 1.5 million people died in a 1943 famine, produced enough extra grain to become a net grain exporter in 1977, even though its population doubled between those years.

Productivity continues to rise: World yield of cereal crops per hectare doubled from an average of 1,608

kilograms per hectare during the period 1979–1981 to 3,263 kilograms per hectare in 2003.

Other Technological Advances

The green revolution is just one aspect of the continuing application of science to agriculture that began in the late eighteenth century known as the **scientific revolution in agriculture.** New pesticides save crops from insects that once wiped out entire harvests, and new fertilizers multiply yields per acre. Parallel scientific research has been directed to livestock. As with crops, the developments include world redistribution, increases in total numbers, improved breeding, and even "engineering" for greater hardiness and higher yield from each animal. In 1950 the average U.S. dairy cow gave 2,339 liters (618 gallons) of milk per year; by 2005 that number had almost quadrupled to 9,275 liters (2,450 gallons). In 1969 the average pig used for breeding in the United States produced 6.7 piglets per year. By 2000 piglet production had risen to 16. So far, however, only a fraction of the world's livestock herds is this productive.

Farm machinery invented since Malthus's day reduces the number of fieldworkers and at the same time increases yields by improving the regularity of plant spacing and the efficiency of harvesting. Heaters rescue many crops threatened by freezes. During the past 200 years, humans have literally re-formed Earth with drainage projects where there was too much water and with irrigation projects where there was not enough. Projects such as these are not new in theory, but they are achieved in a scale beyond anything that Malthus foresaw.

AGRICULTURE TODAY

Humankind has thus managed to prevent the widespread starvation Malthus predicted. Not all of these developments, however, have taken hold equally everywhere. The following pages will first catalog regions of different types of present-day agriculture, and then we will examine the distribution of food supplies. What regions produce surpluses, and where are people hungry? Finally, we will address the question of whether and how we will be able to produce enough food for an increasing human population in the future. The pessimistic neo-Malthusian point of view introduced in Chapter 5 will again be set against the optimistic Cornucopian point of view.

Chapter 6 introduced Varro's theory of the evolution of human societies. The theory was based, principally, on the evolution of agriculture, for humankind's first concern is to get enough to eat. According to Varro, humans were at first hunter-gatherers before they domesticated animals and developed pastoral nomadism. Later the domestication of plants encouraged settled agriculture, which is more productive. That is, settled agriculture yields a greater and more secure quantity of food. **Subsistence agriculture**—agriculture to feed oneself and family—eventually yielded to **commercial agriculture**—raising food as a product for sale.

A few hunter-gatherer bands still do exist and provide evidence of how all humans lived in prehistoric times, before the invention of agriculture. Isolated groups inhabit the periphery of world settlement—the Arctic, the Amazon basin in South America, central and southern Africa, and Papua New Guinea. Examples include the Bushmen of Namibia and the Aborigines of Australia. Anthropologists believe, however, that these few remaining bands will probably abandon their ways or die out within decades. In 2002 the government of Botswana forcibly rounded up and settled the last of the Basarwa, or San Bushmen—about 2,000 people. "Hunting and gathering," decreed President Festus Mogae, "is no longer a viable way of life." By 2006, few were left in the "protected area," and the government stood accused of genocide.

Subsistence Farming Contrasts with Commercial Farming

Chapter 6 discussed how societies evolve from self-sufficiency to participation in trade in general terms by examining three villages connected by a road. Once peoples begin to exchange products, they have an incentive to specialize and increase the production of a good that can be exchanged. Farmers can specialize in one crop and import other foods from elsewhere. Thus, agricultural **polyculture,** the raising of a variety of crops, may yield to **monoculture,** specialized production of one crop.

There is no clear dichotomy between subsistence agriculture and commercial agriculture. They are the endpoints of a spectrum. The most self-sufficient hunter-gatherer may trade extra berries for fish, and the most specialized wheat farmer in Kansas may raise his or her own tomatoes. Nevertheless, many farmers today are called subsistence farmers because they do consume most of what they produce, and they produce most of what they consume. Subsistence farmers may sell some of their output, but surplus production is not their primary purpose, and in years of poor harvests they may not even have a surplus.

The world is too interconnected, however, for there to be many completely self-sufficient subsistence farmers left anywhere. Farmers generally respond to economic incentives, so if a farmer has a market opportunity, he or she will try to produce something for it. In virtually all countries, urban populations offer markets for farmers to satisfy. The day after a dirt road or a railroad reaches the smallest village, many peasants are on their way to market something in the nearest city. The 800 million peasants of China are often referred to as subsistence farmers because most of what they raise is for their own consumption. They do, however, manage to raise and market enough of a surplus to feed China's 500 million urbanites and still have enough left over for China to achieve net exports of food.

Several characteristics differentiate subsistence agriculture from commercial agriculture. Subsistence farming usually relies on enormous investments of human labor, but little animal or machine power. It is usually low in technology and, therefore, in productivity, and most of the food raised is consumed by the farmers themselves.

Modern commercial farming, by contrast, relies on substantial capital investment in machinery, chemicals, improved seeds, and livestock. The high price of farm machinery necessitates that it be used maximally, and this encourages an ever-increasing average farm size. Commercial farm products are seldom sold directly to consumers, but rather to large companies that process, package, store, distribute, and retail food. Individual companies are called agribusinesses—agricultural businesses—and many of them, such as Ralston Purina and General Mills, are household names in North America. Most farms may be owned by individual families, but a high and rising percentage of food is produced on gigantic farms owned by the agribusinesses themselves.

Types of Agriculture

According to the United Nations Food and Agriculture Organization (FAO), in 2005 about 1.5 billion hectares (3.75 billion acres) of Earth's land area were devoted to cropland (11 percent of the total, excluding Antarctica); and an additional 3.4 billion hectares (8.4 billion acres, 26 percent) were devoted to permanent pasture. Cereals (grains) and potatoes are the world's basic foods. Cereals have been grown since the dawn of history, and every major civilization has been founded on them as their principal source of food. Wheat, rice, maize, barley, sorghum, oats, rye, and millet are all members of the grass family of plants. They yield more food, both in bulk and in nutritive value, than most other crops. The preeminence of cereals as a food is also partly explained by the ease with which they can be produced, stored, and transported (Figure 8-7).

Most of Earth's "nonagricultural land" is hot, cold, or dry (compare Figure 8-1 with Figure 2-31). Agriculture has not been productive in these areas, nor have humans settled there in large numbers (compare Figure 8-1 with the map on the rear endpaper). In Figure 8-1, the categories of nomadic herding and low-technology subsistence agriculture cover enormous areas, but these activities support very few people. The category of irrigated agriculture covers a wide range of farming styles, from the intensely worked and productive fields of California to subsistence oases in the Sahara (Figure 8-8).

FIGURE 8-7 The shipment of grain. In the mid-nineteenth century, grain was still stored and transported in individual bags. This required enormous amounts of labor, and a lot of grain was lost to spillage. Late in the nineteenth century, however, methods were devised to store and ship grain as if it were a liquid. These techniques greatly reduced the cost and improved the efficiency of transportation. (Courtesy of the Kansas Wheat Commission)

The nine other categories of agriculture in Figure 8-1 are defined by the way that five variables interact in each place: (1) the natural environment, (2) the crops that are most productive in that environment, (3) the degree of technology used by local farmers, (4) the degree to which local farming is market-oriented,

FIGURE 8-8 Oasis in Algeria. The Grand Erg Occidental Taghit oasis in the Sahara Desert in Algeria supports a small population and feeds an occasional caravan passing through the surrounding sands. (RAGA/Corbis/Stock Market)

and (5) the degree to which crops are raised either for human consumption or to feed animals.

Four of the nine categories are basically subsistence: nomadic herding, low-technology subsistence farming, intensive rice farming, and Asian mixed cereal and pulse farming. The five other categories are predominantly commercial: mixed farming with livestock, prairie cereals, ranching, Mediterranean, and plantation agriculture.

Nomadic herding Pastoral nomads depend primarily on animals rather than crops, and pastoral nomadism may still be a successful low-technology adaptation to arid or semiarid environments. It exists across North Africa, the Middle East, and into parts of Central Asia. The Bedouins of North Africa and Saudi Arabia and the Masai of East Africa are examples of nomadic groups.

The animals upon which nomads depend may include horses, camels, goats, sheep, or cattle. The nomads will consume their animals or else sell them to sedentary farmers in exchange for grains. Sometimes, however, a nomadic group will plant crops at a fixed location and later return to harvest them. Pastoral nomads do not wander aimlessly; they have a strong sense of territoriality.

There are probably no more than 12 to 15 million pastoral nomads today, and many of them are settling. Many governments are forcing nomads to settle in order to control them. The government of Kenya, for example, is forcibly settling the Masai people. Some nomads are today allowed to move about only within limits fixed by governments. In the future, pastoral nomadism will probably be confined to areas that cannot be irrigated or that do not contain valuable mineral resources.

Low-technology subsistence farming The least intensive type of farming is shifting cultivation, called *slash-and-burn*, or **swidden** cultivation, from a word meaning "to singe." It is usually called "cultivation" rather than "agriculture," because the word *agriculture* implies greater use of tools and animals and more extensive modifications of the landscape (Figure 8-9). Nevertheless, swidden cultivation is quite sufficient in tropical areas such as the Amazon region of South America, Central and West Africa, and Southeast Asia, including Indochina, Indonesia, and New Guinea.

Small patches of land are cleared with machetes and other bladed tools, trees are stripped and killed, and fires are set to clear the land for planting. The ashes contain a simple fertilizer, potash (potassium). Various subsistence crops are planted with the use of a simple digging stick or hoe. Polyculture is practiced even in a single swidden, which may contain a large variety of intermingled crops. Soil nutrients are rapidly depleted, so the cleared land can be used to grow crops for only a short time, usually three years or less. Then it

FIGURE 8-9 Swidden burning. This fire was started to clear forest land for agriculture in Brazil. When the soil is exhausted, the site will be abandoned and reclaimed by forest. (James P. Blair/National Geographic Image Collection)

FIGURE 8-10 Rice terraces. These rice terraces in China increase the amount of land available for rice cultivation by turning steep slopes into flat steps. (Keren Su/CORBIS)

must be left for many years to regenerate, although it may be recultivated someday.

Shifting cultivation can support low levels of population without causing environmental damage, but it is being replaced by commercial logging, cattle ranching, and cultivation of cash crops. These activities, however, require cutting down much greater expanses of forest. Wholesale destruction of the rain forests reduces global biological diversity and may be contributing to global warming, as discussed in Chapter 4.

Intensive rice farming

The greatest number of farmers living in the large population concentrations of East, South, and Southeast Asia practice intensive subsistence agriculture. Even tiny, irregular plots are planted, and individual farmers' holdings are often fragmented as a result of subdividing holdings among children through generations. Most of the work is done by hand or with animals, largely because human labor is abundant, but capital for the purchase of equipment is scarce. Livestock are rarely permitted to graze on land that could be used to plant crops, and little grain is grown to feed the animals. The intensive agriculture region of Asia can be divided between areas where wet rice dominates and areas where other crops predominate. Wet rice is characteristic of Asia; dry rice is a grassy staple largely confined to West Africa.

About half the world's population subsists wholly or partially on wet rice. Wet rice occupies a relatively small percentage of the land, but it is the most important source of food in Southeast China, East India, and much of Southeast Asia. It has a protein content of 8 to 9 percent. Successful production of large yields of rice is an elaborate, intensive process. Wet rice should be grown on flat land, because the plants are submerged in water much of the time. Therefore, most wet rice cultivation is located in river valleys and deltas, although additional suitable land may be created by terracing the hillsides of river valleys (Figure 8-10). A farmer prepares the field using a plow drawn by water buffalo or oxen. The plowed land is flooded via dikes and canals that must be maintained to control the quantity of water in the field. Flooded fields are called **sawah** in Indonesian. Rice is either broadcast through the field, or else seedlings that have been nursed elsewhere are transplanted. Rice plants grow submerged in water for approximately three-fourths of the growing period, and they are harvested by hand.

In places with relatively warm winters, such as South China and Taiwan, two rice crops can be harvested per year from one field, a process known as **double cropping.** Most parts of India have dry winters, so double cropping there, where possible, usually involves alternating between rice in the wet summer and another cereal such as wheat in the drier winter.

Asian mixed cereals and pulses

Wet rice cannot be grown where summer precipitation levels are too low and winters too harsh. Therefore, other crops predominate in the interior of India and in northeast China. In milder parts of the mixed cereals and pulses region, crop rotation may allow double cropping. Wheat is the most important crop, followed by barley, which is the world's fourth most important cereal (Figure 8-11). Barley has the shortest growing season of all cereals and can be grown farther north, at higher altitudes, and in more arid regions than any other cereal. It is an important food crop in parts of Asia and in Ethiopia.

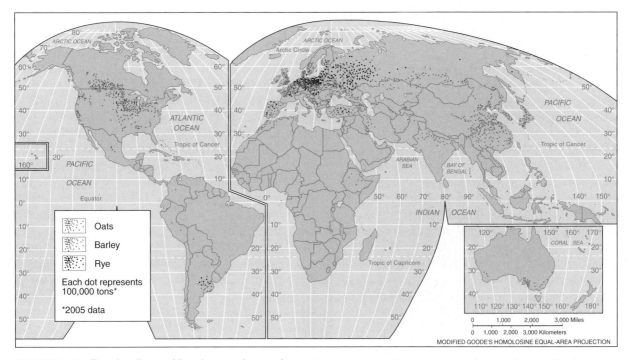

FIGURE 8-11 Production of barley, oats, and rye. Barley, oats, and rye can serve either as human food or as feed for livestock. (Data from FAO)

Pulses is the botanical name for a large family of herbs, shrubs, and trees, also called the pea or legume family. The seeds, which include beans, peas, lentils, peanuts, soybeans, and carob, are rich in protein and may form the principal source of dietary protein in poor regions or in regions where religious belief prohibits the consumption of meat. In China and Japan, tofu, made from soybeans, is a major food source. The root nodules of pulses have bacteria that absorb nitrogen out of the air and release it into the soil (a process called *fixing* nitrogen). This ability makes pulses valuable as crops grown temporarily to reduce erosion and build up a soil's nitrogen content.

Mixed farming with livestock Mixed farming with livestock is usually a commercial undertaking, for most of the crops are fed to animals rather than consumed directly by humans. Mixed farming allows the farmer to distribute the workload throughout the year (fields require less attention in the winter) and also to maintain his or her income by selling animals or animal products throughout the year. Crop rotation allows the farmers to maintain the soil fertility because different crops may deplete the soil of different nutrients but restore others.

Mixed farming is the dominant form of agriculture in much of the world today. Although it is still widespread in North America, farming there is evolving toward more specialized production. Farms that are truly mixed, including both feed production and animal production, remain common; but today corn and beans are increasingly grown on farms that no longer

keep more than a few animals, and the feedstock is sold to feeding operations that produce the meat. Specialization has separated the two components of the system in both ownership and location. In the United States "corn belt," which stretches from Indiana across Iowa to the Dakotas, approximately half of the cropland is planted in corn. Some of this corn is consumed by people directly as oil and other processed foods, but most of it is fed to livestock (either on the farm or concentrated in feed-lots), including pigs, cattle, and poultry. The second most important crop in this region, soybeans, is grown almost exclusively as animal feed. The third most important crop in the United States by area harvested (although not by value) is hay, which is also an important component of mixed farming systems.

This farming is enormously productive. Corn production in the United States represented 40 percent of world total production in 2005; soybeans grown in the United States also accounted for 40 percent of the world total of that crop.

Prairie cereal farming Crops on commercial grain farms are grown primarily for consumption by humans rather than by livestock. Large-scale commercial grain production is found in only a few countries, including the United States, Canada, Argentina, Australia, Russia, Ukraine, and France. This production is mostly, but not exclusively, in regions that are too dry for mixed crops and livestock. Wheat is today the most important cereal in world food production. It supplies about 20 percent of the total calorie consumption of the human species

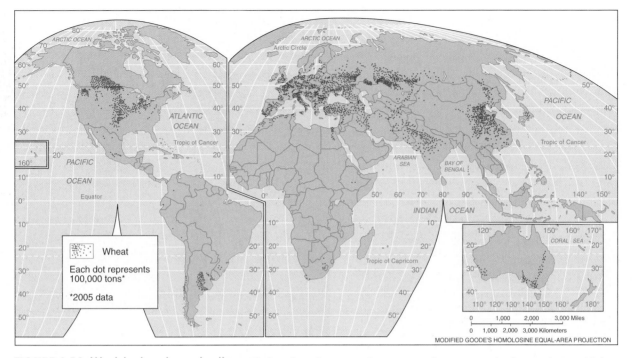

FIGURE 8-12 World wheat production. Today wheat is grown almost everywhere except in the tropics, and it is a major commodity in international trade. (Data from FAO)

(Figure 8-12) and provides a staple for over one-third of the world's population. The protein content varies between 8 and 15 percent.

Wheat generally has more uses as human food than other grains, so it is more valuable. It can be stored relatively easily without spoiling and transported long distances both easily and economically. Hard-kerneled varieties yield flour with a high content of *gluten* (a protein substance) and is used to make breads; flour from soft-kerneled varieties is starchier and is used to make cakes and biscuits. *Durum* wheat is an especially hard-kerneled wheat used in pasta products. Wheat grain, *bran* (the fibrous outer coat of the wheat kernel that remains as a residue after milling), and the rest of the plant are all valuable livestock feed.

In North America, large-scale grain production is concentrated in three areas. In the winter wheat belt that extends through Kansas, Colorado, and Oklahoma, the crop is planted in the autumn and develops a strong root system before growth stops for the winter. The wheat survives the winter and is ripe by the beginning of summer. In the spring wheat belt of the Dakotas and Montana in the United States, and in southern Alberta, Saskatchewan, and Manitoba in Canada, the winter is usually too severe for winter wheat, so spring wheat is planted in the spring and harvested in the late summer. A third important grain-growing region is the Palouse region of eastern Washington State. Large-scale grain production is highly mechanized and conducted on large farms, but the effort required is not uniform throughout the year. Therefore, some individuals or firms may own fields in each belt in order to keep their expensive machinery in use longer. Combine compa-

nies start working in Texas in late May; then move to Oklahoma in early summer and farther north through the season.

Ranching   Ranching is the commercial grazing of livestock over extensive areas. It is best suited to arid or semiarid land, where the vegetation is too sparse or soil too poor to support crops. Cattle raised on ranches are frequently sent for fattening to farms or to local feedlots before being moved to meat processors (Figure 8-13).

Spaniards introduced cattle to the Western Hemisphere in the sixteenth century, and the cattle thrived on the grazing lands in both North and South America. In the late nineteenth century, British capital developed the commercial cattle industry in Argentina in areas convenient for shipping to overseas markets. Today a large portion of the prairies of Argentina, southern Brazil, and Uruguay are devoted to grazing cattle and sheep (compare Figure 8-1 with Figures 2-31 and 4-23).

The interior of Australia was opened for grazing at the same time, although sheep are grazed along with cattle (see Figure 5-28). Ranches in the Middle East, New Zealand, and South Africa are also more likely to raise sheep. Commercial ranching is found in other relatively developed regions of the world, but it is rare in Europe, except in Spain and Portugal.

Mediterranean agriculture   Mediterranean agriculture is a type found in the distinctive Mediterranean climate regions—usually on the west coasts of continents at about 30–40° north and south latitude around the Mediterranean Sea, in Southern California, central Chile, the southwestern tip of South Africa, and southwest

FIGURE 8-14 Grapes in Chile. Chile long exported substantial quantities of grapes such as these, but Chileans now multiply the value of this this rich agricultural harvest by converting the grapes to wine and exporting wine. A wine made entirely, or chiefly, from one type of grape is called a *varietal*; these Malbec grapes produce a fine red wine. (Andy Christodolo/Alamy Images)

FIGURE 8-13 Cattle feedlot in Greeley, Colorado. Cattle that have been grazing over extensive areas are usually fattened at feedlots such as this before slaughter. A cattle-identification system uses a high-speed digital camera to scan the retinas of calves. A GPS signal can then track each animal and record its progress from birth to the slaughterhouse. (Lowell J. Georgia/Photo Researchers, Inc.)

Australia. The Mediterranean climate has hot, dry summers and cool, rainy winters. The lack of summer water and good grazing land hinders livestock raising, so Mediterranean-style farmers traditionally derive only a small percentage of income from animal products. Many of these areas are mountainous, or at least hilly, so some farmers do practice transhumance—keeping sheep and goats on the coastal plains in the winter and transferring them to pastures in the hills in the summer.

Most crops in Mediterranean lands are grown for human consumption, including most of the world's olives, grapes, and other fruits and vegetables. Hilly landscapes allow farmers to practice polyculture, with some crops facing the Sun, while others face away.

Two-thirds of the world's wine is produced in countries that border the Mediterranean Sea, especially Italy, France, and Spain, while other Mediterranean regions elsewhere (such as Southern California, Chile, and South Africa) produce most of the rest (Figure 8-14). About half of the land may be planted in cereals, especially wheat for pasta and bread. Seeds are sown in the fall, and the crops harvested in early summer. Land is periodically left fallow to conserve moisture in the soil.

California has less land planted in cereals than is traditional in Mediterranean regions, because California growers concentrate on profitable citrus fruits, tree nuts, and fruits produced by deciduous trees (apples, peaches, plums, and similar fruit). Rapid urbanization along the coast has driven agriculture inland, where irrigation is required. Competition for water among farmers, urban and recreational areas, and environmental demands constitutes one of the most important political issues in the West.

Plantation farming A plantation is a large farm that specializes in the commercial production of one or two crops. Plantations found in tropical Latin America, Africa, and Asia are often owned or operated by Europeans or North Americans, and they are worked by corporate employees. The crops are grown primarily for sale in the rich countries. Latin American plantations grow coffee, sugarcane, and bananas, while Asian plantations may provide rubber and palm oil (Figure 8-15).

FIGURE 8-15 Tapping a rubber tree in Indonesia.
This woman is tapping a rubber tree on a plantation on the
Indonesian island of Sumatra. (© Vince Streano/CORBIS)

What Determines Agricultural Productivity?

The various types of agriculture we have described all
concentrate on different sorts of crops, but we can learn
something about the great range of productivity in agriculture around the world by looking at yields of grain
(Figure 8-16). This range of output is partly attributable
to variations in the physical environment. For example,
Western European countries enjoy mild climates, and
their agriculture is highly productive, whereas the Sahara
Desert region of northern Africa is generally unproductive. Surprising exceptions, however, can also be found.
For example, several Near Eastern Arab countries record
some of the highest grain yields anywhere on Earth.

When we compare Figure 8-16 with Figure 8-17
and Figure 8-18, however, we see a considerable
correlation between the variation in productivity
and the application of fertilizers and the use of tractors.
The countries that enjoy the highest crop yields rank
high in at least one of these two investments. Investment in agriculture seems to be a principal factor
determining productivity. Many of the rich places
that are able to invest in tractors and fertilizer can also
be assumed to be the places where farmers have the

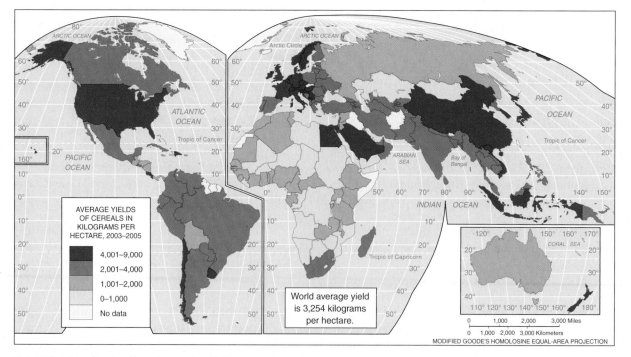

AVERAGE YIELDS
OF CEREALS IN
KILOGRAMS PER
HECTARE, 2003–2005

- 4,001–9,000
- 2,001–4,000
- 1,001–2,000
- 0–1,000
- No data

World average yield
is 3,254 kilograms
per hectare.

MODIFIED GOODE'S HOMOLOSINE EQUAL-AREA PROJECTION

FIGURE 8-16 Agricultural productivity. Yields per hectare vary greatly around the world, and these variations are
not always clearly related to natural environmental conditions. This map does not reproduce the pattern of precipitation
(Figure 2-30), climates (Figure 2-31), or biomes (Figure 4-23). Factors other than natural environmental conditions
must be found to explain agricultural productivity. (Data from FAO)

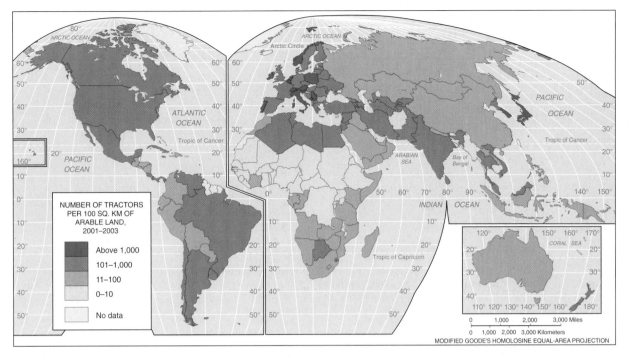

FIGURE 8-17 The use of tractors. Capital inputs in agriculture vary tremendously. The use of tractors is a good measure of modern capital-intensive farming. Does this map reproduce the pattern of Figure 8-16 and, therefore, help explain variations in yield? (Data from FAO)

best hybrid seeds. These are the product of careful interbreeding through generations of the best strains of a crop. These countries also have the best irrigation, pesticides, and all the other modern technological apparatus of agriculture. Therefore, we can conclude that capital input is a principal determinant of agri-

cultural productivity today. Capital investment may be as important as the natural environment, or even more so. If investment in agriculture could be increased in the regions where there is little investment today, total world agricultural output would undoubtedly soar (Figure 8-19).

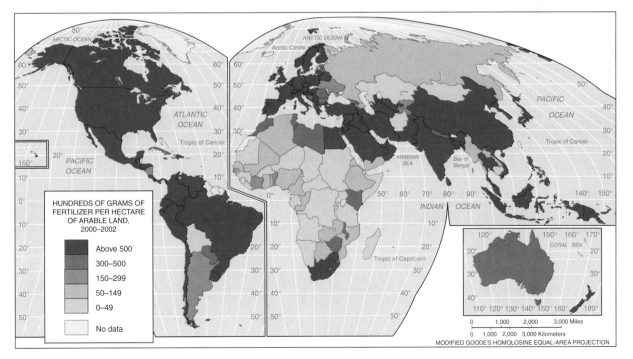

FIGURE 8-18 The application of fertilizer. Fertilizer is another important capital input in agriculture. Its application can override deficiencies in the local soil. (Data from FAO)

(a)

(b)

FIGURE 8-19 Extremes of capital investment. These two photographs illustrate the extremes of capital investment in agriculture. If the Senegalese farmer (a) had as much capital to invest in farm machinery, fertilizer, and improved seeds as the Nebraska farmers (b) do, who knows how much food he could raise? ([a] Carl Purcell/U.S. Agency for International Development and [b] U.S. Department of Agriculture)

LIVESTOCK AROUND THE WORLD

Humans began to domesticate animals about the same time as they did plants, some 8,000 to 10,000 years ago. Dogs were probably domesticated first, independently at a number of places around the world. Dogs helped in hunting, and they were themselves eaten for food. The domestication of a great variety of other animals followed through centuries at various locations (look back at Figure 8-2 again). As humans learned to domesticate plants, they found it convenient to herd plant-eating ani-

mals and to pen them close to settlements. From then on, growing crops and raising livestock advanced side by side.

A wide range of animals have been domesticated as livestock, and their numbers totaled almost 21 billion in 2005—over three times the human population. These numbers included about 17 billion chickens, 3.5 billion ruminants (cud-chewing animals, mainly cattle, sheep, goats, buffalo, and camels), and about 960 million hogs (see Figure 8-20 and also look back at Figure 7-37). Domesticated livestock provide high-quality protein, either as meat or milk as well as supply hides, wool, and other raw materials. Livestock serve as

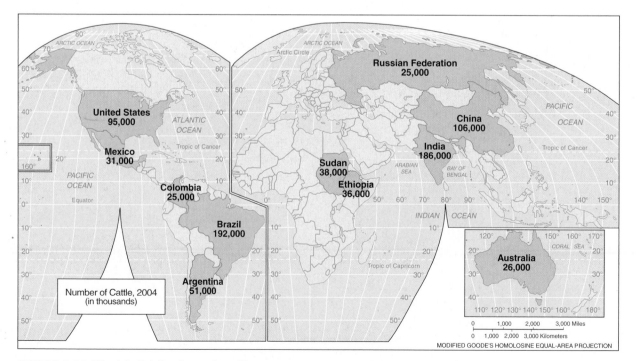

FIGURE 8-20 World distribution of cattle. Cattle are widespread, but these 11 countries counted almost two-thirds of the world's herds in 2004. (Data from FAO)

FIGURE 8-21 Masai cattle. Cattle represent wealth among the Masai people of Kenya. As the government requires these traditionally nomadic people to settle, multiplying cattle overburden the environment. (Photograph by Eliot Elisofon, 1959. Image no. EEPA11002. Eliot Elisofon Photographic Archives, National Museum of African Art, Smithsonian Institution)

draft animals that pull plows and wagons; large numbers of people in poor regions still depend on oxen and buffalo. In India alone, there are over 80 million draft animals. Among pastoral peoples, livestock represents wealth on the hoof (Figure 8-21).

The Direct and Indirect Consumption of Grain

Many domesticated animals eat grass, foliage, and other plant material that humans cannot digest. Some grains, however, called *feed grains,* are fed to livestock. Eventually either the livestock or their products (milk, butter, cheese, and eggs, for example) are consumed by humans, so we say that feed grains are consumed by humans indirectly. Over 40 percent of the world's total grain harvest is fed to livestock. This percentage is higher in the richer countries. In the United States, for example, the figure is approaching 80 percent, whereas in India the figure is about 4 percent.

As a country develops economically, the population consumes a greater amount of grain, but an increasing percentage of that grain is consumed indirectly as meat and dairy products. The people of the United States and Canada, for example, consume as much as 2,000 pounds of grain per person per year, but only about 150 pounds of that amount is consumed directly as bread or cereal. The rest is consumed indirectly. In poor countries, on the other hand, only about 400 pounds of grain are available for each person per year, and it mostly is consumed directly. Meat is the center of a people's diet only in the richest countries, and some nutritionists argue that the consumption of so much meat and dairy products leads to increases in debilitating diseases such as heart disease. Nevertheless, most

people want to enjoy meat and dairy products. Therefore, rising incomes in some countries multiply the populations' appetites for grain consumed indirectly as meat.

Global per capita consumption of meat rose from 26 kilograms (57.2 pounds) in 1970 to 40 kilograms (88 pounds) in 2003. Rapidly increasing quantities of meat consumption in China—mainly pork and chicken—account for much of this. Many observers see this rise in meat consumption as good news, but neo-Malthusians believe that it threatens the food supplies of the poorest people. As long as rich people are willing to pay high prices for meat, grain will be fed to animals instead of being available to poor people. Cornucopians, however, argue that if grain prices rise, triggering a rise in the price of meat, grain production will multiply, increasing the supplies and thus lowering the price of grain again, making more available for human consumption. In any case, improving cereal yields might cause prices to continue to fall. We do not know which of these economic events will occur, but such future price variations will affect the diet of billions of people.

Some kinds of animals transform grain into meat more efficiently than others do, and humankind overall could greatly improve its food supply by concentrating on raising these types of animals. Chickens are the most efficient. They yield 1 pound (45 kilograms) of edible meat for every 4 pounds (1.8 kilograms) of grain they consume. Pigs produce 1 pound for every 7 pounds (3.2 kilograms) of grain, and beef cattle 1 pound for every 15 pounds (6.8 kilograms) of grain. In addition, chickens reach maturity and can be consumed in just seven weeks—much faster than pigs or cattle (Figure 8-22).

FIGURE 8-22 Raising roasting chickens in Florida. Chicken production in advanced countries has moved from the barnyard into factories in which chickens are mass-produced. Hog production is undergoing a similar transformation. Factory-like conditions maintain control over feed, meat characteristics, breeding and reproduction, size, and other characteristics. (Norm Thomas/Photo Researchers, Inc.)

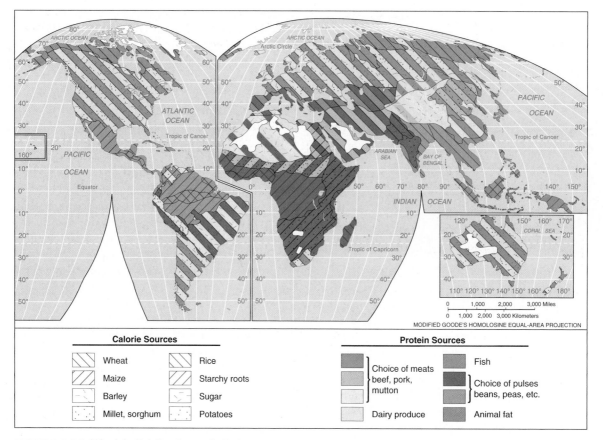

Calorie Sources

Wheat	Rice
Maize	Starchy roots
Barley	Sugar
Millet, sorghum	Potatoes

Protein Sources

Choice of meats beef, pork, mutton	Fish
	Choice of pulses beans, peas, etc.
Dairy produce	Animal fat

FIGURE 8-23 World distribution of diets. Everybody needs both calories and protein. The stripes and patterns on this map convey the predominant local source of calories, and the colors convey the predominant source of protein. The meat and dairy diet of Europe and of Europeans' settlements around the world stands out sharply. East and Southeast Asians rely more on fish, while pulses are important in South America, Africa, and in Southwest and South Asia. (Re-created from *The Times Atlas of the World,* Times Book Division at The New York Times Co., 1983, by John Bartholomew & Sons Limited and Times Books Limited, London)

There are already more than twice as many chickens on Earth as people, and today many countries are concentrating on increasing their chicken populations. China, for example, more than quadrupled its chicken population between 1984 and 2005. In many societies, however, cattle are a status symbol, and this preference delays the switch to more productive livestock.

In some places people avoid meat and dairy products because of religious prohibitions. Figure 7-37 noted the absence of pigs, for example, in Jewish and Muslim regions. These people meet their protein requirements from beef, lamb (mutton), fish, grains, or pulses (Figure 8-23).

Problems Associated with Animal Production

There are about 1.4 billion cattle on Earth today, and some scientists criticize these herds as inefficient and even harmful. The indictment against them is long. In the United States, runoff from feedlots is a serious water-quality problem. In sub-Saharan Africa, cattle contribute to desertification by denuding arid lands of fragile vegetation. In Central and South America, ranchers are felling rain forests for pasturage for cattle.

Some agronomists argue that the millions of acres of land dedicated to feedstocks should be rededicated to growing crops for direct human consumption.

In addition, cattle contribute to greenhouse gases. The bacteria that live in every cow's gut enable the animal to digest cellulose, a tough fiber found in grass and other plants that humans cannot digest. As a by-product, however, these bacteria produce methane, which is one of the most important greenhouse gases. The amount of methane produced by livestock is estimated to be almost double the amount produced by all termites (which have the same bacteria) or from landfills (which give off gas from decomposition).

Other livestock also contribute to environmental devastation. Sheep and goats, for example, have overgrazed substantial areas in the Mediterranean basin, as well as parts of Africa and India. The global populations of hogs and chickens are rising faster than the cattle population, and their proportionate contributions to environmental problems, especially water pollution associated with feedlots, are increasing.

The pollution associated with "industrial" livestock farming is increasing in the United States. Industrial farms housing tens of thousands of hogs generate millions of gallons of waste each year. Hog waste is collected in cesspools because it must be treated before it can be used as fertilizer by being sprayed on fields or into the air in liquid form. Seepage from unlined waste ponds pollutes groundwater, and runoff pollutes local waters, too. Cesspools the size of football fields emit the toxic gases hydrogen sulfide and ammonia. Enormous dairy farms present the same problems. Furthermore, concentration of livestock increases their vulnerability to disease, as well as the chances of transferring diseases to humans. Chapter 5 noted the problems associated with the use of antibiotics in animal feed. These huge facilities are truly factories, but they are not treated as such in the law or subject to the same environmental regulations.

Dairy Farming and the Principle of Value Added

Nearly 65 percent of the world's supply of cow's milk is produced in developed countries. In the past, fresh milk was available only on farms and in nearby towns, but in rich countries urbanization created a demand for large-scale commercial milk production. Milk is heavy and spoils quickly, so the cost of transporting it is high. Therefore, most cities receive their supplies from dairy farms in surrounding regions referred to as the cities' *milksheds.*

Some areas that are remote from urban centers but that have cool, damp climates unsuited to grain farming may specialize in dairy farming. Wisconsin in the United States and Switzerland in Europe are two such locations. Much of the milk from these regions, however, is first transformed into butter, cheese, or dried, evaporated, or condensed milk. These products are not only lighter and less perishable than milk; they are more valuable. The difference between the value of a raw material and the value of a product manufactured from that raw material is called the **value added by manufacturing.** Adding value to raw materials produces wealth. Wisconsin cheese and Swiss cheese and milk chocolate are exported worldwide, providing incomes to those two regions.

There are many steps and activities that transform the milk from a cow into a milk chocolate bar in a child's hand. In economic discussions, it is common to use the image that all products "flow" from their sources as raw materials to their ultimate consumers; the economic activities closer to the consumers are called **downstream activities.** Downstream activities add more value, and therefore they are profitable. Figure 8-14 notes how Chileans add value to their grapes by converting them to wine. We shall see in Chapter 9 and again in Chapter 12 that the possession of either agricultural or mineral raw materials may not necessarily enrich their possessor significantly. The key to wealth is adding value to raw materials and taking a share of that value added as profit. The geography of wealth is largely the geography of adding value to resources. Economists even speak of **psychological value-added,** that is, adding to the sales price of an object not by increasing its actual functionality or usefulness, but increasing its price through design, packaging, or "status" advertising.

FOOD SUPPLIES IN THE FUTURE

All the factors we have discussed have held off the specter of worldwide starvation. Is it possible that humankind is now, at last, at the end of its ability to increase food supplies? The answer to this question is a cautious "probably not." If demographers are correct in their projections of Earth's future population, people can be fed. Humankind has scarcely begun to maximize productivity with the best contemporary technology, and that leading technology has been applied to only a small portion of Earth.

Several of the other factors we have listed that have increased total production since Malthus's time still offer potential for advance. Spreading urbanization is replacing agriculture in many places, but more lands can still be farmed. In 1995 the FAO estimated that developing countries of Latin America and sub-Saharan Africa could more than quintuple their cropland; the countries of East Asia could double their cropland. Irrigation can open new acreage (Figure 8-24), but in most dry areas where irrigation is most effective, water is

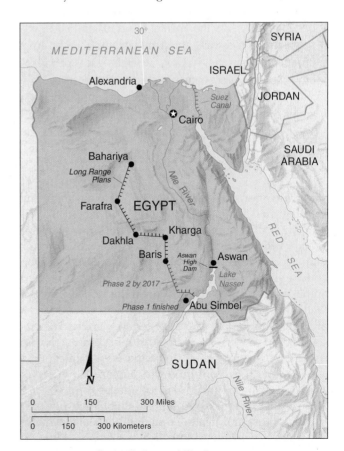

FIGURE 8-24 Egypt's "new Nile." Egypt has begun construction of an irrigation canal that will parallel the Nile River through its western desert.

Focus ON

How Farmers Decide the Ways in Which to Use Their Land: Von Thünen's "Isolated City" Model

Johann Heinrich von Thünen (1783–1850) was a Prussian aristocrat who wondered what product he could grow on his suburban estates and market most profitably. Thinking about this problem led him to question how various crops became distributed across any countryside. To answer his question, von Thünen devised one of geography's earliest models. He began with the idea of an imaginary city market in a perfectly flat plain with absolutely no variations on it. This is called an **isotropic plain.** Von Thünen drew a model pattern for the distribution of different land uses.

Von Thünen noted that the different uses to which parcels of land are put result from the different values placed upon the land, called its *rent value*. On an isotropic plain, transport costs are the only variable, so differences in rent reflect the transport costs from each farm by straight-line distance to the market—as if everything moved by helicopter. The greater the transport cost, the lower the rent that can be paid if the crop produced is to be competitive in the market. In addition, perishable products such as milk and fresh vegetables need to be produced near the market, whereas less perishable crops such as grain can be produced farther away.

Von Thünen deduced that a pattern of concentric zones of land use will form around a city market. The intensity of cultivation—that is, the amount of labor and capital applied, will decline with distance from the market. Perishable crops that have the highest market price and the highest transport costs per unit of distance,

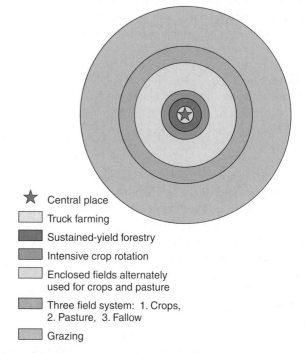

★ Central place

Truck farming

Sustained-yield forestry

Intensive crop rotation

Enclosed fields alternately used for crops and pasture

Three field system: 1. Crops, 2. Pasture, 3. Fallow

Grazing

Land use around an isolated city. Von Thünen's model shows rings of decreasingly intensive land use concentrically outward from the city. The ring of sustained yield forestry may be surprising, but forestry was then an intensive land use. The forest provided grazing for livestock, fuel, and raw materials for building and many other purposes.

---

already in short supply. Greater opportunities lie in increasing the efficiency of irrigation. Only about 37 percent of all irrigation water in the world, for instance, is actually taken up by crops; the rest is lost in runoff, evaporation, or leakage. Technological advances in irrigation could reduce the percentage lost. Drip irrigation, for example, is a method that pipes only as much water as crops need and delivers it directly to their roots. It reduces the average amount of water needed per irrigated acre, boosts crop yields, removes the threat of parasitic disease spread by irrigation canals, and reduces soil salination.

Research is continuing to explore the uses of *halophytes*, plants that thrive in saltwater. Samphire (Salicornia), for example, is a tasty vegetable that not only contains more and healthier oil than soybeans, but also produces a valuable animal feed as a by-product. It thrives in saltwater, and it is now cultivated commercially in Mexico, Arizona, Egypt, Iran, Syria, and around the Arab Gulf (Figure 8-25). Interbreeding halophytes with

FIGURE 8-25 Halophyte cultivation. Samphire is presently cultivated in India. (Courtesy of Planetary Design Corporation)

such as vegetables, will be grown closest to the market. Today a hint of this zone around a city survives in New Jersey's nickname, The Garden State. New Jersey farms historically supplied the cities of Philadelphia and New York. Today New Jersey is heavily industrialized in some parts, yet it also still specializes in greenhouse products, dairy products, eggs, and tomatoes, which are typical market garden commodities. In von Thünen's model, fields farther away from the market will be dedicated to less perishable crops with lower transport costs per unit of distance. In the rings farthest outward from the cities, livestock grazing and similar extensive land uses still predominate.

Von Thünen elaborated his model by adding a navigable river flowing through the town. He noted that the areas along the river's banks would enjoy greater accessibility to the city market. In other words, the cost distance to the city market was lower along the riverbank, so each of the zones would extend out along the river.

Some scholars have tried to apply von Thünen's model to entire countries. They have plotted the use of cropland in the United States, for instance, as if the Eastern Seaboard cities or Chicago were a central market. Trying to apply von Thünen's model to vast territories under contemporary conditions, however, strains the model's usefulness. The model focuses on distance from city markets, but in the real world proximity to the market is only one consideration. Physical environmental conditions, governmental regulations, the economic system, the pattern and the regulation of the transport

system, and still other factors must be considered. Also, transport costs have fallen dramatically relative to other costs of providing food. Most of the fruits and vegetables consumed in New York and Philadelphia, for example, come not from New Jersey but from California, Florida, Chile, and other distant places where production costs are lower and growing seasons longer.

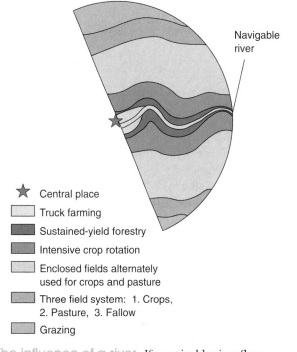

★ Central place

▢ Truck farming

▢ Sustained-yield forestry

▢ Intensive crop rotation

▢ Enclosed fields alternately used for crops and pasture

▢ Three field system: 1. Crops, 2. Pasture, 3. Fallow

▢ Grazing

The influence of a river. If a navigable river flows through the town, the activity characteristic of each zone extends outward along the river.

conventional crops has made these crops more salt resistant, which means that they can grow in more diverse environments. Farmers are today harvesting lands once thought too salt-soaked to support crops in Egypt, Israel, India, and Pakistan. Conventional crops may someday be grown in saltwater.

Techniques of storage and transport could still be improved in many areas. China loses at least one-quarter of its wheat harvest, for example, due to inadequate drying and storage, rats and mice, and poor transport. Improvements have been achieved, but in 1998 the U.N. World Food Program estimated that 40 percent of the world's crops were destroyed as they grew or before they left the field.

New Crops Offer New Potential

The green revolution focused on the improvement of just a few of humankind's most important crops, and the success in improving these has raised concern that humankind is becoming too reliant on too few crops

(Figure 8-26). If a new disease suddenly appeared and attacked any of these, it could destroy a significant

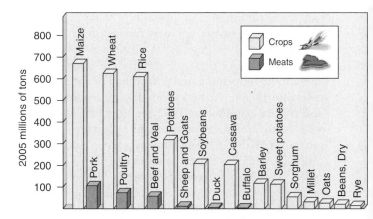

FIGURE 8-26 World food production. The great bulk of the human diet is today based on just a few crops and animal meats, but many others offer food potential. (Data from FAO)

percentage of humankind's total food supply. A model of such a disaster occurred in the U.S. corn crop in 1970. A new fungus suddenly appeared that was well matched to the T-cytoplasm that had been incorporated into 80 percent of the country's seed corn, and U.S. corn production fell 15 percent.

To forestall a catastrophic loss of seed material, Norway, Iceland, Sweden, Denmark, and Finland have joined to build an International Seed Vault to hold seeds from all of the world's farm crops. Placed on the island of Svalbard, far to the North of mainland Norway, the vault will hold samples inside concrete walls 70 meters (230 feet) under the permafrost, to be released only if all other seed sources are destroyed or exhausted.

A diversity of crops provides protection against catastrophe in case any one crop should fail, but modern commercial agriculture is increasingly specialized. Only "primitive" agriculture, which is in fact very sophisticated, has preserved diversity. For example, ethnobotanists have reported more than 50 kinds of potato grown in a single village in the Andes. At least 75,000 plant species have edible parts, and many of them are superior to the plants cultivated today. The New Guinea winged bean, for instance, is entirely edible—roots, seeds, leaves, stems, and flowers. It grows rapidly, up to 4.5 meters (15 feet) in a few weeks, and it offers a nutritional value as high as that of any known crop. Several Brazilian rain forest fruits—including the acerola, camucamu, and açai—are both tasty and extraordinarily rich in vitamins and minerals. The National Academy of Sciences has recommended the cultivation of 36 other crops, including amaranthus, buffalo gourd, tamarugo, guar, mangosteen, and soursop. Most of us have undoubtedly never heard of these, yet each offers enormous potential for food. The potential of raising food from more and different plants yet to be researched is a reason why the extinction of plant species threatens humankind's welfare. The steady loss of genetic diversity before botanists can study each species sacrifices potential new crops or genes for interbreeding.

Agricultural technology offers many alternative foods, but what people eat, or refuse to eat, is to a great degree cultural (Figure 8-27). Culinary imagination could probably make a wide variety of alternative crops and animals appealing and palatable. For example, the menu for the New York Entomological Society's 100th anniversary banquet in 1992 featured live honey-pot ants as an appetizer, cricket and vegetable tempura, mealworm balls in zesty tomato sauce, roasted Australian kurrajong grubs, waxworm fritters in plum sauce, giant Thai waterbugs sautéed in olive oil, and for dessert, chocolate cricket torte. All these insects are very nutritious, and they are common fare in many countries. South Africa and Botswana do in fact annually export hundreds of tons of mopanie worms. A dried worm is 56.8 percent protein, and 15 worms provide a

FIGURE 8-27 Would you eat this? Peasants bring live iguanas into downtown Hanoi, Vietnam, for sale for Saturday night dinner. Many people around the world find nutrition in and enjoy eating things most Americans do not find tempting. (Courtesy of Julia Nicole Paley)

full daily supply of calcium, iron, and riboflavin for an adult. The worms taste like peanuts.

The Scientific Revolution in Agriculture Continues

The scientific revolution in agriculture is not over. Some scientists argue that it has just begun and that new scientific advances will multiply future food yields. The economist Henry George (1839–1897) succinctly contrasted the rules of nature with the multiplication of resources through the application of human ingenuity. "Both the jayhawk and the man," he said, "eat chickens, but the more jayhawks, the fewer chickens, while the more men, the more chickens." This principle is key to understanding and counting all resources.

Biotechnology is the term given to a variety of new techniques for modifying organisms and their physiological processes for applied purposes. One aspect of biotechnology is **gene splicing,** or **recombinant DNA.** Scientists can now join the DNA of two organisms to produce a recombinant (or "recombined") DNA. Scientists can then introduce this recombinant DNA into another organism, thus permanently changing the genetic makeup of that organism and all its descendants. Such products are said to be **genetically modified (GM).** Worldwide plantings of GM crops expanded from 8.1 million hectares (20 million acres) in 1997 to over 81 million hectares (200 million acres) in 2005.

Another technique of biotechnology is **cloning,** the production of identical organisms by asexual reproduction from a single cell of a preexisting organism. Still other techniques of biotechnology develop new microorganisms for diverse purposes ranging from

FIGURE 8-28 A natural insecticide. The rows of potato plants on the left and right were devoured by the Colorado potato beetle. The plants in the center row, however, had a gene introduced into them from a common soil bacteria. Those plants produce a protein that acts as a natural insecticide. Biotechnology offers the promise of protecting many crops in this way. (Courtesy of Pharmacia Corporation)

FIGURE 8-29 A leader in research on GM tropical crops. Dr. Florence Wambugu of Kenya has won world renown for developing a virus-resistant sweet potato that more than doubles yields for subsistence farmers. She herself was raised on sweet potatoes on a subsistence farm. "I wasn't even supposed to be educated," she has said with a laugh. "My mother had to go to a tribal tribunal to get permission. I was a girl who was supposed to be married, and that was it." Today she heads a Nairobi-based scientific foundation and is an advisor to many scientific boards around the world. (Reprinted with permission from *The Globe and Mail*)

pharmaceuticals to cleaning up oil spills. Biotechnology can be applied to both plants and to animals.

Biotechnology offers genetically altered crops that can be custom-designed to fit the environment and produce bountiful harvests. It can even replace chemical pesticides (Figure 8-28). Breeding natural herbicides and pesticides into the plants themselves reduces the enormous quantities of chemicals and energy consumed in the manufacture, packaging, distribution, and application of chemical herbicides and pesticides. For example, a new GM corn significantly reduces the need for pesticides on that crop and a new type of Russet Burbank potato contains its own protection against the Colorado potato beetle. In 1995 an estimated 5,000 barrels of oil went into the manufacture and distribution of insecticides to fight this one insect pest in the United States, and less than 5 percent of the insecticide reached the target insect. By 2001, the use of GM crops was reducing U.S. use of pesticides by an estimated 21 million kilograms (23,100 tons).

The first GM food to reach U.S. groceries was the Flavr Savr tomato, in 1994, engineered to stay firm longer than regular tomatoes, but GM soybeans, cotton, potatoes, corn, and rape soon followed. Most historic advances in the green revolution addressed improving yields of wheat, potatoes, and rice, but today many other traditional crops, including tropical crops such as sweet potatoes and cassava, are winning new attention (Figure 8-29). Today lentils provide one of the world's leading sources of protein, and the GM Mason lentil is up to 18 percent more productive and is hardier than the traditional strain.

Today's food has been improved not only in quantity but in quality. For example, GM pigs produce

omega-3 fatty acids, which help reduce the risk of heart disease. Scientists have doubled the vitamin content of most vegetables and produced soybeans that are easier to digest and that also produce heart-healthy oils. In 2005, White House executive chef Ariel De Guzman, when asked why he used canned and frozen vegetables, replied, "Canned or frozen vegetables have more nutrition than fresh vegetables. It's added from the companies, the manufacturer." (He made no claims, however, for superior flavor or texture.)

GM crops have even been bred to produce pharmaceuticals, in a process called biofarming or farmaceuticals. DNA science has spawned a generation of drugs made from human antibodies, but the antibodies can be produced more cheaply by splicing antibodies into the genetic fabric of plants, growing them in fields, and then extracting and purifying them. Hemoglobin is being bred into corn and soybeans. Corn-bred enzymes stimulate insulin production in diabetics, and safflower plants can even produce human insulin—an advance of increasing importance in a world of overweight diabetic humans. Bananas and potatoes can deliver a dose of vaccine; scientists have increased tomatoes' content of lycopene, which seems to prevent some types of cancer, and have introduced genes that produce beta-carotene, the precursor of vitamin A, into rice grains. A gene from *E. coli* bacteria allows rice to withstand drought, saltwater, and cold. Even a decaffeinated coffee bean has been grown.

The application of biotechnology to livestock itself is yielding hardier and more productive livestock.

Critical THINKING

Continuing Crop Redistribution

Different peoples still harvest different indigenous crops and enjoy different diets, and food crops continue

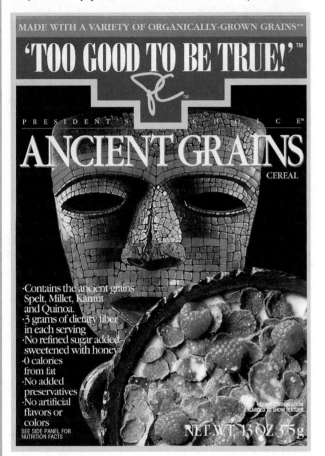

MADE WITH A VARIETY OF ORGANICALLY-GROWN GRAINS**

'TOO GOOD TO BE TRUE!'™

P R E S I D E N T' S   C H O I C E™

ANCIENT GRAINS

CEREAL

·Contains the ancient grains
Spelt, Millet, Kamut
and Quinoa.
·3 grams of dietary fiber
in each serving
·No refined sugar added—
sweetened with honey
·0 calories
from fat
·No added
preservatives
·No artificial
flavors or
colors
SEE SIDE PANEL FOR
NUTRITION FACTS

SERVING SUGGESTION
ENLARGED TO SHOW TEXTURE

NET WT. 13 OZ 55g

Quinoa is gradually being introduced into the North American diet. (Courtesy of Loblaw Brands Limited)

to migrate around the world. *Chenopodium quinoa,* for example, a member of the genus goosefoot, is a native of South America. It is an annual, broadleaved herb usually standing about 3 to 6 feet high. Leafy flower clusters rise from the top of the plant. The dry, seedlike fruit is about 2 millimeters in diameter and enclosed in a hard, shiny, four-layered fruit wall. Quinoa seeds are flat and pale yellow, and they can be steamed, ground into flour, or fermented to make a mildly intoxicating beverage. The leaves are highly edible and often used for livestock feed. Quinoa is sometimes called Inka rice.

Of all the world's grains, quinoa is the highest in protein content, and it has an amino-acid profile that parallels the ideal standard set by the FAO. New varieties of this plant are being bred that are better suited to North American conditions, and quinoa has recently been appearing in North American food products.

Questions

1. Can you think of the first time you tried some new food? How about a kiwi or a Japanese buckwheat noodle?
2. Find exotic foods in local markets. Why were these first imported to your local store? Were they brought for a local immigrant population?
3. New foods from around the world are introduced almost daily into heterogeneous urban centers. How has the change in the source areas of immigrants to North America discussed in Chapter 5 caused the introduction of new foods?

Biotechnologists are even improving animal feed. One product helps animals absorb phosphorous better, thus reducing the ecologically damaging phosphorous in animal waste. Cloning first achieved publicity with the birth of Dolly, a sheep, in 1997 (Figure 8-30). Dolly was the first mammal cloned from an adult cell, but since then mice, goats, monkeys, cats, horses, and rabbits have followed.

Resistance to Biotechnology

The U.S. National Academy of Sciences has found no evidence that GM foods are unsafe to eat, but the Academy does insist that rigorous monitoring of research must be maintained. History demonstrates that

technological solutions to problems can trigger unexpected new problems, so some people fear that GM products may have unforeseen effects in nature (outcompeting and destroying natural species, for example) or on our own bodies.

In 2000 a GM corn that had been approved for cattle but not for human consumption found its way into commercially available taco shells, and the anecdote opening this chapter mentions the GM rice found in 2006. U.S. officals insisted that neither discovery threatened human health or the environment, but the discoveries triggered fear of unknown and unintended side effects.

The cloning of Dolly reminded some people of the novel *Frankenstein,* by Mary Shelley (1797–1851). In

Focus ON

Good-Bye to the Banana?

Without a miracle of genetic engineering, bananas may be extinct within ten years. The banana is one of our oldest crops, the first edible variety having been propagated about 10,000 years ago from a rare mutant of the wild banana, which, with a mass of hard seeds, is virtually inedible. All edible bananas are genetically decrepit sterile mutants—effectively clones of that first plant. Bananas are, therefore, unable to evolve to fight off new diseases.

Today we are seeing the global spread of black sigatoka, a fungal disease that cuts yields up to 75 percent and reduces the reproductive lives of banana plants from 30 years to only two or three. The fungus reduced yields in Uganda 40 percent in one year, and it continues to spread across Brazil.

Genetic engineering may be the only way to save this popular fruit. Scientists hope to sequence the genetic blueprint of the banana by 2008, focusing on the inedible wild banana, which remains resistant to black sigatoka. Some producers fear that consumers will not accept a GM banana, but some 500 million people in Asia and Africa depend on bananas for up to one-half of their daily calories.

FIGURE 8-30 Dolly, the cloned sheep. On February 22, 1997, Scottish embryologist Ian Wilmut stunned the world by announcing that he had created an exact copy—a clone—of an adult Dorset sheep. The historic lamb, created from DNA extracted from a sheep's mammary gland, was named in honor of Dolly Parton. Dolly died February 14, 2003, of a lung infection. (Remi Benali/Stephen Ferry/Getty Images, Inc.—Liaison)

this classic, Dr. Frankenstein's confidence in science leads him to create a living creature that he hopes he can control but that turns on him and kills him. Some critics refer to GM products as *Frankenfood.* The idea survives in our imagination that science can create "monster problems" more serious than the problems that science solves. In 2001 scientists in Australia did, inadvertently, create a modified mousepox virus that destroys the animal's immune system. This scientific breakthrough may lead to a better understanding of AIDS, but it could also inspire new developments in biowarfare. Cloned animals have often exhibited severe and unpredictable health problems, including lung, heart, and immune system defects.

In 2005 Swiss voters supported a five-year-ban on the farming of genetically modified crops, and some food manufacturers, supermarkets, and fast-food chains in several countries have stopped stocking food with GM ingredients. Opponents of GM foods have orchestrated demonstrations and even destroyed fields of experimental crops. Europeans insist that GM foods be labeled, but Americans oppose labeling as unnecessarily frightening. There are, however, no clear definitions of what GM means. For example, are hogs that have been fed GM corn necessarily GM hogs? Lack of agreement means that contradictions in distribution abound: Frito-Lay will not use GM corn in its corn chips, yet its parent corporation, Pepsi-Cola, uses syrup from GM corn in its soda. McDonald's bans GM potatoes, but the company cooks potato fries in vegetable oil made from GM corn and soybeans. Some American opponents of GM foods have begun labeling things that are not GM as "natural" or "organic." To clarify labeling, in 2002 the United States defined standards for "organic" production and processing. Canada has lost potential international food sales by being slow to do so.

Delegates from 130 nations met in Montreal, Canada, in 2000 and adopted a treaty regulating trade in GM products. The Cartagena Protocol (after the city in Colombia where the talks started) is an outgrowth of the Convention on Biological Diversity forged in Rio de Janeiro, Brazil, in 1992. The new treaty went into effect in 2003, having been ratified by the requisite 50 signers. The treaty allows countries to bar imports of genetically altered seeds, microbes, animals, and crops if they deem the product is a threat to their environment. Labeling is required on products only in international shipments, not when the product is on store shelves.

Exporters must obtain permission in advance from an importing country before the first shipment of a particular "living modified organism" is meant for release into the environment—like seeds, microbes, or fish that are to be put into a river. Advance notice and permission are not required for exports of commodities meant for eating or processing. The United States has never ratified the 1992 Biodiversity Convention, so it cannot be a party to the Cartagena Protocol. It will, nevertheless, have to abide by the terms of this treaty when exporting products.

To some degree, popular resistance to biotechnology is not entirely a technological question but also a religious one. Many people feel that nature is immutable or that tampering with it is sacrilegious. Prince Charles of the United Kingdom, for example, has said he will never eat any GM food. "That takes mankind into realms that belong to God, and to God alone," he insists. Monsignor Elio Sgreccia, however, president of the Vatican Bioethics Institute, has said, "We are open to the use of genetic technology in agriculture and with animals, as long as we don't do it with man. We believe that man has a primacy on this planet.... Nature is here for him."

Regardless of whatever happens in the rich countries, many of the developing countries are betting on GM foods to increase their agricultural yields and food supplies. China, for example, is second only to the United States in biotechnology research, and it is forging ahead with productive and pest- and disease-resistant strains of rice, cotton, tomatoes, tobacco, sweet peppers, green peppers, among other types of crops. China, India, and Indonesia are among the leaders in increasing plantings.

Concerns about the application of biotechnology in agriculture are greatly magnified when biotechnology is applied to humans themselves, as in the controversy surrounding research with DNA, human embryos, and stem cells. Therapeutic applications offer great promise, but biotechnology undoubtedly will be used soon for physical enhancement. Augmented athletic prowess, for example, will almost certainly be demonstrated in sports contests in the near future. Some people may find these possibilities literally monstrous.

Other social obstacles remain in the path of technological advances in agriculture. For example, biotechnology is expensive. Will all farmers have access to this technology, or only those with sufficient capital? Some observers worry that the relatively few large corporations that own the rights to enhanced seeds and other advances will exercise increasing economic control over agriculture.

Another threat seems like an ironic paradox: If biotechnology truly unleashes the productivity the optimists foresee, then just a fraction of the farmers on Earth today will be able to feed everyone. Millions of farmers—hundreds of millions—will be unemployed. What will they do for a living? How will humanity be kept busy? Multiplying our ability to produce food threatens to impoverish billions of people, to launch even greater migration to cities that are underequipped to absorb the influx, and thus to trigger civil unrest on scales never before contemplated in history—as well as, perhaps, to worsen the health problems associated with obesity outlined in Chapter 5. Will our political and economic systems be able to transform human societies fast enough to avoid chaos? Chapter 10 examines the pressures on cities already caused by the migration of farmers from the countryside, and Chapter 12 examines possibilities in economic development.

Global Warming

Forecasts of changes resulting from global warming might modify our optimistic scenarios of continuing increases in food production. The FAO has estimated that by the year 2080, the world's poorest 40 countries (determined as of 2000) could lose 20 percent of their ability to raise food because of global warming. Food-growing capability would also lessen in the United States, France, Romania, Hungary, Belgium, the United Kingdom, the Netherlands, the Czech Republic, the Ukraine, all of Central America and the Caribbean region, and India. In Canada, by contrast, agricultural production could double. Other "winners" would be Finland, Norway, New Zealand, Russia, and China.

WORLD DISTRIBUTION OF FOOD SUPPLIES AND PRODUCTION

In 1968 Professor Paul Ehrlich of Stanford University wrote in his bestselling book *The Population Bomb*: "The battle to feed humanity is already lost.... We will not be able to prevent large-scale famines in the next decade." Local famines did occur, but they were not caused by a lack of food on Earth. World excess food stocks relative to consumption have steadily outpaced population growth. Nevertheless, the debate between the pessimists and the optimists, the neo-Malthusians and the Cornucopians, continues.

Figure 8-31 maps the prevalence of undernourishment around the world. A balanced diet includes carbohydrates (derived from staples such as rice, corn, wheat, and potatoes), proteins (from meat, fish, poultry, eggs, dairy products, or pulses), vitamins (from fruits and vegetables, as well as other sources), fats, and minerals. The map suggests that, overall, the food supplies in some countries fall short of per capita needs. Many people in those countries are well fed, but many others are hungry. Even in the countries in which overall supplies are adequate, however, some people are hungry.

No country is completely self-sufficient in food. Most countries both import and export food, and a few

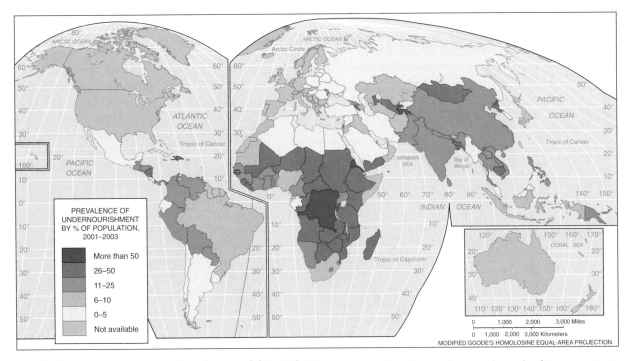

FIGURE 8-31 Prevalence of undernourishment. The amount of food needed per capita varies from country to country, depending on the age distribution of the population, activity levels, health, climate, and other factors. Nevertheless, this map reveals that in many countries, substantial percentages of the populations are undernourished. In virtually all places where data are unavailable, undernourishment afflicts only a very small percentage of the population.

countries are net exporters of food despite the fact that portions of their own populations are undernourished. This imbalance may be due to injustice or civil strife. Political instability contributes to hunger. Several African countries, for example, are environmentally richly endowed, yet a great many of their people go hungry. Food-supply problems are often problems of economics or politics, not problems of geography or technology. In poor countries, rich people have plenty to eat, while poor people are starving. An increase in wealth that allows rich people to eat more meat without improving the standard of living of the poor only increases political instability. For everyone on Earth to be adequately fed today as well as in the future as Earth's population continues to rise, we need both an increase in production and improvements in distribution; these must occur both within countries and among countries.

Peter Rosset, director of the Institute for Food and Development Policy, wrote in 1999, "There is no relationship between the prevalence of hunger in a given country and its population. . . . The world today produces more food per inhabitant than ever before. Enough is available to provide 4.3 pounds to every person every day: two and a half pounds of grain, beans and nuts, about a pound of meat, milk and eggs, and another of fruits and vegetables—more than anyone could ever eat. The real problems are poverty and inequality. Too many people are too poor to buy the food that is available or lack land on which to grow it themselves." These are the problems we must examine in the following pages.

Problems in Increasing Food Production

Technology creates the potential for increasing food production, but it cannot guarantee that this will occur or that the food will be produced or distributed where it is needed. Other potential problems in the effort to increase food supplies include diminishing returns to fertilizer, lack of financial incentives for farmers, and inequitable land ownership or unsuccessful land redistribution.

Fertilizers: Overuse and diminishing returns

Fertilizers can improve crop yields significantly when used in appropriate amounts. For many years, however, the quantities of fertilizer applied in advanced countries have far exceeded the point of **diminishing returns.** Diminishing returns exist when, in adding equal amounts of one factor of production, such as fertilizer or labor, each successive application yields a smaller increase in production than the application preceding it. Table 8-2 applies this concept to a hypothetical example of fertilizer. In this table the point of diminishing returns is 4 pounds of fertilizer. This is true because adding the second and third pounds of fertilizer increased the yield by 11 bushels, but adding the fourth increases the yield by only 10 bushels. At some point, then, the diminishing increases in crop yields no longer justify the application of greater amounts of fertilizer at one place. That fertilizer could more efficiently be applied in another place that has not reached the point of diminishing returns.

TABLE 8-2	Diminishing Returns in the Addition of Fertilizer	
FERTILIZER (POUNDS)	TOTAL YIELD (BUSHELS)	PRODUCTION INCREASE (BUSHELS)
1	10	10
2	21	11
3	32	11
4	42	10
5	51	9
6	59	8

The United States uses about 110 kilograms per hectare (20 pounds per acre) of cropland. If each U.S. farmer's last 5-kilogram bag of fertilizer were given to a farmer in a poor country, the world food supply would increase dramatically. In other words, the same fertilizer that yields only diminishing returns in the United States could be used much more productively in a poor country.

Improvements being made in the techniques of fertilization, however, offer the promise of improving the efficiency of its application. Computers with global positioning systems (GPSs) installed in the cabs of farmers' tractors memorize the coordinates of each field and its characteristics, allowing farmers to target the amount of fertilizer and pesticide appropriate to each small subsection of a field (Figure 8-32). This precision reduces the error of applying more than the optimal amount, thus reducing pollution.

Many farmers lack financial incentives

Chapter 6 referred to the agronomists' dictum: "The market produces the surplus." In other words, if a market exists for any surplus that farmers can produce, and if the political and economic system sufficiently rewards farmers for their effort, production of a surplus is all but assured (assuming adequate technology is available).

In many countries, however, farmers are so heavily taxed that they have little incentive to increase food production, so they remain at the subsistence level. Many governments force farmers to sell their crops to the government at fixed low prices, charge high taxes on food exports, support high exchange rates for their national currencies, and impose high import duties on the tools and agricultural chemicals the farmers need. These are all ways of taxing farming. The wealth that is squeezed from the farmers in these ways is often used to subsidize urban food supplies or to support urban civil service bureaucracies, military spending, or unprofitable state-run industries. In some countries, deteriorating conditions in the countryside trigger migration to the cities (a topic we shall examine in Chapter 10), and swelling urban populations increase the demand for subsidized food. When recent political changes have given farmers incentives to produce more food, the increases in production have been astonishing. India lifted some agricultural price controls in 1991, and since that year grain production has risen about 4 percent annually. This rate is more than twice the rate of population increase, so per capita food supplies have improved markedly. When farmers enjoy higher profits, more rural jobs are created both in agriculture and in rural craft industries.

Systems of land ownership can discourage production

In many countries, systems of land ownership reflect broad political and social inequalities. In Brazil, for example, about half of the country's arable territory is held by only three percent of all landowners. This concentration of land ownership spurs the landless to attempt to farm new claims in the rain forest regions, which degrades the environment while producing only disappointing yields. The government has announced land redistribution schemes, but it has been slow to act. Furthermore, to be successful, newly landed peasants require infrastructure (such as roads), technical support, and even assistance marketing and financing their production. Hundreds of landless peasants have died in clashes with armed landowners and police in Brazil in recent years, but, at the same time, the modern, highly capitalized sector of Brazilian agriculture achieved great success increasing exports of sugar, soybeans, and beef.

In many countries the landholding system is a holdover from its colonial period. Colonial governments replaced communal systems with systems of private property, and the distribution of the property was highly inequitable. This created one class of large estate

FIGURE 8-32 A GPS monitor in a tractor cab. GPS satellite navigation plus GIS field mapping and data analysis allow farmers to distribute seeds and apply fertilizers and pesticides mixed specifically for each square meter of a large field as the tractor drives across it. (Courtesy of Ag-Chem Equipment Co., Inc.)

holders and another, much larger, class of landless poor. The large landowners often moved to the cities, becoming absentee landlords unwilling to make the investments necessary to maximize the return from their own agricultural holdings.

In the Philippines, for example, large blocks of land were granted to the Spanish colonial elite, and the natives on these land grants became tenant farmers. After a few generations the landlords were mostly mestizos (persons of mixed native and colonial ancestry), and their descendants owned the best agricultural land. This land-grant system prevailed in all areas colonized by Spain, and it is at least partly responsible for political unrest in Central and South America today.

Indonesia presents a contrast to the former Spanish colonies. For more than 300 years under Dutch rule, only native Indonesians could own land, except for city lots. Dutch, Arabs, Chinese, and English could rent agricultural land, but they could not acquire title to it. Thus, land ownership among Indonesian peasants remained much more widespread than it was in the nearby Spanish Philippines. Indonesian peasants who have inherited their land and expect to leave it to their children are much less likely to join peasant guerrilla movements than are landless Filipinos. In addition, individual farmers of small personal holdings are likely to maximize their productivity.

When land is leased to those who actually work it, the conditions of tenancy and of payment determine whether the farmers are encouraged to produce significant surpluses. If the landlord takes too large a share, surpluses will be small.

Communist regimes in Eastern Europe collectivized agriculture in large state-owned farms and in cooperative enterprises. Neither provided sufficient incentive to encourage farmers to maximize productivity. Since the fall of Communism, however, private farming is being encouraged. Giving or selling government assets to private individuals or investors is called **privatizing** an activity. Small private market gardens were always allowed in the former Soviet Union, and although these accounted for less than 2 percent of total Soviet farmland, they yielded 60 percent of the country's potato crop and approximately 30 percent of its vegetables, meat, eggs, and milk. Today the government of Russia is privatizing agriculture. Poland ended food subsidies and restrictions on direct marketing in 1989; farmers' markets sprang up overnight, food shortages eased, and prices fell. The other formerly Communist Eastern European countries are privatizing the large communal landholdings. Vietnam ended collectivized agriculture in 1990, and the country swiftly became able to export rice.

China has experimented with several landholding systems in the past 50 years. When Communists came to power in China in 1949, they seized land from landowners (killing thousands of them in the process) and redistributed it to the peasants, but in the 1950s the peasants' holdings were collectivized. In 1978 China relaxed its state-directed agricultural system. The collectives were dismantled, and peasants were allowed decision-making power and profit incentives. Within 10 years farm output rose 138 percent, and China turned from a net importer of food products into a net exporter. The peasants were granted 30-year leases, but not ownership, so they still could not sell it, rent it out, or borrow money to invest in improving productivity. These restrictions on peasant land rights slowed rural development and contributed to a growing gap between rural and urban incomes. To remedy this problem, China has recently abolished a wide range of fees and even the ancient land tax.

Privatizing of industry and of other aspects of national economies is proceeding around the world; we will examine this important development in more detail in Chapter 12.

Agricultural productivity can also be hampered when title to land—legal ownership—is unclear, as is the case in many countries. Generations of peasants may have tilled small plots, but their families have never registered title, and bureaucratic complications make it difficult to establish title now. It has been estimated that in poor countries only about 20 percent of all land has clear title. Lack of clear title discourages the occupants from making improvements on the land. Furthermore, as we have seen, successful farming today requires capital investment in irrigation pumps, chemicals, improved seeds, and machinery. Farmers can borrow money to make these investments only if the land can serve as *collateral* (a thing of value that can be seized by a creditor in case of nonpayment of a debt). Farmers without clear title to the land lack collateral. The satellite-based GPS and GIS have greatly reduced the cost of land registration, and new programs have been launched around the world. Establishing farmers' title to land triggers investment and increases productivity.

Unsuccessful land redistribution: Mexico

Mexico demonstrates how communal land ownership can restrict productivity because communal holdings cannot serve as collateral. More than one-half of the country's arable land is held in *ejidos,* a form of land tenure in which a peasant community collectively owns a piece of land and the natural resources and houses on it (Figure 8-33). Mexican law states that *ejidos* are "inalienable, nontransferable and nonattachable." They cannot be used as collateral, so banks will not extend loans. This system, combined with a government tradition of paying farmers low prices for their crops while at the same time subsidizing food for urban consumers, has kept farmers poor and productivity low. These conditions in turn intensify the migration of peasants to the cities. Mexico imports 35 percent of its total food supplies, including nearly half of its staples of corn, wheat, and beans, from the United States.

FIGURE 8-33 A Mexican *ejido*. All members of a Mexican *ejido*, or farming collective, gather to make decisions regarding planting, harvesting, and marketing their crops. *Ejidos* represent the traditional Mexican form of communal landholding. The land is inalienable, so the *ejido* cannot borrow money for investment, and thus productivity is low. (Phil Schermeister/CORBIS-NY)

The government is struggling to overhaul the *ejido* system, but official recognition of *ejido* ownership was a principal issue of the Mexican Revolution (1910–1920), in which more than 1 million Mexicans died. Many Mexicans remain attached to the concept of the *ejido* system. In 1992 the government introduced a program of certifying individual land rights, but by 2000 scarcely 1 percent of *ejido* land had been sold. Since the NAFTA trade pact between the United States and Mexico was launched in 1994 (see Chapter 13), Mexican farmers have found it difficult to compete with U.S. farmers. The Mexican government has ended price guarantees without helping farmers market their produce, and it costs about twice as much to get a ton of wheat to Mexico City from Mexicali, near Mexico's northwest corner, as it does to get a ton of wheat to Mexico City from Kansas. Some Mexican farmers are switching to new commercial crops, or they are contracting to agribusinesses. Agribusinesses provide the needed capital investment, as well as an assured market. The Mexican government has also experimented with guaranteeing loans made to the *ejidos*, thus encouraging capital investment and improving agricultural productivity.

Barriers to increasing production in Africa

Many of the problems found in Mexico's agricultural economy can also be found in Africa. Landholding is often communal, so successful farmers cannot expand their holdings or borrow money to invest in greater productivity. Lack of investment in fertilizer has caused widespread soil depletion. Many African governments themselves hold ownership of agricultural land and lease it to farmers. In Zimbabwe, for example, the government nationalized numerous large white-owned private farms that were exporting food. The government relocated black settlers onto the properties but did not transfer ownership, so the farmers cannot borrow to invest in increasing productivity. Productivity has fallen, and Zimbabwe has lost export income.

In addition, farmers in many countries are required to sell their crops to state marketing boards. These boards often pay the farmers very little. In Nigeria, for example, the government owns almost all land, leases it to farmers, and then pays the farmers fixed low prices for their crops. Many African state marketing boards have proved so corrupt that the farmers' returns for their work are virtually stolen from them. Some governments also have failed to invest in a rural infrastructure or in agricultural research to develop technologies or native crops suited to local conditions.

These conditions discourage farmers and reduce agricultural output. Many farmers migrate to the cities for work, but because land tenure is based on occupancy, they must leave their families in the villages. In some parts of rural Zimbabwe, 40 percent of the households have lost the father to the town. The overworked women, left to tend children as well as crops, can scarcely rise above subsistence. The United Nations Children's Fund (UNICEF) estimates that women grow 80 percent of Africa's food. AIDS has also taken a terrible toll on African farming, killing almost 10 million farmers by mid-2006, thus greatly reducing food output.

Civil wars are still another factor that has reduced food production in some areas of the world, again particularly in Africa. The continent was feeding itself in 1960, but today Africa imports about 40 percent of its food supply.

Summing up, the reasons for inadequate food production in many countries are not necessarily environmental or even technological. Food production could be greatly increased if many governments could end their civil wars, change national economic policies that discriminate against farmers while subsidizing urban food supplies, stop discriminating against women, reduce import taxes or quotas on imported tools and chemicals, improve infrastructure (particularly roads and water supplies), and reform landholding systems. All of these factors influence food production.

Commerical crops in the poor countries

Today many farmers in poor countries do not concentrate on growing staple crops for local consumption. Instead, they raise cash crops for sale. Senegal, for example, exports peanuts, the Ivory Coast exports cocoa, Angola coffee, Zimbabwe tobacco, and Kenya tea and flowers, even though food availability in each of these

FIGURE 8-34 Flowers as a cash crop. Some 30 percent of Kenyans suffer undernourishment, yet more than 330 tons of flowers and fresh vegetables are air-freighted out of Kenya to Europe every day. The industry employs some 90,000 Kenyans and earns Kenya about $500 million per year, an amount that exceeds the money earned today from tourism and coffee, which were previously principal exports. (Francesco Broli/The *New York Times*)

countries is low (Figure 8-34). If any country chooses to dedicate its efforts to specialty crops, it can buy basic foodstuffs on world markets. In that way, the country can achieve what is called *economic self-sufficiency* in food. Farmers who choose to depend on cash crops, however, face two risks: the price for their product could collapse because of overproduction elsewhere, or prices for the foods they will have to buy could rise faster than their own incomes.

Raising cash crops may increase farmers' disposable incomes rapidly, while maintaining food production. Guatemalan farmers raising snow peas for New Yorkers, for example, can make 14 times as much per acre as they would raising corn. Even after taking into account the increase in their costs, their return for each day of work more than doubles. A great share of the increased costs are for hired labor, so more people find work.

The rise in income may not, however, improve the lives and health of farm families. Studies have revealed that in many poor countries, women control the production of food crops, but men control the production of cash crops. The men often spend "their" money on things such as alcohol and tobacco, which does not increase family welfare.

Several Caribbean countries have found it profitable to surrender self-sufficiency in food and to dedicate their cropland to grass—that is, to fairways for golf courses that attract tourists. High earnings from tourism allow the countries to buy food on world markets. The role of tourism in national economies will be examined in Chapter 12. Unfortunately, the world's most successful cash crop remains the one most governments try hardest to discourage—illegal drugs (Figure 8-35).

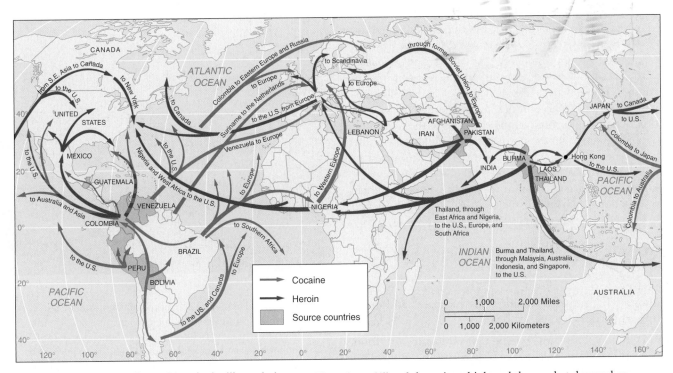

FIGURE 8-35 International trade in illegal drugs. The value of illegal drugs is so high and the market demand so strong that international trade flourishes despite international police efforts to stop it. The flow of opium from Afghanistan has increased since the U.S.-led war ousted the fundamentalist Taliban from rule there (see Chapter 13). (Information from the U.S. Drug Enforcement Agency)

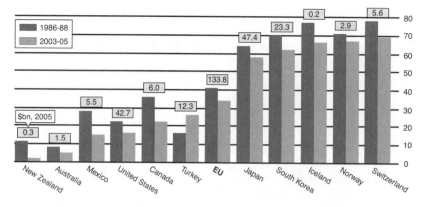

Purple lines: farm subsidies as a percent of total farm income

FIGURE 8-36 Farm subsidies. Farmers in the rich countries are subsidized both by direct government payments and by consumers, who pay more than world prices for their food. (Data from Organization for Economic Cooperation and Development)

The Rich Countries Subsidize Production and Export of Food

Many rich countries have erected figurative walls of high *tariffs* (taxes on imported goods) to protect their markets against food imports from poor countries. This requires the urban consumers in the rich countries to pay inflated prices for food. These inflated prices are indirect subsidies to the rich countries' farmers. Many rich countries also give their farmers direct subsidy payments. In 2005 the European Union spent $134 billion dollars in agricultural price supports, the United States about $43 billion, and Japan about $50 billion (Figure 8-36). As a result, farmers in many rich countries produce food surpluses, which the countries then export or sell on world markets at prices below the costs of production (a practice called *dumping*).

These subsidies distort the world geography of agriculture. They encourage production of agricultural surpluses in many rich countries and discourage increasing production in the poor countries. As long as the rich countries sell their surpluses on world markets at low prices, the governments of many poor countries neglect to invest in their own rural areas. Farmers in the poor countries cannot export to the rich countries or even compete in their own national markets so they remain at the subsistence level, or they are driven off their farms entirely.

Why Do Some Rich Countries Subsidize Agriculture?

Many rich countries subsidize their farmers and erect import barriers for several reasons:

1. Some countries pursue self-sufficiency in food production as a national security measure. The governments want some defense against grain embargoes and crop failures in those countries that normally have surpluses to market. Subsidizing a country's own farmers offers some protection

against these threats. Following U.S. threats to halt grain shipments to the Mideast in the 1970s, for example, Saudi Arabia spent heavily to improve its agriculture. Saudis give farmers subsidies to raise grain, import irrigation equipment, and even raise dairy cattle in air-conditioned sheds where outside temperatures exceed 48°C (120°F) (Figure 8-37).

FIGURE 8-37 Food for national security. The desert in the background is a typical landscape of Saudi Arabia, but Saudi Arabia has invested some of its oil income in irrigating the desert to achieve agricultural self-sufficiency. Large capital investment in agriculture allows Saudi Arabia to achieve such high productivity that it occasionally dumps surplus food on world markets. This bankrupts farmers in other countries with environments more naturally suited to agriculture. Thus, political and economic considerations can outweigh the role of physical conditions for agriculture. (Courtesy of Aramco World Magazine)

2. Many rich countries, particularly in Europe, subsidize agriculture to preserve traditional agricultural communities and keep their farmers from migrating to the cities in search of work. Rural–urban migration can overwhelm the ability of cities to absorb the new labor force.

3. Some rich countries subsidize farming as part of national land-use plans. They enjoy agricultural landscapes and want to preserve green areas around cities, called *greenbelts*. Their urban citizens are willing to subsidize farms in order to preserve them.

4. Many rich countries' political systems favor farmers. The farmers in rich countries constitute a decreasing percentage of the national populations, yet, particularly in Europe and Japan, the electoral districts have never been redrawn to reflect the relative depopulation of the countryside and the urbanization of the populations. The U.S. Senate disproportionately favors farm-state populations, as we shall see in Chapter 11. Thus, farmers enjoy disproportionate political power over urbanites. The contrast is sharp between the rich countries, in which the national governments generally drain wealth from the cities to support farming, and the poor countries, in which governments often heavily tax farmers and subsidize the urban populations.

5. Furthermore, in the rich countries only a small percentage of what urban consumers pay for food actually goes to farmers; most of the cost represents value added by processing and packaging. In 1980 the cost of farm commodities in the United States—the raw material—represented 37 percent of the price of food; by 2000 that had fallen to 19 percent. Food processors and middlemen, including marketing and advertising, take an ever-larger share. For example, the cost of the corn in a box of cornflakes is only about 3 percent of the retail cost of the item; the box is worth more. The rest of the retail price represents processing, packaging, distributing, and advertising the cereal. Advertising rates affect the retail price of cornflakes more than corn prices do. Therefore, most of the increase in urban food costs is attributable to these activities, not to farmers' profits. Most urban consumers remain unaware of how much the prices they pay for food are inflated to subsidize their nation's farmers. They may also be largely indifferent because the cost of food as a percentage of urban incomes in most rich countries is falling anyway, due to increases in agricultural productivity.

The American diet has actually become so full of rich and processed foods that the country has experienced periodic trade deficits in food products since 1998. That is, the United States has been importing more food—by value—than it has been exporting. This is astounding for a country so rich in agricultural resources. The reason for this is that America exports basic foodstuffs (wheat, for example), but Americans are importing food products of ever-higher value added (notably wines and out-of-season fruits and vegetables).

A myth exists that agricultural subsidies in rich countries go to struggling small family farms. In fact, the bulk of subsidies go to large multinational corporations (such as Nestlé, the world's largest food company) and to wealthy or well-connected individuals (such as members of the French Senate and even families of U.S. congressmen). In America, the top 10 percent of subsidized crop producers collected 72 percent of all subsidies from 1994–2004.

The rich countries periodically negotiate to end their subsidies, but they cannot even agree on what constitutes a subsidy. The United States heavily subsidizes water for California farmers but denies that this support is a "farm subsidy." Throughout California's Central Valley, sprinklers irrigate fields of alfalfa, cotton, and rice—crops more suited to wetlands than to a desert—and more than 50 percent of federally irrigated land in California is devoted to crops that are already in surplus. The issue of farm subsidies is complicated, and in each country it is a volatile domestic political issue. The overall result, however, is to increase food surplus in the rich countries and to discourage food production in the poor countries. If the rich countries stopped subsidizing their farmers and opened their markets to crops from developing countries, the farmers of the poor countries might be able to sell more food to the rich countries. This action would benefit the poor countries more than direct foreign aid does.

On July 11, 2003, when U.S. President George W. Bush was touring Africa, Amadou Toumani Touré and Blaise Compaoré, the presidents, respectively, of the African countries of Mali and Burkino Faso, published a joint plea to the American people. "Your farm subsidies," they wrote, "are strangling us." Rich countries' subsides to their own cotton producers totaled almost $6 billion in the production year 2002, "distorting cotton prices and depriving poor African countries" of their own ability to market this crop. During that year, they noted, "America's 25,000 cotton farmers received more in subsidies—some $3 billion—than the entire economic output of Burkino Faso. . . . Further, United States subsidies are concentrated on just 10 percent of its cotton farmers. Thus, the payments to about 2,500 relatively well-off farmers has the unintended but nevertheless real effect of impoverishing some 10 million rural poor people in West and Central Africa." Powerful U.S. congressional interests, however, continue to favor support for U.S. cotton growers, and subsidies actually rose to $4.2 billion in the United States alone in 2005. In that year, the subsidies were ruled illegal under international treaties, but the United States has not changed them. Leaders of many of the world's poor countries have cited the rich countries' agricultural protectionist policies as the reason why the poor countries have balked at progress in international trade agreements (see Chapter 12).

Focus ON

The U.S. Federal Agricultural Improvement and Reform Act of 1996

In 1996 the U.S. federal government attempted to reform and reduce U.S. agricultural subsidies, but the failure of the effort demonstrates how political and economic considerations can overwhelm intentions.

The 1996 act ended a farm policy in which farmers voluntarily accepted certain government restrictions on farm practices in exchange for subsidies for their crops. These programs began as efforts to ensure farmers a living during the Depression in the 1930s. If prices for certain crops dropped below fixed levels, the government made up the difference through deficiency payments. Farmers who agreed to keep land idle and to limit the production of crops could receive government payments. The programs stayed in place through the years, and government payouts for crop subsidies averaged $5.5 billion per year between 1990 and 1996.

The subsidies even redistributed the geography of U.S. farming. Subsidies to the dairy industry, for example, were higher the farther the dairy farm was from Wisconsin, which is the environment most appropriate for dairy farming and the traditional "Dairy State." The intent was to ensure a supply of fresh milk for children everywhere. The subsidies rewarded dairy farming in warm Florida and California. Since the invention of air conditioning for cowsheds, California even has become America's leading dairy producer.

Many of the rules and regulations had desirable results, such as encouraging soil conservation and discouraging water pollution. For example, in order to qualify for subsidies, farmers with land classed as "highly erodible" were required to manage that land in ways that would reduce erosion. Highly erodible lands were also targeted in programs that left some cropland idle. Other provisions preserved wildlife habitats. The 1996 legislation continued some of these programs.

By the 1990s, however, critics argued that some government regulations had become rigid and arbitrary.

By guaranteeing generous prices for certain crops while at the same time removing cropland from production, the supports may have encouraged ever-increasing uses of fertilizers and pesticides. Furthermore, by making it difficult to switch from one crop subsidy program to another, the policies discouraged experimentation with new crops, new methods, and even crop rotation—a good farming practice. Thus, the 1996 program intended to increase productivity, flexibility, and rewards. Skeptics argued that, instead, it endangered valuable environmental safeguards and that the elimination of subsidies would accelerate the trend toward capital-intensive industrial forms, thus eliminating small family-based farming.

The 1996 law covered most major grains but left in place rules to force American consumers to pay prices for tobacco, peanuts, and sugar above international market prices. These three farming interest groups have enormous political power.

The 1996 legislation did not, however, accomplish what its designers had hoped. Prices fell, and government payments to U.S. farmers rose to $23 billion in 1999. Some of the payments provided disaster relief, but huge harvests were a larger factor in the payments. Farmers still get paid for every bushel produced.

In 2002 Congress passed a new bill that guaranteed agricultural subsidies of more than $200 billion over a ten-year period, retained generous subsidies for cotton, rice, wheat, corn, and soybeans, and added price guarantees for lentils, chickpeas, ginseng, onions, apples, and even catfish. Other provisions seek to preserve farmlands, save wetlands, and improve water quality and soil conservation on working farms. Thus, the new legislation retained farmers' freedom to make planting decisions, but abandoned efforts to wean American farmers from subsidies; the "revolution" of 1996 was defeated. In an additional effort to protect American producers from rising imports, the law ordered all meat, fish, and produce to be labeled with its country of origin. Agricultural subsides in 2005 reached $43 billion, some 17 percent of total farm income.

Our study of world food production and availability illustrates many ways that the world's agricultural economy is manipulated and distorted. Thus, we cannot know what quantities of food the world's farmers would be capable of producing if these conditions were eliminated, or what the geography of food production would be.

THE HARVEST OF FISH

Seafood represents about 6 percent of the total human protein intake. Humans harvest four main groups of marine species. *Demersal fish* (bottom dwellers), including cod, haddock, sole, and plaice, tend to concentrate on broad continental shelves, especially of the North

Atlantic. *Pelagic fish* (surface dwellers) include herring, mackerel, anchovy, tuna, and salmon. These two groups together make up about 72 million tons of catch per year. *Crustaceans,* such as lobsters, shrimp, and other shellfish, provide more than 11 million tons. And *cephalopods,* including octopus and squid, yield about 2 to 3 million tons. The Japanese harvest more cephalopods than any other nation, but cephalopods also are important in the diets of Mediterranean peoples and many poor countries. In 2003, world *capture fishery production,* that is, basically, hunting and gathering seafood, totaled 90 million tons.

Traditional Fishing

Traditional, or artisanal fishing, may be distinguished from modern fishing. Traditional fishing is concentrated mainly in the developing countries, particularly in Asia and Oceania. Little capital is invested, and incomes are low, but these activities employ about 80 percent of the world's fishermen. Traditional fishing is important in the lives of the majority of the populations of islands and coastal areas, and it directly supports an estimated 30 million people.

Fishing is frequently a dangerous occupation, and it demands skills and local knowledge. Physical risks are great, and so are the risks in terms of income, which can fluctuate greatly. Many traditional fishing areas are characterized by close-knit communities with distinct customs and rituals. Traditional fishermen from Canada to Madagascar to Indonesia share certain hardships

and experiences. Yields of fish from these traditional methods feed substantial numbers of people, but they are only a small fraction of the global catch. This way of life is rapidly being undermined and replaced by modern, highly capitalized fishing.

Modern Fishing

Concentrations of aquatic species suitable for commercial harvesting are called **fisheries.** The world's major commercial marine fisheries are shown in Figure 8-38. Between 1950 and 1989 the annual harvest from ocean fishing rose from 22 million tons per year to 100 million but then it slowly slipped down through the 1990s. In 2003 China, Peru, the United States, Japan, Indonesia, Chile, and Russia accounted for more than half of the total marine catch.

The demand for fish and fish products continues to rise. Even as people eat more meat, they still create a demand for animal feed supplements, including fishmeal. Almost one-third of the global fish catch now goes into meal and oil, mostly to feed livestock and pets in the rich countries of Western Europe and North America. The Japanese, by contrast, want to consume more fish directly.

Overfishing and depletion of the seas It is difficult to regulate the exploitation of a resource that no individual owns because each country or individual fishing vessel harvests as much as it can. This often leads to overexploitation and rapid depletion of the

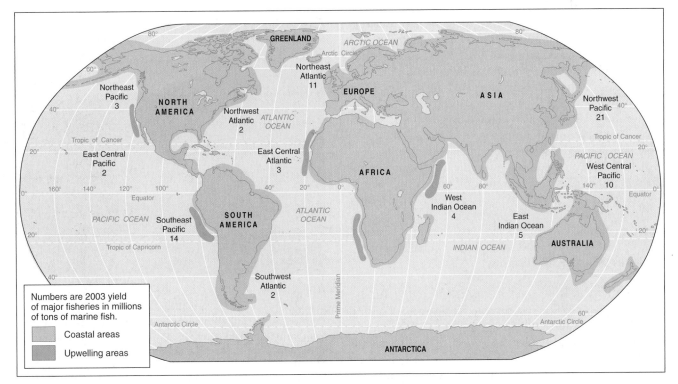

FIGURE 8-38 The world's major fisheries. This map shows each major fisheries' contribution to the total world catch. Many of these are currently fished to their maximum sustainable yield. (Data from the FAO)

resource, a process called *depleting the common*. This has been the case with fish. The United Nations has sponsored periodic conferences on the world's fisheries, and most observers agree that several major fisheries are already fished to their sustainable limits or beyond, and that virtually every fishery is at risk. The major fishing nations have enlarged their fleets and developed new technologies to multiply their harvest: New electronic devices, for example, discover and track wholly new groups of fish. So-called factory ships are equipped with huge drag nets that catch vast numbers of fish and other sea life indiscriminately. In 1999 the world's fishing nations agreed to reduce the size of their fishing fleets, and the World Summit on Sustainable Development held in South Africa in 2002 called for restoring stocks by 2015, but governments have been slow to act.

We do not fully understand fish populations and their migrations. Systematic fisheries research began only in 1902, with the formation of the International Council for Exploration of the Sea, which studies North Atlantic species. Today a great many international fishery commissions and advisory bodies exist. Some only consult, but others set quotas as well as conduct research. Some factors in changing fish populations are caused by human interference, but some are natural. Changes in water temperature and salinity can wipe out certain stocks, especially the small pelagic species such as sardines, anchovies, pilchard, and capelin.

Overfishing is not the only problem. Marine animals rely on coastal wetlands, mangrove swamps, or rivers for spawning grounds, but the world's wetlands and coasts are being destroyed by pollution and overdevelopment. About a third of the world's urban population lives within 60 kilometers (100 miles) of ocean, contributing to the pollution that reaches the seas. Heavy metals have contaminated fish and damaged the health of people who eat them. Sewage, fertilizers, and runoff from agriculture have overfed algae (tiny marine plants), causing them to grow so rapidly that they use up the oxygen that fish need to breathe.

The depletion of atmospheric ozone lets in ultraviolet radiation that harms life in the sea. Ultraviolet radiation may be responsible for the observation that the amount of plankton (microscopic animal and plant life that form the lowest link of the food chain) has decreased by as much as 20 percent near the surface of Antarctic waters, and the surface is where most marine growth and reproduction take place. Global warming might also alter ocean currents.

International law and regulation of fishing

International talks focus on two categories of fish: species like pollock in the Bering Sea and cod off Canada's eastern coast, whose wanderings cause them to straddle territorial and international waters, and fish like tuna, swordfish, and billfish, whose seasonal migrations cover thousands of miles. Coastal states dispute rights to these species with countries that operate long-

range distant-waters fleets. The coastal states claim that factory ships in international waters are destroying the stocks in their territorial waters. Others argue that the crisis stems from coastal countries' mismanagement. Coastal countries want binding rules to regulate catches in international waters and countries that practice distant-water fishing want nonbinding guidelines drawn up regionally.

About 90 percent of the world's marine fish harvest is caught within 370 kilometers (200 nautical miles) of the coasts, and therefore many coastal countries have recently extended their claims out into the sea. The 1982 U.N. Convention on the Law of the Sea authorizes each coastal state to claim a 200-nautical-mile **exclusive economic zone (EEZ),** in which it controls both mining and fishing rights (see Chapter 13). Many countries have already restricted foreign fishing vessels in their waters.

Aquaculture

Global seafood consumption is on the rise, but the catch from the oceans has peaked. Therefore, humankind has begun to shift from fish capture to **aquaculture,** which involves herding or domesticating aquatic animals (Figure 8-39). By 2004 the world yield from aquaculture was 50 million tons, over half as much as from fish capture, and this percentage is rising rapidly. A full two-thirds of the aquaculture yield is in China. Fish farmers can monitor both the purity of the aquatic environment and what the fish are fed. Fish waste is a valuable fertilizer. Aquaculture technology is advancing (computer-controlled feeding is an example), and

FIGURE 8-39 Aquaculture of salmon. These are coastal salmon pens near Chile's salmon-farming center, Puerto Montt. Chile is the second-largest producer of farmed salmon in the world, after Norway. Chile's favorable growing conditions—clean waters, mild climate, long daylight hours, abundant feed, and absence of native salmon species—make it an ideal area to grow salmon. Most of the salmon grown in Chile are exported to the United States and Japan. (Courtesy of the Chilean National Tourist Board)

experiments with different species are revealing which are most productive in aquaculture. Mahimahi, for example, can be grown from fingerlings to 1 to 2 kilograms (2 to 5 pounds) in just five months.

Large-scale experiments have also begun with mariculture, which is fish farming in huge cages in the seas. Aquaculture and mariculture offer promising frontiers for the production of high-protein food.

CONCLUSION: CRITICAL ISSUES FOR THE FUTURE

Technology creates the potential for increasing food production, but it cannot guarantee that this will occur or that the food will be produced or distributed where it is needed. Technology is not spread around the whole world, nor is it distributed in the most economic way. Furthermore, agricultural research and development are occurring overwhelmingly in the private sector—for profit. The fruits of the green revolution emerged almost entirely from public-sector laboratories and national breeding programs, allowing easy access for users. The development of new biotechnologies, however, has come by private companies that have shielded their investments with patents. This may enhance the productivity of rich farmers, but not that of the farmers in poor countries. How can they get this expensive technology? The private corporations that invest in the development of technology cannot give it away. The governments of the rich countries have not indicated willingness to buy the technology in order to provide it to the poor as foreign aid, and the governments of the poor countries are not making significant investments, either. If, then, we rely on technology to boost food supplies, who will pay to transfer the technology to the world's poor farmers?

Chapter Review

SUMMARY

In the 200 years since Thomas Malthus wrote his theory, global starvation has not occurred. This has been because of the farming of more areas of the planet, the transplantation of many food crops, improvements in worldwide transportation and the technology of food storage, the introduction of higher-yielding and hardier strains of crops, new pesticides, and improvements in livestock and farm machinery. These developments are called the green revolution and the scientific revolution in agriculture.

Today ten principal types of agriculture can be identified: four of these are basically subsistence (nomadic herding, low-technology subsistence farming, intensive rice farming, and Asian mixed cereal and pulse farming), and five are basically commercial (mixed farming with livestock, prairie cereals, ranching, Mediterranean, and plantation agriculture). Irrigated farming can be either subsistence or commercial. The richer a nation becomes, the more total grains it consumes, but the higher the percentage of grain consumption that is usually in the indirect form of meat and dairy products. Humankind overall could greatly increase food supplies by concentrating on raising the most efficient animals. Some areas that are remote from urban centers may specialize in dairy farming and add value to milk by producing dairy products.

Variation in agricultural yields worldwide are partly attributable to variations in the physical environment, but today capital input is also important. Biotechnology is providing new possibilities, but some observers fear possible injurious ramifications from tampering with plant and animal biology.

Technology creates the potential for increasing food production, but technology is not readily available throughout the whole world, nor is it distributed in the most economic way. Problems in the effort to increase food supplies include lack of financial incentives for farmers, unsuccessful land redistribution, and inequitable land ownership.

Many rich countries have erected tariffs as protection against food imports, and they subsidize their own farmers. These subsidies contribute to the production of agricultural surpluses in many rich countries and discourage the production in poor countries. We cannot know what quantities of food the world's farmers would be capable of producing or what the geography of food production would be if these conditions were eliminated.

The world's harvest of fish has significantly increased Earth's per capita food supplies, but future increases from fish capture are dubious. Aquaculture offers new possibilities.

KEY TERMS

aquaculture p. 348
biotechnology p. 334
cloning p. 334
commercial agriculture p. 320
diminishing returns p. 339

double cropping p. 322
downstream activities p. 331
ejido p. 341
exclusive economic zone (EEZ) p. 348

fishery p. 347
gene splicing (recombinant DNA) p. 334
genetic engineering p. 316
genetically modified (GM) p. 334

QUESTIONS FOR REVIEW AND DISCUSSION

1. What are the world's principal types of low-technology subsistence agriculture, and where can each be found?

2. What current conditions in many sub-Saharan African countries frustrate attempts to raise more food?

3. How have humans forestalled a "Malthusian" catastrophe for the past 200 years?

4. Why do the rich countries subsidize agriculture? What effect does this have on the urban populations of the rich countries, and what effect does it have on the farmers in the poor countries?

5. Why are countries increasingly clashing over the rights to fishing at sea?

THINKING GEOGRAPHICALLY

1. If fewer and fewer farmers can produce all the food the world needs, what will the rest of the people do for a living?

2. Why is humankind just now domesticating fish thousands of years after we managed to think of domesticating animals?

3. What percentages of the grain you consume do you consume directly? Indirectly?

4. How does your diet differ from that of your parents or grandparents? Compare the amount of meat or fish you eat. How much more processed is the food you buy? Compare the price of the ingredients of a cake with the cost of a cake.

SUGGESTIONS FOR FURTHER LEARNING

Agricultural Statistics. Washington, D.C.: U.S. Department of Agriculture, annually.

Altieri, Miguel, and Clara Nicholls. *Biodiversity and Pest Management in Agroecosystems.* New York: Food Products Press, 2nd ed. 2003.

Clay, Jason, and World Wildlife Fund. *World Agriculture and The Environment: A Commodity-by-Commodity Guide to Impacts tnd Practices.* Washington, D.C.: Island Press, 2004.

Cramer, Gail, Clarence Jensen, and Douglas Southgate. *Agricultural Economics and Agribusiness.* New York: John Wiley & Sons, 8th ed. 2001.

Fedoroff, Nina, and Nancy Marie Brown. *Mendel in the Kitchen: A Scientist's View of Genetically Modified Foods.* Washington, D.C.: National Academies Press, 2004.

Fridell, Ron. *The War on Hunger.* New York: 21st Century, 2003.

Fromartz, Samuel. *Organic, Inc.: Natural Foods and How They Grew.* New York: Harcourt, 2006.

Grigg, David. *An Introduction to Agricultural Geography.* New York: Routledge, 2nd ed. 1995.

Kiple, Kenneth F., and Krienhild C. Ornelas, eds. *The Cambridge World History of Food.* 2 vols. New York: Cambridge University Press, 2000.

Kloppenburg, Jack Ralph. *First the Seed: The Political Economy of Plant Biotechnology.* Madison, WI: University of Wisconsin Press, rev. ed. 2005.

Knutson, Ronald, and J. B. Penn. *Agricultural and Food Policy.* Upper Saddle River, NJ: Prentice Hall, 5th ed. 2003.

Lurquin, Paul. *High Tech Harvest: Understanding Genetically Modified Food Plants.* New York: Westview Press, 2002.

Pierce, Francis J. and David Clay. *GIS Applications in Agriculture.* Vol. 1 (Book/CD Rom). Ottawa, Canada: Communications Research Centre, 2007.

Pringle, Peter. *Food, Inc: Mendel to Monsanto.* New York: Simon & Schuster, 2003.

Sokolov, Raymond. *Why We Eat What We Eat: How the Encounter Between the New World and the Old Changed the Way Everyone on the Planet Eats.* New York: Summit Books, 1992.

Valdes, Alberto, ed. *Agricultural Support Policies in Transition.* Washington, D.C.: World Bank, 2000.

Winston, Mark L. *Travels in the Genetically Modified Zone.* Cambridge, MA: Harvard University Press, 2003.

World Agriculture: Situation and Outlook Report. Washington, D.C.: U.S. Department of Agriculture, Economic Research Service, quarterly.

Zohary, Danile, and Maria Hopf. *Domestication of Plants in the Old World: The Origin and Spread of Cultivated Plants in West Asia, Europe, and the Nile Valley.* New York: Oxford University Press, 1988.

WEB WORK

For information on the Web regarding food and agriculture, you might start with the World Agricultural Information Center of the United Nations Food and Agriculture Organization:

http://www.fao.org/

Its statistical index, called faostat, is at:

http://faostat.fao.org/

FAO even has an interactive world food trade flow map at:

http://www.fao.org/ES/ess/watf.asp

The United States Department of Agriculture has a home page:

http://www.usda.gov/

Information about the Department's National Organic Program is at:

http://www.ams.usda.gov/nop/indexIE.htm

The Iowa State University's College of Agriculture maintains a directory of sites related to agriculture:

http://www.ag.iastate.edu/other.html

For information from an environmental point of view, try the home page of Resources for the Future:

http://www.rff.org/

The World Resources Institute also has information from an environmental point of view:

http://www.wri.org/

The home page of the Cartagena Protocol can be found at:

http://www.biodiv.org/biosafety/default.aspx

For general libraries of information about economics that can help you can focus your research about agriculture, look at the World Wide Web Virtual Library in Economics:

http://netec.wustl.edu/WebEc/WebEc.html

Also look at Econolink:

http://www.progress.org/econolink/

9

EARTH'S
RESOURCES AND
ENVIRONMENTAL
PROTECTION

A policeman directing traffic in Bangalore, India, covers his face with a mask as protection from air pollution. (Aijaz Rahi/AP Wide World Photos)

What Is a Natural Resource?

A natural resource is something that is *useful* to people. Usefulness is determined by a mix of cultural, technological, and economic factors in addition to the properties of a given resource. We often substitute one resource for another as our needs change.

Mineral and Energy Resources

Mineral resources include metals and nonmetals. As mineral resources are depleted, they are usually replaced with substitute materials. Landfills accumulate used materials and may one day be a source of materials for reuse. Our principal energy resources come from three fossil fuels—oil,

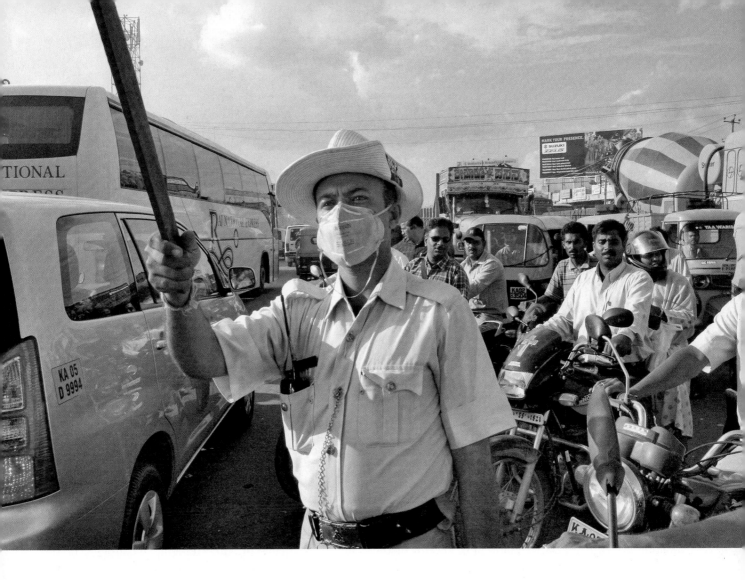

coal, and natural gas. As fossil fuels are depleted, they are likely to be replaced with nuclear and renewable energy.

Air and Water Resources

Pollution results when a substance is discharged into the air or water faster than it can be dispersed or removed by natural processes. Recycling and pollution prevention are growing in importance as methods for solving pollution problems.

*I*t is estimated that worldwide, 3 million premature deaths are caused by air pollution each year. Of these, the greatest number occur in India. Air pollution in India is worst in the major cities, where vehicle exhausts (motorcycles, trucks, and buses in addition to automobiles) are the dominant emission source, but industrial sources are also significant, as is household fuel combustion for cooking. Large numbers of old motorcycles and scooters, as well as three-wheeled motorized rickshaws are powered by polluting 2-stroke engines. Suspended particulate matter, sulfur oxides, and ozone are especially severe problems. The urban population of India has nearly tripled in the last three decades, and improvements in infrastructure have not kept pace with the growth. In addition, an increase in wealth allows a higher percentage of the population to own automobiles. The result is extreme road congestion with slow-moving vehicles emitting pollution. For much of the population mobility and accessibility have actually declined. In most of the major cities government regulations have been established that require vehicles to meet strict emission standards, similar to those in many wealthy countries. But enforcement has been difficult, in part because there are not sufficient low-emission vehicles available to meet the transport needs of these cities. ▶

Everything we consume is extracted from our planet and returned to it in one form or another. As we use Earth's resources of air, water, minerals, energy, plants, and animals, we simultaneously discharge our waste into the environment. As Earth's human population of 6.5 billion grows to almost 10 billion in the twenty-first century, consumption of resources will increase. This expanded population will place tremendous stress on Earth's remaining resources and the ability of the planet's air, water, and land to accommodate human waste.

Consumption and waste vary among cultures and over time. For example, different nations at different times have obtained energy from wood, coal, petroleum, natural gas, running water, nuclear energy, the wind, and the Sun. Issues of natural resource use and environmental quality must be understood in both their physical and human dimensions.

Resource management is exceedingly complex, because each resource varies geographically and physically. Resources also vary in value, depending on human factors: culture, technology, beliefs, politics, economics, and style of government. Many resources are publicly controlled, so resource management is a political process.

The issue of air pollution in India illustrates this complexity. India is experiencing rapid economic growth, and Indian cities are centers of opportunity, which is why people are drawn to them. But the improvement in quality of life that comes with higher incomes is counterbalanced by health problems created by urban population densities. In a wealthy country such population density might be possible without excessive pollution, but that is not possible in India at this time. Managing these problems will require consideration of politics, economics, technology, trade, and cultural attitudes in the context of very rapid social change.

In this chapter, we explore the factors that affect the value of resources. These factors include the physical characteristics of resources and the natural systems in which they exist, the changing technology of resource use, and human value systems. We then consider how changing resource values affect what and how much resources we use. Finally, we look at environmental pollution and resource conflict and management.

WHAT IS A NATURAL RESOURCE?

A **natural resource** is anything created through natural processes that people use and value. Examples include plants, animals, coal, water, air, land, metals, sunlight, and wilderness. Natural resources are especially important to geographers, because they are the specific elements of the atmosphere, biosphere, hydrosphere, and lithosphere with which people interact. Natural resources can be distinguished from human-made resources, which are human creations or inventions such as money, factories, computers, information, and labor.

We often use natural resources without considering the broader consequences. For example, the burning of oil to generate heat or to power an automobile engine pollutes the atmosphere with exhaust gases. The nitrogen oxides emitted into the atmosphere contribute to acid rain, which then pollutes streams. Oil consumption also weighs heavily on international economic and political relations.

Characteristics of Resources

A substance is merely part of nature until a society has a use for it. Consequently, a natural resource is defined by the three elements of society:

▶ A society's *cultural values* influence people's decision that a commodity is desirable and acceptable to use.

▶ A society's level of *technology* must be high enough to use the resource.

▶ A society's *economic system* affects whether a resource is affordable and accessible.

Consider petroleum as an example of a natural resource in North America:

▶ *Cultural Values.* North Americans want to drive private automobiles rather than use public transport such as trains.

▶ *Technology.* Petroleum is the preferred fuel in private automobiles because autos are easily powered by gasoline engines.

▶ *Economic System.* North Americans are willing to pay high enough prices for gasoline to justify removing petroleum from beneath the seafloor and importing it from distant places (Figure 9-1).

The same elements of society apply to the study of any example of a natural resource: rice to the Japanese, diamonds to South Africans, forests to Brazilians, and air quality to residents of Los Angeles. In every case, a combination of the three factors is necessary for a substance to be valued as a natural resource. Differences among societies in cultural values, the level of technology, and economic systems help geographers understand why a resource may be valuable in one place and ignored elsewhere.

Cultural values and natural resources

To survive, humans need shelter, food, and clothing, and make use of a variety of resources to meet these needs. We can build homes of grass, wood, mud, stone, or brick. We can eat the flesh of fish, cattle, pigs, fish, or mice—or we can consume grains, fruit, and vegetables. We can make clothing from animal skins, cotton, silk, or polyester. Cultural values guide a process of identifying substances as resources to sustain life.

A swamp is a good example of how shifting cultural values can turn an unused feature into a resource. A century ago swamps were seen in the United States as noxious, humid, buggy places where diseases thrived instead of a place that provides usable commodities. Swamps were valued only as places to dump waste or to convert into agricultural land. Eliminating swamps was good, because it removed the breeding ground for mosquitoes while simultaneously creating productive and valuable land.

During the twentieth century, cultural values changed in the United States. Scientists and environmentalists praised the value of swamps and documented their importance in controlling floods, providing habitat for

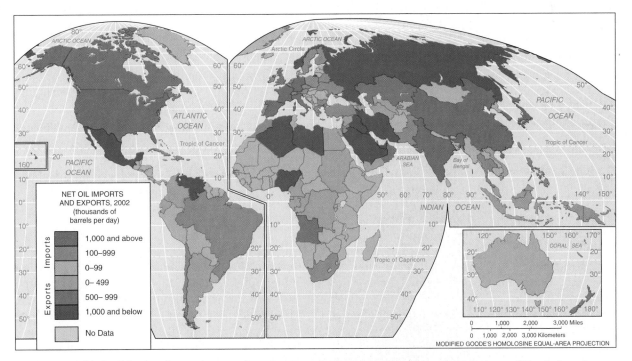

FIGURE 9-1 Net oil imports and exports. The largest oil importers are the wealthy nations of North America and Europe. Many developing nations are also net importers, and although the quantities they import are much smaller, the cost may be a higher percentage of the total national income for poor countries. (Data from U.S. Energy Information Administration)

FIGURE 9-2 Wetlands in southern Louisiana.
Louisiana is losing wetlands at a rate of about 75 square kilometers (29 square miles) per year. In addition to providing important water quality and ecological functions, these wetlands help protect cities like New Orleans from flooding. (Jim Wark/Airphoto)

wildlife, and reducing water pollution. Philosophers increasingly regarded nature as beautiful and praiseworthy. Changing public attitude toward swamps is reflected in our vocabulary: instead of calling them *swamps*, now we use a more positive term, *wetlands*.

As a consequence of cultural change, wetlands now are a valued land resource, protected by law (Figure 9-2). We restore damaged wetlands, create new ones, and restrict activities that might harm them. Even in the Netherlands, where deep-rooted cultural and landscape traditions are based on draining wetlands to convert them into farmland, some of these same farmlands are now being converted back to wetlands to improve water quality and enhance species diversity.

Technology and natural resources

The utility of a natural substance depends on the technological ability of a society to obtain it and to adapt it to its purposes. Metals are elements that are usually heavy, reflect light, can be hammered and drawn, and are good conductors of heat and electricity. A metal ore is not a resource if the society lacks knowledge of how to recover its metal content and how to shape the metal into a useful object, such as a tool, structural beam, coin, or automobile fender.

Earth has many substances that we do not use today, because we lack the means to extract them or the knowledge of how to use them. Things that might become resources in the near future are **potential resources.** As a result of its high level of biodiversity, the tropical rain forest is brimming with plants and animals that North Americans regard as potential resources. New medicines, pesticides, and foods might be developed from these substances. To the indigenous peoples of the Amazon rain forest, some of these plants and

animals are already resources. But deforestation threatens the availability of these resources for indigenous peoples and their potential use by others even while it creates economic opportunities for Brazilians. By destroying the rain forest, we are diminishing Earth's pool of both current and potential resources.

Human needs can drive technological advances. People living in cold climates invented insulated homes and heating technology. The need to increase the supply of food drove people to develop new agricultural technology.

Because human need drives many technological advances, new technologies may emerge when a resource becomes scarce. New technology for reusing materials is being developed in part because space in landfills has become a scarce resource, especially in large urban areas of relatively developed wealthy countries, where consumption is highest. This scarcity is stimulating development of new methods for reusing and recycling materials. Most waste currently is not reusable. But as we deplete the resource of landfill space, we will make more things recyclable and manufacture new products from waste, and waste materials themselves will become resources.

Economics and natural resources

Natural resources acquire a monetary value through exchange in a marketplace. The price of a substance in the marketplace, as well as the quantity that is bought and sold, is determined by **supply and demand.** Common sense tells us some principles of supply and demand.

▶ A commodity that requires less labor, machinery, and raw material to produce (bicycle), will sell for less than a commodity that is harder to produce (automobile).

▶ The greater the supply, the lower the price (corn). The greater the demand, the higher the price (World Series tickets).

▶ Consumers will pay more for a commodity if they strongly desire it (computer) than if they have only a moderate desire (textbook).

▶ If a product's price is low (beer and hamburgers), consumers will demand more than if the cost is high (champagne and prime rib).

In general, natural resources are produced, allocated, and consumed according to rules of supply and demand. Water is a good example. In areas where water is plentiful because of high rainfall and low demand—such as northern Minnesota—consumers pay a low price, little more than the cost of pumping it from the nearest well, river, or lake. But in the arid southwestern United States, water rights must be purchased, and scarce water must be carried hundreds of kilometers through pipelines and aqueducts. Thus, prices are generally much higher. Because of these high costs, the government may subsidize water provision, and prices paid by consumers may not reflect the true cost of obtaining the resource.

Many important natural resource problems result from the inability of markets to account for pollution. **Externalities** are exchanges of commodities that take place outside the marketplace and thus have no price attached at the time of the exchange. For example, a coal-fired power plant generates air pollution in the process of producing electricity. The people who buy the electricity pay for it directly. The price they pay reflects their desire to have the electricity and their willingness to pay for it. But people downwind from the power plant receive the pollution, whether they like it or not. The power plant is not charged directly for the privilege of discharging pollution to the atmosphere. This pollution thus constitutes an externality—a hidden cost. If the power plant were charged for its pollution and the people who receive it were compensated, perhaps the power plant would emit less pollution and thereby have less impact on the people exposed to it.

Example: Uranium

In every society, *cultural values*, *level of technology,* and *economic system* interact to determine which elements of the physical environment are resources and which are not. Uranium is a good example. Until the 1930s, uranium was a resource only because its salts made a pretty yellow glaze for pottery. After German physicists realized that the great energy stored in uranium atoms might be released by "splitting" their nuclei (nuclear fission), uranium gained value as a power of weaponry during World War II.

▶ *Technology.* The United States and its allies, fearing that Germany might develop a nuclear bomb, began what was known as the Manhattan Project, a crash program to develop the bomb technology first. Germany surrendered before the Manhattan Project achieved its goal, but the technology was eventually developed and used to win the war against Japan.

▶ *Economic System.* After World War II, nuclear technology was applied to generating electrical power. Nuclear-generated electricity was slow to gain acceptance because it was more expensive than alternative power sources, but after Middle East petroleum supplies were threatened during the 1970s, more nuclear power plants were built. The cost of nuclear power has had little to do with the cost of uranium, which is cheap and abundant. Recent increases in the price of non-nuclear-generated electricity associated with growing demand, concern about global warming, and rising prices of fossil fuels have renewed interest in nuclear power.

▶ *Cultural Values.* With the construction of nuclear power plants, the public became increasingly concerned about the risks. Following power-plant accidents at Three Mile Island in Pennsylvania

(1979) and Chernobyl in the Ukraine (1986), government agencies regulated nuclear plant safety much more closely. Higher safety standards increased the cost of nuclear power at a time when conservation efforts had succeeded in reducing demand for electricity. Fear of the hazards of nuclear power has prevented growth of the industry, even though the hazards are probably less than those routinely accepted in other aspects of life, such as the risk of death in an automobile or airplane accident. Orders for new power plants ceased in the United States during the 1980s. Renewed concern about security of radioactive materials may also reduce support for nuclear power.

Substitutability

Many natural resources are valued for *specific properties*—coal for the heat it releases when burned, wood for its strength and beauty as a building material, fish as a source of protein, clean water for its healthiness. In most cases, several substances may serve the same purpose, so if one is scarce or expensive, another can be substituted (Figure 9-3). Copper is an excellent conductor

FIGURE 9-3 House construction in the United States. Wood is used in house construction in the United States mostly because it is relatively inexpensive and easy to work with. Here oriented strand board, a manufactured wood product, is being used in new construction. This board is much cheaper than sawn lumber, but has ample strength for covering exterior walls, floors, and roofs. This wood product costs about the same as a foam board made from petroleum that is an alternative to wood-based board in exterior walls. (John Blaustein/Woodfin Camp & Associates)

of electricity, but it may be expensive relative to wire made of other metals that can be substituted. For information transmission, such as in computer networks, using light in a fiber-optic cable is much more efficient than using electrons in a copper wire.

The **substitutability** of one substance for another is important in stabilizing resource prices and limiting problems caused by resource scarcity. If one commodity becomes scarce and expensive, cheaper alternatives usually are found. Such substitution is central to our ability to use resources over extended periods without exhausting them and without decline in our standard of living.

However, many resources have no substitutes. There is only one Old Faithful Geyser and only one species of sperm whale, so if we destroy Old Faithful or force extinction of the sperm whale, we have no substitutes waiting to be tapped. Other geysers and whales exist, but they are not the same as those we now know. The uniqueness of these resources is the essence of their value.

Renewable and Nonrenewable Resources

In thinking about Earth's resources, we distinguish between those that are renewable and those that are not:

▶ **Nonrenewable resources** form so slowly that for practical purposes they cannot be replaced when used. Examples include coal, oil, gas, and ores of uranium, aluminum, lead, copper, and iron.

▶ **Renewable resources** are replaced continually, at least within a human lifespan. Examples include solar energy, air, wind, water, trees, grain, livestock, and medicines made from plants.

Even a renewable resource can be *depleted*, or used to a point where it can no longer be economically used. The only ones we cannot deplete are solar energy and its derivatives: wind and precipitation.

MINERAL AND ENERGY RESOURCES

Mineral resources are substances that we derive from the lithosphere. They are basic materials that we use to construct roads and buildings, manufacture goods, and power transportation. Without minerals, modern industrial societies could not function. Our use of mineral resources in industry and commerce is governed primarily by technology and economics.

We value most minerals for their properties of strength, malleability (ability to be shaped), weight, and chemical reactivity rather than for their aesthetic characteristics. Few car owners care if the engine is made from aluminum or iron—what matters is that it is powerful, durable, and efficient. Few people care very

much whether the roof of their house is made from slate or asphalt shingles, as long as the roof keeps out the rain and doesn't cost too much. Gold is the rare exception of a mineral valued mostly for its beauty in jewelry, although even gold is increasingly demanded for industrial uses, especially electronics.

Because we value a mineral primarily for its mechanical or chemical properties, our use of mineral resources is continually changing as our technology and economy change. As new technological processes and products are invented, demand can suddenly increase for materials that had little use in the past. When these new processes and products replace older ones, demand may be reduced for minerals that were previously important. As a result of changes in consumer demand, remaining supplies, and prices, one mineral may become more favored while another is less desired.

Mineral Resources

The terms *Stone Age, Iron Age,* and *Bronze Age* capture the importance of particular minerals at various times in the past. Minerals are as essential to civilization as plant, animal, water, and energy resources. They are present in virtually every product we manufacture, and though their value may be a small part of the total value of a finished good, those products could not be made without minerals (Figure 9-4).

Earth has 92 naturally occurring chemical elements, but most of Earths' crust is made up of 8 elements: oxygen, silicon, aluminum, iron, calcium, sodium, magnesium, and potassium. These elements, as well as rarer ones, combine to form thousands of minerals, each with its own properties and distribution pattern throughout the world. Each mineral potentially is a resource if people find a use for it.

Metallic minerals, such as copper, lead, silicon, tin, aluminum, and iron, usually occur in ores, from which they must be extracted. Nonmetallic minerals including building stone, graphite, rubies, sulfur, slate, and quartz are generally easier to obtain because they are more plentiful and usually require less processing. Both metallic and nonmetallic minerals must be discovered, mined, transported, refined, and manufactured into useful goods.

Variations in Mineral Use

Historically, the use of particular metals and nonmetals has fluctuated between periods of high demand and price and periods of low demand and price. Discovery of a new resource could create a "rush" of people to the area of discovery. The "Gold Rushes" to California, Colorado, and Alaska are nineteenth-century examples. The period between 1970 and 1985 featured especially volatile mineral prices, as a result of rises and declines of industrial output and inflation. For example, the price of copper doubled between 1973 and 1980, then

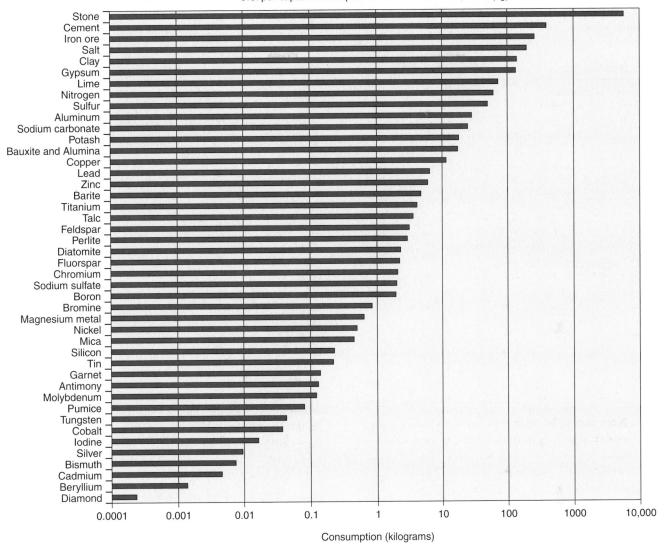

U.S. per capita consumption of selected minerals, 1999 (kg)

Consumption (kilograms)

FIGURE 9-4 Annual per capita consumption of nonenergy minerals in the United States, 2004. (U.S. Census Bureau)

fell nearly 40 percent between 1980 and 1985. Prices of most minerals were relatively stable through the 1990s, but since 2003 prices of some metals have risen dramatically in response to growing industrial production in China and India. For example, the price of copper tripled between 2003 and 2006, and over the same time period the price of steel more than doubled. Historically these spikes in mineral commodity prices have been short-lived, but sustained industrial growth in China, India, and elsewhere may prolong these price trends.

Mineral deposits are not uniformly distributed around the world. Most of the world's supply of particular minerals is concentrated in a handful of countries. For example, five countries—Australia, Brazil, Guinea, China, and Jamaica—produce about three-fourths of the world's total aluminum ore (bauxite). Russia, Canada, Australia, Indonesia, and New Caledonia together produce about two-thirds of the world's nickel, and

China, Indonesia, and Peru, together produce five-sixths of the world's tin (Table 9-1). The United States, Russia, Canada, China, and Australia are especially rich in metal and nonmetal mineral resources.

The concentration of mineral resources and production in a few countries favors the establishment of **cartels.** A cartel is a group of countries that agree to control a particular market by limiting production in order to drive up prices. During the 1970s, when world demand for minerals was strong, a few cartels were able to control world markets for brief periods. But weak demand, falling prices, and political instability have limited the strength of cartels in recent years. In addition, the United States—the largest consumer of most minerals—has accumulated a stockpile of important minerals to protect against short-term reductions in supply caused by high prices, political instability, or hostile foreign governments.

TABLE 9–1   Major Producers of Selected Minerals

Countries that produce at least 5 percent of world total production are listed. Note that some countries are listed for several minerals (Australia, Canada, China, Indonesia, Peru, Russia, USA), indicating that they are rich in several minerals.

COUNTRY	PERCENT OF WORLD PRODUCTION	COUNTRY	PERCENT OF WORLD PRODUCTION
Bauxite and alumina		**Lead**	
Australia	35.6	China	30.5
Brazil	11.6	Australia	20.6
Guinea	10.1	United States	14.3
China	9.4	Peru	9.8
Jamaica	8.4	Mexico	4.5
India	7.1	Canada	2.5
All others	17.8	All others	17.7
Cadmium		**Mercury**	
China	14.9	China	45.6
Japan	11.9	Kyrgyzstan	22.4
Korea, Republic of	11.2	Spain	11.2
Kazakhstan	10.1	Algeria	8.2
Canada	10.0	Finland	4.9
Mexico	8.5	Russia	3.7
All others	33.4	All others	4.0
Copper		**Nickel**	
Chile	37.1	Russia	22.7
United States	7.9	Canada	13.4
Peru	7.1	Australia	12.8
Australia	5.9	Indonesia	9.6
Indonesia	5.8	New Caledonia	8.5
Russia	4.6	Colombia	10.6
All others	31.7	All others	22.4
Gold		**Tin**	
South Africa	14.1	China	42.0
Australia	10.7	Indonesia	25.1
United States	10.6	Peru	15.9
China	8.8	Bolivia	6.4
Peru	7.1	Brazil	4.7
Russia	7.0	Vietnam	1.5
All others	41.7	All others	4.4
Iron		**Zinc**	
Brazil	22.9	China	23.5
Australia	19.4	Australia	13.9
China	13.8	Peru	12.6
India	10.4	Canada	8.2
Russia	7.6	United States	7.7
Ukraine	4.9	Mexico	4.8
All others	21.0	All others	29.2

Source: U.S. Geological Survey

Depletion and Substitution

Fluctuations in the price of a mineral as a result of actions by a cartel, a political dispute, or a limited supply rarely continue for a long period of time. If high prices persist for several decades, technological innovations usually enable the substitution of cheaper minerals for more expensive ones. For example, when the price of copper rose rapidly in the 1970s, plumbers began to substitute polyvinyl chloride (PVC). Today PVC (a plastic made from petrochemicals) has largely replaced copper pipe for plumbing in new buildings.

The substitution of one mineral for another has an important consequence: Even though world supply of a mineral resource may be limited, we will never run out of it. The reason is that if the supply of a resource dwindles relative to demand, its price will rise. The increase in price has four important consequences:

1. Demand for the mineral will decrease, slowing its rate of depletion.
2. Mining companies will have added incentive to locate and extract new deposits of the mineral, especially deposits that might have been neglected when prices were lower.
3. Recycling of the mineral will become more feasible.
4. Research to find substitute materials will intensify, and as use of the substitute increases, demand for the scarce mineral will cease before the supply is exhausted.

Although at current rates of use the world would exhaust its remaining supply of lead in 23 years, zinc in 20 years, and copper in 33 years, you should not be worried about running out of these materials in your lifetime. However, if you know that a company has just invented a product that will replace one of these minerals cheaply, consider investing in it!

Disposal and Recycling of Solid Waste

The average American throws away 2.0 kilograms (4.4 pounds) of solid waste per day, twice as much as his/her parents did in 1960. Paper accounts for one-third of all solid waste in the United States. Discarded food and yard waste accounts for another one-third (Figure 9-5). Relatively developed societies generate large quantities of packaging and containers made of paper, plastic, glass, and metal. We dispose of this solid waste in three ways: landfills, incineration, and recycling. Each of these methods poses significant problems, either in environmental degradation or in costs of disposal. Normally the choice of one disposal method over another means that costs are shifted from one group to another, making conflict inevitable.

Landfill disposal About 55 percent of solid waste generated in the United States is trucked to landfills and buried under earth in **sanitary landfills** in which a layer of earth is bulldozed over the garbage each day. They are considered *sanitary,* because burying the garbage reduces emissions of gases and odors from the decaying trash, prevents fires, and discourages vermin. Unlike air and water pollution, which are reduced by dispersal into the atmosphere and rivers, solid-waste pollution is minimized by concentrating the waste in thousands of landfills. However, landfills have been closed in many communities because they might contaminate groundwater, devalue property, or have been filled to capacity. Opening new landfills is difficult because present environmental regulations are more stringent, and local opposition to new landfills is usually overwhelming. The result in many areas has been a solid-waste crisis. Disposal sites are few and costly and some communities must pay to use landfills elsewhere. San Francisco trucks solid waste to Altamont, California, 100 kilometers (60 miles) away. Passaic County, New Jersey, hauls waste 400 kilometers (250 miles) west to Johnstown, Pennsylvania.

Incineration One alternative to burying waste in landfills is incinerating it. Incineration reduces the bulk of trash by about three-fourths, and the remaining ash requires far less landfill space. Incineration also provides energy. The incinerator's heat boils water, producing steam that can be used to heat homes or to generate electricity by operating a turbine. More than 100 incinerators now burn about 14 percent of the

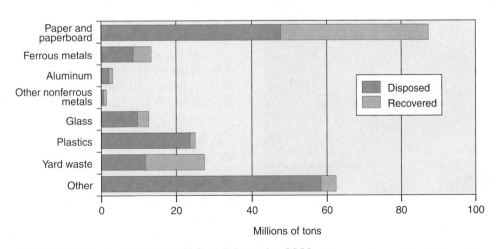

FIGURE 9-5 Composition of U.S. solid waste, 2003. Paper and paperboard form the largest part of solid waste generated in the United States. About one-third of all paper and paperboard waste is recovered and reused; much of this is made into cardboard boxes used for shipping. Recovery rates for aluminum are also about one-third. (U.S. Census Bureau)

trash generated in the United States. However, because solid waste is a mixture of many materials, it does not burn efficiently. Burning releases some toxic substances into the air, while some remain in the ash.

Recycling

Recycling solid waste reduces the need for landfills and incinerators and reuses natural resources that already have been extracted. Recycling simultaneously addresses both pollution and resource depletion. Most U.S. communities have instituted some form of mandatory or voluntary recycling, and about 30 percent of municipal solid waste in the United States is recyled.

Several barriers to recycling must be surmounted: separating waste, consumer resistance, market, cost, and indirect loss.

▶ *Waste Separation.* Solid waste comprises a variety of materials that must be separated to recycle (Figure 9-6). Metals containing iron can be pulled out of the pile magnetically; but paper, yard waste, food waste, plastics, and glass cannot be separated easily from each other. Consequently, many communities require consumers to separate solid waste themselves. Typically, consumers must place newspaper, glass, plastic, and aluminum in separate containers for pickup. Each type is collected on a different day or by a different truck, and is shipped to a specialized processor. The procedure

FIGURE 9-6 Resource recovery. Resource recovery from solid waste requires sorting and separating different types of materials. One approach is to collect them together and separate them at a centralized facility like this one. Alternatively they can be separated eart the source (home, school, work) and transported separately to reprocessing facilities. (Gabe Palmer/Kane/Corbis/Stock Market)

is generally more expensive for the community than picking up all the trash together.

▶ *Consumer Resistance.* Separation for recycling is a nuisance for people in relatively rich developed countries who are used to throwing things away. To encourage recycling, some communities charge high fees to pick up nonrecyclables, but take recyclables at little or no charge. Bottle and can laws requiring a deposit on beverage containers have been enacted in many states to encourage recycling and reduce roadside litter.

▶ *Lack of Market.* To succeed, recycled products must have a market. The lack of an assured market for many recycled products is perhaps the most difficult obstacle to increased recycling. Demand for recycled goods among industries and consumers is uncertain. For example, mixed plastics can be used as a substitute for wood in products such as picnic tables and playground equipment. The price of wood is not very different from that of recycled plastic, however, and for aesthetic reasons most consumers prefer wood. This helps keep the market for recycled plastic small. To increase consumer acceptance, the quality of recycled material must also improve. Poor quality has resulted from the fact that recycling became mandatory before proper manufacturing methods had been developed.

▶ *Hidden Costs.* Recycling involves far more than just "melting down and reshaping." For example, to recycle paper, ink must be removed, which is an additional step compared to conventional papermaking and therefore an added cost. It is difficult to remove all of the ink, so recycled paper is unacceptable for some uses because it is too gray or speckled. Removing ink may create some pollution, though probably less than would be produced in making virgin paper. The relatively high cost of processing and the lower value of the products results in a limited market for the recycled materials.

▶ *Indirect Losses.* Trash burns only if it contains enough combustibles. If paper is recycled and yard waste is composted instead of being thrown in the trash, the trash may be difficult to burn. As recycling has increased during the 1990s, some communities have not had enough combustible waste to operate their incinerators.

Many companies are developing manufacturing methods and packaging that facilitate recycling. To reduce packaging volume, detergent is being sold in concentrated form; refillable containers are available for more products; and toner cartridges for some photocopy machines are built for reuse rather than for disposal. At the same time, however, the low cost of many materials means that recycling depends on cheap labor to separate materials prior to reuse.

One alternative is to require consumers to bear a greater share of the cost of waste generation. If those costs are clearly associated with the products that generate the wastes, then consumers have an incentive to reduce waste. For example, in 2002 the Irish government imposed a tax of about 17 cents on the use of disposable plastic shopping bags. Within a year use of such bags dropped 95 percent, as consumers either re-used bags or carried their purchases in other, reusable containers. The portion of plastic bags in litter dropped from 5 percent to 0.3 percent.

Landfills: An example of changing resource values

Sanitary landfills are building mountains of trash. The Roosevelt Regional Landfill in south central Washington is an example (Figure 9-7). It is the fourth largest landfill in the United States and receives about 35 percent of the State of Washington's solid waste.

Garbage dumps are nothing new, as excavations of ancient cities reveal, but they are growing much more rapidly than in the past. Traditionally, food waste was fed to livestock, iron was used in durable goods, packaging was simple, and plastic products were unknown. Far fewer goods were manufactured. Our "throwaway culture" is a modern invention.

Garbage dumps have been sited in swamps or other low-value land. For years garbage simply was dumped and left uncovered. Rubber tires, deliberately burned in landfills to keep fires going, emitted offensive black smoke. Fires smoldered, rats thrived, flies buzzed, and homes situated downwind rarely enjoyed cookouts. Many coastal cities dumped their garbage at sea.

To reduce fire, vermin, and odor, cities have converted open dumps to sanitary landfills, which reduced

some environmental problems but aggravated others. Landfill toxic materials are leached by groundwater, polluting it. This consequence, plus much publicized problems such as the chemical landfill at Love Canal near Buffalo, New York, heightened public concern. In response to these concerns, the U.S. government has passed several laws and regulations that control handling of toxic substances.

As landfills run out of space, new ones are difficult to create. People may support the creation of a facility, but they may not want it near their own homes. This attitude is sometimes called the NIMBY attitude—Not in my backyard. Most people agree that landfills are needed, but few want them near their homes. Stricter regulations have forced some landfills to close and have prevented others from opening. The number of operating landfills in the United States declined from about 30,000 in 1976 to fewer than 1,800 in 2002. The few remaining landfills continue to grow as they accept the waste that formerly went to other sites.

At the Roosevelt Regional Landfill an innovative project is underway in which garbage is burned to produce electricity. The CO_2 produced in the process is sold to nearby greenhouses that formerly burned propane to increase the CO_2 levels that accelerate plant growth in the greenhouses. This is encouraging: An industry that disposes of unwanted things also sells a commodity that people need. In this case, not only are we producing electricity but doing it in a way that reduces greenhouse gas emissions.

Modern facilities such as this are expensive, as is the process of collecting waste and transporting it to landfills. Unfortunately, in many parts of the world the cost of collecting and disposing of solid waste is too high for society to bear. In our global economy modern mass-produced materials are ubiquitous and inexpensive, so solid waste is becoming a major problem in many poor countries. The most visible example of this is the presence of plastic bags, which have replaced paper bags, baskets, and other biodegradable containers in rich and poor countries alike. Where solid waste collection is absent, plastic bags and other nondegradable waste accumulates wherever people congregate (Figure 9-8).

Today landfills are a scarce resource that must be used carefully. To slow the rate of filling, many landfills no longer accept grass clippings and leaves, which can be composted (decomposed harmlessly) at home, or any other materials that can be recycled.

Energy Resources

Earth has bountiful and varied sources of renewable energy that humans have been able to harness:

▶ Solar energy comes from the Sun.

▶ Hydroelectric power and wind power come from natural movements of water and air caused by solar energy and gravity.

FIGURE 9-7 The Roosevelt Regional Landfill, Washington. This landfill is the site of an innovative program to convert waste to electricity, thereby reducing the need for combustion of fossil fuels for that purpose. In addition, the carbon dioxide produced from the landfill will be delivered to greenhouses that otherwise would burn propane to stimulate plant growth by increasing the CO_2 concentrations in the greenhouses. (AP/Wide World Photos)

FIGURE 9-8 Plastic bags litter a beach in Mumbai, India. These bags are very inexpensive and difficult to recycle, making them a solid waste problem worldwide. (Rajesh Nirgude/AP Wide World Photos)

▶ Geothermal energy comes from Earth's internal heat in volcanic areas, including California, Iceland, and New Zealand.

However, most of the world's energy comes from chemical energy stored in such substances as wood, coal, oil, natural gas, alcohol, and manure. Energy is released by burning these materials. People burn these substances to heat homes, run factories, generate electricity, and operate motor vehicles. Most of our energy comes from fossil fuels, but burning them reduces their supply—they are nonrenewable resources. We can continue to burn them for another few decades, but eventually we must switch to new energy resources if we are to preserve our current standard of living.

Energy from Fossil Fuels

Oil, natural gas, and coal, known as **fossil fuels,** come from the residue of plants and animals buried millions of years ago. Through photosynthesis, plants convert solar energy to the chemical energy stored in their tissues. When plants die, this energy remains in their tissues. The energy may be released promptly if animals or decomposers consume the dead plants or if a fire burns the field or forest. Or the plants may become fossilized as coal, and the stored energy may wait millions of years to be released when the coal is mined and burned.

Oil and gas are also stored sunlight. When we burn these fossil fuels today, we are releasing the energy originally stored in microscopic plants (phytoplankton) millions of years ago. When animals eat plants, they store the energy that was incorporated into the plant tissues. As the sea's countless creatures died over thousands of years, their bodies sank to the bottom, creating an organic sediment. Over time, this was converted to oil, accompanied by natural gas. Coal, oil, and natural gas still are being created, but the processes are so slow that from a human perspective, fossil fuels are nonrenewable resources: Once burned, they are gone forever as useful sources of energy.

From wood to coal, oil, and gas   From the time that humans first lived in North America—probably at the end of the last Ice Age about 18,000 years ago—until the mid-1800s, wood was the most important source of energy. Prior to the arrival of European colonists, North American residents used wood almost exclusively for all of their needs, but because their total population was small, they did not significantly deplete the resource. When Europeans arrived in North America beginning in the seventeenth century, they harvested the forests for fuel and lumber and cleared the land for agriculture. By the end of the nineteenth century, most of the forests near populated areas of the eastern United States had been cut down, and fuel wood became very expensive. It was also inadequate for providing the large amounts of energy demanded by a growing industrial economy. Wood still provides the largest portion of energy in developing countries, though supplies are dwindling in areas of dense, rural populations, such as in East Africa and southern Asia.

Coal served as a substitute for fuel wood during the nineteenth century. Although large amounts of coal have been consumed, abundant supplies still remain in the United States and several other countries. Oil, another fossil fuel, was a minor resource until the diffusion of motor vehicles early in the 1900s. Today it is the world's most important energy resource. A third fossil fuel, natural gas, once was burned off during oil drilling as a waste product because it was too difficult to handle and markets for it were not established. In recent years, however, it has become an important energy source. Today these three fossil fuels provide more than 80 percent of the world's energy and more than 90 percent in relatively developed countries (Figure 9-9).

For U.S. and Canadian industry today, the main energy resource is natural gas, followed by oil and coal. Some businesses directly burn coal in their operations, while others rely on electricity generated primarily at coal-burning power plants. At home, energy is used to generate heat and hot water and to operate diverse electrical devices. Natural gas is the most common source for home heating, followed by petroleum and electricity. Nearly all transportation systems operate on petroleum products, including automobiles, trucks, buses, airplanes, and most railroads. Only subways, streetcars, and some trains run on electricity, much of which is generated from burning fossil fuels.

Distribution of fossil fuels   Fossil fuels are not uniformly distributed beneath Earth's surface (see Table 9-2). Mineral *reserves* are known deposits that can be extracted profitably using current technology. Some

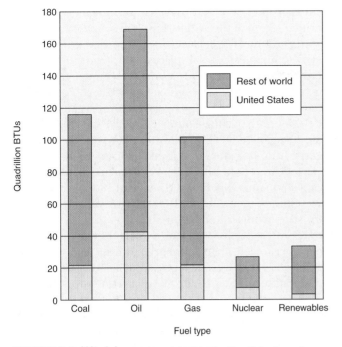

FIGURE 9-9 World energy sources. Fossil fuels make up the major proportion of global energy production; renewable and nuclear energy are relatively minor. (World Resources Institute)

regions have abundant reserves, whereas others have none. This distribution reflects how fossil fuels are formed. Coal forms in swampy areas, rich in plants. The lush tropical wetlands of 250 million years ago are

today the coal beds of the world, relocated to midlatitudes by the ponderous movement of Earth's tectonic plates (see Chapter 3). Oil and natural gas form in seafloor sediment, but Earth's tectonic movements eventually elevate some seafloor above sea level to become land. Today we drill for petroleum on both land and the seafloor.

Because minerals are only available in specific geologic environments their distribution is very uneven. Table 9-2 reveals that, like the minerals listed in Table 9-1, fossil fuel resources are highly concentrated in a few places. For each of the fossil fuels listed, the top five countries hold more than 60 percent of world reserves.

Relatively wealthy developed countries—which comprise about one-fourth of the world's population—possess more than 60 percent of the world's coal and more than 60 percent of natural gas. China is the leading producer of coal, but the United States is also a major coal producer and user. Russia is the leading producer of natural gas. Other than China and India, which have significant coal deposits, most countries in Africa, Asia, and Latin America have few reserves of coal or natural gas.

The distribution of oil is somewhat different. Two-thirds of the world's oil reserves are in the Middle East, including one-fourth in Saudi Arabia. Mexico and Venezuela also have extensive oil fields. In contrast, oil reserves in North America and Europe are relatively small, and production in those regions has passed its peak.

TABLE 9-2	Reserves of Oil, Natural Gas, and Coal, 2004

For each resource only the top 15 countries are listed. The countries listed account for 92 percent of all oil reserves, 84 percent of all gas reserves, and 96 percent of all coal reserves.

COUNTRY	CRUDE OIL (BILLION BARRELS)	PERCENT OF WORLD TOTAL	COUNTRY	NATURAL GAS (TRILLION CUBIC FEET)	PERCENT OF WORLD TOTAL	COUNTRY	TOTAL RECOVERABLE COAL (MILLION SHORT TONS)	PERCENT OF WORLD TOTAL
Saudi Arabia	262.075	24.2	Russia	2,361.053	33.7	United States	267312	26.8
Iran	130.800	12.1	Iran	944.670	13.5	Russia	173074	17.4
Iraq	115.000	10.6	Qatar	913.400	13.1	China	126215	12.7
Kuwait	99.675	9.2	Saudi Arabia	238.500	3.4	India	101903	10.2
United Arab Emirates	69.910	6.5	United Arab Emirates	204.050	2.9	Australia	86531	8.7
Russia	67.138	6.2	United States	192.513	2.8	South Africa	53738	5.4
Venezuela	52.400	4.8	Nigeria	180.000	2.6	Ukraine	37647	3.8
Nigeria	36.630	3.4	Algeria	171.500	2.5	Kazakhstan	34479	3.5
Libya	33.550	3.1	Venezuela	150.500	2.2	Serbia and Montenegro	18288	1.8
United States	21.371	2.0	Australia	128.610	1.8	Poland	15432	1.5
Qatar	20.000	1.8	Iraq	112.600	1.6	Brazil	11148	1.1
China	15.443	1.4	Norway	84.261	1.2	Germany	7428	0.7
Algeria	15.303	1.4	Egypt	66.000	0.9	Colombia	7287	0.7
Mexico	14.803	1.4	Indonesia	63.000	0.9	Canada	7251	0.7
Brazil	11.243	1.0	Canada	60.715	0.9	Czech Republic	6120	0.6
All others	116.473	10.8	All others	1,126.395	16.1	All others	43652	4.4

Source: Energy Information Administration, *Oil and Gas Journal*

North America and Europe have much higher per capita oil consumption rates, so they account for nearly two-thirds of the world's energy consumption. The United States, with less than 5 percent of the world's population, consumes nearly 20 percent of the world's commercial energy. In North America and Europe, the high level of energy consumption supports a lifestyle rich in food, goods, services, comfort, education, and travel.

Because relatively rich developed countries consume more energy than they produce, they must import energy, especially oil, from developing countries. The United States imports roughly half of its needs, Western European countries more than half, and Japan more than 90 percent. U.S. dependency on foreign oil began in the 1950s, when oil companies determined that the cost of extracting domestic oil had become higher than for foreign sources. U.S. oil imports have increased from 14 percent of total consumption in 1954 to about 60 percent in 2005. European countries and Japan increasingly depend on foreign oil because of limited domestic supplies.

Oil production and prices Early in the twentieth century, the United States was an oil exporter, but its needs soon exceeded domestic supplies. Europe has always been a net importer of oil. In their search for cheaper oil, U.S. and Western European companies drilled for oil in the Middle East and sold it inexpensively to consumers in relatively developed countries. Western companies set oil prices and paid the Middle Eastern governments only a small percentage of their oil profits. To reduce dependency on Western companies, the countries possessing oil created the Organization of Petroleum Exporting Countries (OPEC) in 1960. OPEC members include eight countries in the Middle East (Algeria, Iran, Iraq, Kuwait, Libya, Qatar, Saudi Arabia, and United Arab Emirates), plus Venezuela in South America, Gabon and Nigeria in Africa, and Indonesia in Asia.

Angry at the United States and Western Europe for supporting Israel during its 1973 war with the Arab states of Egypt, Jordan, and Syria, OPEC's Arab members organized an oil embargo during the winter of 1973–1974. OPEC states refused to sell oil to countries that had supported Israel. Soon gasoline supplies dwindled in relatively developed countries, and prices at U.S. gas pumps soared. Another oil crisis followed in 1979 as a result of a revolution in Iran. World oil prices went from $3 per barrel (42 gallons, or 160 liters) in 1974 to more than $35 in 1981 (Figure 9-10). To import oil, U.S. consumers spent $3 billion in 1970 but $80 billion in 1980, nearly 3 percent of the Gross National Income (GNI). This rapid price increase

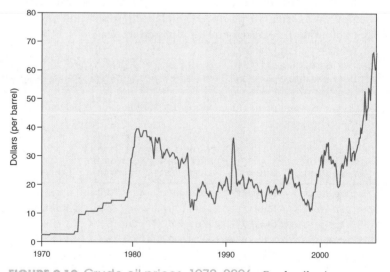

FIGURE 9-10 Crude oil prices, 1970–2006. Crude oil prices vary depending on the characteristics of the oil. The spot price of West Texas Intermediate, a benchmark commodity, jumped sharply in 1973–1974 and 1979–1981, but dropped dramatically in the mid-1980s. There was a brief spike in oil prices during the 1991 Persian Gulf War followed by a dramatic dip resulting from the 1998 financial collapse in Southeast Asia, and another dip resulting from the economic slowdown immediately after the 9/11 attacks. Since about 2003 prices have risen dramatically, primarily as a result of worldwide consumption taking up reserve production capacity, along with sporadic supply problems associated with isolated events such as hurricanes and other disruptions in key producing areas. The prices shown here are not adjusted for inflation; if inflation were taken into consideration the total price rise since 1970 would be much smaller. (Data from the Finincial Forecast Center: http://www.forecasts.org/ index.htm)

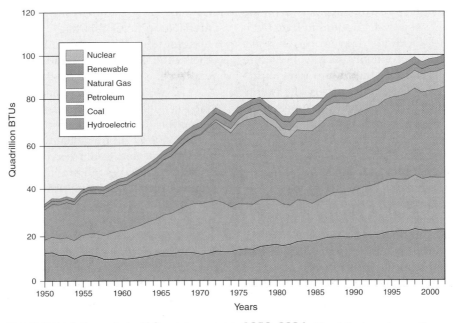

FIGURE 9-11 Trends in U.S. energy use, 1950–2004. Fossil fuels make up the major proportion of energy consumed in the United States. Use of oil and natural gas increased rapidly from the 1950s to the 1970s, but growth has been much slower since the energy crises of 1974 and 1979. (Data from U.S. Energy Information Administration)

abruptly halted the rise in energy use in the United States and elsewhere (Figure 9-11) and caused severe economic problems in relatively developed countries during the 1970s and 1980s.

In each of these oil crises, long lines formed at gas stations in the United States, and some motorists waited all night for fuel (Figure 9-12). In some cases gasoline was rationed by license plate number; cars with licenses ending in an odd number could buy only on odd-numbered days. Some countries took more drastic action; the Netherlands banned all but emergency motor vehicle travel on Sundays.

Developing countries were especially hurt by the price rises. They had depended on low-cost oil imports to spur industrial growth and could not afford higher oil prices. Relatively wealthy developed countries somewhat lessened the impact of higher oil prices on their economies by encouraging OPEC countries to return some of their money by investing it in real estate, banks, and other assets in North America and Western Europe. Poorer nations could not offer this opportunity for reinvesting oil wealth.

But no sooner had the world begun to adjust to high energy prices than the trend was reversed. Energy conservation measures in consuming countries reduced demand. Internal conflicts weakened OPEC's influence: Iraq and Iran fought a war that lasted for eight years, and in the early 1980s some OPEC members broke ranks and reduced oil prices. The price of oil plummeted from more than $30 to around $10 per barrel. Exporting countries flooded the world market with more oil in an unsuccessful attempt to maintain the same level of revenues as they had received during the

FIGURE 9-12 Gas line in 1970s. Cars lined up for gas in Los Angeles during the oil crisis in the 1970s. (Craig Aurness/Woodfin Camp & Associates)

1970s. While supplies increased, demand in relatively developed countries remained lower than before the boycott. Oil prices briefly spiked during the 1991 Persian Gulf War but then quickly dropped back to about $20. In early 1998, following a significant drop in oil prices, OPEC members agreed to reduce production in an attempt to drive prices back up, and by late 2000 prices were again above $30 per barrel. They fell back to about $20 by late 2001, falling particularly sharply after September 11, 2001, but then rose again as tensions mounted in Iraq. The steep rise that began in 2003 was certainly influenced by supply disruptions in Iraq, but it also reflects growing worldwide demand for oil, especially in rapidly growing economies such as India and China. While there had been a significant reserve productive capacity for the last two decades of the twentieth century, that capacity has virtually disappeared so that any significant disruption of supply causes a rise in prices. In this environment prices rose dramatically from 2003 to 2006, peaking at above $70 in mid-2006. Today oil-consuming countries around the world are looking for new sources of oil, hoping to secure and diversify sources of imports. There is considerable potential for new oil discoveries in many areas, including Africa, Central Asia, and Southeast Asia (Figure 9-13). Even if the oil-producing countries of the Middle East remain politically stable in the future, world oil prices are not likely to fall significantly because of increasing worldwide demand.

Future of fossil fuels

How much of the fossil fuels remain? Oil, natural gas, and coal occur beneath Earth's surface, so we cannot see them. Geologists can estimate fairly accurately the *proven reserves* of fossil fuels available in fields that have been explored. Geologists disagree sharply on potential reserves of the fossil fuels—the amount in fields not yet discovered and explored.

If we divide Earth's current proven oil reserves (about 1,000 billion barrels) by current annual consumption (about 24 billion barrels), we get about 40 years of oil supply remaining. Rates of consumption will change and new reserves will be discovered, but proven oil reserves will probably last only a few decades. Thus, unless large potential reserves are discovered, Earth's oil reserves will be significantly depleted during the twenty-first century, possibly in your lifetime.

Every discovery of new deposits extends the life of the resource. But extracting oil is becoming harder—and therefore more expensive. When geologists seek oil, they look first to accessible areas where geologic conditions favor accumulation. The largest, most accessible deposits already have been exploited. Newly discovered reserves generally are smaller and more remote, such as beneath the seafloor, where extraction is costly. Exploration cost also has increased because methods are more elaborate and the probability of finding new reserves is less.

Unconventional sources of oil are being studied and developed, such as *oil shale* and *tar sandstones*. Oil shale is a "rock that burns" because of its tar-like content. Tar sandstones are saturated with a thick petroleum. The states of Utah, Wyoming, and Colorado contain large amounts of oil in shale, more than ten times the conventional oil reserves found in Saudi Arabia. Canada has very large tar sandstone deposits.

Unconventional energy sources such as oil shale are not being used today, because economically feasible and environmentally sound technology is not currently available to extract the oil from it. The cost of conventional oil resources must increase dramatically before unconventional sources could contribute significant portions of our needs. By the time prices have risen that high, different sources of energy may be less expensive to obtain than oil.

How long the world's supply of oil will last depends heavily on how we use it today. We can either continue to use it rapidly, as we are now doing, knowing that oil must get more expensive and that we will eventually have to switch to some other fuel. Or we can conserve the oil that we have and switch to substitute energy sources as soon as possible.

If we increase our rate of consumption without finding more fossil-fuel reserves, we will deplete oil resources with relative speed. In general, since the 1970s, we have made more efficient use of energy. For example, motor vehicles have lighter engines, fewer metal parts, and more efficient transmissions than before the 1970s oil crises. The average fuel efficiency of automobiles in the mid-1990s was about 50 percent greater than it was in the mid-1970s. In the 1990s,

FIGURE 9-13 Oil wells near Hassi-Messaoud, Algeria. Oil exploration and production in Africa and elsewhere in many parts of the world outside the Middle East is likely to grow in coming years. (Parrot Pascal/Corbis / Sygma)

FIGURE 9-14 *Fuel-efficient personal transportation.* Scooters are a common, fuel-efficient form of personal transportation in many cities, such as Naples, Italy, shown here. These scooters typically exceed 43 kilometers per liter of gas (100 miles per gallon) and dramatically reduce parking problems in congested areas. (CORBIS–NY)

however, low gasoline prices helped to increase the popularity of inefficient sport utility vehicles to the extent that the average fuel economy of new vehicles declined in the United States in the late 1990s. The technology to improve fuel efficiency is readily available, however, as evidenced by the introduction in 2000 of hybrid vehicles that combine gasoline engines and electric power. Such vehicles are capable of 60 or more miles per gallon. For those whose transportation needs allow it, personal transportation with fuel efficiency exceeding 100 miles per gallon is readily available (Figure 9-14).

If progress in efficiency continues and alternative energy sources are developed, we probably will not completely exhaust the world's oil reserves. We may be able to continue using oil for those few tasks for which it is best suited. Earth is not in immediate danger of "running out" of oil, but our dependency on this non-renewable fossil fuel will prove to be a remarkably short period of human history.

Natural gas and coal: Short-term oil substitutes

In searching for alternatives to oil, we look first to the other two major fossil fuels—natural gas and coal. Natural gas is important in the United States as the clean-burning fuel of choice to heat more than half of the country's homes. A 1.6-million-kilometer (1-million-mile) pipeline network efficiently distributes gas from the production areas in the Gulf Coast, Oklahoma, and Appalachia to the rest of the country. Demand for natural gas is increasing, in large part due to air pollution considerations.

Natural gas does not offer much of an alternative to oil, because the global distribution of the two resources is similar. At current rates of consumption, natural gas reserves would last for 60 years, although potential reserves may be greater than for oil. Russia and the Middle East have about two-thirds of the world's natural gas reserves; the United States has only a few years of its own proven natural gas reserves. Continued intensive use of gas will depend on increased imports.

Coal reserves are more abundant than oil or natural gas and, at current rates of use, can provide nearly 400 years of proven reserves to the world. Coal can play an especially important role in providing the United States with energy, because the country has a large percentage of the world's proven reserves of coal. Coal is used mainly to produce electricity, and our use of electricity is expanding much faster than our use of other forms of energy. But several problems hinder expanded use of coal:

▶ *Air Pollution.* Uncontrolled burning of coal releases sulfur oxides, nitrogen oxides, hydrocarbons, carbon dioxide, and particulates into the atmosphere. The sulfur and nitrogen oxides are a major component of *acid deposition*. Acid deposition occurs when sulfur and nitrogen oxides combine with water to form acidic precipitation or when these acids are formed in the soil from pollutants that settle from the atmosphere. Many communities suffered from coal-polluted air earlier in this century and encouraged their industries to switch to cleaner-burning natural gas and oil. The Clean Air Act now requires utilities to use low-sulfur coals or to install emission-control devices on smokestacks. Pittsburgh, Pennsylvania, once noted for severe air pollution from burning coal for steel mills and glass factories, today has air relatively free of particulates and sulfur oxides. But coal-fired power plants still pump large amounts of carbon dioxide and sulfur oxides into the atmosphere.

▶ *Land and Water Impacts.* Both surface mining and underground mining cause environmental damage, an externality caused by burning coal. The damage from surface mining is visible: Vegetation, soil, and rock are stripped away to expose the coal. Today surface-mined land must be restored after mining, but restoration may not leave the land as productive as it was before. Underground mining causes surface subsidence and can release acidic groundwater.

▶ *Limited Uses.* Coal is a bulky solid, incapable of powering the internal combustion engines used in cars, trucks, and buses. It is not well-suited to shipment through established pipelines, though it can be mixed with water and sent through pipelines as a muddy *slurry.* Coal-generated power plants can provide the power to run electric vehicles, however, which are expected to become more common in the next few years.

Critical THINKING

Should We Raise the Price of Automobile Fuels?

The history of energy policy in the United States presents a curious paradox of a country that appears to want to do one thing but behaves in a totally contradictory fashion. While most Americans would agree that energy conservation, like Mom and apple pie, is basically a good thing, our actions tell a different story.

Ever since the energy crisis of the 1970s, energy conservation has been a valued goal both as measured by public opinion and in government policy. The initial reaction to the energy crisis was that the crisis demonstrated that oil supplies were perilously short and that immediate and significant conservation measures were needed. We have since learned that the crisis was a short-term phenomenon, that we have substantial flexibility in our energy-use system, and that oil will probably be around for another half century or so. Nonetheless, the negative consequences of our energy use are well known. It increases pollution, strains our foreign-exchange accounts by requiring massive oil imports, and necessitates increased military expenditures, including the possibility of waging an occasional war to protect our access to foreign oil reserves.

In the 1970s and early 1980s, the U.S. government established a wide range of voluntary and mandatory measures to encourage energy conservation and reduce dependence on foreign oil. These measures ranged from tax incentives for conservation-related investments to a nationwide 55 mph speed limit. Most of these programs have since been either abandoned or sharply curtailed in favor of free-market energy choice. When gasoline was cheap in the late 1980s and 1990s consumers acquired a fleet of gas-guzzling sport-utility vehicles equipped with four-wheel drive, which for most of us has true utility only a few hours of the year when plows may have difficulty keeping up with snowfall. Today vehicle manufacturers are struggling, as they did in the mid-1970s, to sell heavy, inefficient vehicles and re-tool to manufacture and sell more efficient ones.

The evidence from past energy prices and subsequent consumption patterns demonstrates that a small increase in energy prices, as occurred during the energy crisis, can bring major improvements in energy conservation. A modest energy tax could generate a new wave of energy conservation while still keeping energy prices well below those in most industrial nations. In most European nations fuel prices are more than double their levels in the United States, with taxes accounting for the higher price. Increasing fuel prices in the United States would have several important long-term benefits, including increasing conservation, reducing oil imports, reducing oil prices, reducing air pollution, and helping balance governmental budgets. There is also a strong argument that says that oil conservation would pay for itself by reducing the need for a major military presence in the Persian Gulf. Between 2003 and mid-2006 the United States spent about $300 billion on the war in Iraq—roughly equivalent to about 1 dollar per gallon of gas consumed in the United States in that time period. But U.S. consumers are used to low prices and unwilling to endure the stresses that would be necessary to change their vehicle-use habits.

Questions

1. Implementation of the Kyoto Protocol (Chapter 13) would probably require an energy tax and other measures to encourage conservation by raising energy prices. Currently this is not politically feasible, as most voters strongly object to tax increases, especially on gasoline. What do you think is the best way to reduce carbon dioxide emissions?

2. Would you support a modest (say, $2 per gallon phased in over 2 years) increase in the price of gasoline if it would bring about a 20 percent decrease in fuel consumption? Would you support a $30-per-month increase in your electricity bill if it allowed us to live without nuclear power, or if it produced a 10-percent decrease in acid rain?

Nuclear and Renewable Energy Resources

Where can we turn for energy that is safe, economical, nonpolluting, widely available, and not controlled by a handful of countries? In the long run, we must look to

energy sources that are renewable—or at least to resources like nuclear energy that are so abundant they are not likely to be depleted. The two most promising energy sources are nuclear and solar. Other alternatives at present include biomass and hydroelectric power. We will now look at each of these sources.

Focus ON

"Peak Oil"

The phrase "peak oil" is commonly heard today. It has become the rallying cry for those who think that the age of oil as the basis of global industry and transport is coming to an end. The basic idea is over a half-century old, however, having been put forth in the 1950s by a petroleum geologist named M. K. Hubbert. Hubbert argued that the life cycle of a fossil fuel was characterized by a bell-shaped curve. The curve begins with a period of rapid rise during which fuel deposits are discovered and exploited. However, as the largest and easiest-to-exploit deposits begin to be exhausted, the rate of discovery declines. Hubbert predicted in 1956 that U.S. production in the lower 48 states would peak between 1965 and 1970, and he turned out to be correct. Although Hubbert's ideas were originally applied to individual oil-producing regions, they can be applied to the world as a whole. The key question, then, is when will world oil production reach a peak and begin to fall? If that happens soon, we will need to plan for technologies that do not rely on oil, but the transition to new energy sources and technologies takes quite a while. In the United States, for example, imagine how long it would take for a majority of the population to switch from automobiles to mass transit as a principal means of getting to work. When we build suburbs that are miles from major employment places, we are making long-term commitments of housing based on the assumption that people will be able to get to work. When we build roads we expect them to last many decades. Should we instead be building rail lines? Those who argue that the peak of oil production is coming soon—say around 2010 or 2015—point to the need to invest now in systems that will rely on other energy sources. On the other hand, those who believe that the peak won't come until 2020 or 2025 would delay such investments until our needs are clearer. Whether the peak comes sooner or later, though, few would doubt that Hubbert's fundamental ideas were correct.

Nuclear energy Nuclear power can be generated either by fission (splitting an atom into two or more parts) or by fusion (joining two atoms together). Most peaceful generation of nuclear power today relies on fission of uranium, although plutonium also can be used. Someday we may use fusion to generate nuclear power, but as yet such technology is unavailable. The big advantage of nuclear power is the tremendous energy that is available from a small amount of material. A kilogram of nuclear fuel contains more than two million times the energy of a kilogram of coal.

The peaceful use of nuclear reactors to generate electricity began in the 1950s, and today about 400 reactors are operating around the world. Nuclear power is used exclusively to generate electricity, a growing part of our energy needs. It supplies about one-third of all electricity in Europe. Japan (which has virtually no fossil fuels), South Korea, and Taiwan also rely on nuclear-generated electricity (Figure 9-15). Nuclear power generates approximately 20 percent of U.S. electricity (Figure 9-16). The United States derives a smaller portion of its energy from nuclear power than other wealthy developed nations, in part because of its more abundant coal reserves.

Like coal, nuclear power presents serious problems. These include potential accidents, the generation of radioactive waste, public opposition, and high cost.

▷ *Potential Accidents.* A nuclear power plant cannot explode like a nuclear bomb, although it is possible to have a runaway chain reaction. The reactor can overheat, causing a meltdown, possible steam explosions, and a scattering of radioactive material into the atmosphere. An explosion did occur in 1986 in a nuclear power plant at Chernobyl, Ukraine, near the border with Belarus (both countries at the time were part of the Soviet Union). The accident has caused over 3,500 deaths, including about 800 among emergency cleanup workers. Cancer rates remain elevated in southern Belarus. For example, childhood thyroid cancer rates in Belarus increased from fewer than 5 per year before the accident to more than 80 by 1994. The impact of this accident extended through Europe: Most European governments temporarily banned the sale of milk and fresh vegetables because of possible contamination by radioactive fallout. In addition to accidental radioactive releases, many people are concerned about potential terrorist attacks involving either nuclear weapons or "dirty" bombs composed of radioactive materials.

▷ *Radioactive Waste.* The uranium-based fuel used in nuclear reactors eventually is broken down to the point that it is no longer usable. However, waste materials in this spent fuel remain highly radioactive for thousands of years. Spent fuel can also be reprocessed to extract plutonium, which can be used in nuclear bombs. Pipes, concrete,

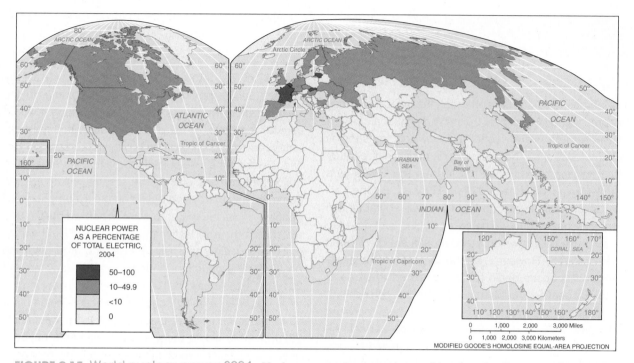

FIGURE 9-15 World nuclear power, 2004. Nuclear power plants are clustered in more developed wealthy regions, including Western Europe, North America, and Japan. Nuclear power has been especially attractive in the United States (which produces about 30 percent of the world's nuclear power) and in some European nations that lack abundant reserves of petroleum and coal. (Energy Information Administration)

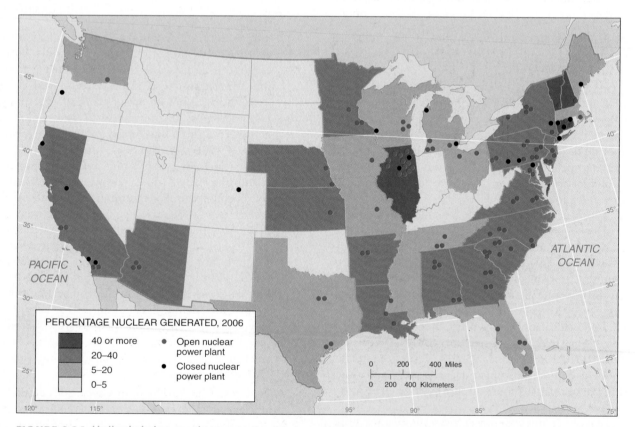

FIGURE 9-16 United states nuclear power. Nuclear power is an important source of electricity in several northeastern and midwestern states. Some locations have more than one nuclear reactor. (Data from Nuclear Regulatory Commission and Energy Information Administration)

and water near the fuel also become "hot" with radioactivity. This waste cannot be burned or chemically neutralized. It must be isolated for several thousand years until it loses its radioactivity. Currently, spent fuel generated in the United States is stored in cooling tanks at nuclear power plants, but these are nearly full. Work is under way to develop a long-term storage facility in Nevada, but many problems remain to be solved before waste can be moved from temporary storage sites to a permanent facility. Kazakhstan, which already has radioactive waste storage and disposal facilities dating from the Soviet era, is building a facility to receive nuclear waste from the European Union.

▶ *Public Opposition.* Public concern about safety has been an obstacle to the diffusion of nuclear power since the technology first emerged. The accident at Chernobyl, as well as less damaging incidents in the United States and other countries, dramatically increased public concern about nuclear power. At Shoreham, Long Island, near New York City, a nuclear power plant was ready to operate and begin generating electricity when, under pressure from worried citizens, the government decided not to allow the plant to open and compensated the electric company for the loss of the plant. In 1992, Italian voters rejected future nuclear power development in that country, and public opposition is similarly strong in Germany and Scandinavia. Even in France, where over 70 percent of electricity is generated from nuclear power, public opposition is a major barrier to new development. In addition, nuclear power plants cannot operate without a reliable source of cooling water, and summer heat waves in 2003 and 2006 (see Chapter 2) caused critical cooling-water shortages in some areas of Western Europe.

▶ *High Cost.* Nuclear power plants cost several billion dollars to build, primarily because of elaborate safety measures. Without double and triple backup systems, nuclear energy would be too dangerous to use. As a result, the cost of generating electricity is much higher from nuclear plants than from coal plants.

Biomass
Biomass fuel derives mostly from burning wood but includes the processing of other plant material and animal waste. Energy is generated by either burning directly or converting substances to charcoal, alcohol, or methane gas. Biomass provides most of the world's energy consumption for home heating and cooking, especially in the developing world. In China, some individual homes have fermentation tanks that convert waste to methane, which is used for cooking and heating.

Forms of biomass, such as sugarcane, corn, and soybeans, can be processed into motor vehicle fuels;

Brazil in particular makes extensive use of biomass to fuel cars and trucks. There has been much discussion in the United States recently of increasing our production of biofuels using corn, biodiesel, or other crops as the feedstock. Unfortunately, as President Bush stated, "ya gotta eat some." In other words, it takes a significant amount of land to produce the feedstock for biofuels, and a substantial increase in biofuel production would require much more agricultural land than is currently available. Also, the amount of fuel energy that is produced to manufacture ethanol from corn or biodiesel from soybeans is small, after one accounts for the amount of fuel consumed for tractor fuel and other needs in producing the corn or soybeans. Some analyses even indicate that more fuel is used to grow the crops than is produced when the crops are converted to fuel. On the other hand, biomass-to-energy conversion is much more appealing when it also serves to dispose of wastes. In Bangladesh, for example, in 2003 a new pilot plant began generating electricity using the waste from 5,000 poultry farms. The concentration of U.S. meat production in large "factory farms" (see Chapter 8) means that manure is already concentrated and managed instead of being scattered over the ground in barnyards. This increases the potential for using it to produce commercially valuable fuels. If the price of fossil fuels remains high, as seems likely, these new forms of renewable energy production are likely to grow in importance.

Hydroelectric power
Flowing water has been a source of mechanical power since before recorded history. In the past, water was used to turn a wheel that in turn could operate machines capable of grinding grain, sawing timber, and pumping water. Since the early 1900s the energy of moving water has been used primarily to generate electricity, called **hydroelectric power.** It supplies about one-fourth of the world's electricity, which is more than any other source except for coal.

To generate hydroelectric power, water must abruptly change height, as at a dam. The falling water turns turbines that power electrical generators. A hydroelectric plant produces clean, inexpensive electricity, and a reservoir behind the dam can be used for flood control, drinking water, irrigation, and recreation.

Canada, China, Brazil, and the United States are the largest producers of hydroelectric power in the world. Together, North America and Europe generate about half the world's hydroelectric power. Hydropower supplies about 6 percent of total commercial energy production worldwide, but in the United States it supplies less than 3 percent and in Canada 9 percent of commercial energy consumption. Most of the best sites for hydroelectric generation are already in use in the United States and Europe, but in many areas there is considerable undeveloped potential. China, Brazil, Indonesia, Canada, and the Congo have especially large hydroelectric potential.

Opposition to construction of big dams and reservoirs is strong among environmentalists who fear the environmental damage they cause, such as loss of farmland or animal habitat. Hydroelectric dams may flood otherwise usable land and displace the people who lived on it. The dam converts a free-flowing stream to a lake, thus altering aquatic life. Many good sites for generating hydroelectric power remain in the world, but political considerations restrict their use, especially if the river flows through more than one country. For example, Turkey's recently built dam on the Euphrates River was strongly opposed by Syria and Iraq, through which the river also passes. They argue that the dam diverted too much water from the river and increased its salinity.

Despite these problems, hydroelectric power remains an attractive alternative to fossil- or nuclear-fueled power plants, largely because it generates no pollution. In rapidly industrializing countries such as China, increased electric generation can mean an increase in the standard of living not only in industrial cities but in small towns where electricity provides clean energy for indoor cooking and lighting and operates pumps that help provide clean drinking water. The case of the Three Gorges Dam, which has drawn much international criticism for its negative impacts on the environment and human settlements, does have the advantage of providing a clean alternative to dirty coal-generated electricity.

Solar energy

Solar energy—energy derived directly from the Sun—offers the best potential for providing the world's energy needs in future centuries. The Sun is a nonpolluting and perpetual source of energy.

At present, solar energy is used in two principal ways: thermal energy and photovoltaic electricity production. Solar thermal energy is heat collected from sunshine. Collection may be achieved by designing buildings to capture the maximum amount of solar energy. Alternatively, special collectors may be placed near a building or on the roof to gather sunlight. The heat absorbed by these collectors is then carried in water or other liquids to the places where it is needed.

Photovoltaic electric production is a direct conversion of solar energy to electricity in **photovoltaic cells.** Each cell generates a small electric current, but banks of them wired together can produce a large amount of electricity. Solar-generated electricity is now used in pocket calculators and where conventional power is unavailable, such as spacecraft and remote places on Earth. As more photovoltaic cells are produced, and as the technology and efficiency are improved, the cost of solar power will decline. Photovoltaic cells are already competitive with conventional energy sources in many new residential and commercial installations (Figure 9-17).

Solar energy can be generated either at a central power station or in individual homes. Many countries are wired for central distribution, so central generation

FIGURE 9-17 Photovoltaic cells supplying electricity to a home in Hamburg, Germany. Although it is more common to install such devices in new structures older building can be renovated to make use of solar energy. (Mike Schroder/Peter Arnold, Inc.)

by utilities makes sense. But solar power technology now makes feasible individual home systems. An installation costing several thousand dollars provides a solar energy system that provides virtually all heat and electricity for a single home. Because the high installation price is offset by low monthly operating costs, home-based solar energy is economical for consumers who remain in the same house for many years. Individual solar energy users do not face rising electric bills from utilities that pass on their cost of purchasing fossil fuels and constructing facilities. The United States, Israel, and Japan lead in solar use at home, mostly for heating water. Solar energy is likely to become more attractive as other energy sources become more expensive.

Wind generation of electricity is one of the fastest-growing solar technologies today. Significant improvements in turbine efficiency in recent years, combined with a growing demand for clean electricity, are making commercial-scale wind projects much more attractive. Large-scale wind-generating facilities are popping up across the landscape in many areas. These facilities, economical and clean as they are, are not without controversy, however. Environmentalists are divided: some favor wind generation because it is renewable and pollution-free, while others oppose it on the grounds of its adverse visual impacts or its hazards to birds. For example, a proposed array of wind generators off the coast of Cape Cod would produce as much electricity as a small conventional power plant, but it is facing stiff opposition from boaters, owners of recreational property on the coastline, and some environmentalists. Nonetheless the new growth in wind energy production signals a significant trend in the shift toward renewable energy.

Transition to new energy sources

The world contains a variety of energy sources. In addition to fossil fuels, people make use of hydroelectric, biomass, solar, and nuclear power. Other technologies are also in commercial use today, including geothermal, wind, and tidal power.

The emergence of new energy technologies suggests that we are beginning the transition from a fossil-based energy system to a new mix of energy supplies. Oil and gas are likely to remain important for several decades, but they likely will diminish in importance relative to other resources, especially renewable alternatives.

At present, the emerging energy sources are less versatile than oil and gas and are likely to find only specialized uses. Solar thermal energy might be used to heat buildings, whereas photovoltaic cells can power small electrical appliances. Centralized electric power plants—coal, nuclear, or hydroelectric—will probably remain the major source of energy for heavy users, such as large factories and shopping malls.

If history is a reliable guide, the mechanism that will drive this transition to new energy supplies is the market. If the price of oil rises significantly, alternative technologies that are currently uncompetitive would become attractive. As these new technologies are more extensively used, their prices will decrease, further encouraging a shift away from oil.

Conservation is equivalent to discovering a new energy source, and in all likelihood the greatest source of new energy resources will be conservation. More efficient use of energy means that we can produce more goods, operate more motor vehicles, and heat more buildings with the same amount of energy. Conservation was one factor in the slow growth of energy consumption since the 1970s. Sluggish demand for petroleum contributed to the decline in oil prices in the 1980s; similarly, slow growth in electricity demand contributed to the demise of the nuclear power industry.

The slow growth of electricity consumption in the 1980s and resulting lack of investment in new generation capacity contributed to new problems at the end of the 1990s. In 2000 a crisis of electricity supply in California arose. Electricity demand grew steadily in California in the 1990s, increasing by 11 percent between 1990 and 1999. In that same time period California's generation capacity decreased by about 2 percent, largely as a result of the closure of aging generating facilities. The state deregulated the electric energy system in the hope of stimulating production, but maintained controls on prices paid by consumers. Utilities were unable to meet rising demand, and they were forced to buy electricity from other states at prices that were higher than they were permitted to charge consumers. When the utilities were unable to afford to pay for added power, they cut supply to consumers, causing blackouts.

The electric energy crisis in California and the 2003 blackout in the northeastern United States and Canada are believed by many to be signs of problems that may be experienced with increasing frequency in the coming years, as electric energy demand grows while the production capacity and distribution infrastructure have remained relatively unchanged. One important factor contributing to the growth of electricity demand is the rapid expansion of the Internet. The computers that host websites and transmit data across the Internet consume large amounts of electricity. In addition, more and more homes and businesses have computers and other electric devices. While individually these consume relatively little power, their expansion collectively increases demand. It is likely that this, along with increasing population, will create the need for many new power plants and transmission lines in the coming years.

The steep increases in oil prices of 2005 and 2006 are prompting the United States to seek new sources of energy. Energy independence is virtually impossible, but we may find more secure sources, albeit at higher prices. Are we willing to abandon the unstable Middle Eastern Islamic regimes and pay this price? Our energy dependence has previously affected our foreign policy. In 1991, then-Secretary of State James K. Baker said that the 1991 war against Iraq (which was triggered by Iraq's invasion of Kuwait) was really about oil and American jobs. As long as the United States imports oil from the Persian Gulf the American economy and American foreign policy will remain hostage to these governments. In addition to urging exploration for new energy sources within American territory in places such as the Alaska National Wildlife Reserve, instability in the Middle East has increased U.S. interest in Africa and Central Asia—areas in which there is potential for new oil development.

The United States has several options for meeting its energy needs. It might burn more coal, perhaps installing improved devices to protect against air pollution. Nuclear energy might win new proponents. Conservation and renewable energy sources also offer much potential. But none of these alternatives to coal and oil will grow significantly without the incentives provided by much higher energy prices. We can provide those incentives through government policy, or we can wait for increasing global demand for oil to drive prices up, forcing us to find alternatives. In any case, oil is certain to become increasingly expensive, and less attractive as the basis of our economy.

AIR AND WATER RESOURCES

Consumption of resources is half of the resource equation—waste disposal is the other half. Burning fossil fuels pumps carbon dioxide into the atmosphere. The sulfur from burning oil and coal also enters the atmosphere and returns as acid deposition. Thus, the geography of resource consumption is also the geography of environmental pollution.

Air and water share two important properties: They are critical to human and other life on Earth, and they are useful places to deposit waste. An extreme environmentalist may argue that we should not discharge *any* waste into air and water, but in practice we need to rely on air and water to remove and disperse *some* waste. Not all human actions harm the environment, for the air and water can accept some waste. For instance, when we wash chemicals into a river, the river may dilute them until their concentration is insignificant. **Pollution** (elevated levels of impurities in the environment caused by human activity) results when more waste is added than a resource can accommodate.

Pollution levels generally are greater where people are concentrated, especially in urban areas. When many people are clustered in a small area, the amount of waste they generate is more likely to exceed the capacity of air and water to accommodate it.

Air Pollution

The purity of air is paramount to life on Earth. Some air pollutants come from natural processes unrelated to human actions, such as dust, forest fire smoke, and volcanic discharges. Humans add to this by discharging into the atmosphere smoke and gas from burning fossil fuels, incinerators, evaporating solvents, and industrial processes.

The atmosphere is constantly stirred by temperature and pressure differences that mix vertically and horizontally. As it moves from one place to another, air carries with it various wastes. The more waste we discharge to the atmosphere, or the less the air circulates, the greater the concentration of pollution.

Average air at the surface contains about 78 percent nitrogen, 21 percent oxygen, and less than 1 percent argon. The remaining 0.04 percent of air's composition includes several *trace gases*. *Air pollution* is a human-caused concentration of trace substances at a greater level than occurs in average air. In addition to the carbon dioxide (CO_2) emitted by "clean" combustion of fossil fuels, the most common air pollutants include **carbon monoxide, sulfur oxides** (SO_x, where the subscript x stands for the number of oxygen atoms), **nitrogen oxides** (NO_x), **hydrocarbons,** and **particulates** (very small particles of dust, ash, and other materials). Concentrations of these pollutants in the air can damage property and affect the health of people, other animals, and plants (Figure 9-18).

Each pollutant entering the atmosphere behaves differently. For example,

▷ SO_x and NO_x combine with water and fall to Earth as acid precipitation.

▷ *Particulates* in smoke are cleansed from the atmosphere quickly by gravity and precipitation.

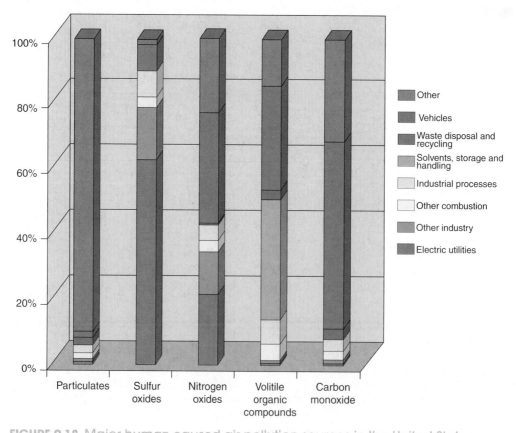

FIGURE 9-18 Major human-caused air pollution sources in the United States, 2002. Nationwide, particulates come mostly from soil erosion, but in urban areas they are derived from vehicles and industrial sources. Sulfur oxides are mostly derived from industrial sources. Transport, mostly cars and trucks, is the most important source of carbon monoxide as well as a major source of nitrogen oxides and hydrocarbons. (U.S. Census Bureau)

▶ *Photochemical smog*—the product of *hydrocarbons*, as well as NO$_x$ and sunlight—is created by chemical reactions that occur in the atmosphere itself.

▶ *Chlorofluorocarbons* (CFCs), chemicals used as refrigerants and in a variety of industrial applications, remain in the air long enough to be widely dispersed and carried into the upper atmosphere, where they damage Earth's protective ozone layer.

We are polluting our air in many ways today. We will focus on two important air pollution issues: acid deposition and urban air pollution.

Acid deposition More commonly known as acid rain, **acid deposition** occurs when sulfur oxides and nitrogen oxides, produced mainly from burning fossil

fuels, are discharged into the atmosphere. The sulfur and nitrogen oxides combine with oxygen and water in the atmosphere to produce sulfuric acid and nitric acid. When dissolved in water, the acids may fall as *acid precipitation,* or they may be deposited in dust. We use the term *acid deposition* to include both types of pollution (Figure 9-19).

Acid deposition seriously damages lakes and kills fish and plants. But the most severe damage of acid deposition is to the soil. The acid deposits harm the soil in a number of ways. Some of the acid deposited in soil is neutralized by calcium, magnesium, and other naturally present chemicals, but the amount of acid deposited can exceed the capacity of the soil's chemicals to neutralize the acid. If soil water grows too acidic, plant nutrients are leached away and are unavailable to plants. Acid deposition can dissolve aluminum in the soil, which can be toxic to plants and interfere with their nutrient uptake. Acids may also harm the soil-dwelling worms and insects that decompose organic matter.

Acid deposition is a regional problem, most severe in the densely populated industrial regions in Europe, eastern North America, and eastern Asia. It is a major source of damage to forests, especially in the industrialized and densely populated regions of eastern United States, Central Europe, and eastern China, where high levels of SO$_x$ discharges combine with precipitation to form acids (Figure 9-20). Much of the acid deposition in the northeastern United States is derived from emissions in the Ohio Valley carried eastward on the prevailing west-to-east winds. Damage to individual forests varies widely, depending on its age, tree species,

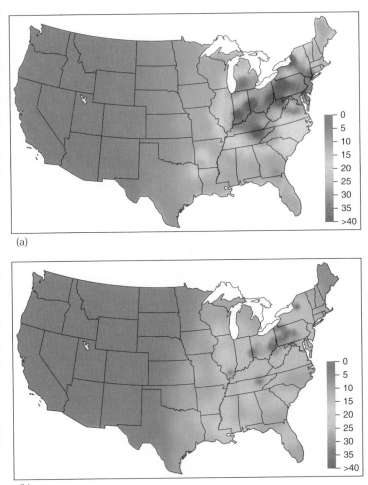
(a)

(b)

FIGURE 9-19 Acid deposition in North America. Wet sulfate deposition rates in the conterminous United States, 1989–1991 and 2001–2003. Acid deposition problems are caused by both nitrate and sulfate. The areas of North America most affected by acid deposition are in the Ohio Valley, eastern Great Lakes, and populated areas of the eastern United States. Acid deposition from sulfate, the larger component of acid deposition, has declined significantly as a result of emissions controls. Acid deposition from nitrate (not shown) has declined less, and has actually increased in some areas. (U.S. Environmental Protection Agency)

FIGURE 9-20 Trees in the Black Forest, Germany, suffering from forest decline probably caused by acid deposition. The Black Forest, in the heart of Europe's industrial region, has been a focus of concern about the effects of acid deposition. (Gernot Huber/Woodfin Camp & Associates)

the capacity of the soil to neutralize acids, and interactions between trees and other organisms in the forest. Because the relationship between tree damage and high discharges of SO_x and NO_x has not been documented precisely, some governments are unwilling to impose the cost of controlling emissions on their industries and consumers.

Nonetheless, significant progress has been made. Since the early 1970s the United States has reduced SO_x emissions about 47 percent. Over this same time period, emissions have been cut by larger percentages in other relatively wealthy developed countries, including 60 percent in Canada, 84 percent in France, 92 percent in Germany, and 92 percent in Sweden. NO_x emissions, which are more difficult than SO_x to control, have remained at about the same level in the United States during the past quarter century. Although precise figures are not available, SO_x emissions have probably increased in developing countries, especially China, which is responsible for nearly one-fourth of the world's coal combustion, the major source of SO_x emissions. China's sulfur emissions are a major source of acid deposition in Japan, again borne by prevailing westerly winds. In an attempt to reduce this problem, Japan is helping pay for the installation of pollution-control equipment in China.

Urban air pollution Urban air pollution results when a large volume of emissions is discharged into a small area. The problem is aggravated in cities when the wind cannot disperse these pollutants. Urban air pollution has three basic components:

1. Proper burning in power plants and vehicles produces carbon dioxide (CO_2), but incomplete combustion produces carbon monoxide (CO). Breathing CO reduces the oxygen level in blood, impairing vision and alertness and threatening persons who have chronic respiratory problems.

2. Hydrocarbons also result from improper fuel combustion, as well as from evaporation of solvents as in paint. Hydrocarbons and NO_x in the presence of sunlight form **photochemical smog,** which causes respiratory problems and stinging in the eyes.

3. Particulates include dust and smoke particles. You can see particulates as a dark plume of smoke emitted from a smoke stack or a diesel truck—not a white plume, which is condensed water vapor. Many particles are too small to see, however.

Three weather factors are critical to urban air pollution: wind, temperature, and sunlight.

1. When the *wind* blows, it disperses pollutants. When the air is calm, pollutant concentrations build up.

2. Air *temperature* normally drops rapidly with increasing altitude. But over cities, conditions sometimes cause **temperature inversions,** in which warmer air lies above cooler air. This limits vertical circulation, trapping pollutants near the surface (Figure 9-21).

3. *Sunlight* is the catalyst for smog formation.

As a result of these three factors, the worst urban air pollution occurs under a stationary high-pressure

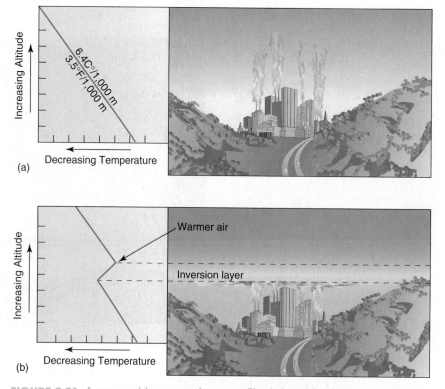

(a)

(b)

FIGURE 9-21 A normal temperature profile (a) and a temperature inversion (b) above a city. The temperature inversion traps pollutants. (From R. W. Christopherson, *Geosystems,* 3rd ed. Upper Saddle River, NJ: Prentice Hall, Inc., 1997.)

cell, where a combination of slight winds, temperature inversions formed by descending air, and clear skies allow pollutants to accumulate. A city that experiences frequent stationary highs, such as Denver, Colorado, has frequent pollution problems.

Mexico City is notorious for severe air pollution, especially in winter, when high pressure often dominates, and the surrounding mountains discourage dispersal of pollutants by wind (Figure 9-22). In the eastern United States, pollution problems are worst in summer and autumn, because stationary highs are most common then. In West Coast cities such as Los Angeles and San Francisco, the pollution "season" is also summer and autumn because inversions and bright sunshine are more persistent then.

Progress in controlling urban air pollution is mixed. In relatively wealthy, developed countries with strict regulations, air quality has improved. Controls on use of coal and improvements in automobile engines, manufacturing processes, and the generation of electricity have all contributed to higher-quality urban air.

To reduce auto emissions in the United States, for example, catalytic converters that oxidize unburned hydrocarbons have been attached to exhaust systems since the 1970s. As a result, carbon monoxide emissions have declined by more than three-fourths and nitrogen oxide and hydrocarbon emissions by more than 95 percent. Gains in relatively developed countries have been offset somewhat, though, by increased use of cars and trucks in recent years.

In many developing countries, urban air pollution is getting worse (Figure 9-23). Although ownership of motor vehicles is less common than in wealthy nations, the cars and trucks that do exist are older and lack pollution controls found on newer vehicles in relatively developed countries. Instead of relying on gas or electricity, many urban residents in developing countries burn wood, coal, and dung for cooking and heating. These smoky fires can create acute air pollution problems in poorly ventilated areas, even in the countryside. In developing countries an estimated 4 million children die each year from acute respiratory problems, for the most part caused or aggravated by air pollution.

Water Pollution

Water is our most immediate resource other than air—it is consumed daily and comprises about 70 percent of our bodies. Oceans occupy 71 percent of Earth's surface. From the sea we obtain fish, shellfish, oil, gas, sand, salt, and sulfur. The seafloor may someday yield manganese and cobalt and may become a burial site for nuclear waste. Countries with inadequate freshwater supplies, such as Saudi Arabia, desalinate seawater.

FIGURE 9-22 Mexico City suffers from a combination of circumstances that all lead to significant air pollution problems. The city lies in a mountain basin that limits dispersion of pollutants. Automobiles are numerous and traffic is heavy, and there are few emission controls. In the early 1990s the problem reached crisis proportions, and a pollution-control effort is now under way. (Wesley Bocxe/Photo Researchers, Inc.)

FIGURE 9-23 Satellite image of air pollution in northeastern India and Bangladesh. The grey-colored haze seen in this image is common in winter, often remaining for weeks at a time. (Jacques Descloitres, MODIS Rapid Response Team/NASA Headquarters)

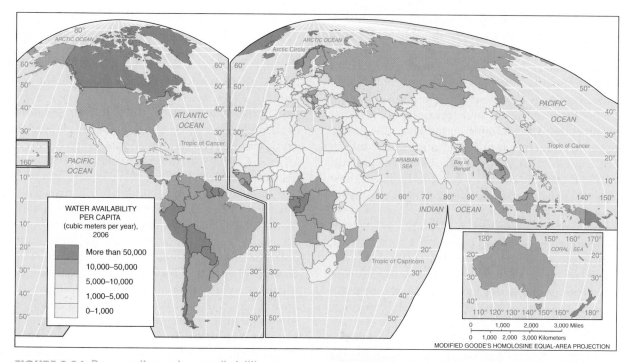

FIGURE 9-24 Per capita water availability. Water is most available in South America, Central Africa, Russia, and Canada. These places have high amounts of runoff and relatively low populations. Major problems of inadequate water supplies occur in India, China, and other densely populated countries, as well as semiarid and arid regions. (World Resources Institute)

In the United States, about 1.5 billion cubic meters (400 billion gallons) of freshwater are pumped from the ground or from rivers and lakes each day. About 12 percent of the water is used in homes and businesses, 53 percent in industry (most of which is for electric power production), and 35 percent in agriculture. About two-thirds of the water we use is returned to rivers and lakes in liquid form, while the remaining one-third is evaporated, mostly from irrigated fields.

Water availability varies greatly around the world. South America has vast water resources (about 18 percent of all the freshwater runoff in the world flows to the sea via the Amazon River) and a relatively sparse population. Asia, on the other hand, is home to three-fifths of the world's people, and large parts of this vast continent are relatively dry, so per capita water availability is limited (Figure 9-24). Use of water for irrigation is especially significant in dry regions because the water is evaporated and not available for reuse (Figure 9-25).

Water is a "universal solvent." It can dissolve a wide range of substances, and it can transport bacteria, plants, fish, sediment, toxic chemicals, and trash of all kinds. As with air pollution, water pollution results when substances enter the water faster than they can be carried off, diluted, or decomposed. Water pollution is measured as the amount of waste being discharged in relation to a body of water's ability to handle the waste.

Pollutants have diverse sources. Some come from a *point source*—they enter a stream at a specific location, such as a wastewater discharge pipe. Others may come from a *nonpoint source*—they come from a large diffuse

FIGURE 9-25 The bed of the Yellow River (Huanghe), exposed at low flow. Diversion of water from the river for irrigation has severely depleted its flow, and today the river barely reaches the sea for much of the year. Water is in critically short supply in much of northern China. (Panorama Images/The Image Works)

area, as happens when organic matter or fertilizer washes from a field during a storm. Point-source pollutants are usually smaller in quantity and much easier to

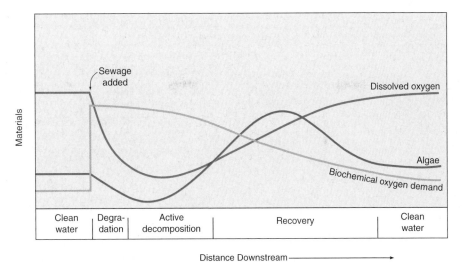

FIGURE 9-26 Water quality varies downstream from a point where sewage is added. Dissolved oxygen is depleted as organic matter is consumed, but it recovers further downstream. Nutrients are released to the water, and algae increase downstream as a result. (Adapted from H. B. N. Hynes, *The Biology of Polluted Waters.* Liverpool: University of Liverpool Press, 1978.)

control. Nonpoint sources usually pollute in greater quantities and are much harder to control.

The *concentration* of pollution, such as sewage effluent, usually declines downstream from where the waste is discharged (Figure 9-26). This reduction occurs because the waste is diluted, and natural processes decompose pollutants and remove them from the water.

Reduction of oxygen in water

Because aquatic plants and animals need oxygen, the **dissolved oxygen** concentration in water indicates the health of a stream and lake. The oxygen consumed in decomposing organic waste constitutes the **biochemical oxygen demand.** If a lake or stream contains an excessive amount of decomposing waste, the oxygen demand is too great and the water becomes oxygen starved, killing fish and other animals living in the water.

This condition often occurs when water becomes loaded with municipal sewage or industrial waste. The sewage and industrial pollutants consume so much oxygen that the water can become unlivable for normal plants, animals, and even fish. Similarly, when runoff carries fertilizer from a farm field into a stream or lake, the fertilizer nourishes excessive production of algae, which consume oxygen when they decompose.

Agriculture is the leading consumer of water, primarily for irrigation, and it is also the leading water polluter. Runoff from agricultural land discharges sediment, fertilizers, animal waste, and small quantities of chemicals into nearby streams. The fertilizer and animal waste, which are rich in nitrogen compounds, can nourish aquatic plants to the point of overproduction. Decomposition of these plants consumes so much oxygen from the water that fish may be unable to survive (Figure 9-27).

FIGURE 9-27 Water quality variations in lakes in the Minneapolis area revealed by remote sensing. This image was created using the relation between light reflectance from water and concentrations of nutrients and sediments. Lakes that are blue have generally good water quality, green and yellow lakes are intermediate in quality, and those that are red have poor water quality. This map was created by a combination of GIS and remote sensing technologies. (Image courtesy Upper Great Lakes Regional Earth Science Applications Center/NASA)

Wastewater and disease

Wastewater comprises about 15 percent of the flow of U.S. rivers. Improved wastewater treatment is critical, especially in a world with a rapidly growing human population. Rich, industrial countries like the United States generate more wastewater than do poorer nations, but they also have greater capacity to treat this wastewater. In the rich countries, strict legislation requires treatment facilities to be upgraded, making their rivers cleaner than a few decades ago. However, many of the ambitious water-quality goals of this country's Clean Water Act, which was first passed in 1972, have not yet been met.

In the developing world, untreated sewage often goes directly into rivers that also supply drinking water. A combination of poor general sanitation, nutrition, and medical care can make the drinking of water deadly. Waterborne diseases, such as cholera, typhoid, and dysentery, are major causes of death in developing countries. Because of improper sanitation, millions of people in Asia, Africa, and South America die each year from diarrhea. As people in these rapidly growing regions crowd into urban areas, drinking water becomes less safe, and waterborne pathogens flourish (Figure 9-28).

Chemical and toxic pollutants

Any waste discharged onto or into the ground may pollute streams or groundwater. Pesticides applied to lawns and golf courses find their way into streams and groundwater. Landfills and underground tanks at gasoline stations can leak pollutants into groundwater, contaminating nearby wells, soil, and streams. Petroleum spilled from ocean tankers contaminates seawater. During times of

flooding, normally secure tanks of chemicals may be dislodged and broken open and their contents mixed with the floodwaters.

Toxic substances are chemicals that are harmful even in very low concentrations. Contact with them may cause mutations, cancer, chronic ailments, and even immediate death. Major toxic substances include PCB oils from electrical equipment, cyanides, strong solvents, acids, caustics, and heavy metals such as mercury, cadmium, and zinc.

During the 1950s and 1960s, toxic wastes were often buried, but by the 1970s this method of isolating toxic wastes had proved inadequate because many waste sites were leaking. One of the most notorious was Love Canal, near Niagara Falls, New York, where several hundred families were exposed to chemicals released from a waste disposal site used by a chemical company. The chemicals had been left by the Hooker Chemicals and Plastic Company, which had buried toxic wastes in metal drums in the 1930s. In 1953 a school and several hundred homes were built on the waste site. Eventually the metal drums were exposed, and beginning in 1976, residents noticed a strong stench and slime oozing from the drums. They began to suffer from high incidences of health problems, including liver ailments and nervous disorders. After four babies on the same block were born with birth defects, New York state officials relocated most of the families and began an expensive cleanup effort.

Love Canal is not unique—toxic wastes have been improperly dumped at thousands of sites. It was, however, very significant in raising public concern about groundwater pollution and stimulating action to combat it. In the United States such polluted sites are slowly being cleaned up, but the cost is tremendous. As safe and legally approved toxic-waste disposal sites become increasingly hard to find, some European and North American firms have tried to transport their waste to West Africa and other areas where pollution regulations are less strict.

The U.S. government has spent well over $100 billion in construction of new sewage-treatment plants since the mid-1960s. Industry has also spent heavily to meet water-quality standards. Many streams and lakes that had become open sewers now are suitable for recreation. But the job is unfinished. Serious efforts are made to control chemical waste in relatively developed nations, but developing nations often give little attention to this problem, and wastes are freely released into streams.

FIGURE 9-28 Women washing and collecting water at a river in India. Less than 25 percent of the wastewater generated in India's 12 largest cities is collected by sewers. Most smaller cities have no wastewater treatment at all. It is estimated that between 5 and 10 million persons die each year from diarrhea caused by waterborne organisms in Asia, Africa, and Latin America. (Macduff Everton/The Image Works)

Reducing Air and Water Pollution

The most often used pollution-control strategy is to remove pollutants from water or exhaust gases before they are discharged to the environment. Control of air pollution in the United States and other industrialized nations

Meat Production and Water Pollution

Farm animals such as cattle, pigs, or fowl produce considerable amounts of waste that can constitute a significant source of water pollution. Many rural streams carry substantial loads of nitrogen, phosphorus, and partially decomposed wastes. In rural areas agriculture is usually the dominant source of water pollution, and feedlots (where animals are kept and fed) are the most obvious and concentrated pollution source in the rural landscape. As just about anyone who lives in a farming region in eastern North America can confirm, pigs are especially noticeable in this regard.

In the late 1980s a radical transformation began in the U.S. pork production industry. Prior to that time, the usual pig production system was a form of mixed agriculture combining grain and meat production. A typical farm may have included perhaps 100–200 hectares (250–500 acres) of land on which grain—primarily corn—was grown. Most of this was fed to a herd of perhaps 100–200 pigs kept in a feedlot, preferably downwind from the farmstead. But a new form of pig production is growing rapidly and replacing the traditional system. Huge industrial-scale feeding operations are springing up across the central and eastern United States. The new farms, if they can be called that, have pig populations in the thousands, fed from grain purchased on the grain market and not necessarily locally produced. These farms tend to be concentrated in specific areas, often supplying a single centralized slaughterhouse. For example, a slaughterhouse in Guymon, Oklahoma, processes 2 million pigs per year, mostly drawn from a few surrounding counties. Large pig-feeding operations enjoy economies of scale that allow them to fatten animals much more cheaply than can be done on a traditional farm.

How does this concentration of pig-raising affect water quality? In many areas where new industrial-scale operations have sprung up, the impacts have been severe locally. An adult pig produces significantly more waste than an adult human. A facility feeding 5,000 pigs produces more sewage than a small city. And while facilities for collecting and treating the wastes are required, standards for design and operation are usually not as stringent as they are for municipal sewage-treatment plants. Wastes are often stored in lagoons, and in a few instances the dams holding these wastes have failed, turning the rivers downstream into open sewers. Pollution from commercial-scale pig and chicken farms has been blamed for outbreaks of a mysterious organism called pfisteria that is killing fish in many streams in Maryland, Virginia, and North Carolina.

Concentration of animal production, however, does not necessarily reduce water quality. If wastes are produced at a few centralized locations instead of many scattered small farms, then it becomes easier to collect and treat. It is also easier—legally and politically as well as technically—to regulate and inspect waste-treatment systems. If industrial-scale feedlots are carefully regulated, they might provide an opportunity to manage a form of pollution that otherwise has been largely unmanageable. Unfortunately, the existing regulations have not been strong enough to prevent significant water-quality degradation in the immediate vicinity of these facilities, and local water quality is suffering.

---

usually means installing equipment that either reduces the amount of waste gases generated or removes wastes from the exhaust. For example, particulate emissions from coal combustion are often controlled with devices called electrostatic precipitators, which use the principle of electric charge to attract particles or other systems that remove ashes from the smoke. SO_x emissions are controlled both by burning fuels that contain less sulfur and by using devices that remove sulfur from exhaust gases after combustion. Similarly, automobiles are fitted with catalytic converters that reduce harmful emissions in the exhaust and with computer-controlled fuel-injection and ignition systems that generate less pollution by increasing the efficiency of combustion.

The largest source of water pollution—sewage—is also controlled by cleaning water before it is discharged. In a typical sewage-treatment plant, large solid particles are screened from the water or allowed to settle and the remaining water is oxygenated to allow bacteria to break down organic matter. The water is then discharged to the environment. In advanced systems, some of the products of that breakdown, such as nitrogen and phosphorus, are removed. Most of the world's population living in developed countries is served by some kind of sewage collection and treatment system. In developing countries, however, sewage systems are less common. In Latin America, for example, only about 80 percent of the urban population is served with sewage collection systems, in Asia about 60 percent have such service, and in Africa the figure is only 53 percent. In most poor countries the sewage collected in urban areas is not treated, but simply piped to rivers or the sea. In rural areas of developing countries, sanitation systems are uncommon.

Instead of using these so-called "end-of-pipe" approaches that remove pollutants from air or water *after* they are produced, some pollution can be controlled by simply not producing it in the first place. This is known as **pollution prevention,** as opposed to pollution *control*. Many opportunities exist for reducing or eliminating industrial pollution without sacrificing the quality of products or increasing manufacturers' costs. For example, toxic cleaning products are used in manufacturing to remove oil, grease, soldering residues, dust, coatings, and metal fragments. Trichloroethylene (TCE) is an excellent solvent, but it is also very toxic. Less toxic cleaning agents can be substituted, such as alcohol or detergents.

Another approach is to recycle polluting substances instead of discharging them into the environment. Some industrial processes, such as molding plastics and stamping sheet metal, generate recyclable scraps. Normally, larger scraps are recycled, but small fragments, like dust from polishing or sanding, instead may be placed in landfills as solid waste or discharged in water. Capturing and recycling these fragments has dual benefits: It reduces pollution while increasing the supply of scrap for making new materials. Health benefits also may accrue if the dust presents a health hazard. In this way, environmental externalities are eliminated without increasing producers' costs.

In the early 1990s a fundamental shift in pollution management began, from pollution control to pollution prevention. This change was spurred by several factors:

1. Despite pollution controls, significant water and air-quality problems remained.
2. Polluters faced legal liability if they discharged substances that were *later* found to cause harm, so reducing discharges to the environment reduces potential later liability.
3. Pollution prevention often is cheaper than pollution control.
4. Publicizing its efforts to reduce pollution can improve a company's image and help it market its products, recruit employees, and negotiate with government regulators.

Pollution prevention has quickly been adopted by many industries, especially in the relatively rich developed countries of Europe and North America. Large corporations—which are especially visible to governments, consumers, and the general public—have led the way in reducing discharges into the environment. Companies increasingly recognize that the best way to prevent pollution is to design a product that can be manufactured with a minimum of toxic chemicals, used without generating pollution, and which can be recycled easily when it is no longer wanted.

While industrialized countries have been successful in reducing pollution during the past few decades, pollution problems are increasing in many poor countries. Most of the differences in pollution between rich and poor countries result from differences in wealth and level of industrial development. Where electricity, clean-burning fuels, and pollution-control devices are widely available—as in Europe, North America, and Japan—urban air pollution mostly has been controlled. Problems are much more acute in places where people cook and heat with wood and drive old vehicles.

As manufacturing expands in developing countries, new facilities could be built with pollution control in mind. But pollution controls can be costly. Reducing the threat of large-scale pollution in developing nations requires development of pollution-prevention methods that reduce—not increase—costs.

FORESTS

As the world population grows and resources are used for more purposes, conflicts over resource use are inevitable. In some cases the conflict concerns *who* has access to a resource in short supply, such as water in a desert region. But increasingly, conflicts concern *how* a resource should be used. Should a valley be dammed to generate hydroelectricity, or should people continue to live there and practice their traditional way of life? Should a wilderness area be protected as a habitat for endangered species, or should it be opened for development?

One resource may have competing uses, some of which may be incompatible with others. Balancing competing and incompatible uses requires careful management, and difficult choices must be made. In this section we will review some of the important uses of Earth's forestlands and the reasons why these lands have long been, and will continue to be, controversial.

About 4.1 billion hectares (10.1 billion acres) or roughly one-third of Earth's land surface (excluding Antarctica) is covered with forest and woodland. The amount of forestland has decreased substantially as a result of clearing land for agriculture, and experts estimate that forests may originally have covered as much as half of Earth's land surface.

A forest is part of an ecosystem in which vegetation plays a major role in biogeochemical cycles. By absorbing carbon dioxide and releasing oxygen, a forest is an essential part of a local climate; it also stores carbon that otherwise would be in the atmosphere as carbon dioxide. A forest reduces erosion, aids in flood prevention, and provides a place for recreation. A forest is drained by rivers that provide drinking water, irrigation, habitat, and waste removal. People may use a forest as a habitat and a place to obtain food, fuel, shelter, and medicine.

Because forests cover such a large portion of Earth's surface, often bordering populated areas, they have become centers of natural resource conflicts. Whereas agricultural land is typically under the control of individuals, corporations, or local political authorities, forest land is more often communally owned. And because of this communal or semicommunal ownership, many different people may have some level of legitimate claim to forest land. This further intensifies resource-use conflicts. So as you think about forests, remember that conflicts occur over other multiple-use resources, especially ones like water and oceans that are commonly owned.

Forests as Fiber Resources

Forests, the wildlife they support, and the water that flows through them are all renewable resources. Conflicts occur between those who wish to harvest timber and those who wish to leave the trees standing (Figure 9-29).

The United States, Russia, China, India, and Brazil are major timber-harvesting regions because their citizens demand wood products (Figure 9-30). Only a relatively small percentage of the world's remaining forest land is protected from cutting. Nearly all forests in the United States (except those in Alaska)

FIGURE 9-29 Environmentalists and forests. These huge karri logs have been cut in Western Australia to be chipped for paper. Environmentalists decry this destruction of old-growth forest, particularly when paper can easily be made from much smaller logs. (Cary Wolinsky/PN/Aurora & Quanta Productions, Inc.)

have been cut at least once during the past 300 years, so very little original (virgin) forest remains. In Canada much more original forest remains, especially in the north and west. The timber industry wants to cut remaining original forests because their large, straight trees yield high-quality lumber (Figure 9-31). Harvesting

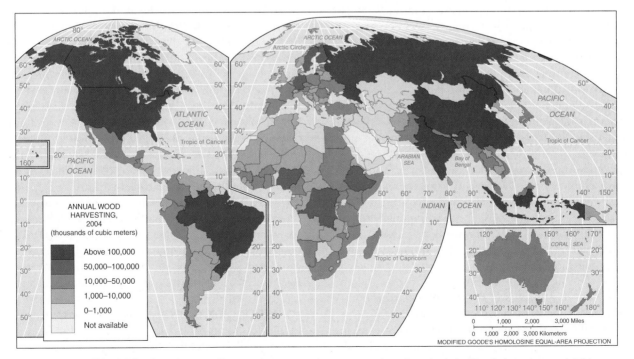

FIGURE 9-30 World timber harvesting. Most timber harvesting takes place in Asia, North America, and Africa, and lesser amounts are harvested in South America and Europe. Data from the former Soviet Union are lumped, with virtually all this harvesting occurring in Russia. (World Resources Institute)

Figure 9-31 Deforestation in Java, Indonesia. Indonesia and Malaysia are the leading exporters of wood products in Asia. (Georg Gerster/Photo Researchers, Inc.)

provides jobs and benefits lumber companies bykeeping the price of timber low. Restrictions on cutting original-growth forests would possibly increase unemployment in regions where these forests are being harvested.

Environmentalists argue that when trees are cut, the forest no longer supports the same wildlife or maintains clean water as effectively as it did before. Cutting down a forest places its inhabitants at risk. During the late 1980s, the spotted owl was endangered in the original-growth forests of Oregon and Washington when timber interests wanted to cut the habitat. Environmentalists point out that U.S. laws prohibit the government from actions that could harm an endangered species. U.S. government attempts to find a compromise between the timber industry and environmentalists has proved elusive, because the two uses of the original-growth forest resource—commercial timbering and as habitat for the spotted owl—ultimately are incompatible.

Sustained yield management is a strategy that can maintain the productivity of a resource even as it is being used. In forestry, sustained yield management means that the number of trees harvested should not exceed the number replaced by new growth. The strategy also emphasizes harvesting in a manner that minimizes soil erosion, and therefore enhances the ability of the forest to regenerate. Sustained yield management is the official policy of the U.S. Forest Service, which manages the national forests. In practice, however, the Forest Service has allowed timber to be cut faster than mature trees are produced in national forests. As a result, we have a shortage of harvestable, mature timber in key areas such as the Pacific Northwest.

Similar conflicts regarding forest use are taking place in developing countries. For example, during the 1980s tropical deforestation became a global issue, primarily because the rate of deforestation in the Amazon basin was increasing rapidly. Environmentalists argued

that once the forest was cut, it could never be restored to its original condition. Species would become extinct, the soil would be ruined, and indigenous cultures would be wiped out. But those involved in clearing the Amazon forests responded that most of the forestland in the United States and Western Europe had already been cleared, some of it many centuries ago. By what right can relatively wealthy developed countries tell Brazilians to save their trees?

Ideally, the solution to conflicting uses of forests is sustainable management. Trees can be cut, but not to the extent that the future productivity of the forest is reduced. Some original forest should be protected to provide habitat for native species and prevent extinctions. Resource managers in northwestern North America and in the Amazon are working toward such compromise solutions.

Other Important Forest Uses

Recreation in forest areas
In many countries, especially wealthier ones, forests are important recreational resources. They are relatively undeveloped and thus have open space available for hiking, camping, and other outdoor activities. They offer a strong contrast to the noise, commotion, crowding, and pollution of cities. Their shade and water availability provide relief from summer heat. Many forest regions are located in mountainous areas that offer other amenities such as skiing, climbing, and the cooler weather of high elevations. And because they serve as habitats for game animals, they can be preferred areas for hunting and fishing.

In the United States and Canada many important recreational centers are located in forested regions. Consider, for example, the coasts of Maine and Nova Scotia, the shores of the northern Great Lakes, or the mountains of the West. The most heavily used national parks, such as Acadia, Great Smoky, Yellowstone, Banff, and Yosemite, are all located in forest regions. Many of these parks are bordered by national forest lands, and tourists drawn to the parks also enjoy the forests, both en route and when the parks are too crowded. In fact, in the United States, recreational use of the national forests (measured in number of visitor hours) is more than double that of the national parks.

The tourists who use these forest areas appreciate them for their natural (or natural-like) landscapes: tall trees, clean rivers, wildlife. They are not interested in seeing swaths of land with only stumps, sharing the road with lumber trucks, or hearing the buzz of chain saws. Recreational use clearly conflicts with timber harvesting.

Forests and biodiversity
As we learned in Chapter 4, human occupation of Earth has resulted in significant threats to global biodiversity. Although damage to biodiversity is probably greater in agricultural lands than in forests, the remaining forests of the world contain some of the most important reserves of species diversity.

The tropical forests are especially important centers of diversity by supporting both larger numbers of tree species and complex vertical habitats. Foremost among these is the Amazon, a vast region of more than 4 million square kilometers (1.5 million square miles), containing literally millions of species. The Amazon has been a focus of attention in part because of the relatively high rates of deforestation taking place there. Since the 1980s, for example, the Brazilian Amazon has been cleared at rates of slightly less than 1 percent per year, an area roughly the size of the state of Connecticut. Worldwide, perhaps 40,000 square kilometers (15,000 square miles) of tropical rain forest have been cleared annually in recent years.

Concern over this loss is based partly on practical considerations, especially the potential medicinal uses of natural substances. Many important medicines are derived from forest species, and as mentioned in Chapter 6, indigenous peoples occupying these lands often have very good knowledge of such medicinal values. When the tropical forest is destroyed, these cultures may also be lost in the process, so both the species that provide medicines and the valuable source of information about them are at risk.

But a more powerful argument in favor of biodiversity preservation may be a moral one: that humans, possessing unprecedented powers over nature and Earth, have a special responsibility to protect the natural environments and the species they contain from destruction by our own hands. Our great numbers and advanced technology do not bestow on us the right to eliminate other species from the planet.

Forests and carbon storage   Another long-term impact from cutting down forests may be the release of stored carbon into the atmosphere, through decay or burning of unusable parts of trees. Deforestation in the tropics may be responsible for as much as 10 percent of the increase in CO_2 in the atmosphere, as discussed in Chapter 2. Left alone, forest growth removes carbon from the atmosphere and stores it in the biomass of living trees and soil organic matter. We must therefore consider both deforestation and reforestation.

In much of the world, the rate of deforestation exceeds the rate of reforestation. Net deforestation is occurring, especially in the tropical rain forest, because people are cutting trees faster than they can be replanted or can regrow naturally. In some areas cleared forests do revert to second-growth woodlands, while in others the land is converted to other uses more or less permanently. The timber industry does replant forests, especially in the midlatitudes. And in some areas, notably the eastern United States, forests now are growing in areas that were farmed or grazed until the early twentieth century, when these activities became no longer profitable. Forests can require more than a century to regrow, however, so these second-growth forests lack some important habitat properties of the old growth

they replace. Nevertheless, as these forests grow, carbon is gradually removed from the atmosphere. The potential for removing carbon dioxide from the atmosphere in this way is considerable if forest growth is promoted worldwide. Some estimates indicate that as much as 15 percent of fossil-fuel emissions in the next half century could be trapped in biomass. Of course, this would preclude using these forests in some other ways.

Balancing Competing Interests

Political and economic relations are the key to any situation where competing interests battle for control of a scarce resource. Decisions on allocation of forests among various uses will be made both by governmental regulation and by the marketplace. In the case of government-owned lands in the United States and Canada, the political arena is where most decisions are made. The important interest groups include the lumber industry, environmentalists, recreational interests, and occasionally ranchers and the mining industry. Each of these has its own power base. Some may have strong allies in government, while others have based more mass support among voters. The fate of the remaining old-growth forest in the northwestern United States and southwestern Canada has been a particularly controversial issue because both of the principal competing interests—the lumber industry and environmentalists—are powerful groups with wide influence.

In market economies the relative values that society places on different forest uses may be expressed through prices of resource commodities. If the price of lumber rises, there is more pressure to harvest timber and sell lumber. If cheaper substitutes are available for lumber, then the forest is more likely to be available for other uses. Similarly, if lumber from one region becomes expensive because of, perhaps, government restrictions on harvesting, then lumber from other regions will be demanded instead. Japanese demand for lumber leads to forest clearing in Southeast Asia and Alaska, while U.S. lumber interests argue that protecting old-growth forest there will only lead to a loss of business as foreign lumber is imported from Canada and elsewhere.

Government regulators sometimes make use of market-based principles, as well as political considerations, in deciding how to manage resources. Those who argue for restricting timber harvesting in favor of recreation and forest preservation, for example, note that the value of timber harvests in U.S. national forests in the early 1990s averaged about $1 billion per year. The national forests also provided about 300 million visitor days of recreation. If each of these visitor days were valued at $3.00, then the recreational value of the forests would be about the same as the value of timber harvested. Of course, both of these activities took place simultaneously in different parts of the forests. But as forest resources become increasingly scarce, such comparisons will likely lead managers to restrict logging further.

Similarly, other values may compete with lumber in economic terms. Since the Earth Summit in Rio de Janeiro in 1992, international negotiations aimed at reducing global carbon dioxide emissions have been under way. Industrialized nations consuming large amounts of fossil fuels are arguing that it would be cheaper to limit CO_2 releases by curtailing deforestation than by reducing fossil-fuel use. They are proposing to establish forest reserves in tropical countries as an alternative to reducing fossil-fuel use at home. In theory, this increases the value of those forests for carbon storage and reduces the supply of timber available for harvesting. We will discuss this further in Chapter 13.

CONCLUSION: CRITICAL ISSUES FOR THE FUTURE

Whether we consider forests, water and air resources, energy, minerals, farmland, or any other natural resource, it is clear that these resources are coming under increasing pressure worldwide, and the environmental impacts of resource use are increasing. As long as population grows and/or per capita resource use increases (and in general these trends have been ongoing for centuries), competition for increasingly scarce natural resources will increase.

In the early 1970s an influential book called *The Limits to Growth,* by D. H. Meadows and others, argued that depletion of Earth's natural resources had placed us on a disastrous course. The report predicted that natural resource depletion, combined with population growth, would disrupt the world's ecosystems and economies and lead to mass starvation. If natural resource protection were not in place within 20 years, the report claimed, environmental systems would be permanently damaged, and everyone's standard of living would decline.

Few contemporary geographers accept the pessimistic predictions made by *The Limits to Growth* a quarter century ago. Use of many resources has declined significantly since then. We may still deplete some resources, but substitutes and other strategies are available. Although pollution continues to degrade natural resources, industrial development has been made compatible with environmental protection in some locations. Nevertheless, many natural resource problems are likely to intensify considerably in the coming decades because of continued population growth and economic development.

The developing countries face the greatest challenges. In places suffering from extreme poverty, sound management of resources for the future is difficult. Careful management of Earth's natural resources is even more difficult where population is increasing rapidly. The desire for economic growth, like that occurring in China, is a powerful incentive for increasing resource use and for valuing industrial uses of the environment above environmental protection. Air pollution in India and China is a result of that industrial expansion. We can only hope that the Chinese are able to reap the benefits of economic growth and use some of that wealth to invest in technology that will help to clean the air and reduce the damaging health effects it is causing.

Chapter Review

SUMMARY

A natural resource is an element of the physical environment that is useful to people. Cultural values determine how resources are used. Technological factors limit our use of some resources by determining the particular applications to which certain materials can be put. Economic factors such as resource prices and levels of affluence influence whether a resource is used, and how much. Renewable natural resources include air, water, soil, plants, and animals. Nonrenewable resources include fossil fuels and nonenergy minerals. Most resources are substitutable to some degree, so that if one resource is less available or more expensive another resource is available to take its place.

We depend on a great many different materials in our daily lives. Wealthy countries use large quantities of resources, causing depletion of some mineral resources. Mineral wastes and other materials accumulate in landfills. Recycling can help conserve landfill space as well as reducing resource use.

Over time society has changed its use of energy resources from wood to coal, oil and gas. At present the world is dependent on fossil fuels for energy. The United States has abundant coal resources. Although the United States is a major oil producer it must import most its oil from other countries. Growing worldwide demand has caused large increases in energy prices. As fossil fuels are depleted, we will need to use energy more efficiently and develop other sources of energy. Nuclear power, renewable electricity, and energy conservation are promising new sources of energy.

Air pollution is a concentration of trace substances at a greater level than occurs in average air. Acid deposition and pollution of urban areas are particularly harmful forms of contemporary air pollution. Acid deposition is a regional problem that is most acute in areas that burn large amounts of coal, or have large numbers of automobiles and fossil-fuel-fired power plants. Air pollution is particularly severe in urban areas where there is a large concentration of pollution sources. Water pollution results from both point and non-point sources. Industrial facilities and municipal sewage plants are important point sources while agricultural and urban runoff are significant nonpoint sources. Pollution prevention is a promising approach to reducing water pollution.

Forests are an example of a resource with many different uses, and conflicts over which uses are most important often arise. Among the important uses of forests are timber products such as lumber and paper, recreation, biodiversity preservation, and carbon storage. Sustained yield management is a strategy that attempts to balance the productive use of a resource while not depleting its supply.

KEY TERMS

acid deposition p. 377
biochemical oxygen demand p. 381
carbon monoxide p. 376
cartel p. 359
dissolved oxygen p. 381
externality p. 357
fossil fuel p. 364
hydrocarbon p. 376
hydroelectric power p. 373

natural resource p. 354
nitrogen oxide (NO$_x$) p. 376
nonrenewable resource
 p. 358
particulate p. 376
photochemical smog p. 378
photovoltaic cell p. 374
pollution p. 376
pollution prevention p. 384

potential resource p. 356
renewable resource p. 358
sanitary landfill p. 361
substitutability p. 358
sulfur oxide (SO$_x$) p. 376
supply and demand p. 356
sustained yield p. 386
temperature inversion p. 378
toxic substance p. 382

QUESTIONS FOR REVIEW AND DISCUSSION

1. What is a resource? How do political–cultural, technologic, and economic factors determine whether substances in the environment become valuable resources or not?

2. What is a renewable resource? What is a nonrenewable resource? Give examples of each. Can any resources be considered either renewable or nonrenewable, depending on their use?

3. How has human use of coal, oil, and natural gas changed over the past 300 years? What changes are

likely in use of these fuels over the next few decades? What are some energy resources that are likely to be substituted for fossil fuels?

4. How do emission rates and rates of pollution dispersal interact to determine the severity of air and water pollution? What weather conditions aggravate air pollution?

5. How has water quality in U.S. rivers changed in the past few decades? What have been the major factors in this change? How does water pollution in wealthy nations differ from that in poor ones?

THINKING GEOGRAPHICALLY

1. Where does the drinking water come from where you live or attend school? What are the most significant potential pollution sources affecting your water supply? Your air supply?

2. If you drive, estimate the number of gallons of gasoline your automobile consumes in a year by dividing the number of miles you drive by your fuel efficiency in miles per gallon. If each gallon of gasoline contains about 2 kilograms of carbon, how many kilograms of carbon were emitted to the atmosphere by your automobile? Compare this with the number of hectares of growing forest it would take to store this carbon in biomass if a hectare of forest accumulates carbon at the rate of 750 kilograms per year.

3. Of the approximately 90 million new people added to the world's population each year, over 80 million are in developing countries. But the per capita resource consumption of people in developing countries is a small fraction of that in rich nations. If population control is important for limiting global use of resources, should population control efforts be focused in developing nations or industrial nations? Why?

4. What are the most important issues of environmental quality where you live? Has environmental quality there improved or deteriorated in the past 30 years? What is your evidence?

SUGGESTIONS FOR FURTHER LEARNING

Allee, D. J., and L. B. Dworsky. "Watersheds: A new picture in an old frame." *Canadian Water Resource* 23, no. 1 (1998): 85–92.

Azzaria, L. M., and R. G. Garrett. "Mercury in the Canadian environment: Current research challenges." *Geoscience Canada* 25, no. 1 (1998): 1–14.

Balirwa, J. S., C. A. Chapman, and L. J. Chapman, and others. "Biodiversity and fishery sustainability in the Lake Victoria Basin: An unexpected marriage?" *Bioscience* 53, no. 8 (2003): 703–715.

Brown, Lester R., and others. *State of the World 2006*. New York: W. W. Norton (see most current edition of this annual book).

Brown, Lester R., and others. *Vital Signs 2006-07*. New York: W. W. Norton (see most current edition of this annual book).

Cannell, M. G. R. "Growing trees to sequester carbon in the UK: Answers to some common questions." *Forestry* 72, no. 3 (1999): 237–247.

Clark, W. C. "Urban environments: Battlegrounds for global sustainability" *Environment* 45, no. 7 (2003): 1.

Clark, W. C., and N. M. Dickson. "Sustainability science: The emerging research program." *Proceedings of the National Academy of Science, USA* 14 (2003): 8059–8061.

Costanza, R. "The ecological, economic, and social importance of the oceans." *Ecological Economics* 31, no. 2 (1999): 199–213.

Dagnall, S., J. Hill, and D. Pegg. "Resource mapping and analysis of farm livestock manures—assessing the opportunities for biomass-to-energy schemes." *Bioresource Technology* 71, no. 3 (2000): 225–234.

Deffeyes, K. S. *Beyond Oil: The View From Hubbert's Peak*. New York: Hill & Wang.

Everard, M. "Towards sustainable development of still water resources." *Hydrobiologia* 395/396 (1999): 29–38.

Farrell, E. P., E. Fuhrer, D. Ryan, F. Andersson, R. Huttl, and P. Piussi. "European forest ecosystems: Building the future on the legacy of the past." *Forest Ecology and Management* 132, no. 1 (2000): 5–20.

Field, Barry C. *Natural Resource Economics: An Introduction*. Long Grove, IL: Waveland Press, 2005.

Gleick, P. H., N. Cain, D. Haasz, C. Henges-Jeck, C. Hunt, M. Kiparsky, M. Moench, M. Palaniappan, V. Srinivasan, and G. Wolff. *The World's Water 2004–2005: The Biennial Report on Freshwater Resources*. Washington, DC: Island Press, 2004.

Gupta, S., and A. Gilman. "Canada's contribution to the international reduction of certain persistent organic pollutants." *Environmental Practice* 2, no. 1 (2000): 71–78.

Harrington, W., V. McConnell, and A. Ando. "Are vehicle emission inspection programs living up to expectations?" *Transportation Research* (Part D: Transport and Environment) 5, no. 3 (2000): 153–172.

Jim, C. Y. "The forest fires in Indonesia 1997–98: Possible causes and pervasive consequences." *Geography* 84, no. 364 (1999): 251–260.

Meadows, D. H., and others. *The Limits to Growth*. New York: Universe Books, 1972.

Oki, T. and S. Kanae. "Global Hydrological Cycles and World Water Resources." *Science* 313 (25 August 2006): 1068–1072

Pimentel, David, J. Allen, A. Beers, L. Guinand, R. Linder, P. McLaughlin, B. Meer, D. Musonda, D. Perdue, S. Poisson, S. Siebert, K. Stoner, R. Salazar, and A. Hawkins. "World agriculture and soil erosion." *BioScience* 37 (April 1987): 277–283.

Redford, Kent H. "The empty forest." *BioScience* 42 (June 1992): 412–422.

Roberts, P. *The End of Oil: On the Edge of a Perilous New World*. New York: Houghton Mifflin, 2004.

Smil, V. "China's thirsty future." *Far Eastern Economic Review*. December 2005.

Smil, V. "Limits to growth revisited: A review essay." *Population and Development Review* 31 (2005): 157–164.

Smil, V. "Peak oil: A catastrophist cult and complex realities." *World Watch* 19 (2006): 22–24.

Szaro, R. C., D. Langor, and A. M. Yapi. "Sustainable forest management in the developing world: Science challenges and contributions." *Landscape and Urban Planning* 47, nos. 3–4 (2000): 135–141.

Thomas, C. E., B. D. James, F. D. Lomax, Jr., and Ira F. Kuhn, Jr. "Fuel options for the fuel cell vehicle: Hydrogen, methanol or gasoline?" *International Journal of Hydrogen Energy* 25, no. 6 (2000): 551–567.

van Kamp I., Leidelmeijer K., Marsman G., and others. "Urban environmental quality and human well-being: Towards a conceptual framework and demarcation of concepts" (a study of the literature). *Landscape Urban Plan* 65, nos. 1–2 (2003): 7–20.

Williams, M., *Deforesting the Earth: From Prehistory to Global Crisis*. Chicago, IL: University of Chicago Press, 2006.

World Resources Institute. *World Resources 2005*. New York: Oxford University Press, 2005.

WEB WORK

The Web is an excellent example of a new technology that can improve our lives and increase wealth while reducing the use of resources. We get information faster, and we don't have to drive our cars to get it. Please resist the temptation to waste paper by printing a copy of every page that contains a few words of potential interest!

Many environmental organizations and academic institutions have lists of links to environmental sites. Envirolink is one good one, plus it also includes an excellent compilation of environmental news items:

http://www.envirolink.org/

U.S. government agencies responsible for resource management are excellent sources of environmental statistics as well as information on government policies and programs. The U.S. Environmental Protection Agency can be found at:

http://www.epa.gov/

The U.S. Energy Information Administration is at:

http://www.eia.doe.gov/

The U.S. Forest Service site is at:

http://www.fs.fed.us/

The U.S. Geological Survey Minerals Information is at:

http://minerals.er.usgs.gov/

The Carbon Dioxide Information Center at Oak Ridge National Laboratory can be found at:

http://cdiac.esd.ornl.gov/

Among the international organizations with information about global environmental issues are the World Bank at:

http://www.worldbank.org/

The Consortium for International Earth Science Information Network at:

http://www.ciesin.org/

The World Resources Institute at:

http://www.wri.org/

And the United Nations Food and Agriculture Organization at:

http://www.fao.org/

10

CITIES AND
URBANIZATION

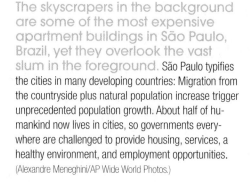

The skyscrapers in the background are some of the most expensive apartment buildings in São Paulo, Brazil, yet they overlook the vast slum in the foreground. São Paulo typifies the cities in many developing countries: Migration from the countryside plus natural population increase trigger unprecedented population growth. About half of humankind now lives in cities, so governments everywhere are challenged to provide housing, services, a healthy environment, and employment opportunities. (Alexandre Meneghini/AP Wide World Photos.)

A Look AHEAD

Urban Functions

Cities may have first developed for cultural reasons, but the economic reasons for their existence rose in importance. Today we can define basic and nonbasic sectors of urban economies and analyze their roles in the primary, secondary, and tertiary sectors of national economies.

The Locations of Cities

Some cities were founded to utilize advantageous sites, such as mines, while others exploit favorable situations, such as a crossroads or a bottleneck on transport routes.

World Urbanization

Urbanization is occurring everywhere, both because of natural increases of the urban populations and because of continuing in-migration of rural people. Some governments are trying to regulate the growth of cities.

The Internal Geography of Cities

Any city's internal geography is defined by economic considerations, social considerations, and government actions interacting in the local culture.

Cities and Suburbs in the United States

U.S. metropolitan areas have exploded across the countryside. Job opportunity and housing continue to grow at the periphery, but some central cities that were once hollowed out started growing again at the end of the twentieth century. The many local governments in metropolitan areas must devise ways of cooperating on common challenges.

Scientists and planners everywhere are looking for ways of reducing cities' impact on the natural environment, and of improving the environments of the cities themselves. Two legal battles in California illustrate the complexities of the issues.

Each year, Los Angeles sends 65 million gallons of sludgy processed human waste to a farm it owns in nearby Kern County to be used as fertilizer. In 2006, however, Kern County residents voted to stop accepting all but a small fraction of the waste.

What can large cities do with their waste? For years, Los Angeles waste flowed from treatment plants into the ocean, but it was choking marine life, so Los Angeles bought the Kern County farm, over which it spread the waste to grow corn, wheat, and alfalfa. About half the nation's human waste is applied to land, and it can be used on crops for human consumption if federal and state rules are followed. The Los Angeles waste, however, is not being used on edible crops. The crops are fed to cows, and the milk they produce is sold. Residents of Kern County, however, fear that the sludge will pollute groundwater.

Meanwhile, in California's Central Valley, developers and air quality regulators have fought over new construction fees intended to reduce the region's chronic smog problem. The fees, which go into a fund for pollution control, require builders of commercial and residential projects to use energy-saving technology and traffic-reduction features in their projects. Local residents hope to make developers more accountable for the explosion in traffic and emissions that typically accompany building. ▶

A **city** is a concentrated nonagricultural human settlement. Every settled society builds cities, because some essential functions of society are most conveniently performed at a location that is central to the surrounding countryside. Cities provide a variety of services, including government, education, trade, manufacturing, wholesaling and retailing, transportation and communication, entertainment, business, and defense and religious services. The region to which any city provides services, and upon which it draws for its needs, is called its **hinterland.**

In ancient Egyptian hieroglyphics—the earliest writing we can read—the ideogram meaning "city" consists of a cross enclosed in a circle. The cross represents the convergence of roads that bring in and redistribute people, goods, and ideas. The circle around the hieroglyph denotes a moat or a wall. Few modern cities have walls, but cities do have legal boundaries, and within those boundaries a degree of self-government is usually exercised. The process of defining a city territory and establishing a government is called **incorporation.**

Sometimes several cities grow and merge together into vast urban areas called *conurbations.* In the northeastern United States, for instance, one great conurbation stretches all the way from Boston to Washington, D.C. This conurbation has been called *Megalopolis,* which is Greek for "great city." The world's largest urban areas are generally called *metropolises,* Greek for "mother cities" (Figure 10-1).

In several countries one large city concentrates a high degree of the entire national population or of national political, intellectual, or economic life. These cities are called **primate cities.** Paris, for example, is the primate city of France, and Bangkok is the primate city of Thailand. Not all countries, however, have a primate city: The United States does not. Whether a country has a primate city depends on its national history and social and economic organization.

Today in all countries urban populations are growing faster than rural populations. This process of concentrating populations in cities is called **urbanization.** The United Nations says that about one half of the world's population lived in urban areas by the year 2000, compared with 30 percent in 1950. Virtually all population growth expected in the coming decades will be concentrated in the urban areas of the world. The degree of urbanization, however, is not the same in all countries (Figure 10-2). One of the purposes of this chapter is to explain why this is so.

The geographic study of cities, **urban geography,** considers three topics:

1. The study of the functions of cities and their economic role in organizing territory;
2. The comparative study of urbanization as it occurred in the past and as it is continuing in different countries today;
3. The study of the internal geography of cities—that is, the internal distribution of housing, industry, commerce, and other aspects of urban life across different cultures.

This chapter will review each of these topics and then examine the growth and internal geography of U.S. metropolitan areas. This examination will

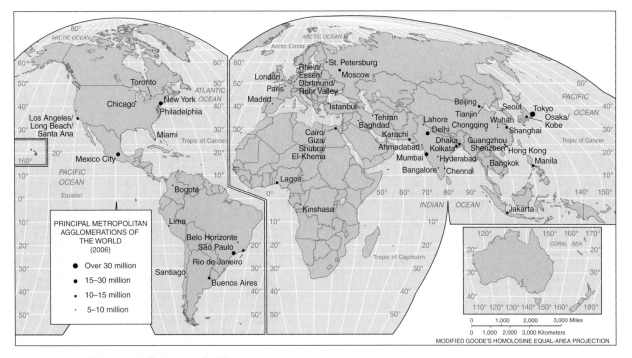

FIGURE 10-1 The world's largest cities. This map shows the 50 largest metropolitan agglomerations. The percentage of people living in urban areas is higher in already developed countries, but most of the world's large cities are in developing countries. The rapid growth of cities in these countries reflects both increases in overall national populations and migration into these cities from rural areas. (Data from United Nations, 2006)

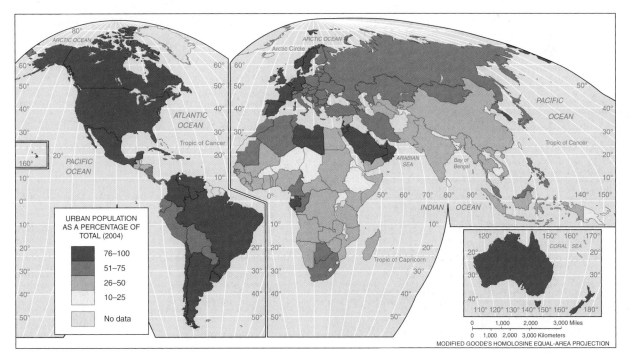

FIGURE 10-2 World urbanization. This map shows the varying degree of urbanization of countries. Different countries use different definitions of "urban area," ranging from settlements as small as 200 persons to as large as 30,000, but most countries use a minimum of 2,000 to 5,000 people. In the United States the minimum threshold population is 2,500; in Canada an urban population is defined as that population living in incorporated places of 1,000 or more and at densities of over 1,000 per square kilometer. (Data from the United Nations)

illustrate many of the principles of urbanization in situations that will probably be familiar to you.

URBAN FUNCTIONS

The first cities appeared in today's Turkey and Iraq about 4000 B.C. Cities developed in Egypt's Nile Valley about 3000 B.C.; in Peru and in the Indus Valley of today's Pakistan by about 2500 B.C.; and in the Yellow River Valley of China by 2000 B.C. Mexico supported cities by about A.D. 500. Some early cities had considerable populations. Ur, which flourished in Mesopotamia by 3500 B.C., may have had 200,000 people, and Thebes, the capital of ancient Egypt, may have had 225,000 in 1600 B.C. Later Rome probably reached a population of 1 million by the second century A.D. Kaifeng, China, housed 1 million people by 1150, and London reached 1 million by the early nineteenth century.

Archaeological evidence suggests that settlements probably originated for cultural reasons rather than economic ones. The first permanent settlements may have started as places to bury the dead, or fixed sites for priests to perform ceremonies. Cities came to be embellished as centers of worship or even as the seats of the gods themselves (Figure 10-3). Many religions teach that the largest house of worship should be the tallest building in a city, and when commercial buildings first overtopped church spires in European and American cities in the late nineteenth century, many observers found it symbolic of an unfortunate reversal in society's values (Figure 10-4).

The earliest settlements also may have served as places to house women and children while the men traveled in search of food. Household objects made by

FIGURE 10-4 Trinity Church at the head of Wall Street in New York City. Trinity Church's 281-foot high spire, erected in 1839, was for many years the highest point in New York City. Today the spire is dwarfed by the skyscrapers of New York's financial district. (Rudi Von Briel/PhotoEdit)

women, such as pots, tools, and clothing, provided a basis for the creation and transmission of a group's values and heritage. Today settlements contain society's schools, libraries, museums, and archives—the repositories of knowledge and the vehicles for passing it through generations.

Early settlements protected groups' land claims and food sources. Palaces arose to house the group's political leaders, and soldiers were permanently stationed there. Many settlements were surrounded by defensive walls. Long after the introduction of artillery, walls could still hold off an attacker until help arrived or the attacker himself ran out of food, so cities still built them. Paris, for example, surrounded the city with new fortifications as recently as the 1840s and did not completely remove them until 1932 (Figure 10-5). Today cities are still the focus of military and political activities, but few retain walls except for historic interest (Figure 10-6). Most city walls were replaced in the nineteenth century by parks or grand boulevards.

FIGURE 10-3 The Todaiji Temple in Nara, Japan. This temple is the world's largest wooden structure. It houses a colossal statue of Buddha 16 meters (53 feet) high. Both temple and statue were originally made in the eighth century, when Nara was Japan's capital city. Today Nara is not a large city nor important in Japanese economic life, but much of it has been preserved as a treasure of Japanese history and culture. (Michael S. Yamashita/CORBIS BETTMANN)

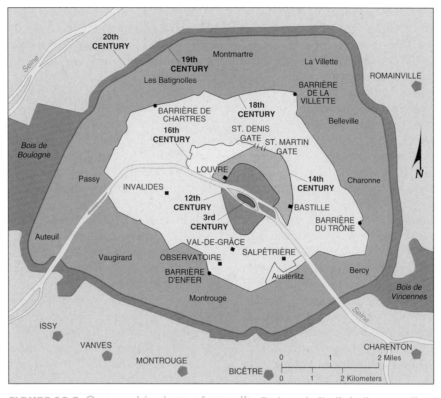

FIGURE 10-5 Concentric rings of growth. Paris periodically built new walls to encompass new neighborhoods as the city expanded, so it built concentric rings of protective walls. The old gates of St. Denis and St. Martin have been preserved, but the walls have been torn down.

FIGURE 10-6 Quebec, Canada. Two-hundred-year-old walls still surround Quebec, the only walled city in North America, standing above the St. Lawrence River. (Courtesy of Luc-Antoine Courier/Andre Caron)

The economic role of settlements may have begun simply as warehousing centers to store food, but as societies develop, production and trade join services as most cities' paramount activities. Cities bring people and activities together in one place for greater convenience. This is called **agglomeration.** Agglomeration promotes the convenient *division of labor,* which is the separation of work into distinct processes and the apportionment of work among different individuals in order to increase productive efficiency. Craft workers flourished within city walls, and specialized occupations emerged, sustained by the peasantry of the hinterland. Settlements serve as convenient sites for trading, and local officials often regulated the terms of transactions, kept records, and created a currency system. Cities thus promote and administer the regional specialization of production throughout their hinterlands. With industrialization, cities become centers of production.

The Three Sectors of an Economy

Economic activities are generally divided into three sectors: primary, secondary, and tertiary. The names used for the three sectors indicate the degree to which each sector is removed from direct involvement with Earth's physical resources.

Workers in the **primary sector** extract resources directly from Earth. Most workers in this sector are usually in agriculture, but the sector also includes fishing, forestry, and mining. Workers in the **secondary sector** transform raw materials produced by the primary sector into manufactured goods. Construction is included in this sector. All other jobs in an economy are within the **tertiary sector,** sometimes called the **service sector.** The tertiary sector includes a great range of occupations, from a store clerk to a surgeon, from a movie ticket seller to a nuclear physicist, from a dancer to a political leader. This sector includes so many occupations that some scholars have tried to split it into a tertiary services sector and a *quaternary sector,* called an *information sector.* There is no consensus, however, as to what this sector includes, and no agencies of the U.S. government or of the United Nations recognize this sector in statistics. Therefore, it is not used in this book.

As economies grow, the balance of employment and output shifts from the primary sector toward the secondary and tertiary sectors. Fewer people work on farms, and more people work in factories and offices. These activities normally require urban settings, and therefore, as an economy's secondary and tertiary sectors grow, its cities grow. This growth of the secondary and tertiary sectors is called *sectoral evolution,* and it will be examined in detail in Chapter 12.

The Economic Bases of Cities

Cities depend on their hinterlands for, at the very least, food. The cities must in turn provide services or "export" something to the outside. Many cities produce and export manufactured goods, but the exports of a city are not necessarily things that leave it. They may be things or services that people come to the city to buy. If people go to Houston for heart surgery, for example, then heart surgery is counted as an export of Houston. Vacations are an export of Miami Beach; gambling is an export of Las Vegas. In this economic sense, capital cities export government.

Some of the workers in a city produce the city's exports, but others serve the needs of the city's own residents. The part of a city's economy that is producing exports is called the **basic sector,** and that part of its economy serving the needs of the city itself is called the **nonbasic sector.** A city's basic and nonbasic sectors may

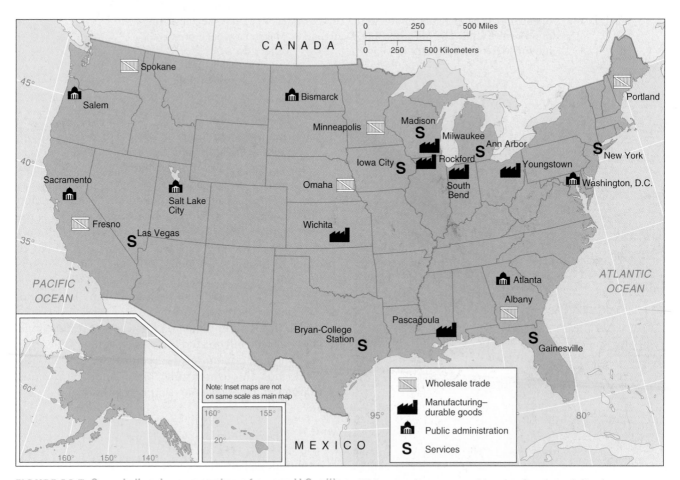

FIGURE 10-7 Specialized economies of some U.S. cities. This map shows some cities that have specialized economies; it does not indicate the cities' individual ranking in specific activities. Los Angeles, for instance, is America's greatest manufacturing center, but it does not appear on this map because its economy is diverse. (Adapted with updating from J. Clark Archer and Ellen R. White, "A Service Classification of American Metropolitan Areas." *Urban Geography* 6 [1985]: 122–51.)

be in the primary, secondary, or tertiary sectors of the economy as a whole.

Jobs in a city's basic sector create jobs in the city's nonbasic sector. For example, if a local factory makes a product that is sold around the world, the factory workers will spend their earnings by shopping locally, getting their hair cut locally, and purchasing other local goods and services. Each job in the basic sector actually supports several nonbasic sector jobs, because earnings from exports circulate and recirculate through the local economy. When a factory worker buys a shirt, the store clerk can get a haircut; the barber in turn might eat at a local restaurant, and so on. Thus, jobs in the basic sector have a **multiplier effect** on jobs in the nonbasic sector.

Cities can be classified economically by examining each city's basic and nonbasic sectors and by comparing these sectors among different cities. New York, Seattle, and Hollywood, for instance, each contains a number of dry cleaners and doctors. These people, for the most part, work within the nonbasic sectors. In terms of exports, however, workers in New York provide specialized financial services, workers in Seattle write computer-software, and workers in Hollywood make movies. A city's employment structure reveals its economic specialization. Any city that has an unusual concentration of workers in a specific job category must be exporting that product or service. A city that has a concentration of autoworkers, for example, presumably exports automobiles (Figure 10-7).

THE LOCATIONS OF CITIES

Today, as in the past, the location of any city depends on a balance of site factors (characteristics of the place itself) and situation factors (its location relative to other places). Choice sites include defensive hilltops, oases, and the locations of mineral resources. Some mining towns virtually sit on top of valuable ore deposits but are otherwise isolated. At the time of the Industrial Revolution, cities often developed at waterfalls to exploit hydropower (Figure 10-8).

The locations of other cities more clearly result from advantageous geographic situations. Cities with the most convenient situations grow (look back at Figure 6-23). Cities frequently grow up at places where two different physical areas meet or along the border between two cultures. In Chapter 6 it was noted that Tombouctou was located on a key trade route, on the border between two physical environments (where the Niger River bends farthest north into the Sahara Desert) and also two cultures (nomadic Arabs to the North and settled black peoples to the South). Many cities spring up as transportation hubs or at bottlenecks, such as at a bridge across a river or where two political jurisdictions funnel trade through border checkpoints. Some cities grow at sites where the

FIGURE 10-8 Paterson, New Jersey. These falls of the Passaic River powered the world's first planned industrial city, Paterson, New Jersey, founded in 1791 by a group of investors lead by Alexander Hamilton. (Peter Bryon/PhotoEdit)

method of transportation necessarily changes. These are called *break-of-bulk* points. A seaport is an example. Another is the head of navigation of a river. If there is a waterfall on a navigable river, cargo has to be unloaded from ships and then reloaded beyond the waterfall up or downriver, or else be shifted to rail or truck.

Louisville, Kentucky, for example, was laid out in 1773 at the falls of the Ohio River. The river is navigable both up and downriver from Louisville. The opening of Louisville's Portland Canal in 1830 allowed ships to pass around the falls, but by then the city was well developed. It provided many services to its rich, developing hinterland and to westward-moving pioneers. The bridge across the Ohio River focused north-south traffic, and later the railroad lines also focused on the city.

If a situation is favorable, a great city may arise on an unfavorable site. Many of Asia's coastal cities were built by European merchants or conquerors at sites that provided access to the sea and that may have been defensible, but they sit on deltas or the swampy foreshores of tidal rivers. To this day, Karachi, Pakistan; Madras and Calcutta, India (today called Chennai and Kolkata); Colombo, Sri Lanka; Yangon, Myanmar; Bangkok, Thailand; Ho Chi Minh City, Vietnam; and Guangzhou, Shanghai, and Tianjin, China, all face formidable problems of drainage, water supply, construction, and health. New Orleans in the United States is a North American example.

One of human geography's great paradoxes is that one of the world's largest cities, Mexico City, is located on one of the world's most unfavorable places to build: a drained lakebed in an earthquake zone in a basin of interior drainage at a high elevation in a dry climate. These conditions combine to cause physical instability, alternating flooding and lack of water, and air pollution. Both human lungs and internal combustion engines are inefficient when high altitude reduces oxygen levels by 23 percent, and air pollution is aggravated by local windstorms. Today's Mexico City, however, was the site of the Aztec capital Tenochtitlán at the time of the arrival of the Spaniards. The Spaniards maintained the site as their capital, and so have modern Mexicans. Thus, this great city testifies to the power of history and geographical inertia.

Central Place Theory

The relationships between cities and their hinterlands have inspired a model of how cities are distributed across territory. Walter Christaller (1873–1969) began with the simplest imaginary landscape—an isotropic plain on which transportation cost is determined according to straight-line distance. Christaller then asked, "If cities are to serve as convenient centers for exchange and other services across an isotropic plain, how will cities be distributed? What will be the pattern of towns and their hinterlands?"

To answer these questions, Christaller developed his **central place theory.** This model has three requirements: (1) the hinterlands must divide the space completely, so that every point is inside the hinterland of some market; (2) all markets' hinterlands must be of uniform shape and size; and (3) within each market region, the distance from the central place to the farthest peripheral location must be minimal. The only pattern that fills these three requirements is a pattern of hexagons—six-sided figures (Figure 10-9). Therefore, market towns will be distributed across an isotropic landscape as foci of a grid of hexagons.

Hexagons appear repeatedly in both nature and in industrial design. Honeycombs are hexagonal, and so probably is the pencil in your drawer. A hexagonal shape allows the greatest possible number of pencils to be cut out of a block of wood.

Urban Hierarchies

Chapter 1 discussed diffusion up or down a hierarchy of cities. There are many small towns, fewer and more widely spaced medium-sized cities, and still fewer big cities. A hierarchy of cities exists because the more specialized a service or product is, the larger the number of potential customers that is needed for that product or service to be offered. No product or service can be offered without a minimum number of customers. This minimum demand is called the *threshold* for that product or service. A coffee shop owner, for example, needs a minimum number of customers to earn a living. Most people periodically visit coffee shops, so the threshold of demand for a coffee shop is low. Each small town or city neighborhood can support one. Similarly, many people frequently buy fresh bread and milk. Therefore, small groceries can be found in each neighborhood.

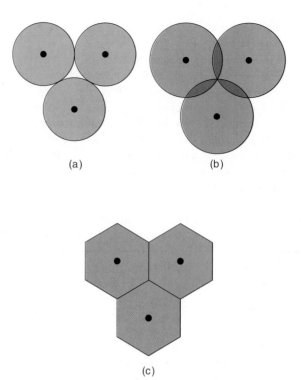

(a)   (b)

(c)

FIGURE 10-9 Hexagons fill space. Circles cannot fill space without overlapping one another, but circles collapse into hexagons that neatly fill space.

Few people, however, take tuba lessons or buy diamond bracelets. The threshold of demand sufficient to support tuba teachers or expensive jewelers is high, so only larger towns will be able to support these enterprises. On an isotropic plain on which people and buying power are equally spaced, coffee shops and grocers will be closely spaced; tuba teachers and jewelers will be widely spaced.

In real landscapes, other factors in addition to distance must be considered. If one area has a concentration of rich people, for example, then the density of jewelers will be greater there. Quantity of purchasing power substitutes for numbers of people. If another area has a high population density, then grocery stores will be more numerous there.

A provider of a service can do one of two things to reach his or her necessary threshold of customers. First, the provider can be *itinerant* (travel from place to place). Alternatively, the provider can set up shop at one convenient place and wait for people to come. Convenience and accessibility are the principal purposes of cities. Agglomeration of services in cities saves travel costs, and it allows thresholds to be reached for more specialized goods and services.

The range of goods and services in a city offers small businesses the opportunity to rent pieces of equipment and to hire services or temporary employees only when they need them. This saves the businesses the cost of investing in equipment or hiring people full time. These available services or goods are called **external economies,** and their availability lowers initial costs for new business ventures. Thus, cities are incubators of new businesses. If a company grows large enough to justify buying its own equipment or hiring full-time employees, then the company has achieved **internal economies.**

The Patterns of Urban Hierarchies

Christaller's model of evenly spaced market towns with hexagonal hinterlands can be elaborated to represent a hierarchy of cities. In the isotropic plain the larger cities that are "above" the small market towns in the urban hierarchy must be more widely spaced than the small market towns, which also must be evenly spaced, and their hinterlands must also be hexagonal. Therefore, the distribution of the larger cities is represented by a grid of larger hexagons superimposed on the grid of small hexagons (Figure 10-10).

Many factors in the real world disrupt Christaller's model because the world is not an isotropic plain. The model has proved useful, however, in planning new

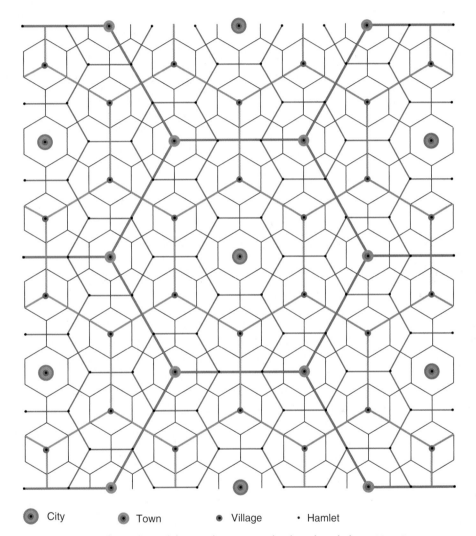

● City   ● Town   ● Village   · Hamlet

FIGURE 10-10 An urban hierarchy on an isotropic plain. If hamlets are the smallest central places, then the distribution of the larger villages could be represented by a grid of larger hexagons (red lines) superimposed upon the grid of the small hexagons that represent the hinterlands of the hamlets (black lines). Geometry defines at least three ways of doing this. The pattern illustrated here is called the K3 model because in it the hinterland of each village includes a hinterland that the village serves as a hamlet plus one-third of the hinterlands of each of the six surrounding contiguous hamlets. When a resident of a hamlet needs a higher-order service available only in a village, he or she finds each of two villages and a town equally convenient. Therefore, that person might visit each of these three settlements occasionally, so news travels rapidly across a landscape organized in this way. Continuing up the hierarchy of central places, town hinterlands (blue lines) are superimposed upon village hinterlands, and city hinterlands (green lines) upon town hinterlands. (From Walter Christaller, *Central Places in Southern Germany*, trans. Carlisle Baskin. Englewood Cliffs, NJ: Prentice Hall, 1966, p. 61.)

cities in countries with unsettled lands. The land that the Dutch have claimed from the sea behind new dikes is similar to an isotropic plain. There, Dutch geographers have planned new market towns on hexagonal grids. Brazil has adapted the model to settle territories in Amazonia. The government built what it calls an "Agrovila" every 10 kilometers (6 miles), with a school, a health-care center, and a post office. Every 40 kilometers (25 miles) the government placed an "Agropolis" that offers the services of an Agrovila, plus sawmills, stores, warehouses, banks, and other commercial services. Each 136 kilometers (85 miles) in Brazil a "Ruropolis," a city, was established for light industry. The ratio of 10, 40, and 136 kilometers cannot be reconciled with the proportions of a "pure" geometric grid, but they have been found useful by the Brazilian planners. The towns serve as foci for economic development. Venezuela has adopted a program similar to Brazil's.

When improvements in transportation allow people to travel farther to obtain services or goods, the smallest central places may lose their reason for existence and disappear. This is happening in the North American farming region from Kansas north into the Canadian prairies, where most small towns developed as commercial centers for farmers. They offered grocery stores, banks, hardware stores, farm implement dealers, automobile dealers, and feed stores. Technological advances, however, starting with the tractor, as well as advances in transportation and communication and other economic forces, have increased the size of farms but reduced the number of farmers. As a consequence, there is less need for the small-scale central places. Between 1960 and 1990 alone, over 70 percent of the 600 towns, villages, and hamlets in Canada's province of Saskatchewan lost the basic commercial functions necessary to sustain the communities. Farm areas are today punctuated by wholly abandoned towns, and the number of rural towns is expected to continue to decline. Only small towns near metropolitan areas can survive or even prosper. These become **exurbs,** the name given to settlements that make up the outermost ring of expanding metropolitan areas.

WORLD URBANIZATION

Urban geographers compare urbanization as it occurred in the past and as it is continuing today. The rapidity of worldwide urbanization today presents many nations with both challenges and opportunities for the welfare of their populations.

Early Urban Societies

In the seventeenth century Holland became the first modern urban country, with more than half its population living in towns and cities. The principal economic functions of Dutch cities were tertiary-sector functions.

The cities arose as administrative and commercial foci of Holland's global shipping, banking, and trading activity, and the urban populations could be supported by Holland's highly productive agriculture. This urbanization occurred before the Industrial Revolution.

In Britain, a larger country with a more varied economy, urbanization occurred with industrialization. More than half of Britain's people lived in cities and towns by about 1900. Several developments over the previous 200 years had resulted in the concentration of Britain's population. These included the following:

1. Improvements in agricultural technology—part of the Agricultural Revolution—reduced the need for the number of agricultural workers. Landowners found it profitable to release employees and to evict tenants. The resulting rural depopulation is described in Oliver Goldsmith's poem "The Deserted Village" (1770), which laments that "rural mirth and manners are no more."

2. Displaced workers migrated to the cities. There, many were absorbed by the concurrent **labor-intensive activity** stage of the Industrial Revolution. An activity is labor intensive if it employs a high ratio of workers to the amount of capital invested in machinery. Other newcomers to the city found work in the tertiary sector. The largest class of urban workers was actually domestic servants—a tertiary-sector job. In 1910 they formed more than one-third of the total British labor force. A similar pattern could be found in the United States and other rich Western countries.

   Rural–urban dislocation caused appalling hardship. The history of urban life in nineteenth-century England records overcrowding in dreadful slums, malnutrition, starvation, crime, and early death. The descriptions of the miseries of England's slum populations in Charles Dickens's novels still haunt our imaginations. Cities have never easily absorbed all those who have flocked to them.

3. Population pressures were somewhat relieved by emigration, or else by the forcible exportation of criminals and debtors throughout the British Empire. The colonies of Georgia and Australia absorbed many of these deported people.

This British experience provided the world's first model for modern urbanization, and most of today's developed countries experienced similar histories. In 1800 the 21 European cities with populations of 100,000 or more held about 4.5 million people, or just under 3 percent of the total European population. By 1900 there were 147 such places with a total population of 40 million, or about 10 percent of the total population. Today about 30 percent of the European population lives in cities with populations of 1 million or more.

The word *model* as used here does not mean that the British experience of urbanization was perfect or that it should be imitated. It imposed terrible hardship on millions of people. Britain provides a model only in the sense that Britain experienced these forces of urbanization before anywhere else, so its experience can be compared with the current situation. In British history the push of rural displacement and the pull of urban job opportunity were not coordinated. Governments have learned how to manipulate both forces in order to ease the process of urbanization, yet some aspects of contemporary urbanization still cause as much hardship as the experience did in Britain.

Urbanization Today

Today urbanization is occurring in many places without concomitant economic development, especially in the world's poor countries. Burgeoning populations overwhelm the cities' ability to absorb the people and to put them to work. The incoming populations overload the cities' infrastructures of housing, education, internal transportation, water supply, and sewerage. By one estimate 90 percent of sewage from urban areas in the developing world pours untreated into streams and oceans today. Reliable water, electricity, and telephone services are rare. From the tops of new skyscrapers in many modern cities, the view presents vast shantytowns of the desperately poor (Figure 10-11). Living conditions in these teeming cities are no better for the majority than those that existed in Europe in the nineteenth century. The stresses of contemporary urbanization have been blamed for the breakdown of family life, recourse to drugs or religious extremism, and the spread of AIDS. The United Nations predicts that by 2025, two-thirds of the world's population will live in cities, most of them in poor countries where infrastructure is inadequate to the population's demands.

Rapid urbanization presents challenges, but challenges also bring opportunities. New urban populations present concentrated labor forces that might be put to productive work in ways that could raise the living standards of all people.

Rapid urbanization is often caused by deteriorating conditions in the countryside, and in many poor countries the problem is exacerbated by sharp cultural differences between rural and urban populations. Chapter 6 noted that many cities in today's poor countries evolved as outposts of international commerce grafted onto the local societies. Chapter 5 noted that their populations may even be cosmopolitan mixes of ethnic groups that are not native to the region. Throughout Latin America, for example, the cities are predominantly white or mestizo, and measures intended to restrict urban migration are interpreted, often correctly, as racist discrimination against Native Americans. In East Africa urban populations were often Asian, and in Southeast Asia urban populations were Chinese and Indian. These cultural

FIGURE 10-11 La Paz, Bolivia. The tall buildings mark the city's downtown core. It occupies a steep canyon in which Spaniards founded this city in 1548 in order to avoid the chill winds of the plateau. The modern city has grown up and down the canyon to the east and west of the historic colonial core. Native American migrants to the city have settled up on the plateau, which is now a distinct city of over 1 million in population, called El Alto. The construction of housing and infrastructure has not been able to keep pace with El Alto's annual 10-percent population increase, so it is a vast slum. Such slums are typical of the growing cities in the poor countries. La Paz is also one of the highest cities of the world, at 3,636 meters (2.3 miles) above sea level, so internal combustion engines are inefficient, and air pollution is a frequent problem. (John Metelsk/U.S. Agency for International Development)

and ethnic contrasts make it harder to deal with rapid urbanization.

At least seven other circumstances differentiate urban growth today from that of the historic British experience.

1. The commercialization and mechanization of agriculture has accelerated, rapidly increasing the displacement of rural workers. For example, the introduction of mechanization to Brazilian sugar plantations reduced the labor force from

1.2 million in 1990 to fewer than 700,000 in 2000. Furthermore, modern agricultural technology requires capital investment, so the income gap between rich and poor widens.

2. National radio and television penetrate the countryside, spreading an image of the city as a place of opportunity. Many rural people feel pulled to the city.

3. In the past, disease and starvation among the poor kept urban death rates above urban birthrates. Without a steady influx of rural people, the town populations would have decreased. Today, however, the introduction of medical care and hygiene lowers urban death rates and triggers natural population increases. These factors compound the population growth that results from in-migration from the countryside.

4. Urban economies have changed in ways that make cities less able to employ the displaced rural population. Historically, urban economies offered entry-level job opportunities for the unskilled—jobs available for anyone with a strong back and a willingness to work. These jobs concentrated in domestic servitude, manufacturing, and construction. Today, however, domestic servitude has declined, and many of the jobs in manufacturing and construction demand skills. Manufacturing and construction are less labor intensive than they used to be; they have become increasingly **capital intensive activities.** Machinery has replaced workers. Tertiary-sector jobs other than domestic servitude have multiplied, but most of these require job skills.

5. For most societies the safety valve of emigration has been stopped. Migration opportunities for the unskilled and for rural workers are decreasing.

6. The world economy is increasingly interdependent, so local governments have less control over local economies. Factory openings or closings in Bangkok are determined in Tokyo. Worker opportunity in Lima, Peru, is regulated from New York, and even job opportunity in New York is subject to international economic forces, as we will study in Chapter 12.

7. Many governments continue to favor urban projects over rural projects. One reason for this may be status. A shiny new hospital in the capital city, for example, may seem more impressive than a thousand new water pumps in poor villages. Another reason may be that the government fears urban rioting. Urban food riots have occurred in several African and Latin American countries. As noted in Chapter 8, many governments subsidize urban food supplies while discouraging their own farmers. These actions enhance the perceived opportunity in the cities without providing real opportunity.

Government Policies to Reduce the Pull of Urban Life

Governments could regulate the migration of people from the countrysides to the cities either by reducing the attractiveness of the cities or else by improving the quality of rural life.

Forceful measures to limit urbanization are not new in human history. The Russians have been required to have internal passports since the days of Czar Peter the Great 300 years ago. Recently the world has even witnessed brutal instances of compulsory "ruralization." When North Vietnam absorbed South Vietnam in 1976, the new rulers relocated millions of urbanites to rural areas to raise food using labor-intensive methods. The government argued that the cities had bloated on U.S. financial assistance. The government of Cambodia similarly relocated urban populations in the late 1970s.

Some governments use media to discourage migration to the city. Ghana, for instance, broadcasts to the villages films suggesting that life is better there than in the cities.

Many cities try to discourage newcomers by restricting housing and economic opportunity. They pass building codes, for example, that ban substandard housing. These codes are often ineffective in stopping the growth of slums, but they make squatters' settlements illegal. Therefore, the squatter-residents do not get city services and will not risk investing to improve the property. The city in turn cannot collect property taxes. In 2006, the United Nations Human Settlement Program estimated that these "illegal" communities, overwhelmingly slums, held 1 billion people—72 percent of the total populations in the cities of sub-Saharan Africa, 57 percent in Southern Asia, 35 percent in East Asia, 31 percent in Latin America, and some 25 percent elsewhere—70 to 95 percent of urban newcomers everywhere. Cities also try to restrict small businesses in residential areas, but backyard workshops may thrive anyway. Cities may discriminate against new urbanites by restricting education, housing permits, business licenses, or other job opportunities. Hawkers are chased from city centers, for example.

In some countries frustrated city authorities have even bulldozed squatters' settlements after giving only one or two days of warning. This has happened in major cities throughout Latin America, Africa, and Asia, leaving hundreds of thousands of people homeless. In 1990, for example, bulldozers flanked by army troops, with air-force planes sweeping overhead, leveled the district of Maroko in Lagos, Nigeria, leaving about 300,000 people homeless. One survivor reported that the soldiers brought truckloads of coffins and said that if anyone was ready to die, the army was ready to bury him. In 2002, the government of India bulldozed shacks along the main airport road into central Kolkata so that World Bank delegates would not see slums on their way to a meeting in the city center. During 2005–2006, the Zimbabwean army virtually declared war on squatters and settlers in Harare, demolishing shanties and markets. Some 1.5 million people were scattered; over 10,000 street urchins were rounded up and dispatched to rural areas. The urban poor were perceived as a threat to the 25-year tyranical rule of dictator Robert Mugabe, and the action was officially called "Operation Murambatsvina" ("drive out rubbish").

China has long enforced a household registration system under which people are not supposed to move anywhere without getting their registration changed, but people have moved anyway. As many as 200 million Chinese peasants have drifted into the cities, where they are called *floaters*. Their residence in cities is technically illegal, so they have no right to education, medical care, and other government services. The floaters serve useful functions, such as bringing produce to the city or providing services not provided by central planning, but their continuing influx threatens to overtax China's urban infrastructure. The government has relaxed the registration system in order to create a unified national labor market, but the city governments themselves introduced new exclusionary techniques, such as granting residency permits only to those who can afford homes or to skilled workers.

Improving Rural Life

Instead of trying to drive people out of the cities, governments might invest in rural health care and education, housing, roads, and other infrastructure to raise rural standards of living and keep people satisfied living in the countryside. This might raise national food production. In 2006, the Chinese government significantly shifted its national budget towards rural roads, water and power supplies, schools, and hospitals.

A government can also organize labor-intensive investments in rural infrastructure (Figure 10-12). During the Depression in the 1930s, even the U.S. government Civilian Conservation Corps, Civil Works Administration, Public Works Administration, and Works Progress Administration all used labor-intensive methods to alleviate unemployment and at the same time build or repair national infrastructure.

The Economic Vitality of Cities

Most of the previous paragraphs about urbanization in developing countries may have seemed pessimistic. They presented urban growth as a problem for which solutions were needed. There is another side to the story, however: It is possible to view the backyard shops, the street hawkers, and the Chinese floaters as examples of opportunity and growth. These activities hint that urban migration, balanced between management and liberty, can provide a reservoir of vitality that can be harnessed for national growth.

The Peruvian economist Hernando De Soto, who is presently Chairman of the U.N. Commission for the Legal Empowerment of the Poor, has long noted that in

FIGURE 10-12 Labor-intensive construction.
Construction of even massive infrastructure projects such as this dam in India can begin with workers carrying baskets of dirt on their heads. In the rich countries, however, labor-intensive construction methods such as this have yielded to capital-intensive methods. (Courtesy of U.S. Agency for International Development)

many cities the productive activities of a substantial share of the population do not appear in official accounts. The people may not have licenses to do what they are doing, they may be avoiding taxes, or for some other reason their activity escapes official notice. These activities make up the **informal or underground sector** of an economy. Every city in the world has such a sector, but the informal sector is particularly important in the cities of the poor countries. The International Labor Organization has estimated that informal employment is a full 72 percent of nonagricultural employment in sub-Saharan Africa, 65 percent in Asia, and 51 percent in Latin America.

Life in the extra-legal world is a constant risk. People lack title to property, so they build housing poorly; therefore many die in earthquakes or other natural disasters. An estimated 4 billion people cannot create wealth or recuperate from disaster because of the lack of legal records. Neither capital nor credit will venture where there are no clear property rights, and seizure of property by the politically well-connected or powerful is a constant threat.

In an early study of Lima, Peru, De Soto found that the informal economy employed fully 60 percent of the population and produced 40 percent of all goods and services. The poor owned and controlled a public transportation network of private taxis and vans, plus land and housing worth billions of dollars. None of this, however, was legal, so it could not be taxed by the government or used as collateral by business owners. If the government simply legalized these activities, these assets would have liquidity and could provide collateral for investment and business enterprise. The government's refusal to recognize what was happening handicapped the country's economic growth and vitality.

In Peru from 1995 to 2001 more than 1.2 million households, including 6.3 million people, received title to the properties they were inhabiting. Title reform enabled more people to work outside the home and more family members to join the labor force, because now no one had to stay home to guard the property. The values of the newly registered properties have soared, and mortgage and consumer credit markets have developed. Studies in the Philippines found that 60 percent of Filipinos are holding real estate assets worth tens of billions of dollars outside the law, and thus illiquid. In Egypt 85 percent of the population lives in homes without property titles. De Soto estimated that in 2000 only 25 countries in the world have genuine contractual urban societies; the rest are informal.

A modern market economy cannot develop unless property rights are acknowledged and protected, and economic growth will probably occur most rapidly in the developing countries that ensure property rights. Therefore, De Soto has argued, governments must formalize the spontaneous emergence of informal property. Chapter 8 already noted how productivity in agriculture can be raised by guaranteeing property rights. The same guarantees could develop in the cities. China has privatized landowning in the cities faster than in the countryside (see Chapter 8), and the privatization has contributed to faster economic growth in the cities. It created a middle class that is using its property as collateral to borrow money to launch enterprises. In both city and countryside, GIS and GPS greatly enhance the ability to record and register land holdings, so the spread of these techniques may spark economic growth and vitality around the world.

If governments view urban migration as a problem, they cannot see how urban immigrants' industriousness could be an asset for economic growth. Millions of people continue to choose to migrate to the cities, where they do survive or even thrive. There is something terrifically dynamic going on, and geographers, economists, and government officials at all levels are challenged to understand and measure it, and to harness the energy.

THE INTERNAL GEOGRAPHY OF CITIES

Urban geographers study not only the distribution of cities in the landscape but also the distribution of activities and of housing within cities. This distribution may be caused by economic forces, social factors, or deliberate actions of the government. In different countries and cultures, each of the three factors carries different relative weight.

Models of Urban Form

A 1945 book by Chauncey Harris and Edward Ullman, *The Nature of Cities,* proposed three models of the internal patterns of North American cities: the concentric zone model, the sector model, and the multiple-nuclei model. Fifty years later, Professor Harris suggested a fourth model, the peripheral model, to describe how American metropolitan areas developed in the second half of the twentieth century.

Figure 10-13 illustrates the *concentric zone model* of urban growth and land use. The core of the city, called

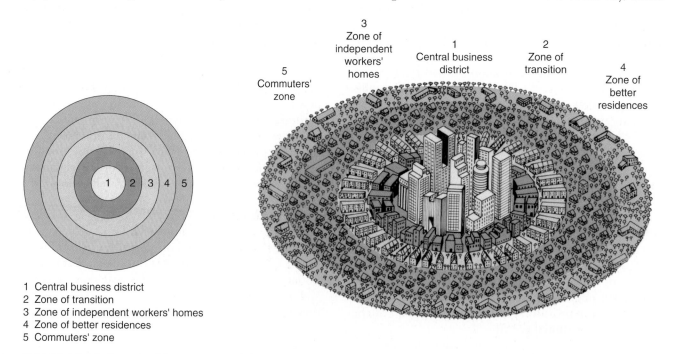

1 Central business district
2 Zone of transition
3 Zone of independent workers' homes
4 Zone of better residences
5 Commuters' zone

FIGURE 10-13 Concentric zone model. According to the concentric zone urban model, a city grows in a series of rings around the central business district.

the **central business district (CBD)**, concentrates office buildings and retail shops. For businesses, accessibility is usually a principal determinant of a location's rent. The success of a department store, for instance, will depend partly on whether customers can reach the store easily. A city's most convenient and busiest intersections are most valuable for commerce. Land owners usually maximize the density of use on this valuable land by building up, so a traditional CBD is identifiable by tall buildings as well as crowded streets. Even within the CBD, clusters of functions appear. Lawyers' offices, for instance, cluster near courts or near the offices of their client firms. Retail stores of one type, such as jewelry stores, may cluster so that consumers can comparison shop. The CBD is surrounded by less intensive business uses such as wholesaling, warehousing, and even light industry—that is, nonpolluting industries that require relatively small quantities of raw materials. Residential land use surrounds this urban core.

The concentric zone model can be modified by considering the effect of transportation routes, which affect accessibility. New means of transport—historically canals, then railroads and tramways—spread out radially from the heart of the city, although their paths are modified by topography. Industrial and residential growth take place in ribbons or fingers along these radial routes, and wedges of open land are usually left between these radial routes.

The second model proposed in 1945, the *sector model* (Figure 10-14), assumes that high-rent residential areas expand outward from the city center along new transportation routes such as streetcar and suburban commuter rail lines. Middle-income housing clusters around high-rent housing, and low-income housing lies adjacent to the areas of industry and associated transportation, such as freight railroad lines.

The third model, the *multiple-nuclei model* shown in Figure 10-15, recognized the development of several nodes of growth within an expanding city area. These nuclei may each concentrate on a different special function.

The *peripheral model*, defined in 1995 (Figure 10-16), shows how radial and circumferential highways continue to draw activities out of the central city and to disperse them around the region. The story of how this happens will be told below.

Social factors in residential clustering

Social considerations play a role in urban residential clustering. Some people want to live with people like themselves, which causes a clustering called **congregation.** Ethnic groups or immigrants of a common background, for example, may want certain services, such as grocery stores offering their traditional foods.

In other cases, however, people live together because discrimination forces them to do so. These people suffer **segregation** from others. In any specific case, it may be difficult to determine the degree to which people are congregating or are victims of segregation. Chapter 7 noted that in the past, Jews were legally segregated in ghettoes, but today the word *ghetto* means any residential concentration of any one kind of people. The factors causing residential clustering of any group—of Chinese Americans in "Chinatowns," of Italian Americans in "Little Italys," of Hispanic Americans in "barrios," or of African Americans as will be discussed later in this chapter—must be evaluated carefully in each case. Sometimes the descriptive words placed in quotation marks in this paragraph

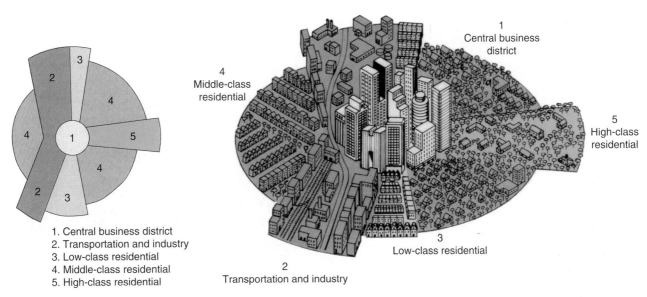

1. Central business district
2. Transportation and industry
3. Low-class residential
4. Middle-class residential
5. High-class residential

FIGURE 10-14 Sector model of urban development. In this model of urban form, a city grows out from the central business district in wedges, or corridors, of various land uses.

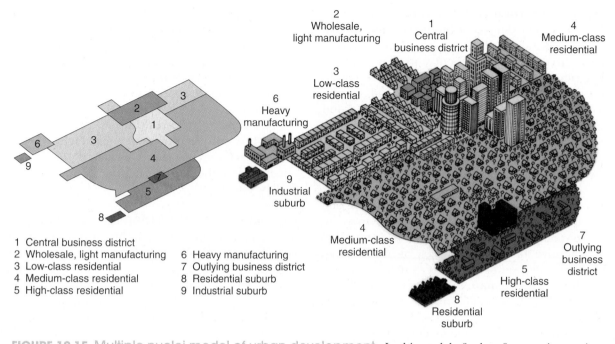

1 Central business district
2 Wholesale, light manufacturing
3 Low-class residential
4 Medium-class residential
5 High-class residential
6 Heavy manufacturing
7 Outlying business district
8 Residential suburb
9 Industrial suburb

FIGURE 10-15 Multiple-nuclei model of urban development. In this model of urban form, a city consists of a collection of individual nodes or centers around which different types of activities and people cluster.

are considered insulting, but they usually are not meant to be.

Religion is another social consideration that frequently causes clustering. People who share a religious faith may cluster around their house of worship, and immigrants of that faith will seek that neighborhood. Language communities frequently form, as do communities of the elderly and communities of gays and lesbians. Almost any factor of social bonding can encourage the creation of an identifiable residential neighborhood.

Urban environments Chapter 2 listed ways in which urban microclimates may differ from those of their surrounding countrysides: greater heat, lack of natural vegetation, paved surfaces, air pollution, and more. Large cities are also noisy, and noise interferes with communication and sleep, raises blood pressure, triggers muscle tension, and precipitates migraine headaches, elevated cholesterol levels, gastric ulcers, and psychological stress. It diminishes people's ability to concentrate, and if it is loud enough for long enough periods of time, it can damage hearing capability. Cities face enormous

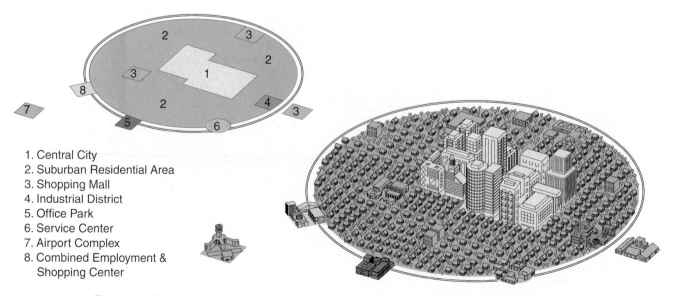

1. Central City
2. Suburban Residential Area
3. Shopping Mall
4. Industrial District
5. Office Park
6. Service Center
7. Airport Complex
8. Combined Employment & Shopping Center

FIGURE 10-16 The peripheral model of urban areas. In this model, describing the growth of U.S. metropolitan areas in the last half of the twentieth century, activities disperse throughout a broad region.

challenges in creating livable, healthy environments, as well as in disposing of their garbage and solid waste, problems that were discussed in Chapter 9.

The role of government   Government may determine land use. Restricting or prescribing the use to which parcels of land may be put is called **zoning.** In U.S. history, local government has often been averse to planning, so zoning has more often been *restrictive* (dictating what cannot be done) than *prescriptive* (dictating what should be done). Industrial and commercial districts, for example, are usually kept away from residential neighborhoods. Each incorporated jurisdiction across vast conurbations exercises independent zoning power. This creates problems that will be discussed shortly.

Today most urban planners believe that the separation of land uses has been overemphasized. It may have been desirable to separate industry from housing when all industry was noisy, polluting, or smelly, but today separating homes from jobs may require excessive commuting. Most new planned communities emphasize the integration of residential and commercial activities, even including some light industry. They offer apartments above downtown shops and offices, for example, as in traditional small towns. The state of New Jersey subsidizes landowners who renovate downtown properties if the renovations create residential space on upper floors. This revitalizes downtowns and encourages the use of public transit—or even walking.

The Western Tradition of Urban and Regional Planning

The process of urban and regional planning applies many of the principles of urban geography to specific situations. The concept of designing an "ideal city" has challenged the best minds for centuries. The founder of Western city planning was probably the Greek thinker Hippodamus of Miletus. He laid out that city in today's Turkey according to a grid plan as early as 450 B.C.

The first modern attempt to formulate the needs of a city as a whole was the work of the British visionary Sir Ebenezer Howard. In *Garden Cities of Tomorrow* (1898), he outlined a plan to stop the unbounded growth of the industrial city and to restore it to a human scale. He wanted to relocate population into new medium-sized garden cities in the outlying countryside. These regional cities would be ringed by greenbelts of farmland and parks. All land would be municipally owned, and each town and its surrounding region would be planned as an interlocking whole (Figure 10-17).

Howard built two "garden cities" just north of London: Letchworth (1904) and Welwyn Garden City

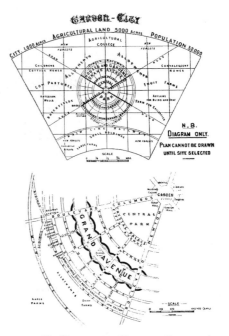

FIGURE 10-17 Sir Ebenezer Howard's garden city. Howard envisioned planned "garden cities" to disperse the concentrations of population in nineteenth-century cities. (Reprinted with permission from Ebenezer Howard, *Garden Cities of Tomorrow.* Cambridge, MA: The MIT Press, 1965.)

(1919). These inspired the Regional Planning Association of America, a private nonprofit organization, to construct two planned communities in the New York City area: Sunnyside Gardens, Queens (1924); and Radburn, New Jersey (1928) (Figure 10-18). Neither is a complete garden city, but both are harmoniously designed and have greatly influenced urban planning in the United States and Europe.

Probably the most important city planner of the twentieth century was the Swiss architect Charles Édouard Jeanneret-Gris (1887–1965), better known by his professional name, Le Corbusier. In a celebrated plan of 1922, he proposed to bulldoze the crowded, rundown historic core of Paris, preserving only the central monuments. In its place he wanted to build a *Radiant City* of tall glass offices and apartments, spaced so far apart that each tower would be surrounded by green space and have a fine, wide view (Figure 10-19). The concentration of facilities within high-rise slabs would liberate the city from its environment. It could be placed anywhere. Le Corbusier brought together two conceptions: the machine-made environment, standardized, technically perfect to the last degree and, to offset this, the natural environment, treated as open space, providing sunlight, air, greenery, and views.

Paris was never torn down and rebuilt as a Radiant City, but Le Corbusier planned Chandigarh, a new capital for the Punjab State in India in 1950. The world's supreme Radiant City, however, is Brasília, the capital of Brazil, designed by Lucio Costa in 1957

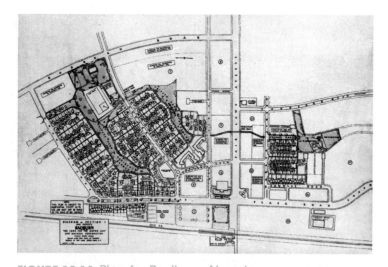

FIGURE 10-18 Plan for Radburn, New Jersey. This planned community, built in 1928, is still a desirable residential settlement. Key elements of the plan include plenty of park space and the separation of pedestrian walkways from car traffic and parking. (Courtesy of the Radburn Association)

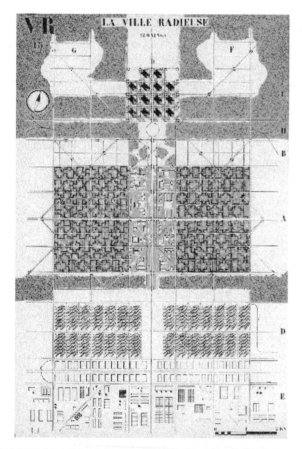

FIGURE 10-19 La ville radieuse (Radiant City). Le Corbusier's plan for his Radiant City featured high-density residential areas strictly segregated from other land uses. (Le Corbusier, *La ville radieuse*. Paris: Editions de L'Architecture D'Aujour'hui, 1933, p. 170/French Government Tourist Office)

FIGURE 10-20 Brasília, Brazil. Brasília, Brazil's capital since 1960, arose on a largely unpopulated open plateau 970 kilometers (603 miles) northwest of Rio de Janeiro. It was an effort to shift the nation's political, economic, and psychological focus toward the interior and away from the former colonial cities on the coast. As this picture reveals, it sometimes lacks water sufficient to keep the grass green or to operate the elaborate fountains. Slums surround the show-place buildings for 50 kilometers (30 miles) in every direction. Architect Lucio Costa admits, "Of course half the people in Brasília live in [slums]. Brasília was not designed to solve the problems of Brazil. It was bound to reflect them." (Getty Images Inc., Stone Attstock)

(Figure 10-20). Unfortunately, the city's gigantic scale demands a completely motorized population. That is the problem with excessive openness—Corbusier's "city in a park" can become a city in a parking lot. The Australian capital at Canberra, planned by Walter Burley Griffin of Chicago, has less openness. Its layout allows walking, and is generally considered superior (Figure 10-21).

In 1930 Le Corbusier planned the town of Nemours in Algeria. It has a geometric grouping of buildings that look like dominoes, and this plan set the

FIGURE 10-21 Canberra, Australia. Canberra became home to Australia's Parliament in 1927, but foreign missions and federal government departments did not move here until the 1950s. (Courtesy of Embassy of Australia)

FIGURE 10-22 Low-income housing in Newark, New Jersey. The abandoned high-rise public housing in the background (soon demolished) has been replaced by the low-rise development in the foreground. Low-rise housing can achieve equivalent density, and it has been demonstrated that it fosters more successful communities than high-rise projects do. Since 1992, the federal government has torn down more than 100,000 apartments nationwide in high-rise buildings and replaced them with low-rise housing. (Keith Meyers/*The New York Times*)

international fashion for high-rise slabs for the next 50 years. These ideas were disseminated worldwide by the 1933 Athens Charter of the International Congress of Modern Architecture (CIAM). In the charter's codification, largely by Le Corbusier, the functions of the city—housing, work, recreation, and transport—provided the city's framework. The charter called for separation of high-rise development, industrial zones, parks and sports fields, and streets of different widths spaced for traffic at different speeds. These ideas diffused to dominate urban planning around the world.

Many people believe today that the widespread adoption of Le Corbusier's ideas produced a half century of monotony—not merely of detail and style but of insensitivity to essential differences between one place and another. The high-rise slabs ringing every big city in the world from Mexico City to Singapore look much alike.

Many urban planners worldwide have come to criticize the concept of high-rise living. Low-rise dwellings can achieve the same density of habitation as Le Corbusier's "towers in a park" can, and many people feel more content living in low-rise dwellings. High-rise public-housing projects, it turned out, can breed a sense of alienation and helplessness, and many have been abandoned and razed across the United States.

The riots in Parisian suburbs in 2005 (see Figure 5-34) occurred in "radiant city" housing projects. Le Corbusier designed new towns on the outskirts of several French cities. The government called them

"modern," but the housing was cheaply made, grim, and boring. When French people refused to live in these projects, the projects were dedicated to housing France's immigrant population. The immigrants' residential separation from traditional French life, as well as the problems of cultural assimilation discussed in Chapter 5, triggered repeated riots among residents. After a 1983 riot in such a project, President Francois Mitterand toured it and wrote that places like that "must disappear from our country." He insisted that it be torn down, and several such projects were dynamited in 1986, in 2000, and more in 2004. Nevertheless, many still surround France's largest cities.

In 1961 a group of younger architects broke away from CIAM and proclaimed that architecture was more than the art of building. It was the art of transforming people's entire habitat. The School of Architecture at the University of California at Berkeley was reconstituted and renamed the School of Environmental Design. Today we still work to design more humane urban environments (Figure 10-22).

Other Urban Models in Diverse Cultures

The concentric zone, sector, multiple-nuclei, and peripheral city models discussed earlier in this chapter were all devised to describe North American experience, but other models have been sketched to describe characteristics of other cultures. The three interacting processes identified earlier—economic factors (including transportation facilities), residential clustering or segregation, and government decree—affect all urban settlements, old and new, in all known cultures. These processes sort out the population and the land uses into distinct patterns that can be identified within any city.

The four models just described were first identified in North America, but they do not fully describe even all modern Western cities. The governments of the Western European countries, for example, have always been concerned with preserving the vibrancy and amenity value in their central cities. Therefore, they severely restrict suburbanization, as will be discussed shortly.

Latin American cities offer still another contrast to North American models. There, central business districts thrive (Figure 10-23). This is partly a result of continuing reliance on public transit and partly because high-income populations choose to live close to the central business district. A commercial spine such as a boulevard extends out from the central business district, and amenities such as opera houses, chic stores, and elegant parks follow this spine. Zones of more modest housing and value surround this elite zone, and the periphery is dominated by squatter slums.

Western forms often overlie indigenous forms Chapter 6 noted that distinct settlement patterns are landscape footprints of distinct cultures (look back at Figure 6-20), but it also noted that many of the world's great cities came into being as a result of trading or political contact between the native peoples and Europeans. The cities that Europeans initiated and founded were planned and built according to European notions of city planning. Their internal geography still shows port zones; enclaves of former European settlements; barracks; colonial government buildings; and racial, religious, and ethnic ghettoes based on the role each group played in the city's founding and during the colonial period (Figure 10-24 and look back at Figure 6-30).

In other cases, Western or modernizing interests built a new city alongside a preexisting native city. This was particularly common in those parts of Latin America (particularly today's Mexico and Peru), Asia, and North Africa where the local peoples had achieved significant urbanization before the coming of the Europeans. Today the two cities often contrast sharply: Historic cores in forms traditional to the local culture stand beside districts of modern commerce, retailing, industry, and associated residential areas. The modern or Western cities typically demonstrate grid layouts. In some cases a third element—vast slums of newcomers—surrounds both cities. The city of Fez, Morocco, for example, is clearly two cities (Figure 10-25): A modern Western-style city of broad, straight avenues lies on the plain below an older, traditional Islamic city. Western

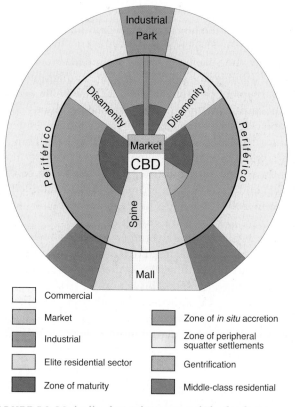

Commercial

Market

Industrial

Elite residential sector

Zone of maturity

Zone of *in situ* accretion

Zone of peripheral squatter settlements

Gentrification

Middle-class residential

FIGURE 10-23 Latin American model of urban growth. This model contrasts with the North American models. In Latin America an elite residential sector often follows a spine of high-value land use stretching out from the central business district. (Adapted from Larry R. Ford, "A New and Improved Model of Latin American City Structure," *Geographical Review* 86 (1996): 438. Used by permission of the publisher.)

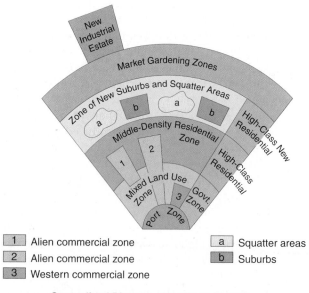

1   Alien commercial zone

2   Alien commercial zone

3   Western commercial zone

a   Squatter areas

b   Suburbs

Generalized Diagram of the Main Land Use Areas of the Large Southeast Asian City

FIGURE 10-24 Model of a Southeast Asian port city. Most Southeast Asian port cities are the product of Western influence, so they reflect Western design forms. (Adapted from T. G. McGee, *The Southeast Asian City.* Reprinted with permission of Greenwood Publishing Group, Inc. Westport, CT. Copyright © 1967 by Greenwood Publishing Group, Inc.)

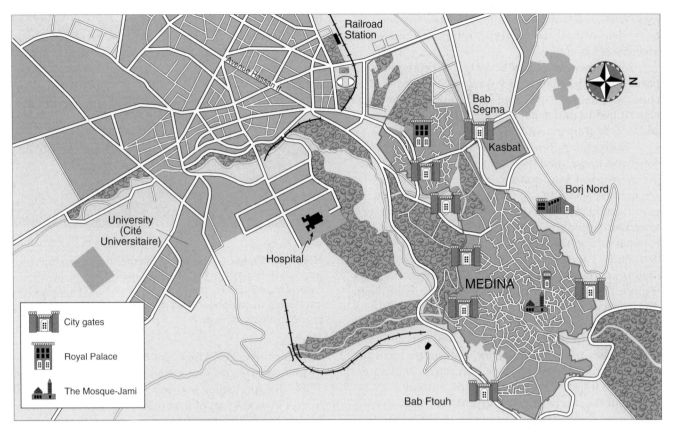

FIGURE 10-25 Fez, Morocco. The Moroccan city of Fez is really two contrasting cities. A modern Western city lies on the plain to the southwest (left in this figure) of an old Islamic city, the *Medina*. The new city has straight formal avenues, a railroad station, university, and a modern hospital. The Medina contains old mosques and narrow constricted streets, and it is surrounded by city walls with great gates.

forms continue to be stamped on the world's cities today. Western architects, engineers, and urban planners, or non-Western individuals educated in Western schools, continue to transform even the older traditional built environments.

Individual cities, however, emerged at different times, for different reasons, and within different cultural contexts. Many great cities still boast historic cores that illustrate indigenous principles of urban planning, and they cannot all be squeezed into three or four simplified models.

Islamic urban form

The non-Western culture with the oldest and most articulate urban planning tradition is the Islamic culture. Traditional Islamic cities illustrate the role of culture in urban form. There are regional differences in cities across the Islamic realm, but most nevertheless show surprising similarity. These cities may seem chaotic to Westerners at first glance, especially to those of us accustomed to grid patterns, but they present an entirely rational structure. The structure develops from the basic premises that religion is the most important consideration in life and that houses of worship are the most important elements of urban design. The resulting design characteristics can be identified from Seville, Granada, and Córdoba in Spain to Lahore

in Pakistan, and elements of these principles can be found from Dar es Salaam in Tanzania to Davao in the Philippines.

The logic of traditional Islamic urban planning is announced in the Koran and has been codified by various schools of Islamic law. Certain basic regulations govern individual rights and the pursuit of the virtuous life in a densely crowded urban environment. For example, Islamic urban planning recognizes the need to maintain personal privacy; it specifies responsibilities in maintaining urban systems on which other people rely, such as keeping thoroughfares or wastewater channels clear; and it emphasizes the inner essence of things rather than their outward appearance. This last principle applies as much to the decoration of houses as to purely spiritual issues.

Take another look at Figure 10-25. At the heart of the traditional Islamic city stands the main mosque, the *jami*, which is typically the city's largest structure. Close to it are the main *suqs*, the street markets and enclosed shopping arcades. These arcades prefigured urban *galleries* in Europe and enclosed shopping malls in North America. Within the suqs, trades are diffused in relation to the mosque. The tradespeople who enjoy the highest prestige, such as booksellers and perfumers, are closest. Farthest away are those who perform the noxious and noisy trades, such as coppersmiths, blacksmiths, and cobblers. The neutral tradespeople, such as clothiers and jewelers, act as buffers.

An immense fortified *kasbat* is attached to the ramparts, on which are located several towers or gates. The *kasbat* was the place of refuge for the governor or sovereign. It had its own small mosques, baths, and shops, in addition to government buildings and barracks.

Everywhere else the city is filled in with cellular courtyard houses tied together by winding lanes. Housing is grouped into quarters, or neighborhoods, that are defined according to occupation, religious sect, or ethnic group. The widest streets usually radiate outward from the core to the gates in the city wall. Slightly narrower streets serve the major quarters and define their boundaries, and still narrower third-order streets are used primarily by people who live in the neighborhood (Figure 10-26). Narrow streets provide vital shade, keep down dust and winds, and use little building land.

Interior courtyards of homes, often with trees and fountains, provide shade in hot climates, but, more important, they provide an interior and private focus for life sheltered from public gaze. This is true in Mediterranean architecture. The outside of a house may be plain, but the interior and courtyard may display lavish wealth and decoration. The interior vividness parallels the Koranic emphasis on the richness of the inner self compared to a more modest outward appearance.

In the United States, New Orleans is the city closest to exemplifying these values (Figure 10-27). In the old Spanish sections of the city (misleadingly called today the French Quarter), houses generally present plain fronts to the street, but many enclose beautiful

FIGURE 10-27 New Orleans's "French Quarter." Here, as in Islamic countries and Mediterranean architecture in general, houses are not placed in the centers of gardens; courtyard gardens are placed in the centers of houses, not impressing the neighbors and passersby but providing cool private retreats. (© Robert Holmes/CORBIS)

private courtyards. When Anglo-Saxons began to move to the city after its annexation by the United States in 1803, the Anglo-Saxon rich preferred to build grand homes in a new part of town called the Garden District. In the French Quarter the courtyard gardens are inside the houses, but in the Garden District the houses sit in the middle of their gardens. This difference is one of cultural preference.

CITIES AND SUBURBS IN THE UNITED STATES

The dominant feature of the metropolitan form in the United States has been the explosive growth of cities across the countryside. Growing cities have spilled over their legal boundaries into areas called *suburbs*. Some suburbs are entirely residential, but others offer services for the surrounding residential population. In some cases, the suburbs are older cities that have been engulfed by the growth of a larger neighbor, but others are newly incorporated settlements. What defines an area as a suburb is its economic and social integration with a larger population nucleus nearby. *Town* and *village* are inexact terms that generally designate settlements smaller than cities, but the settlements may be incorporated.

Many of the developments described in the following discussion are now occurring elsewhere around the world, but they occurred first in the United States—largely because of the nation's prosperity.

The Growth of Suburbs

Early suburbs Most large U.S. cities included manufacturing districts by the late nineteenth century. These were noisy and dirty, and they often attracted a

FIGURE 10-26 Casablanca, Morocco. In many traditional Islamic cities—as here, in Casablanca, Morocco—occasional straight streets provide views of minarets. People are constantly reminded of the importance of religion. (Courtesy of Moroccan National Tourist Office)

working class, largely made up of immigrants, whom many long-established residents found to be unpleasantly "different." These biases pushed those who could afford it to move to the suburbs.

At the same time, a cultural preference for rural or small-town life pulled many people out of the city. Many Americans fell in love with the idea of "the country," and they favored a return to nature, to the land, or to open spaces—even if only a suburban yard. Therefore, when the railroads made older rural communities within commuting distance of the city, many people who had the time and the money necessary to commute to work from a home outside the city began to do so. In other cases, the wealthy built new towns (Figure 10-28). "Streetcar suburbs" sprang up when streetcar transportation was devised. Some of these planned suburbs eventually became completely built up, merged into other settlements, and lost their identities.

The automobile ultimately opened the nation's landscape to suburban growth. For those who disliked urban life, the suburb was the solution. "We shall solve the city problem," wrote Henry Ford in 1922, "by leaving the city."

FIGURE 10-28 Riverside, Illinois. The designers of New York City's Central Park, Frederick Law Olmsted and Calvert Vaux, planned this real estate subdivision 14 kilometers (9 miles) from the center of Chicago in 1869. The plan included two straight business streets paralleling the railway into the city, but all residential streets were curved to slow traffic (before cars!). Open spaces contribute to the sense of breadth and calm enjoyed by "the more fortunate classes" for whom Riverside was designed. In 1992 Riverside residents refused federal financial assistance for traffic control because federal regulations would have required traffic intersections to be reengineered to 90° angles. (Courtesy of the National Park Service, Frederick Law Olmsted National Historic Site)

Government policies and suburban growth

The dispersion of housing to suburbs was slowed by the Great Depression in the 1930s and by World War II in the early 1940s. Following the war, however, government policies established a new balance of push-and-pull forces that encouraged the movement of investment, residents, and jobs out of the central cities into the suburbs.

The Federal Housing Administration (FHA) guaranteed loans, so down payments shrank to less than 10 percent of the house price. Suddenly thousands of families could afford new houses. FHA benefits did not, however, apply equally everywhere or to everyone. The FHA favored the construction of new single-family houses in the suburbs over the rehabilitation of older houses or apartment buildings in the central cities. Also, the FHA opposed what it termed *inharmonious racial or nationality groups.* In some places the presence of one non-White family on a block was enough to cut the entire block off from FHA loans. Thus, government policy helped segregate the suburbs.

The government also granted tax and financial incentives to homeowners, including the deductibility of both mortgage interest payments and of local property taxes from gross taxable income. These two benefits alone often made buying a new house cheaper than renting. In addition, the government protected homeowners from capital gains taxation—that is, taxes on any increase in the value of the house. By fiscal 2007, the tax loss to the U.S. Treasury of these three benefits totaled well over $100 billion per year. These subsidies to homeowners far exceed government spending on public housing. Furthermore, until 1980 the government allowed savings-and-loan institutions to pay higher interest on savers' money than commercial banks, because money in savings and loans was directed into housing.

The Veterans Administration Housing Program, begun in 1944 and called Homes for Heroes, pumped billions of additional federal dollars into housing programs. By 1947 the Levitt Company was completing 30 new single-family homes in Levittown, formerly a Long Island potato field, each day, and similar developments were springing up on the outskirts of every major U.S. city (Figure 10-29). Nationwide housing starts jumped from 114,000 in 1944 to 1,696,000 by 1950.

The suburbs brought home ownership to an increasing share of U.S. families. The percentage of U.S. housing that was owner-occupied rose from 44 percent in 1940 to 62 percent in 1960 and 69 percent in 2006, signaling middle-class status for a rising share of the population. Expanding home ownership has increased the number of citizens who have profited from the many homeowner subsidies, so it also has reduced the political possibility of rescinding them. Some scholars argue that the tax concessions were never necessary. Canada, Australia, and other countries achieved comparable levels of homeownership without offering such concessions.

FIGURE 10-29 Levittown, Long Island. In building their first Levittown, 40 kilometers (25 miles) east of Manhattan, the Levitt family changed U.S. homebuilding techniques. The land was bulldozed and the trees removed, and then trucks dropped building materials at precise 19-meter (60-foot) intervals. Construction was divided into 27 distinct steps. At the peak of production more than 30 houses were completed each day. Through the years owners have personalized their homes so much that few visitors today can see that the houses were originally identical. (UPI/CORBIS BETTMAN)

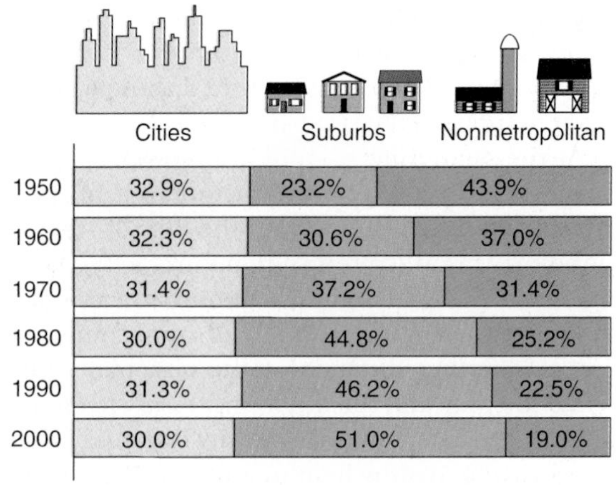

	Cities	Suburbs	Nonmetropolitan
1950	32.9%	23.2%	43.9%
1960	32.3%	30.6%	37.0%
1970	31.4%	37.2%	31.4%
1980	30.0%	44.8%	25.2%
1990	31.3%	46.2%	22.5%
2000	30.0%	51.0%	19.0%

FIGURE 10-30 Distribution of U.S. population. By 1970 a plurality of the U.S. population (37 percent) lived in the suburban portions of the MSAs, and by the century's end, a majority of Americans were suburban. Canada's 2001 census counted two-thirds of Canadians living in 25 "census metropolitan areas," with over 50 percent of the total national population in just four great conurbations (Toronto, Montreal, Vancouver, and Calgary/Edmonton). The Canadian population became predominantly urban in the mid-1920s and predominantly metropolitan in the mid-1960s, about a decade later than the United States. This is explained by Canada's lower population, relatively low density and greater dispersion, and the importance of Canada's resource industries.

At the end of the twentieth century and beginning of the twenty-first century, housing prices in the United States rose rapidly during a period through which interest rates were low. At the same time, income and sales taxes rose, and property taxes fell from 40 percent of local tax receipts in 1970 to under 30 percent by 2005. These factors combined to encourage many people to consider their homes as investments, that is, to buy the most expensive home on which they could possibly meet mortgage payments on the assumption that the value of the property would rise. Mortgage debt rose from 15 percent of gross domestic product in 1945 to 90 percent in 2005. Many economists were alarmed that, if housing prices fell or even leveled, if interest rates rose sharply, or if Americans' incomes fell, many homeowners would not be able to meet the payments on their houses.

As the population spread out, new suburbs incorporated, and the Census Bureau devised a term for these sprawling conurbations: **metropolitan statistical area (MSA).** The Bureau defined a metropolitan area as "an integrated economic and social unit with a recognized large population nucleus." Thus, MSAs are the principal central cities and their suburban counties (except in New England, where the definitions are in terms of cities and towns). When two or more MSAs are contiguous, each is called a primary metropolitan statistical area

(PMSA), and the conurbation is called a **consolidated metropolitan statistical area (CMSA).** By 2000 the nation's 274 metropolitan areas contained 80 percent of the total population. These MSAs covered only 20 percent of the country's land surface (Figure 10-30).

In 2005 the Census Bureau first defined and collected data for 573 **micropolitan areas**, defined as developed regions with a core city of fewer than 50,000 people. These suburban areas offer a middle ground between rural and metropolitan living.

The suburban infrastructure The sprawl of single-family homes is expensive. It first requires roads for individualized transportation, which, in turn, demands energy. Heating and cooling individual homes is also energy-intensive. Dispersed housing also requires enormous investment in sewerage, water pipelines, telephone lines, and electrical wiring. The cost of providing infrastructure for 100 people in an apartment building is much less than the cost of providing it for the same 100 people spread out in 20 single-family homes over several hectares.

Infrastructure costs are further inflated by the fact that U.S. suburbs have not expanded contiguously outward from the city, like the waves from a stone tossed into a pond. Each developer wants to buy land as cheaply as possible, and thus buys land beyond the edge of growth. This is called *leapfrogging*. The infrastructure network cannot be advanced in a regular pattern. Thus leapfrogging increases initial costs. Later the leapfrogged areas are filled in, but then some initial infrastructure has to be rebuilt. Many suburbs, for example, originally relied on individual-home wells for water, and their sewage was treated in individual-home septic tanks. As the suburbs matured and density increased, water supplies became polluted. Homeowners had to pay for wholly new public water mains and sewers. Suburban development also increased stream runoff, so financial losses from flooding have risen virtually every year.

The high infrastructure costs necessitated high property taxes, and these eventually provoked voter backlashes. The most notable was California's Proposition 13 of 1978, which limited property-tax increases. This has hobbled the government's ability to provide services no matter how much they are needed.

Suburbs take up a lot of space, and since 1945 U.S. urban areas have spread at the rate of about 405,000 hectares (1 million acres) per year. This growth has required that a good share of the country's most productive farmland be paved over, such as the Long Island potato fields covered by Levittown. The best agricultural hinterlands of many cities disappeared, so a rising percentage of the nation's food is now grown in conditions requiring expensive fertilizer or irrigation. In addition, the food has to be transported farther, consuming still more fuel, perhaps requiring refrigeration or special handling, and further boosting food prices.

Farmers virtually have been forced to sell their land to developers because property taxes are calculated on land's potential value, not its current-use value. A farmer who was making a small profit by farming could not afford to pay property taxes calculated on the land's potential value as housing. Even if a farmer could somehow meet the annual property-tax bill, the farmer's heirs would eventually have to sell the farm to pay inheritance taxes, which are also based on land's potential value. Only recently have these tax laws been changed to preserve greenbelts around some cities. In some cases, farmers now pay use-value taxes. Other farmers have sold development rights to conservation groups or to local governments, thus guaranteeing that the land will remain as farmland.

In the 1950s and 1960s, U.S. citizens did not worry about the costs of creating new suburbs. Between 1950 and 1973 median family income doubled in real terms. Demands for housing continued to mushroom with the baby boom and the splintering of families into separate households, partly as the result of a rising divorce rate. The average number of occupants of a U.S. household shrank from 3.67 in 1940 to 2.6 in 2000.

Today a greater share of U.S. wealth is invested in housing and the necessary infrastructure than in any other nation. A high proportion of Americans enjoy private ownership of spacious, free-standing, well-equipped homes, and this investment has succeeded in bringing a sense of well-being to many Americans. It is becoming clear, however, that this development has brought with it both high economic costs and steep social costs.

The Social Costs of Suburbs

The suburban lifestyle has imposed social costs not only on those who enjoy it but on many others, also. Americans generally sort themselves out residentially in such a distinct manner that sociologists and mass marketers can confidently construct an astonishingly accurate profile of people on the basis of their address alone (Figure 10-31). This residential segregation of the American people—racial, economic, and in many ways cultural—allows advertisers to target potential customers geographically. A great deal of demographic data on people's income, education, and other characteristics has been sorted by ZIP code, in order to fine-tune the selection of junk mail that arrives in mailboxes. GIS mapping and analysis of neighborhoods and metropolitan areas offers fast-growing career opportunities. Retailers want to know where the most probable potential customers for their product or service reside, what paths they take for work or relaxation and, thus, where the retailer might best locate a new store or service. One marketing consulting firm has categorized the country's ZIP codes and census tracts into 62 "lifestyle types," giving them catchy names such as Gray Power to describe communities of affluent retirees in Sunbelt

FIGURE 10-31 You are where you live. Suburbs may differ greatly from one to the next, but individual suburbs often congregate people of similar income and lifestyle. This clustering allows retailers and advertisers to use demographic data stored in GISs to locate potential customers. (© *The New Yorker Collection.* 1993 Henry Martin from cartoonbank.com. All rights reserved.)

cities; Towns and Gowns to describe college towns; and Latino America to describe neighborhoods of Latino middle-class families.

In low-density suburbs local racial and social homogeneity may have caused conservatism and conformity. In addition, property owners in many suburbs established *restrictive covenants*, which were legal agreements that the land would never be sold to people of a designated race or religious group. Such covenants are no longer legal, but they were common as late as the 1970s. Homeowners' associations, however, which regulate some aspects of property ownership, covered 20 million homes in 2003 (of the nation's 106 million) that house 50 million people, and continue to grow. Most of these associations not only require that homeowners maintain their yards, but some dictate even when owners can put up holiday decorations or park their cars in their own driveways. Major changes in a home's structure or exterior or appearance must be approved by the association, which can fine homeowners or even foreclose the offending homes—without the residents' knowing until they are evicted. Courts have accepted the argument that the regulation of "visual pollution" is analogous to the regulation of noise pollution, even though there may be greater disagreement on what constitutes visual pollution. In Portland, Oregon, for instance, the front of a house must be at least 15 percent windows and doors, and a garage door may take up no more than half the façade. In Arizona, a homeowner was sued for installing solar panels on his roof; a suburban couple in Washington State was sent to jail for painting their house mauve, and in December 2006, a couple in suburban Denver was fined for hanging a holiday wreath in the shape of a peace symbol on their front door. The purpose of these rules is to maintain neighborhood property values, but some Americans find them stifling.

The suburbs overall are diverse—there are poor suburbs as well as rich ones, and there are suburbs of every ethnic and racial group. Individual suburban communities, however, are usually homogeneous.

The movement of jobs to the suburbs

The suburbs first expanded as bedroom communities for the middle-class workers who left their suburban families each morning to go to work in the city. Soon, however, the interstate highway system (authorized in 1956) and similar limited-access highways not only joined cities but also provided peripheral bypasses around them. These peripheral arteries provided access among suburbs and reduced the geographical advantage of a city's central business district. Developers put up suburban office buildings, and corporations built spacious office parks. Retailers soon began to build giant shopping malls at highway crossroads.

Manufacturing establishments abandoned the central city, too, for several reasons. New light-industrial facilities proliferated in the suburbs because new technology favored horizontal buildings rather than the vertical ones characteristic of central cities (called *lofts*). In addition, light industries relocated to escape central-city congestion, as well as the higher costs for energy, taxes, wages, and rent. Warehousing also relocated out from the inner-city railroad yards to the suburban highway interchanges. Metropolitan airports in the suburbs grew to provide jobs in both freight and passenger services.

In the 1950s, suburban growth was fed by young married couples who wanted to raise children away from the cities. By the early 1970s, however, the proliferation of jobs in the suburbs became the driving force for new housing. As the suburbs surpassed the central cities in employment, job opportunity became a pull factor for continuing suburbanization (Figure 10-32). Employment growth in U.S. suburbs has been greater than in central cities for every year since 1965. As jobs and housing have continued to expand outward, the extent of any metropolitan area has had to be continually redefined.

Some of the new exurbs, called *satellite cities* or *edge cities*, boast greater retail sales and contain more office space than the old central cities. They even offer amenities formerly found exclusively in the central cities: art galleries, theaters, sports teams, and fine restaurants.

Changing commuting patterns and problems

When suburban workers commuted in and out of the city, radial mass-transit systems that focused on the central business district could serve transport needs tolerably well. Today, however, suburb-to-suburb commutes account for more than 50 percent of all U.S. metropolitan traffic. Only individualized transportation can serve populations that are spread out at low densities, and, as a matter of government policy, gasoline has never been taxed in the United States at rates comparable to those in other rich countries. Already in 1969, 82.7 percent of workers drove to work, but by 2000 over 90 percent did. More family members have been going to work, so the number of cars has increased faster than the local overall populations.

The Texas Transportation Institute has been studying traffic in the 75 largest U.S. metropolitan areas. The Institute reports that from 1982 until 2003, the annual delay per peak period (rush hour) grew from 16 hours to 47 hours. The number of urban areas with more than 20 hours of annual delay per peak traveler rose from 5 to 51. The total amount of delay reached 3.7 billion hours, and the amount of wasted fuel lost to engines idling in traffic jams rose to 2.3 billion gallons (at increasing prices per gallon). This accounting does not include the costs of air pollution or of highway accidents. A similar study in Canada in 2006 estimated costs of traffic jams in the nine largest cities there alone as some $3.7 billion. Commuting costs now consume more than 20 percent of the average U.S. household budget—as high as 23 percent in metropolitan Atlanta and Houston—and the percentage is rising. Older people often have difficulty driving,

(a)

(b)

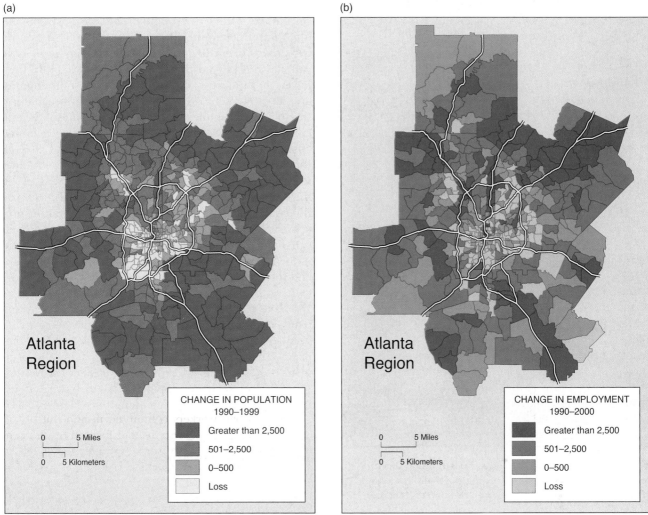

CHANGE IN POPULATION
1990–1999

Greater than 2,500

501–2,500

0–500

Loss

CHANGE IN EMPLOYMENT
1990–2000

Greater than 2,500

501–2,500

0–500

Loss

Source: Atlanta Regional Commission

FIGURE 10-32 Population and employment growth in metropolitan Atlanta. These maps show that in metropolitan Atlanta, Georgia, as in most U.S. metropolitan areas, the greatest population growth (a) and job growth (b) are still occurring in the periphery. The city of Atlanta is at the center of the ring highway visible on the map. Central-city neighborhoods began losing population 40 years ago, and inner suburbs already had begun to lose population 30 years ago.

so as the American population ages, older people may become isolated or traffic accidents may increase.

The U.S. Department of Transportation has reported that the average American household used its cars and trucks for 496 shopping trips in 2001, with each trip averaging 7.02 miles (for a total of almost 3,500 miles). This is up considerably from 1990, when the average household took just 341 shopping trips with an average length of just 5.1 miles (totaling 1,700 miles). People are now taking more shopping trips than trips to and from work. No one knows how may of these trips are in fact discretionary, but suburban Americans seem to have little alternative to driving.

As driving has increased, concerns about energy consumption are rising. America consumes about 25 percent of the world's production of oil, most of it for transportation. Of the 520 million cars in the world in mid-2006, 200 million were driven in America. In the early 1970s, the United States imported about one-third

of the energy it consumed, but by 2006 it imported 60 percent, and the Department of Energy forecast that it will have to import 65 percent by 2025. The United States has only 3 percent of the world's petroleum reserves, and domestic production has been falling for decades. Meanwhile, worldwide demand for energy will rise rapidly, as China and India increase their consumption, thus keeping up the prices American businesses and consumers pay for gasoline and home heating oil, as well as jet fuel and other petroleum products such as plastics and pharmaceuticals.

Potential solutions to increased problems from traffic include mass transit, ramp metering, reserved high-occupancy lanes, improved methods of broadcasting information on traffic jams, better signal timing, and improved clearance of accidents (Figure 10-33). Stockholm, London, Singapore, and other world cities have reduced congestion with fee systems that monitor traffic and automatically charge drivers according to

FIGURE 10-33 Road rage. As the frustration of sitting in traffic jams has increased on U.S. highways, the National Highway Traffic Safety Administration estimates that one-third of all fatal car crashes, in which almost 50,000 people die per year, could be attributed to "road rage," which is defined as "aggressive and even violent behavior by drivers caused by frustration or by the actions of other drivers." Road rage may win entry into the American Psychiatric Association's official *Diagnostic and Statistical Manual of Mental Disorders*. (Vince Streano/Corbis/Stock Market)

the time of day they are using busy roads and the fuel-efficiency of their vehicles. The sophistication of some in-vehicle GPS guidance systems would almost seem to obviate the need for drivers, as well as offer the possibility of collision-free highways, but the realization of such a possibility is years away. Satellite, GPS, and GIS technologies that are currently available only to the military and to security forces would allow tracking of every vehicle and could improve traffic flows, but citizens might be wary of such scrutiny of their travel habits.

In an attempt to stimulate experimentation, the federal government has allowed states to use federal transportation funds for any means of transportation they choose. Salt Lake City, Utah, continues to build highways and to develop 400 hectares (1,000 acres) of land per month. The built-up area is expected to double in size, reaching 800 square miles by 2020, while the population is expected to increase by only 50 percent. Milwaukee, Wisconsin, by contrast, is using federal

funds to tear down highways in an effort to revitalize the central city. Other metropolitan areas are experimenting with traffic monitoring and new methods of informing drivers of traffic jams. The technology for electric cars has not proved practicable. Hybrid cars, however, having both electric and internal combustion engines, are already seen on America's highways, achieving significant economies and lowering pollution. Several cities are rediscovering mass transit. Los Angeles built a 60-mile subway and light-rail system in the 1990s, although many observers think that the system is ill-suited to such a dispersed city with no central district. In 1999 the state of Georgia created a new Regional Transportation Authority with power over all transportation projects in the 13-county Metropolitan Atlanta region. It can finance and operate public transport, take state money away from localities that refuse to enact its plans, and veto major projects such as new malls.

The "New Urbanism"   As suburbs continue to expand, several new developments have tried to re-create small-town community life, with walkable streets, compactness, downtown shopping and services, and even front porches. This effort has been called *The New Urbanism*. Most such developments try to free residents from dependence upon the automobile (Figure 10-34).

FIGURE 10-34 A new planned town. Valencia, California, is a new town 48 kilometers (30 miles) north of downtown Los Angeles, where it occupies a 15,000-acre portion of an historic ranch owned by Newhall Land since 1875. The company has been planning and developing Valencia since 1965, and Valencia currently is home to approximately 48,000 residents in 17,000 homes. Its master plan balances residential uses with employment, education, recreation, open space, shopping, and services, so many Valencia residents live, work, play, and shop entirely within the community. There are 48,000 jobs within Valencia. Its residential villages, business parks, neighborhood retail and recreation centers, schools, parks, and shopping center are all interconnected by a 45-kilometer (28-mile) system of landscaped pedestrian walkways, bike trails, and bridges called *paseos*. (Carol Maglione, Newland Land)

Critical THINKING

The Question of Public Places on Private Property

One significant difference between traditional downtowns and most new villages and suburban malls is that the latter are private property. People cannot be banned from a traditional downtown, and the right to petition on a public sidewalk is constitutionally protected. People may, however, be banned from private property such as malls, and constitutional rights, such as political pamphleteering, may be restricted. In many communities today, malls, although private, are the only public gathering places, so if the mall owners are allowed to decide who may speak in them, mall owners can determine the public's access to ideas. Candidates for political office have been banned from busy malls owned by their opponents, and so have labor union organizers. In 2003, during the U.S. war against Iraq, a shopper was arrested for trespassing at a mall in Albany, New York, when he refused to take off a T-shirt reading "Peace on Earth."

Some communities are even locating schools in malls. The Landmark Shopping Center in Northern Virginia, for example, includes the Fairfax County Mall School. The government insists that the students "feel comfortable" coming to the mall; they can shop during class breaks, and the mall offers opportunity not only for after-school employment but for "career-training programs" in retail sales. Mall restaurants replace costly school cafeterias, and heating and air-conditioning expenses are also reduced.

Private environments are patrolled by private security officers, one of the country's fastest-growing occupations. (Security guards constitute almost 2 percent of the nation's total labor force, which is triple the number of public police officers.) The privatization of space is a new form of economic and social segregation that carries profound influence throughout U.S. political and social life.

U.S. law recognizes that private properties can perform functions traditionally associated with government. This is called the *public function doctrine*. The U.S. Supreme Court has ruled that the U.S. Constitution does not protect citizens' access to shopping centers against the wishes of the owners (*Pruneyard* v. *Robins*, 1980). Some states, however, have upheld the

A security guard in a shopping mall. Does he enforce the law? Or the mall owner's rules? Can he arrest people? Evict them? What for? (L. Mulvehill/The Image Works)

right of public access under their state constitutions. In 2000, for example, New Jersey struck down malls' limits on political leafleting. Each state has balanced public and private rights differently.

Questions

1. Are political candidates allowed to campaign at any malls where you shop? Would you be allowed to set up a table and gather signatures to run for mayor yourself? Where else could you go in your community?

2. Are beggars or charity fundraisers found in your downtown? At the mall? Might their presence distract you from shopping?

3. Does the public function doctrine cover malls in your state?

Seaside, Florida, a small Victorian town on Florida's Gulf Coast featured in the 1998 movie *The Truman Show*, was constructed in the 1980s, but Seaside is just a vacation resort. Celebration, Florida, was developed by the Disney Corporation in the 1990s. It tries to re-create small-town American life: Oak trees and white picket fences line the town's quiet streets, and all buildings are in traditional styles of architecture. Living in the town means obeying the rules of the Celebration Company, a Disney subsidiary created to run the community instead of elected officials.

Some of these suburban centers form self-contained villages but offer mass transit into the central city, thus reproducing the advantages of the original planned suburbs such as Riverside (see Figure 10-28). Metropolitan areas in Oregon, Washington, Minnesota, Colorado, and elsewhere are laying new light rail tracks. Across America, those suburbs that enjoy rail service into the central city are growing, and in many cases rail service is reviving (Figure 10-35). For example, in the 1990s the Massachusetts Bay Transit Authority restored the public rail service system within 50 miles of Boston that was terminated in the 1970s. The result was a boom in each community affected—both in new developments and in historic towns such as Newburyport, Massachusetts. Newark, New Jersey, is restoring a network of trolleys that honeycombs the city and reaches out into the suburbs. The system was abandoned in favor of private automobiles 60 years ago. The 1988 classic film *Who Framed Roger Rabbit?* is a work of comic fiction, but it tells the story of how automobile companies bought and deliberately destroyed Los Angeles's mass-transit network to boost car dependency. Auto interests did the same in many U.S. cities until courts declared such purchases illegal.

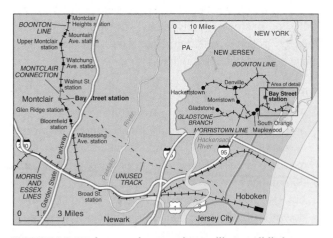

FIGURE 10-35 Improving metropolitan rail links.
A crucial link between two suburban rail stations in the New York metropolitan region that was opened in 2002 shortened commuting time, increased housing prices, and stimulated commercial development, with new stores and businesses opening all along the line. The convenience of suburban rail service—introduced in the nineteenth century—is enjoying new appreciation.

Suburbs at Century's End

Advances in information technology, rising incomes, population growth, and spending to develop infrastructure continue to drive residential and business development to the fringes of metropolitan areas. America's suburban population continues to grow faster than the central city population, so that by 2000 central cities housed only 30 percent of the national population. The number of jobs continues to grow faster in suburban areas than in cities, too, so about 60 percent of metropolitan jobs are now in the suburbs. Furthermore, the rate of development of suburban land actually increased in the 1990s. The national population is growing at 1 percent per year, but land occupied by single-family houses is rising at 2 percent per year.

Suburban airports are multiplying and expanding to serve personal and corporate aircraft, but this activity is launching new battles over land use, air, and noise pollution. When Washington Dulles International Airport opened in 1962 in rural Virginia, it spawned a high-tech corridor that is now the fastest-growing county in the country. Dallas-Fort Worth Airport opened between the two metropolitan areas in 1974. The immediate airport area has all the facilities travelers expect—car rental, hotels, and cargo storage, as well as the company headquarters of American Airlines—with warehouses on all sides. Amazon is building a huge distribution facility in Irving, 15 minutes away, and a new shopping mall is opening. Denver International Airport, covering a full 34,000 acres, opened 40 miles out of town in 1995, and by 2025 it is expected to be the center of a community of 500,000 people—almost as many as live in Denver itself. Working and living in the shadow of an airport has problems—height restrictions, noise, and traffic—but many businesses and people today feel the need to be near a runway.

The geography of retailing Highway congestion, more family members going to work, fear of crime, toll-free telephone numbers, credit cards, television and Internet shopping, and other factors are changing the geography of U.S. retailing. Many Americans are shopping from home by telephone, catalog, or computer. Elaborate catalogs fill mailboxes, and today almost two-thirds of the adult population buys at least occasionally from home. These developments introduce *virtual shopping* (see Chapter 6), which reduces the length of time before orders are filled, the quantity of goods sitting in inventories, and the number of middlemen standing between a product and the consumer.

Electronic commerce between businesses reached $2.4 trillion by 2003, and consumer e-commerce reached $95 billion. Stand-alone music retailers and travel agents are among the types of retailers being replaced by e-commerce, but Amazon is an example of a company that succeeded by lowered marketing, inventory, and warehousing costs (Figure 10-36). Its revenues reached almost

FIGURE 10-36 Amazon's computerized warehouse. Amazon's warehouse in Fernley, Nevada, a quarter-mile long and 200 yards wide, holds over three million different books, CDs, toys, and housewares. Computers send orders to workers' wireless receivers, weigh and sort packages, and apply shipping labels, so the facility can box 11,000 orders per hour. Consumers can order any two items out of the three million and have them delivered quickly in one box—usually at a discount price. (Macduff Everton Studio)

$9 billion in 2005 and have been growing at 20 percent per year. New technology significantly lowers warehousing and delivery costs. Company founder Jeff Bezos insists, "In the physical world it's the old saw: location, location, location. [But] the three most important things for us are technology, technology, technology."

U.S. retailers doubled shopping space per capita between 1970 and 1996, but as shopping at home increases, stores have begun to suffer. Almost 20 percent of the shopping malls in the United States that existed in 1990 were out of business by 2000. Malls are today being rebuilt as destinations for a wider range of activities than just shopping. Today malls offer movies, skating rinks, games, and a greater variety of restaurants and other attractions (Figure 10-37). Even with such diversions, many of the 16 million sales jobs in the United States may be vulnerable to elimination.

Telecommuting It is easier to move information than to move people, so more people are working at home via the use of computer terminals. This is called **telecommuting.** At IBM, for example, at the end of 2006, a full 40 percent of the workforce had no official office. About 15 million employees in the United States work from home on computers, and several state governments have telecommuting programs. Some government bureaus have distributed telecommuting centers throughout the suburbs, complete with office equipment.

Telecommuting eases the pressure on transport facilities, saves fuel, reduces air pollution, reduces the demand for office space, reduces absenteeism, lowers the fixed costs of a business enterprise, and has been shown to increase workers' productivity. It also allows

FIGURE 10-37 New attractions at malls. A new aquarium and marine science center draw visitors to the Mall of America—the largest in the United States—in Bloomington, Minnesota. (Courtesy of the Mall of America)

employers to accommodate employees who want flexible work arrangements, thus opening employment opportunity to more people, such as women who are still homemakers. Telecommuting tends to reduce central-city employment, and we do not yet know what other ramifications increased telecommuting will mean for metropolitan geography. Americans' fear of terrorist acts may also stimulate both virtual shopping and telecommuting, if people fear crowded public spaces and prefer to participate in dispersed activities.

Developments in the Central City

America's central cities suffered economic decline through most of the second half of the twentieth century. The drain of jobs from the central city and the concomitant development of the suburbs hollowed out many U.S. metropolitan areas. Many central business districts lost their purpose. Commercial, professional, and financial offices relocated to the suburbs, followed by upscale retailing. Many U.S. downtowns came to consist only of a government center, a convention center, and a few hotels; the streets were deserted after 6 P.M. The total populations of many central cities fell, particularly of older cities in the Northeast and Midwest. Between 1970 and 2000, population fell by the following amounts in major U.S. cities: Chicago, 14 percent; Philadelphia, 24 percent; Detroit, 41 percent; Baltimore, 30 percent; Washington, 27 percent; Boston, 13 percent; Cleveland, 40 percent; St. Louis, 43 percent; Pittsburgh, 36 percent; and Cincinnati, 27 percent. Visitors to the United States are astounded at the urban infrastructure that Americans seem simply to have abandoned: housing, sewerage, water pipes, roads and streets, industrial buildings, and more.

Economic decline Central cities can thrive as long as (1) their economies offer a complete range of job opportunities, ranging from entry-level jobs for the unskilled up to specialized jobs for skilled workers, and (2) family stability, education, and other social systems help urbanites ascend the socioeconomic ladder. In other words, cities do not have to retain their middle classes, but they have to offer the lower classes opportunity to become middle class. Unfortunately, the departure of the middle classes for the suburbs occurred at the same time as two other developments.

First, urban economies were transformed by the out-migration of entry-level jobs, particularly in manufacturing, construction, and warehousing. This out-migration broke the rungs of the ladder of upward mobility. In 1968 economist John Kain first suggested that the removal of manufacturing jobs to the suburbs and the concentration of the poor in the central cities created a spatial mismatch between the suburban job opportunity and central-city low-income housing. This spatial mismatch, he argued, could explain the high unemployment found in the central cities. The **spatial mismatch hypothesis** has until today dominated analyses of inner-city unemployment—and therefore the formulation of potential solutions. We will discuss a new alternative hypothesis later in the chapter.

Second, at the very time that many unskilled jobs were leaving the central cities, new waves of unskilled workers were pouring into them. The stream of new migrants to the cities included African Americans from the rural South, Latinos, and other immigrants. This influx continued long after the numbers of entry-level job opportunities began to shrink.

Much of the inner-city housing stock began to deteriorate. This was because government incentives still have made it profitable to give up a house in the city and move to the suburbs. Policymakers had thought that if new suburban houses were available to middle-class people from the cities, the urban poor could move into the older city dwellings. This succession, called *filtering*, would solve the housing problem for lower-income families. Much of the central-city population remaining behind, however, was financially incapable of maintaining the inherited housing stock. Therefore, many central-city neighborhoods deteriorated. The term *inner-city neighborhood* became a euphemism for slum.

The service economies At the end of the twentieth century, however, some central cities began to enjoy new growth in financial, information, and specialized technical services. In these areas skyscrapers replaced rusty factories (Figure 10-38). New York's leading export for decades was garments, for instance,

(a) (b)

FIGURE 10-38 Pittsburgh has been transformed. No U.S. city illustrates a transformed economy better than downtown Pittsburgh, which was transformed from a dirty and smoky industrial area (a) into a new park and gleaming service center, a "Golden Triangle" (b). This is where the Allegheny River (at left) meets the Monongahela (at right) to form the Ohio. ([a] UPI/CORBIS BETTMAN; [b] Chuck Savage/Corbis/Stock Market)

but today it is legal and financial services. In 1975 Baltimore's leading employer was the Bethlehem Steel Company; today it is Johns Hopkins University Medical Center. In 2002 almost 10 percent of all jobs in the Megalopolis from Boston to Washington, D.C., were in health care. This shift in job opportunity exemplifies a switch from *blue-collar jobs,* performed in rough clothing and usually involving manual labor, to *white-collar jobs,* which are salaried or professional jobs that do not involve manual labor.

Analysts of urban economies suggest that cities increasingly compete for jobs and growth on the basis of lifestyle. Surveys have found, for example, that growth in the biotechnology and computing industries is clustered in a few metropolitan areas, including Austin, San Francisco, and Boston. These cities do not enjoy traditional economic advantages such as proximity to raw materials, cheap energy, or low costs of living. They do, however, share two features: a thriving arts scene (reflected statistically by a high number of artists, writers, and other arts workers) and a dense, highly diverse, and tolerant social character portrayed by, among other things, a high number of immigrants and gays and lesbians. Business journals periodically rank cities' relative attractiveness, and although the mix of criteria chosen is not scientific, a high ranking can boost a city's fortunes (see Table 10-1). Analyst Richard Florida identified "the three T's" of an economically successful city: tolerance, talent, and technology. These features attract the people who are crucial to economic success: creative workers, engineers and scientists who develop new products and industrial processes, and creative businesspeople, financiers and other workers who start new businesses and improve the old. Such people have the skills and the means to live wherever they choose, and they are attracted to cities that offer the amenities and broad quality of life they desire. *Business Week* magazine has reported that two-thirds of college-educated adults aged 25 to 34 today decide first where to live, and then where to work.

In some cities the holders of the new white-collar jobs—sometimes called *yuppies,* young urban professionals—triggered a rediscovery and revival of urban life. They first occupied and restored select older residential neighborhoods in a process known as **gentrification,** but they soon began to convert even former industrial and warehouse buildings into residences. These conversions were usually in owner-occupied forms (condominiums or cooperatives) so that owners could profit from the tax advantages afforded to suburban homeowners. Some observers argued that the conversion of lofts to residences damaged the cities' chances for industrial resurgence, while others argued that the industries were never coming back, so the conversions were beneficial. Gentrified neighborhoods mix old and new architecture and a vibrant street life with upscale commercial activities such as book shops, gourmet food and wine shops, and art galleries. Yuppies have been joined by *empty nesters*—that is, older people who had raised families in large suburban homes but found urban attractions and activities more appealing in later life. Historic preservation movements have assisted central cities by winning tax advantages for the reuse of older buildings. New zoning for mixed use revives the cities' days and nights. Even some businesses that left the central cities in the period 1950–1980 are moving back downtown, discouraged by suburban traffic tie-ups and attracted downtown by relatively low rents and the availability of high-quality workers. Thus, rising educational levels, shrinking family size, and aging all contributed to some central cities' revival.

TABLE 10-1	Different Ways of Ranking Cities' Desirability

In 2002, Professor Richard Florida of Carnegie Mellon University ranked the following "city-regions" as the most "creative" in the United States:

San Francisco, CA	Houston, TX
Austin, TX	Washington, D.C./Baltimore, MD
San Diego, CA, and Boston, MA (tie)	New York, NY
Seattle, WA	Dallas, TX and Minneapolis-St. Paul, MN (tie)
Raleigh-Durham, NC	

By contrast, in 2003 the American Association of Retired Persons listed the following 15 cities as being America's best places to settle in retirement. The empty nesters referred to in the text choose to move into central cities and enjoy their cultural attractions, but most of these smaller cities offer a relatively low cost of living and a relaxed pace of life. The increasing numbers of retired persons may initiate a revival of many smaller towns. Young upwardly mobile people might find some of these cities relatively boring, but note that two cities are on both lists!

Loveland/Fort Collins, CO	San Diego, CA
Bellingham, WA	San Antonio, TX
Raleigh/Durham/Chapel Hill, NC	Sante Fe, NM
Sarasota, FL	Gainesville, FL
Fayetteville, AR	Iowa City, IA
Charleston, SC	Portsmouth, NH
Asheville, NC	Spokane, WA

Most job and population growth is still in the suburbs, but signs of central-city revival are widespread. The 2000 census revealed that Chicago, New York, Atlanta, and many other central cities enjoyed population growth for the first time in decades. Violent crime dropped 34 percent in America's 10 largest cities in the 1990s, and young people increasingly choose the urban lifestyle. In the newest editions of guidebooks to the best colleges, most of the top-rated schools are urban institutions. Higher education is a mainstay of many central cities' economies, and frequently the cities retain the graduates. Among the largest 100 U.S. cities, the 25 that began the 1990s with the highest percentage of college graduates ended the decade with even greater concentrations. These 25 cities saw the college-educated share of their population jump 6 percent, twice the average growth of the other 75 cities. Cities compete to attract and retain these educated workers.

Additionally, many cities have cultivated the tourist sectors of their economies and now attract tourists with historic quarters, fine cuisine, shopping opportunities, performing arts, and art exhibitions and festivals (Figure 10-39). Memphis, Tennessee, for example, attracts 8–10 million tourists each year, about 14 percent of whom are foreigners. The city has borrowed exhibitions of artworks from around the world, and Memphis's role in American musical history draws many tourists to exhibitions and performances. Many cities are building new museums and performing-arts centers just to pull in suburbanites.

The role of immigrants

As noted in Chapter 5, immigrants to both the United States and Canada concentrate in major cities, giving those cities a cosmopolitan sophistication. Immigrants have also played an important

FIGURE 10-39 Graceland. The home and grave of musician Elvis Presley remains one of the most popular tourist attractions in Memphis, Tennessee. Tourists coming principally to visit Graceland spend an estimated $40 million per year in Memphis. That sum equals $65 per city resident, and tourists' dollars have a high multiplier effect. (Liz Gilbert/Corbis Sygma)

role in the cities' economic rejuvenation. Many bring capital or job skills, so for them a city's traditional advantages of agglomeration and external economies confirms the city's function as incubator of new businesses. In Los Angeles County, for example, corporations that employ fewer than 100 people offer more than half of all the county's jobs. Virtually all of the firms making clothing and textiles, toys, processed foods, furniture, and biomedical supplies are owned by foreign-born individuals. Furthermore, many immigrants have moved into declining neighborhoods and repaired deteriorating homes themselves—a process called investing *sweat equity* rather than money.

Latinos today outnumber Blacks in seven of the ten biggest cities in the United States—New York, Los Angeles, Houston, San Diego, Dallas, Phoenix, and San Antonio. In Los Angeles, Houston, and San Antonio, Latinos outnumber non-Latino Whites as well. (Table 10-2). The changing demographic mix in major U.S. cities carries political ramifications, as Latinos, Asians, and other minorities form new political alliances. The growing diversity of cultures and interest groups tends to submerge America's historic Black–White dialogue in a new chorus of voices demanding political expression.

The shortcomings of service economies

Upscale urbanites compose only a fraction of the total inner-city population. The new white-collar jobs being created do not always equal the number of blue-collar jobs being lost. Although most high-paying jobs are in the service sector (doctors, lawyers, executives, sports stars, etc.), most service sector jobs are not high-paying. Creative website designers, for example, earn higher incomes than janitors do, but usually janitors outnumber website designers. In New York City between 1989 and 1999, a time of national prosperity, the only net growth in the numbers of jobs was in jobs paying less than $25,000 per year, which was just about the national median for a full-time worker. The average weekly wages of workers in New York City's finance and insurance industries in 2006 was $8,323. This average figure conceals an extreme concentration of this income among the highest-paid bankers and stock brokers, but it contrasts vividly with the average weekly wage of $594 of workers in the fast-growing accommodation and food services industry and $803 in retail trade. Furthermore, workers who lose their blue-collar jobs often require retraining or education before they can capture one of the new opportunities.

The median household income in U.S. central cities hovers between 70 and 75 percent of the figure for the suburbs, and central-city unemployment rates hover about one-third above those in the suburbs. Furthermore, the percentage of jobs in most central cities held by commuters is rising—especially the percentage of the best jobs. Some 60 percent of the salaries earned in Washington, D.C., for example, go to suburbanites, and federal law protects them from D.C. taxes. In 1999 the New York

TABLE 10-2 A Decade of Racial Change in the Cities

A new analysis of 2000 census data shows that non-Hispanic Whites are now a minority of the total population of the 100 largest cities, and 18 of those cities became majority non-White over the decade.

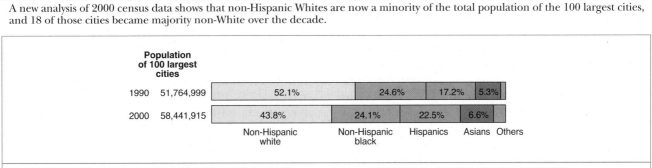

Population of 100 largest cities		Non-Hispanic white	Non-Hispanic black	Hispanics	Asians	Others
1990	51,764,999	52.1%	24.6%	17.2%	5.3%	
2000	58,441,915	43.8%	24.1%	22.5%	6.6%	

FROM WHITE MAJORITIES TO NONWHITE MAJORITIES

Eighteen of the 100 most populous cities that were majority white in 1990 became majority nonwhite in 2000.

	PERCENTAGE NON-HISPANIC WHITE		DECLINE
CITY	1990	2000	1990–2000
Anaheim, Calif.	56.6%	35.9%	**20.8%**
Riverside, Calif.	61.3	45.6	**15.7**
Milwaukee	60.8	45.4	**15.4**
Rochester	58.3	44.3	**14.0**
Sacramento	53.4	40.5	**12.8**
Fort Worth	56.5	45.8	**10.7**
Augusta-Richmond, Ga.	54.0	43.7	**10.3**
Philadelphia	52.1	42.5	**9.6**
Boston	59.0	49.5	**9.5**
San Diego	58.7	49.4	**9.3**
Mobile, Ala.	58.9	49.8	**9.2**
Montgomery, Ala.	56.1	47.1	**9.0**
Columbus, Ga.	57.3	48.6	**8.7**
Norfolk, Va.	55.6	47.0	**8.5**
Albuquerque	58.3	49.9	**8.4**
Baton Rouge, La.	52.9	44.7	**8.1**
Shreveport, La.	53.6	45.9	**7.7**
St. Louis	50.2	42.9	**7.3**

THE IMPACT OF THE HISPANIC INFLUX

Nineteen cities would have lost population were it not for the growth in the Hispanic population.

CITY	GROWTH IN TOTAL POPULATION, 1990–2000	GROWTH IN HISPANIC POPULATION, 1990–2000
Anaheim, Calif.	61,608	69,619
Boston	14,858	23,134
Chicago	112,290	207,792
Corpus Christi, Tex.	20,001	21,029
Dallas	181,703	212,347
Des Moines	5,495	8,509
El Paso	48,320	76,206
Grand Rapids, Mich.	8,674	16,424
Hialeah, Fla.	38,415	39,891
Jersey City	11,518	12,557
Kansas City, Mo.	6,399	13,587
Long Beach. Calif.	32,089	63,673
Los Angeles	209,422	327,662
Miami	3,922	14,387
Minneapolis	14,235	21,275
Oakland, Calif.	27,242	35,756
Riverside, Calif.	28,661	38,489
Santa Ana, Calif.	44,235	65,714
Yonkers	8,004	19,376

Source: Analysis of Census Bureau data by the Center on Urban and Metropolitan Policy at the Brookings Institution.

state legislature, responding to suburban voting power, ended New York City's income tax on commuters. This gesture crippled the city's ability to provide services to them. The labor force participation rates of central-city populations—that is, the percentage of the population that is currently employed or even looking for a job—fall behind those of the nation and of the entire local metropolitan areas. For example, the labor force participation rate for the city of St. Louis in 2005 was 56 percent, whereas for the St. Louis metropolitan area it was 69 percent. This suggests that a smaller percentage of the city's population was working and paying taxes, and also that a higher percentage was receiving public assistance.

Twenty-first century security concerns also distress central cities. Central cities are the most likely targets of terrorism, their populations suffer increased stress lev-

els, and, for the most part, the individual cities must bear the burden of high security costs.

U.S. cities are losing the middle class. Defining "middle" as between 80 and 120 percent of the median, the Brookings Institution found that the percentage of middle income neighborhoods in the 100 largest metropolitan areas had dropped from 58 percent in 1970 to 41 percent in 2000. Middle-income families made up 28 percent of all families in the metropolitan areas in 1970, but only 22 percent in 2000. These figures highlight the overall shrinking of the U.S. middle class.

The urbanization of African Americans and segregation by race and income

Chapter 5 recounted how some 1.5 million African Americans migrated from the South to the cities of the North between

1910 and 1945, and an additional 6.5 million migrated north and west between 1945 and 1970. That second wave arrived just as the central cities were losing their ability to provide entry-level job opportunities. The migrants also often faced discrimination in employment and segregation in housing. Deteriorating conditions in new African-American ghettoes eventually triggered civil unrest. A 1965 riot in the black Watts section of Los Angeles left 34 people dead and more than a thousand injured and required military occupation of 119 square kilometers (46 square miles) to halt the violence. The Watts riot was followed by 150 major riots and hundreds of minor ones in cities across America that summer and the next three summers.

Since the 1960s, civil rights have made great advances, and a great many of the black urban in-migrants have achieved success. Others, however, have been left behind. If a family's first urban generation failed to find employment, skills, and upward mobility, the second and third generations may have failed also. Los Angeles erupted again in April 1992, for example, and in 2001, black neighborhoods in Cincinnati erupted in three days of rioting after the fatal shooting of the fifteenth unarmed black man by police since 1995. In 2003, 20 houses burned in Benton Harbor, Michigan, during riots that broke out after the death of a black motorist killed during a police chase.

Residential racial segregation continues. Statistics on America's 8.2 million individual residential blocks disclose that about one-third of Blacks and more than half of Whites live in blocks that are at least 90 percent of their own race.

During the 1990s, Blacks' overall proximity to jobs improved slightly, but the 2000 census revealed that no demographic group remained more physically isolated from jobs than Blacks. Some decline in the spatial mismatch occurred between 1990 and 2000, and that decline was principally due to the residential movement of Black households within metropolitan areas. Nevertheless, in nearly all metropolitan areas with significant black populations, the separation between residences and jobs was much higher for Blacks than for Whites.

The 2000 census revealed one more changing pattern: Whereas one of America's most dramatic trends from 1960 to 1990 was the movement of the American poor into urban neighborhoods of concentrated poverty, that trend reversed itself between 1990 and 2000, and poverty became less concentrated. Concentrations of poverty magnify the problems associated with poverty in general: crime, delinquency, joblessness, drug trafficking, breakdown of the family, and low-performing local schools. Furthermore, when a neighborhood is a known concentration of poverty, middle- and working-class people often see it as "dangerous." They avoid the neighborhood, which becomes increasingly isolated, socially and economically.

In 1990, 10.4 million people, which was 15 percent of all poor people at the time, lived in census tracts in which at least 40 percent of the households were in poverty. By 2000, however, the number had declined to 7.9 million people (down 24 percent), about 10 percent of all poor people. Among Blacks the concentrated poor population fell from 4.9 million people (30 percent of poor Blacks) to 3.1 million people (19 percent of poor Blacks). The total number of poverty tracts fell from 3,417 in 1990 to 2,510 in 2000. Observers believe that the concentrations of people in poverty dispersed because public housing projects were torn down, central cities enjoyed some economic resurgence, immigrants revived some urban neighborhoods, and millions of poor people left urban slums for other neighborhoods.

People living in poverty had not disappeared between 1990 and 2000; their numbers had in fact increased, but some of them had moved to the inner ring of suburbs. Today fewer central-city neighborhoods resemble the nightmarish scenes portrayed in the popular films *Fort Apache, the Bronx* (1981) or *Boyz N the Hood* (1991). Today's metropolitan reality has been captured better by rapper Eminem's fictionalized biography *8 Mile* (2002), in which poverty is located in the dreary inner suburbs of Detroit—the American city that in fact saw the most dramatic reduction of concentrated poverty 1990–2000, down by almost 75 percent. Suburban poverty is not pleasant, but, as a general rule, a mix of income and ethnic groups in suburbs reduces the isolation of the poor.

Only in pockets of the central cities are conditions at their worst, and even the worst slums of New York City, Chicago, Detroit, and Los Angeles do not compare with the conditions of life for many in Mexico City, Lagos, Nigeria, or Kolkata, India. The inner-city second or third generation living in deprivation, sometimes called the *underclass*, numbers less than 3 million people, which is less than 1 percent of the national population. Still, the conditions of deprivation and the lack of opportunity contrast starkly with the national self-image.

The network hypothesis Scholars are formulating a new alternative to the spatial mismatch hypothesis to explain some urban unemployment. They have learned that the primary qualification employers seek in new unskilled workers is reliability. The best way to find reliable new employees is to ask current employees for recommendations. Therefore, networks of working family or friends provide unskilled urbanites seeking entry-level positions with information about jobs, sponsorship for jobs, and role models for work. Hiring is referential, not residential. Many urban black communities, after a generation or two outside the mainstream of labor opportunity, may lack these crucial links because the social networks conducive to upward mobility have deteriorated. Philip Kasinitz, a sociologist, has written, "The primary reason for ghetto unemployment is not the lack of nearby jobs but the absence of social networks that provide entry into the job market." The theory that it is the lack of these networks that causes unemployment may be called the **network hypothesis.** Scholars investigating this hypothesis note that significant numbers of inner-city jobs, even those in manufacturing, have often been captured by groups whose residences are distant, even though local residents cannot get work. For example, Kasinitz found that jobs in a manufacturing district of Brooklyn, New York, were captured by Hispanics who commuted from New Jersey, even though unemployment stayed high in contiguous Black neighborhoods. Geographer Thomas Cooke concluded that "it is not possible to argue that census tract African American male unemployment rates are related to the number of local job opportunities."

These studies do not conclusively replace the spatial mismatch hypothesis, but they introduce a new partial factor of explanation. Assumptions about what causes a situation will determine approaches to altering that situation. In this case our understanding of the causes of inner-city unemployment will determine what solutions we propose.

Efforts to Redistribute Jobs and Housing

The spatial mismatch hypothesis has dominated thinking about urban unemployment since the 1960s. Therefore, several government programs designed to deal with the problem of inner-city unemployment have addressed the spatial mismatch. Three approaches have been suggested: (1) Bring new blue-collar jobs to the cities, (2) move inner-city residents to the suburbs, or (3) transport inner-city workers to suburban jobs.

1. Bringing new blue-collar jobs to the cities involves efforts to reindustrialize the central cities by establishing **urban enterprise zones,** where manufacturers receive government subsidies. The Federal Budget Act of 1993 called for the designation of both urban and rural enterprise zones. Several states and cities had already designated zones and granted manufacturers assistance in their poorest communities. Some of these zones have enjoyed success, exploiting the cities' advantage of existing infrastructure (water, sewerage, and so forth), but the factors that historically have driven industries from the central cities are difficult to overcome.

In 2000 Congress granted tax credits for commercial projects that create jobs in low-income areas. Tax credits reduce development costs, thus encouraging riskier projects. Private investors pay lower taxes, and the developer passes the savings on to the community by, for example, lowering rent.

Many central cities are scarred by abandoned industrial facilities called **brownfields.** There are an estimated 400,000 brownfield sites across the United States, which include industrial properties, old gas stations, vacant warehouses, former dry cleaning establishments, and abandoned residential buildings that may contain lead paint or asbestos. Developers have shunned these deserted industrial sites because of the liability for buried wastes and other pollution-related risks. It has been easier to buy undeveloped sites at the edges of cities. The most difficult brownfields to clean up, however, are often within two miles of the downtown core, so they offer the greatest potential for reclamation and rededication to new land uses. The U.S. Environmental Protection Administration provides grants for cleanups, and a 2003 survey of 244 cities found that federal help to renovate brownfields had cleaned 19,000 acres, boosting local tax revenue by as much as $1.5 billion and adding up to 570,000 jobs (Figure 10-40).

As many central-city districts were virtually abandoned, some scholars recommended that urban density shrink in a planned manner. In New York City's South Bronx and in parts of Detroit, Cleveland, Houston, and other central cities, decrepit apartment buildings have been replaced by new single-family dwellings that resemble typical suburban homes (Figure 10-41).

2. The second approach is to move poor people from the central city out into new subsidized suburban housing. The federal government actually built suburbs for low-income people in the 1930s: Greendale outside Milwaukee, Greenhills outside Cincinnati, and Greenbelt outside Washington, D.C. Today, however, most suburbs are zoned to exclude subsidized housing or even private apartments with young families. Suburbs prefer expensive single-family homes, because only these pay property taxes sufficient to cover the costs of the services that they and their residents require. Owners of undeveloped property, however, usually want to maximize development on their land, so they battle such *exclusionary zoning.*

FIGURE 10-40 A brownfield reclaimed. South End, in Charlotte, North Carolina, was formerly a blighted area of abandoned textile mills and warehouses; it now boasts new restaurants, stores, and condominiums. (Courtesy Steven Little, Little Photography)

FIGURE 10-41 Central-city transformations. The new detached single-family homes in a neighborhood of New York City's South Bronx, shown here being visited by former president Bill Clinton, resemble typical suburban homes even more than do the homes in Figure 10-22. Here in the Bronx, low-income high-rise apartment buildings had been devastated by fires and virtually abandoned for 20 years. (Stephen Crowley/*New York Times* Pictures)

A clause in the Fifth Amendment to the U.S. Constitution, called the *takings clause*, states that "Private property shall not be taken for a public use, without just compensation." In other words, governments may take private land—this is called the right of **eminent domain**—but the governments must pay for it. Courts have long upheld takings in order to build public infrastructure such as parks and highways. In 1984, however, the Supreme Court extended the right of eminent domain to take land for any project "rationally related to a conceivable public purpose" (*Hawaii Housing Authority* v. *Midkiff*), and in 2005 the Court extended the meaning of "public purpose" still further to rule that fostering economic development is an appropriate use of eminent domain (*Kelo* v. *City of New London, Ct.*). Many people feel that local governments have abused eminent domain powers. The city council of Riviera Beach, Florida, for example, condemned 1,700 houses and apartments housing 5,100 people for a new development of shops, a hotel, a conference center, and yacht slips. Cypress, California, prevented a local church from building an annex in order to give the land to Costco. Such actions have roused complaints of political favoritism. The Supreme Court rulings have thrown the consideration over takings back to the state legislatures and courts, and most states have acted to restrict local governments' power of eminent domain.

Today property owners are insisting that even short of complete seizure of land, exclusionary zoning restricts the use to which owners may put their land so severely that the zoning is in effect an unconstitutional taking. They are protesting that both environmental

and historic preservation legislation are essentially takings. In 2004, Oregon voters approved a law that insists that when land rules reduce the value of property, the government must compensate the owner or waive the regulations. The takings clause is a fine point in the interpretation of law, but in recent years several state supreme courts have banned exclusionary zoning. Several have actually required local communities to zone in order to provide all sorts of housing: high-density apartments and low-density single-family homes, as well as housing affordable to people of all income levels.

An alternative is to allow people to subdivide their suburban houses into *accessory apartments* if their homes are bigger than their needs. Most suburbs, however, ban this for fear of lowering overall property values. This political battle is becoming more common.

For years, researchers have debated whether the poorest Americans would improve their personal situations if they were simply relocated from the worst neighborhoods and housing projects and dispersed among middle-class neighborhoods somehow to absorb a different set of cultural norms. The discussion has been largely theoretical, because an extensive resettling of the poor would be expensive, intrusive, and racially charged—in short, politically impossible.

Recent federal programs tried to break up the concentrations of subsidized housing in the central cities and build more in the suburbs. The government encouraged some suburban areas to accept a greater share of public housing by offering grants for other community facilities. The first experiment, known as Gautreaux, occurred as part of a court-ordered settlement of a racial discrimination lawsuit against the Chicago Housing Authority. From 1976 to 1998, the housing authority moved nearly 25,000 poor African Americans from crumbling city public housing to subsidized housing in suburbs where no more than 30 percent of the population was African American. Studies showed that heads of households were more likely to be employed than their inner-city counterparts, although they still earned poverty-level wages and their families received little in the way of extra counseling or social services. Their children did even better. They were much more likely than their inner-city peers to graduate from high school, go to college, and get good jobs.

In the 1990s, the federal government inaugurated "Moving to Opportunity," which gave vouchers to thousands of residents in poor city neighborhoods. The participants were required to move to neighborhoods with lower poverty levels. Nevertheless, they tended to choose to live near one another. So far, studies have found no statistically significant changes in employment rates for adults or educational attainment for their children. Most scholars agree that the lesson of the two programs is that unless some limits are placed on housing vouchers, the poor will band together elsewhere and simply recreate mini-enclaves with the same problems of crime, single motherhood, and low educational expectations.

The devastation of New Orleans by Hurricane Katrina in 2005 has, unhappily, provided another chance for social scientists to collect evidence on whether relocating the poor is effective policy. The very poor make up a sizable portion of the nearly half-million evacuees spread across hundreds of towns and cities who are recipients of a large, focused package of federal and state aid. The evacuees have received rental vouchers and relocation allowances that they can spend on everything from job retraining to child care. Two relatively small-scale relocation programs suggest, again, so far, that the poor will better their lives as long as they do not end up together again.

3. The third approach to solving the spatial mismatch is to provide inner-city poor people with transportation to jobs in the suburbs. The American Automobile Association puts the average cost of car ownership at over $7,000 per year, which is more than many working Americans can afford. Therefore, several city governments, private employment agencies, and even suburban employers have instituted dedicated bus services for this purpose. Wisconsin and California are experimenting with interest-free loans to allow working poor to buy or maintain cars, and a federal-level program is called Bridges to Work.

These three efforts to solve the problem of inner-city unemployment build from an acceptance of spatial mismatch as the cause of central-city unemployment. The network hypothesis has only recently been offered as an alternative or complementary explanation. Therefore, few government programs have yet been devised to address the lack of social networks. Perhaps we will see proposals in the future.

Still another potential solution to inner-city unemployment among the unskilled is to educate and train the central-city population so that they can capture the skilled tertiary-sector jobs that do open in the central city. The quality of education in the central cities, however, is generally discouraging. Schools in upper-income suburbs usually have more money to spend on modern facilities than inner-city schools do. The latest experiments in public education suggest that the integration of students of varying economic statuses raises student achievement. In Wake County, North Carolina, for example, minority students have made rapid strides in standardized test scores since school officials began to use income as a prime factor in assigning students to schools, with the goal of limiting the proportion of low-income students in any school to no more than 40 percent. Other education districts across the country are experimenting with this model.

Governing Metropolitan Regions

Political geography is the subfield of geography that studies the interaction between political processes and the distributions of all other activities and transformations of the landscape. Political geography can be studied at any scale from local community politics to international boundary disputes and international law. Conurbations present special problems of interest to political geographers.

The legal boundaries of most U.S. cities were originally drawn to include some surrounding land for future growth, and when the city outgrew those boundaries, it annexed suburban areas. By the 1920s, however, the suburban populations had begun to incorporate themselves to avoid annexation (Figure 10-42). Suburbanites argued that their action upheld the U.S. tradition of local self-government, and many may have felt that by incorporating their own communities, they were escaping the problems (and people) of the old city, including high social costs and politics that were often corrupt. Only a few central cities—including Austin, Texas; Charlotte, North Carolina; and Oklahoma City—can still expand by annexing new suburban areas.

Today autonomous municipal units form a legal retaining wall around almost every large city in the United States. Metropolitan areas cover a great number

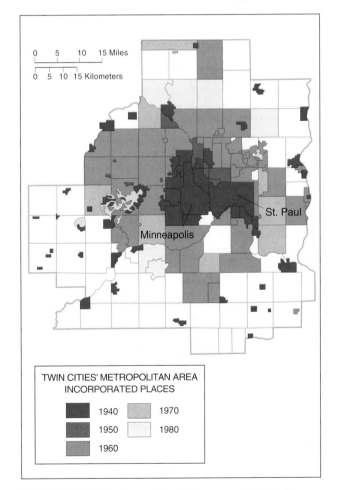

FIGURE 10-42 Metropolitan Minneapolis and St. Paul. Minneapolis and St. Paul, Minnesota, were surrounded by tiers of independently incorporated municipalities, like the rings of growth of a tree. This pattern of suburban incorporation typifies U.S. metropolitan expansion.

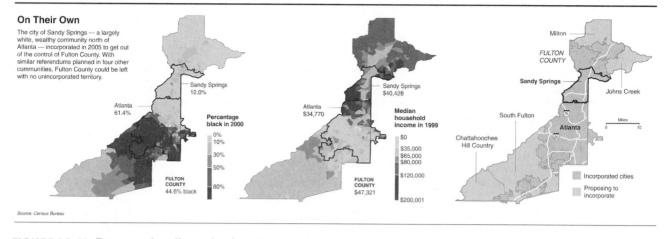

On Their Own

The city of Sandy Springs — a largely white, wealthy community north of Atlanta — incorporated in 2005 to get out of the control of Fulton County. With similar referendums planned in four other communities, Fulton County could be left with no unincorporated territory.

Source: Census Bureau

FIGURE 10-43 Remapping the suburbs. Questions of suburban incorporation continue to concern metropolitan politics. In 2005, Sandy Springs, Georgia, in suburban Atlanta, decided to incorporate itself to escape governance by Fulton County. That action encouraged residents of other parts of the county to incorporate their territories. If each succeeds and the county is left with no unincorporated territory, county government responsibilities will virtually disappear. Some observers have suggested that Sandy Springs residents were motivated by demographics of race and income. (New York Times Agency)

of municipalities and also myriad special district governments, which are incorporated to deal with specific problems. For example, Nassau and Suffolk counties in suburban New York include two cities, 13 towns, 95 villages, 127 school districts, and more than 500 special districts, most of which exercise taxing powers for services from garbage collection to hydrant rental. The five-county Los Angeles metropolitan region contains 160 separate governments. Los Angeles County alone has 82, and even within the boundaries of the city of Los Angeles, there are seven independent city governments.

The boundaries among the many jurisdictions are not obvious to everyone, but each government has its own agenda and marks the landscape. These governments jostle for authority and tax dollars. Property taxes soar, but many metropolitan area residents have no idea which government is responsible for which service.

The governing bodies of many of the special districts are not chosen in elections in which each citizen exercises an equal vote. Instead, they are either appointed or else chosen in elections in which votes are weighted in terms of payments for a service, use of a service, or by some other measure. As a result, the percentage of public funds spent by officials who are directly responsible to the voters shrinks. Furthermore, the boundaries of special districts may not conform to those of general-purpose governments but instead may overlap them. Overlapping boundaries multiply the difficulties in coordinating the provision of services. All these factors discourage voter turnout.

Decisions regarding metropolitan land use and the location of industries, recreation facilities, transport facilities, or new housing cannot be made in the best interests of the entire metropolitan population.

Instead, they are made on the basis of competition among the local governments, each of which wants to enhance its own property-tax base by attracting commercial developments that pay high property taxes but demand little in the way of local services. For example, fast-growing Santa Clara County, California, has zoned for 250,000 new jobs but only 70,000 new homes. Each community hopes to let surrounding towns cope with the additional costs of schooling, pollution, and congestion (Figure 10-43).

Many metropolitan areas have created councils of governments (COGs). These are committees of officials representing each of the local governments in the region. COGs, however, exercise limited powers, but each local government can veto any proposed area-wide action. Some U.S. metropolitan regions are creating regional governments to address area-wide problems. State governments in Washington, Minnesota, Connecticut, Virginia, and elsewhere are devising new schemes to redistribute or equalize both the revenues and the costs of housing and schooling throughout metropolitan areas. Dade County, Florida, has assumed responsibility for many public services for Miami and the 29 other cities in the county, and county voters chose to rename the county Miami-Dade. In 2005 metropolitan Chicago created the Chicago Region Environmental and Transportation Efficiency Project (CREATE). The body plans transport facilities for the entire region, but it must deal with 272 independent municipalities. Other state governments are assuming increasing direct responsibility for governing metropolitan areas. Some states are even drawing and applying state-level land-use plans (Figure 10-44).

The governing of metropolitan areas in Canada presents many of the same problems as in the United States. The province of Ontario created a government

REGIONAL *Focus* ON

Metropolitan Portland, Oregon

Oregon has long championed laws that restrict urban sprawl. Metropolitan Portland, Oregon, has a "Metro" government that controls land use, conservation, and transportation for 24 cities and three counties holding 1.3 million people. Metro requires cities and towns of a certain size to bound their urban areas, thus keeping forests, farmlands, and open space free of development. The law has permitted new housing while protecting a rural character at the edges of cities. As a result, Portland is perhaps America's most "European" city, with a fine mass transit system, compact neighborhoods, and protected forests and farms outside the city limits. Critics point out that the law has pushed up housing prices by reducing the land available for building, but voters have turned down repeated initiatives that would have weakened the government's powers since the law was first passed in 1973.

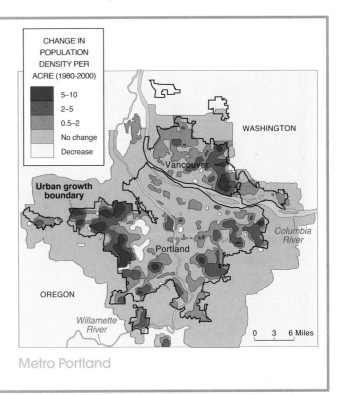

Metro Portland

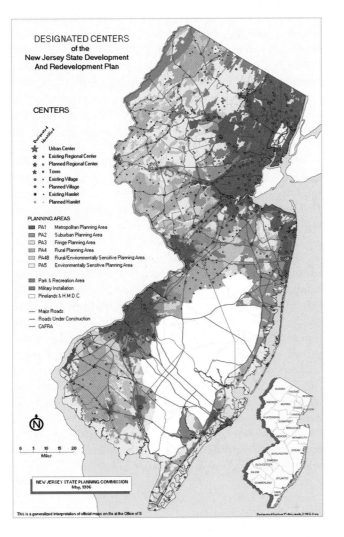

FIGURE 10-44 New Jersey's state land-use plan.

New Jersey is among the states that are preparing land-use plans and realizing them through zoning and purchase. (Courtesy of New Jersey Division of Travel & Tourism)

for metropolitan Toronto in 1953 that has since served as a model throughout North America.

Today many city and suburban governments—backed by downtown business executives, environmentalists, farmers, and church leaders—are banding together to fight continued sprawl by pressuring state and federal governments to end the many subsidies that encourage suburbanization. The government of Ohio, for example, has cut back on building new roads, begun repairing old ones, and prevented new communities from offering tax breaks that might draw businesses from older core metropolises. In 1997 Maryland legislated to confine state spending on infrastructure to existing municipalities. Similar political initiatives have been launched in Minnesota, Pennsylvania, Indiana, and Oregon.

The weakness of metropolitan regional governments in North America contrasts sharply with models elsewhere (Figure 10-45). In most European and Latin American countries, either strong metropolitan governments exist or else the national governments themselves oversee land use and the growth of metropolitan regions. Those governments generally concentrate activities in city centers and regulate land use very strictly to prevent urban sprawl. Also in contrast to the

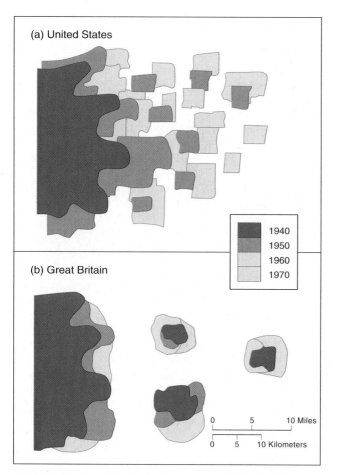

FIGURE 10-45 Suburban growth patterns in the United States and Great Britain. This schematic drawing contrasts (a) the leapfrogging typical of U.S. suburban growth with (b) the tighter, more controlled suburban growth typical of Great Britain. The British system saves open space and allows for better and cheaper planning and installation of infrastructure. (Adapted from Marion Clawson and Peter Hall, *Planning and Urban Growth: An Anglo-American Comparison,* 1973. Reprinted by permission of Resources for the Future.)

North American model is the fact that homeownership is relatively low. Europeans seem more content to rent their dwellings.

CONCLUSION: CRITICAL ISSUES FOR THE FUTURE

Today the majority of people, for the first time in history, lives in cities. Rapid urban growth, driven by demographics and economics, has transformed the face of the planet. Chapter 1 diffentiated a natural landscape, one without evidence of human activity, from a cultural landscape, one that reveals human modification of the local environment. That chapter noted that evidence of human activity is so ubiquitous on Earth that we might be justified in saying that all Earth is today a cultural landscape.

Rapid world urbanization has required us to rethink our common definition of "environment." Up to now, in many people's minds the word has meant Earth's natural landscapes. The phrase "environmental preservation" triggered images of struggles to preserve forests and wetlands, beaches and grasslands. Today, however, the transformation of Earth's surface and the actual circumstances in which most people live force us to direct more of our attention to urban environments. The environmental problems of settlements are appearing on the political agendas of most countries and even of international agencies. Urban environmental problems increasingly impact the health and livelihoods of people in both the rich and the poor countries.

For example, unexpected recurrent epidemics of cholera have focused attention on health threats in human settlements. It had been assumed that the threat of cholera had been eliminated through improvements in water supplies, sanitation, sewage treatment, and food safety, but urban crowding and overloading of the urban infrastructure have brought the problem back. Other urban environmental problems of growing concern include poor or inadequate housing, safe water supplies and adequate sewage treatment, air quality, exposure to chemical hazards in industrialized areas, physical hazards including road accidents, the collection and management of solid waste, and noise. Even education becomes an environmental issue because hygiene education is an essential safeguard of public health.

Cities transform natural landscapes not only within the built-up areas but for considerable distances around them. Ecologist William Rees has called the total area of land required to sustain a city its *ecological footprint*, and the United Nations has singled out four regional impacts of particular concern: uncontrolled and unplanned city expansion, liquid-waste disposal, solid-waste disposal, and acid precipitation. Cities concentrate demands for the products of fertile land, watersheds, and forests. Mining and the extraction of bulky, low-value substances that go into building materials, roads, foundations, and other parts of the built environment also disrupt local ecology, as does the waste matter dumped as a result of excavations. The United Nations notes that there is considerable potential for creating new jobs in reducing resource use and wastes and in recycling or reusing the wastes that are generated.

The challenge is great. We must transform our increasingly "cultural" landscape into a healthy and sustainable physical environment.

Chapter Review

SUMMARY

Every settled society builds cities because some essential functions of society are most conveniently performed at a central location for a surrounding countryside. The region to which any city provides services and upon which the city draws for its needs is its hinterland. Some cities' locations are the result of site characteristics; others are more the result of convenient situations. Urban environments often suffer from heat, noise, air pollution, and water- and waste-management problems.

In many traditional societies, cities' ritualistic and political importance has outweighed their economic importance. As societies develop, cities' economic role usually becomes paramount. Cities offer agglomeration for the division of labor, and they promote and administer the regional specialization of production. Cities can become centers of industrial production and tertiary-sector functions. Some workers in cities produce basic sector exports, whereas workers in nonbasic sectors serve the needs of the cities' residents. Jobs in the basic sectors multiply jobs in the nonbasic sectors. Cities can be classified economically by examining each city's basic and nonbasic sectors and by comparing these sectors among different cities. Central place theory models how cities and hinterlands are distributed across an isotropic plain.

Britain provides the first model for urbanization, but urbanization imposed hardship on millions of people. Today urbanization is occurring in many places without economic development. People migrate from the countryside because of the balance of push and pull factors. Governments can regulate this migration both by reducing the attractiveness of cities and by improving rural life. Urban migrants' industriousness could be an asset for economic growth.

Urban geographers also study the distribution of activities within cities, which may be caused by economic forces, social factors, or government actions.

The concept of designing an "ideal city" has challenged planners for centuries. Sir Ebenezer Howard's garden cities plan tried to stop the growth of the industrial city and to restore it to a human scale. Le Corbusier proposed cities of tall office buildings and apartments spaced far apart and surrounded by green space.

Many of the world's cities were planned and built according to European notions, and this is still reflected in their internal geography. Western forms continue to be stamped on the world's cities. Many great cities' historic cores illustrate indigenous principles of urban planning. Islamic cities present one alternative to Western forms.

In the United States, growing cities have spilled into suburbs. This dispersion has resulted from government policies, cultural choices, and other economic and social forces. Both housing and jobs have dispersed, and commuter traffic has been redirected. The central cities suffered economic decline, but some are reviving with new tertiary-sector activities. Attempts have been made to coordinate governmental activities across conurbations.

KEY TERMS

agglomeration p. 397
basic sector p. 398
brownfields p. 429
capital-intensive activity p. 404
central business district (CBD) p. 407
central place theory p. 400
city p.394
congregation p. 407
consolidated metropolitan statistical area (CMSA) p. 416
eminent domain p. 430
external economies p. 400
exurbs p. 402

gentrification p. 425
hinterland p. 394
incorporation p. 394
informal or underground sector p. 405
internal economies p. 401
labor-intensive activity p. 402
metropolitan statistical area (MSA) p. 416
micropolitan areas p. 416
multiplier effect p. 399
network hypothesis p. 429
nonbasic sector p. 398
political geography p. 431

primary sector p. 398
primate city p. 394
secondary sector p. 398
segregation p. 407
service sector p. 398
spatial mismatch hypothesis p. 424
telecommuting p. 423
tertiary sector p. 398
urban enterprise zones p. 429
urban geography p. 394
urbanization p. 394
zoning p. 409

QUESTIONS FOR REVIEW AND DISCUSSION

1. Define and explain the three sectors of an economy.
2. What are the basic and nonbasic sectors of an urban economy?
3. How are cities distributed across a landscape?
4. In what ways does world urbanization today differ from the model provided by the historical experience of England?

5. How can governments slow urbanization?
6. In what ways does U.S. tax policy subsidize people who buy houses?
7. Contrast modern Western urban-planning ideas with the ideas of traditional Islam.

THINKING GEOGRAPHICALLY

1. What are the site characteristics of your city? Why was the site selected?

2. What are the situational relationships of your city? What principal transport routes converge on your city?

3. What range of external economies is available in your town's Yellow Pages?

4. Who are the biggest employers in your town? Which enterprises do the biggest dollar volume of business? What are your town's basic economic activities?

5. Are any planned suburbs or exurbs located around your town? What forms of transportation do their residents depend on?

6. How many local general-purpose governments and special-purpose governments are in your metropolitan region, or the metropolitan region nearest to you? Study a map of all the forms of independent government in the area. How are the special-district government ruling boards chosen? What kind of regional planning body or government does the region have?

7. Some people enjoy living in cities; some people flee cities. What are the benefits and the drawbacks of living in either cities or suburbs?

8. What would you do to stop rural–urban migration in any given poor country?

9. Quality-of-life measurements can affect a city's prosperity. What are the basic measurements of urban or regional quality of life? How would your community score?

SUGGESTIONS FOR FURTHER LEARNING

Abrahamson, Mark. *Global Cities*. New York: Oxford University Press, 2004.

Allen, John, Doreen Massey, and Michael Pryke, eds. *Unsettling Cities: Movement/Settlement (Understanding Cities)*. New York: Routledge, 2000.

Beauregard, Robert A. *When America Became Suburban*. Minneapolis: University of Minnesota Press, 2006.

Brenner, N. *The Global Cities Reader*. New York: Routledge, 2006.

Bruegmann, Robert. *Sprawl: A Compact History*. Chicago: University of Chicago Press. 2005.

Brunn, Stanley D. *Cities of the World: World Regional Urban Development*. Lanham, MD: Rowman & Littlefield, 3rd ed., 2003.

Davis, Mike. *Planet of Slums*. New York: Verso Press, 2006.

De Soto, Hernando. *The Mystery of Capital: Why Capitalism Triumphs in the West and Fails Everywhere Else*. New York: Basic Books, 2000.

Florida, Richard. *The Rise of the Creative Class*. New York: Basic Books, 2002.

Hubbard, P. *The City*. New York: Routledge, 2006.

Jenkins, P. *Planning and Housing in the Rapidly Urbanising World*. New York: Routledge, 2006.

Keiner, Marco, Martina Koll-Schretzenmayr, and Willy A. Schmid, eds. *Managing Urban Futures: Sustainability and Urban Growth in Developing Countries*. Burlington, VT: Ashgate, 2005.

Kruse, Kevin, and Thomas Sugrue, eds. *The New Suburban History*. Chicago: University of Chicago Press, 2006.

Orfield, Myron. *American Metropolitics*. Washington, D.C.: The Brookings Institution, 2002.

Portney, Kent E. *Taking Sustainable Cities Seriously*. Cambridge, MA: MIT Press, 2003.

The State of the Cities 2000: Megaforces Shaping the Future of the Nation's Cities. Washington, D.C.: U.S. Department of Housing and Urban Development, June 2000.

Storey, Glenn, ed. *Urbanism in the Preindustrial World: Cross-Cultural Approaches*. Tuscaloosa, AL: University of Alabama Press, 2006.

Texas Transportation Institute. *Urban Mobility Report*. College Station, TX: Texas Transportation Institute of Texas A&M University, annual.

UN-HABITAT. *An Urbanizing World: Global Report on Human Settlements 1996*. New York: Oxford University Press for the United Nations Centre for Human Settlements (HABITAT), 1996.

UN-HABITAT. *State of the World's Cities Report 2006/7*. New York: United Nations, 2006.

Von Hoffman, Alexander. *House by House, Block by Block: The Rebirth of America's Inner Cities*. New York: Oxford University Press, 2003.

Whitfield, Peter. *Cities of the World: A History in Maps*. Berkeley: University of California Press, 2006.

WEB WORK

The United Nations Human Settlements Programme is at:

http://www.unchs.org/

A special website provides a study of the most successful efforts to solve certain urban problems common around the world:

http://www.bestpractices.org/

The United States Department of Housing and Urban Development and its Office of Policy Development and Research maintain home pages that provide data and useful links:

http://www.hud.gov/

and

http://www.huduser.org/

Several organizations studying urban development maintain home pages. Among the more interesting is the Center for Urban Policy Research at Rutgers University:

http://policy.rutgers.edu/cupr/

The Planning Commissioners Journal page of this website provides a large and useful array of information and articles about contemporary urban planning problems:

http://www.plannersweb.com/

You may generate maps of urban poverty in the United States at a site operated by The Bruton Center at the University of Texas at Dallas and the Brookings Institution Center on Urban and Metropolitan Policy:

www.Urbanpoverty.net

The Sprawl Watch Clearinghouse was formed in 2005 to monitor issues of American urban sprawl:

www.sprawlwatch.org

A New York City website demonstrates how GIS and GPS technologies can offer information about city facilities and services:

http://gis.nyc.gov/doitt/cm/CityMap.htm

11

A WORLD
OF STATES

The Canadian flag flies over the Canadian Parliament in Ottawa. Canada and the United States are close allies, sharing a great deal of their history, culture, democratic beliefs, and economies. Nevertheless, each organizes its territory differently, apportions power differently between the central government and lower levels, distributes activities differently within its territory, and maintains a national culture. The partition of Earth into sovereign countries affects almost all human activities. (Radius Images/Photolibrary. Com)

A Look AHEAD

The Development of the Nation-State Idea

A nation-state is a sovereign territory that includes a group of people who want to have their own government. The nation-state idea originated in Europe and diffused from there—or was applied—to virtually all the rest of Earth.

Efforts to Achieve a World Map of Nation-States

Relatively few states on today's world political map are nation-states. A map of nation-states could be achieved only by redrawing the map, by expelling people from some countries into others, or by forging national identities in the populations of the existing states.

How States Demarcate and Organize Territory

States maintain their borders and subdivide their territory for governmental purposes. Democratic states have a variety of ways to design representative government, but these techniques can be manipulated for political ends.

Measuring and Mapping Individual Rights

States vary greatly in the degree to which they guarantee all of their citizens equality and rights.

*I*n 2006, makers of world political maps continued to set lines around Somalia and to color it one homogeneous hue (see Figure 1-6). The United Nations General Assembly had one seat for Somalia. In fact, however, no real functioning Somali political unit had existed for 15 years. Somalia had come into existence in 1960, when two former colonies (British Somaliland and Italian Somaliland) were yoked together and granted independence. Both colonies had been entirely artificial creations of the colonial powers, and the new configuration did not represent the wishes of the local populations. A ruthless dictator named Siad Barre held "Somalia" together from 1969 until 1990, but the country dissolved in civil war after his fall. By 2006, one leading protagonist in the civil war was the Somali Supreme Islamic Council (SSIC), a fragile coalition of Islamist warlords. The SSIC intended to create a fundamentalist Sharia state and was receiving assistance from Eritrea. The SSIC's principal opponent in civil war was a Transitional Federal Government (TFG) that was recognized by the United States and the United Nations, but which was literally on the run. Neighboring Ethiopia invaded Somalia in 2006, as it

had earlier in 1993 and again in 1996, announcing that it acted to restore order. In fact, Ethiopia probably has feared the creation of a united Somalia, because such a neighbor might lay claim to the Ethiopian province of Ogaden, which has an ethnic Somali population. Religion plays a role in regional animosities, too, for Ethiopia, the world's oldest Christian state, probably does not want a fundamentalist Sharia state as a neighbor. The SSIC used the presence of Ethiopian troops to inflame both ethnic and religious passions among Somalis, despite the fact that long ago the Emperor of Ethiopia had granted asylum to early Muslims when they were being persecuted in Arabia, before Muhammad's triumph. In 2006, Italy and Yemen were selling arms to whoever would pay for them (according to the United Nations), and, meanwhile, anarchic pirate ships in East African seas endangered global shipping. Two northern provinces of Somalia, Somaliland and Puntaland, had declared their individual independence. This brew of ethnic and religious animosities, conflicting territorial claims, and complicating outside interventions is terrifically complex, but such a situation is not unique. ▶

The world political map is probably the most familiar of all maps, because Earth's division into countries, or states, is the most important territorial organizing principle of human activities. **States** are independent political units that claim exclusive jurisdiction over defined territories and over all of the people and activities within them. The governments are not always able to exercise this jurisdiction completely, but states can encourage or even force patterns of human activities to conform to the political map. Chapters 6 and 7 noted how many aspects of culture are affected by the political partitioning of the world. Patriotism, which is a strong emotional attachment to one's country, is itself a powerful cultural attribute. A country can be in many ways a fixed culture realm.

Several of the countries shown on the world political map (Figure 1-6) are not, however, effectively organized. A great many people do not accept that pattern of countries, and much of the territory is not ruled by the central governments. The existing governments try to consolidate their control over all their territory, but not all are able to do so.

The idea that the whole world should be divided up into countries seems natural to us, but it is a relatively new concept in human history, and many of today's countries are young (Figure 11-1). This chapter will explain how the idea originated in Europe and how it was diffused worldwide as part of European conquest. The neatness of the units on today's world political map nevertheless exaggerates the degree to which all peoples accept the current pattern of countries. The map suggests that all the borders are clearly demarcated and that they divide the activities on their two sides, but in fact some activities overlap the borders. The map also suggests that the areas within those borders are politically homogeneous, which is also false. Governments are only more or less successful in organizing their territory, and no territory can be sealed off. Border wars among countries continue to afflict the world, and civil wars within countries are even more common.

This chapter also reviews the political geography of states, analyzing how they maintain their borders and subdivide their territories. In addition, it tries to measure and map the degree to which various governments either place restrictions on their people's freedom or encourage full development of their people's potential.

The ways in which countries try to organize their territories economically—that is, to integrate and build national economies—will be the subject of Chapter 12.

THE DEVELOPMENT OF THE NATION-STATE IDEA

The idea that a state claims exclusive sovereignty over a demarcated space and all the people and resources within it originated in medieval Europe. Under the Roman Empire, the Roman Catholic Church was geographically organized into dioceses. When the Empire fell, the Church and its system survived. As the Church converted people to Christianity, it also converted them to the idea of territorial political organization. Before conversion, the kings of the peoples in Europe had not ruled over fixed territories, but over groups of followers, wherever they wandered. Government over a group of people rather than a defined territory is called *regnum*. The Church taught the principle of rule over a defined territory, which is called *dominium*. The Merovingian kings (fifth through eighth centuries), for example, called themselves Kings of the Franks, but the later Capetians (tenth through fourteenth centuries) settled down and called themselves Kings of France. Most scholars mark the 1648 Treaty of Westphalia, which ended Europe's Thirty Years' War, as establishing in Europe the general principles of territorially sovereign states and of noninterference by rulers in states other than their own.

In the nineteenth century, anthropologists insisted that conversion from regnum forms of government to dominium is an evolutionary step in human society. This argument, however, might be interpreted as an excuse for nineteenth-century imperialism. The argument that native forms of government everywhere were "backward" compared to European forms offered a justification for European conquest. At least it served as a rationalization after the conquests. British sociologist Herbert Spencer (1820–1903) extended Darwin's theory of evolution to insist that "Nature's law" called for "the survival of the fittest," even among individuals, cultures, and whole peoples. Spencer's theory is called **social Darwinism.**

The Idea of the Nation

A state is a territory on a map, but a **nation** is a cultural entity. It is a group of people who want to have their own government and rule themselves. The feeling of nationality may be based on a common religion or language, but it does not have to be, as the Swiss and many other nations demonstrate. A group sharing a sense of nationalism may or may not share any other attributes. Nationalism is a cultural concept in its own right.

Nationalism is one expression of **political community,** which is a willingness to join together and form a government to solve common problems. Today, in most places, the state is the most powerful level of political community, but Chapter 13 will ask whether worldwide concern over environmental pollution or terrorism may nurture a global political community. The evolution of nationalism is part of the story of the evolution of *legitimacy*, the question of who has the right to rule any group. Upon what sort of consent of the governed, if any, is rule based? Even the most totalitarian states today claim to represent the people.

Chapter 7 noted that Christianity explicitly distinguishes religion from secular life. European emperors and kings signified the state. They ruled over earthly matters with the sanction of the Church—by divine right—but they required only obedience, not loyalty or personal identification with the state. Kings could often assign and reassign thrones among themselves and redraw the political map without significant protest by the people. France's King Louis XIV (reigned 1643–1715) insisted flatly, "I am the state." The Church, however, as protector of people's souls, demanded a greater degree of personal commitment than the sovereign did.

The Protestant Reformation challenged this traditional church-state accommodation. Martin Luther preached that every person was his or her own priest and therefore carried individual responsibility for his or her own soul. This belief sabotaged the divine right of consecrated priests and, by extension, that of consecrated kings. In 1581 the Dutch, who were then subjects of the king of Spain, adopted an Act of Abjuration that renounced (abjured) the theory of divine right and argued that a king had an obligation to rule for the welfare of the people. If he did not, then the people could abjure his rule over them. This turned the notion of divine right upside down. It suggested that the king, or any government, served the people.

The U.S. Declaration of Independence of 1776 includes a list of charges against King George III. Prior to that time, even the English themselves already had risen up against their king, Charles I; defeated his forces in battle; tried him for crimes against "his" people; and executed him in 1649. The subsequent English Bill of Rights (1689) recognized that sovereignty did not lie in the king but in the people.

The Nation-State

The Swiss philosopher Jean Jacques Rousseau (1712–1778) laid the foundation for allegiance to the state as the people. Rousseau believed that in nature people were merely physical beings, but when they united in a *social contract*, they were capable of perfectibility. For Rousseau, politics was a means to moral redemption. Rousseau's ideas swayed France, and the French Revolution gave birth to the French nation. The 1789 Declaration of the Rights of Man stated: "The

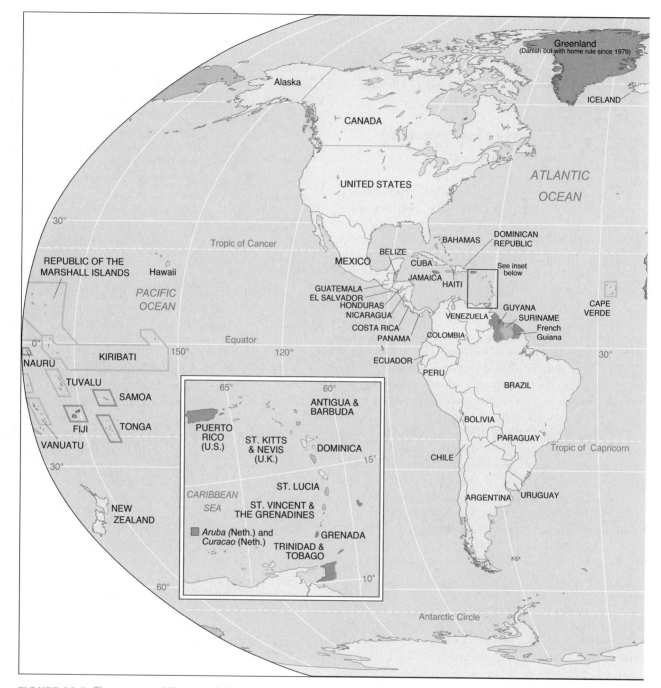

FIGURE 11-1 The ages of the world's states. The current political partitioning of the world is very recent. In many states, citizens remember the day their country achieved independence.

principle of all sovereignty resides essentially in the nation; no body nor individual may exercise any authority that does not proceed directly from the nation."

The nation demanded personal dedication and allegiance from its citizens. Therefore, the perfect state was a **nation-state,** a state ruling over a territory containing all the people of a nation. The theory of the nation-state assumed that nations develop first and that each nation then achieves a territorial state of its own.

Some scholars have argued that several of the nation-states had historic **core areas,** or historic homelands. France, for example, has long been focused on the region of Paris, and Ethiopia on the highlands of the horn of East Africa. In many other cases, however, core concentrations of settlement or activity developed only after the state had come into existence.

Constitutions and laws never fully explain how any government works, because each political community has a unique **political culture.** A political culture is the set of unwritten rules or the unwritten ways in which written rules are interpreted and actually enforced. Political communities differ widely, for example, in whether they honor elders, rich people, or religious leaders; in their tolerance of bribery; and in the rigidity or laxity with which they

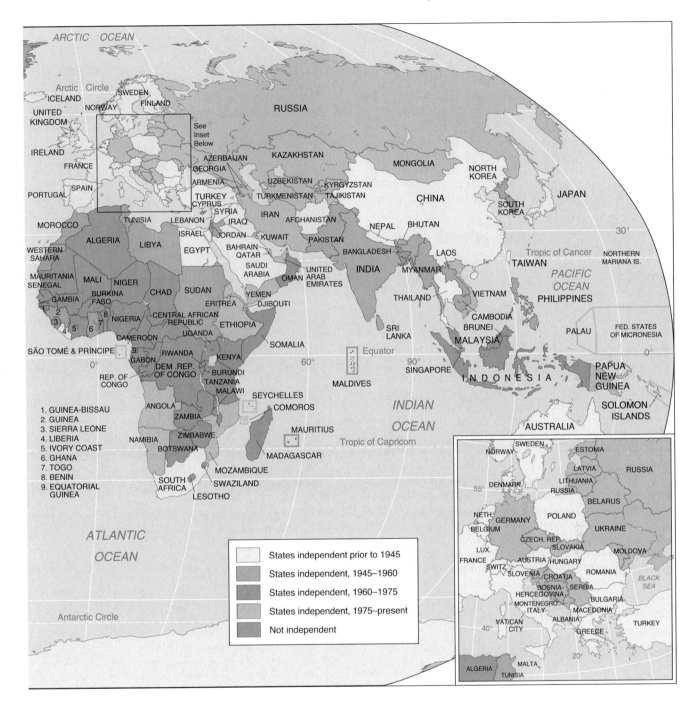

ARCTIC OCEAN

Arctic Circle

ICELAND
UNITED KINGDOM
NORWAY
SWEDEN
FINLAND
IRELAND
FRANCE
PORTUGAL
SPAIN

RUSSIA

See Inset Below

AZERBAIJAN
GEORGIA
ARMENIA
TURKEY
CYPRUS
SYRIA
LEBANON
ISRAEL
JORDAN

KAZAKHSTAN
UZBEKISTAN
TURKMENISTAN
KYRGYZSTAN
TAJIKISTAN

MONGOLIA

NORTH KOREA
CHINA
SOUTH KOREA
JAPAN

MOROCCO
TUNISIA
WESTERN SAHARA
ALGERIA
LIBYA
EGYPT
MAURITANIA
SENEGAL
MALI
NIGER
GAMBIA
BURKINA FASO
NIGERIA
CHAD
SUDAN
CENTRAL AFRICAN REPUBLIC
ETHIOPIA
CAMEROON
UGANDA
GABON
REP. OF CONGO
DEM. REP. OF CONGO
RWANDA
BURUNDI
TANZANIA
MALAWI
KENYA
SOMALIA

IRAQ
IRAN
KUWAIT
BAHRAIN
QATAR
SAUDI ARABIA
UNITED ARAB EMIRATES
OMAN
YEMEN
DJIBOUTI
ERITREA

AFGHANISTAN
PAKISTAN
NEPAL
BHUTAN
INDIA
BANGLADESH
MYANMAR
LAOS
THAILAND
VIETNAM
CAMBODIA
SRI LANKA

30°
Tropic of Cancer
NORTHERN MARIANA IS.
TAIWAN
PACIFIC OCEAN
PHILIPPINES

SÃO TOMÉ & PRÍNCIPE

BRUNEI
MALAYSIA
SINGAPORE
INDONESIA
PALAU
FED. STATES OF MICRONESIA

Equator

MALDIVES

60°   90°

0°

PAPUA NEW GUINEA

1. GUINEA-BISSAU
2. GUINEA
3. SIERRA LEONE
4. LIBERIA
5. IVORY COAST
6. GHANA
7. TOGO
8. BENIN
9. EQUATORIAL GUINEA

ANGOLA
ZAMBIA
ZIMBABWE
NAMIBIA
BOTSWANA
MADAGASCAR
MOZAMBIQUE
SWAZILAND
SOUTH AFRICA
LESOTHO

SEYCHELLES
COMOROS

MAURITIUS

INDIAN OCEAN
Tropic of Capricorn

SOLOMON ISLANDS
AUSTRALIA

ATLANTIC OCEAN

Antarctic Circle

States independent prior to 1945
States independent, 1945–1960
States independent, 1960–1975
States independent, 1975–present
Not independent

NORWAY
SWEDEN
ESTONIA
LATVIA
RUSSIA
DENMARK
LITHUANIA
RUSSIA
BELARUS
NETH.
GERMANY
POLAND
BELGIUM
UKRAINE
LUX.
CZECH. REP.
SLOVAKIA
MOLDOVA
FRANCE
AUSTRIA
HUNGARY
SWITZ.
SLOVENIA
CROATIA
ROMANIA
BOSNIA-HERCEGOVINA
SERBIA
BLACK SEA
MONTENEGRO
ITALY
BULGARIA
MACEDONIA
VATICAN CITY
ALBANIA
TURKEY
GREECE
ALGERIA
MALTA
TUNISIA
55°
40°
20°

enforce laws. Political culture reflects other aspects of a people's culture, such as their religion. This is true everywhere and at every level of government, from a city council district up to the United Nations General Assembly.

For example, one scholar examined political life in Italy and found distinct regional differences in political culture, which he defined as "civic community"— patterns of social cooperation based on tolerance, trust, and widespread norms of active citizen participation (Figure 11-2). In Italy, the distribution of civic community among the regions was already evident as long ago as the thirteenth century. Economists have translated these findings into an economic theory: The more citizens trust one another, the better off their society. These findings suggest that patterns of civic community explain both a region's capacity for democratic self-government and its capacity for economic growth (compare Figure 11-2 with Figure 12-22). Political leaders in regions lacking civic community may lack the fundamental building blocks upon which stable democracy can be built, and civic community may be difficult to create where it does not exist. If this hypothesis is correct, similar studies of whole countries could indicate which ones will most probably enjoy economic growth and democracy in the future.

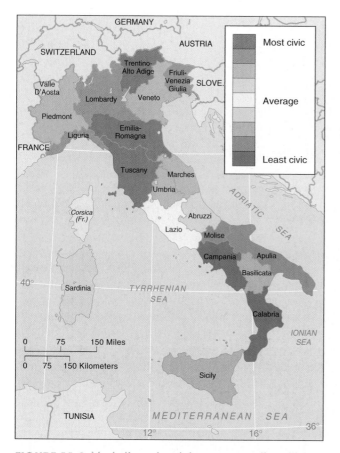

FIGURE 11-2 Variations in civic community within Italy. This map reveals the sharp differences in *civic community* that the scholar Robert Putnam found in each of the regions of Italy. He has hypothesized that economic development and welfare can come about only where local political culture is based on a strong civic community. In the regions north of Tuscany, a sense of political community has threatened to split these regions off from Italy into a new state. (From Robert Putnam, Robert Leonard, and Raffaella Nanetti, *Making Democracy Work: Civic Traditions in Modern Italy.* Princeton, NJ: Princeton University Press, 1992.)

The European Nation-States

The Napoleonic Wars that followed the French Revolution carried the idea of nationalism across Europe. Armies had formerly been composed of hired professionals, but now entire male populations had to serve, and armies were called "the school of the nation." Napoleon was defeated in 1815, but the code of laws that he had imposed left a widespread legacy. It swept away aristocratic privileges and strengthened the middle class (Figure 11-3).

The idea of nationalism matured in Europe during the nineteenth century, as new nations struggled to emerge from old empires and feudal states. In some cases this produced a competitive nationalism, which has caused many wars through the nineteenth and twentieth centuries. For example, several countries claimed the maximum extent of territory over which their people had ever wandered or ruled. These assertions led to overlapping territorial claims.

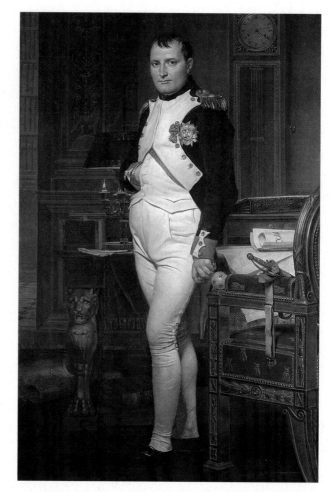

FIGURE 11-3 Napoleon Bonaparte. This 1812 painting by Jacques-Louis David was Napoleon's favorite portrait of himself. He is shown neither in imperial robes nor astride a horse in battle, but as a lawgiver. The pen and scattered documents, the hour on the clock, and the dying candles reveal that the emperor has worked all night on composing the Law Code. The Napoleonic Code remains today the basis of law in 30 countries in Europe and beyond, including Quebec and the U.S. state of Louisiana. Differences occur in many areas. In French civil law, for example, it is a crime not to help a person in need of assistance when help can be provided at no risk to oneself. English Common Law, by contrast, does not compel active benevolence. (Jacques-Louis David (French, 1748–1825) "Napoleon in His Study," 1812, oil on canvas, 2.039 × 1.51 [80¼ × 49¼]; framed: 2.439 × 1.651 × 152 [96 × 65 × 6]. National Gallery of Art, Washington: Samuel H. Kress Collection. © Board of Trustees, National Gallery of Art, Washington. Photo by Lyle Peterzell.)

After World War I, U.S. President Woodrow Wilson advanced the ideal of the nation-state, which he called **national self-determination.** The victors redrew the map of Europe to break up the defeated German, Austro-Hungarian, and Ottoman Turkish empires (but not their own empires) and to grant self-determination to several new European nation-states, including Poland (Figure 11-4). Russia's new communist ruler, Vladimir Lenin, had criticized the czarist Russian empire as "a prison-house of nations." He and his successor, Stalin, reorganized the empire under a new totalitarian

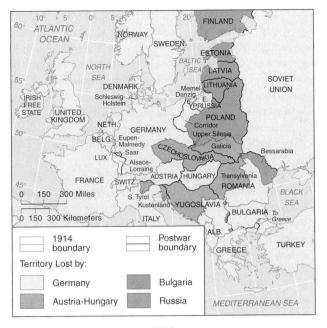

FIGURE 11-4 Europe in 1920. After World War I, the defeated German, Austro-Hungarian, Russian, and Ottoman empires were dismembered, and several new states appeared. Some of them, such as Yugoslavia and Czechoslovakia, did not represent nations but were composed of diverse populations. The new Poland was unsatisfied with the eastern border it was originally awarded, and it seized more territory in a war against Russia.

FIGURE 11-5 Europe in 1946. After World War II the Soviet Union expanded considerably. It retook the territory that Poland had won in the earlier war and gave Poland some German territories, thereby effectively shifting Poland about 240 kilometers (150 miles) to the west. It gave the southern half of German East Prussia to Poland, but kept the northern half (with the important city of Königsberg, which the Russians renamed Kaliningrad), and it swallowed up the three Baltic states. They had been part of the old Russian Empire but had enjoyed independence since 1919. The U.S.S.R. also detached Ruthenia from Czechoslovakia and incorporated it into Ukraine, and Bessarabia from defeated Romania. Bessarabia was combined with part of Ukraine to form a new Republic of Moldavia—today's Moldova. The Soviet Union also seized territory from Finland.

Yugoslavia took territory at the head of the Adriatic Sea from Italy, but the port of Trieste, which Yugoslavia wanted, remained Italian. Compare this map with the current political map (Figure 11-1) to see the changes that have occurred since the end of World War II.

government disguised as a union of nations—the Union of Soviet Socialist Republics, or Soviet Union, but that Union split apart in 1991 (see Chapter 13).

It can be argued that several European nations did exist before they achieved their own independent territory and governments—that is, their own states. Even in Europe, however, national governments conscientiously inculcated patriotism in their citizens. Few states, however, have ever achieved a clean match between people and territory. The European map was redrawn again after World War II (Figure 11-5), but dissident groups in several states still claim parts of their neighbors' territories. Territorial claims on a neighbor are called **irredenta,** from the Italian word for "unredeemed." For example, many Austrians still claim the South Tyrol from Italy, and many Hungarians claim the province of Transylvania from Romania (see Figure 11-4). In both these cases, however, the state governments officially accept existing borders. By contrast, the Bolivian government still formally claims the coastline that Chile took from it after an attack in 1879. George Bernard Shaw well-described the emotional tug of irredenta: "A healthy nation is as unconscious of its nationality as a healthy man is of his bones. But if you break a nation's nationality it will think of nothing else but getting it set again. It will listen to no reformer, to no philosopher, to no preacher, until the demand of the Nationalist is granted. It will attend to no business, however vital, except the business of liberation and unification."

The Formation of States Outside Europe

At the same time as the idea of nationalism was maturing in Europe, the Europeans were actually enlarging their empires. They were not willing to recognize that their colonial subjects had national rights. They argued that non-Europeans were inferior or "not yet ready" for political independence. After World War I the Europeans did not offer national self-determination to their subject peoples outside Europe. The winners just took the losers' colonies while retaining their own colonies. In fact, individual colonies seldom represented political or cultural communities. When the European imperialists carved up the world among themselves, they drew borders that ignored any existing political organization among the native peoples. Some of these new **superimposed boundaries** split native political communities, whereas others combined two or more in one colony (Figure 11-6).

The Europeans often used native rulers as intermediaries between themselves and the people, especially if a colony included several groups. This form of government was called **indirect rule**. In many cases the imperialist powers even amalgamated groups into new tribes and appointed new kings or, in India, maharajahs. These new groupings soon gained the force of "tradition," leaving today's scholars and citizens a bewildering challenge to define what were genuine precolonial communities. Indirect rule hindered the native peoples from uniting against the imperialists: Later the only feeling that brought native groups together in a sense of unified nationalism was often a shared indignation against the colonial power. It has been argued that still today the principal basis of some countries' nationalism is hatred of foreigners, of "traditional enemies," or even of internal minorities: Some countries' rulers skillfully manipulate these hatreds.

The colonies were administered by bureaucracies made up in some cases of Europeans, in other cases of acculturated natives, and in still others of foreign peoples imported by the Europeans. When the colonies received their political independence, these bureaucracies had a vested interest in maintaining the existing units and borders. Therefore, today's world political map is not a map of nations that have achieved statehood. It is a vestige of colonialism.

The threshold principle As late as the 1920s, most diplomats and statesmen insisted that a nation had to have some minimal population and territory to merit self-determination. This principle, called the **threshold principle,** was responsible for yoking together, for instance, the Czechs and the Slovaks, who had never formed a political community and who eventually split apart.

The threshold principle was abandoned after World War II, so today the world map reveals many tiny, independent nation-states (Table 11-1). Some, such as

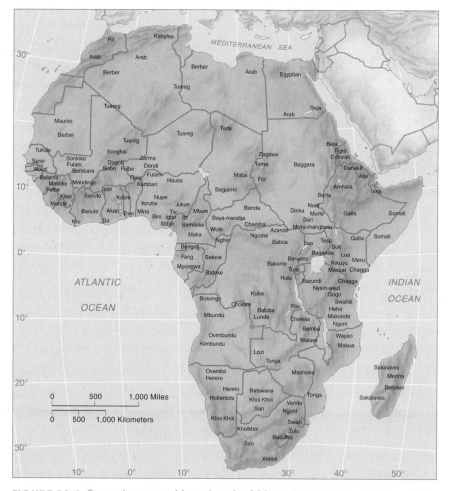

FIGURE 11-6 Superimposed borders in Africa. This map superimposes the borders of today's African states upon the territories of just a few of Africa's many indigenous peoples, just as the European colonial powers superimposed these borders at the Berlin Conference of 1884–85. When the colonies attained independence, the new states seldom contained unified nations.

TABLE 11-1	A Few of the World's Smallest Independent States			
	AREA (KM^2)	AREA (MI^2)	POPULATION (IN THOUSANDS)	GDP (IN BILLIONS OF U.S. DOLLARS)
Bahrain	620	239	699	11.0
Barbados	430	166	280	3.0
Grenada	340	131	90	0.3
Maldives	300	116	359	0.5
Monaco	1.9	0.733	33	0.8
Nauru	21	8.1	13	0.1
San Marino	60	23.16	29	0.5
Tuvalu	26	10	12	0.008
Vatican City	0.438	0.169	0.860	NA
NA – Not available.				

Source: U.S. Central Intelligence Agency.

REGIONAL *Focus* ON

Civil War in Sri Lanka

Approximately 75,000 people have died in a long civil war in Sri Lanka between the 75 percent of the population that is Sinhalese and Buddhist and the 18 percent that is Tamil and Hindu. The 60 million Tamils in India sympathize with Sri Lanka's Tamils, so India has intervened militarily in the civil war. The war has continued so long that the Tamil area has evolved into a shadow state, called Eelam, complete with a flag, national anthem, police, army, taxes, and courts. Talks are complicated by the fact that the U.S. government has labeled the Tamils "terrorists"; not even the government of Sri Lanka can get that designation withdrawn.

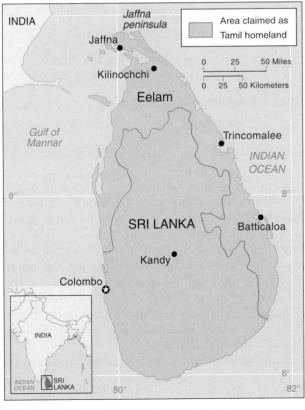

The Tamils call the area they rule Eelam.

This billboard in the Tamil-controlled area of Sri Lanka praises a woman suicide bomber for killing government soldiers. (Photo by Sriyantha Walpola for *The New York Times*)

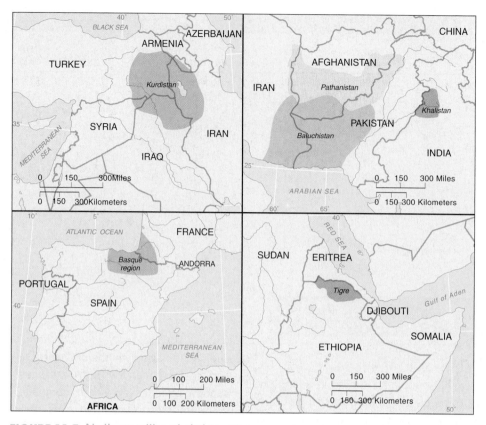

FIGURE 11-7 Nations without states. These are a few of the nations that do not have states of their own but may achieve statehood in the future. The present states, however, may succeed in dissolving these nations. Note that most of these nations overlap two or more existing states, so pressures by these submerged nations to achieve statehood are international issues.

Neither the population of Afghanistan nor that of Iraq constitutes a nation, and many international observers insist that there is no reason for either to exist as a country. After defeating both areas in war early in the twenty-first century, however, the United States committed itself to building nations.

There is very little of a Pakistani nation, either. Pakistan was founded as a faith-based country—Muslim in contrast to secular India—but the civil war that split East Pakistan (today's Bangladesh) from West Pakistan revealed that the religion could not bind the nation. For many Pakistanis—perhaps most—tribal and clan loyalties still today surpass loyalty to the "Pakistani nation."

Bahrain, thrive economically, but others rely on foreign subsidies. Few are militarily defensible. Small, weak, or poor states often negotiate defense or economic treaties among themselves or with more powerful neighbors, but usually they jealously guard their ultimate sovereignty.

Cultural subnationalism When the entire population of a state is not bound by a shared sense of nationalism but rather is split among several local primary allegiances, then that state is said to suffer **cultural subnationalism.** In some states many people grant their primary allegiance to traditional groups or nations that are smaller than the population of the whole new state. These traditional identifications may be strong enough to trigger civil war. In other cases, a group's bonds of affinity may extend beyond the state's borders, and this

may inspire international disputes. For example, the Kurds were promised a national homeland at the end of World War I, but instead they were split among several new countries (Figure 11-7). After the United States defeated Iraq in the First Iraqi War in 1991, it allowed the Kurds in northern Iraq to establish a de facto independent state there. This action has frustrated the Turks, who are trying to squelch demands of the Kurds in Turkey for independence. After the Second War against Iraq in 2003, the United States refused Turkish offers of troops to occupy northern Iraq.

Subnationalism is among the forces, called **centrifugal forces,** that tend to pull states apart. A strong sense of nationalism shared throughout the whole state is among the competing **centripetal forces** that bind a state together. Multinational empires have been held together by military force throughout history, but states

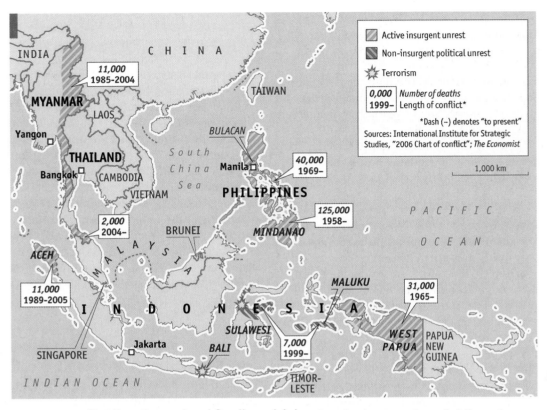

FIGURE 11-8 Fighting throughout Southeast Asia. Ongoing insurgencies and civil wars in Southeast Asia in 2006 illustrate continuing inability to build nation-states in the region.

containing several nations—**multinational states**—usually suffer divisive politics.

Civil wars demonstrate that many countries on the world map are not nation-states (Figure 11-8). At the same time, however, there are many groups that are politically self-conscious and, therefore, arguably are nations, but these groups are politically submerged and do not show up on the map. Hundreds of African and Native American groups might claim national self-determination. Submerged nationalities often present threats of local violence, as events in Southeast Asia, the Mideast, Africa, the former Yugoslavia, and throughout the former U.S.S.R. have demonstrated. The idea of national self-determination retains its powerful attraction, so such local violence may continue.

Rebels in most civil wars claim to be fighting to correct religious, ethnic or political wrongs, but the World Bank has suggested that those claims are often either coincidental or ex post facto justifications for war. The Bank has concluded that the most probable risk factor for civil war is actually economic dependence on commodities that are available for plunder. Prime examples of wars of plunder include the wars in Nigeria (over oil), Colombia (drugs), and Sierra Leone (diamonds, as dramatized in the 2006 film *Blood Diamond*). Therefore, the way to end such civil wars is for countries to develop away from dependence on commodity production, or for international markets to embargo sales of commodities by rebels. The United

Nations did embargo the sale of diamonds from Sierra Leone on world markets in 2000.

Many observers fear a world in which some nations' internationally acknowledged governments cannot claim a monopoly on force within their territories. Colombia, for example, suffered a civil war that took over 35,000 lives and displaced over 1.5 million people in the 1990s. By the end of the decade, the government admitted that it did not exercise sovereignty over its total area, and it recognized guerrilla insurgencies' control over two areas. The United States, however, fearing increases in drug supplies shipped from those areas, supplied forces and arms to encourage Colombia to reconquer them.

Subnationalism and civil wars in Africa

Subnationalism has plagued many African states; some have suffered almost endless civil strife since receiving independence after World War II. Independence released hostilities that had been suspended by foreign rule, and superimposed boundaries cut across historic relationships. Outside powers continued to intervene. As noted in Chapter 8, civil wars have been partly responsible for recurring famines. The Democratic Republic of the Congo, for example, dissolved in war soon after receiving independence in 1960. Eventually a dictator, Mobutu Sese Seko, seized power and held it for 32 years. (The country was called Zaire from 1971 until 1997.) He was overthrown by a rebellion aided by troops from Burundi, Rwanda, Uganda, and Angola, all of which became involved because of complicated loyalties of peoples that overlap all of these countries' superimposed borders. These countries' armies soon turned against one another, and the fighting eventually involved troops from Congo, Rwanda, Uganda, Burundi, Sudan, Zimbabwe, Zambia, South Africa, Namibia, Angola, and Tanzania. It was called "Africa's First World War." Laurent Kabila seized the presidency in 1997, but in 2001 he was assassinated and replaced by his son Joseph, who in 2006 won an election for the presidency.

In the Sudan, the central government, controlled by Muslim Arabs in the North, persecuted a Black animist and Christian population in the South that fought for independence. A ceasefire was reached in 2005, according to the terms of which the South could secede if the Southerners choose to do so by referendum in 2011. The South, however, holds the nation's great untapped oil wealth, so it remains to be seen whether the referendum will peacefully occur. Meanwhile, resentment has grown among the population of Darfur, a once-independent Islamic sultanate in Western Sudan: The central government recruited Arab militias called the janjaweed ("evil horsemen") to put down these non-Arab African tribes. The people of Darfur are Muslim, but the janjaweed are lighter-skinned and speak of killing "Blacks." Tens of thousands died in Darfur between 2003 and 2007, and about three million people were displaced. Janjaweed attacks even crossed the border into neighboring Chad. In 2006, troops sent by the African Union (see Chapter 13) patrolled the region, but peace had not been achieved by year's end.

The Republic of the Congo, Sri Lanka, Somalia, and the Sudan are all examples of what we today call **failed states** or **collapsed states,** that is, countries that have proved incapable of providing their citizens with either economic development or even peace and security.

Subnationalism helps explain the number of authoritarian governments in Africa. Authoritarian rulers argue that iron rule is the only alternative to the tribalism that could tear their countries apart, but this argument is at least partly a rationalization for them to hold on to power. Several have been accused of deliberately fomenting and manipulating subnational strife.

The existence of many failed states in Africa, as well as in Asia and in Latin America, provides breeding grounds for civil wars, guerilla wars, and the multiplication of independent rogue military forces. These have proved capable of projecting violence into the heart of the developed world, as the attacks on the United States on 9/11 demonstrated. (The agents of those attacks were introduced in the Preface and will be examined in greater detail in Chapter 13.) Therefore, since the attacks, the United States has demonstrated greater involvement in the problems of nation-building and state-building.

EFFORTS TO ACHIEVE A WORLD MAP OF NATION-STATES

The territorial state is the highest level of political sovereignty. The United Nations organization promotes cooperation among states and in Antarctica, which is the only portion of Earth not divided into an independent state or states (both of these topics are discussed in Chapter 13). The complete division of the world into stable and peaceful territorial nation-states may or may not be an ideal goal, but the ideal holds powerful attraction.

States use three major strategies to make the map of nations fit the map of states: (1) to redraw the international political map; (2) to expel people from any country in which they are not content, or to exterminate them; and (3) to forge nations in the countries that exist now.

(1) Redrawing the World Political Map

Theoretically, the world political map could be redrawn until everybody is content being in the state in which he or she resides. Unfortunately, this would open endless disputes and provoke new wars. Some countries would split apart, and numerous territories would have to be transferred from one country to another. No satisfactory solution could exist for the states in which different national or ethnic groups intermingle or in which the cities are populated by people of one group, and the countryside by people of another group.

Disputes would also arise over the distribution of natural resources. Each group would claim the most generously endowed territory as its own, or at least a "fair share" of what had been the entire state's endowment. The Ogoni people of Nigeria, for example, occupy one of the world's richest oil-producing regions, yet the central government claims all mineral rights and distributes profits among all regions of the state. The Ogoni profit little from the oil but live instead in toxic pollution and poverty. Ogoni independence movements have been squashed quickly.

Despite inevitable problems, political maps continue to be redrawn. In Europe in recent years East and West Germany joined into one country in 1990. Czechoslovakia split into two countries—the Czech Republic and Slovakia—in 1993. Yugoslavia broke up in 1991 (Figure 11-9a). Montenegro split off from Serbia in 2006, and Kosovo may achieve independence, too (Figure 11-9b).

Few borders have been redrawn in Africa or Asia since decolonization (Figure 11-10 on page 454). African countries could have chosen to redraw their borders at the time they received their independence, but at a meeting of the Organization of African Unity (OAU) held in Cairo, Egypt, in 1964, the African states pledged

Geopolitics

Chapter 10 defined political geography as the study of the interaction between political processes and the distributions of all other activities and transformations of the landscape. The term was coined by German geographer Friedrich Ratzel as the title of a book he wrote in 1897. Ratzel's studies focused on the political geography of states, and he suggested that each nation needed *Lebensraum*, room to live. As national populations increased, he wrote, nations might need more Lebensraum.

Ratzel's idea of Lebensraum was adopted by some of his contemporaries who were at that time drawing analogies between states and living things. They argued that states are a form of biological organism. To explain why some states thrive while others perish, these writers combined Ratzel's idea of Lebensraum with Spencer's theory of social Darwinism, and they concluded that competition among states for territory resulted in "the survival of the fittest." In 1916 the Swedish political scientist Rudolf Kjellen published *The State as a Form of Life*, in which he wrote of "the natural and necessary trend towards expansion as a means of self-preservation." Kjellen proposed a new "science" to study states' competitive strategies. He called it **geopolitics.** In 1924 a German professor named Karl Haushofer founded an Institute for Geopolitics in Munich, where he taught that it was natural for strong states to expand at the expense of the weak. These ideas impressed Adolf Hitler, and when Hitler became ruler of Germany, he quoted geopolitical theories to justify Nazi aggression in World War II.

The term *geopolitics* is still loosely applied to studies of global military strategies, and it is often confused with political geography. Chapter 12 will demonstrate, however, that any state's possession of extensive territory—or even of natural resources—does not strictly determine the welfare of its people. Already in 1899 the British scientist Sir William Crookes had refuted geopolitical theories. His book *The Wheat Problem* suggested that technological progress can replace territorial aggression in raising a nation's standard of living. Crookes coined the metaphor "scientific frontiers," but even he would probably be amazed to learn of the technological advances in agriculture described in Chapter 8.

---

themselves to respect the existing international borders. They made this pledge even though they resented the borders as a colonial legacy and found it difficult to govern within them. In 1994, however, Rwanda's president, Pasteur Bizimungu, demanded that neighboring Congo (then called Zaire) surrender territories occupied by the Banyamulenge, a people related to the Tutsi of Rwanda. President Bizimungu called for "a Berlin II, so that we can reflect on the disorders created by Berlin I [where colonial powers divided Africa]" (see Figure 6-29). As we previously noted, three years later Rwandan troops helped bring down the Congolese government. In 2000 Edem Kodjo, former Prime Minister of Togo and Secretary General of the OAU, said, "A certain number of principles that date from the independence era are already being shattered. . . . The intangibility of borders is likely to be next, because many of our states are inviable, either for economic or political reasons, or because of their population or ethnic composition." Thus, 40 years after the Cairo Conference, African leaders began considering redrawing the map. Some African states may either break up (e.g., Somalia) or else recognize subnational communities by adopting federal forms of government (discussed later in this chapter).

Some remapping might occur in Asia as well: Possible developments include independence for Palestine, re-unification of the Koreas, the break-up of Iraq, Afghanistan, Pakistan, or even China (as will be noted later in this chapter). Other potential new countries that may be created in the coming years include Bermuda (a British territory), Greenland (a division of Denmark), New Caledonia (a territory of France), and Western Sahara (under the control of Morocco, but in dispute).

(2) Mass Expulsions or Genocide

A second strategy used to make a nation "fit" a state territory is to expel from the state those people who are not accepted as members of the nation, or to exterminate them. These policies are tragic and abominable, but both have actually been implemented when groups have attempted to carry the logic of territorial nationalism to its illogical extreme.

Mass expulsions and genocide occurred in Southern Europe during and after World War I. The Turks massacred Armenians in 1915 and later expelled Greeks from Asia Minor. Greece responded by expelling Turks. During the 1930s and 1940s, Germany attempted to bring into the country many Germans who lived in other countries and to eliminate Jews within Germany. After World War II, Germans were expelled from Poland and Czechoslovakia. Millions of refugees fled from Pakistan to India or vice versa at the partition in 1947.

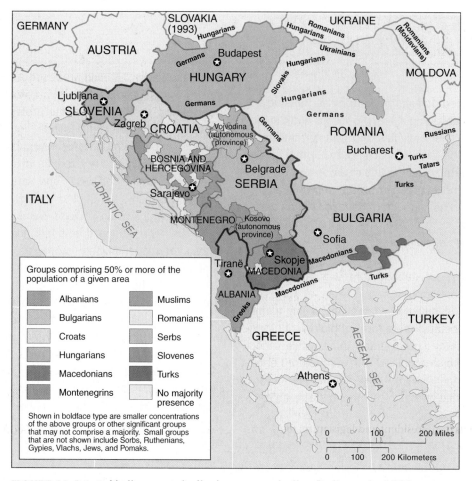

FIGURE 11-9 (a) Nations and ethnic groups in the Balkans in 1990. This pattern has defied attempts to delineate nation-states since the nineteenth century. World War I began in 1914 when a Serb assassinated an Austrian archduke in Sarajevo, which was then an Austrian possession but was claimed by Serbs. After World War I, Yugoslavia was formed, composed of six constituent republics. That state dissolved in 1991, and the constituent republics of Slovenia, Croatia, Bosnia-Herzegovina, and Macedonia emerged as independent countries. Serbia and Montenegro stayed together for several years, but Montenegro broke off in 2006. (For the borders current in 2006, see Figure 11-1.)

Beginning in 1991, Serbian residents of Bosnia-Herzegovina (pink on this map), backed by troops from Serbia, began corralling the Bosnian-Herzegovinin Muslims (green) into ever-shrinking territories or else driving them out of Bosnia-Herzegovina altogether. Croatia has driven the Serbs out of its territory and resettled the area with Croats. In 1998 Serbs opened an offensive to drive ethnic Albanians out of Kosovo, the province of Serbia that has an Albanian population. The United Nations assumed control of Kosovo in 1999, and Kosovo may ultimately win independence or a merger with Albania. In 2001 Macedonia, by contrast, peacefully granted Albanians cultural rights.

Mass expulsions continue. In 1989 Bulgaria expelled about 100,000 ethnic Turks. In 1991 a million Kurds fled or were driven out of Iraq. In 1993 tens of thousands of people of Nepalese origin were driven out of Bhutan. Serbians referred to the slaughter or exile of Muslims in Bosnia-Herzegovina as *ethnic cleansing;* this policy extended to a *cultural cleansing* of the landscape. The Serbs destroyed mosques, libraries, and museums and even bulldozed cemeteries (Figure 11-11). In Tutsi-dominated Rwanda in 1994, Hutus rose up and slaughtered more than 500,000 Tutsis. The next year the Tutsis regained the upper hand, driving several hundred thousand Hutus into Congo. Next door, Burundi has suffered ethnic violence ever since it won independence from Belgium in 1962. The Tutsi minority (15 percent) traditionally ruled the Hutu majority (85 percent), and in 1993 the first democratically elected Hutu president was assassinated by a Tutsi military uprising. Civil war raged, and over 200,000 people were killed and an additional million internally displaced. The Arab majority of Mauritania continues to expel the country's minority Blacks into Senegal. In the Caucasus

FIGURE 11-9 (b) Bosnia and Herzegovina today.

An accord concluded at Dayton, Ohio, in 1995 provided that: (1) Bosnia and Herzegovina would be a single country divided in two parts, a Bosniak (Muslim)-Croat Federation and a Serb Republic with a central government. (2) Each entity would have its own president and legislature; the central government would have a collective presidency; and the president and parliament of the central government would be chosen in free elections. The central government in Sarajevo is responsible for foreign and economic policy, citizenship, immigration, and other issues. (3) Refugees would be allowed to return to their homes, with free movement guaranteed. (4) The peace is implemented by a NATO force. (5) All parties cooperate with an international war crimes tribunal.

region through the 1990s, Abkhazia, Ossetia, and Nagorno-Karabakh all suffered ethnic killings and expulsions involving more than a million people combined.

Censuses taken in the Soviet Union in the late 1980s counted some 60 million "displaced Soviets" living outside the regions of their ethnic identity. When the Soviet Union broke apart, not all of the newly independent states guaranteed full civil rights to their minorities. Ethnic Russians, especially, were subject to discrimination in some countries. Millions of people have migrated into the homelands of their own ethnic origins.

Probably the most ethnically homogeneous major states today are Japan and the two Koreas; they are 99 percent homogeneous. China is 92 percent Han. These countries exist more or less within their historical frontiers. Japan may be the closest realization of the concept of a nation-state. The Japanese virtually closed their islands to foreign trade and cultural exchange for hundreds of years and, as noted in Chapter 5, allow very little immigration today.

(3) Forging National Identities

Today many existing states are struggling to weld their populations into nations. This reverses the theoretical order in which the formation of a nation precedes the achievement of statehood. The significance of this transition for world cultural geography can scarcely be exaggerated. The states are struggling to make the maps of other human activities conform to their pattern. Chapter 7 noted that governments can affect the geography of religion and language. In addition, governments can affect the geography of political community, of law, of land use, and as will be seen in Chapter 12, of economic activity. Countries can encourage the circulation of people, goods, and ideas within their territories, and they can restrict or discourage circulation across their borders. Countries can never control these activities entirely, but modern governments exercise incomparably greater power over human activities than governments did in the past.

We have already emphasized that many units on today's world political map are relatively new. No one can predict how long any world political map will remain fixed, but existing governments try to stabilize it. The history of the United States recounts many struggles to weld one nation out of many diverse groups. In Europe a few nation builders faced their task self-consciously. The Polish hero and later president Joseph Pilsudski (1867–1935) said flatly, "It is the state that makes the nation, not the nation the state." Italy was politically unified in the 1860s, and after the unification the statesman Massimo D'Azeglio observed: "We have made Italy; now we must make Italians." As illustrated by Figure 11-2, Italian politicians have not completely succeeded.

European experiences were echoed by Julius Nyerere, the first president of Tanzania (served 1961–1985). He commented on the African states: "These new countries are artificial units, geographical expressions carved on a map by European imperialists. These are the units we have tried to turn into nations." Many Asian states face the same difficulty, and even many Latin American countries that have been independent for over 150 years still have not welded their populations into nations.

The destruction of subnational loyalties

Before some states can forge their populations into nations, they often try to undermine the cultural subnationalisms within their borders. This destruction of traditional communities can be profoundly destructive psychologically to at least the first generation of the new citizens. Typical is Kenya's campaign to abolish the ancient traditions and distinctiveness of the Masai people, once one of Africa's most powerful nations, and to integrate them into the modern state. The Masai must go to school and conform to a new style of life dictated by the government. They must surrender their age-old lifestyle of nomadic cattle herding, settle down, and take up farming (Figure 11-12).

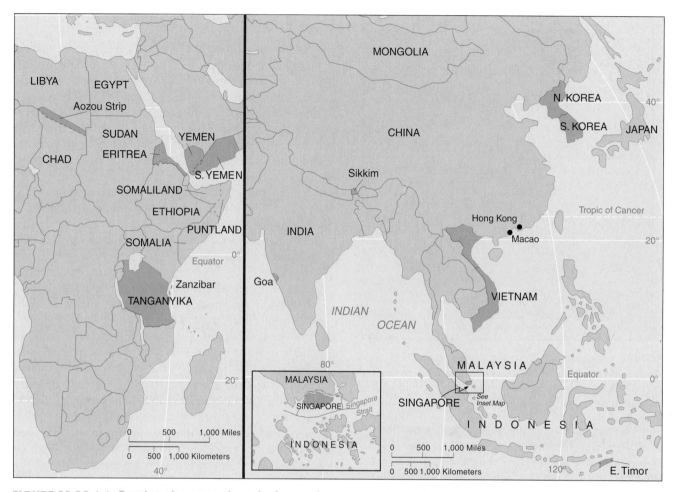

FIGURE 11-10 (a) Border changes since independence. In Africa, Tanganyika and Zanzibar formed a union as Tanzania in 1964, although the terms of that union may be renegotiated. Zanzibar's historic ties link it to the Arab world; Zanzibar was for many years the capital of Oman on the Arabian Peninsula, over 2,200 miles (3,600 kilometers) away. Eritrea achieved independence from Ethiopia in 1993, but, after long fighting, the two countries' border was still being demarcated in 2006. The Aozou Strip between Libya and Chad, plus the borders between Egypt and Sudan, have been disputed, but not redrawn. Nigeria and Cameroon engaged in a long-running dispute over the oil-rich 400-square-mile Bakassi peninsula in the Gulf of Guinea (not shown). In 2002 the International Court of Justice in The Hague awarded the territory to Cameroon, but Nigeria did not surrender it until 2006. Somaliland and Puntaland seized their independence early in the twenty-first century, but neither is yet recognized by the United States or by international bodies.

A few colonial borders have been redrawn in Asia. In 1965 Singapore separated from Malaysia. In 1976 North and South Vietnam joined into one Socialist Republic of Vietnam. Korea, which had been annexed by Japan in 1910, was divided by victorious Russian and American troops in 1945, and it remains split into South and North Korea. North Korea tried to conquer the South, but South Korea rebuffed it with U.N. assistance (1950–1953). Yemen, which had received independence from Turkey in 1918, and South Yemen, which received independence from the United Kingdom in 1967, merged in 1990. India annexed the Portuguese colony of Goa in 1961 and the independent country of Sikkim in 1975. In 1975 Indonesia seized East Timor from Portugal, but East Timor achieved independence in 2002. China reabsorbed Hong Kong from the United Kingdom in 1997 and Macao from Portugal in 1999.

Some countries have absorbed the leaders of their traditional subnational groups into the new political structure (Figure 11-13). Cameroon, for example, still allows the 17 traditional kings within its borders to exercise judgment over minor crimes, and it defers to their ceremonial importance. In Zimbabwe local chiefs appoint 10 members to the 150-seat parliament. In the 1990s South Africa's president Nelson Mandela tried to educate and train the country's 300 Zulu chiefs in law and administration. His efforts were criticized as a centralization of power, but the chiefs' powers had in fact been created by the British colonial authority. This struggle provides a fine example of the confusion of precolonial tradition with imperialist legacies. Members of many African royal families retain positions of power and influence in the modern states. Mandela himself was a traditional prince, and Ghanaian diplomat Kofi Annan, U.N. Secretary General from 1997–2006, is the prince of the Fante people, and through his mother he is heir to the paramount chieftancy of the Akwamu. His father was a thoroughly modern man, working as an executive of a European global corporation. In Malaysia a paramount ruler is elected every five years from among the nine traditional Malay

FIGURE 11-10 (b) After the 1950–53 Korean War, South Korea developed into a prosperous democratic capitalist state, and North Korea remained a poor communist dictatorship. The two states have, however, opened negotiations to achieve reconciliation. As a symbol of that hope, their teams entered the opening ceremonies of the 2002 Olympics in Sydney, Australia, under a single flag, which was the flag of neither country. Their teams did not compete as one, but in 2005 they announced their intention to compete as one country in Beijing in 2008. (Chang Lee/NYT Pictures)

sultans. The curtailment of these sultans' power in 1993 symbolized Malaysia's emergence from a traditional monarchical country to a modern democratic country.

India abolished the privileges of its many maharajahs in the 1970s, although many of them subsequently won elective offices. India remains the world's largest experiment in bringing diverse groups of people together under democratic political structures. India is afflicted by centrifugal forces based on religion, ethnicity, caste, and language. Several of these distinctions are regional and might encourage separatism, but the Indian civil service, a legacy of British rule, remains a strong centripetal force.

Urbanization, which is occurring rapidly everywhere, often dissolves traditional rural-based identities. This development may reduce subnational loyalties.

FIGURE 11-11 The ruined national library of the Sarajevo, Bosnia, University. "Cultural cleansing" of a landscape includes the deliberate destruction of libraries and other cultural monuments. (Noel Quidu/Getty Images, Inc.—Liaison)

Instruments of nation building   Different countries rely on different instruments of nation building, and the choice reflects differences in other aspects of their cultures. Among the instruments most often used are religion, the armed forces, education, symbols, media, political parties, and labor unions.

FIGURE 11-12 Masai people.   The westernized Kenyan on the left appear to regard these Masai men visiting downtown Nairobi with bewilderment (and, it seems, hostility). Kenya is forcibly acculturating the Masai to modern norms, and Masai are prohibited from dressing like this on the streets of Nairobi today. (Courtesy of Robert Caputo)

Nigeria

Nigeria may be the world's most spectacular—and dangerous—example of a failed state.

In the late-19th century, the British established the Royal Niger Company to exploit resources in the Niger River delta. Expanding inland, the British found themselves ruling some 250 ethnic groups that had never before coexisted in a single state. The British divided Nigeria into administrative zones along ethnic and religious lines. After Nigeria achieved independence in 1960, the Muslim north, which is poor, but with half of the country's total population, gained supremacy over the army. Through a succession of military dictatorships, it dominated and plundered the fertile and oil-rich but disunited South, whose largest ethnic groups—the Yoruba in the West and Igbo in the East—together represented just 39 percent of the population. The Igbo declared their region to be the independent state of Biafra in 1967, and during the three years that the central government fought to restore its authority, fatalities were estimated to be more than one million. The workings of the civilian government that came to power in 1999 have been characterized by wholesale bribery and murder of government officials and candidates.

From 1980 to 2005, Nigeria earned more than $300 billion in oil revenues, but annual per capita income plummeted from $1,000 to $390. Billions of dollars have been squandered or stolen outright by the country's leaders. The country appears to be *de*-developing, and terrorist Osama bin Laden has called it "ripe for liberation."

Nigeria has the largest petroleum reserves of any African country, but its oil industry is characterized by crime, neglect, pollution, and corruption. Tapping into pipelines and siphoning oil into tankers hidden in the swamps of the Niger River delta is widespread, causing losses, pollution, and catastrophic fires. Owing to the abysmal state of Nigeria's few refineries, the country must import gasoline. Fuel shortages are endemic.

Nevertheless, Nigeria is the fifth-largest supplier of oil to the United States, and in 2002, the U.S. Executive declared the oil of Africa to be of "strategic national interest," meaning that the United States would use military force, if necessary, to protect it.

A hastily erected façade of modernity is disintegrating and leaving city-dwellers, in particular, struggling to survive in desolation. Lagos, Africa's largest city, with 13 million people, is a chaotic jumble of crumbling roads, falling power lines, electricity blackouts, and crime.

From 1999 to 2006, at least 14,000 Nigerians died in ethnic, religious, or communal fighting, and approximately 3 million people were internally displaced. Nigerian Nobel Prize–winning author Wole Soyinka has joined those Nigerians calling for a national conference that would bring together representatives of the country's groups and debate a new constitution. Only such a debate, Mr. Soyinka argued, would enable Nigerians "to make sense of our remaining together." "What do we want as a nation?" Mr. Soyinka wondered aloud. "We have never been able to decide for ourselves."

International investment. Royal Dutch Shell's pristine oil plant in Bonny, Nigeria, presents a sharp contrast to the squalid poverty of the neighboring traditional fishing village. (Michael Kambler/New York Times Agency.)

Religion   Chapter 7 noted countries in which a national church was a building block of nationalism: Ireland, Ukraine, and Poland. The Orthodox churches are traditionally national churches, and some Protestant denominations are also rooted in specific nations: Lutheranism in Germany, Presbyterianism in Scotland, and Anglicanism in England. The Roman Catholic Church is transnational, but the 1965 Vatican Council (called Vatican II) encouraged countries to form their own national councils of bishops. These councils have begun to reflect cultural differences among countries that are significant enough to surprise and upset the Vatican. The U.S. National Council of Bishops, for example, reflecting U.S. culture, has expressed willingness to allow women a greater role in church services than the Vatican will accept. Chapter 7 noted that the Emperor of Japan renounced his claim to divinity after World War II, but in 2000 Japanese Prime Minister

FIGURE 11-13 The Alake (king) of Abeokuta (seated) and his son. These two men represent the transition from tradition to modern statehood in Africa. The son of the Alake of Abeokuta was educated in Britain and returned home to serve as the Chief Justice of the new state of Nigeria. (Photo by Eliot Elisofon, 1959. Image no. EEPA 2016. Eliot Elisofon Photographic Archives, National Museum of African Art/Smithsonian Institution)

Yoshiro Mori described Japan as "a divine nation with the emperor at its core." When the remark provoked a storm of criticism, the Prime Minister explained that he had meant only that the nation should be more spiritual. Many Russians view Orthodox Christianity as an essential element of Russian identity, and Russia severely restricts missionary activity of other faiths.

Armed forces In many countries the armed forces are still "the school of the nation" and claim a high proportion of the people for service (Figure 11-14). The armed forces may consume a large percentage of the national budget, but defense against external enemies is not always their principal function. The army holds the state together, either by force or just by training and socializing young people. When Hutus were slaughtering Tutsis in Rwanda in 1994, then-U.N. Secretary General Annan suggested that African governments were not

contributing troops to peacekeeping efforts because "they probably need their armies to intimidate their own populations." The remark stirred outrage in many African countries.

In many cases the armed forces provide a disciplined labor force for building infrastructure. The military forces may even exercise an independent role in the government. In Chile, for example, the armed forces are guaranteed their own seats in the national legislature. Chapter 7 noted that the Turkish constitution grants the army a special role in preserving the government. Even the U.S. armed forces have been assigned nation-building tasks. In 1947 President Harry Truman ordered their racial integration. This was brought about successfully decades before the society at large was ready to integrate.

Education National school systems are another tool with which states create nations. Rousseau wrote: "Education must give souls a national formation, and direct their opinions and tastes in such a way that they will be patriotic by inclination, by passion, by necessity." Today in France, it is said, the Minister of Education in Paris can look at his watch at any time of day and say precisely what all French schoolchildren are studying. In 2003, the United Nations administrator of Bosnia and Herzegovina mandated the integration in local high schools of Serbs, Croats, and Muslims, children of the groups who had fought so hard to live apart. It was hoped that integration in school could dispel age-old antagonisms.

Most countries have national curricula and textbooks. The interpretation of national history in those texts is a political issue, and a change in the government can change that interpretation. Mexico, for example, introduced in 1992 a new edition of the history book for fourth graders. After 20 years of teaching that Porfirio

FIGURE 11-14 "The School of the Nation." This young member of the army of the Congo uses a bag printed with cartoon figures to store his grenades. Many national armies—as well as guerilla armies in civil wars—recruit or kidnap child soldiers, despite international convenants against children in warfare. Many countries draft high percentages of their people just to keep them under control. (Michael Kamber/*The New York Times*)

Díaz, the turn-of-the-century dictator, "was very bad for the life of Mexico, because the people were not given the chance to elect their leaders," the new version emphasizes that Díaz achieved stability, tolerated the Catholic Church, built railroads, fostered industrial growth, and attracted foreign investment. These are all programs supported by the current government. The 1992 text also introduces criticism of Emiliano Zapata, a revolutionary hero who defended the ejido agricultural system (see Chapter 8). When Hong Kong was returned to Chinese sovereignty in 1997, new schoolbooks shifted away from British geography and British interpretations of history to a focus on Chinese geography and Chinese interpretations of history. Later, in a further repudiation of British influence, China replaced all English-language textbooks from Britain with English-language textbooks from Canada. Today Chinese students of English learn about Canadian holidays, weather, cities, the flag, hockey, and environmental values. In 2003, U.S. agencies began rewriting Iraqi schoolbooks in which maps depicted Kuwait as Iraqi territory. Education in Japan was patriotic before World War II, but the U.S.-imposed Education Law of 1947 minimized the inculcation of patriotism and emphasized individual rights. Today, however, nationalism is on the rise, and teachers are being punished for refusing to sing the Japanese national anthem or to stand before the flag.

Schools instill in youngsters the society's values and traditions, its political and social culture. This is called the process of **enculturation,** or **socialization.** National schools propagate national culture, literature, music, and artistic traditions. National history and geography lessons leave the rest of the world outside the focus of concern. Classroom maps often promote hostility to neighbors by claiming territory held by a neighboring country.

The United States has traditionally relied on the school system to forge a nation, yet it allows a diversity of programs and standards across the country. Education is not mentioned in the Constitution, so responsibility for it remains with the individual states, which have generally delegated the task down to local school districts. International comparisons revealing the weakness of U.S. elementary and secondary schools have spurred demands for nationwide standardization of curricula and testing. Local districts could still experiment in designing school programs, but standardization could end one of the nation's oldest and most revered traditions of local government.

Manipulating symbols Each country has its own set of national symbols, called an **iconography,** and these items are emphasized to schoolchildren. The national flag is one of the most important of these icons (Figure 11-15), and several countries have laws against its desecration. A nation's national anthem is another icon. In a unique instance of international solidarity, however, the nonpolitical prayer for God's blessing,

FIGURE 11-15 The flag of Montenegro. Montenegro adopted a new flag in 2004, even before full independence was achieved in 2006. The gold border contains a red field, centered in which is the nation's coat of arms. The coat of arms was adopted from that of the 19th century Montenegrin royal family.

"God Bless Africa," serves Namibia, Tanzania, Zambia, Zimbabwe, and South Africa. It was written by a South African schoolteacher and taken up by the African National Congress as its anthem.

The map is another icon of great power. Children can be taught to respect—even to cherish—the size, shape, topography, resources, and variety of their land. The U.S. weather map has demonstrated appeal on television, and news editors know that the weather map is one of any paper's most popular features. Few people actually read the map or even care about the weather in other areas of the country, but the map itself is an icon that wins allegiance.

Iconography also includes the pomp and circumstance of national ceremony—the celebration of holidays, national costumes, the designs on national postage stamps, and international sports competitions. In 2003 the U.S. Treasury introduced a new $20 bill redesigned to foil counterfeiters, but also specifically to add "symbols of freedom," meaning more eagles and flags. Many countries ban public displays of the iconography of past regimes, including fascist symbols in Germany, Hungary, and Romania and Communist symbols in several formerly Communist countries.

Media National media usually can reach and sway more of a national population than any other shared experience, even the army or the school system. Half of India's people, for example, are illiterate, but 80 to 90 percent of them can be reached by state-controlled radio and television. Government media monopolies may inform the population, educate it, win its allegiance, or command it. The United States is the only country in the world in which commercial television broadcasting preceded public programming, and today only the United States, Sri Lanka, and Norway do not require television stations to donate air time to political candidates.

Forms of media that spill over international borders can exert a powerful centripetal force. For example, one reason that East Germany never created an East German nation, despite 40 years of trying, was that the government could not prevent East Germans from watching West German television.

Critical THINKING

Can a Country Insulate Its Borders and Isolate Its People?

Many countries try to limit the information that world citizens—or even their own people—are able to learn about events inside the country. Such censorship, however, is becoming increasingly difficult. New technologies of communication sabotage governments' efforts to control the flow of information across borders.

Today hundreds of dissident political movements in various countries have access to their own satellite transmission facilities or even host their own Internet sites. They can reach the whole world audience with their interpretation of political developments. Refugees in exile can communicate with resistance movements back home and disseminate information that is subversive of a repressive government. No government can completely guarantee the security of its borders against the flow of information and ideas. The English-language Egyptian newspaper *The Cairo Times*, for example, is censored in Egypt, so its publishers in Cyprus put the censored material on a website for Egyptian readers.

Even the United States has laws and governmental directives that are intended to keep out of the country the presentation of opinions that the government considers subversive. Many foreign artists, writers, and political figures have been prohibited from entering the United States personally, even though they are interviewed abroad and featured on U.S. television and newsmagazine covers. Legislation requires schools and libraries with Internet access to install software that blocks access to materials that censors consider "pornographic, obscene, or harmful to minors." A 2003 study found that even at their least restrictive settings, however, these filters blocked 24 percent of all health education sites and 50 percent of safe-sex information sites.

American software makers and service providers have been criticized for helping repressive regimes censor their people's access to the Internet. Saudi Arabia, Iran, Singapore, China, Myanmar, Sudan, and Tunisia are among the countries that regularly use American companies' censoring or filtering products. In China only four state-controlled entities can legally connect with the Internet, and commercial websites cannot hire their own reporters or publish news not already published by the state news agency. By 2006, China had an estimated 110 million Web surfers—second only to the number in the United States. China also had an estimated 35,000 officers patrolling the Web, deleting phrases such as "free speech" and "human rights" from online bulletin boards. If a searcher on any of China's search engines enters a sensitive political term such as *Tiananmen Square*, the computer will crash or simply offer a list of censored websites. Microsoft, Yahoo!, and Google all cooperate with the Chinese government. In 2005, Yahoo! supplied information that helped it track and convict a political dissident who sent an e-mail with forbidden thoughts from a Yahoo! account. "Business is business; it's not politics," said Jack Ma, Yahoo!'s director of operations for China.

Questions

1. What methods do governments use to prevent the influx of information and foreign ideas?
2. Can these methods be entirely successful, given modern technology?
3. Do governments have the right to restrict the flow of information? Why or why not?

Political parties Political parties can politicize and mobilize populations, recruit people, and give them a sense of participation. They can broaden the government's base of support, but they can also be divisive. If people identify themselves more closely with their tribe or faith than with their country, political parties can become the means of promoting narrow communal interests, or, worse, of fomenting ethnic grievances. In Uganda, President Yoweri Museveni, who seized power in 1986, long banned political parties because he considered them divisive, but he allowed them to form in 2003.

Political parties play different roles in different countries, and these roles reflect local political cultures. For example, most states have adopted one of two forms of government that grew out of European political experience, but the forms do not actually work the same way everywhere. In one model, originated by the United States, the chief executive is a president elected independently of the legislature. The chief executive and the legislature can check one another's power. This model was copied for the most part throughout Latin America and later in France's former African colonies. In the alternative model, called the parliamentary form, the chief executive (usually called prime minister) is elected from among the members of the legislature. This form was set up in and retained by Britain's former colonies, including Canada.

Both forms of government assume the existence of competing political parties. Multiparty democracy may be appropriate in some cultures, but in others a one-party government does not necessarily mean totalitarianism. Traditional African political techniques, for instance, are based on building consensus through dialogue. This technique is called *existential democracy*, as distinct from the *adversarial democracy* that Westerners practice. Existential government conforms with traditional African legal systems and religion, which emphasize the maintenance of social harmony rather than retribution. Asian politics are also notably existential, in keeping with traditional Confucian values. The Japanese constitution, for instance, was imposed during the U.S. occupation of Japan, yet the Japanese govern themselves under that constitution according to traditional Japanese values. Consensus on most issues is reached before any vote is taken, so formal votes are seldom contests or measures of power. Japan has been ruled by the same party since 1955. The Philippines is moving to discard the presidential system that it inherited from the United States and adopt the parliamentry form by 2010.

Labor unions Labor unions may serve as still another building block of nationalism. "Labor" or "workers'" political parties can even govern. In totalitarian states the government might sponsor official labor unions and curtail the formation of independent unions. In Poland, under Communist Party rule, the independent Solidarity labor movement nevertheless overwhelmed the Communist Party candidates for parliament and deprived the Communists of any pretense of legitimacy. Independent unions also played a key role in bringing down the Communist government in the former Soviet Union.

Each of these six institutions—religious institutions, armed forces, schools, media, political parties, and labor unions—may be either a centrifugal force or a centripetal force in any country. Geographers study their presence or absence, the mix or balance among them that is unique to each country, and their interaction with other aspects of each country's culture. Furthermore, each institution has a geography within each country: One or more institutions may be equally influential across the entire territory, be concentrated in only one region, or spill over international borders. In many countries, for example, including the United States and Canada, the political parties show regional strengths. In several African countries the army draws disproportionately upon one ethnic group or region for troops.

Nationalism can be threatened either by regionalism or by internationalism. Whereas regionalism tends to split a country into many parts, internationalism tends to dissolve a country's external borders. Today many global social and economic developments exert centrifugal force against the centripetal forces of nationalism. Chapter 12 and 13 will examine those global developments.

Democracies and False Democracies

"Democracy is not like instant coffee, where you can just add water and stir," insisted France's foreign minister, Hubert Védrine, as he declined to sign a U.S.-sponsored international declaration committing the signers to the principle. In 2000, 107 other nations—democratic or not—did sign. Virtually all governments today govern in the name of the people, but some are more genuinely representative of the people than others. The world has many imperfect or just plain fake democracies. For instance, some countries have "elected" presidents, but media (often government-owned) cover only the incumbent; opposition candidates' supporters (or even the candidates) are beaten, arrested on dubious charges, or even killed; and constitutions are routinely rewritten to prolong the terms and powers of presidents. In many countries authoritarianism is tempered by populism. Popularly elected presidents of both Peru and Venezuela, for example, actually seemed to rise in popularity when they unconstitutionally closed their countries' parliaments and fired their Supreme Courts.

People living in genuinely democratic countries assume that most governments represent their people and work for their welfare. A great many governments, however, do not represent the people. Government by an elite privileged clique is called an **oligarchy.** The governments of many countries have even been called **kleptocracies**—government by theft. The national wealth may be funneled off and even taken out of the country by a privileged few, while most of the people remain in poverty. In these cases foreign aid, or any system of providing assistance to the poor through their governments, cannot guarantee that the needy will receive the assistance. For example, General Sani Abacha ruled Nigeria from 1993 until 1998, and in 2000 the Swiss government found $670 million in one Swiss bank account in his name, $600 million in another account in Luxembourg, and similar sums in other accounts in France and Germany. In 2002 American oil companies revealed that Kazakhstan's President Nursultan Nazarbayev regularly accepted large bribes in connection with dispensing his country's oil concessions. In 2003 the Supreme Court of the Philippines awarded to the government $700 million in Swiss bank accounts of the late president Ferdinand Marcos, whose heirs could not justify ownership of the money. The presidents or rulers of many of the poorest countries in Latin America, Africa, and Asia are among the world's richest individuals.

The rich democratic countries might impose fiscal responsibility in the poor countries, but some observers argue that monitoring corruption in the poor countries is a form of neo-colonialism. British Prime Minister Anthony Blair's Extractive Industries Transparency Initiative encourages oil and mining companies to disclose royalty payments to governments, thus, it is hoped, helping ensure that the money shows up in the countries' budgets. Governments, however, have refused to cooperate.

Berlin-based Transparency International nevertheless publishes an annual "Corruption Perceptions Index" that reviews both the poor and rich countries. In the 2006 report, Haiti, Myanmar, and Iraq and Guinea (tied) were rated most corrupt; Iceland, Finland, and New Zealand least corrupt. Canada ranked 14th least corrupt; the United States 23rd least corrupt.

Authoritarian rulers often insist that their people "are not ready for democracy." Eritrean president Isaias Afewerki said in 2003, "We have not yet institutionalized social discipline, so the possibility of chaos is still here. . . . No one in Africa has succeeded in copying a Western political system, which took the West hundreds of years to develop. Throughout Africa you have either political or criminal violence. Therefore, we will have to manage the creation of political parties, so that they don't become the means of religious and ethnic divisions. . . ."

Are there, in fact, preconditions for democracy, and are there states where the people are not ready for it? Without those conditions, do authoritarian rulers have the right to rule, presumably for the good of the people, to guide the people toward democracy? Farouk Adam Khan, today the Prosecutor-General of Pakistan, led a military coup that failed in 1973. During the following five years he spent in prison, he read the great works of Western political philosophy, including *The Federalist Papers* and John Stuart Mill's *On Liberty*. Later he said, "Every single ingredient that the authors of those books say is required for a civil society—education, a moral code, a sense of nationhood; you name it, we haven't got it! Just look at our history. . . . We need someone who will not compromise in order to build a state."

The French observer Alexis de Tocqueville analyzed the young United States in his classic *Democracy in America* (1835). He noted that one of American democracy's great strengths was the number of nongovernmental organizations among the people: clubs, youth organizations, political parties, sports groups, business executives' leagues, church groups, and so forth. In totalitarian states, by contrast, rulers feel threatened by any organization that brings people together under any auspices but the state's own.

China is an example of a state that severely represses independent citizen groups. Falun Gong is a Buddhist faith-healing sect that has spread across China. Its estimated 70 million members outnumber the 60 million members of the Chinese Communist Party. One morning in April 1999, 10,000 Falun Gong members suddenly appeared in Beijing's Tiananmen Square for a silent demonstration. They sat quietly for a couple of hours, and then the crowd melted away as mysteriously as it had appeared. To have brought together so many people required a tremendous amount of planning and coordination, all of it completely unknown to the government. This demonstration so frightened the government that it has cracked down on the organization, tried to infiltrate it, and arrested its identifiable leaders. Western democracies consider

such organizations harmless or even beneficial to a healthy civic life, but totalitarian governments feel threatened by any organization that has its own internal information network and the ability to organize people.

In some Western democracies, however, citizen participation rates have fallen so low that observers wonder whether elections are valid. Italian courts invalidated seven national referenda in 2000, because only 32 percent of potential voters had turned out. The Serbian Supreme Court invalidated three successive presidential elections in 2002 because none of them achieved the constitutionally required 50 percent voter turnout. In the United States an average of just 44.1 percent of the voting-age population turned out for national legislative elections through the 1990s, putting the United States in 139th place among the 163 countries in which turnouts were tabulated. In the elections of November 2006, U.S. voter turnout fell to 40 percent. British Prime Minister Tony Blair has argued that "failure to vote is the mark of a satisfied citizen," but many observers insist that this remark interprets the death of democracy as a vote for the status quo.

The early twenty-first century has demonstrated that there are areas in which most people simply do not want democracy. Many Arab rulers in the Near East, for example, are autocratic, corrupt, and heavy-handed, but they are nevertheless more liberal, tolerant, and pluralistic than any form of democracy that would likely replace them. Elections in many Mideast countries would produce undemocratic totalitarian or fundamentalist results. Thus, paradoxically, a more democratic country can be a more repressive one. A majority of Iraqis, for example, probably favor restrictions of the rights of women and minorities. Achievement of a democratic state is more than just holding elections. Unless majority rule is accompanied by legal protections, tolerance, and respect for minorities, the result can be populist oppression.

Furthermore, in any democracy, the demographic description of those who actually vote may differ from a description of the people at-large. In the United States, for example, the higher one's age and income, the more likely one votes. Popular public opinion polling, by contrast, usually includes all ages and income levels. Therefore governmental activities that respond to the voters' preferences do not always reflect the population's wishes as reckoned in opinion polls.

The detailed analysis of governments is the field of political science, but a geographer comparing the world's countries must be careful never to be deceived by appearances and must instead look deeper for processes at work.

HOW STATES DEMARCATE AND ORGANIZE TERRITORY

Each state maintains its borders and subdivides its territory for political representation or administration. This section will examine the methods used and some of the results achieved.

The Shapes of States

The physical shape of a state may affect its ability to consolidate its territory and control circulation of people and goods across its borders (Figure 11-16). A circle would be the most efficient shape on an isotropic plain because a circular state would have the shortest possible border in relation to its territory, and that shape would allow all places to be reached from the center with the least travel. States with shapes the closest to this model are sometimes called *compact states*. Bulgaria, Poland, and Zimbabwe are examples. *Prorupted states* are nearly compact, but they have at least one narrow extension of territory. Namibia and Thailand are examples. If these extensions reach out to navigable waterways, the extensions are called *corridors* (corridors' special importance to landlocked states is discussed in Chapter 13). *Elongated states* are long and thin, such as Chile or Norway, and *fragmented states* consist of several isolated bits of territory. *Archipelago states*, made up of strings of islands, such as Japan or the Philippines, are fragmented states, but Azerbaijan is also fragmented, having a territory, Nakhichevan, separated from the rest of the country and enclosed by territory of Armenia. Nakhichevan is called an *exclave*. Still other states, called *perforated states*, are interrupted by the territory of another state enclosed entirely within them. South Africa, for example, is perforated by Lesotho, and Italy is perforated by the Vatican and by San Marino.

The shape of a state's territory may influence the government's ability to organize that territory, but this is not always true. A topographic barrier such as a mountain chain may effectively divide even a compact state. Bolivia and Switzerland, for example, are compact in shape, but mountain chains disrupt their interiors. For some of their regions, trade across international borders is easier than trade with other regions of their own country. The people throughout an archipelago state,

by contrast, may be successfully linked by shipping. Before drawing any conclusions about political control from the shape of a state alone, one must consider the distribution of topographic features, of the state's population and resources, and whether any centrifugal forces such as economic or cultural ties straddle the state's borders.

International Borders

Many people think of rivers and mountain ranges as "natural" borders, perhaps because on a map these features often resemble lines. This simple assumption, however, is false, as shown by the numerous wars through history triggered by attempts to enforce it. Rivers bind the peoples on their two banks as much as they divide them, and mountain borderlines can be drawn either from one peak to the next or else up and down the mountain valleys that reach between the peaks. Mountain valleys can actually provide foci for political organization. The early Swiss united the valleys on either side of mountain passes. The passes did not divide peoples, but gave them a common interest.

State borders within the United States illustrate the various types that can be drawn. Some are rivers (as the Columbia River between Washington and Oregon, and the Mississippi River), and some follow mountains (Idaho–Montana, Virginia–West Virginia, North Carolina–South Carolina–Tennessee). Most, however, are geometric lines imposed across the landscape.

Some international borders are defended with minefields and watchtowers. The United States has built sections of high wall along the border with Mexico to keep undocumented migrants from sneaking across, and Chapter 7 discussed Israel's new wall barrier.

Some borders are matters of dispute. Guatemala recognized the independence of Belize in 1991 but claimed most of its territory until a treaty was signed in 2002. In 1992 El Salvador and Honduras accepted a World Court decision about their border, over which they had argued since independence in 1861. India and China have fought over their border several times, and they still have not agreed (see the map in the box on page 302). Other international borders, by contrast, are relatively open and free. The location of an international border can be clear to anyone flying overhead, however, even if there are no border structures. Borders may be evident by contrasting land uses on the two sides, by different types of landholdings, or by discontinuity in the transportation networks (Figures 11-17, 11-18, and 11-19).

An international border across a sparsely populated area may not be marked or supervised at all, and its exact location may become a matter of dispute only if valuable resources are discovered in the region. This has happened along the inland borders of South American countries. In 1941 Peru attacked Ecuador

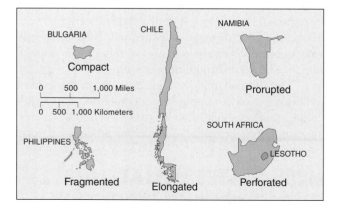

FIGURE 11-16 Shapes of states. Examples are shown of states that are compact, prorupted, elongated, fragmented, and perforated. The five states are drawn to the same scale. In general, compactness is an asset, because it fosters good communication and integration among all regions of the country.

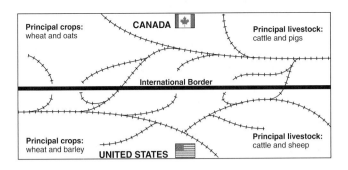

FIGURE 11-17 The U.S.-Canadian border. This schematic drawing illustrates two effects of different policies on the two sides of the U.S.–Canadian border. One feature is that the railroad network frequently approaches the border but seldom crosses it. Countries often try to restrict circulation and trade across national borders. The second feature illustrated here is that different crops and livestock predominate on the different sides of the border. This difference does not result from different environmental or market conditions but from the two countries' agricultural support policies.

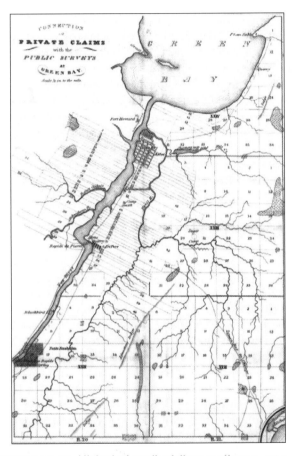

FIGURE 11-18 Historic landholding patterns. This early map of the region of Green Bay, Wisconsin, demonstrates principles of both political geography and historical geography. By juxtaposing two patterns of landholding, it reveals the history of the region through two different legal systems. The first Europeans who settled here were French. They depended on the river for transportation, so they defined the landholdings that you can see extending inland from the riverbank. This system of landholding is called the *long lot system*. Later this territory came to be part of the United States, so it was surveyed, and landholdings were assigned according to the system defined in the U.S. Ordinance of 1785. This system, called the *township and range system*, is a purely geometric pattern imposed regardless of natural features. U.S. surveyors recognized the preexisting long lot claims, but they subdivided all land not already claimed into a pattern of regular squares. Such regular squares of property holdings are one of the most distinctive features of the U.S. landscape and give a "checkerboard" look to U.S. agricultural regions. (Map by S. Morrison, E. Swelle, and J. Hathaway. Courtesy of the State Historical Society of Wisconsin. Neg. #WHi[X3]39431)

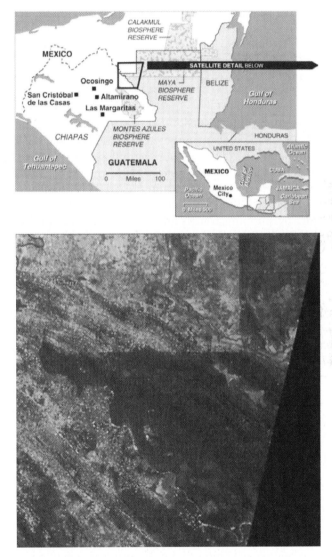

FIGURE 11-19 The border between Mexico and Guatemala. The photograph reveals the devastation of the rain forest on the Mexican side of the border. This devastation has been caused by extensive commercial logging and by chopping down the forest and then trying to farm the poor soils too intensively. Food production has not kept up with population growth. (Map courtesy of NYT Graphics/New York Times Pictures; photograph by John C. Stennis Space Center/NASA Headquarters)

and annexed 55 percent of Ecuador's territory. The discovery of oil in the former Ecuadorian territory of what is today northern Peru triggered renewed fighting in 1995, but the states agreed on a border demarcation in 1998. Venezuela claims about two-thirds of Guyana, and Suriname claims another 10 percent. Both Suriname and Guyana claim the mouth of the Corentyne River, where oil explorations look promising. The jungle border between Brazil and Venezuela is also disputed. Similar disputes have occurred in Middle Eastern deserts. Saudi Arabia and Kuwait share jurisdiction and oil revenues from a 14,763-square-kilometer (5,700-square-mile) "neutral zone" between them. Russia and Japan have not signed a treaty to end World War II, because they dispute possession of four islands. The Russians claim that the four islands in question are part of the Kurile chain occupied by Russian troops during the war and surrendered by Japan; the Japanese claim that they are not. Russia and China finally agreed on the demarcation of their 4,350 kilometer. (2,700 mile) border only in 2005. Canada and Denmark dispute possession of tiny Hans Island (the size of a football field) in the Arctic. The island is uninhabited, but global warming is allowing increased nearby shipping traffic, and local fish stocks are also important. Ownership of the island was not settled when borders between Greenland and Canada were drawn in 1973.

Borders may attract complementary economic activities on the two sides. The zone along the U.S.–Mexican border has developed complementary manufacturing, as we shall see in Chapter 13.

States often want to control the passage of information as well as of people and of goods over their borders. Today, however, direct people-to-people links via telephone lines or satellites have proliferated and destabilized the distinction between what is inside and what is outside a country. Telephone connections have been used to break into computers in foreign lands and to read, steal, or tamper with privileged information. This action challenges international law: In which country was the crime committed? In 2005 a computer virus called "Mytob" caused extensive damage in computers around the world; the virus had been created by a 23-year-old Turk and his 19-year-old Moroccan friend, but neither of the men's home countries would prosecute the action as criminal. Newspaper and magazine editors and publishers are being held liable to legal action in any country where the Web version of their publication can be accessed. Publishers may obey the laws of their home country, but if "publication" of defamation or of a libel is considered to take place where the story is read, rather than where it was written, then the laws of the most censorious and autocratic governments can be applied to material originating in countries with free speech. Furthermore, as geolocation technology improves, territory can be quarantined and censorship applied to the Internet.

Governments can regulate only from within their own borders, and they cannot regulate broadcasts into a country if those broadcasts originate outside its borders. Electronic broadcasts into a country can be blocked, but efforts to do so are expensive and often unsuccessful. Many countries try to ban or regulate the ownership of satellite dish antennas, but these regulations are often ineffective. Globally, radio is still more important than is television, and U.S. broadcasts of Radio Free Europe/Radio Liberty, Voice of America, and Radio Free Asia blanket the world.

Some countries welcome international transmissions. Since Pakistan began to allow transmission of U.S. Cable News Network (CNN) into Pakistani homes, Pakistanis have been astonished to see the range of debate allowed in the United States on every issue. Programs always contain messages that are not obvious to the programmers. People in some countries may watch a U.S. crime show, for example, and notice that the captured criminals were read their rights. Such direct international dissemination of U.S. political style might represent a triumph for democracy and sow the seeds for democratic awakenings.

U.S. Border Security and Internal Security

Congress responded to concern over American security after 9/11 by creating a new cabinet-level Department of Homeland Security. It exercises control over 46 federal agencies, including the Customs Bureau and the Coast Guard. The new Department's responsibilities include both border security and internal security.

It is difficult to monitor borders as open and free as those of the United States. On an average day some 1.3 million people, 348,000 private vehicles, 38,000 trucks and railcars, 2,600 aircraft, and nearly 16,000 freight containers arrive at U.S. borders. Many Americans are demanding improved inspection and tracking of this massive influx of people and goods (Figure 11-20). The United States has imposed new restrictions on the granting of visas to foreign visitors, tourists, and students, and on international shipping: Every one of the world shipping industry's 1.2 million merchant mariners will be required to carry a new biometric identity card. American ports may install scanners that can scan every container for radiation and also gamma-ray screeners, which check for odd-sized objects. The systems can help identify suspicious cargo while keeping detailed records of what passes through the port. The United States has stationed more customs inspectors and migration officers in other countries to inspect ships and planes bound for the United States. In 2006, a corporation owned by the government of Dubai attempted to purchase a corporation that operates several U.S. port facilities.

FIGURE 11-20 "Invaders" of the United States. On February 7, 2003, when United States was officially on "high risk" for terrorist attacks, this team of armed security agents of a hostile foreign government landed on American soil. Their 30-foot Cuban government patrol boat, flying the Cuban flag, docked at the marina of the Hyatt Hotel in Key West, Florida, a short distance from the U.S. Coast Guard Station. Carrying assault rifles, the men marched into the middle of the resort looking for someone to whom they could present a request for asylum. Their story illustrates the difficulty of policing the United States's almost 161,000 kilometers (100,000 miles) of coastline. (Courtesy © Nuri Vallbona/*The Miami Herald*)

Many U.S. ports were already operated by foreign-owned companies, and the government had approved this sale, but public clamor halted it. Perhaps the possibility of Arab ownership stampeded public concern.

Within the country, the 2001 U.S.A. Patriot Act (Uniting and Strengthening America by Providing Appropriate Tools Required to Intercept and Obstruct Terrorism Act), which was renewed almost without change in 2006, and other legislation allow law-enforcement agents a wide range of activities. They can investigate, detain, and deport people suspected of links to terrorist activity or other crimes, as well as hold suspicious individuals, obtain wiretaps and search warrants, monitor computers and bank transactions, and tap telephone conversations. Authority may be granted to use cameras equipped with facial recognition technology to scan crowds for suspects; to monitor chat rooms, cell phones, and e-mail; and even to establish a national identity card.

Cases have been documented in which over-zealous government agents overstepped the bounds that America has traditionally put on police investigative powers. In 2005, for example, agents of the Department of Homeland Security called at the home of a student at the University of Massachusetts who had borrowed to

use for research on a term paper a book written by China's one-time Communist dictator Mao Zedong. In 2006, the American Library Association awarded its Intellectual Freedom Prize to a Connecticut librarian who chose to remain anonymous; he or she had refused to comply with a "national security letter" from the Federal Bureau of Investigation demanding patron records. The government has admitted to inappropriate (and even arguably illegal) wiretapping of Americans' telephones. Some new laws tend to shift the focus of investigations from responding to crimes to trying to prevent them, that is, to preventive measures such as have never before been practiced in America.

John Ashcroft, U.S. Attorney General from 2001–2005, tried to silence criticism of government actions by warning: "To those who scare peace-loving people with phantoms of lost liberty, my message is this: Your tactics only aid terrorists, for they erode our national unity and diminish our resolve. They give ammunition to America's enemies, and pause to America's friends." Democracy, however, requires a strong skepticism of letting the government do whatever it thinks is best.

Territorial Subdivision and Systems of Representation

All countries subdivide their territory, and each country has a unique balance of powers between its local governments and its national government. We generally call those countries in which the balance of power lies at the center **unitary governments,** and those in which the balance of power lies with the subunits **federal governments.** Today most national governments are unitary. Few newly independent and poor countries have adopted federalism, perhaps because they fear that if centrifugal forces were officially recognized, they could pull the country apart. Ethiopia, however, adopted a new federal constitution based on regional self-determination for 11 ethnic groups in 1994.

The definitions of unitary states and of federal states represent models, but in fact countries with federal constitutions may be highly centralized through government by a one-party political system or a military regime. India's federal constitution allows the prime minister to impose direct rule from New Delhi when law and order have broken down in a state. Even the U.S. Constitution obligates the federal government to "guarantee to every State . . . a Republican form of government . . ." (Article IV, Section 4). All governmental systems actually leave much open to improvisation and continual redefinition.

Unitary government In unitary states the central government theoretically has the power to redraw the boundaries of the subunits. This offers flexibility to accommodate geographic shifts in national population distribution or economic growth. It may be useful to redraw internal boundaries from time to time in response to changing needs, for example, by extending the boundaries of cities as they expand into the surrounding countryside. Interests become vested in any given pattern, however, so unitary governments seldom redraw internal boundaries. For example, the 50 U.S. states are each internally unitary, but they seldom redraw the boundaries of their counties. This is largely because political parties are organized at the county level, and they resist change. Also, changes may threaten the existing balance of power among groups, so potential losers struggle to maintain existing patterns.

China is constitutionally "a unitary multinational state." Five principal minorities and about 50 smaller national groups exercise some degree of local autonomy, and Hong Kong has been guaranteed special political and economic privileges for 50 years after 1997 (Figure 11-21). The Turkish-speaking Muslim Uighur people of Xinjiang have rebelled periodically; they may be envious of the political autonomy enjoyed by their ethnic relations in Uzbekistan, Kyrgyzstan, Turkmenistan, and Kazakhstan. China has encouraged neighboring countries to repress Uighur independence movements in their own countries. Meanwhile, settlement of ethnic Chinese in Xinjiang has reduced the Uighurs to a minority there.

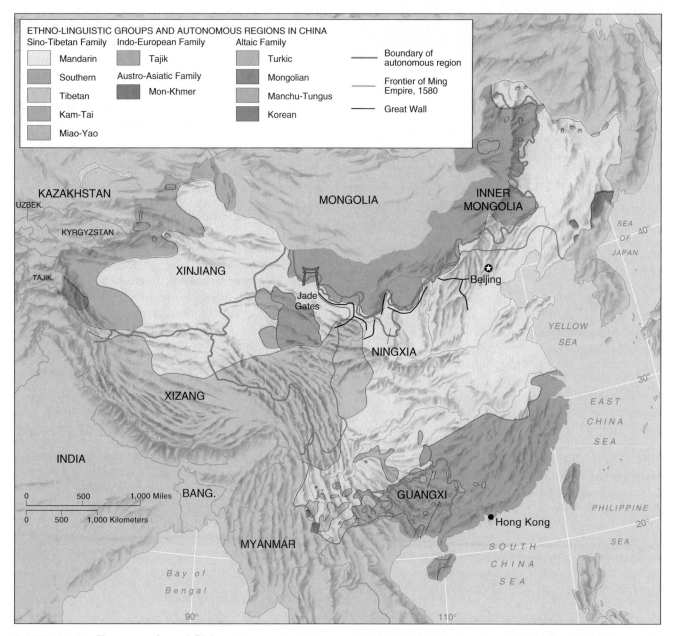

FIGURE 11-21 The peoples of China. The traditional limits of Chinese territory were the borders of the Ming Empire—the Jade Gate on the route to Central Asia and the Great Wall against the Mongols. Through the centuries, however, Chinese rule has intermittently expanded beyond these borders to include non-Chinese peoples. China's contemporary constitution guarantees some political and cultural autonomy to minority peoples in the five peripheral regions that are labeled plus Hong Kong. These regions constitute 42 percent of China's territory but hold only about 8 percent of the Republic's population.

Similarly, the Mongols of Inner Mongolia (a part of China) may envy the Mongols in Mongolia (an independent country) and wish to secede from China, but here too Chinese settlement has reduced the Mongols to a minority in "their own" territory. Nevertheless, political instability may cause the Chinese multinational state to break apart.

Federal government
A federal government assumes that diverse regions ought to retain some local autonomy and speak with separate voices in the central government. Federalism also allows each area to serve as a laboratory for legislation that can be adopted elsewhere if it works. Wisconsin, for instance, might devise a successful school program that other states might copy, New Mexico might create a highway program, or Maine might develop environmental legislation. The protection of diversity, however, may perpetuate economic or social inequality. State education systems usually inculcate state history and geography, just as education systems do in sovereign nation-states. Each state boasts an extensive iconography: state flags, birds, flowers, and so forth. Texas (where schoolchildren pledge allegiance to the state flag) and Hawaii have histories as independent countries. If Puerto Rico were to become a state, it would be the most distinct state culture region.

The U.S. federal government has increasingly preempted state action in recent decades. Examples include federal laws governing consumers, food labeling, class-action lawsuits, and the "No Child Left Behind" education law. The Supreme Court has held that federal anti-marijuana legislation overrides the referenda in 12 states that approved marijuana's medical use. Such federal preemption has stimulated individualized responses from the states. Massachusetts offers near-universal health care coverage and allows same-sex marriage. California imposes limits on the emissions of greenhouse gases. California, New Jersey, Maryland, and Connecticut have allocated money for stem cell research. On the other end of the political spectrum, Idaho, Georgia, and 17 other states have constitutional amendments banning same-sex marriage, and Florida, Mississippi, and Utah even ban adoption by same-sex couples. Historically, issues that enthrall the states eventually get sorted out nationally in federal legislation.

Many observers have suggested redrawing the map of the states. A new pattern could improve the states' conformity with physiographic or economic regions or make each state more compact or equal in population or in economic strength (Figure 11-22). The U.S. Constitution, however, protects the states against restructuring against their will (Article IV, Section 3), although Virginia split into two states during the Civil War in 1863. Strong political currents within California have proposed breaking it into two or three states.

At the level of the federal government, the United States may be merging into one political arena. Several

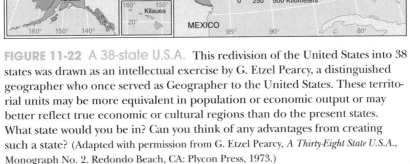

FIGURE 11-22 A 38-state U.S.A. This redivision of the United States into 38 states was drawn as an intellectual exercise by G. Etzel Pearcy, a distinguished geographer who once served as Geographer to the United States. These territorial units may be more equivalent in population or economic output or may better reflect true economic or cultural regions than do the present states. What state would you be in? Can you think of any advantages from creating such a state? (Adapted with permission from G. Etzel Pearcy, *A Thirty-Eight State U.S.A.*, Monograph No. 2. Redondo Beach, CA: Plycon Press, 1973.)

facts provide evidence. First, voter turnout is much higher for presidential elections than it is for House or Senate contests or even for elections for state and local officials. This suggests a weak sense of identification with local representatives. Second, in 1994 Republican candidates for the House of Representatives offered one platform, a "contract" to the American people. This contract effectively nationalized elections to the House of Representatives. Both parties have since offered such contracts. Third, members of Congress are raising increasing percentages of their campaign funds outside their own constituencies. More than half of the U.S. senators, for example, raise more than 50 percent of their funds outside their home states—several raise more than 90 percent. This inspires the question of whom, or what constituency, these senators represent. Widespread fundraising arguably corrodes the federal character of U.S. government.

Some countries' federal patterns reflect polyethnicity (for example, Russia, Myanmar, and India). Some countries' internal units are older than the federal framework, and they participate in the federal system only on the condition that they may retain certain powers (for example, the original 13 United States, Germany, and Canada). Federalism may also serve countries that are relaxing central control. Spain's culturally distinctive peripheral provinces, for example, have never been welded into one nation, and in recent years they have demanded and received considerable autonomy. The Basques have long struggled for autonomy, and in 2006 a full 77 percent of Catalonians voted for semi-independence for their province. A new Italian Constitution adopted in 2001 devolves considerable power from the central government to the provinces. Belgium is decentralizing into three parts: Flanders, Wallonia, and the city of Brussels. The United Kingdom established new parliaments in Scotland and in Wales in 1999, and France has granted some autonomy to the local government of Corsica. In these European examples, some powers of the national governments are devolving down to the regions at the same time as others are being surrendered up to the multinational European Union (discussed in Chapter 13), so the national governments retain few powers. In states torn between two religions, such as Nigeria, the adoption of Sharia law in some of the states can aggravate antagonisms, and if loosening the federal ties proves insufficient, a federal state may dissolve in civil war, as happened in the former Yugoslavia.

Canada's federal system

Canada was formed as much to preserve British institutions in North America as in response to any indigenous nationalism. The British North America Act of 1867 created the Confederation of Canada, and other provinces joined the confederation later (Figure 11-23). The 1931 Statute of Westminster removed all legal limitations on Canadian legislative autonomy, and in 1982 the Canadian Constitution was "patriated," meaning that Canadians obtained the right to amend the Constitution in Canada.

In Canada's parliamentary federal government, executive authority is vested in the sovereign (Queen Elizabeth), who appoints a governor general. That office, however, exercises little authority or influence. Legislative authority resides in the Canadian Senate and House of Commons. The 105-member Senate is appointive, and it too exercises little power. It cannot, for example, initiate legislation, although it can delay legislation. Therefore, power is concentrated in the House of Commons, whose 308 members represent districts based on population.

This concentration of power has allowed the two most populous provinces, Ontario and Quebec (about 62 percent of the total population), to dominate the national agenda. Much of Canadian politics has traditionally been concerned with reassuring the French-speaking (Francophonic) population of Quebec that their culture would be secure in a country with an English-speaking (Anglophonic) majority (see Chapter 7). In 2006 Canada's House of Commons even voted to recognize Quebec as "a nation within a united Canada." Many residents of Western Canada have felt disadvantaged, but newly formed political parties or those strongest in the Canadian West have recently risen to power. The voices of new immigrant groups are also making themselves heard.

Canada's provinces have no significant institutionalized recognition at the level of the national government, but many observers are wondering how national politics will evolve when Ontario has more than one-half of the total population, as is expected by mid-century. Canadian individual provinces do, however, retain more powers than do the individual states of the United States. For example, the Canadian Constitution lacks a "commerce clause" such as the one in the U.S. Constitution that gives Congress the power "to regulate commerce with foreign nations and among the several states" (Article One, Section 8). Trade barriers exist among the Canadian provinces. The provinces signed an Agreement on Internal Trade in 1995, but they have failed to meet agreed deadlines to implement it. Alberta and British Columbia nevertheless forged ahead with a Trade, Investment and Labour Mobility Agreement in 2006.

The people of Quebec have often debated seceding from Canada. The 2000 Clarity Act provides that a province can secede only after a referendum in which the option of independence is clearly stated and secessionists win a clear majority. This has been suggested as a model for other countries that may dissolve.

Canada's Native Americans (also called First Nations) are also gaining political significance, in two ways. First, courts are increasingly ruling in favor of their claims to territories and resources. Second, by arguing that they should get any special privileges that Quebec does, the First Nations have become power brokers in the Anglophone/Francophone dialogue.

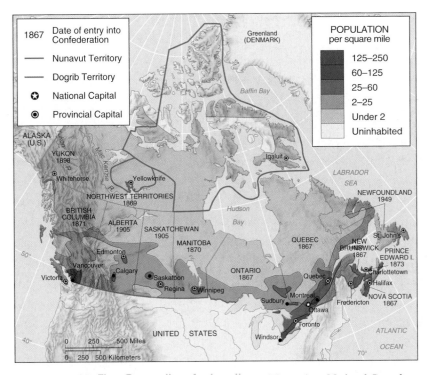

FIGURE 11-23 The Canadian federation. The units of federal Canada slowly came together into the country's present configuration. The 1867 British North America Act united the provinces of Upper Canada (Ontario), Nova Scotia, New Brunswick, and Lower Canada (Quebec). In 1869 the Northwest Territories were purchased from the Hudson's Bay Company; Manitoba was carved from this territory and admitted into the confederation in 1870. British Columbia joined in 1871, and Prince Edward Island joined in 1873. Alberta and Saskatchewan were formed out of previously provisional districts and admitted in 1905. Newfoundland, previously an independent country, joined as late as 1949, and took the name Newfoundland and Labrador in 2001.

The territory of Nunavut formally came into existence April 1, 1999. It covers one-fifth of Canada's land area but contains a population of only 27,000, of whom 85 percent are Inuit. The city that had been Frobisher Bay changed its name to Iqaluit as the new capital. In 2003 the Canadian government granted the Dogrib people local self-government in a territory of 39,000 square kilometers, and three additional self-government treaties are under negotiation.

Special-purpose territorial subdivisions Several countries divide their territory into economic-planning regions, which can gather almost as much power as the legal constituent subunits. This has happened in France. In the United States several federal agencies, including the Federal Power Commission, the National Labor Relations Board, the Bureau of Reclamation, the Federal Trade Commission, the Federal Communications Commission, and the Federal Reserve System, have each subdivided the country into a different pattern of districts for its own uses. Few of these patterns conform to the pattern of the states, so when two districts of a federal agency enforce different policies in two parts of one state, the state government's own powers can be confused or undermined. When Vladimir Putin became president of Russia in 2000, one of his first official acts was to divide Russia into seven new federal zones, each with an administrator appointed by himself, and

to transfer power to these zonal administrators. Thus, Russia's 89 elected regional governors lost much of their power overnight.

The United States is also subdivided by 91 federal district courts and, above them, 11 federal courts of appeal (Figure 11-24). Legal efforts to protect local differences may carry unexpected consequences. Two of today's most controversial topics, abortion and marriage, illustrate this. In 1973 the Supreme Court upheld the legality of abortions, but it allowed the states considerable leeway in defining the conditions under which each state would allow abortions (*Roe* v. *Wade;* upheld in *Planned Parenthood* v. *Casey*, 1992). Wealthy women are able to travel across state lines to get abortions, but poor women have fewer choices. Dissenters cross state lines to protest abortions in states other than their own. The question remains on the national agenda, and legislation may yet be standardized throughout the country.

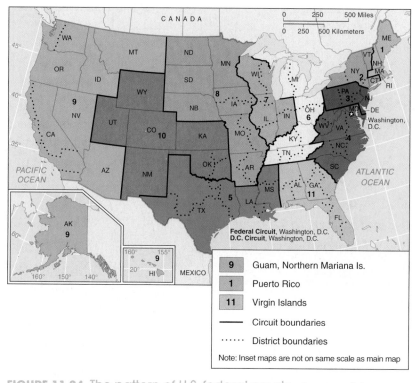

FIGURE 11-24 The pattern of U.S. federal courts. Any two of these courts may interpret federal law differently. Therefore, federal law is not uniform throughout the country until the disagreement is resolved by the Supreme Court.

Several states are now wrestling with conflicts in family law. Massachusetts allows same-sex marriages, and most states are redefining civil unions and partnerships in ways that approach or even approximate "marriage." Conflicts among state laws present formidable challenges to American federalism. Marriage is portable. Couples married in one state move to another, where they may get divorced and fight over child custody. These acts require one state to recognize the validity of other states' laws.

Federal territories Most federal states contain territories that are not included in any of the subunits but that are administered directly from the federal center with, perhaps, some form of local government. These territories may constitute a significant share of a state's total area, including capital districts, colonies, strategic frontier areas, and federal territories.

Territories usually have only limited local government until some presumed future time when they will be ready for full statehood. If the territories lack resources, they may remain territories indefinitely. Most of the land area of the United States was once federal territory, but the territories were organized, populated, and finally admitted into the Union as states. By contrast, today some 40 percent of Canada remains in territorial status. Mexico includes two territories that may become states upon reaching popula-

tions of 120,000 and demonstrating "the resources necessary to provide for their political existence."

The U.S. government still owns about one-third of the total national land area (Figure 11-25). The 1976 Federal Land Policy and Management Act requires the federal government to receive full value for any lands traded away, and Congress still wrestles with the question of how the federal government should exploit or preserve its lands.

Capital cities In some countries the federal capital is also the capital of a component state (Bern is capital of both its own canton and of Switzerland). Austria's federal capital, Vienna, is a federal state. Some federal capitals are governed directly by the federal government (Australia's capital territory, Mexico's Distrito Federal). Residents of Washington, D.C., complain that they have no voice in the federal government; if the city were a separate state, it would rank 44th in population.

Several countries have moved their capitals because they believed that their old capitals were international cities grafted onto national life rather than truly representative of national life. A new capital, it is believed, especially if built inland, will symbolize a rebirth and rededication of a national spirit. In 1918 the Russians moved their capital from St. Petersburg inland to Moscow; in 1923 the Turks moved theirs from

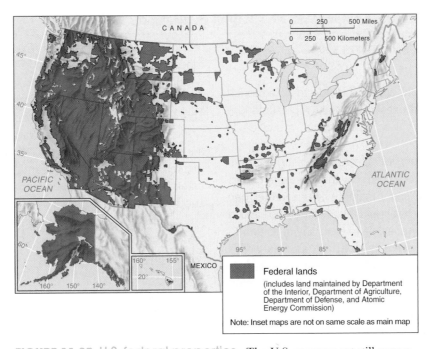

FIGURE 11-25 U.S. federal properties. The U.S. government still owns a substantial share of the total national territory. The largest land trade in U.S. history occurred in 1998. The state of Utah surrendered to the federal government 152,580 hectares (376,739 acres—an area almost half the size of Rhode Island) that had originally been given to the state to finance education; the land was scattered through national forests, parks, and monuments. In return the federal government gave Utah $50 million in cash plus 58,725 hectares (145,000 acres) outside of national parks and areas proposed for wilderness.

Istanbul to Ankara; Brazilians moved theirs inland to Brasília in 1960; Tanzanians moved theirs inland to Dodoma in 1975; Nigerians moved theirs to Abuja in 1991. Newly independent Kazakhstan moved its capital from Almaty (which was thought to be too Russian) to Akmola (today called Astana) in 1997. The Japanese, by contrast, moved their capital from inland Kyoto to coastal Tokyo in 1868 to symbolize Japan's opening to the world. Capitals are frequently designed as showplaces of national pride to impress both citizens and foreigners. In 1999 the offices of Malaysia's prime minister were moved to the newly built city of Putrajaya, next to the country's new scientific- and computer-research center, Cyberjaya. Kuala Lumpur will, however, remain the official capital until 2012. In 2005, the Burmese government announced its intention to move the capital to Pyinmana, about 250 miles north of the current capital, Yangon.

How to Design Representative Districts

The subfield of political geography that studies voting districts and voting patterns is called **electoral geography.** In district systems of political representation, the boundaries of the districts can determine the outcomes of the ensuing elections. This can occur in either one of two ways. First, if the electoral districts are unequal in population, then the ballots cast by some voters outweigh those cast by others. Second, district lines can be drawn in ways that include or exclude specific groups of voters so that one group gains an unfair advantage. This is called **gerrymandering** (Figure 11-26).

The electoral systems in many countries tolerate inequalities in numerical representation. Chapter 8 noted that many countries grant rural populations a disproportionate amount of representation in the government. In the United States this was long true at the level of the state and local general-purpose governments, but in 1962 the Supreme Court ruled that for these governments, the number of inhabitants per legislator in each district must be "substantially equal" (*Baker* v. *Carr*). In 1962 Tennessee was still electing state legislators on the basis of a 1901 apportionment. The population had urbanized, so that by 1962, one rural vote was equal to as many as 19 urban votes. Many other states' legislatures contained similarly disproportionate representation, but nationwide redistricting now follows each decennial census.

After each census, the state legislatures have traditionally also redistricted for federal representation. In

Critical THINKING

Nonterritorial Systems of Representation

Not all governments at all levels divide their territories into districts for purposes of political representation. There are at least two nonterritorial systems of representation.

In *at-large systems of representation*, all members of a governing body are chosen by the entire electorate. In such a system, a slight majority of one group can elect a government that is entirely of its group, and a minority may find it impossible to elect any of its candidates. In a *district system*, by contrast, an area with a minority population can elect at least its local representative. Therefore, courts in the United States have often required local governments to switch from at-large systems to district systems of representation.

Another alternative to territorial districts is government by representatives of interest groups. This is called **corporativism.** The representatives may represent labor unions, agricultural cooperatives, chambers of commerce, private or nationalized industries, pro-

fessionals, and other groups. Corporativism has been tried by Germany, Italy, Ireland, several Latin American countries, France, and several African states, but no countries are governed by corporativist systems today. The U.S. Congress, however, does much of its real work in committees that specialize in various topics: agriculture, labor, and so forth. Members of these committees are lobbied by representatives of groups interested in that topic, and the members receive campaign support from those interested groups, whether the groups are based in the representative's district or not. Some observers have suggested that this system, in fact, imitates corporativism.

Questions

1. Should U.S. representatives and senators be restricted to raising campaign funds within the geographic districts they represent?
2. How would that affect government?
3. Would such a plan hinder free political expression?

FIGURE 11-26 The original gerrymander. The original gerrymander was a district created in Massachusetts in 1812 to concentrate the Federalist Party vote and thus to restrict the number of Federalists elected to the state senate. The district configuration was at first likened to a salamander, but later it picked up the name of Governor Eldridge Gerry, the Anti-Federalist who signed the districting law. ("It looks like a salamander." "No, by golly; it's a Gerrymander!") (Reproduced from James Parton, *Caricature and Other Comic Art.* New York: Harper Brothers, 1877, p. 316.)

2003, however, Republican legislators in Texas redistricted that state shortly after winning the majority in the 2002 elections, thus opening the possibility of redistricting after each election in every local government all across America. The U.S. Supreme Court upheld the legality of the Texas legislature's action (*League of United Latin American Citizens et al. v. Perry, Governor of Texas*, 2006).

The Supreme Court has extended the one-person/one-vote rule to all levels of political representation in the United States except the governing boards of the special-district governments in metropolitan areas.

Representation in the U.S. federal government

The U.S. federal government itself does not achieve the one-person/one-vote ideal. Neither the House of Representatives, nor the Senate, nor the presidency is based on population count. The Constitution specifies that "each State shall have at least one Representative" (Article I, Section 3). Several states have so few residents that without this constitutional protection, they would have to share a representative. There are 435 seats in the House of Representatives. Therefore, assigning one representative to each state leaves 385 seats to be allocated among the states by population. The Constitution's mandates cannot be reconciled to a one-person/one-vote ideal. After the 2000 reapportionments, the 483,000 people of Wyoming had a representative, but each representative from

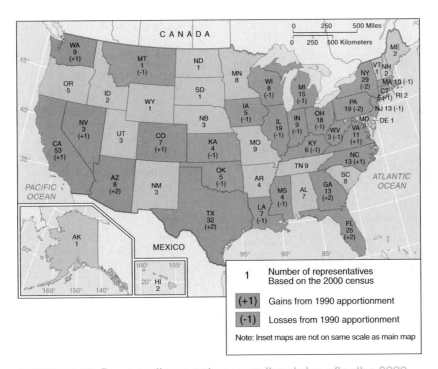

FIGURE 11-27 Reapportionment among the states after the 2000 Census. After the 2000 Census, the seats in the U.S. House of Representatives were reapportioned as shown on this map. Fast-growing states, mainly in the South and West, gained seats, but those with proportionately declining populations, mainly in the North and East, lost representation. Some observers feel that the United States has grown so populous that people are alienated from the Congress. The House of Representatives capped its own number at 435 in 1910, when the U.S. population was 90 million, but if the ratio of members of the House today were the same as it was in 1790, the House would have over 9,000 members.

California represented a district of about 651,000 people, 1.3 times as many (Figure 11-27).

Each state is also represented by two senators, no matter how few residents it has. Therefore, each of the senators from Wyoming represented about 246,000 people, whereas each senator from California represented about 17 million people, 69 times as many.

The president, in turn, is elected indirectly in the Electoral College, in which each state has as many electors as it has total seats in Congress. The candidate who wins a majority of a state's popular vote wins all that state's electoral vote. In 1888 Grover Cleveland won the popular vote over Benjamin Harrison, but Harrison won the electoral vote, and Harrison became president. The presidential election of 2000 repeated the experience of 1888. The Democratic candidate, Al Gore, swept the popular vote by a plurality of over 500,000 votes, but, after the U.S. Supreme Court intervened to stop a recount of votes cast in Florida, the Republican candidate, George W. Bush, won Florida and with it the vote in the Electoral College. In 2004, the Electoral College almost torpedoed the results of the popular contest again. George Bush's margin of victory in the popular vote was 2 percent out of over 122 million votes cast. If fewer than 60,000 people in Ohio had changed

their votes, John Kerry would have won that state's 20 electoral votes, and he would have entered the White House in January 2005 as G. W. Bush did in January 2001—having won the votes of fewer Americans than had the man he defeated.

The United States' national population is becoming increasingly geographically concentrated, so the system of representation is becoming increasingly arithmetically undemocratic. By 2000 California's population was greater than that of the 22 least populous states combined, yet those states had twice California's representation in the Electoral College. It is projected that by 2030, a full quarter of all Americans will reside in just California, Texas, or Florida. Eliminating the Electoral College would require the approval of the low-population states that profit from it, so the possibility is slim. In 2005, when U.S. President Bush told Russian President Vladimir Putin that Russia should have a more democratic system, Putin asked, "Do you mean like the Electoral College?" President Bush abruptly changed the subject.

The fact that the government is not truly numerically democratic causes inequitable distribution of federal spending. As a rule, the more-populous states subsidize the less-populous.

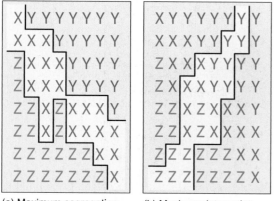

(a) Maximum segregation   (b) Maximum integration

FIGURE 11-28 Models of political districts. If we want to draw three districts of roughly equal populations, and if we have three population bases (*x, y, z*) defined by race, income, or any other characteristic, we could draw the boundaries in many different ways: (a) illustrates maximum segregation of the populations among the districts; (b) demonstrates maximum integration of the populations within each district.

How gerrymandering is done

The following discussion of the techniques of gerrymandering will draw mostly on U.S. examples because they will be familiar to most readers. These techniques, however, apply wherever government is established with representation by territorial districts. Figure 11-28 illustrates the problem of drawing district boundaries in an area of heterogeneous population. It is easier to define the problems of districting than it is to define equitable solutions. Substantial equality of population count in each district is legally required, and contiguity and

compactness of districts is desirable because it eases communication within them. The question still arises, however, of whether homogeneity or heterogeneity of population within a district is preferable. Some people feel that districts should have a common social or economic characteristic, whereas others feel that balanced or integrated districts are preferable. Gerrymandering to achieve one kind of homogeneity (for example, racial homogeneity) may link people who in fact share no other attributes, problems, or political community.

GISs make it easier to gerrymander, because any legislator can call up any district on a screen, shift a boundary, and get an instantaneous readout of what the total population, voting behavior, racial composition, and other characteristics would be of the newly drawn district (Figure 11-29). Some observers fear that easier gerrymandering threatens democracy, as incumbents gerrymander to secure their own reelection. Voters are polarized; widening percentages of victory are interpreted as evidence of gerrymandered districts. In the 2004 congressional elections, only 5 percent of the 435 contests were decided by fewer than 10 percentage points, and only 4 incumbents were defeated. Early analyses of the election of 2006 also suggest that while gerrymandering does not guarantee incumbents their positions, partisan gerrymandering continues to frustrate democracy nationwide. Nevertheless, only Iowa and Arizona have established nonpartisan redistricting commissions.

Gerrymandering and civil rights

A 1982 amendment to the Federal Voting Rights Act of 1965 banned any redistricting that would have a negative impact on the political representation enjoyed by minority

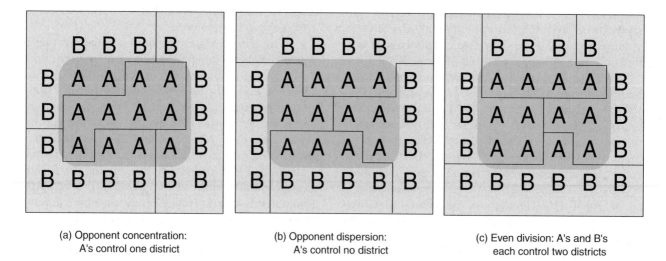

(a) Opponent concentration: (b) Opponent dispersion: (c) Even division: A's and B's
A's control one district   A's control no district   each control two districts

FIGURE 11-29 Types of gerrymandering. In this figure, the B's have an absolute majority in each case, so they have the power to draw the district lines. Drawing (a) illustrates concentration of the opponents (the A's), so the A candidate wins that one district with an unnecessarily large majority, but A's cannot win anywhere else. Drawing (b) splits up the A's so that they do not form a majority anywhere and cannot elect even one representative. In (c), each group controls two districts.

Critical THINKING

Garza v. County of Los Angeles

In 1990 federal courts ruled in *Garza* v. *County of Los Angeles* that the Los Angeles County Board of Supervisors had used a combination of unconstitutional gerrymandering methods in drawing their voting districts in order to exclude Hispanic people from representation. Hispanics made up 35 percent of the county's total population, but none had ever sat on the five-member board.

The court recognized that the lack of a Hispanic on the board did not prove that unconstitutional discrimination had occurred. Alternative hypotheses can partly explain the lack of Hispanic representation. Hispanics may be underrepresented because

- a high percentage of them are aliens
- a high percentage fail to register
- a high percentage are children too young to vote
- a high percentage simply are not interested in politics.

Each of these hypotheses has some validity, but the court ruled that their combined explanatory power did not explain Hispanics' total exclusion unless intentional discriminatory gerrymandering had also taken place.

Board district lines were redrawn, and a district was created in which Hispanics made up 71 percent of the population, although only 51 percent of registered voters. Voter turnout in the subsequent 1991 special election was only 21 percent, but a Hispanic won the seat on the board. In the following elections for that seat, however, non-Hispanic individuals have defeated Hispanic incumbents and other Hispanic candidates.

Questions

1. Does the fact that a particular group is underrepresented or not represented in a governmental body necessarily mean that discrimination has occurred?
2. How can discrimination be proved?
3. How much power should the courts or other institutions have to remedy such situations?
4. What does the defeat of Hispanic candidates in a Hispanic majority neighborhood by a non-Hispanic individual suggest about a contrast between identity politics and the voters' interpretation of their own best interests?

---

communities. The law insists that when any court is examining a case of redistricting, the court must consider "the extent to which members of a protected class have been elected to office." Several states interpreted this to mean that they were mandated to create districts in which the majority of the voters belonged to racial minorities. These districts are called *majority-minority districts*, and they are drawn by practicing the opponent concentration technique illustrated as (a) in Figure 11-29. Such majority-minority districts have been drawn at every level of government across the nation, from city council seats to congressional districts. In 1986 the Supreme Court defined a three-part test for assessing voter dilution that seemed to encourage the formation of majority-minority districts. The ruling stated: "First, the minority group must be able to demonstrate that it is sufficiently large and geographically compact to constitute a majority of a single-member district. . . . Second, the minority group must be able to show that it is politically cohesive. . . . Third, the minority must be able to demonstrate that the white majority votes sufficiently as a bloc to enable it . . . usually [to] defeat the minority's preferred candidate" (*Thornburg* v. *Gingles*).

Throughout the country, however, these majority-minority districts were challenged as representing racism—favoring a designated group rather than oppressing it, but racism nonetheless. Some people, however, argue that the United States must be color-conscious for a while as a way of making up for past discrimination and segregation. In 1993 the U.S. Supreme Court ruled that creating legislative districts with people of the same race who are otherwise separated "bears an uncomfortable resemblance to political apartheid. . . and reinforces the perception that members of the same racial group—regardless of their age, education, economic status or the community in which they live—think alike, share the same political interests and will prefer the same candidates at the polls" (*Shaw* v. *Reno*). Grouping voters by their ascribed identities and trying to win their votes is commonly called *identity politics*, and contrasted with *issue politics*, which argues public matters, but racial stereotyping in districting is legally impermissible. In examining one gerrymandered district, North Carolina's Twelfth Congressional District (Figure 11-30), the Supreme Court ruled that a conscious concentration of black voters did not automatically make a district unconstitutional, as long as the state's primary motivation was

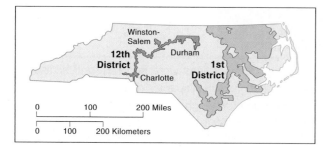

FIGURE 11-30 Efforts to gerrymander districts in North Carolina. In the early 1990s, North Carolina legislators designed these two congressional districts in order to elect black representatives to Congress. The U.S. Supreme Court upheld the constitutionality of the First District, but, in 1996, it ruled that the Twelfth District was drawn in a manner that was too "race-conscious." In 1997 the North Carolina legislature redrew the Twelfth District in a way that would scarcely be distinguishable on a map at this scale. That new pattern was finally approved by the U.S. Supreme Court in 2001, just as North Carolina set about redrawing all of its congressional districts on the basis of the new data from the 2000 census!

political (to create a Democratic district) rather than racial (to create a black district). In other words, party-conscious gerrymandering is constitutional, but race-conscious gerrymandering is not (*Easley* v. *Cromartie, 2001*).

When in 2006 the Supreme Court upheld Texas's right to have redistricted in 2003, it did, however, criticize the redistricting scheme for having violated the Voting Rights Act, thus showing that the Voting Rights Act (extended for 25 years in 2006) remains a tool for minorities who can show that their right to equal participation in the political process has been impaired. Nevertheless, Chief Justice Roberts wrote a strongly worded dissenting opinion, which Justice Alito signed, insisting "It is a sordid business, this divvying us up by race." Arguments over gerrymandering will continue in America's legislatures, and in 2003 the U.S. Supreme Court upheld the federal courts' right to assume the activist responsibility of stepping in to draw legislative boundaries when state processes fail (*Branch* v. *Smith* and *Smith* v. *Branch*) (Figure 11-31).

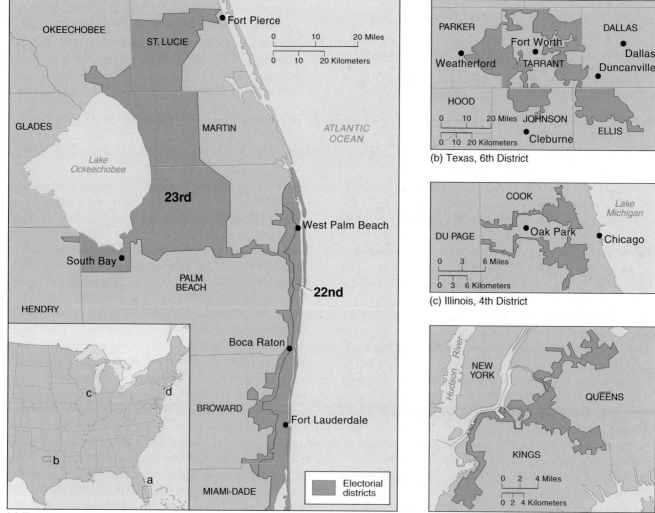

(a) Florida, 22nd & 23rd Districts

(b) Texas, 6th District

(c) Illinois, 4th District

(d) New York, 12th District

FIGURE 11-31 Some U.S. congressional districts in 2003. These are a few "gerrymanders" who sent representatives to the U.S. House of Representatives in 2006. In each case, the districts were designed to manipulate the outcomes of elections held in the districts.

MEASURING AND MAPPING INDIVIDUAL RIGHTS

Countries differ greatly in the restrictions they place on their citizens' freedom and in the degree to which they encourage and achieve full development of their citizens' potential. Predominant attitudes toward some issues, such as women's rights, may be determined by the underlying culture that may cross international boundaries. Arab culture, for example, as noted in Chapter 7, seems relatively insensitive to women's rights. Nevertheless, sovereign states are the units within which laws promote or restrict individual rights, so we can meaningfully analyze and compare the policies of different countries.

A society in which the most capable people can rise to the top on merit alone is called a **meritocracy,** which most countries claim to be. Rigid social stratification can be as unhealthy in a society as hardening of the arteries is in a human body. Hurdles to individual advancement may include stratification of the population by caste, race, or employment restrictions. The modern labor movement has won important victories for working people, but in some countries labor unions may protect their members' jobs at the expense of opportunity for other individuals. In the United States, for example, some labor unions have been bastions of racism.

Many countries of diverse populations have *affirmative action* programs designed to lift to national standards of achievement those segments of the population that have suffered an historic lack of opportunity. Laws in both Malaysia and Indonesia, for example, favor ethnic Malays over the more entrepreneurially successful Chinese. Nigerian law recognizes efforts of school and job-development programs to "reflect the federal character." This is a code for federally mandated ethnic quotas. Some European countries try to achieve diversity in groups in terms of members' age, as well as ethnic composition and sexual orientation. India has an elaborate system of preferences for lower-class "scheduled" castes and tribes, and Brazil enforces new race-based preferences in universities, the civil service, and the private sector. Affirmative action programs have been crafted in the United States in both the public and private sectors. In upholding race-consciousness among other admissions criteria for U.S. universities, the Supreme Court replaced the traditional appeal to historical justice as a rationale for affirmative action with diversity in itself as a key value in American life (*Grutter* v. *Bollinger* and *Gratz* v. *Bollinger*, 2003).

The most pervasive form of discrimination in the world, however, is sexism, and the most powerful boost to individual achievement is education. Individual freedom is a value sought by all individuals. International variations in these three social characteristics—sexism, educational opportunity, and individual freedom—are examined in the paragraphs that follow.

Sexism

Almost everywhere, women are worse off than men. They have less power, less autonomy, less money, more work, and more responsibility. National variations in sexism can be approached through a number of quantitative measures. In Chapter 5 we saw that discrimination against women can begin even in the womb, as reflected in higher rates of abortion of female fetuses than male fetuses, and that female-to-male ratios in national populations may reveal discrimination against women. Discrimination against female farmers, discussed in Chapter 8, lowers agricultural productivity and the levels of nutrition in many countries.

U.N. researchers have measured and mapped striking differences in women's welfare around the world. Women in rich countries are generally better off than women in poor countries, but fewer data focus on the contrasts between the welfare of women and that of men in individual countries. Those are the data that must be isolated to compare sexism among countries. One useful statistic might be the measure of female to male enrollments in primary and secondary school (Figure 11-32). A country's failure to educate females matters not just in itself but also because improving female literacy is one of the most effective ways to control population growth and to reduce infant mortality.

Some issues are almost exclusively women's issues, such as the female genital mutilation that is common across Sahelian and East Africa and in the Near East, and maternal death rates. Other issues affect both sexes, but principally women: abortion rights, birth rates, equal pay, rape, domestic violence, maternity care or maternity leave, availability of contraception, child welfare and rearing practices, and even marriage and divorce rights. By 2000 some 44 of the United Nations' 189 members had enacted laws against domestic violence, but these laws were not always enforced.

In many countries, the presumptions of the law are biased against women. In Brazil, for example, the definition of "rape" is "coercive sex with an honest woman," defined as someone else's wife or an unmarried virgin. An "unchaste," that is, nonvirgin, unmarried woman has no recourse to the law in case of rape. In Pakistan, to prove rape, a woman must have at least four male witnesses. Without them, the woman is herself imprisoned for adultery, a situation in which almost 5,000 women found themselves in mid-2006. The fact that Rio de Janeiro, Mexico City, and Tokyo have "women-only" subways cars may be interpreted not as victories for women's rights in those places, but as evidence that men normally feel entitled to grope or harrass women in those countries.

Some researchers have suggested using the percentage of the labor force that is female as one index of women's rights. The problem with this statistic, however, is that statistical measurements of the labor force measure only paid occupations, and in many countries women are not paid for their labor—particularly homemaking,

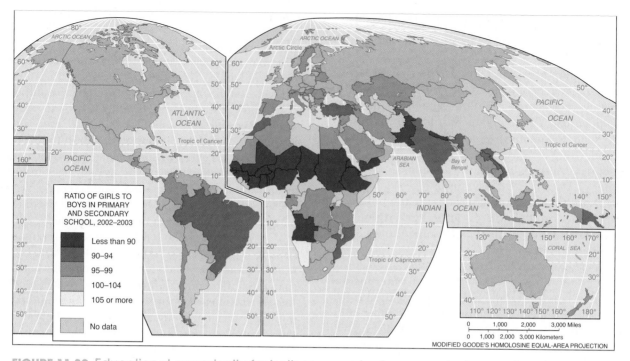

FIGURE 11-32 Educational opportunity for both sexes varies from country to country. Where a lower percentage of females than males are in school, we may hypothesize that females do not receive all of the opportunities that males do in that society. (Data from United Nations)

farming, or even small-scale domestic manufacturing. Women's participation rate in the labor force is everywhere below that of men. It is about 40 percent in North America; slightly less in Japan, Europe, Australia, and New Zealand; and only 20 percent or even less in South Asia and the Middle East. The figures for Africa are even lower, suggesting that very few African women work. Observation of African life, however, reveals that the women are indeed working; they just are not getting paid. We need better statistical measures of women at work.

Women's participation in politics might serve as another index of sexism. In most democratic countries men achieved the vote before women, but when Kuwait gave women the vote in 2006, no countries remained where men can vote but women cannot. Women have formed political parties in several countries, including Holland, Iceland, Belgium, Spain, and Sweden. Women are proportionately underrepresented in legislatures everywhere, even though several countries reserve legislative seats for women. Ghana, Tanzania, Eritrea, Kenya, Rwanda, and Uganda, for example, all have percentages of their parliaments reserved for women. India adopted a Constitutional Amendment in 1993 that set aside one-third of all seats on village councils and of all village mayoralties for women. A certain percentage of those positions must go to low-caste women. This has caused revolutionary changes in local government concerns. The 2005 Constitution of the Democratic Republic of the Congo requires equal representation for women and men in all local provincial and national institutions. France and Portugal require that a minimum percentage of candidates be female.

In 2005 the World Economic Forum ranked 58 countries according to a "gender gap" measure of economic opportunity, economic participation, political empowerment, educational attainment, and health and well-being (Table 11-2). The gap measured narrowest in 4 Nordic countries and widest in 4 Muslim Near Eastern countries.

Regardless of national constitutions and laws, men's and women's relative welfare can be determined by their traditional roles in national culture. Many countries formally provide for sexual equality in the law, but few protect specific job, inheritance, property, or marriage rights or protect women from domestic violence. To learn more about the world geography of sexism, statistical and analytical tools will have to be improved.

The World Geography of Education

The more advanced a society, the larger is the required investment in its human resources, specifically education. The U.S. educator Horace Mann wrote in 1846, "Intelligence is a primary ingredient in the wealth of nations."

A country's wealth and the education levels of its citizenry usually rise together. In the rich countries of Western Europe and Asia and in the United States, adult literacy is high. In the rich countries that attract immigrants, significant numbers of people may be literate in a language or languages in addition to, or other than, the official language of the country. The percentages of literacy drop only slightly in Latin America but drop markedly in Africa. Few African countries boast literacy rates much higher than 50 percent, even for the

TABLE 11-2	Women's Empowerment: Measuring the Global Gender Gap				

THE GENDER GAP RANKINGS

COUNTRY	OVERALL RANK	OVERALL SCORE*	COUNTRY	OVERALL RANK	OVERALL SCORE*
Sweden	1	5.53	Colombia	30	4.06
Norway	2	5.39	Russian Federation	31	4.03
Iceland	3	5.32	Uruguay	32	4.01
Denmark	4	5.27	China	33	4.01
Finland	5	5.19	Switzerland	34	3.97
New Zealand	6	4.89	Argentina	35	3.97
Canada	7	4.87	South Africa	36	3.95
United Kingdom	8	4.75	Israel	37	3.94
Germany	9	4.61	Japan	38	3.75
Australia	10	4.61	Bangladesh	39	3.74
Latvia	11	4.60	Malaysia	40	3.70
Lithuania	12	4.58	Romania	41	3.70
France	13	4.49	Zimbabwe	42	3.66
Netherlands	14	4.48	Malta	43	3.65
Estonia	15	4.47	Thailand	44	3.61
Ireland	16	4.40	Italy	45	3.50
United States	17	4.40	Indonesia	46	3.50
Costa Rica	18	4.36	Peru	47	3.47
Poland	19	4.36	Chile	48	3.46
Belgium	20	4.30	Venezuela	49	3.42
Slovak Republic	21	4.28	Greece	50	3.41
Slovenia	22	4.25	Brazil	51	3.29
Portugal	23	4.21	Mexico	52	3.28
Hungary	24	4.19	India	53	3.27
Czech Republic	25	4.19	Korea	54	3.18
Luxembourg	26	4.15	Jordan	55	2.96
Spain	27	4.13	Pakistan	56	2.90
Austria	28	4.13	Turkey	57	2.67
Bulgaria	29	4.06	Egypt	58	2.38

*All scores are reported on a scale of 1 to 7, with 7 representing maximum gender equality.

Source: The World Economic Forum.

preferred males (Figure 11-33). School enrollment figures tell the same story. The children in the world's rich countries are in school; those in the world's poor countries are not.

The World Geography of Freedom

Some people in the world live in freedom; others do not. But it is difficult to quantify freedom. Surprises appear when comparing statistics. Many civil libertarians, for example, ask why the United States ranks first among all countries in the percentage of its population that is incarcerated (over 2.2 million in 2006). This has been largely because of the "war on drugs." Does this mean that the United States is a totalitarian country? Or does it mean only that if you break the laws in the United States you are more likely to go to jail than anywhere else? Many observers might respond that, in many countries, just living there is the equivalent of being in jail. How are such statistics to be interpreted?

Freedom House is a research institute that monitors political and civil freedom around the world. In the measurement of political rights, the institute assigns the highest scores where elections are fair and free and where the people who were elected actually do rule. The

institute defines civil liberties as including freedom of expression, assembly, demonstration, religion, and association. It rates highest those states that protect individuals from political violence and from harm inflicted by courts and security forces. These states have free economic activity and strive for equality of opportunity. Freedom House assigns each country a score of 1 to 7 in each of these two categories and then sums up the scores in an overall rating "Free," "Not Free," or "Partly Free." Freedom House findings for 2006 are reproduced in Figure 11-34. These rankings change rapidly as governments rise and fall, but the attempt to rate countries reveals wide disparities in freedom. We shall see in Chapter 13, however, that the very definition of freedom may not be the same around the world.

CONCLUSION: CRITICAL ISSUES FOR THE FUTURE

The complete division of the world into stable and peaceful territorial nation-states may or may not be an ideal goal. At present, however, the state remains the ultimate level of absolute sovereignty on Earth. The difficulties and hardships resulting from trying to make nations fit

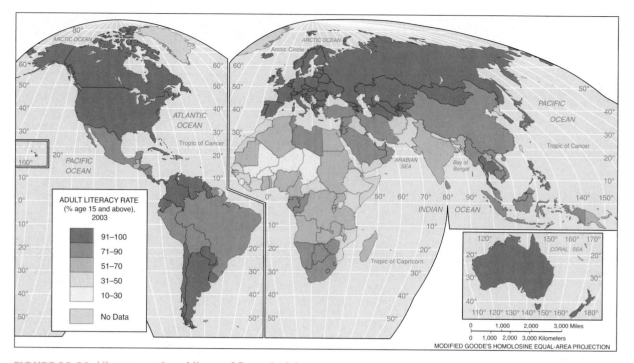

FIGURE 11-33 Literacy rate of those 15 and older. Literacy is virtually universal in the rich countries, but this map reveals considerable variations in national literacy rates. (Data from United Nations)

states, or vice versa, should remind us that the territorial nation-state is only one theory of what might be the best political–geographical division of Earth. Perhaps territorial nationalism is not a positive ideal. A return to regnum forms of government is improbable, but it might be a desirable goal to achieve some universal guarantee of individual rights in any political-territorial framework.

Furthermore, the number of failed states demonstrates that the current political division of Earth is at least partly responsible for continuing war and poverty among hundreds of millions of people. Assuming that we do not want to choose the theoretical option of massacring or expelling the hundreds of millions of people who are unhappy with the current territorial pattern or

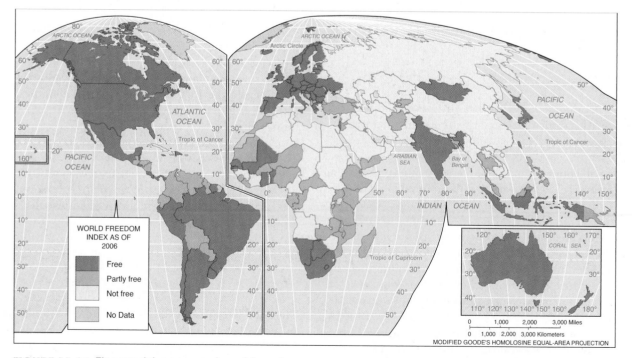

FIGURE 11-34 The world geography of freedom. This map results from an effort to quantify the degree of freedom exercised by the population in each country. (Rankings courtesy of Freedom House)

who are ill-served by it, should we instead try to "enforce" the current political boundaries? Who has the authority to do that? (Who is the "we" in the previous sentences?) Who has the power to do that? Alternatively, who has the authority or power to redraw the map? These questions threaten to trouble the world for the forseeable future.

Chapter 13 will investigate whether a genuine world political community exists and whether the world community might assume the responsibility or exercise the power to act in defense of global peace and universal human rights.

Chapter Review

SUMMARY

Earth's division into sovereign states is the most important territorial organizing principle of human activities. The idea that the whole world should be divided into countries originated in Europe and diffused worldwide with European conquest. The neatness of the units on today's world political map nevertheless exaggerates the degree to which all people accept the current pattern.

A nation is a group of people who want to have their own government and rule themselves. The ideal state is theoretically a nation-state. Several European nations existed before they achieved their own states. The political patterns of the colonies, however, were the conquerors' creations. When independence came, only large units were at first granted independence, but later this threshold principle was abandoned.

Many states suffer from cultural subnationalism. States might employ three strategies to fit the nation to the state territory: redraw the political map; expel people from any country in which they are not content or exterminate them; or forge nations in the existing states. Redrawing the international map presents many difficulties, and expulsion or genocide are abominable, so six national institutions—a country's church or churches, armed forces, schools,

media, political parties, and labor unions—become key centrifugal or centripetal forces in any state. Geographers study the presence or absence of these institutions in any country, the unique balance or mix among them in each country, and their interaction with other aspects of each country's culture.

Each state demarcates its territory's borders, and the terrorist attacks on the United States in 2001 brought new concern to matters of both secure borders and of internal security. Countries also subdivide their territory for political representation or administration. Each country has a unique balance of powers between its local governments and its central government. Special districts can undermine the powers of local governments, and most federal states contain territories that are administered directly from the federal center. Capitals are symbolically important. Electoral geographers note that in any system of representative government, voting-district boundaries can be gerrymandered to determine the outcomes of the ensuing elections.

Countries vary widely in the degree to which they liberate and train their populations. The major form of discrimination is sexism, and the most powerful boost to individual achievement is education. Individual freedom is a value sought by all individuals. The incidence of each of these three social characteristics varies from country to country.

KEY TERMS

centrifugal forces p. 448
centripetal forces p. 448
collapsed states p. 450
core area p. 442
corporativism p. 472
cultural subnationalism p. 448
electoral geography p. 471
enculturation p. 458
failed states p. 450
federal government p. 465

geopolitics p. 451
gerrymandering p. 471
iconography p. 458
indirect rule p. 446
irredenta p. 445
kleptocracy p. 460
meritocracy p. 477
multinational state p. 449
nation p. 441
nation-state p. 442

national self-determination p. 444
oligarchy p. 460
political community p. 441
political culture p. 442
social Darwinism p. 441
socialization p. 458
state p. 440
superimposed boundaries p. 446
threshold principle p. 446
unitary government p. 465

QUESTIONS FOR REVIEW AND DISCUSSION

1. Differentiate a nation, a state, and a nation-state. Give examples of countries that are not nation-states.

2. What important decision was taken at the 1964 conference of the Organization of African Unity (OAU)?

3. How do governments try to forge their populations into nations?

4. What are the two principal methods of gerrymandering?

5. What evidence do you have that people enjoy different degrees of rights and liberties in different countries?

THINKING GEOGRAPHICALLY

1. When and how was the pattern of local governments in your state drawn? What responsibilities do the various local governments have? Could you improve the pattern, reassigning responsibilities to new districts or different levels of government?

2. Read the constitutions of a few countries that you believe to be totalitarian and centralized. Do the constitutions reveal that? Does the U.S. Constitution accurately reveal the division of power between state and federal governments that exists today?

3. How much land in your state is owned by the federal government? What is its legal status?

4. Is your state overrepresented or underrepresented in the Electoral College?

5. Do you think regions of a country ought to be able to secede? If you favor a vote, do you think that majority approval should be required throughout the entire country or only in the areas seeking to secede?

6. Make a list of several mountain chains or rivers around the world that serve as international borders. Which act as effective barriers? Which do not? Why?

7. Investigate the symbolism of the flags of a few countries. Look at Australia, the Comoros, Liberia, Mongolia, Namibia, Venezuela, and South Africa.

8. Why might a unitary government be more successful in Japan than in the United States?

SUGGESTIONS FOR FURTHER LEARNING

Anderson, Benedict. *Imagined Communities: Reflections on the Origins and Spread of Nationalism.* New York: Verso Books, 2nded. 2006.

Barone, Michael, and Grant Ujifusa. "The Almanac of American Politics." Washington, D.C.: *The National Journal,* published biannually.

Boone, Catherine, ed. *Political Topographies of the African State: Territorial Authority and Institutional Choice.* Cambridge, UK: Cambridge University Press, 2003.

Country Reports on Human Rights Practices. Washington, D.C.: U.S. Department of State, annual.

Fitzgerald, Valpy, Frances Stewart, and Rajesh Venugopal eds. *Globalization, Violent Conflict and Self-Determination.* New York: Oxford University Press, 2006.

Galderisi, Peter, ed. *Redistricting in the New Millennium.* Lanham, MD: Lexington Books, 2005.

Hastings, Donnan, and Thomas M. Wilson. *Borders: Frontiers of Identity, Nation and State.* New York: Berg Publishing, 2000.

Hooson, David, ed. *Geography and National Identity.* London: Blackwell (The Institute of British Geographers Special Publications), 2002.

Marx, Anthony W. *Faith in Nation: Exclusionary Origins of Nationalism.* New York: Oxford University Press, 2003.

Monmonier, Mark. *Bushmanders and Bullwinkles: How Politicians Manipulate Electronic Maps and Census Data to Win Elections.* Chicago: University of Chicago Press, 2001.

Roshwald, Aviel. *The Endurance of Nationalism: Ancient Roots and Modern Dilemmas.* New York: Cambridge University Press, 2006.

Sassen, Saskia. *Territory, Authority, Rights: From Medieval to Global Assemblages.* Princeton, NJ: Princeton University Press, 2006.

Smith, Anthony D. *Nationalism: Theory, Ideology, History.* Cambridge, UK. Polity Press, 2001.

Statesman's Year-Book. New York: Macmillan, annual since 1864.

World Boundaries (series coordinated by the International Boundaries Research Unit at the University of Durham, North Carolina). vol. 1, *Global Boundaries,* Clive Schofield, ed.; vol. 2, *Middle East and North Africa,* Clive Schofield and Richard Schofield, eds.; vol. 3, *Eurasia,* Carl Grundy-Warr, ed.; vol. 4, *The Americas,* Pascal O. Girot, ed.; vol. 5, *Maritime Boundaries,* Gerald H. Blake, ed. New York/London: Routledge, 1994.

WEB WORK

Basic information about each country may be found at:

http://www.countrywatch.com/

or at the Central Intelligence Agency World Factbook:

https://www.cia.gov/cia/publications/factbook/index.html

Political news of the world is available at many websites. Most major daily newspapers are posted, including:

http://www.washingtonpost.com

http://www.latimes.com

http://www.nytimes.com

http://globeandmail.com

Some services provide news from specific world regions. Two examples are:

http://www.allafrica.com/

http://www.askasia.org/

The U.S. Department of State provides travel information (often warnings) about affairs in foreign countries at:

http://travel.state.gov/index.html

A good guide to the U.S. government is:

http://www.fedworld.gov/

The U.S. president and each house of Congress have home pages with numerous valuable links:

http://www.whitehouse.gov/

http://www.senate.gov/

http://www.house.gov/

The Canadian government's official website is:

http://www.canada.gc.ca/main_e.html

Information about borders and border disputes can be found at:

http://www-ibru.dur.ac.uk/index.html

The United Nations maintains an Internet Gateway on the Advancement and Empowerment of Women:

http://www.un.org/womenwatch/

Freedom House monitors political developments around the world:

www.FreedomHouse.org

as does the International Institute for Democracy and Electoral Assistance:

http://www.Idea.int

12

NATIONAL PATHS
TO ECONOMIC GROWTH

This suburban high-technology office park could be almost anywhere in the world. In fact it is just outside Bengalooru, India (formerly Bengalore). Here, Wipro Technologies, an Indian giant in the field of information technology, devises computer programs for clients around the globe. By exporting this technology, India earns much greater returns than it could by exporting raw materials or even many manufactured goods. Providing valuable goods and services to global markets is one economic development strategy that some countries employ to raise living standards for their own peoples. (Mahesh Bhat/India Picture)

A Look AHEAD

Analyzing and Comparing Countries' Economies

Gross domestic product and gross national income are the principal measures of nations' wealth, but they must be supplemented by noneconomic measures to compare the quality of life in different countries. Sectoral analysis reveals how countries' economies may be described as preindustrial, industrial, or postindustrial. National wealth is affected by many things, including natural resources, but natural resources are not the only determinant of national wealth.

Adding value to resources or providing valuable services builds wealth. Technology, communications, and government policies constantly change the locational determinants of economic activity.

The Geography of Manufacturing

The world geography of manufacturing is the result of identifiable locational determinants, but as these change, they cause some manufacturing to relocate continuously.

National Economic-Geographic Policies

Countries have different systems to organize and regulate domestic economic activity in order to increase wealth and distribute it around the territory.

National Trade Policies

Some countries try to build their domestic economies without trade, while others participate in international trade to attract capital and win markets.

The Formation of the Global Economy

International economic links have reduced the barriers to exchange and communication, and many activities have expanded their scale of organization to cover the whole globe. Economic globalization has far outpaced cultural or political integration.

On June 6, 2006, Samuel Palmisano, Chairman of IBM, the world's largest computer services company, stood together with Indian President Abdul Kalam in Bengalooru, India, (formerly Bengalore) and announced a three-year plan to invest over $6 billion in India. Three years earlier IBM had faced a public relations disaster when a disgruntled employee had publicized corporate plans to accelerate the relocation of high-paying white-collar jobs out of the United States, but the relocations had proceeded. In 2003, IBM had 9,000 employees in India; by mid-2006 it had approximately 43,000, and its Indian employees earned, on average, 12 percent of what U.S. employees doing similar work had earned. IBM had cut 14,5000 U.S. jobs. IBM's Bengalooru command center today monitors 16,000 computer servers and 10,000 applications around the world. What is more, IBM in India not only maintains services, but devises new technologies. For example, Indian research labs have created a warranty management system for U.S. carmakers. Other technology multinationals, such as Microsoft, Intel, and Cisco Systems also have announced multi-billion-dollar investments in India.

The movement of jobs overseas is the inevitable result of not only corporations' efforts to lower costs but also of the technology that knits the world economy—the phenomenon of globalization. The first jobs to leave the United States and other rich countries were the blue-collar factory jobs, which went to new factories in low-wage countries; China has captured the lion's share. The relocation lowered costs to manufacturers and prices to consumers, but at the cost of job opportunity in the rich countries. Many former factory workers retrained and found good new jobs, but most suffered from decreased incomes, benefits, and pensions.

Now service sector jobs are following the factory jobs. Everything from relatively low-paying call center jobs to software development, accounting, income tax preparation, and even legal and medical analysis can be performed by lower-paid workers in China, India, the Philippines, Russia, or elsewhere because technology allows the out-of-country work to be communicated to the rich countries instantaneously. Three million service jobs are expected to leave the United States alone by 2015, including half of the jobs in packaged software and information technology services. An IBM executive said, "Our competitors are doing it, and we have to do it." Will wages in today's rich countries have to collapse to the levels in China and India to compete?

How will economic globalization redistribute good jobs, or wealth, and how will it affect development in the countries where living standards are low today? The answer will determine future job opportunity in today's rich countries. ▶

Most people agree that one of the principal tasks of any government is to promote the welfare of its people, including their economic welfare. Each country devises a program to build its national economy and also to ensure that economic well-being is distributed across the national territory. Therefore, the world political map is also a map of local economies and economic policies. Many states need economic growth not only to raise standards of living, but also because economic growth may convince the population that the state works for them, that it is at least a success economically.

ANALYZING AND COMPARING COUNTRIES' ECONOMIES

Some countries have achieved high incomes and standards of living for their people. Others have not. Throughout this book we have generally referred to the rich countries and the poor countries. The terms *rich* and *poor* are fairly straightforward and descriptive. Other terms often contrast the *industrial* or *developed countries* and the *undeveloped, underdeveloped,* or *developing countries.* This chapter emphasizes that it is difficult to compare countries' levels of wealth and standards of living. There are many ways to measure economic status and welfare, and these measures do not always rank countries in the same order.

This chapter is divided into five sections. The first section examines several ways of measuring countries' wealth. Then it maps the rich and the poor countries and examines reasons for their wealth or poverty. The second section of the chapter examines the world distribution of manufacturing and reasons for that distribution. The third section of this chapter investigates national economic-geographic policies. Each country has a unique way of organizing its economy and managing the distribution of economic activities within its territory. The fourth section looks at how countries regulate their participation in world trade. We will investigate the meaning of the popular term *the Third World*. The closing section examines the formation of a global economy. World trade has accelerated so fast and multiplied to such a degree that international management consultant Peter Drucker (1909–2005) wrote, "The world economy has become a reality, and one largely separate from national economies. The world economy strongly affects national economies; in extreme circumstances it controls them."

Measures of Gross Product and Their Limitations

The two most commonly used measures of a country's wealth are its gross domestic product and its gross national income. **Gross domestic product (GDP)** is the total value of all goods and services produced within a country. **Gross national income (GNI)** is the GDP plus any income that residents receive from foreign investments, minus any money paid out of the country to foreign investors (Figure 12-1). Either GDP or GNI can be measured *per capita*—per person in the country. GDP provides a better guide to analyzing a country's internal economy, because it measures exactly what activities are occurring and what wealth is being created inside the country. GNI is best used when we want to compare and contrast different countries' total incomes as a guide to the total wealth available to the people in that country.

For most countries there is little difference between the two statistics, but the differences that exist are important to geographers. GDP describes the economic activity inside a country, whereas GNI reveals the impact that foreign investments have on a country's economy. Some countries either receive large payments from abroad or send money abroad, so these countries will record a significant difference between their GNI and their GDP. For example, as Kuwait sells oil from domestic oil wells, the income is counted as part of Kuwait's GDP. Through the years the Kuwaitis have profited from oil sales and invested profits abroad. Today the profits from Kuwaitis' foreign investments that are returned to Kuwait are part of Kuwait's gross

FIGURE 12-1 Lukoil service station. The Russian oil company Lukoil has bought thousands of Getty and ConocoPhillips service stations throughout the United States, so profits from this station in Manhattan are part of Russia's GNI. They are not, however, part of Russia's GDP, because they are not generated inside Russia. Many oil-producing states have purchased refineries and distribution systems in order to control supplies from the ground to the ultimate consumer. Thus, they capture profit from value added at each step. (Robert Azzi/Woodfin Camp & Associates)

national income, but they are not part of Kuwait's gross domestic product. Kuwait's GNI regularly amounts to 110 to 125 percent of its GDP. This reveals that Kuwaitis have enormous investments abroad. In 2004, the GNI of the United States was 104 percent of GDP. Other countries, by contrast, pay out profits to owners in other countries, so their GNIs are smaller than their GDPs. In 2004 Canada's GNI was 8 percent smaller than its GDP, Argentina's GNI was about 6 percent smaller than its GDP, and Ireland's GNI was fully 25 percent smaller than its GDP. These statistics suggest that significant shares of these countries' economies are owned by foreigners.

In 2004, world GDP totaled $41 trillion. The United States generated $11.7 trillion; Japan $4.6 trillion; Germany $2.7 trillion; the United Kingdom $2.1 trillion; France $2 trillion; Italy $1.7 trillion; and China $1.6 trillion. Thus, these seven countries accounted for two-thirds of the measured economic activity on Earth.

Limitations of the GNI and the GDP statistics

Both GNI and GDP are deceptive. Both begin with the idea that most goods and services are of little or no value if they cannot be exchanged. They have no value in their simply being used. Therefore, these measuring techniques underestimate the activities of hundreds of millions of people who provide entirely for their own needs or who exchange very little except through barter. For example, many of the world's farmers buy little food because they eat food they raise themselves. Therefore, farmers' total output and income are probably both undercounted. A peasant farmer in China whose total annual income is recorded as $250 may eat food that would cost a resident of Chicago well over

$1,000 in the local supermarket. Farmers usually make up a greater percentage of the population in the poor countries than in the rich countries, so undercounting farmers' production underestimates the material welfare of many people in the poor countries. By noting this, however, we do not mean to exaggerate the quality of life those people may have or to suggest that they live well. They are still poor by our Western standards.

Because statistics undercount subsistence areas, they exaggerate the degree to which modern areas, especially cities, dominate national economies. In extreme cases, a single city can provide most of a country's measurable output. Abidjan, with 15 percent of the Ivory Coast's population, accounts for 70 percent of all economic and commercial transactions in the country. São Paulo, with 10 percent of Brazil's population, contributes a quarter of that country's measurable economic activity. Statistics particularly overlook the work of women (Figure 12-2).

Government statistics cannot measure activity that is illegal and therefore clandestine. Colombia's profitable drug exports do not seem to exist. Neither do illegal drugs in the United States, yet the federal government admits that the drug trade is one of the nation's largest industries. Nor does gross product include the activities in a nation's informal sector (discussed in Chapter 10).

Statistics examined per capita over time reveal a relationship between population growth and changes in the statistic measured. A country's GNI can grow, but if the population grows faster than the GNI, GNI per capita will fall. For many countries of sub-Saharan Africa, per capita incomes rose in the 1960s, flattened

out in the 1970s, and fell from the mid-1980s through the rest of the century. The per capita GNI of Zambia fell at an annual rate of 1.9 percent from 1975 to 2004. That of the Democratic Republic of the Congo fell at an annual rate of 4.9 percent. Falling per capita incomes in many African countries were due in part to the general deterioration of African economies, to civil wars, and in part to rapid population growth.

The geography of exchange rates   Attempts to compare economies are further complicated by the fact that there are no common measures of value. Gross products are measured in local currencies. This complicates comparisons among them because the exchange rates among currencies are changeable and may be manipulated. This is of special concern to geographers not only because it hinders comparison of one place with another but also because variations in exchange rates affect the flows of goods, investment, and people that geographers study. Between 1985 and 1990, for example, the U.S. government deliberately lowered the value of the U.S. dollar by 50 percent relative to the currencies of Japan and the Western European countries. This had extensive geographic results. American goods were cheaper for foreigners to buy, so exports of U.S. manufactured goods rose 80 percent; new exports boosted the output of the nation's manufacturing belt (the Northeast–Midwest industrial region; look ahead to Figure 12-18). Foreigners invested billions of dollars in the United States, and millions of foreign tourists came to visit. In 1989, for the first time ever, foreign visitors spent more money in the United States than U.S. citizens spent abroad. As the dollar fell, goods flowed out, but tourists and investment flowed in.

These new flows of people and goods diffused culture. Tourists took away impressions of American life, and Americans received new impressions of foreigners. Many more Europeans and Japanese could afford U.S. clothes, cassettes, compact discs, videos, and other cultural exports when those goods' prices fell 50 percent. Millions more people saw U.S. movies and television shows.

At the same time, the exports to the United States from foreign trading partners sagged, their economies suffered, and the internal economic geography of their countries changed as their export-dependent regions suffered most. Billions of dollars that U.S. manufacturers had planned to invest in factories abroad were invested in the United States instead, and communities from the Philippines to Brazil went without new vocational training programs, roads, and schools. Places dependent on U.S. tourists suffered recessions.

From 1990 to 2001, the exchange value of the dollar generally rose, and all these flows reversed. From 2001 through 2006, the dollar fell again, and the flows reversed once more. Geographers monitor shifting exchange rates such as we have described to understand changing patterns of travel and trade and of local prosperity.

FIGURE 12-2 Subsistence farming. Chapter 11 noted that official statistics in many countries do not record women as working at all, but Chapter 8 noted that women raise most of Africa's food. Nor does subsistence production count in a country's GDP. Therefore, only if this woman sold her produce to her neighbors would its value or her work appear in the country's GDP. (Getty Images, Inc.)

Furthermore, exchange rates differ from what economists call *purchasing power parity (PPP)*. The currency values established in foreign exchange markets often do not accurately reflect purchasing power, but PPP attempts to measure what goods or services a certain amount of money will buy at different places. When you read that annual incomes in a certain country are a certain amount of money, do not assume that the people there live at the same level of material welfare that that amount of money would buy where you live. For example, in 2005, at the international exchange rate of Indian rupees to U.S. dollars, India's per capita GNI was only $620 dollars. In India, however, food, clothing, housing, and other necessities were much cheaper than in the United States. Therefore, when the Indian income was adjusted for purchasing power, India's GNI per capita jumped to $3,300. China's per capita GNI in 2005 was only $1,692, but, adjusted for purchasing power, that figure soared to $6,800. Those two countries alone contain about 37 percent of the total human population, so adjusting income for PPP tremendously raises our measures of humanity's standard of living.

Gross Product and the Environment

Measures of gross product fail to assess the environmental damage that may result from growth. In standard techniques of bookkeeping, machines and buildings are counted as capital assets. As they age, their declining value is deducted from income. When a country's natural resources are exploited, however, annual depreciation is not deducted. Many countries sell off their timber and minerals, destroy their fisheries, mine their soils, and deplete their water resources, and national accounting treats the proceeds as current income. The loss of the natural resources is not deducted as a depreciation of national assets.

If natural resources were treated as assets, then statistics might demonstrate that protecting the environment is sensible economic policy. The United Nations held the world's first conference on sustainable development in 1988, and since then many economists have been working to calculate statistics that would subtract from gross product the value of natural resources that are destroyed or depleted. The World Bank's 2003 *World Development Report* analyzes several variations. Economists have estimated that when resource loss or depletion is considered, many countries' wealth has been growing more slowly than their GDPs would suggest. Many of the countries that are exporting their raw materials, such as some oil-exporting states, may not be growing at all, but actually depleting their national wealth.

The Pacific Ocean island-nation of Nauru illustrates the extreme of unsustainable development (Figure 12-3). The people of this 21-square-kilometer

FIGURE 12-3 Phosphate mining on Nauru. After years of phosphate mining, the island of Nauru may have to be abandoned. (Michael Friedel/Woodfin Camp & Associates)

(8-square-mile) island allowed strip-mining of phosphate for fertilizer exports to ravage their homeland. This was immensely profitable, and the government trust fund once held over $1 billion for the national population of just 13,000 people. By 2006, however, much of that money seems to have been squandered, and the entire population may even have to abandon the island. Nauru would then have a GNI, but absolutely no GDP.

Many economists doubt the value of trying to measure sustainable development. They point out that the countries that are rich today had exploited their environments in the past at unsustainable rates, but after these countries had achieved wealth, they were able to repair much environmental damage (Figure 12-4). Perhaps sustainability is irrelevant to a poor population struggling to raise its standard of living to some minimum level, but once the population attains that level of welfare, it can afford to maintain or even repair the environment. Many poor countries resent rich countries' suggestions that they slow their economic activity to a sustainable rate. Imposing conservation measures on poor countries has been derided as *green imperialism.*

Nevertheless, when the world's banks are financing infrastructure projects such as dams, power plants, and pipelines, they are increasingly accepting the *Equator Principles,* a set of environmental and social-impact standards drawn up by The World Bank. These principles were designed to prevent large construction projects from poisoning air and water, denuding forests, and destroying the livelihoods of local residents in poor countries, where government regulators are often ill-equipped or unwilling to mitigate potential side effects of foreign-financed initiatives. The principles are strict but voluntary, and there is no enforcement mechanism except negative publicity.

FIGURE 12-4 Restoring a forest. It was a natural disaster—the eruption of Mount Saint Helens on May 18, 1980—that felled these trees in Washington State, but human replanting and care is restoring the forest as a productive asset for human use. (Courtesy of the Weyerhaeuser Corporation)

The Gross National Product and the Quality of Life

Figure 12-5 is a world cartogram, with each country drawn in proportion to its GDP. It is strikingly different from a territorial map. Some of the smallest countries—Belgium, for instance—have the largest GDPs. Figure 12-5 also differs from the population cartogram pictured in Figure 5-2. Some countries with small populations have enormous economies. Clearly, neither population nor territorial size explains wealth.

Figure 12-6 establishes categories of per capita GNI on a regular world map. What we really want to know, however, is what percentage of *people* live at various levels of income. Therefore, Figure 12-7 colors the various levels of income on the cartogram of world population. By this device, we can see that about 40 percent of the world's population lives in countries in which per capita GNI is less than $826. Many rich people, however, do live in each of these countries, and many poor people live in some of the richest countries.

For most people, incomes are rising. The proportion of people living on $1 per day has fallen from 41 percent of the world's population in 1981 to 19 percent in 2006, while the percentage living on only $2 per day has fallen from 67 percent to 50 percent. This is progress—agonizingly slow—but progress.

Noneconomic measures of national welfare

Per capita GNI is only one measure of national welfare. A picture of world standards of living and welfare can be improved by combining it with other statistics. We have already examined infant mortality rates (Figure 5-12), food supplies (Figure 8-31), literacy (Figure 11-33), freedom (Figure 11-34), and the appendix figures on energy consumption. There is a great deal of coincidence among these statistics. The countries with the lowest infant mortality rates generally have the best food supplies and education, and their citizens consume the most energy. These are all measures of the standard of living, and the highest-ranking countries are much the same as those that have the highest per capita GNIs.

Countries' rankings according to per capita GNI, however, do not always correspond exactly with their rankings according to all other measures of welfare. The concentration of wealth, government spending choices, and other factors will affect the population's quality of life. Therefore, the United Nations has devised a statistical attempt to compare the quality of life among countries. This is called the **human development index (HDI).** The United Nations argues that living a long and healthy life, being educated, and enjoying a quality standard of living are measures of human development, so the HDI combines statistics of life expectancy, school enrollment, literacy, and income (Figure 12-8). The statisticians weigh the relative importance of each of these statistics in order to arrive at one final HDI figure for each country. HDI rankings differ from those based on any one criterion alone. Rankings by HDI and by GNI per capita can differ, showing that high levels of human development can be achieved without high incomes, and that high incomes do not guarantee high levels of human development. For example, Saudi Arabia has a far higher income than Thailand, but a similar HDI ranking. Guatemala has almost double the income of Vietnam, but a lower HDI ranking. Large gaps between wealth and HDI rankings usually indicate deep social and structural inequalities. In the 2005 report, HDI values ranged from a high of 0.963 (Norway) to a low of 0.281 (Niger). Canada tied with Luxembourg and Sweden for third place (0.949), and the United States ranked tenth (0.944).

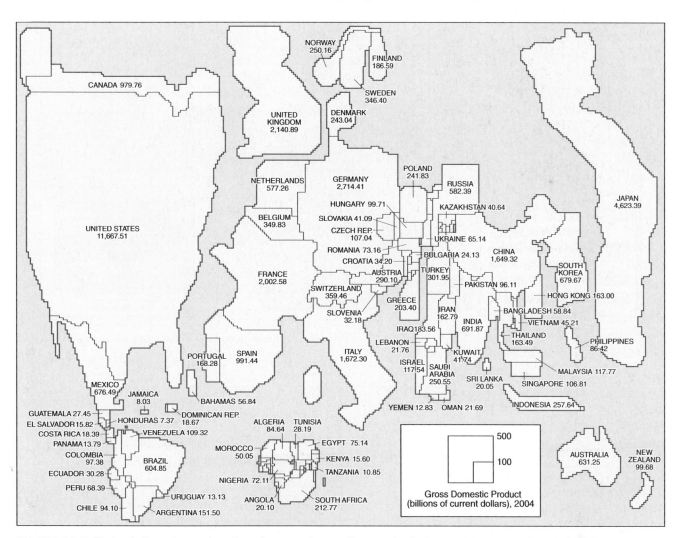

FIGURE 12-5 The relative sizes of nations' gross domestic products. The size of the countries on this cartogram corresponds to their total gross domestic product. The United States, Europe, and Japan grow considerably beyond the sizes of their relative territories. Africa, in contrast, shrinks. This cartogram contrasts sharply with the world population cartogram (Figure 5-2). The world's most populous countries are not necessarily those with the greatest economies. (Data from The World Bank)

HDI figures echo world GNI per capita figures in that both numbers are rising for humanity as a whole, but both are nevertheless falling in some individual countries. HDI has fallen in 18 countries since 1990, reflecting particularly the impact of AIDS in many African countries, and the result of social and economic turmoil in Russia since the fall of communism (see Chapter 13). In most countries, however, both GNI and HDI are advancing. The steady advance of both HDI and per capita GNI in China and in India significantly improves the statistical averages for all humanity.

Any index that weighs and compares different criteria is, admittedly, comparing apples and oranges. Chapter 10 described how popular magazines periodically rank cities based on the quality of life they offer, and we noted then that we may not agree with any magazine's criteria or weighing method. Baseball teams, good hospitals, and mild winters may be important to me; golf courses, art museums, and good schools may be more important to you. Therefore, we may not all

agree with the ranking. We may not agree that the best criteria were chosen for the Human Development Index, or we may not agree with the weights assigned to each criterion in the final tabulation. Whatever criteria are chosen, however, statistics reveal that conditions of life vary greatly. Furthermore, for most of humanity, conditions seem to be improving, but conditions of life in some countries are still poor and even falling.

Preindustrial, Industrial, and Postindustrial Societies

Geographers want to know what people are doing for a living and which activities are producing the wealth in each country. To do this, they must analyze more than just the total GDP. Chapter 10 noted that, over a long period, most countries that are rich today experienced shifts in the distribution of jobs among what are defined as their economies' primary, secondary, and tertiary

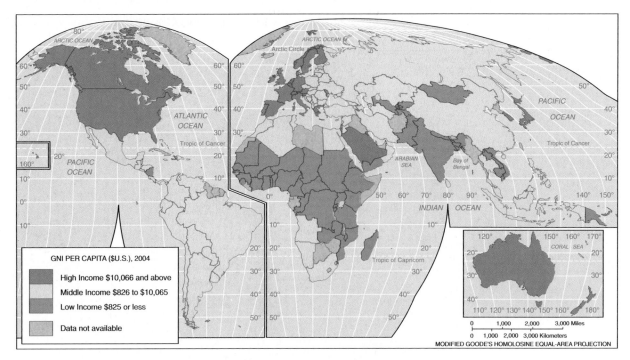

FIGURE 12-6 Per capita gross national income. Per capita national incomes in many countries in Asia and Africa are a fraction of those in Europe and North America. (Data from The World Bank)

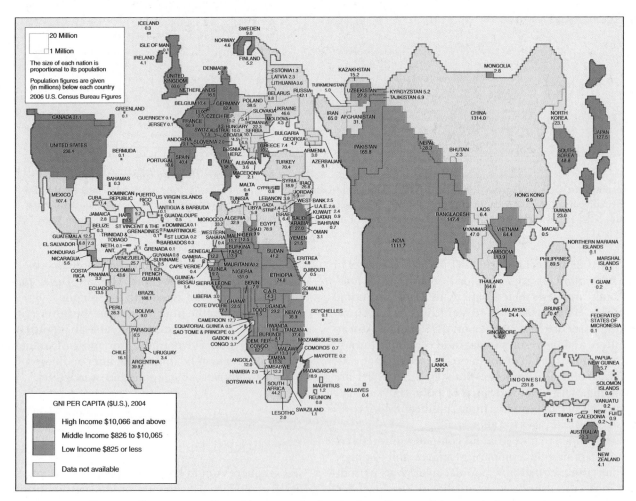

FIGURE 12-7 Per capita gross national income indicated on the world population cartogram. This figure improves on Figure 12-6 by placing the categories of income on the population cartogram that we first saw as Figure 5-2. Therefore, this figure clarifies how many people live at each income level. The colors correspond to those in Figure 12-6, so we can see that 40 percent of Earth's population lives in low-income countries, 44 percent in middle-income countries, and 16 percent in high-income countries. (Data from The World Bank)

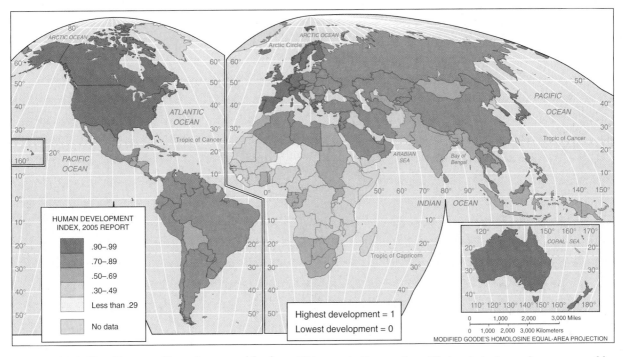

FIGURE 12-8 The Human Development Index. This map indicates where life is relatively good, as measured by the United Nation's Human Development Index, which rates countries on the basis of a statistic that balances life expectancy, school enrollment, literacy, and income. (Data from *Human Development Report 2005*)

sectors. This shift was defined as **sectoral evolution.** The best way to examine any country's economy is to study the relative importance of each sector in its GDP.

Sectoral evolution changes national employment

Before industrialization, most of a country's labor force is occupied in the primary sector (Figure 12-9). Societies with the bulk of their

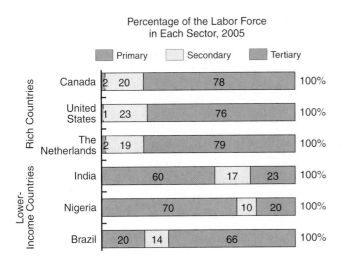

FIGURE 12-9 Sectoral analysis of the labor force.
This figure shows the proportion of the labor force in each of six countries that works in each of the three sectors of that country's economy. Workers in the rich countries (exemplified by the top three countries) work in the secondary and tertiary sectors, whereas higher percentages of workers in the poor countries (exemplified by the bottom three countries) are concentrated in the primary sector. (Data from The World Bank)

employment in the primary sector are called **preindustrial societies.** Many societies still today are preindustrial, and even though they have high percentages of their workers in agriculture, many can barely feed themselves (Figure 12-10).

As some countries industrialized, many workers found employment in factories, and the proportion of the labor force employed in the primary sector declined. This is true despite the fact that both the total number of people in the primary sector as well as the value of the primary sector's output may have risen. The workers in the sector usually became more productive. The primary, extractive activities remain crucial to many economies, but they provide a diminishing share of jobs.

Many jobs lost in agriculture were initially replaced by new opportunities in industry, and the proportion of workers in the secondary sector increased until, at no precisely defined point, certain societies came to be called **industrial societies.** In the United States, for example, manufacturing employment overtook farm employment during World War I, and by 1925 manufacturing workers had become the largest single occupational group.

Continuing evolution of some countries' economies has drawn a higher percentage of workers into the tertiary sector, producing services instead of goods. Services accounted for almost 25 percent of all jobs in the United States by 1929. Sometime in the 1940s the proportion first exceeded 50 percent, and the United States became the world's first **postindustrial society.** By 2000 tertiary employment represented three quarters of

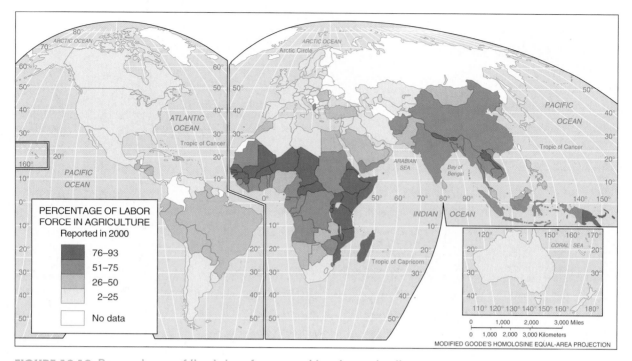

FIGURE 12-10 Percentage of the labor force working in agriculture. If you compare this map with Figure 8-31, which illustrates the prevalence of undernourishment, you will see that many countries with high percentages of their labor forces in agriculture are barely able to feed themselves, whereas several countries that have only small percentages of their labor forces working in agriculture enjoy food surpluses. What can explain this paradox? The explanation is that economic and technological development have allowed some countries to invest in agriculture, raise agricultural yields, and at the same time release workers for manufacturing and service activities. Thus, a high percentage of workers in agriculture reveals technological backwardness and poverty. (Data from The World Bank)

the nation's jobs. The leading job categories in the United States in 2005, by numbers, were teachers, retail salespersons, secretaries and typists, truck drivers, farmers and farm laborers, janitors and cleaners, waiters and waitresses, nurses and home health aids, and freight and stock handlers. All advanced nations have seen the percentage of jobs in the tertiary sector increase. This trend has also been noted in a few poor nations, whose tertiary sectors are well integrated into the international economy—tourist destinations, for example.

At first in the United States both the secondary and the tertiary sectors increased their shares of total employment as the primary sector's share dropped, but in recent years the proportion of workers in the secondary sector has begun to fall as well. The nation began losing manufacturing jobs as a percentage of all employment about 1960, and even the absolute number of manufacturing jobs has been declining since at least 1970.

The "new economy"   The production of computer hardware, software and services, telecommunications, and of products on and for the Internet are all collectively referred to as the *new economy*. This term, however, like the term *quaternary sector* discussed in Chapter 10, is not specific or agreed upon. Sectoral evolution and technological evolution proceed so rapidly that economic statisticians cannot always agree on how to

measure developments. For example, some economists believe that the production of computer software should be counted as a manufacturing activity, and they have ranked it as the third-largest manufacturing industry in the United States (after automobiles and electronics). Silicon Valley, the 80-kilometer (50-mile) corridor along Highway 101 from San Francisco to San Jose, California, is home to thousands of computer and software companies (Figure 12-11). The Standard Industrial Classification System (SIC) that originated in the 1930s, however, classified the production of computer software as a service. The new North American Industry Classification System (NAICS), which has been developed jointly by the United States, Canada, and Mexico, is continually revised to sharpen our ability to monitor economic evolution. Professor Rudi Dornbusch of MIT wrote in 1998, "In the 19th century, the great effort was to draw up a map of each country. The 20th century equivalent is to get a good set of statistics." Even into the twenty-first century, rapid change frustrates achievement of that task.

Measuring how each sector contributes to GDP   The share that each sector of an economy contributes to GDP is not necessarily the same as the share of the national labor force that sector employs (Figures 12-12 and 12-13). Some countries have high proportions of their workers in the primary sector, but

FIGURE 12-11 Silicon Valley, California. By some measures, California's Silicon Valley is the manufacturing center of the United States, yet no smokestacks are visible. Clearly, technology has progressed from the "dark Satanic mills" of industrialization described by poet William Blake (1757–1827). (George Hall/Woodfin Camp & Associates)

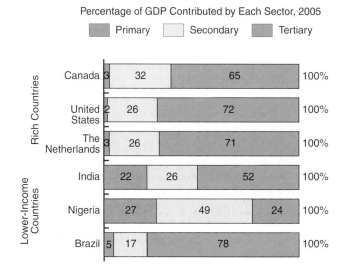

Percentage of GDP Contributed by Each Sector, 2005

Primary   Secondary   Tertiary

Rich Countries

Canada	3	32	65	100%
United States	2	26	72	100%
The Netherlands	3	26	71	100%

Lower-Income Countries

India	22	26	52	100%
Nigeria	27	49	24	100%
Brazil	5	17	78	100%

FIGURE 12-12 Sectors' contributions to GDP. This figure shows the origin of the GDP, by sector, for the same six countries that were shown in Figure 12-9. The percentage of a country's GDP that is produced in the secondary sector almost always exceeds the percentage of the country's labor force working in that sector. This means that value added per worker is usually highest in the secondary sector, so average workers' incomes are usually highest in that sector. The large percentages of the labor forces that work in the primary sectors of the poor countries produce relatively small shares of GDP. (Data from The World Bank)

those workers produce relatively little of the measurable national output. In the secondary sector, by contrast, output per worker is usually high, so the secondary sector usually contributes a larger share of GDP than it employs of the labor force.

The continuing evolution of the U.S. economy toward the tertiary sector may explain a widening gap in incomes. Many of the occupations that pay the highest incomes are usually in the tertiary sector (sports star, heart surgeon, lawyer, etc.), but most of the

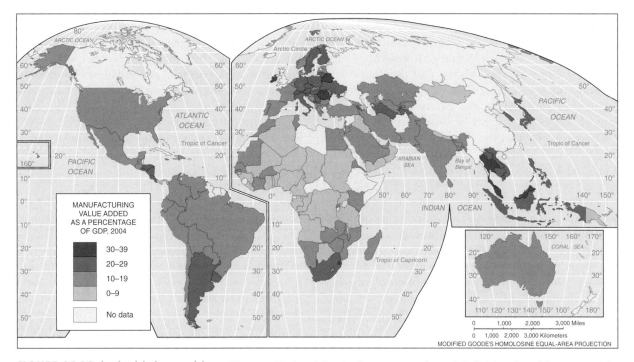

FIGURE 12-13 Industrial countries. The term *industrial society* has no exact formal definition, but this map reveals the countries in which industry—the secondary sector—contributes the highest share of GDP. The industrial sector is relatively small in some of the richest postindustrial countries and high in a few relatively poor countries that do process some mineral ores. (Data from The World Bank)

tertiary-sector jobs do not pay as well as factory jobs. The jobs listed above as most numerous in the United States—teacher, retail salesperson, secretaries and typists, and so forth—are not generally the highest-paying occupations. The inflation-adjusted median income of U.S. workers has been virtually flat since 1973. Median *family* and *household* incomes have risen only as more members of each household have gone to work; first the women, and then the youngsters, and in the twenty-first century, median incomes of households headed by someone under the age of 66 have actually been falling. Those people at the very top of the income scale, however, have enjoyed substantial increases. At the 90th percentile of income, inflation-adjusted median wage and salary income rose 58 percent between 1966 and 2001; at the 99th percentile, 121 percent; and at the 99.99th percentile (representing America's 13,000 highest-paid workers), 617 percent. Generally speaking, those people with more education are doing better, while the incomes of those with less education are falling. Educational requirements, however, are rising for almost all occupations. Today many college graduates hold jobs that once required only a high school diploma.

We do not know whether the widening wage gap in incomes is inevitable in a postindustrial society or whether this phenomenon is the result of uniquely American tax laws or other policies. International comparative studies are not conclusive.

Why Some Countries Are Rich and Some Countries Are Poor

If you compare the map of per capita GNI (Figure 12-6) and the map of HDI rankings (Figure 12-8) with the data on location of world natural resources in Chapter 9 (Table 9-1), you can see that the countries most richly endowed with raw materials are not necessarily enjoying the highest per capita GNIs and the highest standards of living. Conversely, some of the richest peoples who do enjoy the highest standards of living do so in environments with meager natural resources. The geography of resources alone does not explain the geography of wealth.

Environmental determinists (see Chapter 6) would find this inexplicable, but the study of geography teaches that the key to wealth is not *having* raw materials, but *adding value* to raw materials. The principle of adding value was first introduced in Chapter 8, where we read of dairy farmers converting milk into cheese. Some resource-poor countries are able to import raw materials, process them, manufacture items from them, and enjoy their ultimate use.

The value added to raw materials downstream in the secondary and tertiary sectors surpasses the value of the original raw materials, whether the raw materials are mineral or agricultural. The value added by refining copper ore and manufacturing things out of it, for ex-

ample, quickly surpasses the value of copper ore. The value added by grinding wheat into flour is greater than the value of the wheat. This principle holds true no matter how valuable the original raw material may be. Diamonds are valuable, but the value added to them by diamond cutters and polishers is greater than the value of the original uncut diamonds. The greater the value added, the greater the potential for profit.

The places that add the greatest value to raw materials prosper, while the places that export raw materials without adding any value to them do not enjoy the same economic growth. They may even have to buy goods manufactured out of raw materials that they originally exported themselves. For example, Jamaica exports bauxite to the United States, where that bauxite is made into aluminum and then into consumer goods, which Jamaica then imports back. As long as this continues, the United States will grow richer than Jamaica.

Even within the secondary sector, there is a hierarchy of value added. The manufacture of textiles, for example, adds less value than does the manufacture of clothing, so places that manufacture textiles try to develop clothing manufacture. Other generally low-value-added manufactured goods include shoes, toys, sporting goods, simple television sets, and inexpensive cars. Countries try to progress from production of these items to production of higher-value luxury cars, computers, chemicals, and ever more complex electronic items. An agricultural example is to note how Chile began exporting grapes, then evolved to exporting moderately priced jug wine, and today is exporting high-value fine wines. **Economic development** is a process. It consists of progressively increasing the value of goods and services that a place is able to produce in order to enjoy or to export (Figure 12-14). This process has no foreseeable end, but we do commonly speak of the rich countries as "developed" and the poor countries as "underdeveloped," or optimistically as "developing." In fact, weekly visits to your local electronic-goods store will demonstrate that the complexity of the products available in even the richest countries continues to increase.

Many of today's poor countries still export only unprocessed raw materials or materials with little value added, such as agricultural goods, basic metals, or textiles, for example (Table 12-1). Only a small percentage of their exports are manufactured goods (Figure 12-15). The welfare of their people rises and falls with world prices of commodities over which they have no control, and which have slowly declined (Figure 12-16). In 2002 the world market prices of the 33 leading non-fuel commodities in world trade were only 55 percent of what they had been in 1970. Thus, the export earnings of countries exporting these items had dropped by as much as 45 percent.

Exporters of some raw materials have tried to form cartels, such as OPEC (see Chapter 9), but few have succeeded. If production of a commodity can easily be increased, then any nation can gain a short-term

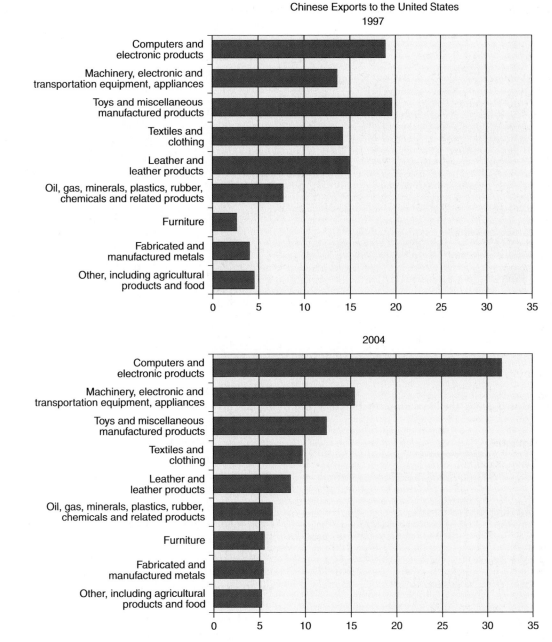

FIGURE 12-14 The developing economy of China. China exemplifies a fast-developing economy. It has reduced the share of its exports that are low-value-added items (notably toys and miscellaneous textiles, clothing, and leather products) and progressively increased the share of its exports that are higher-value-added items, such as electronic products and machinery. (Data from The U.S. Department of Commerce)

TABLE 12-1 A Few Countries' Merchandise Exports 2004 (as percentages; data from The World Bank)

	FOOD	AGRICULTURAL RAW MATERIALS	FUELS	ORES AND METALS	MANUFACTURES
Low-income countries					
Bolivia	27	2	38	19	14
Cameroon	19	24	47	5	5
Colombia	17	5	38	1	38
Kazakhstan	4	1	65	14	16
High-income countries					
Austria	6	3	3	3	84
Japan	0	0	0	2	93
The Netherlands	15	3	9	3	70
United States	7	2	2	2	82

497

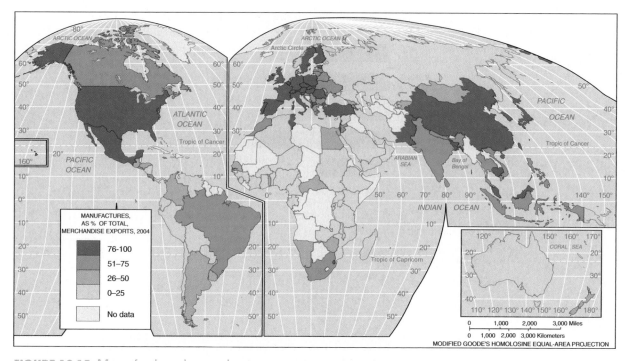

FIGURE 12-15 Manufactured exports. As a general rule, the higher the percentage of a country's exports that are manufactured goods, the more industrialized that country is. (Data from World Bank)

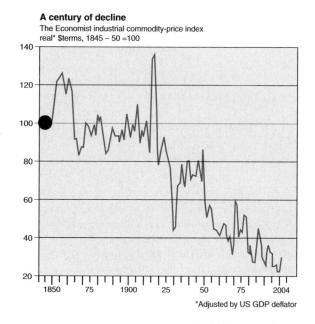

FIGURE 12-16 Declining values of commodities.
Economic development and increasing demands on resources may yet cause commodity prices to rise on world markets, but they have been declining for over 150 years. (Data from *The Economist*)

advantage by dumping stockpiles on world markets, causing prices to collapse. The Association of Coffee-Producing Countries tries to keep coffee prices high, but they have not been able to cooperate as a united group. The price of coffee on world markets periodically collapses, thus in the countries dependent on coffee exports, suddenly there is no money for chil-

dren's schoolbooks; national roads go unpaved; and hospitals run out of medicines. Erratic swings in commodity prices can harm the economies of countries dependent on them as much as steadily falling prices do.

The production of commodities usually employs only a small fraction of the national labor force, and this can cause problems of distributing the national wealth. The export of oil, for example, can be very profitable, but the task requires so few workers that having oil to export is often referred to as the "petro curse." About 90 percent of global oil and gas resources are owned by wholly or partially government-owned entities, as in Saudi Arabia, Iran, Russia, Qatar, and Kuwait. Economists agree that the wisest course is for a government to "sow the oil" (in the words of Venezuelan writer Arturo Pietri) by investing in education, infrastructure, and developing industries. The temptation is great, however, to spend the income on current consumption. About 80 percent of Libya's national budget, for example, comes from oil exports, but few workers are needed, so the government "makes work" for others. The National Oil Company employs probably twice the number it needs, and another 20 percent of the Libyan working population is employed by the civil service, where most are admittedly unnecessary. Government subsidies for food and housing allow citizens to refuse menial jobs, so heavy labor is done by sub-Saharan Africans and more skilled labor is handled by Egyptians. Thus, paradoxically, the official unemployment rate is 30 percent, yet over two million foreigners are working. Libya's Mininster of Finance, Abdulgader Elkhair, has said, "If we hadn't had oil, we would have developed. Frankly, I'd rather we had water."

Norway is practically the only oil-exporting nation to have achieved balanced economic growth. When Norway first pumped crude oil from the North Sea in 1975, it had the advantages of being a stable democracy with honest civil servants, a well-established legal system, and a large middle class. Nevertheless, oil wealth at first seduced Norwegians: The national welfare system was expanded, and the work week declined. In the 1990s, however, Norway segregated the oil business and diversified the rest of the economy. The government set up a Petroleum Fund to finance Norwegian retirements.

As Figure 12-1 illustrates, some countries that export commodities capture the value added downstream by buying processing plants, fleets, refineries, or even retail networks to reach ultimate consumers in rich countries. Russia's President Vladimir Putin himself cut the ribbon to open the first U.S. Lukoil service station in Manhattan. The Venezuelan National Oil Company (PDVSA) owns Citgo, with stations across the United States, and which even, in a gesture of "foreign aid" to America's poor, has supplied home heating oil to low-income households in Boston and New York at discount prices. Mideast oil-producing states have built facilities for manufacturing petrochemicals and pharmaceuticals. Diamond-producing countries in Africa are requiring that raw stones be cut and polished before export, so local citizens are encouraged to study diamond cutting and jewelry design.

Can culture cause poverty?

Some peoples are rich and others are poor, and we have seen that this is not necessarily related to the country's possession of natural resources. If we know the reasons for the distribution of wealth, and if we want to achieve high levels of welfare everywhere, then two paths of action are open to us. Either we could redistribute world wealth (in other words, give foreign aid from the rich to the poor), or else we could teach or encourage the poor places to generate their own wealth locally. In order for the poor areas to create their own wealth, however, they might have to change local government policies (such as taxation or budgeting) or even to change aspects of their local culture (such as women's rights or education). Culture does have consequences, and some scholars argue that some cultures hinder economic development. It has even been said that there are "cultures of poverty." In 2002 the United Nations issued a special report on economic development in the 22 member nations of the Arab League. The report, written largely by Arab economists, concluded that the principal barriers to Arab development were shortages of three essentials for development: freedom, knowledge, and womanpower. All of these are matters of domestic politics and culture.

If a people's culture does hinder their economic development, the people may nevertheless resent suggestions that they change their culture in order to improve their standard of living. They would rather just be given material assistance. At the U.N. World Conference on Poverty, held in Copenhagen in 1995, the rich countries argued that poor countries could develop if they would just end corruption, guarantee workers' rights and property rights, and help women and children. Representatives of the many poor countries responded that such suggestions were unacceptable intrusions into local customs. They instead demanded immediate financial aid.

The debate on whether it is necessary to change culture in order to achieve wealth engages not only diplomats but all social scientists—economists, sociologists, anthropologists, as well as geographers. Many of the ideas presented throughout this textbook may help you form your own opinions.

THE GEOGRAPHY OF MANUFACTURING

Most countries strive to industrialize themselves so they can capture the high value added by manufacturing. One of the earliest scholars to identify the locational determinants of manufacturing was Alfred Weber (1868–1958). Weber's analysis focused on the role of transport costs, and it included models that differentiate material-oriented manufacturing from market-oriented manufacturing.

Material-oriented manufacturing is located close to the source of the raw material for one of two reasons. The first reason is that the raw material is heavy or bulky, and manufacturing reduces that weight or bulk. The steel industry is an example. The value added in manufacturing steel is low relative to the cost and difficulty of transporting the raw materials (iron ore, coal, and water), so steel mills generally are located where the raw materials can be found or cheaply assembled. The second reason why material-oriented manufacturing may be located near the raw material is because some raw materials are perishable and need immediate processing. Examples include the canning and freezing of foods, and the manufacture of cheese from milk.

Market-oriented manufacturing, Weber's second category, is located close to the market, either because the processing increases the perishability of the product (baking bread, for instance) or because the processing adds bulk or weight to the product. A can of a soft drink, for example, consists of a tiny amount of syrup, plus water, in a can. The syrup can be transported easily and cheaply, but canned soda is heavy and expensive to move. Therefore, the water should be added and the beverage canned close to the market.

Weber elaborated his models of the location of manufacturing by adding additional considerations, such as the availability of a labor force. In that three-factor model, manufacturers locate factories to minimize the cost distance from three points: the location of the raw materials, the labor force, and the market. The optimum location for any specific manufacturer depends on the balance of these costs in his or her business.

Several examples demonstrate that consideration of Weber's factors is still important in many industrial location decisions. Wood furniture making typifies an industry that may locate near its source of raw materials. In the United States, Grand Rapids, Michigan, became a center for wood furniture manufacturing in the nineteenth century, drawing on Michigan's extensive forests. When those forests were depleted, furniture makers relocated to their new sources of wood: Georgia and North Carolina. The copper industry exemplifies a bulk-reducing industry. In the United States most copper ore is smelted (separated into its metallic ingredients) at the mines in New Mexico and Arizona, because smelting removes a high percentage of the original ore as waste. The largest copper refineries and manufacturing facilities, by contrast, are found in Baltimore, Maryland; Norristown, Pennsylvania; and Perth Amboy, New Jersey. These locations are closer to markets and labor forces, and they are also situated to draw on imported copper supplies. The automobile industry is an example of a fabricated metal industry, which brings together a number of previously manufactured parts and assembles them into a more complex product. In the United States several new automobile assembly factories are located in the center of the country (Figure 12-17). In addition to geographical convenience, this region is attractive to manufacturers because labor conditions and terms are more favorable to manufacturers in this region than they are in the regions of older, more heavily unionized industry.

Locational Determinants for Manufacturing Today

Weber's studies still help us understand many industrial location decisions, but several things have changed since Weber's time. Transportation costs have steadily fallen (look back to Table 6-1), and at the same time the value added in manufacturing has increased as manufactured products have become more and more complex. The value of the iron, copper, and other raw materials that go into a mobile telephone is only a minuscule percentage of the value of the completed unit. As transport costs fall and the value added in manufacturing rises, high-value-added manufacturing can more easily relocate; it is, as we learned in Chapter 6, increasingly footloose. Furthermore, today trade binds world regions, so locational determinants are no longer tied to local, regional, or even national scales. The entire globe must be examined as one theater of interconnected operations.

Weber's models considered the locations of the raw materials, the labor force, the market, and the transportation costs in determining the best site for a manufacturing operation. Since Weber's time the balance among locational determinants for many industrial processes has tipped away from what Weber considered. For example, the value of the labor input is generally a shrinking percentage of the value of manufactured goods. In the consumer electronics industry, for example, it is no more than 5 to 10 percent. At least five other considerations, however, must be added to Weber's (Table 12-2). Each of these considerations is an *input*, or ingredient, for manufacturing. The first is *capital*, because manufacturing is increasingly capital

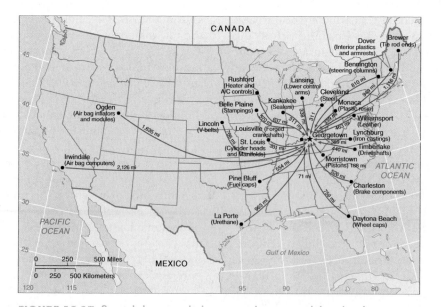

FIGURE 12-17 Supplying parts to an auto assembly plant. Georgetown, Kentucky, was chosen as the location of a Camry assembly plant partly in order to draw conveniently on supplies of parts from around the United States.

TABLE 12-2	Locational Determinants for Manufacturing	
FACTORS CONSIDERED BY ALFRED WEBER	**ADDITIONAL FACTORS THAT ARE IMPORTANT TODAY**	
1. Raw materials	1. Capital	
2. Labor force	2. Technology	
3. Market	3. Governmental regulations, including environmental legislation, labor rights, and labor union power	
4. Transport costs	4. Political stability	
	5. Inertia	

intensive. The second is *technology.* Third, industries seek places with *hospitable governmental regulations;* that usually means low taxes, little environmental regulation, and restraints on labor rights and unionization. *Political stability* is the fourth consideration. Manufacturers hesitate to invest in volatile political environments. The *inertia* of facilities and suppliers that are already in place is still another consideration in the geography of manufacturing. Factories are major investments, and networks of suppliers and trained labor forces develop around them; therefore, they are not quickly abandoned.

Figure 12-18 is a map showing the world's most important manufacturing regions. A map of manufacturing measured by value added, however, would show that the greatest value added is in those countries that dominate as shown in Figure 12-5, the cartogram of the nations' gross domestic products. Much of the world's manufacturing takes place in Europe, for historical reasons. Europeans brought about the Industrial Revolution first, giving Europe a technological lead which enabled it to overwhelm most of the rest of the world

politically and economically. Europeans generally located the manufacturing in their own countries in order to retain the value added. It was not their intention to enrich their colonies. In some cases they actually intended to retard the development of manufacturing in their colonies. For example, from 1651 until 1776, at the very beginning of the Industrial Revolution, Britain's Navigation Acts prohibited Americans from manufacturing products from local raw materials. Raw materials had to be shipped to Britain, where value was added by manufacturing, and then the products were shipped back to the colonies—at higher prices. Even iron from New Jersey had to be shipped to Britain, where it was made into nails, and the nails were then shipped back to New Jersey. The colonists' resentment of this exploitation helped trigger the U.S. War of Independence. Many colonies in Africa and Asia later suffered the same deliberate restraint on their industrialization.

When manufacturers did locate factories in countries other than their own, they generally chose other rich countries where they could find markets and labor

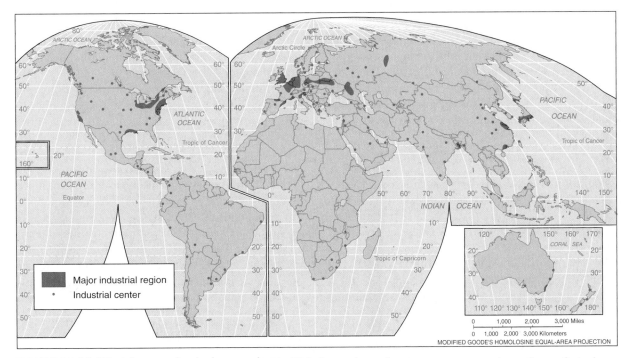

FIGURE 12-18 World manufacturing regions. This figure shows the greatest concentrations of manufacturing today. (Adapted from Paul L. Knox and Sallie A. Marston, *Human Geography: Places and Regions in Global Context.* 4e. Upper Saddle River, NJ: Prentice Hall, 2006.)

forces. Today the rich countries still have the technological lead and the best-trained labor forces. They usually are politically stable, and they provide the eventual market for most goods. This geographical system has prevailed for so long that it has a tremendous inertia. Raw materials continue to flow from the poor countries to the rich countries, where they are transformed into more valuable goods. Still today, for example, Ghanaian cocoa beans may enter Europe freely, but Ghanaian cocoa powder faces a stiff tax. This pattern has been labeled *economic colonialism* or *economic imperialism*. It is noteworthy, however, that neither Japan nor Switzerland, two countries with wealth but without substantial natural resources, ever had a major political empire.

In their efforts to industrialize, today's poor countries suffer from a number of problems that reinforce one another. Markets in poor countries are not as important as those in the rich countries. In addition, cheap manufactured goods from the rich countries can flood the poor countries' domestic markets and discourage competitive domestic manufacturing. The countries' poverty increases their political instability. They need capital, and their people need education.

Poor countries can, however, offer three locational determinants: raw materials, inexpensive labor, and hospitable regulatory environments. Workers in poor countries receive relatively low wages, and they will often work in conditions worse than will workers in the rich countries. Poor countries' governments seldom complain about pollution coming from a new factory that provides jobs. Today's poor countries can exploit these advantages to attract new industry, and many corporations from the rich countries are today placing manufacturing operations abroad or buying from suppliers abroad to take advantage of these factors. **Outsourcing** is the name given to the delegation of any of a firm's operations to an external subcontractor who specializes in that activity. If the outsourcing is to a foreign country, it is often called **offshore outsourcing.** Thus, some poor countries are industrializing by attracting the offshore outsourcing from rich countries.

Locational Determinants Migrate

The geography of locational determinants is not fixed. The locational considerations for manufacturing make it possible for manufacturing to migrate around the world as the mixes of attributes among places change and attract it.

There are, in fact, reasons why the geography of locational determinants for industry—and thus the geography of industry—will continually change. Chapter 6 already suggested relevant factors: new products, new technologies, new raw materials, new sources of traditional raw materials, new methods of manufacture, new governments with new policies, growing and shrinking labor supplies, and the opening and development of new markets.

If a place first attracts manufacturing because it offers a low-wage labor force, its local standard of living and local wages will rise. As local wages rise, the industries that developed there to take advantage of the low wages will be driven away. They will migrate to still poorer places, where wages are lower. Thus, ideally, as each location develops, it surrenders low-value-added activities to poorer regions. These regions in turn begin to climb the ladder of economic development. WestPoint Stevens, Inc., for example, America's largest textile maker, outsourced production offshore and reduced its U.S. payroll by half between 1988 and 2000. The U.S. company is concentrating on marketing and on adding psychological value by signing licensing agreements with designers Ralph Lauren and Martha Stewart.

The manufacture of men's dress shirts for the U.S. market offers another example. Shirt making is labor intensive, so manufacturers are continuously searching for cheaper labor. In the 1950s, U.S. shirt manufacturers located in Japan to use the low-paid workers there. When the costs of labor and real estate rose in Japan, the companies moved to Hong Kong. As Hong Kong's factories gave way to offices, the shirt makers moved again, first to Taiwan and Korea and then, in the 1970s and 1980s, to China, Thailand, Singapore, Indonesia, Malaysia, and Bangladesh. Toward the end of the 1980s and into the early 1990s, Costa Rica, the Dominican Republic, Guatemala, Honduras, and Puerto Rico became shirt-manufacturing centers. The Nien Hsing Textile Company, for example, which was founded in Taiwan in the 1970s, moved its jeans sewing factory to Managua, Nicaragua, in the 1990s. Its 1,800 Nicaraguan workers earn from $65 to $124 per month; the labor input in a pair of jeans that retails in the United States for $25 (under several different brand names) is 50 cents. The most expensive jeans manufacturers maintain high prices for their garments largely by adding psychological value through status advertising.

The Caribbean basin countries offer cheap labor and U.S. tax advantages that were granted under a U.S. government program, the Caribbean Basin Initiative, which was designed to help these countries industrialize. These countries' proximity to the United States offers several additional advantages. For one, U.S. managers can oversee operations more easily. Proximity also promises speedier delivery, and fashion demands the fast introduction of new styles. Also, renting a ship for an additional week's sailing time from the Far East is more expensive than the shorter trip from the Caribbean. Still another advantage of great concern to business executives depends on interest rates, or what we call the *cost of money.* Shirts in the hold of a ship in transit are a form of capital, and that capital is not earning any profit. One shipload of shirts may be worth $15 million. The interest on that capital each day that the ship is on its way to the United States adds a few cents to the retail price of each shirt. "Time is money," and distance is money, too.

Manufacturing in the United States

No other country yet matches America's total manufacturing output of $2 trillion (value added in 2004). In the United States, however, the share of the labor force in manufacturing and the sector's contribution to GDP are falling—from 31 percent of jobs and 27 percent of GDP in 1960 to 10 percent of jobs and 15 percent of GDP in 2004. American corporations are building new factories abroad, and Americans are increasingly buying imported goods. By 2003, about half of the manufactured goods that Americans bought were made abroad, up from 31 percent in 1987. Economists debate whether this economic evolution away from the secondary sector is good for the United States.

In 2002, most U.S. steel makers were uncompetitive with global competitors. To salvage what was left of the industry, the United Steelworkers Union agreed to cut wages, benefits, and jobs. The steel companies ended retiree health plans and transferred the costs of pensions to the federal Pension Benefit Guaranty Corporation (which offers lower pensions). The result was to clear billions of dollars off the firms' books and out of steelworker retiree's pockets. Today, America's airlines and auto firms appear to be headed down that same path.

In 2005, Delphi auto parts company filed one of the largest bankruptcies in U.S. history, and in 2006 it announced it would sell or close 21 of its 29 plants in the United States. The last eight plants would reduce their workforces, and remaining workers would suffer cuts in benefits and wages. Chief Executive Officer Robert S. Miller said, "We've got 60,000 workers in Mexico. Labor there is paid $7,000 a year. Before long, it's going to be very difficult for someone with a manual-labor job in the U.S. to have the kind of lifestyle they've enjoyed up to now. It ain't going to happen." A preview of the competition that U.S. automakers can expect appeared at the 2006 Detroit Auto Show: The Chinese Geely 7151 CK sedan, which is scheduled to retail in the United States in 2008 for about $10,000.

The importation of manufactured goods, notably cars, apparel, electronic goods, computers, and furniture, has lowered their cost to American consumers. Falling prices is called *deflation*. The deflation of prices of manufactured goods comes at the cost of American jobs, and, as we have seen, the substitution of service jobs for manufacturing jobs contributes to the spreading gap in incomes in the United States. Manufacturing jobs pay an average of 20 percent higher wages than tertiary sector jobs. In addition, manufacturing jobs have a higher multiplier effect than service jobs: The average production-sector job produces three times as many additional employment opportunities as does the average service job. Manufacturing also disporportionately employs non-college-educated workers, and as of 2006 more than 60 percent of U.S. workers lacked college degrees. We may applaud that the incomes of female workers, traditionally associated with service occupations, have risen relative to men's wages. Full-time female workers' wages held steady at about 60 percent of men's for decades after World War II, but by 2006 they had risen to 78 percent of their male counterparts', but this does not help the nation as long as median wages do not advance.

Some observers are disturbed that an increasing share of U.S. military procurement comes from foreign suppliers. They find this arrangement upsetting even if the suppliers are foreign subsidiaries of U.S. corporations.

Can this evolution away from manufacturing continue? Americans will be able to import manufactured goods only as long as Americans can offer the rest of the world services plus a few sufficiently high-value goods. In 2003 David Heuther, chief economist at the United States National Association of Manufacturers, stated flatly that "you have to assume that manufacturing will continue to disappear," but he argued that America's high-tech advantage and its ingenuity will sustain a small high-value manufacturing base. Nobel prize–winner George A. Akerlof, by contrast, argued that the value of the U.S. dollar will continue to fall against other currencies. This will make manufacturing operations relatively cheaper in the United States, and that event will entice manufacturing operations back to the United States. How much would the dollar have to fall, and what would be the additional ramifications of that event? We do not know.

While prices of imported items are deflating, prices of goods produced locally and of services such as health care, education, and such day-to-day activities as trash collection and auto repair, rise (*inflate*). In the early years of the twenty-first century, the deflation in the costs of manufactured goods has slowed the inflation of the overall cost of living.

The Economy of Japan

Japan is the most astounding and paradoxical success story in economic geography today. The Japanese islands have few natural resources, but on this meager natural geographical endowment, the Japanese have developed a great economy. How?

The answer lies in Japan's cultural resources and in its trading patterns. Japan's economic success demonstrates how a place can prosper by importing raw materials and adding value in manufacturing, or by providing other downstream services. The key cultural resources include the people's education and labor skills, technology, and the country's style and degree of cooperative organization. All these factors are carefully managed by the government. Japan's participation in international exchange is also carefully regulated. The Japanese import raw materials, transform them into manufactured goods, and export these finished goods, plus an array of valuable services, around the world. In 2005, Japan's merchandise imports were 13 percent food, 3 percent agricultural materials, 20 percent fuels, 5 percent ores and metals, and 57 percent manufac-

Vietnam

In 1975, North Vietnam conquered South Vietnam, and the United States withdrew, imposing an economic embargo that lasted until 1994. Even without U.S. trade, however, Vietnam adopted market-oriented reforms and attracted foreign investment. First international oil companies bid for licenses to search for oil, but eventually manufacturers built factories. The first were simple assembly plants for apparel companies such as The Limited Brands and Nike; by 2006, Nike subcontractors employed 130,000 Vietnamese workers. Foreign investment has climbed the ladder of value added. The Intel Corporation is building a computer processor factory in Ho Chi Minh City (formerly known as Saigon) that will employ 1,200 workers. Canon announced that it was investing in an ink-jet printer factory in Hanoi. The government aims to increase exports of electronics by 27 percent per year. Vietnam is an increasingly attractive alternative to China, where factory workers today earn as much as five times what Vietnamese factory workers earn. Furthermore, in Vietnam land is cheap, and shipping is cheaper than from Thailand or Indonesia. Vietnam continues to invest in roads and airports and to improve its legal system, banking system, and regulations, making it a safer and more reliable business environment.

Vietnamese students of technology. These students of a Hanoi technical university greeted Bill Gates like a rock star during his 2006 visit. (Richard Vogel/AP Wide World Photos)

Standing 108th in HDI, Vietnam outranks its regional neighbors Indonesia (110), India (127), Myanmar (129), and Cambodia (130). The Vietnamese people, half of whom have been born since the end of the war in 1975, are looking toward the future, not the past. In 2006, thirty-one years after the end of the war, the United States and Vietnam signed a bilateral trade pact.

---

tures. Japan's merchandise exports, however, were 97 percent manufactured goods.

In the 1950s Japan was a relatively poor country, but it began to manufacture products of relatively low value added. These products included cargo ships, toys and games, sporting goods, inexpensive clothing and cars, and simple electronic appliances. Japan built its prosperity by steadily adding value to its exports.

As Japan developed, it surrendered the manufacture of the items that are low in the value-added scale. Japan invested in factories to manufacture these products in less-developed countries that are dominated economically by Japan. Through the 1990s, Japan's production of lower-technology consumer electronic equipment began to fall as the production of these items was relocated by Japanese firms to Vietnam, Thailand, and Indonesia. Japan's own production of the highest-technology digital compact cassettes, electronic components and devices, and industrial electronic equipment, however, continued to rise. For example, 18 to 25 percent of the cost of a laptop computer is in its liquid crystal display (LCD)—almost twice the percentage of value that is in its microprocessor chip. By 2000 the manufacture of LCDs was essentially a Japanese monopoly.

Entering the twenty-first century, the Japanese economy continued to evolve. By 2002, with about $2 trillion in assets invested overseas around the world and

$500 billion in foreign reserves in its vaults, Japan for the first time earned more money from its investments abroad than it did from exporting manufactured goods. Some observers have described this as stagnation. Japan, a nation with a fertility rate below the replacement rate, an aging population, and now beginning to live off its savings, has been called "A nation in retirement."

Technology and the Future Geography of Manufacturing

Because the balance among the inputs to manufacturing is continuously changing, maps of the world geography of wealth, of manufacturing, and of economic development will continue to change. Each nation must tailor its economic policies in order to compete. For example, the importance of technology is increasing. When Japan was industrializing, it bought technology. By one estimate the Japanese paid foreign manufacturers a total of only $10 billion for patents and licenses between 1950 and 1980. That must rank as the shrewdest investment any nation ever made. Today Japan holds almost 50 percent of all U.S. patents granted to non-U.S. residents, and it has joined the United States and the other most advanced countries among recognized holders of key patents. Japan has continued to achieve advances in industrial productivity. In 2004,

Japan's 356,000 installed robots outnumbered those in the United States by more than three-to-one.

Technology and capital contribute the most rapidly increasing shares of the final value of most manufactured goods, and technology and capital are exceptionally footloose. In the manufacture of computer disk drives, for example, the raw materials are an insignificant fraction of the drives' final value, so computer disk drives can be easily manufactured and distributed worldwide from almost anywhere. Why then are more than half manufactured in Singapore? They were not invented there. The decisive locational determinants for this high-value product are the availability of inexpensive skilled labor, technology, capital, and political stability in Singapore. None of these factors, however, is unique to Singapore. How long will Singapore continue to dominate world manufacture of disk drives?

Tomorrow's world geography of manufacturing may be guessed by mapping investment in research today. Some analysts feel that U.S. investment in research may be falling behind that of Japan and other international economic competitors, measured both as a percentage of GDP and in actual money terms. Singapore recognizes that its current dominance in the manufacture of electronics is uncertain, and the government has bet that pharmaceuticals and biotechnology can replace that industry. It has opened a research center called Biopolis, which has attracted some leading U.S. scientists. Dr. Neal Copeland and Dr. Nancy Jenkins left the U.S. National Cancer Institute to take positions in Singapore, saying, "We wanted to be in a place where they are excited by science and things are moving upward."

NATIONAL ECONOMIC-GEOGRAPHIC POLICIES

Each country organizes its domestic economy, manages the distribution of economic activities within its territory, and regulates its participation in world trade and investment.

Political Economy

Each country defines a set of principles to organize its economic life. The study of these principles is **political economy.** In all economic and political systems, the government usually provides those services that are unprofitable to the provider but that diffuse benefits throughout the economy, called *positive externalities* (see Chapter 9). These services include education, transportation, water and sewerage, and other aspects of the infrastructure.

Each country also manipulates its national budget and expenditures, interest and exchange rates, and money supply in order to promote growth and high employment. The tools for manipulating these are

imperfect, but national governments' economic planning and management, their revenue and expenditure, and their role in redistributing income have made national economic policies more central to most people's lives than ever before.

Direct government participation in the economy ranges theoretically from a *communist system,* in which both the natural resources and the productive enterprises are **nationalized**—that is, owned by the government in the name of the people—to a **capitalist system,** in which the state defers to private enterprise and a stock market raises and allocates investment capital. The economy of each country lies between these two extremes: Most countries have mixed economies.

Even among those countries generally considered capitalist, there is a great range of government involvement in the economy. Each country's political economy usually reflects other characteristics of that country's culture. The United States, for example, generally favors a system that minimizes the government's role. This is called **laissez-faire capitalism,** from the French for "Leave us alone," supposedly said by the French economist Vincent de Gournay (1712–1759) to a government bureaucrat. The U.S. government has never developed an explicit industrial policy, although its regulations, research funding, and defense budget have greatly influenced the national economy. The United States also designs its tax system to allocate capital. For example, Chapter 10 noted how the tax code favors investment in housing.

The attacks of 9/11 caused some reevaluation and readjustment of America's laissez-faire stance. Many people came to see the federal government as uniquely capable of performing some urgent tasks: fighting enemies abroad, providing security at home, rescuing bankrupt airlines, protecting Americans from bioterror, and making the skies safe again. The *liberal* political philosophy generally favors government action to solve problems. These sentiments caused a move toward greater governmental regulation and even governmental assumption of some responsibilities that had been privatized. For example, in 2001 Congress created a new federal force responsible for airport security, a responsibility that previously had been surrendered to private security forces.

The *conservative* political philosophy, however, favors privatization, arguing that the private sector can deliver many public services at higher quality for lower cost than government bureaucracies can. Therefore, the conservative presidential administration of G. W. Bush privatized many government services. By 2006, some 40 percent of federal discretionary funding went to private companies doing everything from cleaning federal parks to feeding the U.S. troops in Iraq. Some privatized programs revealed corruption, price inflation, and incompetence, but others defeated bureaucratic inaction and inspired new approaches to problem solving. Case-by-case analysis is required to examine this argument in full.

In several Asian countries, the government plans and regulates the economy, although it does not own many companies outright. This is called *state-directed capitalism*, and it conforms with these countries' Confucian bureaucratic cultural traditions (discussed in Chapter 7). South Korea, for example, is considered a capitalist free-market economy, yet the government proposes national economic plans and subsidizes new industries. Until the late 1990s, industry was dominated by combinations of corporations, called *chaebols*, which were affiliated by interlocking directorships and ownership. Daewoo, Hyundai, and Samsung were examples familiar around the world. The economies of Japan and Taiwan are similarly organized, but in the United States such combinations would be broken up as illegal trusts.

Western European countries have been privatizing their economies through recent decades, although nationalized companies still account for about 25 percent of GDP in France, 11 percent in Germany, and even 3.5 percent in Britain, which began the privatizing trend in the early 1980s. Eastern European countries are rapidly privatizing as they abandon communism, and the special case of China will be discussed shortly. Some African countries long have insisted on national ownership of natural resources as a point of national pride, but even they have begun to privatize and invite foreign capital. In 1997, Zambia's then-president Frederick Chiluba said, "We don't care who buys the mines in Zambia so long as the mines make money and contribute to the exchequer [national treasury]." Many countries are today creating new national stock markets (Figure 12-19). The combination of privatization, reduction of government regulation, opening to international investment, and increasing participation in international trade is often called a *neoliberal* approach to political economy.

Most Latin American countries have traditions of outright government ownership of the nation's assets and industries, but in the 1980s and 1990s, these governments began privatizing their assets. Purchasers included domestic investors and corporations, but most purchasers were foreign. Early in the twenty-first century, however, nationalist politicians fanned public resentment of foreign ownership of resources, and, encouraged by rising energy prices, several Latin America governments began to assert greater control over their energy resources (Figure 12-20).

National political-economic policies also dictate the conditions of employment. The United States favors a flexible labor market. Employers are quite free to hire workers as needed, but also to release workers they no longer need, which means that Americans change jobs at a high rate. Europeans' jobs, by contrast, have traditionally been more protected. Employers cannot so freely release unneeded workers, and therefore, it has been argued, economic growth is hampered because employers are reluctant to hire workers in the first place. For example, through most of 2005, the unemployment rate of people 15–24 years of age in France was 22 percent; in the United States it was 12 percent. The next year the French government attempted to increase the flexibility of the labor market and encourage the hiring of young people by allowing employers more easily to release workers under 26 years of age. The initiative triggered destructive riots

FIGURE 12-19 An emerging stock market. The new Ugandan Stock Market lists stock shares of only seven Ugandan corporations, and it is open only a few minutes per week. Nevertheless, the government hopes that its existence will stimulate Ugandans to save and invest in Uganda's growth, and to incorporate new business ventures. (Vanessa Vick)

FIGURE 12-20 Nationalizing resources. "Nationalized: Property of the Bolivians," reads the banner. In May, 2006, Bolivia's newly elected president Evo Morales ordered the military to occupy his country's oil and gas fields. "The time has come," he declared, "the awaited day, a historic day, in which Bolivia retakes absolute control of our natural resources. The looting by the foreign companies has ended." Bolivia's resources had in fact been privatized unconstitutionally in the mid-1990s, without the approval of the Bolivian Congress. (AP Wide World Photos)

across the country; the young felt that they were being asked to sacrifice job protections that older workers had traditionally enjoyed. Some individuals in any country may be described as *opportunity-seeking*, whereas others are more *security-seeking*. Whole cultures may tend toward one or the other of these two types, and some economists argue that rigid job protections for workers seeking security hamper an economy's ability to compete.

For private enterprises to succeed, the government must in any case perform certain basic functions. Economists regularly attempt to measure and compare countries' "business environments," asking, for example, whether a country's government secures property rights, whether the judiciary is independent and reliable, how long it takes to obtain licenses necessary to start a new enterprise, and whether corruption and bribery are endemic. For example, The World Bank has reported that it requires 2 procedures and 2 days to start a business in Australia, but 7 procedures and 32 days in Ethiopia. It takes 15 procedures to register property in Brazil, but only 3 in Denmark. An average sub-Saharan business executive takes 50 days to get the 21 signatures on 9 forms necessary to export an item, and those procedures may require several bribes. An Indian exporter needs to collect 22 signatures on 10 documents. Such studies indicate that the business environment in some countries significantly retards economic growth.

The results of privatization

If a country privatizes some assets, it must devise a policy for investing the windfall of money that it receives. This challenge faces many countries that are privatizing assets today. If the country is so poor that it spends the money on immediate needs, such as food, then it will soon be worse off than it was before privatization. The assets will have been sold and the money spent. In other cases, some countries' corrupt rulers have virtually stolen the money from privatization and deposited it in personal accounts in foreign banks. Our present state of economic theory and experience suggests that if a government invests the income from privatization in education, infrastructure, or health care, however, then the country might achieve economic development.

The privatization of national assets or industries ramifies across a country's economy and politics. This is because nationalized activities fulfill responsibilities beyond the merely economic ones. Nationalized enterprises might be subsidized in order to maintain high employment, to sustain the economies of poor regions, or just to win votes. These political purposes may provide positive externalities across the society, but they may not be economically justifiable. The principal goal of a private enterprise, by contrast, is to maximize profits for shareholders. This usually requires the enterprise to be cost efficient in order to compete in world markets. Today, as more countries embrace privatization

and capitalism as the surest route to economic development, thousands of workers are being laid off from economically inefficient nationalized industries or from enterprises that are being privatized. In many countries from Argentina to China these workers are heading for the ballot box (or even taking to the streets) in protest. National leaders are gambling that privatization will release economic dynamism and stimulate economic growth fast enough to create new job opportunities for these workers.

Crony capitalism

Crony capitalism is the name given to situations in which financial markets are not free and fair, but are manipulated by politicians or insiders. Crony capitalism sometimes occurs in countries that practice state-directed capitalism, but are run by oligarchies. In this case a political crisis can immediately become an economic crisis. In the autumn of 1997, for example, an international loss of confidence in the stability of the government of Indonesia triggered a reevaluation of that country's financial strength, and the ensuing turmoil endangered the financial structures of several other East Asian countries. Political and economic restructuring required adopting regulatory infrastructures and legal safeguards for markets, democracy, and the rule of law.

Events early in the twenty-first century revealed that the United States is not immune to crony capitalism. Corporations were found to have been creating hundreds of millions of dollars in fake profits (Enron), disguising operating expenditures as capital investments (WorldCom), booking nonexistent revenues to inflate the stock price (Global Crossing, Lucent, and many others), and overstating earnings and backdating accounts to inflate executives' rewards (United Health Group, Fannie Mae, and many others). These latter activities were called "stealing, pure and simple" by *Fortune* magazine. Numerous large corporations have been virtually pillaged by their boards of directors and executives, while one of Wall Street's most trusted stock analysts urged his clients to buy stocks that he privately called "crap." America's most prestigious accounting firms and banks colluded.

Multi-billion-dollar corporations went bankrupt, and hundreds of millions of dollars were levied in fines against corporations, accountants, and banks. Several corporate executives have been jailed, and new regulatory legislation was passed to rectify these faults in America's economic system. Stockholder and legal suits and criminal proceedings continue. These events, however, have significantly damaged the prestige that the American capitalist model has held for 50 years, and the ramifications of this damage will continue in world trade negotiations for decades. The fact that other countries felt freed by the situation to develop alternative models may, however, in the long run, produce innovations in economic development theory that will benefit all countries.

REGIONAL *Focus* ON

South Korea

South Korea demonstrates that a country can grow under one political economy and then, if continued growth is threatened, it can virtually reverse its course and adopt a new form. South Korea has few natural resources, and in 1962 its economy was still in shambles from the Korean War (see Chapter 13). Nevertheless, for the next 35 years the South Koreans achieved an annual rate of economic growth of over 9 percent, the highest ever achieved anywhere. South Korea followed the Japanese model of state-directed capitalism. The Japanese had occupied Korea from 1910–1945, and Japanese-stye institutions were in place. The government owned the banks, which loaned no money to consumers or to small businesses, but only to the *chaebols* at artificially low interest rates. The *chaebols* could capture world markets for South Korean goods because neither capital nor labor cost were high. South Korea exported many goods that South Koreans themselves could not afford to enjoy. Workers were not satisfied with low wages, but enjoyed lifetime job security—famously called an "iron rice bowl". The national tax structure encouraged savings and discouraged consumption. The country was governed by virtual dictators.

During two shattering weeks in late 1997, Japan devalued its currency so Japanese goods were cheaper than South Korean goods on world markets. South Korean exports halted, and because there was little domestic demand, the national economy collapsed. The new president, Kim Dae-jung, saw the necessity of reversing economic policies. He privatized the banks and opened the country to foreign investment. The *chaebols* were broken up, and jobs were no longer secure. Economic growth and welfare took off. By 2005 South Korea held almost $250 billion in foreign reserves and was enjoying an annual per capita GDP growth rate of over 4 percent. Its HDI had steadily improved, placing it in a tie for 23rd place among countries.

The South Korean experience demonstrates that different types of economic organization may be best for a country at different stages in its own economic development, perhaps even at different times in the development of the global economy, or at different times in each country's own social, cultural, or political development. We do not know the answers to questions such as these.

Yoo Jung Hwan, the leader of the labor union at South Korea's Jinro distiller, left, and Kim Seon Joong, the corporation's chief executive, console each other after a South Korean court ordered the company into bankruptcy in 2003. Foreign creditors had demanded the action. Under the directed-capitalism economic system that South Korea had employed until 1997, the government probably would have stepped in and saved the company. Under the laissez-faire system of global capitalism, however, the inefficient distiller went broke, and many workers lost their jobs. Nationalist outrage over the closure of Jinro triggered new Korean defensive economic strategies in 2006, when American financiers made a bid to take over Korea's biggest cigarette maker, KT&G. Koreans demanded that the government protect Korean corporations from other foreign corporate investors, all of whom were seen as marauders. (Seokyong Lee/Bloomberg News/Landov LLC)

Variations in Wealth Within States

Countries try to boost their economic growth by organizing their populations, territory, and resources for production. A strong sense of nationalism can encourage the population to work together, but the population also needs education and training. The territory must be at peace, and the country should constitute a single market for raw materials, goods, and labor.

Internal mobility of the population is another factor that may induce growth. Mobility assists each individual in realizing his or her own potential, and it also allows employers to draw on the potential of the entire national population. We would expect workers to move from areas of high unemployment to areas of low unemployment, so variations in regional unemployment can provide a measure of labor mobility and national cultural homogeneity. Great variations in unemployment suggest that something is discouraging people from moving. Factors preventing mobility may include people's attachment to place or family; racial or ethnic animosity among different localized groups; lack of nationwide availability of certain cultural products and services (religious services, foods, or other consumer goods); regional variations in unionization and difficulty in joining unions; variations in workers' pay, rights, or benefits; nationwide acceptance of degrees or certification for professionals; and a lack of knowledge of distant opportunities. Italy exemplifies a country in which unemployment rates in one region, the South, are usually two or more times as high as in the North. This reflects continuing cultural differences between the regions.

States with frontiers

Not all states occupy and utilize their full territory. The world population map (on the rear endpaper) reveals that many countries have core areas of dense population and development but also **frontier** areas, which are undeveloped regions that may offer potential for settlement. Environmental limitations may hinder exploitation of frontiers. Brazil, for example, contains vast sparsely settled areas in which the government has encouraged settlement, but the results have often been economically disappointing and ecologically disastrous. Similarly, the island of Java is only 7 percent of Indonesia's national territory, but it is home to 60 percent of the population. The Indonesian government has tried to relocate settlers to other islands, but many new settlements have had difficulty supporting themselves.

Some African states seem overpopulated relative to their current ability to feed themselves, yet they are only sparsely populated. With an end to civil wars and with greater investment in agriculture, some of these territories could support greater populations. Canada's northern regions are poorly suited to agriculture (compare Figure 4-23 with Figure 11-23), but they are exploited for mineral, timber, and hydropower resources. Expanses of the Siberian region of Russia have never been densely settled or even fully explored for resources, yet most of Russia's population remains west of the Urals (Figure 12-21). Japan might become a partner with Russia in developing Siberia's potential.

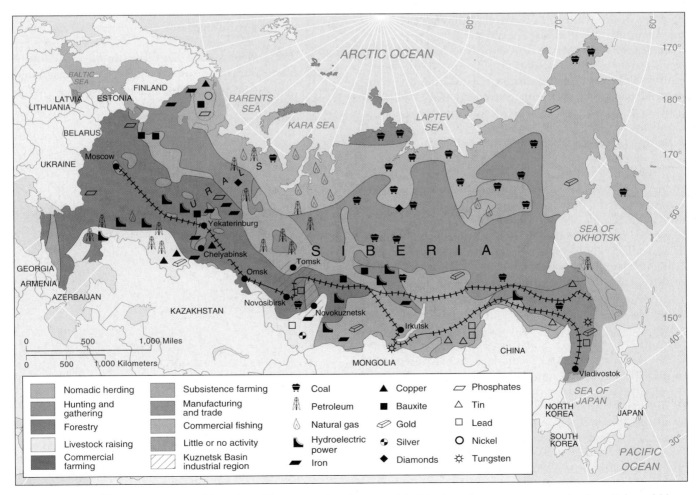

FIGURE 12-21 The resources of Russia. Siberia remains almost uninhabited, and enormous amounts of capital would be needed to overcome its inhospitable climate and great distances to exploit its raw materials.

How Do Governments Distribute Economic Activities?

Most countries try to maintain a fairly equal standard of living throughout their territory, although many have biases that favor some regions over others. Great disparities of wealth from region to region are a centrifugal force that may pull the state apart (Figure 12-22). Economic competition among regions may dominate national politics. Politicians weigh the domestic regional impact of every program and devise new ones to reduce disruptive imbalances. We have already seen the cultural differences between Italy's northern and southern provinces, which may be either a cause or an effect of regional variations in income (compare Figure 11-2 with Figure 12-22). Norway builds bridges and tunnels to reach even the most isolated parts of its territory. This keeps the population in these outlying farms and villages. The apportionment of the Norwegian Parliament even deliberately underrepresents the main cities in favor of rural areas.

The geography of any country's resource endowment may favor some regions over others. The relative fortunes of regions may change, depending on the discovery of new resources, shifting trade patterns, or patterns of innovation. In the area of today's Belgium, for example, the lowland coastal Flemings grew rich from trade and commerce during the medieval and Renaissance periods, and they dominated the highland Walloons. Later the discovery of coal and the industrialization of Wallonia brought the Walloons prosperity. Today the decline of coal and steel manufacturing and the development of a trade and service economy has swung the pendulum of prosperity back to the Flemings.

The national distribution of prosperity can also change as the sectors of the national economy evolve. As employment shifts from sector to sector, job opportunity shifts from place to place. The expansion of the secondary and tertiary sectors, for example, brings urbanization. These shifts might occur faster than workers can be retrained or relocated, so certain regions of a country may suffer while others thrive.

Countries often devise special development programs for poor regions, just as cities or states designate urban enterprise zones (discussed in Chapter 10). For example, the United States created the Tennessee Valley Authority in 1933, an Appalachian Regional Commission in 1964, and a Lower Mississippi Delta Development Commission in 1988 (Figure 12-23). Income and welfare were significantly below national levels in these three regions, so federal programs attempted to boost their economies.

A government may locate its factories or other enterprises in poor regions, or it may distribute its offices, research institutes, and military installations. Governments may lure private enterprises by offering subsidies, tax waivers, free development sites, or loans.

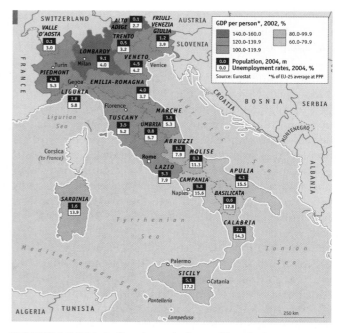

FIGURE 12-22(a) Regional disparities in income. Italy (a) and Indonesia (b) exemplify countries in which great disparities in regional incomes threaten to pull the country apart.

The Spanish government, for example, subsidizes international businesses that invest in Spain's poorer provinces, and Italy subsidizes investment in its south. India encourages firms to locate in officially designated "backward areas" by offering low-interest loans, tax reductions, and low freight rates on government railroads for products from the designated areas.

No matter what economic policies a government devises, some regions may dominate a national economy

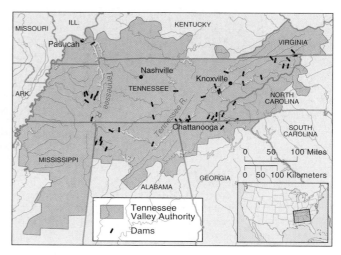

FIGURE 12-23 The Tennessee Valley Authority region. The Tennessee Valley Authority was created to bring electric power and jobs to this economically deprived seven-state region in 1933. Despite the authority's many dams, 60 percent of the electricity it produces today is generated in its coal-fired plants. The modernization of these plants is an issue of current political debate.

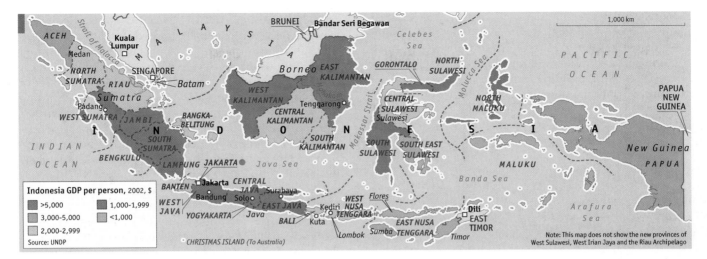

FIGURE 12-22(b).

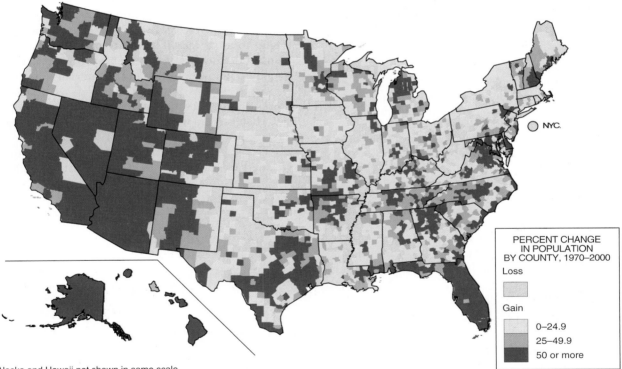

Alaska and Hawaii not shown in same scale.

FIGURE 12-24 Population change in the United States by county. Some counties in the central U.S. farming region and some old cotton-farming centers in the South counted their maximum populations before 1900. The populations of many other areas across the Great Plains and in Appalachia peaked by 1940, when the prosperity that had come from farming and mining was past. The national population today seems to be concentrating at the margins of the country. (Reprinted from *Scientific American* August 2000, © Rodger Doyle 2000)

while others suffer population losses. As a whole, the central regions of the United States and Canada have been losing population since World War II. Kansas alone has over 2,000 officially registered ghost towns (Figure 12-24). While U.S. Plains states have suffered population declines, the South and Southwest, known as the Sunbelt, have absorbed most U.S. population growth.

National Transportation Infrastructures

Political considerations Some regions in Southeast Asia and in parts of South America and Africa are in open rebellion against their national governments, and the central governments invest in roads specifically to "occupy" the territory. Colombia, for example, has recently discovered oil deposits beneath its eastern plains, but exploiting these deposits will require the central government to seize control of this region from drug dealers and independence movements. China has been accused of building railroads to strengthen its grip on peripheral minority regions (Figure 12-25).

FIGURE 12-25 China's new high-altitude train to Tibet. China opened a new railroad to Tibet in 2006. The difficulties of its engineering made the train one of the wonders of the world. High-tech systems stabilize the tracks over permafrost, and passenger cabins are enriched with oxygen to help passengers cope with the high elevations. Critics, however, argue that the influx of tourists and of Chinese influence will destroy the Tibetan culture. (Xinhua, Chen Xie/ AP Wide World Photos)

Transportation and economics Economic growth, however, is usually the explanation for investment in transportation. Transportation and communication allow different territories to specialize their production and to trade. This territorial division of labor is comparable to the division of labor among workers. The design and regulation of a country's transport infrastructure are two important influences on internal geography. In some countries transportation is slow and difficult. Some regions are practically inaccessible. In other countries virtually every place is easily and cheaply accessible from every other place (Figure 12-26).

Whenever anything is in transit, the value it represents is idle. The owner is therefore losing an opportunity to invest that value in some other activity that might pay a return every minute. A financial return sacrificed by leaving capital invested in one form or activity rather than another is called an **opportunity cost.** Goods in transit cost their owner opportunity costs as well as transportation charges, and that is why people want things to move fast. If a country has a poor transportation infrastructure, then much of the country's capital is tied up in raw materials or in materials in transit. This principle applies on a small, local scale: Your local stores do not need to carry large inventories if they can get whatever you want quickly. Furthermore, if factories can receive regular supplies of the materials they need, they can reduce their inventories of parts. Minimizing inventories is the key to the modern manufacturing

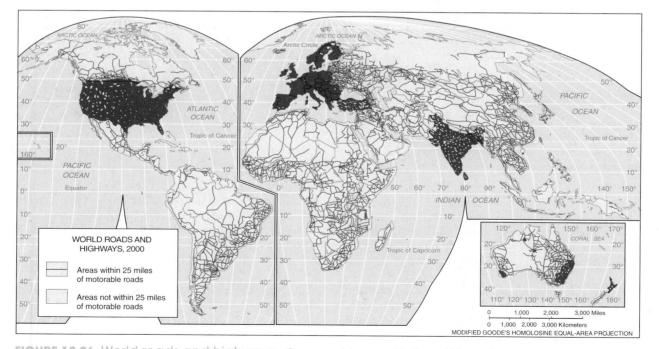

FIGURE 12-26 World roads and highways. Compare this map with the world population distribution shown in the map on the rear endpaper. Many densely settled regions, such as Eastern China, would benefit from improved road networks. The African continent is three times the size of the United States, yet it contains only one quarter of the mileage of paved roads. (Adapted from *The New Comparative World Atlas.* Maplewood, NJ: Hammond Inc. 1997.)

REGIONAL *Focus* ON

India's New Roads

In 2005, the Indian government announced a 15-year project to widen and pave approximately 65,000 kilometers (40,000 miles) of narrow, decrepit national highways. This is the most ambitious infrastructure project since the country's independence in 1947. The new roads will save billions per year in transportation costs, and, perhaps, further unify the multicultural country.

The first stage of the highway project has been dubbed the "Golden Quadrilateral." This rough quadrilateral's 5,840 kilometers (3,625 miles) run through 13 states and India's four largest cities: New Delhi, Kolkata (formerly Calcutta), Chennai (formerly Madras), and Mumbai (formerly Bombay).

The project has experienced delays in the process of acquiring the land along the highway—8,326 hectares (20,574 acres). The government has the power of eminent domain, but it must compensate for land taken and must rely on cumbersome regulations and local officials. Some observers contrast the Indian process with that of China, an absolutist country is which land is simply taken. China has built more roads faster.

method called *just-in-time manufacturing*. In the early 1990s, for example, Dell Computer Company stored 30 days of inventory and components in warehouses. By 2004, however, the company had no warehouses. Instead, it required suppliers to stock 8 to 10 days' worth of goods no further than 90 minutes from its assembly plants. Its de facto warehouses, therefore, were the lines of semitrailers lined up in the truck bays of its plants.

Fast and efficient transportation releases capital for productive investment, so developing a country's transportation system is key to economic growth. In the United States, deregulation lowered both truck and rail freight rates between 1981 and 2002, and new computer systems helped businesses track inventory better. As a result, the share of national GDP accounted for by transportation and logistics fell from 16.2 percent to 8.7 percent—an enormous national saving.

Just-in-time manufacturing saves money, but it is vulnerable to disruption, especially considered together with outsourcing. Parts come to Dell Computer factories from all over the world, yet on any day, one missing shipment can stop the entire operation. Disruptions can be natural or can be caused by humans. For example, on September 21, 1999, an earthquake measuring 7.6 on the Richter scale struck Taiwan, where 90 percent of the world's scanners and over

50 percent of computer motherboards (over 50 percent of those in one industrial park in Hsinchu) were manufactured. Production was restored at most factories within a week, but another such disruption could do more lasting damage to global supply lines. On September 12, 2001, the day after the attacks on the World Trade Center, much of U.S. manufacturing activity came to a sudden halt when the United States closed its border and grounded all flights. Terrorist threats anywhere in the international transportation system threaten the development of global just-in-time manufacturing.

India exemplifies a country whose economic growth is held back by an inadequate road system. Traffic in passengers and goods on India's roads has increased thirtyfold since independence, but the total road mileage in India has increased only fivefold. Roads throughout India are overcrowded, and thousands of villages lack all-weather roads. The government hopes to have every village of more than 500 people connected to an all-weather road by 2010. Accessibility would allow regional specialization of production and cash cropping, especially of high-value perishable foods, about 40 percent of which spoils during transit delays today. Better transportation could also improve the Indian diet.

China is investing heavily in new roads and railroads. The government has created an extensive

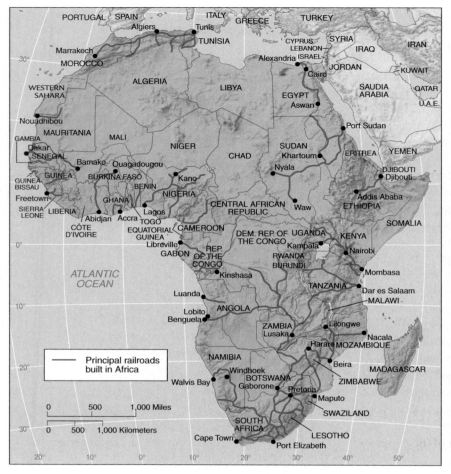

FIGURE 12-27 Railroads in Africa.
Europeans built these railroads to penetrate Africa and to haul out agricultural products and minerals, and still today railway lines provide the only freight service in many roadless areas. Africa's coastline offers few natural harbors, and many of its rivers drop off the continental interior through rapids and waterfalls only a few miles inland from the coast (compare this figure with the map on the front endpaper, world topography). Thus, the continent is relatively impenetrable by water, which presents a major impediment to African economic development. Many lines shown on this map, however, are disrupted by civil strife or else they are inadequately maintained, so they offer only intermittent service. Deterioration of the rail network increases transport costs and lowers income for many African states. Lines in Sudan, for example, and the important line reaching into central Africa from Benguela, Angola, have not operated for 30 years.

network of multiple-lane highways, but its plans include the construction of an additional 85,000 kilometers (53,000 miles) of intercity highways and urban ring roads within 30 years. The government also intends to construct 5,400 kilometers. (3,355 miles) of high-speed rail track by 2010, plus to build or refurbish 40,000 kilometers (25,000 miles) of its railways. Plans include a 1,320-kilometer (820-mile) link from Beijing to Shanghai capable of speeds up to 354 km per hour (220 mph), and a second 177-kilometer (110-mile) magnetic levitation line from Shanghai to Hangzhou capable of 450 km per hour (280 mph). China will import the technology, but at least 70 percent of all parts will be built in China, thus greatly enhancing China's manufacturing technology.

In many countries the pattern of the transport network is not integrated, but rather consists of lines penetrating into the country from the ports to bring out the exploitable wealth. These routes are often the legacy of colonialism or the product of neocolonialism, and they facilitate getting into or out of a country, but they do not facilitate getting around inside it (Figure 12-27). Australia, too, exhibits such a pattern of railroads, and Australia's rail routes were built at many different *gauges* (widths between the tracks). The Australian government has tried to knit one national rail

system, but Sydney and Melbourne, Australia's two largest cities and only 700 kilometers (435 miles) apart, were not linked until 1995, and the first transcontinental north–south line, promised in 1911, did not open until 2004.

Regulating transportation Mapping a national transport network is only a first step toward understanding what moves, where, and why. Lines on a map are static, but movement on those lines can be manipulated by conditions and charges that are "hidden" behind the map. Most countries blend public and private transportation services, and they manage freight rates on government-owned railroads, for example, as India does, to distribute industry or other activities.

In all countries, trucks, railroads, and airlines compete for traffic. They may also compete for government subsidies, and these subsidies can in turn affect domestic economic geography. In Kenya, for example, the trucking industry has successfully prevented national investment in railroads. In most European countries, in contrast, government-owned railroads receive generous subsidies. In the United States the trucking industry, the oil industry, the highway construction industry, and the car-driving public make up a lobby powerful enough to overwhelm the railroad interests. As a result, trucks pay only a fraction of their real highway-use costs, and they do not pay for the many negative externalities that they impose, such as air pollution, traffic accidents, and costs due to congestion. This subsidy to the trucking industry enables it to capture much long-haul traffic that is actually best suited to railroads. Railroads still handle 42 percent of U.S. bulk traffic, but trucks have captured 31 percent, causing railroad trackage to be abandoned. Thus, in the United States as in Kenya, political competition distorts economic competition in the transport system and locks inefficiency into the economy.

U.S. freight railroads are nevertheless enjoying a revival. Computers improve tracking of shipments and timing of schedules, and highway congestion (and the concomitant pollution and accidents) encourages the rail alternative. State and local governments are helping finance improvements in the network. For example, the six major North American railroads announced in 2003 a multi-billion-dollar project to untangle their lines around Chicago, speeding freight across the country.

Countries may also manipulate domestic freight rates to direct traffic to favored seaports, and the port facilities themselves may be subsidized to attract business and create jobs. The Netherlands, for example, organizes its national transport network and fixes prices on it to capture an international hinterland for its port of Rotterdam.

These examples provide only a hint of the manipulation of charges built into each country's transportation infrastructure. When geographers investigate economic activities, they often discover these "hidden" policies.

National communications infrastructures

The dissemination of information can be as important as the dissemination of goods and materials. The degree to which citizens of a country enjoy access to sources of information can reveal its economic development and political liberty, and it can also promote these two values.

The availability of telephone lines is a good measure of access to information (Figure 12-28). Today fax machines, personal computers with Internet access, and other modern communications devices are largely restricted to the rich countries, but we can hardly guess how their widening availability will transform the geography of the twenty-first century.

All aspects of infrastructure—pipelines, electricity generators, highways, telephone facilities, and more—require continuous maintenance and updating. As societies become increasingly dependent on these facilities, and as the facilities become more complex, both maintenance costs and repair costs in the event of failure rise dramatically. In 2005, however, the American Society of Civil Engineers released a "report card" on the state of America's infrastructure. The overall grade was "D," with the highest mark (C+) for solid-waste handling. The most dramatic underinvestment noted involved the nation's roads. Simply maintaining existing roads would require increasing annual spending by 50 percent;

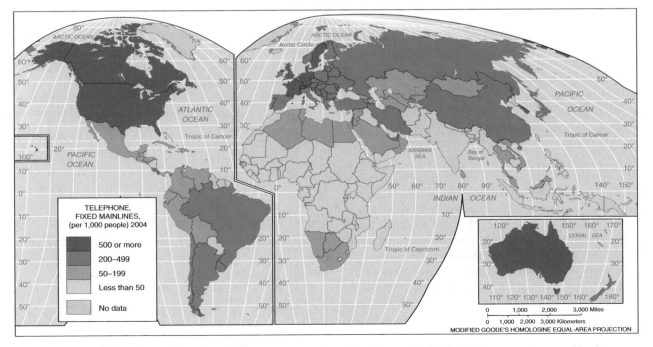

FIGURE 12-28 Telephone fixed mainlines. A few places that have never had telephone service are skipping mainline technology and jumping right to mobile telephone technology. The existence of telephone mainlines, however, still provides a good measure of the availability of information and communications in a society. (Data from The World Bank)

Focus ON

Socks and Politics

In 2005, two cities called themselves "The Sock Capital of the World:" Fort Payne, Alabama, and Datang, China. In 2004 Fort Payne produced 728 million pairs of socks; Datang, 9 billion pairs. Workers in Fort Payne averaged $10 per hour; workers in Datang, 70 cents per hour. When the U.S. Congress considered a new trade pact with Central America in 2005, Fort Payne's representative, Robert B. Alderhol, expressed concern that such an agreement could unleash a new flood of imported goods into the United States. Nevertheless, to the surprise of many observers, Representative Alderhol voted for the trade pact. Five days after the vote, the U.S.

government announced that it would extend limits on imports of socks from China. This action raised the prices of socks for all Americans and perhaps temporarily saved some jobs in Fort Payne.

Questions

1. Do you think that Representative Alderhol's vote may have been part of a deal?
2. Can the jobs of sock manufacturers in Fort Payne be saved forever? Should they?
3. Who would profit and who would suffer if socks could be imported freely and if any subsequently unemployed sock workers in Fort Payne received job training or simply cash pensions?

improvements would require doubling spending. The engineers described America's infrastructure as, quite simply, "falling apart." Electricity blackouts are increasingly afflicting major urban areas. These events are warnings about infrastructure maintenance.

NATIONAL TRADE POLICIES

Countries differ in the degree to which they participate in world trade and circulation. Each country wants to develop its own economy, to ensure its national security, and to some degree, to protect its national culture or identity. Chapter 8 noted how some countries subsidize domestic food production as a national security measure. Countries also protect national security industries—armaments, for example.

The Import-Substitution Method of Growth

A government may protect a national industry from imported competition as only a temporary measure. This is done in the belief that newly developing industries, called **infant industries,** cannot compete with imports. Infant industries manufacture only small quantities of the product, so manufacturing costs per unit are high. Eventually, however, the industry might build a national market for its product and produce larger quantities of it, thus reducing the cost per unit. This is called achieving **economies of scale.** When economies of scale have been achieved in protected national markets, then imports will be allowed to compete. This policy of protecting domestic infant industries is called the **import-substitution method of economic growth.**

This form of economic protectionism helped some countries industrialize in the past, including, to some degree, the United States, but it involves economic risks. If a country raises tariffs to protect a particular industry, all national consumers are subsidizing that industry. Some protected industries may never be able to survive competition with imported goods, so the national public may go on subsidizing them. This may be an inefficient investment of national resources. For example, the United States long imposed high tariffs on imported apparel; the U.S. Department of Commerce estimated that this protection cost U.S. consumers $145,000 per job saved in the U.S. apparel industry, which was several times what those workers actually earned. The wages apparel workers earned, however, multiplied through the economy, and if the workers had had no jobs, some might have had to go on welfare. The United States agreed gradually to reduce barriers to trade in textiles after 2005, but China's share in the U.S. market for textile and apparel products jumped, so the United States threatened to reintroduce tariffs unless China imposed restraints on its own exports. Chinese eventually imposed export tariffs on its own industries. Overall, it is difficult to balance the costs of protection, but manufacturers and workers still lobby for protection from foreign competition.

EXPORT-LED ECONOMIC GROWTH

The alternative to the import-substitution method of economic growth is the **export-led method.** Countries that adopt this program welcome foreign investment to build factories that will manufacture goods for international markets. In that way, they can achieve economies of scale immediately. Export-led policies rely on global capital markets to facilitate international investment, and they rely on global marketing networks to distribute the products. Export-led development is usually accompanied by low tariffs and export duties, market-based domestic economic policies, privatization of most economic activity, and general laissez-faire openness to economic innovation.

The countries that have grown the fastest in recent decades have generally followed export-led programs. The success of these policies demonstrates again how, increasingly, what happens *at* places is the result of what happens *among* places.

For a country's economy to grow by exporting, other countries must be willing to accept the developing nation's products. Through recent decades the United States has been the major importer of products from countries choosing this method of development, and, in doing so, the United States has accumulated substantial trade deficits. At the end of the 1970s, the United States was a creditor country, with a net stock of foreign assets worth about 10 percent of GDP. Persistent current-accounts deficits, however, turned the country into a net debtor by 1985, and since then the United States has been getting deeper and deeper into debt. The *current accounts balance* summarizes a country's current transactions with the rest of the world, which include trade, income from international investments, and transfers. At the end of 2006, the U.S. current accounts deficit soared over $900 billion, which is an unprecedented 7 percent of GDP. This debt is in addition to the national debt of about $8.5 trillion. Some economists fear that this debt threatens America's prosperity and even independence of action in foreign affairs.

China: How trade policies can affect the national distribution of wealth
International trade and investment sometimes multiply regional economic inequalities within a country. If a country adopts protectionism, all regions of the country may not profit equally, which can intensify regionalism. If a government adopts the alternative export-led growth policies, outside investment might concentrate in favored regions.

China exemplifies on a colossal scale how foreign investment can intensify regional imbalances in economic growth and how the centrifugal force of these imbalances can threaten national stability. China has always had an economic and cultural dichotomy between an outwardly focused southeast coastal region and an inwardly focused central region. Chinese capitals were in the north or center, and the government limited the trading activities of the merchants in the southern coastal cities. In the nineteenth century, however, European powers forcibly "opened" China's southeastern ports to trade, and industrial development, inspired by international trade, clustered along the coast. The Chinese government keenly felt this humiliation.

The Communist government that Mao Zedong brought into power in 1949 set out to boost the inland regions out of their relative economic backwardness by redistributing industry inland. From 1949 to 1979, foreign trade plummeted; China's coastal areas and cities stagnated.

Mao died in 1976, and in 1979 the Chinese government reversed its policies. It designated zones along the coastline where relatively free capitalistic practices were allowed. These areas were soon expanded as international corporations built factories in the zones to exploit the cheap land and labor, and the designated areas leaped to create new wealth (Figure 12-29). China had engaged in very little foreign trade in 1979, but from 1980 to 2006 trade increased by 12–15 percent per year.

As China's seaboard provinces grew in wealth, the inland provinces slipped behind noticeably, and the historic contrast between the interior and the periphery reemerged. The disparity of incomes across China was ameliorated only by workers' remittances from the coastal provinces to their inland home provinces. Today the government is investing in education, health care, and public works in the interior.

The rift in the nation's economic geography duplicated a widening rift between the state and private sectors of the economy. China remained dedicated to Communism in theory, but in practice it continued to privatize the economy. Chapter 8 noted that when agriculture was privatized, food production soared. Similarly, the industries of the state sector declined as the private industries flourished (Figure 12-30). Closures, mergers, and privatizations of state-owned enterprises lowered the state-owned sector of the economy from almost 100 percent in 1980 to less than one-third by 2006. China's 1954 Constitution said that business executives must be "used, restricted, and taught." The 1988 Constitution softened those words to "guided, supervised, and managed," but amendments added in 1999 declare private business "an important component of the socialist market economy."

In August 1997, *The People's Daily*, China's principal Communist Party newspaper, announced, "The words 'market economy' have been writ large on the flag of socialism. . . . We ran a planned economy for more than 20 years and created the foundations of industrialization, but. . . ahead is a new world. There is no way back." Within weeks the Fifteenth Communist Party Congress declared that the government will continue to insist on ownership of only a few key industries, and the definition of ownership will include holding only a minority stake in enterprises. This is privatization in all but name. In 2006 the Chinese government approved an 11th Five Year Plan for future economic development, but for the first time that document retreated from the phrase "plan" (*wu nian ji hua*) and substituted the less-controlling phrase "blueprint" (*wu nian gui hua*).

China's economic evolution exemplifies the ladder of economic development: Wages are rising, and so

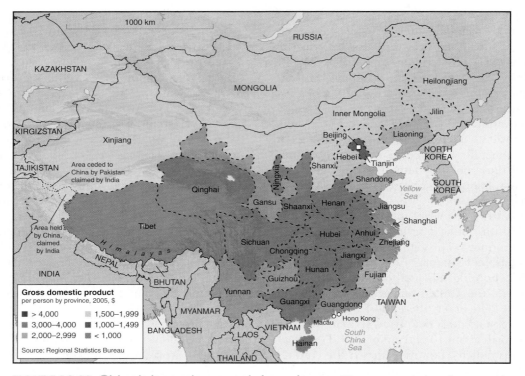

FIGURE 12-29 China's booming coastal provinces. The great variations in per capita gross product among China's provinces result largely from foreign investment and astounding economic development in the coastal regions. China has invited foreign investment since 1979, and capitalist-style economic growth has boomed.

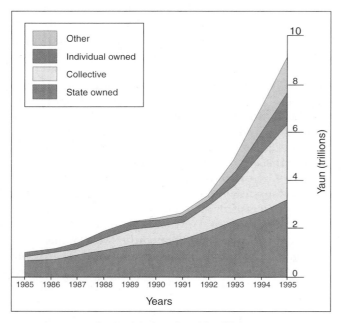

FIGURE 12-30 Industrial output in China, measured in trillions of yuan. The private sector of the Chinese economy soared after economic liberalization. The share of the total economy created by state-owned enterprises fell from about 75 percent in 1975 to about 28 percent by 2000. If we accept the economic definition that Communism is state ownership of industry, then China ceased being a Communist country about 1993. (*Statistical Yearbook of China* 1997)

is the value added to Chinese exports. In 2002 its exports were still 70 percent garments, toys, shoes, and furniture—items of low value added—but other products of increasing sophistication are following: from plastic to better-quality shoes, clothing, ceramic tiles, even autos. In 2004, China surpassed the United States as the world's leading exporter of technology and telecommunications goods—$180 billion against $149 billion. China had earlier pulled ahead of Europe and Japan. China is already experiencing a shortage of skilled workers—both industrial and managerial—which is inflating salaries. Multinational corporations, including General Motors, Intel, Honda, and Motorola, are relocating factories deeper into the Chinese interior, where wages and land costs are lower, or to lower-cost countries such as Vietnam and Indonesia.

Some observers feared that an avalanche of cheaply made goods from China might inundate world markets and virtually bankrupt every other country, but that will not necessarily happen. Chinese exports have indeed been responsible for deflating prices of manufactured goods in the world's rich countries, but, as China develops, the Chinese themselves are absorbing more of their own production and importing more of other countries' products. By the early twenty-first century, Eastern China already boasted a lower-middle-income market of 500 million people. The domestic market for cellphones, cars, cameras and film, and DVD players, was already number one on Earth. Toshiba, the Japanese electronics and industrial equipment giant, once used China as an export base, but by 2002 Toshiba was selling over 60 percent of what it made at its 34 Chinese factories to the Chinese themselves: televisions, refrigerators, computers, and automation equipment. On April 20, 2006, while visiting the United States, China's President Hu Jintao insisted, "China is pursuing a policy of boosting domestic demand, which means that we'll mainly rely on domestic demand to further promote economic growth." Within a week of his return to China, the government changed taxes and interest rates in ways designed to encourage Chinese to spend more and to save less: Thus, Chinese development may lift both their own and the world's standard of living.

Taiwan Taiwan is a principal source of investment funds for China. The 23 million people of this island enjoy a per capita GDP of almost $28,000 per year (PPP), contrasted to about $7,000 in China. This contrast underlines a bitter political division.

In the 1930s a civil war raged in China between the Communists, led by Mao, and the Kuomintang, a political party led by General Chiang Kai-shek (1887–1975). In 1949 the Communists drove the Kuomintang offshore to Taiwan, where it established a separate government. The ethnic Chinese, although a minority among the Taiwanese, continued to control the government until the elections of 2001. Meanwhile, state-directed capitalism achieved phenomenal economic growth. In the 1990s, Taiwanese began to invest and transfer low-wage, low-technology manufacturing to China, but the sophistication of the technology being transferred has steadily risen. In 2003, for example, Taiwan Semiconductor Manufacturing Company, the world's largest contract chip maker, built a huge semiconductor factory in Shanghai. The Chinese government welcomes such investment from its "rebel" province.

Both the government in Beijing and that on Taiwan claim that there is only one China and that they represent it, but at present China and Taiwan are in fact two separate countries. The U.N. General Assembly ousted the Taiwan government and seated the People's Republic in its place in 1971, and the United States recognized the People's Republic as the sole government of China in 1978. China says it will attack Taiwan if—but only if—Taiwan explicitly declares independence; Taiwan says it will declare independence only if China attacks. Some political accommodation will probably eventually be reached between the two governments—perhaps following the slogan explaining China's reincorporation of Hong Kong: "One country, Two systems." China and Taiwan might agree on joint economic ventures and joint representation in foreign affairs, for instance, without completely merging their domestic governments. The economic integration of Hong Kong, Taiwan, China, and the considerable economic vitality of the overseas Chinese is proceeding rapidly.

As a result of China's slowing rate of population growth (see Chapter 5), its surge in food production (see Chapter 8), and foreign investment, China has achieved a rate of economic growth estimated at 10 to 12 percent per year. Overall, the Chinese economy quadrupled in size between 1980 and 2000, and since 2000 China's contribution to global GDP growth has been greater than America's and more than half again as big as the combined contribution of India, Brazil, and Russia, the three next-largest emerging economies. This has come about with repressive government, and any number of factors might slow or halt this development, including political disruption, environmental disruption, an energy supply crisis, or inadequacies in China's infrastructure. If the growth continues, however, China will overtake the United States as the world's largest economy by 2010. The population of China, a full 22 percent of the human population long living in poverty, might achieve considerable material comfort within 50 years. China's dramatic economic and industrial growth will likely transform many aspects of global political, economic, and environmental patterns in the twenty-first century. The growth is taking place over a period of years rather than suddenly, so it does not make the headlines, but in terms of its impact on the geography of wealth and power in the world, it is one of the most important "news items" of our time.

Where Is the Third World?

The term *Third World* was introduced by the French demographer Alfred Sauvy in 1952. It reflected attitudes during the Cold War that followed World War II. An ideological line was drawn between the capitalist countries led by the United States and the Communist-ruled countries headed by the Soviet Union. At the height of the diplomatic pulling and tugging, it was expected that all countries should line up on one side or the other. Many did so, but a few, led by presidents Tito of Yugoslavia, Nasser of Egypt, Nehru of India, Nkrumah of Ghana, Makarios of Cyprus, Sukarno of Indonesia, and Emperor Haile Selassie of Ethiopia, clung to a precarious neutrality. These maverick states came to be known collectively as the **Third World,** to distinguish them from the **First World** of the Western bloc and the **Second World** of the Soviet bloc.

Over the years the terms gained economic and sociological connotations, many of which were false and unfair. The First World, for example, was interpreted as a haven of science and rational decision making: progressive, technological, efficient, democratic, and free. The First World included as allies in the Cold War all reasonably well-to-do non-Communist countries even though some of them were politically repressive, such as South Africa, South Korea, and Taiwan. The Second World was also defined as modern, powerful, and technologically sophisticated, but as dominated by totalitarian Communist governments. In fact, the countries of

Focus Box

The Great Race

International observers have been comparing China and India since the late 1940s, when China became Communist while India received independence as a democracy. The world's two most populous countries were regarded as being in a race: Which would develop faster and bring a higher quality of life to its people? In 2006, the combined populations of the two giants totaled 37 percent of humanity: India's population of 1.1 billion, growing at a rate of 1.4 percent, was projected to overtake China's 1.3 billion population, growing at a rate of only 0.59 percent, within a few years.

Politically, India has long been the world's largest democracy. It has endured rough spots—wars with Pakistan, occasional sectarian strife between its Hindu majority and Muslim minority, even the assassination of Prime Minister Indira Gandhi by Sikh separatists in 1984—but it has held stable. China, by contrast, has remained a totalitarian dictatorship, disrupted by bloody suppressions of periodic riots and demonstrations.

China chose a communist economic system in 1949, but it has rapidly evolved toward economic liberalization and export-led development since 1979, thus becoming a global manufacturing giant. India held to an import-substitute economic development program until the early 1990s. Since then, it has slowly liberalized its internal economic system, opened its markets, and succeeded in capturing a significant share of the world's outsourced tertiary-sector jobs. Analysts have cited insufficient investment in infrastructure and sluggish bureaucracy as holding back Indian economic development. From 1979 to 2005, China attracted over $600 billion in foreign investment; India barely $30 billion.

Observers in the world's press and in political capitals dispute the ongoing measures of the race between these two countries. In 2005, China's per capita GDP (PPP) was $6,800; India's was $3,300. China ranked 85th in HDI; India 127th, but Freedom House rated India "Free," and China "Not Free."

America's mixed attitude. U.S. press coverage of the economic development of India and China alternates between fear of economic competition and featuring opportunity for U.S. investment. (Getty Images)

the Second World never achieved economic growth equivalent to that of the First World, and the collapse of many Communist regimes early in the 1990s exposed their true poverty.

The Cold War rhetoric unfairly characterized the Third World in contrast with the First and Second Worlds, as a world of underdevelopment, overpopulation, irrationality, and political chaos. Much of the Third World had still been organized as colonies in 1939, and had not "emerged" until after World War II. According to the theory, the Third World was poor but struggling to develop. The rhetoric interpreted the Third World as a zone of competition between

the First and Second World models of modernity (Figure 12-31).

Today the terminology of First, Second, and Third Worlds is out of date. A region, by definition, must exhibit meaningful homogeneity. Historically, the unifying element in the Second World was the existence of Communist regimes. When these regimes in Eastern Europe and the former Soviet Union collapsed, however, and the nations individually adopted new political and economic systems (see Chapter 13), the Second World as a political-economic bloc ceased to exist. Furthermore, the term *Third World* carries a racial, or even racist, aspect. Most Third World countries were

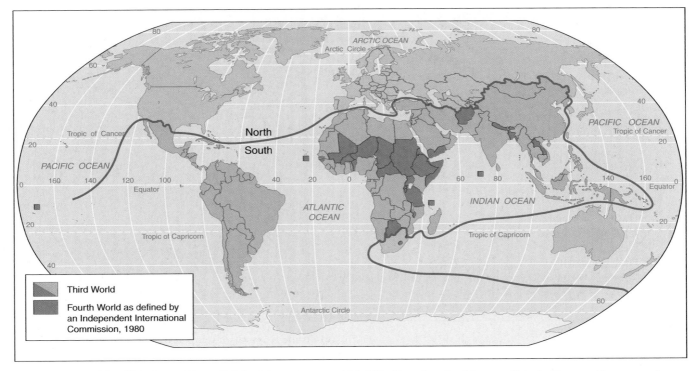

FIGURE 12-31 The Third and Fourth Worlds. The term *Third World* confused cold war political alliance with economic status, so efforts were made to refine it. Some scholars subdivided the Third World and defined a poorest "Fourth World." Others tried to differentiate a rich "North" of the world from a poor "South."

non-White, and therefore the term suggests, that they are somehow, "all the same." The term actually lumped together peoples with extremely diverse histories and cultures who inhabit a broad range of physical environments that possess a great variety of natural and cultural resources. If we consider just the countries whose leaders made up the original political Third World—Yugoslavia (for its successor states today, see Figure 11-9), Egypt, India, Ghana, Cyprus, Indonesia, and Ethiopia—they cannot be homogenized economically, culturally, or sociologically. It is ultimately insulting to refuse to recognize or appreciate their differences. The term *Third World* will eventually be remembered as a condescending relic of a political cold war mentality. It regarded political alignment in a bygone struggle as the most significant distinction among all the world's nations and cultures.

THE FORMATION OF THE GLOBAL ECONOMY

International trade has been carried on for millennia. Trade always triggers regional specialization and increases in productivity, as well as new cultural possibilities and combinations (see Chapter 6). As new means of transportation and communication have linked world regions at lower cost, more goods are produced for trade, so trade multiplies faster than total production.

From 1950–2000 world GDP sextupled, but world exports rose by a multiple of 20, and in 2004 world trade reached a total of over $9 trillion in merchandise and an additional $2 trillion in commercial services (Figure 12-32).

International economic and communication links have "shrunk" Earth in terms of time or cost distance, and economic globalization has far outpaced cultural or political integration. Tensions among these economic, cultural, and political organizing activities generate much of today's international news. The globalization of the economy is already redistributing economic activities in every economic sector, and countries are continuously renegotiating conditions of production, trade, and exchange.

Transnational Investment and Production

Two developments in economic globalization accelerated in the late nineteenth century. One was the evolution of a world market for certain primary products, foods, and minerals. Chapters 8 and 9 traced the results of that development. The second development was an increase in international investments. At first these investments were limited to shares of stocks and bonds representing minority holdings in foreign companies, called *portfolio investments,* but foreign direct investment soon followed. **Foreign direct investment (FDI)** means

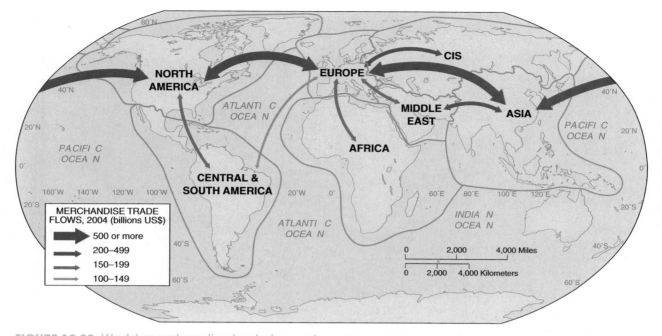

FIGURE 12-32 World merchandise trade by region. Most trade is among developed regions, but some developing regions, such as Latin America, are capturing an increasing share. (Data from WTO)

investment in enterprises that are actually operated by a foreign owner.

Modern transnational manufacturing was launched in 1865, when the German chemical magnate Alfred Bayer built a factory in Albany, New York. Today technology allows the integration of manufacturing processes worldwide. For example, South Korea's Kia Motors Corporation manufactures a popular sport utility vehicle, the Kia Sorrento, named for a town in Italy. The Sorrento is assembled in South Korea, but it is full of components from around the world. The optical pickup units of the vehicle's CD player, which read the CDs, are made in China, for example. Their mechanical structure and electronic components are added in Thailand, and then the units are shipped to Reynosa, Mexico. From there the units are trucked to Matamoros, Mexico, where they are assembled into an audio system. The audio systems are trucked to California and shipped to South Korea, whence the Sorrentos are shipped to the United States.

Enormous enterprises have grown up that own and coordinate production and marketing facilities in several countries. These are called **multinational** or **transnational corporations** (Table 12-3). Transnational companies (as defined by the United Nations) that have international production facilities now number 60,000 and boast 500,000 foreign affiliates, accounting for an estimated 25 percent of total global production.

Their evolution historically followed four stages. In the first stage, the corporation exported products from its home country to meet demand abroad. In the second stage, the corporation established production

TABLE 12–3	Sales of the World's Largest Corporations Compared to a Few Countries' GDPs (in billions of dollars, 2004)
Sweden	346
Exxon Mobil	**340**
Wal-Mart Stores	**316**
Royal Dutch/Shell	**307**
Turkey	303
BP	**268**
Indonesia	258
Poland	242
Denmark	241
General Motors	**193**
Chevron	**189**
DaimlerChrysler	**186**
Toyota Motor	**186**
Ford Motor	**177**
Thailand	162
Argentina	153
Israel	117
Czech Republic	107
Egypt	79

facilities abroad to supply specific markets abroad. Exports from the home country dropped. In stage 3, the foreign production facilities began to supply foreign markets other than their own local markets. In stage 4, the foreign production facilities began to export back into the home country. These stages appeared first in trade in primary materials such as copper and oil. Later the stages were repeated during the evolution of international manufacturing.

Today the evolution of transnational activity can skip the historic first stage of export from the transnational's home country. Transnationals can invest in countries that choose the export-led growth strategy, and they can export globally from those countries (called *export platforms*) before those host countries are themselves markets for the goods they produce. Transnational corporations' ownership and activities are so widespread that it is difficult to tell which corporations are corporate citizens of which countries. Nestlé, the world's largest food company, with corporate facilities around the globe, is generally regarded as a Swiss corporation, yet its shares are traded on the London Stock Exchange. The company's board of directors includes citizens of several countries, but no records reveal Nestlé's actual stockholders. One economist estimates that U.S. citizens own about 14 percent of Nestlé. Gilbert Williamson, president of another global concern, NCR Corporation, has said, "I was asked the other day about United States competitiveness, and I replied that I don't think about it at all. We at NCR think of ourselves as a globally competitive company that happens to be headquartered in the United States."

Multinational corporations challenge individual nations to regulate them or to tax them. Imagine, for example, that a multinational corporation manufactures sneakers for $1 per pair at a subsidiary in Malaysia and sells them to a U.S.-based subsidiary for $40 per pair. The multinational company has kept enormous profits in Malaysia, where tax rates may be lower than in the United States. Trade among different divisions of individual corporations today accounts for an estimated one-third of all international trade, and the corporations avoid any country's taxes by allocating investment and profits wherever they wish. In 2004, 42 percent of all U.S. trade in goods occurred among arms of the same companies.

The U.S. government takes the position that any company that employs and trains U.S. workers and that adds value in the United States is a U.S. corporation, no matter which flag it flies at its international headquarters or where those headquarters are. Accordingly, the U.S. government, acting on the behalf of a Japanese-owned office machine factory in Tennessee, has protested the trading practices of a Singapore corporation that is a subsidiary of the U.S. Smith-Corona Corporation, which is owned by a British corporation.

The International Tertiary Sector

When most people hear the words *foreign trade*, they think of oil and iron, bananas and coffee, cars and clothes being shipped around the world. It is easy to imagine such "things." The international economy, however, has transcended trade in primary products and manufactured goods. Trade in tertiary sector services is expanding, and so is international investment in the tertiary sector.

The United States is the world's leading exporter of services (15 percent of the total of global service exports in 2004), followed by the United Kingdom, Germany, and France. The United States's net surplus in trade in this sector partly offsets the country's annual deficit in merchandise trade. In education, for example (which students come from abroad to receive, but which is counted as an export), the United States has enjoyed an export surplus of almost $5 billion per year, although new restrictions on visas to the United States instituted since 9/11 have deflected the global stream of students to Canada and to Europe, so U.S. education exports have declined. This has harmed America's ability not only to educate these students, but to retain them as skilled immigrants. Other U.S. tertiary exports include real estate development and management, accounting, medical care, business consulting, computer software development, legal services, advertising, security and commodity brokerage, and architectural design (Figure 12-33). U.S. entertainment productions dominate the world's airwaves and movie screens. Movies, music, television programs, and home videos together account for a significant annual trade surplus.

In many cases corporations that provide services to manufacturing corporations in their homelands

FIGURE 12-33 Design is a U.S. export. U.S. architect Frank Gehry (born in Canada) designed this spectacular new Guggenheim Museum that opened in Spain in 1997. Architectural design is a significant U.S. export. This building carries a political message, for it was entirely planned, built, and paid for by Basques, a minority group within Spain, in Bilbao, their largest city, to assert their presence in the world (see Chapter 11). On the façade of the building, huge letters proclaim "Museoa," which is not Spanish, but the Basque language. The shimmering metal used to coat the petal-like steel panels is titanium. (Eric Vandeville/Getty Images, Inc.—Liaison)

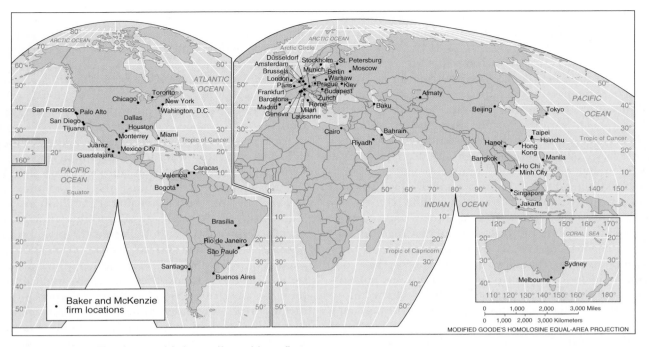

FIGURE 12-34 The largest international law firm. The Chicago-based law firm of Baker & McKenzie can provide legal advice on every area of business law in 61 offices in 40 countries. (Courtesy of Baker & McKenzie)

have followed these out into the world arena. This partly explains the internationalization of legal counsel, business consulting, accounting, and advertising (Figure 12-34). Many professionals, such as architects and physicians, market their skills around the world. Financial services, entertainment, and still other services are produced for global markets. For many nations, welcoming tourists and foreign students to their lands have become major elements of their economies. These activities count as exports of services.

International trade in services demonstrates the same trends as trade in goods. Countries are losing their autonomy to an increasingly global market. The costs of global electronic transmissions are virtually negligible, so transnational corporations treat the world as "one place" and exploit resources, locate facilities, and market their products accordingly. The global market for tertiary labor skills was originally restricted to such back-office work as key punching and standardized data processing, but today global clients can draw on providers of even the most specialized skills and services anywhere, and in many places, these services cost much less than they would if they were provided by providers based in the rich countries. For example, local accountants in India and the Philippines working for the firm Ernst & Young calculate taxes for the firm's U.S. and global clients. American and Canadian law firms send legal work to be done in India, with almost one million English-speaking lawyers trained in common law who can research cases and draft contracts, patent applications, wills, and even divorce papers. X-rays taken in hospitals in the United States are today routinely transmitted to India where they are analyzed

by capable doctors at a fraction of the price that service would command at home. A California-based company offers tutoring services to U.S. students by Indian tutors. American parents pay $20 per hour, and the Indian tutors earn $230 per month, which is twice local teachers' salaries. Even several U.S. states's welfare benefits and food stamps "Help lines" are located in Mumbai, India, which is connected to toll-free help lines (Figure 12-35). White collar workers in the rich coun-

FIGURE 12-35 An Indian telephone "help" line for U.S. welfare recipients. Manisha Martin in Mumbai, India, is answering questions telephoned in by welfare recipients in New Jersey. Manisha's monthly $200 salary is five times the per capita income in India, but less than that of the welfare recipients whose questions she answers. (Amy Waldman/The New York Times)

tries enjoyed the deflating prices of manufactured goods when domestic blue collar jobs migrated to the newly industrializing countries. Today, however, many of even the most skilled white collar workers see their own jobs departing.

In the short run, the transfer of jobs abroad puts deflationary pressure on wages for people doing these jobs in the rich countries; in the long run, many jobs may disappear from the rich countries. For the populations of the rich countries, the crucial question is: Will the white-collar workers in the rich countries be able to add ever-higher value to services, and thus advance into jobs paying ever-higher incomes than they now enjoy? Or will they fall into lower-paying service-sector jobs, as most former manufacturing workers have? In other words, how many current or would-be data processors or computer programmers in the rich countries will capture positions as movie stars, lawyers, top athletes, or surgeons? How many, by contrast, will find jobs only flipping fast-food hamburgers or greeting customers at discount stores at miniumum wages? We do not know the answers to these questions. As more and more jobs become footloose and the world becomes one labor market, each worker must compete with every other worker with his or her skills in the world. The only workers protected against the globalization of the labor market are those whose work requires personal contact, but much personal contact is already being replaced in many contexts by videoconferencing and other innovations in communications. As *Fortune* magazine, one of America's leading business periodicals, editorialized colloquially in May 2006, "The hard truth is, we ain't seen nothin' yet."

The early-nineteenth-century English economist David Ricardo wrote that "Nothing contributes so much to the prosperity and happiness of a country as high profits." Globalization, however, has broken down the historic relationship between corporate profits and national prosperity. Today the profits of multinational companies based in the developed economies can soar even as the home-country economies stagnate and workers' real incomes stay flat or fall. This is because the firms make profits in countries other than their homelands. The world's 40 biggest multinationals now employ on average 55 percent of their workforces in countries other than their homelands, and they earn 59 percent of their revenues abroad. If profits and executive pay soar while ordinary employees' wages stagnate and pensions and health benefits erode, workers might demand higher taxes on profits, restrictions on overseas investment, import barriers, or guaranteed employment. These actions, however, might be suicidal, because the companies would just move their headquarters to friendlier jurisdictions. Many have already done so, as some large companies and financial institutions have established their headquarters in microstates.

The Geography of Foreign Direct Investment

The amount of FDI is today several times the amount of foreign aid being given around the world, and FDI has been much more successful in triggering economic growth. Developing countries used to seek aid from foreign governments and international lending institutions, but today they are more likely to court foreign companies to invest in their countries. Since the 1980s, global FDI has grown three times faster than world trade and four times faster than total world output, reaching $920 billion in 2005. This flow of FDI has greatly increased and significantly redistributed the world's productive capacity. No world agency tracks portfolio investment, but including portfolio investment would multiply the total amount of capital invested internationally.

Companies from an increasing number of countries have become active investors in other countries. Transnational corporations from developing countries, including Brazil, Thailand, South Korea, and Mexico, have reached around the world, but three sources—the United States, the combined European Union countries (see Chapter 13), and Japan—still account for about 80 percent of FDI. The United Nations refers to these three sources as the Triad.

U.S. firms account for about 30 percent of global FDI, and the United States also has attracted the highest accumulation of FDI from other countries. By mid-2006, FDI in the United States totaled about $1.5 trillion, and U.S. FDI abroad totaled over $2 trillion. Additionally, foreign investors owned $1.4 trillion worth of U.S. stocks and $900 billion worth of corporate bonds. These foreign investments and purchases testified to foreign confidence in the American economy and political stability, and they supported U.S. prosperity. Foreign firms carry on research, introduce new technology and management techniques, and change Americans' daily lives in the workplace as well as their consumption habits. Nevertheless, since 9/11, the United States has occasionally vetoed foreign investment in "sensitive" or "strategic" industries. Chapter 11 noted the public uproar that stymied Dubain investment in U.S. ports, and protests in 2005 stopped the Chinese oil company CNOOC from buying U.S-based Unocal.

FDI in the developing countries   Three trends characterize the geography of FDI. First, the greatest share of global FDI—about 75 percent—is investment from one rich country to another. For example, U.S. firms have their greatest investments in Europe.

Second, FDI in the developing countries has been geographically selective. The countries that have attracted the most investment are countries that have chosen the export-led method of economic growth. Their economies have boomed. Four Asian nations—Singapore, South Korea, Taiwan, and Hong Kong (reunited with China in 1997, but still often considered separately in international trade statistics)—attracted FDI and grew so fast during the 1980s that they earned the nickname The Four Tigers. These areas reached European levels of prosperity within a decade and quickly became sources of capital and technology for the next tier of Asian developing lands: Thailand, China, Malaysia, and Indonesia. Since 2002, China has even surpassed the United States as the world's number-one recipient of new FDI each year.

The areas that have not attracted FDI have been slower to develop. Africa has attracted only about 2 percent of total FDI, and almost all of that investment has been in extractive activities—oil and mining. Political instability has been a major factor discouraging international investors there.

A third characteristic of the pattern of investment is that each Triad member has a majority share of the FDI in a cluster of countries that have become its economic satellites. Japanese investment dominates in East and Southeast Asia, U.S. investment in Latin America, and Western European investment in Eastern Europe and Central Asia. These clusters help explain world patterns of trade. For example, in recent years the United States has had a trade deficit in electronic consumer goods with Thailand. Thai electronics factories are mostly owned by Japanese corporations. Thus, the consumer electronics that the United States imported from Thailand profited the Japanese corporations as well as the Thai economy. In a similar fashion, many of Europe's imports from Latin America are the products of U.S. corporations. This pattern of global investment carries political ramifications: Trade sanctions against any individual country with which the United States runs a large trade deficit (such as China) may not necessarily succeed in reducing total U.S. imports because the transnational corporations simply move the production of the lower-cost goods from the sanctioned country to another low-cost country and ship the goods to the United States from there. Thus, the U.S. trade deficit with the sanctioned country falls, but the deficit with another country rises.

Geographers watch capital flows carefully, because the pace of development within individual countries is increasingly determined by their ability to attract foreign capital. The world map of wealth continues to change kaleidoscopically.

Some observers have hypothesized broad political ramifications of the rise in global FDI. To attract private investment, governments must reveal national economic and social conditions they previously may have kept secret. Furthermore, no country can attract investment unless it guarantees consistency in the rule of law. Arbitrary government actions—often typical of totalitarian regimes—will isolate a country from international capital flows. For example, several major international corporations have decided against investing in Malaysia out of concern that Malaysia might try

to regulate Internet access. International capital cannot be attracted without acceding to international rules, and local political leaders realize that the authority they exercise over their own people cannot be exercised over international investors. The process of compromising and accommodating investor's concerns is often called the *discipline of the market*. Therefore, the substitution of private investment for foreign aid may have a liberalizing and democratizing influence worldwide.

The Globalization of Finance

The FDI that is redrawing the world economic map is only a fraction of the vast amounts of capital moving around the world. Multinational banking and telecommunications make it possible to monitor and trade instantaneously in national currencies, stocks, and bonds listed anywhere in the world. Today over $2 trillion is traded daily on the world's foreign exchange markets—over 100 times the average daily merchandise trade. An additional $10 trillion worth of stock changes hands on the world's major stock exchanges. This activity is monitored by the International Organization of Securities Commissions (IOSC), but as noted previously, no international agency records ownership of global assets, even by national source.

Nations are expected to regulate their own stock exchanges, and the home countries of global banks are responsible for supervising the banks' operations, but countries do not agree on how to do these things. Many giant banks have located their international headquarters in places renowned for lenient banking regulations. Tiny Monaco (population 30,000) counts over 340,000 bank accounts holding almost $50 billion. The 260-square-kilometer (100-square-mile) British Caribbean colony of the Cayman Islands (population 36,000) is one of the world's greatest financial centers, with assets of $700 billion held in over 600 chartered banks. By comparison the state of New York has only 154 chartered banks. The Cayman Islands, however, has almost no vaults, tellers, security guards, or even bank buildings. Assets are held electronically in computers. Many banks consist of one-room offices, and the confidentiality of their transactions is protected by local laws. For example, international agencies cannot trace the profits of the international drug trade. Shortly after 9/11, the United States and its allies passed laws designed to curtail money laundering as a means of financing international terrorist activities, but tracing the global migration of money remains a difficult challenge.

Today Japan is one of the world's principal sources of capital, so the regulations governing the Tokyo stock market affect investments and jobs throughout the world. The Arab oil states are another source of capital, so a costly Mideast war raises the interest rate on U.S. college tuition loans. Bank policies in Frankfurt, Germany, and Johannesburg, South Africa, affect the checking-account charges of U.S. banks. Financial manipulations in Rio de Janeiro, Brazil; Bangkok, Thailand; and Mumbai, India, affect the security of many U.S. elderly people's pensions, which have been invested abroad. And the interest rates set at the Federal Reserve Bank in Minneapolis affect business taxes in Indonesia. The explanations of the "where" and "why there" of local economic affairs may lie in global finance.

Tourism

Tourism is by some accounts the world's largest industry. Every year a greater proportion of the world population travels, and in most countries tourism is one of the fastest-growing sectors of the economy. International tourist arrivals numbered over 800 million in 2004, and international tourist receipts totaled approximately $750 billion, excluding air fares, which is equivalent to 7 percent of global exports. Tourism is the top earner of foreign exchange for many countries, and the World Tourism Organization (a U.N. organization headquartered in Madrid) believes that tourism will continue to expand despite the threat of terrorism.

Citizens of the rich countries account for the largest tourist expenditures, and the rich countries are also the most popular destinations. Measured by tourist arrivals, the leading destinations in 2004 were France, Spain, the United States, China, and Italy. When tourism income is measured as a share of total national income, however, the importance of tourist income to the poor countries becomes clear. Tourism made up more than 10 percent of the exports of almost 50 countries in 2004; 20 percent of the exports of Cambodia, 28 percent for Egypt, and 44 percent for Jamaica.

The tourist potential of any spot depends upon its "three A's":

▶ *Accessibility*, usually via convenient air routes
▶ *Accommodations*, at all levels from simple pensiones to grand resort hotels

FIGURE 12-36 Cancún, Mexico. This resort was carefully planned. The government of Mexico mapped many relevant locational determinants, including annual days of sunshine, temperature ranges, attractiveness of local beaches, temperature of the water, water currents, variety of sea life, scenery, and accessibility of nearby archaeological ruins. When these factors all came together at one place, the government invested millions of dollars to develop Cancún as a new vacation destination. (Jose Fuste/RAGA/Corbis/Stock Market)

▶ *Attractions*, which can include a pleasant climate, as well as beaches, museums, architecture, historic sites, festivals and performances, shopping opportunities, and more (Figure 12-36).

Areas spectacularly endowed with wildlife may feature **ecotourism**, which is travel to see distinctive examples of scenery, unusual natural environments, or wildlife (Figure 12-37). The rich bird life on the islands of Lake Nicaragua, the lions of South Africa, and the tropical vegetation in many countries are important tourist destinations. Ecotourism today provides the money needed to save many countries' natural environments as well as many species of rare animals. Overall, the countries best endowed for tourism have both natural and cultural attractions, pleasant climates, good beaches, and reasonably well-educated populations. Political stability is a necessity. Therefore, despite Africa's wealth of ecological and cultural attractions, political instability and lack of accommodations have restricted its income to only 2 percent of global tourist dollars, and those were divided almost exclusively between the countries of Mediterranean North Africa and the country of South Africa.

FIGURE 12-37 Ecotourism. "Shoot with a camera, not with a gun," insist these intrepid sightseers in Africa. (Martin Harvey/Peter Arnold, Inc.)

Tourism creates many jobs and offers many multiplier effects. For example, a new hotel creates demand for a host of related goods and services. Tourism may also stimulate the local agricultural economy, and it is generally a high-value-added and labor-intensive industry that trains local labor. Tourism may, however, corrupt local culture or provoke inflation, and too many tourists may overwhelm the natural environment.

International Regulation of the Global Economy

The rules regulating business vary so widely from nation to nation that international trade requires cooperation or even political regulation. Under United Nations auspices, a General Agreement on Tariffs and Trade (GATT) was signed in 1947, but it was superseded by the creation of the World Trade Organization (WTO) in 1995. WTO establishes panels to investigate trade disputes, and countries that are found to have broken WTO rules must change their own policies or laws or else offer compensation; if they do not, they can face trade sanctions.

In mid-2006, 149 countries belonged to the WTO. Russia and Iran still stood outside, accused of avoiding the WTO's free-market principles. Membership in WTO brings nations a significant advantage. Each member receives from every other member the terms of trade that that nation grants its most favored partners—called *most favored nation* status.

Negotiations aimed at continuing reductions in global trade barriers was optimistically begun in Doha, Qatar, in November, 2001. These negotiations, however, collapsed in failure in the summer of 2006. The poorer countries would not further open their markets to goods and services from the rich countries unless the rich reduced the subsidies to their own agriculture, which the rich refused to do. The collapse of the "Doha Round" may cripple the usefulness of the WTO and multiply bilateral agreements.

Other international organizations oversee specific aspects of international commerce, such as the International Organization of Securities Commissions mentioned earlier. The London-based International Accounting Standards Board works to standardize corporate accounting procedures, and the International Labor Organization (ILO) defines labor standards with the goal of improving working conditions.

The Geneva-based International Standards Organization (ISO) determines and publishes standards for manufacture of items in world trade. Virtually all items in international trade must conform to ISO standards, so items manufactured even for nations' domestic markets do, too. For example, the specifications of your tennis racket are covered by eight pages in ISO 11415: 1995. A new series ISO 14000 establishes environmental standards for manufacturing processes and products. It is hoped that this series will reduce global pollution and environmental degradation.

CONCLUSION: CRITICAL ISSUES FOR THE FUTURE

The meeting of the World Trade Organization in Seattle, Washington, in November 1999 was expected to be a festival of celebration for the principles of global trade. Instead, however, the meetings were disrupted by violent demonstrators. Demonstrators have since regularly targeted meetings of international bodies that regulate the global economy. The demonstrators have included a mixture of groups, holding a mixture of opinions, some of them even contradictory. Some demonstrators oppose what they see as the destruction of the world's many cultures by homogenization. Others represent trade unions in the rich countries, objecting to the export of jobs to poor countries. Others object to the "exploitation" of workers in the poor countries. Still others fight to save the global environment from over-exploitation for economic development or from pollution by economic development.

Some observers have asked who elected these street demonstrators to extort agreements from corporations or to change policies of democratically elected governments. In 2000 Mexico's president Ernesto Zedillo condemned the "globophobes" for opposing free trade. "Free trade," he said, "gives developing nations a leg up, because with lower wages and less generous social policies, it costs them less to produce products for export. Over the long run, that raises incomes and living standards." In other words, only development allows the poor countries to raise their wages, social policies, and environmental standards. To require the poor countries to observe the same labor and environmental standards that exist in the rich countries would short-circuit this process. Some critics fear that comments such as these constitute a global "race to the bottom," in other words, competition to lower wages, environmental standards, taxes, or any other factor that might affect international locations. Furthermore, as this text has noted, increasing trade or interchange does not guarantee that all countries involved will benefit equally. Some countries whose economies are most completely integrated into the world economy are some of the poorest.

For the future, uncertainty threatens the global economy. Political instability frightens capital investment. If terrorists succeed in causing more deaths, damage, or fright, and if civil or international wars are protracted, then corporations will worry about the safety of their staff and facilities, and the free flow of people and goods will slow, which will hurt economic efficiency. The cost of capital will rise because investors will demand higher returns for their risk. Higher capital costs lower the rate of new investment.

Globalization could suffer from insecurity. At the end of the twentieth century, open world markets led to a boom in trade. Globalization helped disseminate capital, technology, and entrepreneurial ideas, as well as products. It lowered prices for consumer goods, enabled manufacturers to boost productivity with just-in-time supply chains, and opened pools of technical talent. The world's markets may be too integrated to break them apart now, but continuing globalization could become slower and costlier. Higher risk means higher cost. Companies will probably have to pay more to insure and provide security for staff and property. Heightened border inspections could slow movements of cargo, forcing companies to stock more inventory. Tighter immigration policies could curtail the liberal flows of labor. Also, companies obsessed with political risk would narrow their horizons for making new investments. The only nations that could attract new FDI would be the most stable and growing. Any place that suffers political instability would be out of consideration for FDI by multinational corporations.

Critics of globalization argue that the gap between rich and poor is widening, but this is not clear. Gaps in income between the richest and poorest countries may be widening, but rising incomes in China offset declining incomes in many countries with low populations. China and India have done well in recent decades. Income gaps do seem to be widening within countries, but in most cases the rich got richer faster than the poor did, but the poor did not get poorer; they just got rich slower.

Certainly advances in global communication are increasing awareness of gaps in living standards—awareness of gaps within countries and awareness of gaps among countries. This increasing awareness intensifies pressures for migration (discussed in Chapter 5) and may be responsible for political restlessness in many countries.

Poverty among many people is a pressing moral, political, and economic issue, but whether turning back globalization is the answer to this problem is less definite. International economic integration is not inevitable. Governments and large corporations have it in their power to slow, or even reverse, the economic trends of the last 20 years, but economists doubt whether this would be a victory for the poor or for the citizens of the rich countries.

Chapter Review

SUMMARY

Every country devises a program to build and regulate its national economy and to distribute economic well-being across the national territory. Gross national income, gross domestic product, and gross sustainable product are three ways of measuring wealth, although each undervalues subsistence production and unpaid labor. Maps of these values reveal great differences in global incomes. The Human Development Index reveals standards of living. Welfare may vary greatly within countries, and statistics examined per capita and through time may reveal a relationship between population growth and economic growth.

Sectoral evolution reveals that some countries are preindustrial, some industrial, and some postindustrial. Various sectors' contribution to national product is not the same as their share of employment.

The world distribution of wealth is not the same as the distribution of raw material resources. Downstream secondary- and tertiary-sector activities add value and produce wealth. Today's map of global manufacturing is a result of historical developments modified by new balances among relevant locational determinants, including capital and technology, as well as labor and raw materials. Several locational determinants are themselves footloose, so manufacturing will probably be redistributed continually.

Each nation's political economy usually reflects other aspects of local culture. Nations try to knit and develop the national territory and to maintain an equal standard of living throughout it. Transportation networks allow regional specialization and trade, and their improvement releases national resources for production. Communication infrastructures both reflect and boost economic development as well as, potentially, political liberty. The term *Third World* combines economic measures with Cold War political values.

Import-substitution policies protect national growth behind tariff walls, whereas export-led policies encourage participation in international trade as a source of capital and as markets. Both methods entail both economic risks and political risks.

International trade has multiplied faster than total world production. Capital has been invested internationally, and world markets for primary products, foods, and minerals evolved. Today manufacturing industries are achieving globalization, and flows of FDI are redistributing world productive capacity. China is today the largest recipient of FDI. A global capital market has formed, and international trade in tertiary-sector services is growing. Tourism, including ecotourism, is today arguably the world's largest industry. The World Trade Organization regulates international commerce.

KEY TERMS

capitalist system p. 505
economic development p. 496
economies of scale p. 516
ecotourism p. 527
export-led economic growth p. 516
First World p. 519
foreign direct investment (FDI) p. 521
frontier p. 509
gross domestic product (GDP) p. 487
gross national income (GNI) p. 487

human development index (HDI) p. 490
import-substitution economic growth p. 516
industrial society p. 493
infant industries p. 516
laissez-faire capitalism p. 505
market-oriented manufacturing p. 499
material-oriented manufacturing p. 499
multinational corporation p. 522

nationalized p. 505
opportunity cost p. 512
outsourcing and offshore outsourcing p. 502
political economy p. 505
postindustrial society p. 493
preindustrial society p. 493
Second World p. 519
sectoral evolution p. 493
Third World p. 519
transnational corporation p. 522

QUESTIONS FOR REVIEW AND DISCUSSION

1. Define and explain the three sectors of an economy. What is sectoral evolution?

2. What locational determinants for manufacturing did Alfred Weber identify? What new determinants are important today?

3. What is the origin and meaning of the term, the Third World?

4. Differentiate laissez-faire capitalism from state-directed capitalism.

5. Why is transportation important to economic development?

6. What is the concept of gross sustainable product? Why do poor nations dispute the idea?

7. Aside from GNI or GDP, what are some alternative noneconomic measures of welfare in different places?

THINKING GEOGRAPHICALLY

1. Americans have difficulty understanding privatization because Americans live in a country in which most activities are and always have been private. Can you, however, imagine selling the national parks to the Disney Corporation? What would be the advantages and disadvantages? Would the corporation run the parks better?

2. What physical and cultural attributes are most conducive to a nation's economic development?

3. Compare Figures 11-33 (literacy) and 11-34 (freedom) with Figure 12-6 (GNI per capita). Can we hypothesize any cause-and-effect relationships?

4. Latin America is rich in natural resources, and yet it has long suffered economic and political problems. Many of these problems stem from class and ethnic structures and from political history. Why?

5. At the opening of international currency markets on August 11, 2006, the following values were quoted as the value of each currency in U.S. dollars:

Australian dollar	0.7670
Brazilian real	0.4624
Euro	1.2784
Indonesian rupiah	0.000110
Japanese yen	0.008675
South African rand	.1470

Check these values in today's newspaper. Calculate the percentage each has risen or fallen against the dollar or against one another. What flows of goods, of investment, or of tourists should these fluctuations have triggered?

6. Did the United States develop in a sustainable way, or at a sustainable rate, between 1870 and 1910? Does it today?

7. Could foreign ownership of businesses hurt the economy of a developing country?

8. What countries have frontiers? What barriers exist that are retarding development of those areas? How might these barriers be overcome? Investigate and compare the real development potential of the frontiers of Canada, Russia, and a South American or African country.

9. How does the quotation from Peter Drucker at the opening of this chapter (p. 487) echo that of Clifford Geertz in Chapter 6 (p. 221)?

10. Find the latest edition of the *Human Development Report* in your library and read the authors' justification for what they have done. If you disagree with their methods, find any one of the other statistical yearbooks published by the United Nations or The World Bank, browse among the quantified descriptions of life in each country, and devise your own overall "ranking of the quality of life" among the countries.

SUGGESTIONS FOR FURTHER LEARNING

Ginsburg, Norton. "Natural Resources and Economic Development." *Annals of the Association of American Geographers* 47, no. 3 (1957): 197–212.

Grove, Richard. *Green Imperialism: Colonial Expansion, Tropical Island Edens, and the Origins of Environmentalism 1600–1860.* New York: Cambridge University Press, 1996.

Held, David. *Global Inequality: Patterns and Explanations.* Cambridge, UK: Polity Press, 2007.

Holliday, Charles O. *Walking the Talk: The Business Case for Sustainable Development.* New York: Berrett-Koehler, 2002.

Lopez, Ramon, and Michael Toman, eds. *Economic Development and Environmental Sustainability: New Policy Options.* New York: Oxford University Press, 2006.

O'Meara, Patrick, Howard Mehlinger, and Matthew Krain, eds. *Globalization and the Challenges of the New Century: A Reader.* Bloomington, IN: Indiana University Press, 2000.

Rees, Gareth, and Charles Smith. *Economic Development.* New York: Palgrave Macmillan, 2006.

Simon, David. *Fifty Key Thinkers on Development.* London: Routledge, 2005.

United Nations Conference on Trade and Development. *Handbook of International Trade and Development Statistics.* New York: United Nations, annually.

Winters, L. Alan, and Shahid Yusuf. *Dancing With Giants: China, India, and the Global Economy.* Washington, D.C.: The World Bank, 2006.

The World Bank. *World Development Indicators.* Washington, D.C.: The World Bank, annually.

The World Bank. *World Development Report.* Washington, D.C.: The World Bank, annually.

Yergin, Daniel, and Joseph Stanislow. *The Commanding Heights: The Battle Between Government and The Marketplace That Is Remaking the Modern World.* New York: Simon & Schuster, 1998.

WEB WORK

In Chapter 8 we cited two general libraries of information about economics that are useful for exploring concepts developed in this chapter. You might also want to begin a search for information about economic development at the Yahoo site at:

http://www.yahoo.com/Social_Science/Economics/

The United Nations hosts a Center for Sustainable Development at:

http://www.un.org/esa/sustdev/

The International Political Economy Network offers links to studies of economic development and economic policies from a political point of view at:

http://www.isanet.org/sections/ipe/

The World Trade Organization has a home page that can be the source of much useful information on the world economy and world trade:

http://www.wto.org/index.htm

The World Bank Group home page offers innumerable studies, plus links to still others:

http://www.worldbank.org/

The United Nations Conference on Trade and Development offers numerous studies and updates on economic matters:

http://www.unctad.org/

The World Tourism Organization home page may be found at:

http://www.world-tourism.org/

and that of the International Labor Organization at:

http://www.ilo.org/

POLITICAL
REGIONALIZATION
AND GLOBALIZATION

Haiti's presidential palace gleams in the background, but the blue helmets on the two Brazilian soldiers in the foreground identify them as troops under the command of the United Nations Stabilization Mission. In 2004, the United Nations Security Council became persuaded that civil unrest in Haiti constituted a threat to international peace and security, so it has stationed international troops in Haiti since 2004. The United Nations is not a world government, but it has often played a key role in defusing world tensions and combating terrorism. (Ariana Cubillos/AP Wide World Photos)

A Look AHEAD

The Collapse of Empires

The collapse of empires through the twentieth century gave birth to many new states. The U.S.S.R. fragmented into 15 countries in 1991.

New Unions of States

New organizations of states are forming to achieve common ends. The European Union and the North American Free Trade Agreement are particularly significant, but numerous

other international organizations coordinate policies among countries. Cooperation among states can result in redistributions of populations and activities within and among them.

Global Government

The United Nations was formed at the end of World War II as an international peacekeeping organization. Its functions have broadened to include economic and social

development. Political management of unoccupied portions of Earth's surface, such as Antarctica and the oceans, has been established through international treaties.

Protecting the Global Environment

Global economic development and population growth have raised concern about global-scale environmental problems. Rapid industrialization in poorer countries is increasing pol-

lution. International negotiations about environmental issues focus on balancing the need for economic development with concerns for environmental protection.

Chicago, a global center for many commodities markets, is now home to a market in greenhouse gases. It is called the Chicago Climate Exchange (CCX), and was established in 2003 to serve as a marketplace for greenhouse gas emissions allowances. The idea of a market for pollution allowances has been around for some time, and markets for sulfur emissions allowances have been very effective in reducing sulfur emissions in the United States. The European Union has a carbon emissions trading system in place. These markets operate on the principle that a unit of pollution controlled in one place is as good as a unit controlled somewhere else, and that the most effective way to control pollution is to do so where it can be accomplished at the lowest cost. Emitters either are required to buy permits, or allowances, or they voluntarily agree to do so. Suppose, for example, an allowance for emitting 100 tons of carbon costs $4.00 (about the current price on the CCX). A company that can control its emissions for $3.00 per 100 tons would be better off reducing emissions, while one for which the cost of reducing emissions is $5.00 per 100 tons would be better off simply buying additional allowances.

Much of the impetus for reducing carbon emissions comes from the Kyoto Protocol, an international treaty that went into force in 2005. The United States has not ratified the Kyoto Protocol, and at the present time the U.S. government has allowed companies' participation in the CCX to remain voluntary. But a similar program has been set up in Europe, which has ratified the Kyoto Protocol, and in 2006 California passed a carbon-emissions control law that includes a carbon-emissions trading program. These schemes could ultimately result in polluters in one part of the world paying governments or other entities in distant lands for permission to pollute our global atmosphere. ▸

Individual countries remain the highest level of absolute territorial sovereignty, but frameworks do exist to regulate and coordinate activities among them. Chapter 12 introduced the World Trade Organization, the International Organization of Securities Commissions, and the International Standards Organization. International monitoring of control over weapons of mass destruction (WMDs) may necessitate greater cooperation or enforcement of agreements. The International Court of Justice, a branch of the United Nations, sits in The Hague, in the Netherlands. It hears cases according to international law and issues rulings, but the litigant states must choose to come before it, and the Court has no power to enforce its decisions.

Another topic of increasing concern is potential international clashes over water rights, because ensuring the continuing availability of fresh water is a long-range problem in many regions. Many of the world's major river systems are shared by two or more countries, so resolving disputes over international waters may be an important function of international courts in the future.

Organizations that coordinate activities among two or more countries are called **international organizations,** whereas organizations that actually exercise power over countries are called **supranational organizations.** Countries choose to surrender some powers to supranational organizations, but each country can escape that control—by making war, if necessary. Therefore, the allocation of powers in either international or supranational organizations is actually flexible and open to continual renegotiation. A *commonwealth* is another kind of international organization that exercises limited powers over its members, and a **bloc** is a group of countries that presents a common policy when dealing with other countries. Most blocs are regional.

International and supranational organizations are not new in history. Independent governments have always made alliances and coordinated policies. Today some international organizations coordinate military defense, but the most active organizations work for economic or broader cultural goals. They are trying to achieve across several countries many of the same things that sovereign states try to achieve within their individual borders. These goals include creating unified markets for materials, goods, and services; knitting area-wide transport networks; ensuring free travel throughout the territory; and even guaranteeing common standards of political or civil rights. These goals enlarge the territorial scale of organization and thus redistribute people and activities.

This chapter first reviews how the collapse of great empires through the twentieth century gave birth to many new states. More than half of today's countries containing about half of the Earth's population and covering about half of Earth's surface are less than 50 years old (Figure 13-1). The political fragmentation of space triggered economic and even cultural breakup, redistributing and remapping activities.

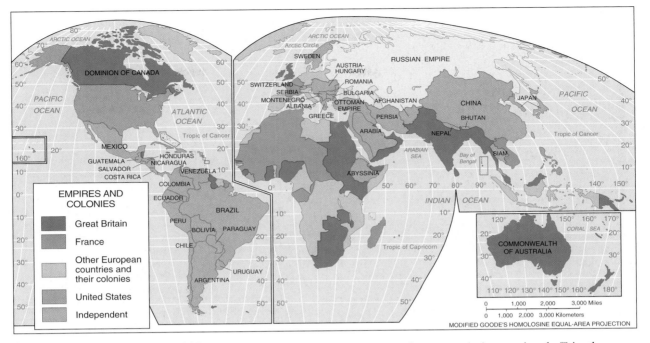

FIGURE 13-1 The world in 1900. In 1900 there were far fewer independent actors in international affairs than there are today. Compare this map with today's political map, Figure 1-6, p. 10.

Even as the number of states has increased, however, the states' genuine independence of action has been undermined by international economic links and other transactions and travels. Therefore, the second part of this chapter examines instances in which nations are joining together to achieve common ends. The European Union and the North American Free Trade Area (comprised of those countries participating in the North American Free Trade Agreement) are particularly significant because the participants have great economic and military power and because in each case the members have surrendered considerable autonomy to a central body or at least to a common policy. Numerous other international organizations coordinate policies among countries, although their goals may be limited.

The third section of this chapter discusses the instruments and institutions through which international law may be evolving toward global government. At present no effective global government exists, but at least systems of global agreement are needed to regulate conduct among nations. International agreements also regulate activities in the areas of Earth that all nations hold in common, such as the seas. The fourth section of the chapter notes the ways in which humankind is beginning to agree on rules to protect the global environment.

THE COLLAPSE OF EMPIRES

The twentieth century saw the fragmentation of empires that Europeans had built up over hundreds of years. The few remnants of empires can be seen on

Figure 13-2. Since the breakup of the U.S.S.R., China is the greatest remaining multinational state.

British Empire to Commonwealth

At its peak in 1900, the British Empire covered over one-quarter of Earth's land surface. This empire has gradually been transformed into a loose association of independent states called the Commonwealth of Nations. Most of the United Kingdom's former colonies have chosen to join, and today the Commonwealth has 53 members. Queen Elizabeth remains head of state in 15 countries in addition to the United Kingdom, including Canada, New Zealand, and Jamaica, but most former colonies have become republics. The British legal system, language and education system, the Anglican Church, and other cultural traditions linger in each of the countries. The Commonwealth offers a framework for consultation and cooperation for the achievement of common ends, where they exist; and its achievements have enticed Mozambique, a former Portuguese colony, to join.

Britain retains possession of the Falkland Islands ("Malvinas" in Spanish) against claims by Argentina, which even invaded the islands in 1982; and Britain retains Gibraltar against claims to it by Spain (Figure 13-3). Bermuda is a Crown Colony with its own internal representative government. All three colonies have repeatedly voted to maintain their current status.

Northern Ireland's status is open to question. The United Kingdom gave independence to 26 of Ireland's 32 counties in 1921, but the six counties of Northern Ireland (Ulster) elected to remain within the United

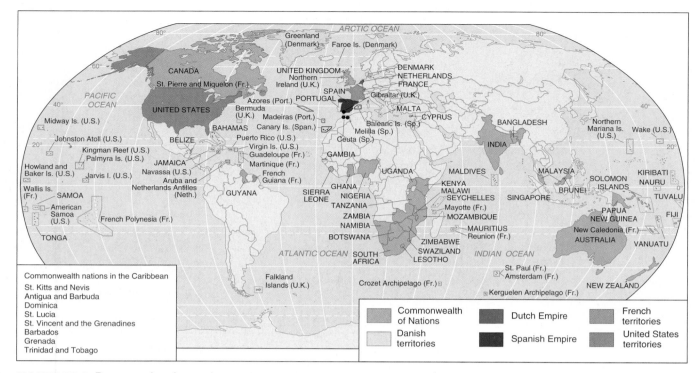

FIGURE 13-2 Remnants of empires. A few empires linger, whereas the breakup of others caused the creation of cooperative organizations. Zimbabwe, ruled by Robert Mugabe, has been suspended from the Commonwealth for lack of democratic government, and Fiji was suspended after a military coup d'etat in 2006.

FIGURE 13-3 Gibraltar. Britain's 6.5-square-kilometer (2.5-square-mile) colony of Gibraltar offers a fine harbor strategically located at the entrance to the Mediterranean Sea. Spain claims the colony, and the United States even fought in alliance with Spain to wrest it away (1779–1783), but Britain will not surrender it unless its 30,000 residents vote to change its status. (Topham/The Image Works)

Kingdom. Great Britain and Ireland agree that Ulster can remain a province of Britain for as long as most of its people want it to, but the province has been wracked by almost nonstop violence since the 1960s. The province's Roman Catholics complain of discrimination by the majority Protestants, and both Catholic and Protestant paramilitary groups are guilty of murder and terrorism. Several government schemes have been

attempted, but none has yet been able to guarantee peace. In 2000 Ulster was 57 percent Protestant, but birthrates are so much higher among the Roman Catholics that they might soon form the majority.

The French Empire

France built up an empire in the seventeenth and eighteenth centuries but lost most of its colonies during an eighteenth-century rivalry with Britain. During the nineteenth century France built a second empire, beginning with the occupation of Algeria in 1830 and culminating in 1919 with the assumption of rule in Syria, Togoland, and Cameroon. The French believed that their subject peoples would mature not to independence but to full representation in the government in Paris. The four colonies of Martinique, Guadaloupe, Réunion, and Guiana were organized as Overseas Departments of France, and each was granted representation in the French National Assembly in 1946. Algeria won independence in 1962 after a long and bloody war. France granted independence to most of its other colonies in the 1950s, but it continued alliances. France's former African colonies receive aid and still even host French military forces.

The present French Republic encompasses, in addition to mainland France and the four Overseas Departments, two Territorial Collectivities and four Overseas Territories. France also possesses a number of islands, including St. Pierre and Miquelon off the south coast of Newfoundland, a remnant of France's once-vast North American holdings. New Caledonia in the southwestern Pacific has major nickel reserves, and in 1998 its people agreed to defer a vote on independence for at least 15 years. France's other territories are generally poor and sparsely populated.

The Successor States of the Ottoman Empire

When the Ottoman Turkish Empire collapsed and a new Turkish nation was born in the 1920s, most of the Turkish Empire in the Middle East was divided between France and Britain. New states were eventually granted independence, but the map of today's states results from the interests of the colonial powers rather than the sentiments of the local people. Several borders have been disputed for 90 years. Israel was carved out to provide a homeland for the Jewish people, and bitter confrontation over this land—historic Palestine—continues. Furthermore, the region has remained a cauldron of competing loyalties. Many Arabs bestow

their loyalty on units that are much smaller than states (clans, tribes, or families) or on ideas that are much bigger than states. These ideas include *pan-Arabism*, the notion that there exists some bond among all Arabic-speaking peoples that is more legitimate than the modern Arabic states, and a universal bond of Islam. The Middle East is plagued by border disputes; rulers whose power is based on foreign interests rather than popular support; religious animosities; and political and economic intervention by international oil companies and banks. Several wars have troubled the region since World War II, frustrating many hopes for democracy and economic development.

The Russian Empire, Revolution, and Reorganization

At the same time that the Western European countries were building overseas empires, Russia was building an empire across Asia (Figure 13-4). During World War I, the empire suffered internal rebellion as well as external attack. The Communists, led by Vladimir Lenin

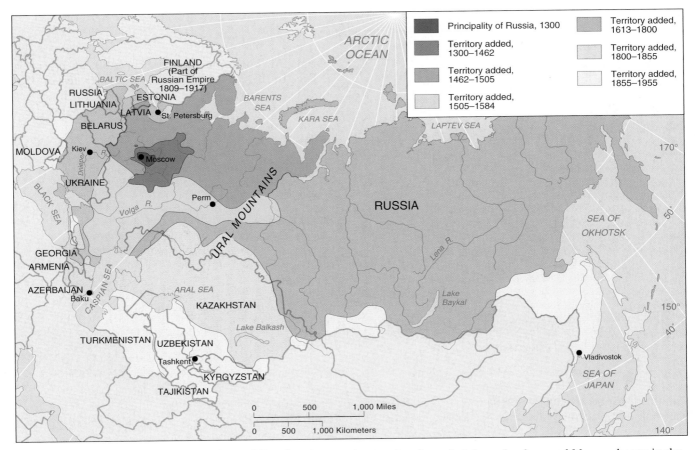

FIGURE 13-4 The historic expansion of Russia. Russians' expansion from their homeland around Moscow began in the mid-sixteenth century, and by 1647, Russian cavalrymen had reached the Pacific. During the seventeenth and eighteenth centuries, Russians pushed south to the Black Sea. Russian explorers crossed over the Bering Strait to incorporate Alaska (which Russia sold to the United States in 1867) and to locate settlements as far south as California. In the nineteenth and early twentieth centuries, czarist power advanced still further into Eastern Europe and Central Asia.

(1870–1924), seized control in 1917 and offered a new propagandistic ideal of a federal union among the former subject nations. This union, the Union of Soviet Socialist Republics (U.S.S.R. or Soviet Union), came into being in 1922.

The configuration and internal organization of the U.S.S.R. changed through the years, but in 1990 the Soviet Union covered 17 percent of Earth's land area (excluding Antarctica), and its population of about 290 million made it Earth's third most populous country, after China and India.

The Soviet Union was subdivided into 15 union republics, each of which was theoretically the homeland of a national group and therefore was named for that group (Figure 13-5). Several republics contained within them the homelands of less numerous ethnic groups, and these groups enjoyed lesser territorial autonomy. Russia was the largest republic, occupying fully 76 percent of the total territory of the U.S.S.R., and Moscow was the capital of both the Russian Republic and also of the U.S.S.R. Within Russia itself, 88 constituent units, 20 of which were ethnically based republics, enjoyed varying degrees of autonomy (Figure 13-6). Russia is, therefore, still technically referred to as the Russian Federation.

The individual Soviet republics ostensibly enjoyed considerable autonomy, but in fact all power was centralized in Moscow. Russian Communists dominated the Union government and even in the non-Russian republics. Russians attempted to acculturate the other peoples. The Russian language, for example, was preferred for central government affairs, and it was one of two official languages in all non-Russian republics. Migration of Russians into the other homelands was yet another form of Russian imperialism.

The Soviet Union contained 55 million Muslims in 1990—concentrated in Kazakhstan, Turkmenistan, Uzbekistan, Tajikistan, and Kyrgyzstan in Central Asia plus Azerbaijan in the Caucasus. Birthrates were high among these peoples, and many of them were politically dissatisfied. Their dissatisfaction threatened the Union's political stability.

The economic geography of the U.S.S.R.

The two principal characteristics of Soviet economic development were economic self-sufficiency for the Union as a whole and centralized state control. Natural resources were the property of the state, and no one was allowed to own "means of production" in such a way as to

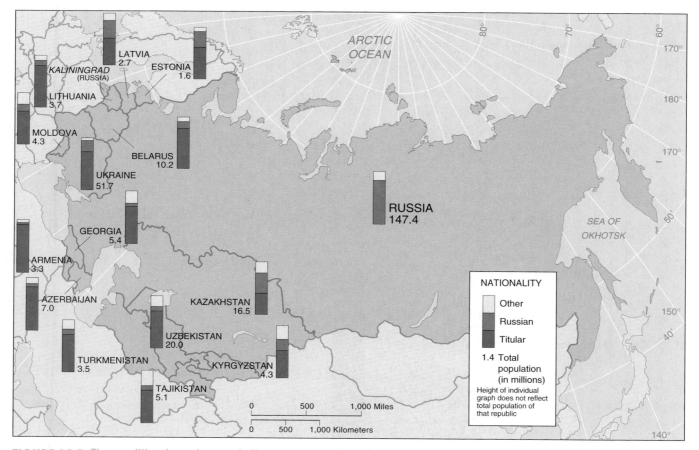

FIGURE 13-5 The political and population geography of the U.S.S.R. in 1990. Most of the states of the former Soviet Union are dominated by nationalities for which the republic was named, here shown as "titular" nationalities. The map illustrates that Kazakhs were a minority in "their own" republic, and in several republics immigrant peoples—often Russians—formed substantial minorities.

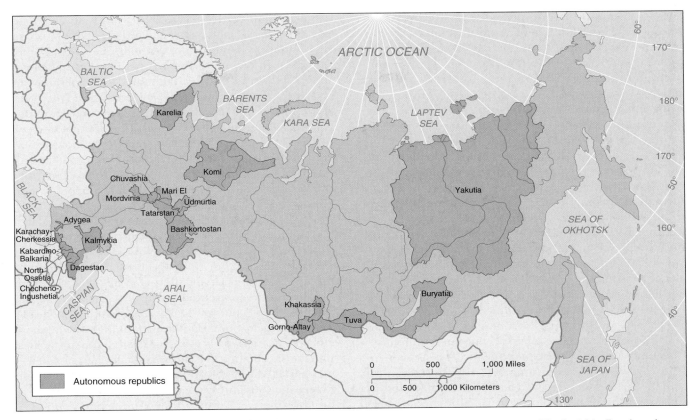

FIGURE 13-6 Russia's autonomous republics. These 20 ethnically based "autonomous republics" within Russia enjoy varying degrees of autonomy from the government in Moscow.

profit from the labor of others. Only a little private property and a few private-sector services were allowed.

Russia's treasury and supplies of raw materials subsidized inefficient industries in other republics. Agriculture was brutally collectivized. New lands were opened to agriculture, notably the steppes of northern Kazakhstan, but conditions across much of the U.S.S.R. territory—particularly northern Russia and the desert regions of central Asia—remain unfavorable for agriculture. Agriculture also suffered from mismanagement, lack of worker incentive, lack of investment, poor transport, and government policies that kept farm prices low. Heavy industry, guided by planning bureaucracies, made immense strides, but the Soviet Union lagged behind the West in light industry, in the production of consumer goods, and in high-technology products. Both production and internal trading were highly centralized, and extremely inflexible. These three factors—centralization, rigidity, and geographic concentration of production—bound the republics together in trade.

Environmental protection was sacrificed for industrialization. In 1990 Alexei Yablokov, director of the Institute of Biology at the Academy of Sciences, estimated that 20 percent of Soviet citizens lived in "ecological disaster zones" and 35 to 40 percent more in "ecologically unfavorable conditions."

The breakup of the U.S.S.R. Mikhail Gorbachev assumed power as Union president in 1985, and he tried to boost the Union's economy by launching two liberalizing initiatives: *glasnost*, a loosening of restraints against freedom of speech and the press, and *perestroika*, a restructuring of the economy and of politics. Gorbachev gambled that his policies could unleash individual initiative and productivity while retaining the centralized control of the Communist Party and of the central Union government. Gorbachev lost. The tensions unleashed brought down the old governmental system before Gorbachev could bring a new one into place.

During 1991 much was done to reform or overthrow the government of the U.S.S.R., but by the end of the year the individual republics had declared their independence. On December 26, 1991, the Soviet Parliament formally voted the U.S.S.R. out of existence. Representatives of the three Slavic republics (Belarus, Russia, and Ukraine) created a new Commonwealth of Independent States (CIS), which was eventually joined by nine more former republics.

Estonia, Latvia, and Lithuania (called the Baltic Republics) were independent democracies after World War I, but the Soviet Union forcibly reincorporated them in 1939. They achieved relative prosperity under the Union government, but since the collapse of the

Soviet Union they have turned their hopes toward the European Union, discussed later in this chapter.

Developments in Russia

After the fall of communism and the dissolution of the Soviet Union, Russia suffered a virtual implosion of calamitous proportions from which it seems to be emerging only now, early in the twenty-first century. Every measure of income, health, or welfare still reveals a society facing enormous challenges. The economy has shrunk, the birthrate and the total fertility rate have dropped, and the death rate has risen. Life expectancy has shortened, and Russia's population, which was 147 million in 1999, fell to 143 million in 2006 and is projected to fall to 121 million in 2050. Neither attempts to achieve democracy nor to achieve capitalism seem to have succeeded.

Vladimir Putin was elected president in 2000 in elections would not have been considered open and fair in any Western state. State-owned media, for example, relentlessly attacked Putin's opposition. Most of the basic governmental framework of the Russian Federation has been maintained, but one ethnic unit, Chechnya (formally, Checheno-Ingushetia as seen in Figure 13-6) launched a long and bloody struggle to achieve full independence. The 20 million Muslims in Russia constitute the majority in 7 of the autonomous republics, and fertility rates among them are high, so in 2003 Russia joined the Organization of the Islamic Conference. Russia's Muslims, said President Putin, "have every right to feel part of the Muslim world."

Meanwhile, the privatization of the state-owned economy meant seizure of many assets by the members of the old Communist government bureaucracies, by organized crime, or by a few ruthless oligarchs. *Privatizatsia* (privatization) turned into *prikhvatizatsia* (to grab). Hundreds of billions of dollars were siphoned out of Russia and deposited in foreign banks during unscrupulous privatizations of state-owned companies between 1900 and 2000. Only history will decide how many leaders or profiteers in Russia in the 1990s were naïve, incompetent, corrupt, or all three.

Russia still has immense natural resources, a large skilled and educated population, and scientific institutes second to none. Soaring energy prices early in the twenty-first century returned enormous profits to Russia, but it may be many years before it fully reassumes a place as a great power.

The other formerly Soviet republics

Experience in most of the other formerly Soviet republics (aside from the Baltics) has echoed the Russian example. Politically, they have renegotiated their relationships with the former autonomous regions within them, although several newly independent republics, including Moldava, Tajikistan, and Georgia have experienced civil wars. Fifteen years after winning independence, most guaranteed few freedoms and continued to be ruled by autocrats or oligarchs. The highly centralized command economy of the U.S.S.R. left a legacy ensuring that none of the successor states has a balanced economy. Most are dependent on other former republics for essential supplies and rely on enterprises in other former republics as their sole markets. Russia continues to supply most of them with natural resources at prices below those on world markets. It is proving very difficult to unravel this economic interdependence, and the process of unraveling has imposed considerable hardship. Most of the new countries are privatizing their economies, but not all are privatizing at the same rate. As in Russia itself, the process in several has been tainted by corruption. Several of the states have skidded toward worsening poverty, instability, and international isolation. Russia still maintains military bases in several of them.

Ukraine was the second most powerful of the Union republics, but since the breakup, its industrial might has proved largely illusory. Many factories were obsolete and represented more of a liability than an asset to a new economy. The government has been slow to reform the economy and reluctant to privatize, so inefficient state-owned industries still pile up losses. Income in Ukraine fell approximately 50 percent between 1990 and 2000 and it remains dependent upon Russia for fuel and raw materials. Ukraine, known as the "breadbasket" of the former Union and potentially one of the richest agricultural regions on Earth, has even had to import food.

Many former republics want to diversify away from economic dependence on Russia by giving priority to trade and investment with other partners in the West, Asia, and the Middle East. The newly independent Central Asian Islamic countries retain economic links to Russia, but religion and other cultural ties pull them away. These countries are landlocked, and the countries surrounding them compete for influence. Turkey offers a democratic and secular model, and it is strengthening its links with its co-Turks across the region, but the Iranians, who are related to the Tajiks in language and culture and who share Shiite Islam with the Azerbaijani, have subsidized new mosques and schools in Azerbaijan and are initiating joint economic ventures (Figure 13-7). To the east, China has cultivated economic and political contacts, and even Japan and South Korea have invested in Kazakhstan and Kyrgyzstan.

Some of the Central Asian Republics—particularly Kazakhstan—may prove to be rich in oil, so the questions of who will take the lead in developing that resource and by what routes the oil will reach world markets have launched diplomatic initiatives among many concerned countries. Each possible route carries political and economic complications. In 2005, the China National Petroleum Corporation bought Kazakhstan's largest oil company and built a pipeline to the Chinese border.

Even the United States has assumed a role in central Asia. The United States has occupied military bases in Tajikistan and in Uzbekistan in exchange for long-term commitment to friendship and assistance.

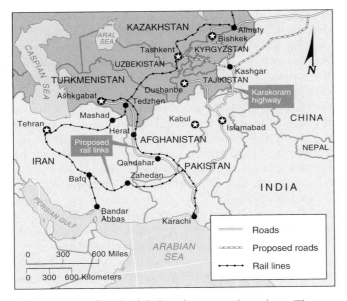

FIGURE 13-7 Central Asian transport routes. The Iranian government completed a rail link from Mashad to Tedzhen in 1996, so today Iran's port of Bandar Abbas is central Asia's principal outlet to the sea. Completion of a railroad from Mashad to Bafq will shorten that route by 900 kilometers (560 miles). When the line is completed from Bafq to Zahedan, goods from central Asia will be able to bypass war-torn Afghanistan to reach Pakistan and India. A pipeline is also under construction from oil-rich Turkmenistan around the southern shore of the Caspian Sea through Iran to Turkey.

This was partly to protect these governments from their own Islamic fundamentalist organizations. These agreements triggered concern among many regional governments that the United States intends to establish a permanent military presence in central Asia—perhaps to obtain the oil.

One of the most volatile issues among the new central Asian republics is their competitive demands for the limited regional water supplies. Problems associated with the unequal water resources of mountainous Tajikistan and Kyrgyzstan and their low-lying neighbors Uzbekistan, Kazakhstan, and Turkmenistan broke up a U.N.-sponsored conference on regional environmental issues in 2003 and have even raised fears of armed conflict.

The dissolution of the Union of Soviet Socialist Republics created new units on the world political map and released powerful new political, religious, cultural, and economic forces. These forces continue to cause global redistributions of cultural, political, and economic activities.

The Empire of the United States

The United States organized most of its territories and admitted them to the Union fairly quickly. With the granting of statehood to Alaska and Hawaii in 1959, the United States even incorporated overseas territories, as France had.

The United States took Cuba from Spain in 1898 but granted it independence in 1934 and today retains only a naval base in the island's Guantánamo Bay. In 2002, President Bush authorized the holding of non-U.S. citizens in indefinite detention and the shipment of "unlawful enemy combatants" to that naval base without the prisoners having recourse to either U.S. courts or the protections of the Geneva Convention, which prohibit "mutilation, cruel treatment and torture." The subsequent mistreatment of prisoners at Guantánamo Bay disturbed many Americans and friends. In 2004, however, the U.S. Supreme Court ruled that the prisoners were not beyond the reach of the U.S. court system (*Rasul* v. *Bush*), and in 2006 the Court ruled that the military tribunals created to try suspects there violated both American military law and the Geneva Convention (*Hamdan* v. *Rumsfeld*). Congress must redefine the reach of U.S. law over Guantánamo.

The United States also took the Philippines from Spain in 1898, but it then had to defeat a powerful national independence movement there. The Philippines were granted independence in 1946.

The United States provoked Panama's uprising for independence from Colombia in 1903, and then by prearrangement leased the Canal Zone from the new country (Figure 13-8). The United States invaded

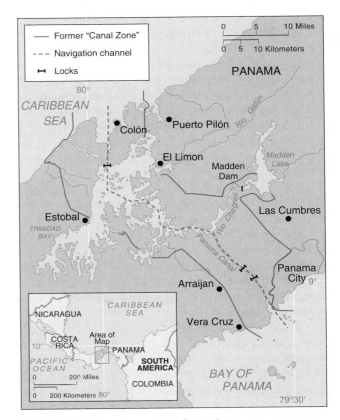

FIGURE 13-8 The Panama Canal. The United States turned the Panama Canal over to Panama in 1999. The Canal is too small to accommodate many of today's ships, so in 2006 the government announced a plan to double the capacity of the canal by 2014.

Panama in 1989 to interdict drug traffic to the United States and occupied Panama until the end of 1999, but U.S. officials report that narcotics traffic through Panama has not been reduced. Panama is one of the 11 countries that accept the U.S. dollar as legal tender in an effort to stabilize their own currencies. Some 14 percent of the world's merchant ships are registered in Panama. This is an astonishingly high percentage for such a small, poor country, but many multinational corporations register their ships in Panama because of its low taxes and regulation. For this reason the Panamanian flag is known as a *flag of convenience.*

Liberia was founded as a haven for freed U.S. slaves and received independence in 1847. Inequities in wealth and political power fuel antagonism between the dominant Americo-Liberians (about 5 percent of the total population of 2.1 million) and the indigenous Africans. The indigenous population is itself split among ethnic groups. Civil war broke out in the 1980s, and even intervention by forces from a consortium of West African states has been unable to stop the bloodshed. The Liberian flag is another flag of convenience. Liberia registers about 6 percent of the world's merchant marine fleet, representing 13 percent of total world gross tonnage. The threat of terrorism has stimulated efforts by the London-based International Maritime Organization to clarify true ownership of ships flying flags of convenience.

The United States still retains a modest empire of islands. Their total population is about 5 million, most of whom are citizens of the United States. American Samoa, Baker Island, Guam, Midway Islands, the U.S. Virgin Islands, Wake Island, and a few other spots are dependent areas. The Northern Mariana Islands is a commonwealth in political union with the United States, while Palau, the Republic of the Marshall Islands, and the Federated States of Micronesia are independent states with Compacts of Free Association with the United States.

Puerto Rico   The United States granted Puerto Rico territorial status in 1917 and elevated it in 1952 to the status of a free commonwealth. Puerto Rico enjoys internal autonomy, but Puerto Ricans are U.S. citizens also under the jurisdiction of the federal government. They do not vote in presidential elections, and their resident commissioner in Congress can vote only in committee meetings. The United States and Puerto Rico share a common market and monetary system, but Puerto Ricans pay no federal taxes, and some federal customs and excise duties are paid back into the island's treasury. Puerto Ricans are divided over their political status. In a nonbinding 1993 plebiscite, 48 percent of voters opted to continue the commonwealth status, 46 percent preferred statehood, and 4 percent voted for independence.

Most arguments in the political debate are economic. The densely populated, resource-poor island enjoys a standard of living higher than that of any other Caribbean or Latin American nation, so it may be considered to be either a rich Caribbean country or, conversely, a poor part of the United States. Puerto Rico's 4 million people would rank it twenty-fifth among the states in population, but its per capita income is less than half that of Mississippi, which ranks last among the states. Both the federal government's role and that of the local government is large: Federal transfer payments (i.e., welfare) make up more than 20 percent of the island's personal income, and about 30 percent of the island's jobs are in the public sector. The United States long granted tax benefits to U.S. corporations that located in Puerto Rico, so today manufacturing accounts for 45 percent of Puerto Rico's economic output and for 20 percent of employment. Agriculture accounts for only 1 percent of output and 3 percent of employment.

Puerto Ricans are Spanish speaking, and many wish to retain their linguistic and cultural traditions. This might make it difficult for Puerto Rico to assimilate fully into U.S. life. If Puerto Rico were a state, it would send 2 Senators to Congress, plus 6 or 7 Representatives, costing 6 or 7 states one Congressman each. If, on the other hand, Puerto Ricans ever choose independence, terms of continuing financial assistance plus leases on U.S. military bases (a full 13 percent of the territory) will have to be renegotiated.

NEW UNIONS OF STATES

The breakup of empires has fragmented and reduced the scale of political organization of territory and, often, the cultural and economic organization as well. New organizations of states, however, are binding activities across regions. The European Union is the international organization with the most ambitious goals—nothing less than eventual political unification. The North American Free Trade Agreement has more limited goals.

The European Union: Nations Knitting a Region Together

World War II ended in Europe in 1945 with the surrender of the Axis powers: Germany, Italy, Romania, Bulgaria, and Hungary. Japan had been an Axis Asian ally. The principal victorious Allied powers were the United States, Canada, Great Britain, France, the U.S.S.R., and China. In Europe the Allies had squeezed the Axis powers from West and East, and their victorious armies met in the middle of the continent.

The Allies themselves, however, were already divided by mistrust, and Europe soon split into two competitive blocs. British Prime Minister Winston Churchill said, "From Stettin in the Baltic to Trieste in the Adriatic, an Iron Curtain has descended across the continent." To the east, Soviet armies installed satellite regimes in Poland, Hungary, Romania, Czechoslovakia, and

Bulgaria. Local Communist leaders seized power in Yugoslavia and Albania without Russian help. The constitutions of all these countries were amended to guarantee Communist Party dominance in national politics, and their economies were largely collectivized.

Germany itself was split into four zones of military occupation. Already by 1949, however, the United States, French, and British zones were united functionally, and a new Federal Republic of Germany (West Germany), with Bonn as its capital, gained sovereignty in 1955. The Soviet Union reacted by granting its occupation zone independence as the German Democratic Republic (East Germany). Germany's capital city of Berlin had also been jointly occupied, so it too was split between West and East Germany, even though it lay about 201 kilometers (125 miles) inside East Germany. Its western sector was recognized as a part of West Germany, and East Berlin became the capital of East Germany. West Berlin showcased Western prosperity in the middle of East Germany. As a result, so many East Berliners fled into West Berlin that in August 1961, East Germany built a wall across the city to seal in its citizens. Austria, which had been annexed by Germany, also had been split into four occupation zones, but they were reunited in a newly independent state in 1955.

The Iron Curtain across Europe followed fairly closely the far older cultural division that Chapter 7 demarcated in orthography and religion: The West was defined by Roman Catholicism or Protestantism and the use of the Roman script, and the East by Orthodox Christianity and the Cyrillic script. The Iron Curtain deviated from this line, however, in that Estonia, Latvia, Lithuania, Poland, Eastern Germany, Czechoslovakia, Hungary, Croatia, and Slovenia, which had never been "Eastern European" culturally, were east of the Iron Curtain.

The division of Europe into two competing blocs was furthered by Soviet seizure of territory during and after World War II (look back at Figure 11-5). In 1975, 35 countries—mostly European, but including the United States and other World War II participants—agreed to accept European borders as inviolable, but they did not rule out revision by peaceful agreement. Some European borders have been redrawn since then. For example, in 1990 the Federal Republic of Germany and the German Democratic Republic peacefully united and recognized Berlin as the capital of the united country (Figure 13-9). Eastern Germany still lags behind the Western part, but standards of living are rising. Unification has been judged a political success overall, despite continuing difficulties in absorbing the East. In 1991 Yugoslavia split apart into warring new states, and Montenegro separated from Serbia in 2006. In 1993 Czechoslovakia peacefully split into two countries.

Military pacts The postwar political division of Europe was accompanied by its division into two military alliances. In the West the **North Atlantic Treaty Organization (NATO),** established in 1949, united Belgium,

FIGURE 13-9 Germans celebrate the opening of the Berlin Wall. At 7 o'clock on the evening of November 9, 1989, an East German official mistakenly handed a television speaker a draft of a new regulation suggesting that East Germans could go to the West without a visa at once. By 10 o'clock, hundreds of thousands stood at the wall, shouting, "Open the Gate!" Confused guards did, and East Germans surged through, to be met by West Germans and doused with beer and champagne. The Berlin Wall had fallen, and East Germany was soon swept away by history. (Anthony Suau/Getty Images, Inc.—Liaison)

Denmark, France, Great Britain, Iceland, Italy, Luxembourg, the Netherlands, Norway, and Portugal with Canada and the United States. Greece, Turkey, West Germany, and Spain joined later. In 1955 the *Warsaw Pact* linked the U.S.S.R. and its satellites. The U.S.S.R. used this treaty to justify the continued occupation of Eastern European countries by Soviet troops, which suppressed occasional anticommunist uprisings. The Western and Eastern blocs faced one another in a protracted cold war: Although the two blocs never engaged directly in combat, they competed for economic growth and for influence among the new countries that gained their independence after World War II.

The pattern of alliances in Europe dissolved in 1989. The U.S.S.R. directed its attention to its own internal problems and released its grip on the countries of Eastern Europe. These countries, in turn, repudiated the privileged role of the Communist Party in their own countries and within a few months scheduled free elections. The Warsaw Pact formally dissolved in 1991. NATO eventually expanded to include most of its former putative adversaries: Poland, the Czech Republic, Hungary, Latvia, Lithuania, Estonia, Bulgaria, Slovakia, Slovenia, and Romania. In 2002 NATO signed an agreement with Russia on terrorism, arms control, and international crisis management, so today even Russia joins NATO conferences. Croatia, Macedonia, Albania, and Georgia have expressed their wish to join.

Despite the end of the cold war, NATO has become increasingly active in international affairs. In 1999 NATO for the first time in its history attacked a sovereign nation: It engaged in Yugoslavia to end fighting there. In September 2001, NATO invoked, for the first time, a clause that binds the members to recognize an attack directed from abroad on any member as an attack on all members. This clause was designed to protect Europeans from the U.S.S.R.; few observers would have ever suspected that it first would be invoked to protect the United States from terrorists. NATO forces helped defeat the Taliban in Afghanistan and have remained there. The organization's expansion and its redirection of effort so far afield have caused many observers to wonder whether NATO must redefine its focus and purpose.

Economic blocs

In 1947 U.S. Secretary of State George C. Marshall (Figure 13-10) announced a plan of financial aid to war-shattered Europe. Aid was offered to the Soviet Union and to the countries of Eastern Europe, but Soviet dictator Joseph Stalin (1879–1953) rejected it. The **Marshall Plan** became a program for the economic rehabilitation of Western Europe.

FIGURE 13-10 George C. Marshall. George C. Marshall (1880–1959) is shown here (at right) with Mrs. Marshall and President Eisenhower in 1957, at ceremonies commemorating the tenth anniversary of the Marshall Plan. During World War II, Marshall served as army chief of staff and rose to five-star rank, as illustrated in the painting behind him. As secretary of state (1947–1949), he devised the Marshall Plan for the recovery of Europe, for which he won the Nobel Peace Prize in 1953. He also served as secretary of defense (1950–1951). During World War II Dwight D. Eisenhower (1890–1969) was supreme commander of the Allied Expeditionary Force in Europe, and then Eisenhower served as U.S. president (1953–1961). (George C. Marshall Foundation, Lexington, VA, GCM Photograph Collection #307)

The Soviet Union and its Eastern European satellites formed a competing economic pact, the Council for Mutual Economic Assistance (COMECON), that achieved a partial integration of their economies. Some of the East European economies grew under their communist regimes, enough to surpass the Soviet Union itself in per capita measures, but they did not keep up with Western Europe. COMECON formally disbanded in 1991, and the East European countries began to privatize their economies and to negotiate new economic links with the more prosperous countries of Western Europe.

The European Union today The Marshall Plan encouraged economic cooperation among the European countries, and in 1952 six of the Western European nations united their industrial economies to form the European Coal and Steel Community: Belgium, the Netherlands, Luxembourg, West Germany, France, and Italy. The success of this union encouraged further cooperation, and the European Economic Community (EEC) and European Atomic Energy Community (Euratom) came into existence in 1957. All three merged into one European Community (EC) in 1967. Denmark, Ireland, the United Kingdom, Greece, Spain, and Portugal soon joined, too, and in 1993 these 12 countries changed the name of their organization to the **European Union (EU).** Sweden, Finland, and Austria joined in 1995; the Czech Republic, Estonia, Hungary, Latvia, Lithuania, Poland, Slovakia, Slovenia, Malta, and Cyprus in 2004; and Bulgaria and Romania on January 1, 2007 (Figure 13-11). Norway has voted not to join.

There are three kinds of economic groupings. A **free-trade area** has no internal tariffs, but its members are free to set their own tariffs on trade with the rest of the world. A **customs union** enforces a common external tariff. A **common market** is a still closer association of states; it is a customs union, but in addition, common laws create similar conditions of production within all members. The EU is forming a full common market, and it even declares political merger to be an eventual goal. It has been noted that the United States developed a political community before merging as an economic union, whereas the European states are trying to reverse that process.

The members of the European Union share many broad cultural patterns, which provide a basis for their successful cooperation. Common traditions include democratic ideals and parliamentary institutions, civil rights and legal codes, Judeo-Christian ethics, respect for scientific inquiry, artistic traditions, and humanism and individualism. It may be difficult to extend Union membership to some states, such as Turkey, that are pressing for entry but do not share these European traditions.

The European Union has several goals. One is to end the wars that have plagued the continent. Another is economic growth, and a third reason for European

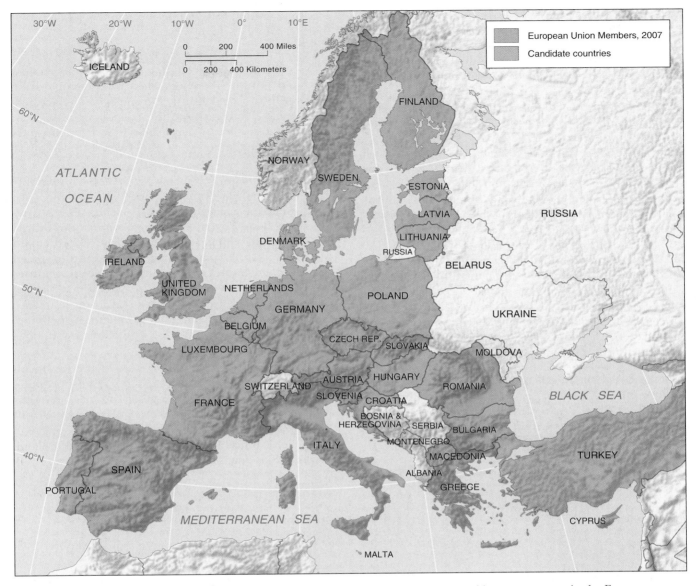

FIGURE 13-11 The European Union. Germany has both the largest population and largest economy in the European Union.

unification is the urge to play a role in world affairs. Economic strength has largely replaced colonialism or military might as a measure of prestige, and no European nation alone can command prominence on the world stage. United Europe, however, can assume a leading position.

European Union government

The Union is governed by four institutions: the European Commission, the Council of Ministers, the Parliament, and the European Court of Justice. The Commission, with one member from each country, proposes legislation and is responsible for administration. It sits in Brussels, Belgium, which has enjoyed considerable growth and prosperity as de facto capital of the Union. Commissioners are not permitted to take instructions from the government of the country that appointed them, but are supposed to represent the interests of the citizens of the EU as a whole. The legislative body is the Council of Ministers. Seats in the Council of Ministers are awarded roughly according to the populations of the EU members, but no decision can be taken by a coalition representing less than 62 percent of the total EU population. The council decides many issues by majority vote, thus considerably reducing individual countries' sovereignty.

The European Parliament sits in Strasbourg, France, and in Brussels. The apportionment of its seats compromises between numerical democracy and equality among nations. Its members are elected directly by the citizens of the member countries, and the members who represent affiliated political parties from different countries join together as international voting blocs. Therefore, the Parliament and to a degree the commission are genuinely supranational. The

European Parliament is not equivalent to a national legislature. It can scrutinize proposed Union laws, offer amendments, vote against directives, veto the actions of any branch of the Union, or dismiss the European Commission, but it cannot initiate legislation, select the executive, or levy taxes. Nevertheless, Parliament is the most democratic of Union institutions, and it has been gaining power.

European law is adjudicated by the European Court of Justice. In cases of conflict between Union law and national laws, the supremacy of Union law is being steadily affirmed. Union institutions have become so important to the member states that about 60 percent of each member's domestic legislation is drafted in Brussels and simply translated into national law. Final adoption of the European Convention on Human Rights in 2000 profoundly changed government in several European states that, unlike the United States, had no Constitution or independent judiciary to enforce guarantees against official abuse. After 2000, for the first time, national judges were bound to measure acts of their national legislatures against the EU Convention. This is truly supranational law.

European Union integration   The EU largely achieved the creation of one economic community by the end of 1992. The following six criteria, however, demonstrate that the Union is much more than an economic union. It is uniting its members and binding their territories in many ways.

1. *The Creation of One Market for Goods.* The Union established a common agricultural policy, and Union bureaucracies have standardized business regulations. European manufacturers would like to achieve the economies of scale in one European market that U.S. manufacturers have long enjoyed in the huge U.S. domestic market.

2. *The Creation of One Market for Capital.* Thirteen EU members have adopted one currency, the euro (Belgium, Germany, Greece, Spain, France, Ireland, Italy, Luxembourg, the Netherlands, Austria, Portugal, Finland, and Slovenia). Coordination among the countries that use the euro has been strained because rules require individual countries that use the euro to keep down national budget deficits. This rule may force cuts in social welfare programs and other governmental spending. Nevertheless, the euro has won increasing favor as an international currency.

3. *Transport System.* The members of the European Union are redesigning and rebuilding Europe's transport system to change it from national webs into a single coordinated web (Figure 13-12). All railroads, for example, must use the same gauge

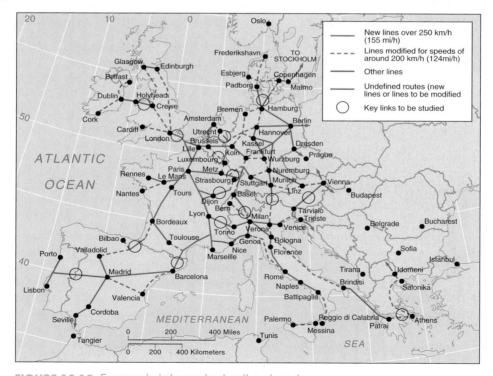

FIGURE 13-12 Europe's integrated rail network. An integrated network of high-speed trains has been superimposed on the existing national networks. It required laying thousands of kilometers of new track and improving much of the rest. In an area as compact and densely populated as Europe, city-to-city train travel is a sensible alternative to air travel at little or no additional time cost. (Data from the European Commission)

and electricity systems, and all highways, bridges, and tunnels must have the same construction specifications. One air-traffic control system has evolved and was extended in 2006 to ten European countries outside the EU.

4. *Regional Policy.* The Single Europe Act commits the countries "to strengthen economic and cultural cohesion." Living standards and quality of life ought to be equalized throughout the Union area, just as nations strive to equalize incomes throughout their territories (Figure 13-13). The Union tailors policies for poor agricultural regions and declining urban regions and furnishes funds for needed infrastructure, such as transport and telephone service.

5. *Social Policy.* The Union members have tried to strengthen cultural ties. International committees are rewriting school history books, for example, to soften nationalistic antagonisms. EU policies equalize worker health and safety standards, the international transfer of pension rights, the guarantee of rights to migrant workers, and mutual recognition of technical qualifications. The Workers' Charter regulates the workweek, overtime pay, holidays, and other terms of employment. The European Court of Justice has enforced standard civil rights and antidiscrimination legislation. Union citizens may travel, settle, and work anywhere within the Union, whereas non-Union citizens face common immigration and visa requirements. EU citizens can vote in any EU country in which they reside, and noncitizens can even hold offices throughout the member countries. The concept of "European citizenship" is being made real.

6. *Environmental Policy.* The EU has become a leader in international efforts to protect the global environment. The EU has established policies of protecting the quality of the environment, protecting human health, ensuring prudent use of natural resources, and promoting international measures to deal with regional or global environmental problems. These policies have led to new standards for air and water quality, restrictions on pollutant discharges, and procedures for environmental impact assessment.

The Communist governments in most of the EU's new Eastern European members had compelled managers to meet industrial production targets regardless of the cost or hazards to people's health or to the environment. Unbridled industrial spewing and spilling carried disastrous environmental consequences that have not all been stopped, and the degradation will have to be remedied.

Through the years, statistical descriptions of life in the EU member countries have converged. These include statistics as diverse as median educational attainment of the population, degree of urbanization, percentage of females in the labor force, crude birthrates, and total fertility rates. This convergence demonstrates that life in the member countries is growing measurably more alike. At the same time, the various European national cultures are growing less distinct than they were just decades ago. Styles of dress, cuisine, architecture, and other artifacts are merging. Almost two-thirds of the Union's citizens report speaking at least one European language in addition to their native tongue.

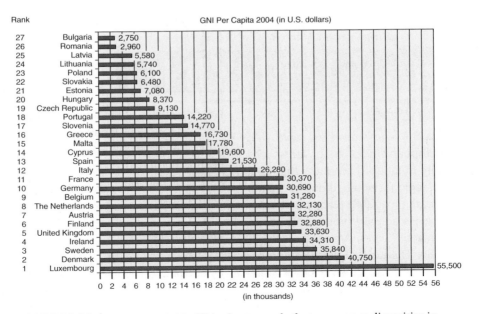

FIGURE 13-13 Income gaps. This chart reveals that enormous disparities in GNI per capita exist across the EU, but the Union hopes to achieve high living standards for all.

The challenges of expansion and international relations Extensions of the Union have presented enormous challenges. Since 2004, the Union has struggled to "digest" many new members, several of which have large populations that are economically far behind the earlier members. The Czech Republic, the Baltic states, and Hungary seem to have evolved successfully into capitalist democracies, with Slovenia not far behind, but other countries have had only limited success in privatizing, and several have seen their economies crumble. Today the boundary between West and East Europe traces an economic divide almost as deep as that between the United States and Mexico. Eastern European countries generally have lower per capita incomes, shorter life expectancies, higher percentages of their populations in agriculture, and lower measures of HDI. Eastern Europeans are migrating to the West. Migration from the new Eastern European EU members to Ireland, for example, has been so great that Ireland's population reached a modern high of 4.2 million in 2006. In that year 400,000 people in Ireland were foreign-born, double the figure from 2002. The EU has begun to assist economic restructuring and development in the East.

The accession of new members has also presented political and even cultural challenges. For example, in the government of the EU, power has slowly been shifting to the democratic European Parliament, in which poorer Eastern European countries wield substantial political power. This has caused frictions within the wider Union. Poland, for example, has balked at granting rights to women and gay people and at protecting natural habitats and endangered species. These are matters to be watched carefully.

Croatia, Macedonia, and Turkey remain candidates for membership, and Switzerland enjoys close links without membership.

The European Union negotiates trade pacts with outside countries and other blocs, but it lacks an overall foreign policy. This reveals how far the countries remain from true union. Nevertheless, since 1997 the EU has had a secretary general, who represents the foreign policy decisions of the Council of Ministers on the international stage. The Union has also formed a Defense Force, which took over some NATO peace-keeping responsibilities in Macedonia in 2003. The EU Defense Force was more important for its symbolism—320 soldiers serving under the Union flag—than for its military might.

Many U.S. corporations do business in Europe or export to Europe, so European regulations increasingly affect U.S. economic affairs, and European governments are much less laissez-faire than is the U.S. government. The European Commission has intervened in everything from U.S. antitrust policy to food safety. For example, Europeans regulate what Internet companies may do with personal information collected from consumers far more strictly than the U.S. government does. European governmental review of corporate mergers is also much stricter than in the United States, so it was Europeans who in 2001 blocked the proposed merger of the American giant corporations General Electric and Honeywell. American corporations' acceptance of European regulations may make it harder for these corporations to fight the introduction of similar regulation back in America.

The existence of the euro currency also threatens U.S. economic welfare. Many nations have long had such confidence in the United States that they have kept their own national reserves in U.S. dollars—hundreds of billions of them. This has allowed the United States literally to print money to pay its debts and to run high current accounts deficits. If many countries gain sufficient confidence in the euro that they shift even a fraction of their national reserves into euros,

Critical THINKING

Geographical Indicators

In negotiations among its own members and with other countries, the European Union has become concerned with the integrity of labeling products that take their names from the regions where those products originated. For example, the following place names can now be used only for wines actually produced in these regions: Bordeaux, Chianti, Claret, Madeira, Malaga, Marsala, Medoc, Moselle, Burgundy, Rhine, Sauternes, Chablis, Champagne, Port, and Sherry. Other geographical indicators include Parma ham and Stilton, Parmesan, Asiago, Roquefort, and Gorgonzola cheeses. Many of Europe's trading partners, including

Canada, have agreed to stop using these European place names to label their domestic imitations. In 2006 the United States agreed to limit its winemakers' use of European place names such as Port, Sherry, and Chianti, but only when the U.S. wines are exported.

Questions

1. How many of the places named in the products above can you find on a map of Europe?
2. How many U.S. products can you find in your local grocery store or wine shop that use the European geographical indicators?

or if global trade in oil came to be priced in euros rather than dollars, as has been suggested by Iran and Venezuela, the United States could be challenged to meet its international financial obligations.

The Formation of a North American Trade Bloc

On January 1, 1994, the North American Free Trade Agreement (NAFTA) came into effect, linking the United States, Canada, and Mexico. The agreement does not propose a full customs union or common market, nor are there any plans for a supranational governing body. NAFTA concerns only trade and investment. At the beginning of 2007, the North American partners outstripped the 27-member European Union in total area (20 million square kilometers to 4.5 million), but not in total population (440 million to 487 million). Each bloc produced a total GDP of about $14 trillion.

The Canada–U.S. free trade agreement

Canada and the United States signed a Free Trade Pact in 1988, by the terms of which all barriers to trade in goods and services were eliminated between the partners by 1999. The treaty also removed or reduced the hurdles to cross-border investment, government procurement, agricultural sales, and the movements of employees.

Canada and the United States have long been by far the world's two leading trading partners (Figure 13-14), as well as major investors in each other's economies. About two-thirds of foreign investment in Canada comes from the United States, representing one-third of all U.S. investment abroad, and Canada is the fourth largest investor in the United States. About 10 percent of the chief executives of the largest U.S. corporations are Canadian-born.

U.S. corporations were generally strong supporters of the pact, because Canadian tariffs were two to three times higher than U.S. tariffs, but Canada's labor unions fought the pact. The unions pointed out that employment fringe benefits in Canada are generally much more generous than they are in the United States, and they feared a drop to U.S. standards. In Canada, labor unions represent 40 percent of the labor force, whereas in the United States unions represent only 19 percent. Canada's farmers also fought the pact. Canadian agriculture is at a competitive disadvantage because of the harsh climate, vast distances, and small scale of regional markets. Canadian small producers suffer competition from big food-processing firms that can buy agricultural goods more cheaply in the United States.

Some Canadians feared that strengthening economic ties with the United States would weaken Canadian nationalism. If you look back on Figure 11-23, you will notice a thin line of Canadian settlement along the border and an almost empty empire stretching away to the north. Some 75 percent of Canadians live within

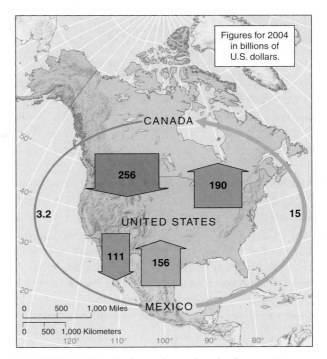

FIGURE 13-14 North American trade flows. The United States buys about 85 percent of Canada's and of Mexico's exports each year, whereas Canada takes about 20 percent of U.S. exports and Mexico just over 10 percent. Trade between Canada and Mexico is limited, but increasing rapidly.

241 kilometers (150 miles) of the U.S. border, and they absorb U.S. news and culture. Canadian means of transportation and communication have always been designed to foster east–west linkages, as illustrated in Figure 11-17. Nevertheless, the international border to the south is punctured by more than 75 million border crossings per year, as well as by common telephone lines, pipelines, computer links, power lines, contracts, cross-ownership, and even by common professional sports allegiances (Figure 13-15). A 2006 survey found that 30 percent of Americans think that Canada is a state of the United States.

America's security concerns after 9/11, however, have caused a tightening of this border. For 189 years, the U.S.–Canadian border was famously, "the world's longest undefended border," but in 2006, the United States mounted armaments on Coast Guard ships in the Great Lakes. The militarization of the border had required an addendum to the 1817 Rush-Bagot Treaty. The addendum allowed the weaponry but restricted its use to the U.S. side of the lakes. Another U.S. initiative will require all travelers between the United States and Canada to have passports. Since only 20 percent of Americans and 30 percent of Canadians held passports in 2006, it is expected that these new regulations will strangle border crossings. The United States has suggested erecting a joint security cordon around the two countries, but Canada resists the appropriation of Canadian sovereignty that step would entail.

What Is Canadian Music?

A Canadian law insists that 35 percent of the daytime playlist on Canadian radio be "Canadian." But what is Canadian music? The nationality of the singer, of the composer, and of the lyricist must all be taken into account. Some results of the government directive may seem bizarre. For example, Lenny Kravitz's 1999 remake of the 1970s song "American Woman" qualifies as Canadian. Kravitz isn't Canadian, and the song isn't about Canada, but the song was written by Guess Who, a Canadian band. Celine Dion, a Canadian singing "My Heart Will Go On," a song by a U.S. composer and lyricist, does not count as Canadian. Rod Stewart's "Rhythm of My Heart," however, is Canadian because both the music and lyrics are by the Canadian Mark Jordan.

Similar problems of definition plague other cultural products. *Top Cop*, a U.S. television program filmed in Canada, is Canadian, but the Disney Corporation's movie *Never Cry Wolf*, a 1983 film version of a Canadian classic book by (Canadian) Farley Mowat, is not Canadian because it was produced in the United States.

Questions

1. Should Canada feel that it is necessary to protect its culture from the United States?
2. How can the determination be made as to what is really Canadian?
3. Should the United States somehow identify and label which pop stars and film stars are Canadian?

The trade pact provides protection for Canadian broadcasting, publishing, and related cultural industries, but it may be difficult even today to distinguish Canadian cultural products from those of the United States. The majority of Canadians do not watch Canadian television, read Canadian books, listen to Canadian music, or go to Canadian movies and plays. Moreover, many U.S. citizens are probably unaware of the Canadian nationality of authors Robertson Davies and Margaret Atwood, reporter Morley Safer, and entertainers Nelly Furtado, Rich Little, k. d. lang, Jason Priestley, Michael

J. Fox, Celine Dion, Shania Twain, Alanis Morissette, Sarah McLachlan, the Tragically Hip, Avril Lavigne, Carolyn Dawn Johnson, Remy Shand, Chad Kroeger, Diana Krall, Cathy Fink, Marcy Marxer, Jane Bunnett, Walter Ostanek, the Barenaked Ladies, Mike Myers, David Cronenberg, Atom Egoyan, James Cameron, Norman Jewison, and Jim Carrey. Also Canadian is Alias/Wavefront, which developed the animation software prominent in virtually every special-effects-based feature film in recent years (*Spiderman, Ice Age, Hollow Man, The Matrix, The Perfect Storm,* and the *Lord of the Rings* and *Star Wars* series of films).

The degree of cooperation between the United States and Canada may be represented symbolically by the fact that Canada is the only country that the United States has allowed to build an embassy on Pennsylvania Avenue in Washington, D.C., America's grand ceremonial boulevard; and in October 1999 President Clinton traveled to open a new U.S. embassy in Ottawa. This was the first time that a U.S. president has ever traveled abroad for this purpose.

Mexico A trade agreement among the United States, Canada, and Mexico went into effect on January 1, 1994, to bring down tariffs and other barriers to trade in four, 5-year steps.

The integration of Mexico into a North American trade pact might be more difficult than the integration of the U.S. and Canadian economies. The United States and Canada are both rich, postindustrial societies with populations that are growing slowly. Mexico is a less-developed nation with a population still growing rapidly—from 25 million in 1950 to 107 million by 2006. To provide work for this rapidly growing labor force, Mexico needs economic growth, which might reduce the push

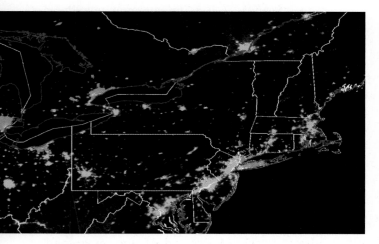

FIGURE 13-15 The blackout of August 14, 2003. The power grids of the United States and Canada are so interconnected that the border itself could not be identified on this satellite image of the international blackout of August 14, 2003. This image was created from the difference between two other images, one collected before the blackout, the other during it. The areas with power outages are either dark or red. (Christopher D. Elvidge, Ph.D./NASA Headquarters)

Focus ON

Food Safety and International Food Trade

Chapter 8 noted the opportunities that the continuing scientific revolution in agriculture plus international trade in food offer. It also noted, however, that biotechnology may carry serious risks. These risks are multiplied by increasing international trade in foods. The explosion in world trade, travel, and tourism is enabling diseases to move around the world, and many of these diseases threaten both humans and other animals. Alien animal and plant pests represent such a broad set of dangers that the U.S. National Intelligence Council has termed foreign animal diseases a threat to national security.

Many of these diseases are transmittable to humans. One of the most worrisome is Mad Cow disease (technically BSE, *bovine spongiform encephalopathy*). In the 1980s and 1990s BSE spread through Great Britain and later much of Western Europe via protein feed that contained infected animal parts, requiring the slaughter of almost four million cattle. Britain imposed a domestic ban on the tainted feed in 1988, but the feed continued to be sold all over the world. The spread of BSE severely damaged the beef industry in some European countries, and it seems to appear in humans as the related new variant Creutzfeldt-Jakob Disease (nvCJD). By 2003 over 100 people had died of nvCJD, and about 50 new cases were reported in each of the following years. In 2000 three flocks of sheep in Vermont that had been imported from Belgium were killed for fear of BSE. It was later determined that they did not have BSE but did have scrapie, a related disease. BCE cows have occasionally been found in Canada early in the twenty-first century, triggering embargoes against Canadian beef, cattle, and animal feed, but each animal was found to have been born before Canada implemented restrictions on potentially dangerous feed in 1997.

BSE is caused by a type of malformed protein called a *prion*, which is also believed to cause a contagious, invariably fatal disease called chronic wasting disease (CWD). This disease occurs in wild and captive deer and elk in the western United States. CWD has not been observed in wild animals in Canada, but it has been diagnosed in Saskatchewan game-farm elk that were imported from the United States. This prompted the Canadian Food Inspection Agency to slaughter thousands of elk on Saskatchewan ranches.

Foot and mouth disease (FMD) is another disease that, while rare in humans, is very dangerous among some meat animals. Japan and Korea have had to slaughter hundreds of thousands of cattle that suffered FMD, which may have come from China. In 1997 Taiwan slaughtered 3.8 million pigs with FMD, and in 2001 an outbreak of FMD in Britain devastated the meat industry. Millions of cattle, hogs, and sheep had to be slaughtered. FMD also appeared in several other European countries. Chapter 5 noted the early twenty-first century threat to poultry and humans from the H5N1 virus.

These events have raised major concerns worldwide about the safety of our food supplies. Such concerns can often rise out of proportion to the actual risks involved, just as many people fear flying on airplanes but are very willing to ride in an automobile, even though cars are much more dangerous than planes. Intense public concern about food safety can lead governments to place restrictions on international trade in food. Cultural and political differences between countries mean that perceptions of food hazards may be very different from one country to another. If a few countries that import significant amounts of food place restrictions on the way such food is produced, then there is pressure on major producers around the world to modify their production techniques. Pressures such as these can lead to disputes between trading nations, and are likely to place major strains on international food trade in the coming years.

for Mexicans to migrate to the United States. Mexico's Hispanic culture and language contrast with the British traditions and English language held in common in the two countries to the north (except in Quebec), but that cultural contrast may not be as troublesome in building a free trade pact as it would be if the three countries were building a full common market.

Signing the trade pact was part of an evolution of Mexican policies toward export-led economic development. Mexico's economy had been state directed and centralized through most of the twentieth century, but in 1986 Mexico began privatizing its state sector, selling off factories, shopping centers, mines, and public utilities. Only PEMEX, the state oil monopoly created when Mexico's oil industry was nationalized in 1938, was viewed as politically untouchable. PEMEX supplies 9 percent of Mexico's GDP and more than one-third of government revenue, but economists view it as inefficient, with too many employees and too little reinvestment of earnings in exploration or adding value. Although Mexico

Critical THINKING

Intellectual Property and Globalization in the Drug Industry

As the U.S. population ages, the cost of prescription drugs has become a political issue. Drugs are expensive even for Americans, but in poor countries such as India, where the per capita GNI is just $620, such drugs are simply out of reach of all but the richest citizens. India does, however, possess the ability to manufacture these drugs, and some brand-name products sold by European and U.S. companies are actually wholly or partly manufactured in India under contract. Generic equivalents of brand-name drugs are also made in India, and one Indian company, Cipla Ltd., makes approximately 400 important drugs sold elsewhere under patents owned by Western pharmaceutical companies. These are sold in India, under different names, at a fraction of the prices charged by the patent holders. Cipla claims to make an equivalent of every product of America's giant Pfizer pharmaceutical corporation.

On the other hand, the large pharmaceutical companies, based primarily in wealthy countries, insist on the need to protect their investments in research and development to produce new drugs. It costs many millions of dollars and years of research to produce a new drug, and unless those companies can charge high prices for their products, there is no way to recover that investment. Without patents that protect the interests of drug developers, there would be little incentive to develop new drugs. The pharmaceutical companies are concerned about what they see as "pirating" of their property by Cipla and similar companies. That property is indeed valuable; pharmaceutical companies as a group are one of the most profitable sectors of the U.S. economy.

India has long pursued an import-substitution development strategy after gaining independence. The spinning wheel at the center of the Indian flag symbolizes Indians' determination to manufacture at home products that their former colonial rulers forbade them to make; during colonialism, for instance, the British had required Indians to purchase British-produced textiles. Still today, Indian patent and trade law is designed to inhibit Indian companies from purchasing patented technology from abroad; it encourages local production, therefore drugs produced locally can be sold at prices dramatically lower than imported drugs. If Indians paid U.S. prices for these drugs, they would pay perhaps 20 times the price of the Indian equivalents. This reduction in cost brings the drugs within reach of many people who would otherwise suffer without them. It can also be argued that foreign drug companies would not sell many drugs in India at the high U.S. prices, so the companies' losses may not be as large as they might claim. Improved health care in India has contributed to an increase in life expectancy from only age 38 in 1956 to 64 today. Cipla has offered to supply AIDS-fighting drugs to African countries at prices well below those charged in the United States.

When Americans faced the threat of biological warfare with the spread of anthrax spores at the end of 2001 (see Chapter 5), they stampeded to buy Cipro, a patented medicine that provides the best protection. Not enough Cipro was being made to meet the new demand, so the U.S. government found itself questioning the protection of the patent, and it even considered turning to the Indian producer of a generic version for new supplies. Americans were outraged to learn that while Cipro cost $350 per month in the United States, the generic costs only $10 a month in India.

The summit conference of the World Trade Organization held in Doha, Qatar, in 2001, agreed that, in cases of national emergency, governments could override the patents on life-saving drugs and produce the drugs themselves or purchase generics on international markets. Within weeks, 10 African countries signed agreements with international pharmaceutical companies that slashed the cost of AIDS drugs in those countries by up to 85 percent. Negotiations continue on providing drugs to poor countries at prices they can afford while still preserving drug companies' rights and encouraging research on new drugs.

Questions

1. Should pharmaceutical companies and other technology-developing industries in rich countries allow poor countries to use their intellectual property free of charge, so that those who would otherwise have no access to these goods may enjoy at least some modest improvements in the quality of life?

2. Should we insist that patents and other intellectual property be rigorously protected, so that corporations have an incentive to create new technology?

is the fifth-largest producer of oil in the world, it imports 20 percent of its refined petroleum products. Mexico's second-largest source of foreign exchange, after oil, is workers' remittances from the United States, which total about $16–18 billion per year.

Many U.S. companies were active in Mexico even before the trade pact was signed. They had moved labor-intensive operations to Mexican factories called **maquiladoras** concentrated along the U.S.–Mexican border. Mexico imported components duty free; these were assembled and the products re-exported to the United States. The United States charged tariffs only on the value added, which was low because Mexican wages are low. The existence of these factories by 1994 had already significantly redistributed Mexican population and urbanization toward the border, but maquiladoras multiplied after the signing of the pact (Figure 13-16). New FDI in Mexico, about $2 billion per year in the 1980s, rose to $12 billion per year in the 1990s. Exports of manufactured goods rose from 43 percent of Mexico's merchandise exports in 1990 to 80 percent by 2004. In 2000 about 800,000 *maquiladora* jobs existed in the border region.

Already early in the twenty-first century, however, Mexico's *maquiladoras* began to suffer competition from China, where wages were lower. In the 1990s Mexico had risen to the rank of number two among nations exporting to the United States (after Canada), but China seized that position by 2003. The number of *maquiladoras* began to fall, at a cost of hundreds of thousands of jobs, and Mexico realized that continuing economic growth depended on lowering the cost of energy and improving the transportation infrastructure and education. The government encouraged the study of engineering, and the number of students rose to 451,000 by 2006 (compared to 370,000 in the United States). Multinational corporations responded by moving higher value-added activities to Mexico. For example, General Electric established facilities to design and test jet engines and turbines in the city of Querétaro, where engineers earn about one-third of U.S. salaries. Similarly, Tijuana is the world capital of television manufacture, and many companies that started manufacturing television sets are now assembling computers and other higher-value electronics.

Many Americans feared that the pact with Mexico would trigger a migration of U.S. manufacturing jobs to Mexico, where wages, working conditions, and environmental protection laws are all below U.S. standards. A principal reason that Mexican wages are lower than U.S. wages, however, remains that Mexican productivity is lower than U.S. productivity; that is, each Mexican worker's output is much lower than the output of each U.S. worker. A worker's productivity is usually determined by the amount of capital investment that has been made per worker, and that investment is much higher in the United States than in Mexico.

Some of the new FDI in Mexico comes from European and Asian corporations that re-export goods from Mexico into the United States. This occasionally causes trade disputes. Negotiators of trade pacts must agree what percentage of the total value of a good entering one country must have been added in the second country for that product to qualify as a product of the second country. This is called a **local content requirement.** Arguments over local content requirements typically arise whenever two countries negotiate free-trade pacts that are not customs union agreements. Disagreements have arisen between the United States and Mexico about whether goods assembled in Mexico out of components made in other countries should enjoy free access to the U.S. market. Such disputes have also arisen with Canada.

The *maquiladoras* are not the only border phenomena that geographers have noticed. Wherever two cities—one Mexican, one U.S.—face each other across the border, these so-called "twin cities" are being compelled to integrate and often internationalize their infrastructures. In 1998, for example, Mexican and U.S. officials opened the International Wastewater Treatment Plant, which treats Tijuana's excess sewage on the San Diego side of the border, the first international facility of its kind in the world. A unified air-quality district covers El Paso, Texas, and Ciudad Juarez.

Mexican politics
From 1929 until 1997, Mexico was ruled by one political party, the Institutional Revolutionary Party (PRI), whose political apparatus made little distinction between the party and the state. Accusations of corruption and repression abounded. Some Americans hesitated to join in a pact with Mexico because they doubted Mexico's democracy, but others hoped that closer ties to the United States would strengthen it. In 1997, the PRI lost its majority in the lower house of Mexico's Congress, and by 2006, the PRI had slipped in status to the third largest party in Congress. In 2000 the PRI

FIGURE 13-16 A new Mexican industry. These workers are assembling Web televisions in a maquiladora in Guadalajara. (Veronica Garbutt/Panos Pictures)

lost the presidency to Vicente Fox, of the National Action Party (PAN). Many observers attribute this political change to the economic changes that have come with NAFTA. President Fox himself later regretted that he had not been able to achieve all of the "historic transformations our times demand," but he insisted that democracy and the rule of law had made progress under his presidency. Felipe Calderón, of PAN, won the presidency in 2006, to succeed President Fox.

After the attacks on the United States in September 2001, President Fox offered all possible assistance. Mexico has traditionally remained neutral in international affairs or even expressed distrust of U.S. motives, but it has recently been stepping forward on the international stage: In 2001, Mexico won a two-year seat on the U.N. Security Council (discussed below), but Mexico's refusal to support the U.S. second war against Iraq angered some Americans.

President Fox suggested that the United States and Mexico seal the two countries' borders against outsiders but open the borders to freer movements between the two countries; approximately ten percent of Mexico's population already lives in the United States. Whether these people will eventually retire in the United States or return to Mexico is one of the most important questions in both countries' future.

Agriculture remains a sore point among NAFTA nations, especially between the United States and Mexico. U.S. subsidized products can overwhelm the farm economies of either neighbor, bankrupting Canadian farmers and impoverishing millions of Mexican peasants. Agriculture produces only about 5 percent of Mexico's GDP, but roughly 20 percent of the labor force is directly involved in it. Mexico has been flooded with imported corn. In 2006 one-third of the corn used in Mexico was imported from the United States, and the price of corn in Mexico had fallen over 70 percent since 1994, thus reducing the incomes of the 15 million Mexicans who depend on producing corn for their livelihood. Canada, meanwhile, imposed a tariff on subsidized corn from the United States. Mexico's exports of fruits and vegetables to the United States, however, had increased so much as to double Mexican agricultural exports to the United States overall from 1994–2004. Therefore, in 2007 U.S. fruit and vegetable growers launched a lobbying campaign in Congress to add their products to the list of agricultural products allready subsidized in the United States. Tariffs were removed on all agricultural products traded between the United States and Mexico on January 1, 2003, except beans and white corn, which Mexico must stop protecting as of January 1, 2008.

Just as the U.S.–Mexican pact took effect, a native American guerrilla organization calling itself the Zapatista National Liberation Army, in honor of Emiliano Zapata (discussed in Chapter 11), rose up in Mexico's southern state of Chiapas. The rebels were motivated in part by fears that imports of U.S. corn and other market pressures would wipe out their traditional agricultural economy. "The free trade agreement," said the leader of the group, "is a death certificate for the Indian peoples of Mexico." Chiapas is one of the poorest states in Mexico, and largely Indian in population. The insurrection has received support from local Roman Catholic priests, who are followers of liberation theology discussed in Chapter 7. The Mexican government negotiated land tenure and other issues with the rebels, but the uprising serves as a reminder of how far Mexico has to go to achieve the goals of economic welfare and political equity for all of its peoples. Some Mexicans have demanded the renegotiation of agricultural trade rules, but in 2006 U.S. under-secretary of agriculture J. B. Penn insisted, "We have no interest in renegotiating any parts of the agreement."

The geography both of manufacturing and of agriculture in Mexico highlight a growing split between the country's North and South. New manufacturing plants have concentrated in Northern Mexico, most of the new productive commercial agriculture is in the North, and the North has rapidly urbanized, resulting in a regional economic growth rate over 4.5 percent. The South, by contrast, retains the traditional *ejido* agriculture system (see Chapter 8) and has enjoyed little industrialization. Growth there has occurred at less than 1 percent per year. PAN generally represents the capitalist, globalized economic interests of the North, where it won its first governorship in the late 1980s, and whence it has grown to national power. President Calderón, for example holds a master's degree in public administration from Harvard University.

Results of the NAFTA pact

Since the formation of NAFTA, the individual economies of the three signatories have experienced many changes, so it is difficult to isolate the impact of NAFTA. Trade and investment among the three partners have risen substantially. Imports into the United States from both Mexico and Canada have surged even more strongly than U.S. exports, so that net value of trade between the United States and the two partners has swung from a net surplus to a net deficit.

U.S. labor unions argue that the United States has lost jobs, that wages and living standards are down, and that workers' rights and bargaining power weakened in all three signatory countries. Both U.S. and Canadian labor unions have launched strong recruitment and organizational training drives in Mexico. Some factories are undoubtedly being closed and moved to Mexico. In February 1997, for example, the RCA Corporation announced its intention to close the world's largest TV assembly site, a plant in Bloomington, Indiana, and to move 1,100 jobs to Mexico. The company cited wage costs as the principal reason for the move. RCA is not a U.S. corporation, but is owned by Thomson, a French multinational conglomerate.

International manufacturing and marketing have begun to reorganize in scale from national to North American scale planning. The Ford Motor Company,

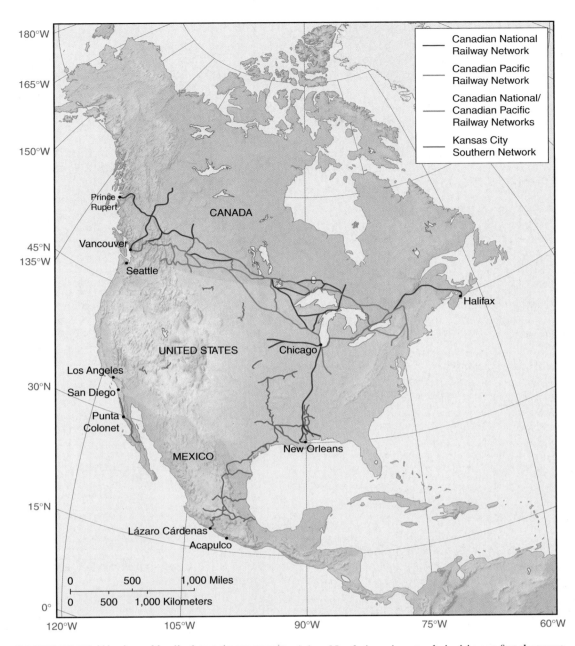

FIGURE 13-17 Western North American ports. Asian–North American trade is rising so fast that west coast ports in both Mexico and Canada are expanding and integrating with the U.S. transport net. In 2005, the Los Angeles–Long Beach port still handled 80 percent of U.S. imports from Asia, but in Mexico, Lázaro Cárdenas, which is served by the Kansas City Southern Railroad, is being upgraded and expanded. Developers want to build a huge new port facility in Punta Colonet, a bay on the Baja Peninsula 250 kilometers (155 miles) south of the U.S. border. Canadian National Railways is expanding port facilities at Prince Rupert on Canada's west coast. Lower port fees and fear of terrorist threats on U.S. soil increase these ports' attractiveness to U.S. importers.

for example, began to make Escorts and Tracers for the eastern U.S. market in Michigan and for the western U.S. market in Hermosillo, Sonora, Mexico.

The pact's effects on Canada are difficult to ascertain. Total U.S. exports to Canada have risen, and some Canadian industries seem unable to compete, but overall Canada's trade surplus with the United States has doubled, creating many new jobs in Canada.

The transport infrastructures of the three countries are being knit (Figure 13-17). In 1998 the Canadian National railroad, with a main line stretching across Canada from Halifax, Nova Scotia, to British Columbia, bought the Illinois Central Railroad, giving the Canadian line a route south to the Gulf of Mexico ports of New Orleans, Louisiana, and Mobile, Alabama. Canada's deepwater port at Halifax was already winning trade from New York, Boston, and other U.S. East Coast ports. The Canadian National cooperates with America's Kansas City Southern Railroad, whose line to Mexico City carries 40 percent of Mexico's rail traffic. The United States has promised to allow Mexican trucks to haul goods throughout the United States, but the question of harmonizing truck inspections and driver qualifications has slowed implementation of this agreement.

Trade between Canada and Mexico rose at an average annual rate of 12 percent from 1993 to 2006, but it is still a small fraction of either country's trade with the United States.

Many environmentalists remain critical of NAFTA. NAFTA's Commission on Environmental Cooperation issues an annual report called *Taking Stock*, in which it has criticized the worst polluters in North America, but the Commission has no powers to enforce cleanups.

Expanding Western Hemispheric Free Trade

The United States continues to sign free-trade pacts with individual countries. Between 2001 and 2006, the United States signed free-trade agreements with 15 countries with a total population of 230 million and combined GDP of $2.2 trillion.

The United States has suggested the creation of a Free Trade Area of the Americas, excluding only Cuba, and negotiations have continued on that initiative. A number of trade blocs formed in the Western Hemisphere in the early 1990s. Venezuela, Colombia, Peru, Bolivia, and Ecuador formed the Andean Pact. The English-speaking countries of the Caribbean united in a Caribbean Union and Common Market (CARICOM). In 2006, Venezuela joined Argentina, Brazil, Paraguay, and Uruguay in a customs union called MERCOSUR; Chile and Bolivia are associate members and may join. Citizens of member countries may live and work in any country and be granted the same rights as citizens of those nations. Perhaps a genuine United States of South America is in the works.

The unification of transport infrastructures again reveals the expanding territory being organized (Figure 13-18). A new railroad tunnel and new gas and oil pipelines have pierced the Andes Mountains between Chile and Argentina and now link Bolivia and Brazil. New roads join Brazil and Venezuela, and a new power line brings electricity from Venezuela's Guri dam to Brazil's Amazon cities. Argentina uses Chilean ports for its exports to Asia. Brazil and Argentina have reconciled the gauges of their railroad systems, and new highways link several countries. A new inland waterway allows barges to travel 3,424 kilometers (2,140 miles) down the Paraguay and Paraná Rivers from eastern Bolivia to the sea.

Trade within MERCOSUR has soared, but in 2005 only 23 percent of total MERCOSUR trade was with other MERCOSUR countries. This is a small amount compared with the level of trade typical between industrialized neighbors, and it suggests that the various Latin American states did not produce items they could exchange to mutual profit. The new trade agreements, however, encourage specialization of production, increases in productivity, and intraregional trade and growth. The United States is the principal supplier of

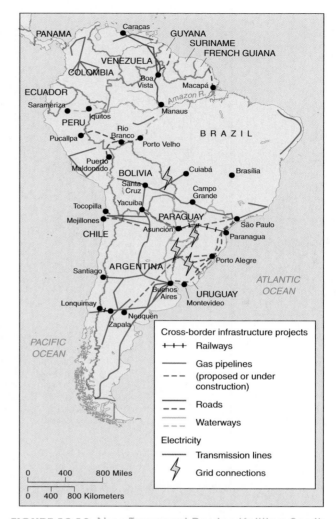

FIGURE 13-18 New Transport Routes Knitting South America. These roads, railroads, pipelines, and electrical grids are pulling together the population of South America, which the map on the rear endpaper shows to be, for the most part, scattered around the fringes of the vast continent.

imported goods to most Latin American countries, so growth in Latin America boosts the U.S. economy.

Other Regional International Groups

The European Union and the Western Hemisphere trade areas are the most important regional international organizations today, but many others exist (Figure 13-19). Some of these are primarily military-defensive. Since the end of the Cold War, however, they have been turning into general economic and cultural associations. For example, the Association of Southeast Asian Nations (ASEAN) began as a military alliance, but it became a trading organization and hopes to achieve free trade among its ten members by 2015: China may join. Other organizations, such as the Southern African Development Coordination Conference (SADCC), are economic. The purpose of each international organization is simply to provide a

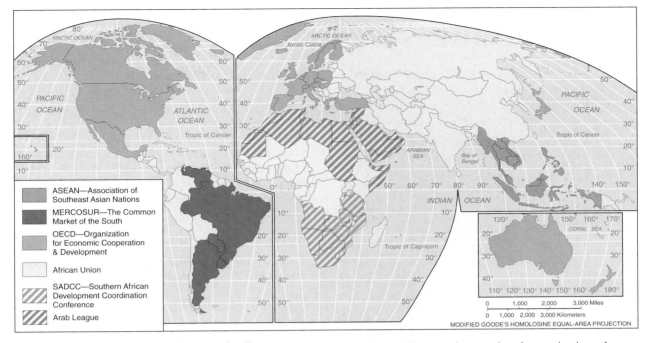

FIGURE 13-19 International organizations. These are just a few of the many international organizations that provide frameworks for cooperation.

framework for consultation so as to reach agreement whenever possible in areas of mutual concern. In some cases, however, the members of these organizations can agree on very little and have even gone to war with one another. The African Union, for example (called the Organization of African Unity before 2002), and the Arab League serve few functions, although both are negotiating trade agreements.

GLOBAL GOVERNMENT

The Dominican priest Francisco de Vitoria (1486–1546) earned recognition as "The Father of International Law" by arguing that each people has the right to its own ruler. Therefore, de Vitoria argued, relations among nations would have to recognize mutual rights. Subsequent centuries, however, have shown only token respect for this principle.

Discussions of world order are generally based on principles of state sovereignty and self-determination, but as the number of states has multiplied, the number of actual and possible conflicts among states has grown. Causes include aggressive ambitions, as with Iraq in the early 1990s; border disputes or rival claims on territory, as between Ethiopia and Eritrea; and domestic crises or policies that have effects abroad, causing other states to threaten to intervene, as with the former Yugoslavia. The necessity for common action in areas of common concern, such as the oceans or the global environment, is another pressure for global government. International society as a whole, however, has neither the centralized government, judicial system, or police that

characterize a state, and consensus on what constitutes an international crime is only slowly evolving.

The first lasting international political structure was the League of Nations. The league was founded by the Allies after World War I, but several nations, including the United States, refused to join. The League provided a forum for discussion of world problems, and it carried out many humanitarian projects. It exercised little real power, however, and was unable to resolve the disputes that eventually led to World War II.

The United Nations

In 1942, 26 states that were allied against the Axis powers joined in a Declaration by the United Nations, pledging to continue their united war effort. A meeting of Allies in San Francisco in 1945 drew up a charter for a United Nations organization to continue in existence after the war, and by the end of 1945 the new organization had 51 members. As more colonies received independence and joined the United Nations, membership rose to 192 by 2006—virtually all Earth's sovereign states.

The U.N. General Assembly serves as a rudimentary legislature of a world government, but it has no power of enforcement (Figure 13-20). The power that the United Nations does exercise is vested in the 15-member Security Council. Five members of the council (the United States, China, France, the United Kingdom, and Russia—originally the U.S.S.R.) hold their seats permanently, and the remaining 10 members are elected by the General Assembly for two-year terms. This allocation of seats reflects the allocation of

FIGURE 13-20 United Nations headquarters in New York City. The United Nations found a permanent home on the east side of Manhattan Island, where its headquarters buildings were designed by an international committee of architects, including Le Corbusier. The high slab of the Secretariat dominates the group, with the domed General Assembly to the north. The land legally is not part of the United States of America. (Rafael Macia/Photo Researchers, Inc.)

world power in 1945, but recent discussions have suggested that granting additional seats to, perhaps, Germany, Japan, India, and Brazil might better reflect the present world balance of population and power. The U.N. Secretariat, with the Secretary General at its head, handles all administrative functions. On January 1, 2007, Kofi Annan ended two 5-year terms as Secretary General and was replaced by the South Korean diplomat Ban Ki-Moon.

The Security Council has occasionally voted for the United Nations to field armies. United Nations forces fought in Korea from 1950 to 1953. In 1990, a Security Council resolution gave the United States and its allies sanction to force Iraq to retreat from Kuwait,

and on September 28, 2001, the United States obtained a Security Council resolution obliging U.N. members to cooperate in combating terrorism. In a greater number of cases, the United Nations has sent observer troops to patrol world trouble spots, and these incidents have been increasing. Examples include Kashmir; the Golan Heights; Lebanon; Cambodia; Yugoslavia; the Mediterranean island of Cyprus (Figure 13-21), split by feuding Greeks and Turks; and Haiti still today. U.N. troops have attempted both *peacekeeping* and *peacemaking* efforts. Peacekeeping involves patrolling a situation in which combatants have agreed to a peace. Peacemaking, however, is trying to impose a cease-fire or peace on combatants, and United Nations forces have had less success in these efforts.

In many cases the United Nations has assumed responsibility for monitoring the freedom of national elections. This idea that national governments can be legitimized by international standards represents a reversal of international law, which presumes the priority of national governments.

The International Court of Justice in The Hague is another branch of the United Nations that on the surface works like a branch of government. The litigant states, however, must choose to come before it, and the court has no power to enforce its decisions.

Perhaps the greatest accomplishment of the United Nations is when nothing happens; in other words, when countries are at peace. States that are at least talking to each other are seldom making war on each other. In September 2000 the United Nations hosted the greatest meeting of heads of state ever—more than 150 came together. U.S. Secretary of State Madeleine Albright pointedly attended a speech by Iranian President Mohammad Khatami, and he attended hers. Two days later U.S. President Clinton found himself standing next to Cuban President Fidel Castro. They shook

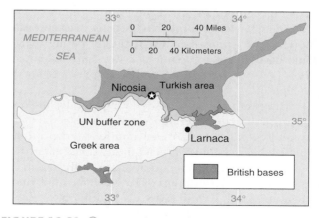

FIGURE 13-21 Cyprus. Since 1974 Cyprus has been divided into Greek and Turkish portions, with little mingling between the two groups. The Turkish sector has declared itself to be the Turkish Republic of northern Cyprus, but only Turkey recognizes it as an independent country. The Greek portion joined the EU in 2004. Continuing antagonisms may stymie Turkey's efforts to join the EU.

hands and spoke a few words. This was the first time since Castro came to power in 1959 that a sitting U.S. president recognized him as a leader. Such small civilities mark steps toward reconciliation between countries officially "not talking to each other."

U.N. special agencies

The United Nations has more than 40 specialized agencies. Some facilitate communications among member states: the Universal Postal Union, the International Civil Aviation Organization, the World Meteorological Organization, and the Intergovernmental Maritime Consultative Organization. The U.N. International Telecommunications Union, for example, has allocated frequencies on the radio spectrum, thus allowing international expansion of mobile telephones, satellite computer transmissions, and digital radio broadcasting. Other U.N. agencies encourage international cooperation toward specific humanitarian goals: the Food and Agriculture Organization; the World Tourist Organization; the United Nations Educational, Scientific, and Cultural Organization (UNESCO, Figure 13-22); the International Atomic Energy Agency; the International Bank for Reconstruction and Development; and the International Monetary Fund. The U.N. Conference on Trade and Development (UNCTAD) works to assist the economic development and trade of the poor countries. UNCTAD has, for example, developed a standardized computer customs collection system that is in use in over 50 countries. In 2000 the United Nations sponsored the Millennium World Peace Summit of Religious and Spiritual Leaders; about 800 attended from around the world.

The United Nations is also helping define principles for multinational business policies. Under U.N auspices, leading multinational corporations have joined labor associations and nongovernmental organizations in a pact that commits all to support human rights, eliminate child labor, allow free-trade unions, and refrain from polluting. This was a declaration of principles, not a legal code, but it sets guidelines for international standards in an area of growing international concern.

Nongovernmental Organizations and Terrorist Organizations

Today states are no longer the only players on the international stage. Initiatives are increasingly taken by **nongovernmental organizations (NGOs).** These are organized international interest groups that cross state boundaries and carry out direct actions to further their goals as well as put pressure on the existing governments of states. They include multinational corporations (p. 522), religious and charitable organizations, environmental organizations, investors able to move their money almost instantly from one country to another and thus destabilize national economies

FIGURE 13-22 Yellowstone National Park.

Yellowstone Park was the first World Heritage Site. UNESCO supervises the Convention Concerning the Protection of the World Cultural and Natural Heritage. Signatories to the convention—virtually all of the world's countries today—have agreed that certain areas have such unique worldwide value that they should be treated as part of the heritage of all humanity. Designated sites range from Sinharaja Forest Reserve in Sri Lanka to the ancient city of Hatra, Iraq; Canterbury Cathedral in the United Kingdom; Dinosaur Provincial Park in Alberta, Canada; and Bolivia's historic capital of Sucre. Yellowstone National Park was the world's first national park and its 100th anniversary in 1972 inspired the global convention. (Ray Mathis/Corbis/Stock Market)

(p. 526), but also drug cartels, Mafia, and terrorists. The backlashes against biotechnology (p. 336) and globalization (p. 528) are led by NGOs. (It is ironic that the worldwide struggle against globalization is one of the world's most globalized activities.) Few activities remain either purely international or purely domestic anymore.

The global civil society is unmanageable and hard to study. A world stage with millions of private empowered actors means a world of unlimited vulnerability. It is paradoxical that this is especially frightening for the United States, because the United States has done so much to destroy borders and walls, to shape a world market, and to promote freedom of communications, information, and movement. This modern reality demonstrates again what nineteenth-century critics of communications—starting especially with the telegraph—already argued: Accelerated and increased communication among peoples is not sufficient to guarantee good will or even an understanding of one another.

Few governments interfere with the work of the charitable NGOs, such as, for example, Doctors without Borders, which provides emergency medical help in all situations around the globe. Other NGOs, however, have more militant or aggressive purposes. Some have demonstrated the willingness—and the ability—to use violence in the pursuit of their ends. The Earth Liberation Front, for example, founded in England in 1993, has engaged in acts of violence against people or activities it deems are harming the environment. Groups against GM foods have destroyed laboratories and crops in fields, and animal-rights groups have destroyed laboratories that use animals in scientific research.

Today even the ability to make war is no longer confined to state governments. Many world terrorist groups, such as Al Qaeda, are not recognized governments, nor do they represent the governments of recognized states. In much of Africa and parts of central Asia, centralized state power has effectively collapsed, and wars are fought by irregular armies commanded by political and religious organizations, often clan-based and prone to savage internecine conflicts. Rich societies cannot be insulated from the repercussions of collapsing states and the new forms of war. The attacks on the United States on 9/11 proved that the ragged, irregular armies of the world's most collapsed state—Afghanistan—can reach to the heart of its richest and most powerful state.

Governments' difficulties in fighting NGOs were illustrated by the United States' first efforts to strike back at Al Qaeda. On Friday, September 14, 2001, the U.S. Congress gave President Bush the right to "use all necessary and appropriate force against those nations, organizations, or persons he determines planned, authorized, committed, or aided the terrorist attacks." The United States has fought several military engagements since World War II—in Korea, Vietnam, Nicaragua, Panama, and still other countries—but Congress has passed no declarations of war since entering World War II. The 2001 Congressional Resolution reminded many observers of the 1964 Gulf of Tonkin Resolution that had authorized U.S. attacks on Vietnam. In U.S. and international law, a declaration of war would define new sets of circumstances regarding presidential prerogatives, federal spending, civil rights, and many other domestic and international matters. In this case, however, it is unclear against what government or organization the U.S. could legally declare war. Resolutions such as the Tonkin Gulf Resolution and this new resolution create confusing legal situations in, for example, the treatment of terrorists captured abroad or even at home.

Not all states in which terrorists operate are their willing accomplices. Some states are too weak to exert control (e.g., Lebanon), and others are insufficiently vigilant (even the United States itself). A U.S.-determined project to rid the world of all rogues and terrorists would be interpreted around the world as imperialism.

Some observers fear that authoritarian rulers in some countries may use the generally approved war against terrorism as an excuse to repress any dissidents in their own countries. The United States often defends international stability, even when many peoples resent or struggle against present circumstances or political structuring. The map of states from Morocco to China, for example, was not drawn by the indigenous peoples, but by European imperialist powers, and the governments of each of these states, when each first gained independence, were installed by imperialists. In some cases, the original governments or even forms of government were overthrown by the populations. The United States, as leader of the Western alliance, has not always welcomed what may have been genuine democratic revolutions. The original revolutions may have been democratic even if the replacement rulers turned out to be autocratic themselves, as in Libya, Iran, Iraq, Egypt, and Syria. In other countries, autocratic kings, emirs, and sheiks are still supported by the United States. America cooperates with these regimes and even defends them contrary to America's own values, so the local peoples see American forces not as symbols of freedom but as the support of a corrupt order they seek to replace.

America must, however, respond to the groups who have declared war on America. Some of these are religious fanatics who think the United States or all of the West is a machine for cultural subversion, political domination, and economic subjugation. This kind of conviction feeds on experiences of despair and humiliation, which are issues that can be understood and addressed.

Is National Sovereignty Inviolable?

National sovereignty means that international borders are inviolable. No matter how monstrous any regime may be, no matter how much it persecutes its own people or oppresses minorities, no outside state or international agency has the right to interfere. This is true even when internal persecution triggers international flows of refugees. The U.N. Charter insists, "Nothing contained in this present Charter shall authorize the United Nations to intervene in matters which are essentially within the domestic jurisdiction of any state." This is the legacy of the Treaty of Westphalia (see Chapter 11).

Disarray in several states, however, and the formation of powerful NGOs may have signaled the beginnings of change in international law. The first war against Iraq in 1991 did not overthrow Iraq's government, but the United Nations did subsequently protect some Iraqis from their own government (see the discussion of Iraq on page 563). In 1992 the Security Council voted to send a U.S.-led force to Somalia, which was wracked by civil war, to guarantee the distribution of food-relief supplies. The purpose of that mission was defined as "humanitarian," but the mission expanded to restoring central civil government to Somalia. Several states contributed troops (notably Canada and Pakistan), but Somalia still has not achieved peace.

In 1999 then-U.N. Secretary General Kofi Annan, reviewing cases of civil war and slaughter in Africa, argued, "Nothing in the Charter precludes a recognition that there are rights beyond borders," and referred to a "developing international norm in favor of intervention to protect civilians from wholesale slaughter...." The secretary general pressed the United Nations formally to establish the principle that massive and systematic violations of human rights must be stopped by international intervention. Many world leaders, however, agree with President Abdelaziz Bouteflika of Algeria, who has argued that "interference can only occur with the consent of the state concerned." Many countries fear that U.N. intervention would serve as a cloak for Western interference. Russia's Foreign Minister Igor Ivanov has said bluntly, "Human rights are no reason to interfere in the internal affairs of a state." China's President Jiang Zemin has emphasized, "Dialogue and cooperation in the field of human rights must be conducted on the basis of respect for state sovereignty... So long as there are boundaries between states, and people live in their respective countries, to maintain national independence and safeguard sovereignty will be the supreme interests of each government and people." Global debate over the "absoluteness" of state sovereignty will probably intensify.

The globalization of justice After World War II, the victorious Allies tried the leading German and Japanese officials for "crimes against humanity" and for "genocide." Since then, international law has affirmed that these crimes are subject to universal jurisdiction. They transcend the province of the state where the crimes occurred, and any state may prosecute. Few states, however, have been willing to do so. Therefore, in 1993 the United Nations created a War Crimes Tribunal to try Serbian leaders for "crimes against humanity" that had occurred during the fighting that accompanied the breakup of Yugoslavia. During the rule of Slobodan Milosevic (1987–2000), Serbia fought and lost five wars, bringing upon itself terrible destruction, charges of international crimes, and international disgrace. Milosevic was eventually overthrown by nationwide rioting and brought to trial for crimes against humanity, but he died of heart ailments before the end of his trial. Other U.N. tribunals have investigated African wars, and another is trying the leaders of the

Khmer Rouge that killed millions in Cambodia during its rule 1975–79.

In 1998 the United Nations voted to create a permanent International War Crimes and Genocide Tribunal, and two years later delegates from over 100 nations agreed on a catalog of acts that constitute international crimes. The required 60 signatory nations ratified the treaty by 2002, so 21 justices were sworn in, and the Court began operating in The Hague, the Netherlands. The Court must defer to national courts; its prosecutors can issue indictments only when national courts are unwilling or unable to deal with atrocities such as war crimes, crimes against humanity, and genocide. The Court has investigated atrocities in Africa and issued indictments against those responsible.

The United States originally signed the treaty to establish the Court, but it never ratified the treaty, and in 2002 the U.S. government repudiated even having signed the treaty. American officials remember that in 1967 an International War Crimes Tribunal sitting in Stockholm found the United States guilty of terrorism and genocide in Vietnam. The United States has more troops engaged in more actions in more countries all around the world than any other country, and the U.S. government fears that groundless charges would frequently be brought against U.S. officers and officials.

The "Axis of Evil"

In January 2002 U.S. President George W. Bush labeled Iran, Iraq, and North Korea as an "Axis of evil, arming to threaten the peace of the world." The term *axis* was coined by Italian dictator Benito Mussolini to describe the World War II alliance of Italy, Germany, Romania, Bulgaria, and Japan. The three countries named in 2002, however, were not allied. Iran and Iraq were at the time bitter enemies.

Iran Iran's 70 million people make it one of the Middle East's most populous countries, and its vast reserves of oil and natural gas make it one of the richest. It is strategically located, bordering several powerful or volatile states. As a Shiite nation, it was a natural enemy of the Taliban Wahabists in Afghanistan, so it absorbed millions of refugees during the years of fighting in Afghanistan. As noted in Chapter 7, Iran is today a theocracy against which the United States encourages internal rebellion.

Iranians angrily remember that the U.S. Central Intelligence Agency engineered the removal of a democratic government there in 1953 and replaced it with the tyrannical Shah, who was himself toppled by the religious revolution in 1979. During that revolution, Iranian groups took 52 Americans hostage and kept them for 444 days. The confiscated former U.S. embassy in Tehran has been converted into an anti-American museum. Iran's leaders refer to the United States as "The

Afghanistan, the Taliban, and Al Qaeda

Afghanistan is an undeveloped, landlocked state in the middle of Asia that came into existence in the nineteenth century as a buffer between the imperial power of Britain in India to the south, and that of the Russian Empire spreading its power from the north. The population includes many ethnic groups, clans, and tribes that are quite independent of one another. Few observers would argue that there is an "Afghan nation." In 1979 the Soviet Union invaded Afghanistan because Soviet rulers feared that Afghans were encouraging political unrest in the nearby Islamic Soviet republics.

Osama bin Laden, a native of Saudi Arabia, joined thousands of other Arabs who went to Afghanistan to help the Afghans in their *jihad*, or holy war, against Soviet occupation. Most of these Arabs were followers of Wahabism, the fundamentalist Islamic sect that is the official religion of Saudi Arabia (see Chapter 7). Bin Laden founded a group, Al Qaeda ("the foundation" in Arabic), which allied with an indigenous Wahabist Afghan group called the Taliban (meaning "student"), the members of which were largely Pathan. These united fighters called themselves *mujahedeen*, or "holy warriors." Arabs remained the backbone of Al Qaeda.

The United States assisted the Taliban and Al Qaeda in the war against the Soviet Union, although the exact degree of collaboration is a U.S. government secret. In 2006, however, Russian President Vladimir Putin noted, "When the Soviet Union was present [in Afghanistan], the whole Western community was creating bin Ladens there in large numbers, and spared no money and efforts for that."

The Soviet Union failed to subdue Afghanistan completely, but it did install a Communist government before it withdrew in 1989. When this Communist government fell in 1992, the United States withdrew its support for Afghans—military, political, food, and other aid. The country fell into a many-sided civil war. The Taliban, together with its Al Qaeda allies, eventually gained control of about 90 percent of Afghanistan. They imposed upon the people the most severe restrictions of life and rights by following extreme fundamentalist teachings. The Taliban repudiated religious toleration spectacularly in 2001, when it destroyed colossal statues of Buddha that had been carved into rock cliffs in Bamyan province 2,000 years ago. These sacred Buddhist statues counted among the greatest of humankind's monuments.

The Taliban failed to win international diplomatic recognition. Saudi Arabia recognized the Taliban government, and Pakistan provided it with considerable assistance, but most other countries and organizations, including the United States and the United Nations, recognized another government, called the Northern Alliance. This group, dominated by minority Uzbeks and Tajiks, held onto only about 10 percent of Afghanistan, in the northeast.

The Taliban and Al Qaeda, emboldened by their triumph over the Soviet Union, then began a struggle against the United States and other countries that they believed were preventing the creation of true Islamic government throughout the Middle East. Bin Laden was a man of enormous wealth, and he used his wealth to finance several terrorist attacks. Through the 1990s, Al Qaeda was "credited" with bombings in Saudi Arabia, Kenya, and Tanzania, and with the ramming of the naval destroyer *U.S.S. Cole* by suicide bombers in the harbor of Yemen in 2000. New York's World Trade Center Towers were bombed in 1993 by a group of terrorists that may have been affiliated with bin Laden. Fifteen of the 19 men identified as carrying out the attacks on the United States on 9/11 were Saudi Arabs.

Many Islamic governments in other countries had supported the Taliban, but these governments eventually learned that they had been "playing with fire" when the Taliban and Al Qaeda extremists later turned against these governments as insufficiently religious. Pakistan supported extremists, partly in order to recruit guerrillas who would be committed to fight Indian rule in Kashmir (see Chapter 7). Afghanistan became an impoverished nursery for zealots and an exporter of governmental instability to many countries. Al Qaeda formed alliances with other international terrorist groups long identified as dangerous sources of terrorist activity. Islamic fundamentalist volunteers have appeared in the Chinese province of Xinjiang, for example, trying to detach it as an independent Islamic republic, as well as in the Russian provinces of Chechnya and Daghestan, seeking to carve out independent Islamic states there.

From October until December 2001, a U.S.-led coalition invaded Afghanistan and overthrew the Taliban regime, although Osama bin Laden was not captured. Allied forces established a new government, which, under the continuing protection of U.S. and allied forces, attempts to build a nation. At the end of 2006, however, regional warlords still ruled without interference from the capital city, Kabul; the Taliban had reformed and showed new strength; and exports of opium had rebounded to record levels. The Taliban even claimed to have created a new independent Islamic state in isolated parts of Afghanistan.

Great Satan" (see Chapter 6). The United States insists that Iran supports Hezbollah and other regional terrorist groups. Iran is in fact a multicultural country; half of its population is non-Persian. Azerbaijanis account for one-third of the national population, but significant Arab, Baluch, and Kurdish populations also exist, and relations among the groups are not always friendly. These antagonisms constitute centripetal forces because each of these groups laps over Iran's international borders and, if unhappy, looks to its members over the borders for support.

Iran has made several friendly gestures toward the United States: After the attacks on the United States in 2001, Iran observed moments of silence for the American victims, and when U.S. attacks on Afghanistan began, Iran said that it would rescue any American military personnel in distress in its territory.

Other incidents in U.S.–Iranian relations still rankle Iran, however. In September, 2001, Iran's leader Ayatollah Khamenei criticized U.S. support of Israel, and then he commented, "[Americans] divide terrorism into good and bad. In the skies of the Persian Gulf they down an Iranian airliner with hundreds of passengers on board, but then they give the commander of the warship an award." In 1988 the *U.S.S. Vincennes* did shoot down an unarmed passenger liner. The U.S. government claimed it was an accident and awarded the ship's commander a medal for "exceptional conduct." The Ayatollah continued, "Evidence shows that the American government intends to repeat in Central Asia what it did in the Persian Gulf. They intend to come and establish themselves under the pretext of a lack of security here." Since 2003, U.S. troops in Iraq and Afghanistan have surrounded Iran, and Iranians are distressed at the continuing U.S. presence in the region.

Iran is developing nuclear installations. The government insists that these are only to generate electricity—they were commenced under the Shah—but the United States insists that Iran is developing nuclear weapons. U.N. inspectors regularly visit Iranian nuclear sites, but the International Atomic Energy Association has voted to declare Iran in violation of the Nuclear Non-Proliferation Treaty, to which Iran is a signatory. The U.N. Security Council may apply sanctions against Iran, even a war.

On May 8, 2006, Iranian President Mahmud Ahmadi-Nejad sent U.S. President Bush an 18-page open letter on Iran's positions on international affairs. A great deal of the letter was arrogant bombastic rhetoric, but many observers nevertheless regretted that President Bush did not respond in any way. Former U.S. Secretary of State Henry Kissinger pointed out that the letter was "the first direct approach by an Iranian leader to a U.S. president in more than twenty-five years," and a response to the letter would have presented an opportunity for a clear statement to the world of U.S. intentions. President Ahmadi-Nejad sent another open letter to the American people on November 28, 2006, but he has received no response to that, either.

Iraq   Saddam Hussein was an officer in the Iraqi army who became Iraq's president and virtual dictator in 1979. The next year he launched a war against neighboring Iran, which was then a U.S. antagonist. In 1983 U.S. President Ronald Reagan sent Donald Rumsfeld (later U.S. Secretary of Defense in the George W. Bush administration) personally to affirm U.S. support for Saddam Hussein. The amount and type of military support the U.S. gave Iraq remains a secret. Iraq and Iran fought until both were exhausted in 1988.

In 1990 Iraq invaded Kuwait, claiming (with considerable justification) that the territory of Kuwait was in fact historically Iraqi, and that existence of the Kuwaiti state was an illegitimate relic of colonial rule. In 1991 the United States and several allies, fighting under U.N. auspices, forced Iraq to retreat from Kuwait, but they neither conquered Iraq nor brought down Hussein's government. The United Nations did, however, establish "secure zones" within which the Iraqi government was prevented from attacking its own citizens who had cooperated with the U.N. allies (Figure 13-23). Shiites were protected south of the 33rd parallel, and the Kurds enjoyed a sanctuary north of the 36th parallel; there, through the following years, the Kurds created a virtual Kurdish free state. This was much to the annoyance of the Turks, who feared Kurdish separatism in Turkey's southeast (see Chapter 11).

The 1991 war devastated Iraq's economy and infrastructure, and the United Nations subsequently

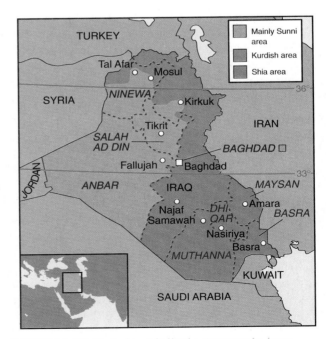

FIGURE 13-23 Religious/ethnic groups in Iraq.
Terrible strife accompanies continuing U.S. efforts to forge an Iraqi nation.

Is the United States the World's Policeman?

Recent U.S. diplomatic initiatives have stimulated consideration of America's evolving view of its own role in world peacekeeping.

Early in U.S. history, statesmen maintained an *isolationist* diplomatic position, that is, one that turns inward and is concerned only with its own domestic affairs. President George Washington's farewell message (1797) emphasized, "It is our true policy to steer clear of permanent alliances with any portion of the foreign world," but America could have "temporary alliances for extraordinary emergencies." President Thomas Jefferson's inaugural address (1801) similarly promised "no entangling alliances." On July 4, 1821, then-Secretary of State John Quincy Adams spoke against sending the U.S. Navy to interfere in the affairs of Spain's former colonies Colombia and Chile: "America does not go abroad in search of monsters to destroy. She is the well-wisher to the freedom and independence of all. She is the champion only of her own."

The United States took a more active role in 1823 with the proclamation of the Monroe Doctrine. It stated that whereas the United States would not interfere with existing European colonies in the Western Hemisphere, it would allow no new colonial claims. Thus, the United States defined itself as the "defender" of the Western Hemisphere.

When President William McKinley asked Congress to declare war on Spain in 1898, he injected a new consideration into U.S. foreign policy: a concern for human rights elsewhere. Referring to alleged Spanish mistreatment of Cubans, McKinley said: "It is no answer to say this is all in another country, and is therefore none of our business. It is especially our duty, for it is right at our door." The Spanish-American War involved fighting not only in the Caribbean, but on the other side of the world in the Spanish colony of the Philippines. It was this war that provoked satirical writer Ambrose

Bierce's wry observation, "War is God's way of teaching Americans geography." After defeating Spain, the United States took four more years to conquer the Philippines in order to hold it as its own colony (until 1946). The Philippines provides a humbling perspective on nation building, because although U.S. troops finally left the Philippines in 1990, they returned in 2002 to help hunt guerrillas tied to Al Qaeda, and democracy remains chronically insecure in the Philippines.

McKinley's successor President Theodore Roosevelt proclaimed the 1904 "Roosevelt Corollary" to the Monroe Doctrine, according to which "chronic wrongdoing...[may]...require intervention by some civilized nation, and in the Western Hemisphere the adherence of the United States to the Monroe Doctrine may force the United States, however reluctantly, in flagrant cases of wrongdoing or impotence, to the exercise of an international police power." This is a clear proactive stance.

After World War I, President Woodrow Wilson (who served 1913–1921) repudiated isolationism and announced U.S. readiness to join in a permanent association of nations working for peace. His inability to compromise with Congress, however, led to the failure of the United States to join the League of Nations. After World War II, the United States helped found the United Nations, and the 1947 Truman Doctrine (President Harry Truman served 1945–1953) pledged U.S. assistance to any country anywhere fighting a takeover by Communists. It was to be presumed that communism emanated exclusively from the U.S.S.R., and was, therefore, a foreign attack upon an independent nation. This policy came to be called one of "containing" Communism during the Cold War. The degree to which individual Communist movements around the world were in fact initiatives of the U.S.S.R. or else were homegrown nationalist movements will long be debated.

President Jimmy Carter (who served 1977–1981) accepted the necessity of accepting as our allies in the

allowed Iraq to export oil only under international supervision and only in exchange for food and medicine. Saddam Hussein awarded those supplies to his loyal soldiers rather than to Iraqi children, and the deaths of thousands of Iraqi children brought Iraq considerable world sympathy. Much wrath was directed at the U.N. sanctions rather than, perhaps more properly, at Saddam Hussein.

The United Nations also insisted that Iraq allow U.N. monitors to guarantee that Iraq was not producing nuclear, chemical, or biological weapons of mass destruction (WMDs). After Iraq drove out those inspectors in 1998 and repeatedly frustrated efforts to ensure that it had no WMDs, the United States invaded Iraq again in 2003. This time acting without U.N. sanction, the United States was joined by a few allies

Cold War some countries that were by no means truly "free." Part of his legacy, however, was renewed concern with the protection and encouragement of human rights in other countries.

The 9/11 attacks on the United States changed many Americans' isolationist worldview and ended American willingness to respect the sovereignty of states that harbor terrorists. On September 20, 2001, President George W. Bush told Congress, "We will . . . drive [terrorists] from place to place, until there is no refuge or no rest. And we will pursue nations that provide aid or safe haven to terrorism. Every nation, in every region, now has a decision to make. Either you are with us, or you are with the terrorists. From this day forward, any nation that continues to harbor or support terrorism will be regarded by the United States as a hostile regime." Shortly thereafter, the president emphasized: "Every nation has a choice to make. In this conflict, there is no neutral ground. If any government sponsors the outlaws and killers of innocents, they have become outlaws and murderers themselves. And they will take that lonely path at their own peril." Many observers feel that with these words, President Bush repudiated even the 1648 Treaty of Westphalia (see Chapter 11).

The world has shrunk, and the United States no longer stands relatively safe and isolated behind ocean barriers as in George Washington's day. President Bush argued that if other countries cannot control terrorists who operate from within their borders, then the United States does not have to recognize the territorial sovereignty of those countries, because it has the right preemptively to destroy enemies of America. If we accept that preemptive strikes against enemies may ever be necessary, then we might have to accept that secrecy may be needed: How can such an issue be debated in a democracy? Who should make the decision to attack first, and to whom should the action be explained or justified? Could it even be necessary for the government to withhold information or sources in order to protect the sources? Questions such as these prompted the young Congressman Abraham Lincoln to write in 1848, when he was protesting the U.S. war against Mexico, "Allow the President to invade a neighboring nation whenever he shall deem it necessary to repel an invasion . . . and you allow him to make war at pleasure. If today he should choose to say he thinks it necessary to invade Canada to prevent the British from invading us, how could you stop him? You may say to him, 'I see no probability of the British invading us,' but he would say, 'Be silent: I see it if you don't.'" Lincoln's own later experience as president, however, convinced him of the necessity for presidential secrets. Questions such as these have always plagued diplomats and political theoreticians.

The debate continues within the United States. In 2002, Patrick Buchanan, senior advisor to three Republican presidents and the Reform Party's presidential candidate in 2000, wrote, "What happened on September 11, 2001, was a direct consequence of an interventionist U.S. policy in an Islamic world where no threat to our vital interests justifies our massive involvement. We are a republic, not an empire. And until we restore the foreign policy urged upon us by our Founding Fathers—of staying out of other nations' quarrels—we shall know no end of war and no security or peace in our own homeland." Today, however, in contradiction to the isolationist tradition in American foreign policy, perhaps the United States must "go abroad in search of monsters to destroy" in its own defense.

Another official U.S. government document, "The National Security Strategy of the United States of America," released in October 2002, emphasizes that "Our forces will be strong enough to dissuade potential adversaries from pursuing military buildup in hopes of surpassing, or equaling, the power of the United States." This document suggests to many readers that the United States has defined for itself a role as preeminent global policeman.

(notably the United Kingdom). Three justifications were given for the war: (1) that Iraq had WMDs; (2) that Iraq assisted international terrorist organizations; and (3) that Saddam Hussein was persecuting many of his own people. The world learned later that only the third of these assertions was true, and many observers wondered whether it was sufficient cause to have invaded.

The Second War brought down the Hussein government within weeks, but American efforts to establish a new government in Iraq met considerable difficulty. Few nations were willing to help rebuilding and nation-building efforts without U.N. sanction. By the end of 2006, Iraq had created a "national unity" government dominated by the majority Shiites, who had been suppressed by Sunnis under Hussein; but that government's real power was limited. The war had become a terrible civil war in which U.S. forces had suffered almost 3,000 fatalities and thousands of casualties, and tens of thousands of Iraqis had died. Conditions of the fighting and of the U.S. occupation had eroded the perception of U.S. power and moral authority in the world. Already in April 2006, the official consensus of America's intelligence agencies, as stated by the Director of National Intelligence, was that "The Iraq conflict has become the 'cause celebre' for jihadists, breeding a deep resentment of U.S. involvement in the Muslim world and cultivating supporters for the global jihadist movement." Thus America's earlier conjectural fears about Iraq's possible collaboration with terrorist groups had attained a terrible reality.

The United States struggled to midwife an Iraqi nation, but perhaps Iraq will inevitably split into separate Kurdish, Sunni Arab, and Shiite Arab states. Whether Iraqi Shiites continue to dominate in a unified state, or whether Iraq breaks apart, Iraqi Shiites' links to neighboring Shiite Iran will continue to concern the United States. Evidence from early national elections, the Shiite constitutional program, and from Islamic rule already in place in the South of Iraq suggest that most Shiites want an Islamic theocracy, with many features borrowed from neighboring Iran. Only time will tell which cultural forces are stronger among Iraqi Shiites: the desire for democracy or for theocracy and an Iraqi secular patriotism or Shiite solidarity with Iran.

Territorial breakup would unleash a struggle over Iraq's oil wealth, which comprises the world's third greatest reserves after Saudi Arabia and Iran. Some 80 percent of the reserves are in southern Iraq, where Shiites, who constitutue 60 percent of the total population, predominate. Another 15–18 percent of the oil is in northern Iraq, where the Kurds, who constitute 20 percent of the total population are concentrated; less than 5 percent of the reserves are in Al-Anbar, Nineveh, and Salahuddin provinces, where most of the population that is Sunni lives.

Korea
After the 1950–1953 war in which United Nations forces defended South Korea against North Korea, South Korea prospered while North Korea stagnated as a Communist dictatorship under Kim Il Sung (who ruled 1948–94) and then his son Kim Jong Il. North Korea has remained uniformly hostile to the United States, perhaps partly as a way of binding its own people, and it has created the most paranoid, totalitarian state in history while millions of its people have starved to death (Figure 13-24). As late as 2006, the United States maintained almost 40,000 troops in South Korea in case of another North Korean invasion. Nevertheless, as South Korea prospered, many South Koreans began to resent the American presence and domination and to yearn for reconnection with the North (see Figure 11-10b). South Korea has invested in North Korea and provided it with financial aid;

FIGURE 13-24 "Ruthless punishment to U.S. imperialism." Posters such as these are found everywhere in North Korea, exhorting the people to sacrifice in "defense" of their homeland, although the attack on the U.S. Capitol here seems offensive. (AP Wide World Pictures)

thousands of North Koreans now work in "special economic enclaves" in the North. Cho Yong-nam, Director-General of South Korea's Unification Ministry, said in 2006 that South Korea invests in North Korea in order to raise Northern living standards so that, in a unified Korea, North Koreans would not constitute "a displaced, misfortunate minority group." Ko Gyoung-bin, another Ministry official, said, "It's de facto unification. It's already under way." A railroad from South Korea across North Korea, currently being negotiated, would provide South Korea with a much-desired direct land link to China and even to European markets. Trying to unite and harmonize South Korea's economy, culture, and politics with North Korea, however, would be immeasurably more difficult even than it was for the two Germanies to merge.

North Korea, in contrast to both Iran and Iraq, does have biological weaponry, nuclear bombs, and the capability to deliver these weapons by missile to it neighbors and possibly even to the United States. It has provided weaponry to many global terrorist organizations and hostile states. The United States and North Korea agreed in 1994 that, in exchange for regularization of diplomatic relations, trade, and aid, North Korea would end its nuclear programs. Neither side, however, fulfilled the terms of that agreement, and negotiations have continued fitfully. China has long defended and aided North Korea (providing up to 80 percent of its food and oil at concessionary prices), and it does not want the United States to attack, but it also fears continuing mass migration into China of desperately poor North Koreans. Japan fears North Korean nuclear arms. Many states fear that Japan may be provoked to develop its own nuclear capability. Japan was disarmed after World War II, and, despite Japan's contribution of peacekeeping forces in Iraq, Japan has little military capability.

Human Rights

Global government presumes global acceptance of fundamentals of a political culture, which rests in turn on fundamentals of human culture. Two hundred years ago Thomas Jefferson could confidently write that certain truths regarding human rights were "self-evident," despite the fact that he himself owned slaves.

The U.N. General Assembly adopted a Universal Declaration of Human Rights very early, in 1948. At that time, however, most U.N. members were Western countries, and they were still holding most African and Asian societies as colonies. As these societies have achieved political independence, they have begun to challenge Western ideas of human rights. They deny that any human rights are in fact universal and "self-evident." Human rights, they insist, may vary from culture to culture.

Two questions have arisen in international organizations. First, what is the balance between the rights of individuals and those of the whole society? Even Western countries disagree on this issue. Canada, for example, is more authoritarian than the United States. Canada's equivalent to the United States' Bill of Rights is the Charter of Rights and Freedoms, passed only in 1982. It reflects a commitment to gender equality and social equality far beyond that of U.S. documents, but its commitment to civil liberties is arguably less. It guarantees "freedom of thought, belief, opinion, and expression," but these are subject "to such reasonable limits prescribed by law as can be demonstrably justified in a free and democratic society."

The second question is the balance between political rights and what some people call "economic rights." In 1997 China's President Jiang Zemin insisted, "As a developing country of 1.2 billion people, China's very reality determines that the right of subsistence and development is the most fundamental and most important human right in China. Before adequate food and clothing is insured for the people, the enjoyment of other rights would be out of the question."

At a 1993 U.N. Conference on Human Rights, many countries were concerned that Western countries would use their definitions of human rights as excuses for withholding aid or other economic benefits from developing nations that may not accept the same definitions. Singapore's Foreign Secretary Kishore Mahbubani insisted, "Too much stress on individual rights over the rights of the community will retard progress.... In the future, these agreements will assert the rights of society over the rights of individuals." Indonesia's foreign minister stated flatly that "Human rights exist as a function of history, culture, value systems, geography, and phases of development." Several Asian countries jointly offered a statement that fairness and justice should be measured against "regional particularities and various historical, cultural, and religious backgrounds." An African statement demanded that observers must take account of "the historical and cultural realities of each nation." Not all cultures recognize the equality of women with men, and the assembled nations could not even agree on a condemnation of torture. They insisted, however, that "the right to development is inalienable." The United States hesitates to accept a statement that development is a "right," because that could lead to demands that the rich countries redistribute their wealth among the poor countries.

After many days of debate a final document was adopted, stating that while differences of "historical, cultural and religious background must be borne in mind, it is the duty of states, regardless of their political, economic and cultural systems, to promote and protect all human rights and fundamental freedoms." In other words, the states agreed to disagree.

The United Nations long had a Commission on Human Rights, the chairmanship of which was awarded not by merit, but by rotation among members. Thus the chairmanship periodically fell to U.N. members whose own records on human rights were not good. Therefore, in 2006, the U.N. replaced the Commission with a new Human Rights Council with restrictions on membership. When the new Council first met, however, Iran, China, Cuba, Pakistan, Russia, and Saudi Arabia were counted among its members, thus continuing to illustrate the difficulty of reaching international agreement on definitions of "human rights" and who enjoys them.

Jurisdiction over Earth's Open Spaces

One of the most important issues requiring international agreement is the adjudication of the rights that individual states have in Earth's open spaces. Open spaces include the Arctic, Antarctica, and the world's seas.

The Arctic and Antarctica   Eight countries have territory north of the Arctic Circle: the United States, Canada, Denmark, Finland, Iceland, Norway, Sweden, and Russia. In 1996 these countries formed an Arctic Council that pledged to cooperate in monitoring Arctic pollution and in protecting the region's plant and animal life. Canada claims dominion over the uninhabited Arctic islands north of Canada all the way to the North Pole, but the United States has challenged Canada's claims.

Antarctica has never had any permanent inhabitants, but seven states claim overlapping sovereignty. Australia, New Zealand, Chile, and Argentina, the world's most southerly states, claim sovereignty on the basis of proximity. The claims of Britain, Norway, and France are based on explorations (Figure 13-25). Neither the United States nor Russia, which support most of the scientific research there, make any territorial claims, nor do they recognize other nations' claims.

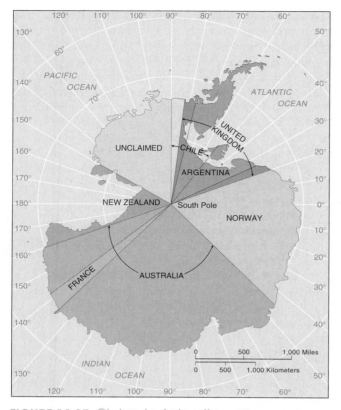

FIGURE 13-25 Claims to Antarctica. The United States does not recognize any of these overlapping claims to Antarctica.

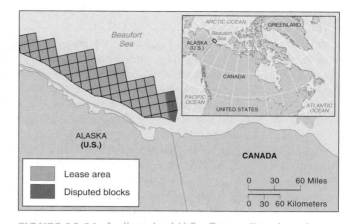

FIGURE 13-26 A disputed U.S.–Canadian border. The United States and Canada dispute the seaward extension of their border here in the Arctic, and the United States has auctioned off the rights to explore for oil in the disputed tracts of the Beaufort Sea.

Forty states are parties to a 1959 Antarctic Treaty that resolves that "Antarctica shall continue forever to be used exclusively for peaceful purposes and shall not become the scene or object of international discord." The treaty prohibits military installations, but it opens the continent to all nations for scientific research and study. The treaty also prohibits signatories from harming the Antarctic environment.

Despite the terms of this treaty, however, some countries, led by Malaysia, demand that exploration should be undertaken for Antarctic resources and that any resources found should be exploited for the good of all nations. Other nations, led by France and Australia, insist that Antarctica should be set aside as a wilderness preserve.

Territorial waters The world's states have divided up all the land on Earth (except Antarctica), but international agreement negotiates the question of how far out to sea the territorial claim of a country can reach.

The Dutch jurist Hugo Grotius (1583–1645) first argued that the world's seas are open space, *mare liberum*, which no political unit can claim. Each coastal state, however, has claimed coastal waters for defensive purposes. A later Dutch jurist, Cornelius van Bynkershoek, accepted this in his book *De Domina Maris*

(1702). The limit of sovereignty was set at three nautical miles from shore, which was generally accepted for more than 200 years. All distances at sea are measured in international nautical miles of one minute of latitude—1,852 meters (6,076 feet). Territorial sea is measured from the low-tide mark, and any disputes between states arising from irregular coastlines or islands are resolved by individual negotiations. A 1958 International Conference on the Law of the Sea set rules for ascertaining the seaward extension of land borders (Figure 13-26).

The United States accepted a 3-mile limit in 1793, but in 1945 President Truman broke that international covenant and claimed sole right to the riches of the continental shelf up to 200 nautical miles out. A **continental shelf** is an area of relatively shallow water that surrounds most continents before the continental slope drops more sharply to the deep-sea floor. The 1958 Sea Law Conference agreed on a water depth of 200 meters (656 feet) as the definition of the outer edge of a continental shelf (Figure 13-27).

The Truman Proclamation claimed shelf mineral rights, but it did not claim control over fishing or shipping in the seas over the shelf beyond the three-mile territorial limit. It nevertheless triggered extended claims to fishing rights by other nations. In 1976 the United States extended its own claims to exclusive fishing rights up to 200 nautical miles out from its shores. As noted in Chapter 8, about 90 percent of the world's marine fish harvest is caught within 200 miles of the coasts.

The delineation of continental shelves in the Arctic is of increasing dispute. Canadian, Danish, and Russian scientists are mapping a topographic feature under the Arctic seas called the Lomonosov Ridge. At stake are the rights to unknown, but potentially rich, deposits of minerals, oil and gas, as well as control over Arctic

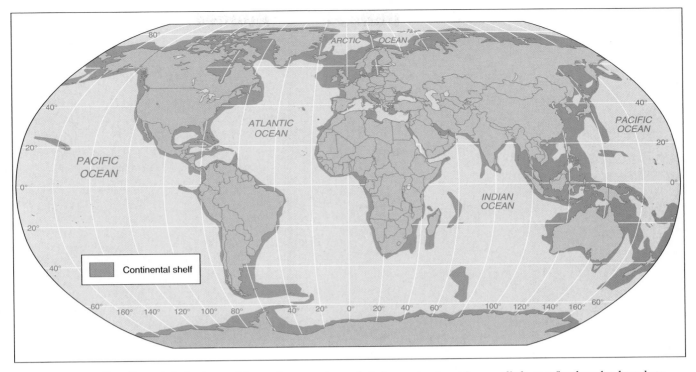

FIGURE 13-27 Continental shelves. The various continental shelves extend out from as little as a few hundred yards to more than 600 miles.

navigation. The ridge appears to run from the area around Ellesmere Island and Greenland all the way under the polar ice cap for a couple of thousand kilometres to Russia. A Canadian and Danish team is trying to demonstrate that the Lomonosov Ridge is an extension of the North American continent, whereas the Russians are trying to prove that the ridge is an extension of the Siberian continental shelf. Canadian professor of international law Dr. Michael Byers has described mapping Canada's claim as "Canada's moon mission. . . . It's just as tough. But we can do this." A settlement may be many years away.

The 1982 United Nations Convention on the Law of the Sea

In 1982 the United Nations proposed a new Law of the Sea Treaty. Sixty countries ratified it by 1993, thus bringing it into force among the ratifying parties in 1994. The Law of the Sea Treaty authorizes each coastal state to claim a 200-mile exclusive economic zone (EEZ), in which it controls both mining and fishing rights. Therefore, possession of even a small island in the ocean grants a zone of 326,000 square kilometers (126,000 square miles) of sea around that island. This might help explain why France retains many small island colonies. France claims a full 10.4 percent of the world total EEZ. The United States claims 8.4 percent, New Zealand 5.8 percent, and Indonesia 4.7 percent. No other state claims more than 4 percent. The potential for discoveries of oil in the world's seas has triggered an international scramble to claim even

tiny uninhabited outcrops of rock. For example, China, Japan, and South Korea lay conflicting claims to areas of the East China Sea. Vietnam, China, Taiwan, the Philippines, Thailand, and Malaysia extend overlapping claims in the South China Sea.

Another clause of the 1982 treaty guarantees ships **innocent passage** through the waters of one state on the way to another. The extension of countries' territorial waters, however, means that countries now claim many of the world's narrow waterways. They remain open in times of peace, but they can be closed in time of war (Figure 13-28).

The United States at first refused to sign the Law of the Sea Treaty in protest against provisions calling for internationalization of seabed mineral resources and the creation of an International Seabed Authority to control mining. The United States insisted on rights for free enterprise. Several of the treaty's other provisions, however, did serve U.S. interests, so U.S. President Reagan announced that the United States would regard all but the seabed provisions as law, even though the country had not signed the treaty. For example, one provision allows nations to declare a 12-mile territorial limit. Therefore, in accordance with the treaty, on December 28, 1988, the United States extended its territorial limit to 12 miles. Given the length of the U.S. coastline, this seaward extension enlarged the territory of the United States by some 478,000 square kilometers (185,000 square miles). Compromises on disputed provisions were eventually

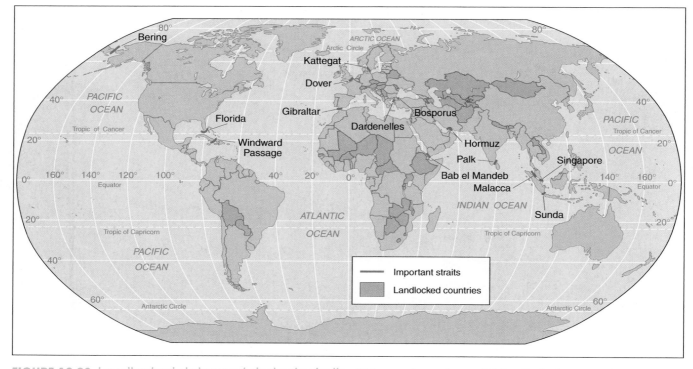

FIGURE 13-28 Landlocked states and strategic straits. This map shows the world's landlocked states and some of the world's most important narrow sea lanes. As states have extended their seaward claims, many of these narrow waterways have been claimed by the adjacent states. A terrorist act in one of these narrow straits could disrupt the world economy. The Straits of Malacca, for example, is at certain points only 2.5 kilometers (1.5 miles) wide, but it carries 40 percent of world trade.

reached, and the United States signed and ratified the treaty in 1996.

Another clause of the 1982 treaty allows countries to claim some powers in a "contiguous zone" of 12 miles further out beyond the 12 miles of territorial waters. Therefore, on September 2, 1999, the United States announced that it would start enforcing U.S. law and would be boarding ships up to 24 nautical miles off the coast. The U.S. proclamation applies to the territorial waters around all U.S. possessions.

Landlocked states The independence of Montenegro from Serbia in 2006 added Serbia to the list of the world's now 43 landlocked states without seacoasts. Several other states are not totally landlocked, but their own coasts are unsuitable for port development, so they must rely on neighbors' ports. Landlocked states must secure the right to use the high seas, the right of innocent passage through the territorial waters of coastal states, port facilities along suitable coasts, and transit facilities from the port to their own territory.

Landlocked or partially landlocked states may gain access to the sea in one of three ways. First, any navigable river that reaches the sea may be declared open to the navigation of all states. Freedom of navigation on rivers that flow through several countries was first proclaimed by France in 1792. The French revolutionary government proclaimed that the freedom of rivers was a "natural law." International commissions regulate navigation on many international rivers, and often these same commissions guard against pollution and regulate the drawing of irrigation waters from the river (Figure 13-29).

Second, a landlocked state may obtain a corridor of land reaching either to the sea or to a navigable river. Several countries have long, thin extensions of land out to seaports (Figure 13-30). Some of these, such as the Congo's corridor to the Atlantic Ocean, are important transport routes, but others, such as Namibia's Caprivi Strip to the international Zambezi River, serve no significant traffic function.

The third way a landlocked state can gain access to the sea is to obtain facilities at a specific port plus freedom of transit along a route to that port. Coastal states have signed international conventions promising to assist the movement of goods across their territories from landlocked states without levying discriminatory tolls, taxes, or freight charges. Chile, for example, helped build a railroad connecting La Paz, Bolivia's principal city, to the Chilean port of Arica, and Chile guarantees free transit. Argentina grants Bolivia a free zone at the Argentine city of Rosario on the Paraná River, and Peru gives Bolivia a free-trade zone in the port of Ilo. The 1998 treaty between Ecuador and Peru guaranteed Ecuador navigation rights in Peruvian ports on the Amazon River. In 1993 Ethiopia joined the ranks of the landlocked states when its coastal province of Eritrea gained independence. Eritrea promised to assist

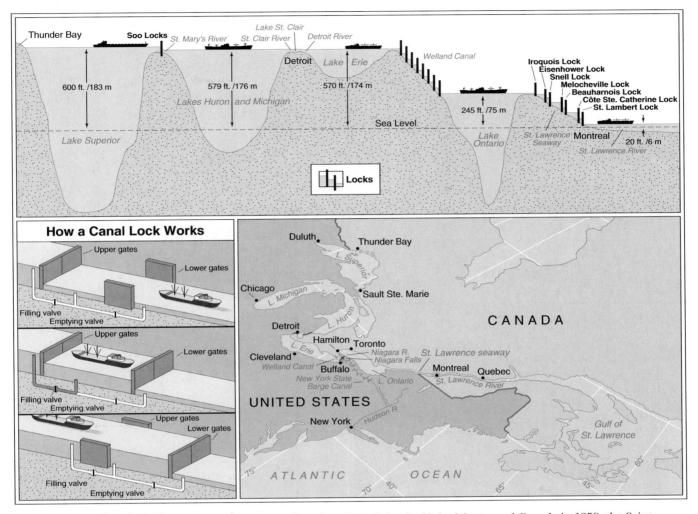

FIGURE 13-29 The Saint Lawrence Seaway. Completed jointly by the United States and Canada in 1959, the Saint Lawrence Seaway is an example of international cooperation on an international waterway. It allows ocean vessels to reach from the Atlantic Ocean into the Great Lakes. The joint project included the construction of hydroelectric power plants.

Ethiopian import and export trade, but in fact most of Ethiopia has long relied on transit via the port of Djibouti (look back at Figure 12-27).

Coastal states can profit by granting transit rights. The coastal state's railroads have a captive customer, and their ports gain extra business and opportunity to serve as break-of-bulk points for processing imported or exported raw materials (see Chapter 10). The fact that a landlocked state misses these opportunities can hinder its economic development. Landlocked states also lose out in the apportionment of the resources of the continental shelves, exclusive economic zones, and fishing opportunities.

The high seas All nations agree that beyond territorial waters are high seas, where all nations should enjoy equal rights. Several problems, however, threaten any area where rights are not specifically assigned. These include depletion of resources (such as overfishing) and pollution. International conventions regulate each of these potential abuses of the high seas. The

1972 London Dumping Convention, for example, signed by 70 countries, originally banned the dumping of radioactive and other "highly dangerous" wastes at sea, but it was extended to ban dumping of all forms of industrial wastes in 1995. The enforcement of international agreements such as these is uncertain, and their effect is limited.

Airspace The question of how far up a nation's sovereignty extends is as complex as the question of how far out to sea it extends. Ships enjoy innocent passage through territorial waters, but airplanes have never been granted innocent passage to fly over countries. Airlines must negotiate that right, and states commonly prescribe narrow air corridors, the altitudes at which aircraft must fly, and even the hours of the day the passages are open. U.S. President Dwight Eisenhower first proposed an "Open Skies" treaty in 1955, but a treaty was not realized until 1992. The treaty establishes conditions for unarmed observation flights over the entire territory of participants. In accordance with the treaty, a

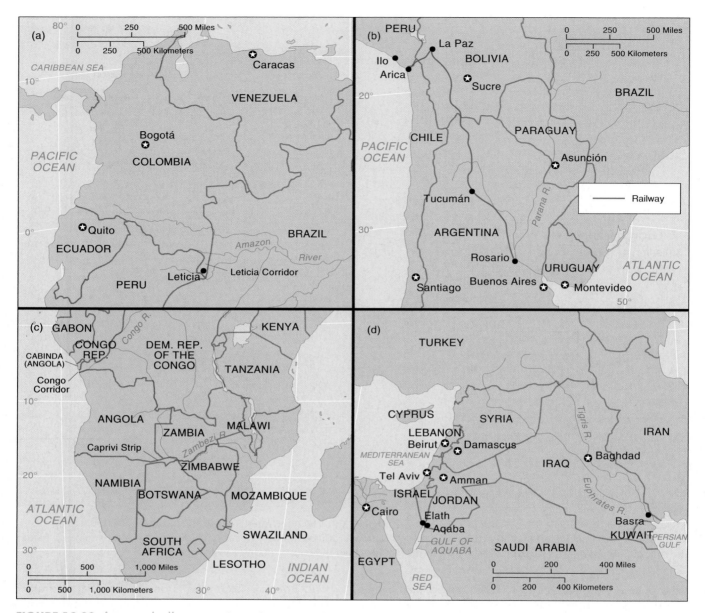

FIGURE 13-30 Access to the sea. Several countries that would otherwise be landlocked have narrow corridors of territory creating access to the sea or to a navigable river. Alternatively, neighboring countries grant the landlocked state special railroad access to ports, as in (b). Sucre is the constitutional capital of Bolivia, although La Paz is the seat of government.

Russian Air Force jet spent five days in August 1997 flying over the United States taking photographs of U.S. military installations. This astonished anyone who grew up during the Cold War.

Diplomats cannot agree on the altitude at which national airspace ends. One possible definition of national airspace would limit it to the lowest altitude at which artificial unpowered satellites can be put into orbit, at least once around Earth. That ranges between 113 and 161 kilometers (70 and 100 miles). The U.N. Office for Outer Space Affairs does try to formulate recommendations for designing and flying space vehicles to reduce the amount of debris they produce and their chances of colliding with one another.

Many observers dread the militarization of space, but suggestions have been made that the eventual parceling-out of space is almost inevitable, as a measure of national security. In 1997 the United States successfully targeted a laser beam on an orbiting satellite, and it spends more than a billion dollars each year on space weapons research. The U.S. government's "National Space Policy" published in August 2006, rejects any arms control agreements that might hinder "freedom of action in space." Countries have so far refrained from extending territorial claims to the Moon or planets, but several have insisted that they reserve the right to do so (Figure 13-31). Our study of geography, therefore, is so far restricted to Earth.

Focus ON

Regulation of Whaling

Whales are an excellent example of a commonly owned natural resource that can only be managed by international negotiation and agreement. Whales have been hunted for thousands of years, but the development of large-scale commercial whaling in the eighteenth and nineteenth centuries reduced the populations of many species to dangerously low levels. By the 1920s whalers began to realize that some species had been overfished, and in 1946 the International Whaling Commission (IWC) was formed to regulate the industry. In its early years the IWC's conservation achievements were modest at best, but by the 1980s widespread support for whale conservation had developed in much of the world. In 1986 the IWC declared a worldwide moratorium on whaling except for purposes of scientific research. The moratorium also allowed continued whaling by indigenous Arctic peoples in northern Russia and Alaska. Member nations retain their sovereignty over their whaling fleets, however, and the IWC cannot enforce its restrictions. Norway and Japan continue to harvest whales, though in fewer numbers than they did before the IWC was formed. In late 1997 the IWC agreed to ban whaling completely on the high seas but to allow commercial whaling in coastal waters. An agreement of this type effectively limits whaling to those areas where national sovereignty can be exercised. If it holds, it will allow more effective management of whaling and reduce the likelihood of overhunting. In 2006, however, Iceland renounced the global moratorium on commercial whale hunting.

International regulation of whaling contrasts with the lack of controls over harvesting other marine species. As described in Chapter 8, there is no effective worldwide regulation of fishing, and only in a few cases (most notably the European Union's Common Fisheries Policy) have nations been able to agree on regulation of fisheries. Even close friends like the United States and Canada have bitter disputes over shared fisheries, and the consequence has been worldwide devastation of fish populations similar to that experienced prior to regulation of whaling.

FIGURE 13-31 Moon exploration. When U.S. astronauts visited the Moon, they planted U.S. flags. The United States denies that the gesture was intended as a claim. (NASA Headquarters)

PROTECTING THE GLOBAL ENVIRONMENT

Just as the world has become more integrated and interdependent in recent decades, so has it become apparent that we share common concerns about environmental quality. While local environmental quality issues such as urban air pollution remain important—and are growing in importance in industrializing regions—global environmental issues have risen in importance. International governmental and nongovernmental organizations are increasingly involved in finding ways to manage concerns about the environment in a changing world.

Changing patterns of economic activities at the global scale mean changing patterns of resource consumption and pollution outputs. Whereas most of the world's industrial activity and most output of industrial pollutants are still concentrated in the rich countries, a significant shift in pollution output to the poorer countries is occurring (Figure 13-32). Two factors are driving this shift. First, heavy manufacturing, and the pollution that accompanies it, is shifting from rich countries to poor ones. Second, populations are growing faster in poor countries than in rich ones, so consumer-generated pollution such as sewage, solid waste, and

(b)

(a)

FIGURE 13-32 Changing locations of heavy industry. (a) An abandoned steel mill in Volkingen, Germany, and (b) a steel mill in Benxi, Liaoning Province, China. The locus of heavy industry in the 1950s and 1960s was in Europe, North America, and Japan, but today much of this industry has shifted to developing nations, and along with it, much energy consumption and pollutant emissions. ([a] Haackenberg/Masterfile Corporation; [b] Michael Wolf/Aurora & Quanta Productions Inc.)

automobile emissions are growing rapidly in poor countries. Unfortunately, the poor countries are less able to afford pollution-control systems, so their populations increasingly suffer the negative effects of living with pollution (Figure 13-33).

Energy consumption In general, poor countries are using rapidly rising amounts of natural resources, whereas the use of natural resources in the rich countries is growing slowly. Rich countries were provoked to decrease their natural resource use and to increase the energy efficiency of their economies by the rise in energy prices in the 1970s. The most obvious example of this increase in energy efficiency is the change in automobile design between the early 1970s and the 1990s. Modern autos are much more fuel efficient: A typical mid-1990s five-passenger sedan travels at least 50 percent farther on a liter of gasoline than a comparable-sized car built 20 years earlier. At the same time, since the 1980s Americans have been buying larger vehicles, especially minivans and sport-utility vehicles, which are less efficient than the sedans and station wagons that were more common earlier. This has caused a reduction in our overall vehicle fuel efficiency that persists, even though rising gasoline prices have slowed sales of inefficient vehicles.

Air-pollution-control regulations are another factor in this increased efficiency. Using the auto as an

example again, one of the easiest ways to reduce emissions is to improve the efficiency of fuel use. The replacement of carburetors and mechanical ignition points with computer-controlled fuel injection and electronic ignition both reduces pollution output and improves fuel efficiency. Similar changes that have occurred throughout all industries have reduced both the amount of resources used and the pollution generated.

At the same time that the rich countries were increasing their efficiency of fossil-fuel use, they were also experiencing a rapid economic transition away from energy- and pollution-intensive heavy industries toward light-industry and service-based economies. Steel manufacturing and related heavy-equipment industries were especially hard hit by the oil price increases of the 1970s. The steel industry in the United States, Canada, and much of Europe was aging, and many mills used less efficient, older technologies. These mills also faced increased competition from other regions. Developing countries built their own steel industries, and global production capacity was greater than demand. The domestic automobile industry in the United States and Canada faced increasing competition from Japan. When fuel prices rose, consumers demanded smaller foreign-made cars, not the large gas guzzlers that the North American plants were producing. Increases in energy costs caused a general slowdown in manufacturing. The combined effects of these factors contributed

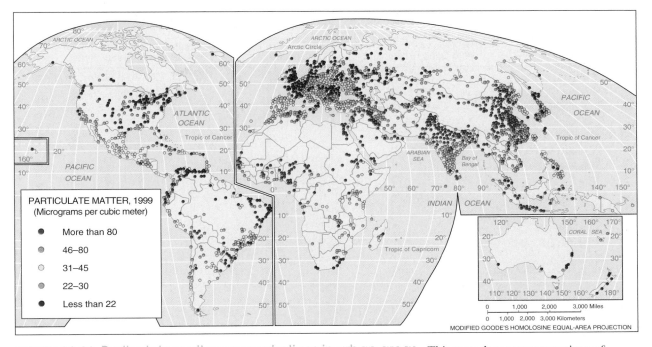

FIGURE 13-33 Particulate matter concentrations in urban areas. This map shows concentrations of particulate matter in urban areas. Particulate matter is a major cause of health problems such as asthma, chronic bronchitis, and acute respiratory problems. (Redrawn from The World Bank Atlas, 2003)

to the decline of heavy industry in the northeastern United States and adjacent Canadian cities, in northern Britain, and to a lesser extent, in Germany. Along with this came a significant drop in fossil-fuel use, which helped to reduce pollution emissions in these regions (Figure 13-34).

The decline of the traditional heavy industrial regions of the developed world was accompanied by the growth of manufacturing in rapidly industrializing regions of East Asia and Latin America. Much of this manufacturing, however, is labor- and capital- rather than energy-intensive, so it has made only a minor contribution to increased energy consumption in these regions. The increase in wealth that this manufacturing brings, especially in urban regions, is much more significant. However, this added wealth raises demand for energy-consuming machines, especially automobiles. The number of autos in use in China, Malaysia, Brazil, and other nations with rapid economic growth rates is rising, so petroleum use and air pollution also rise.

Energy efficiency trends
It takes energy to generate income. Some ways of generating income take more energy than others, however. Making steel uses large amounts of fuel per dollar earned, whereas making televisions requires much less. Typically, manufacturing takes much more energy than providing services. Also, some ways of making a given product are more efficient than others. Modern factories or office buildings usually use much less energy per unit of output or office space than old ones. We can see changes in energy efficiency by

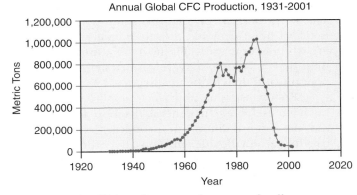

FIGURE 13-34 Chlorofluorocarbon production. Production of chlorofluorocarbons peaked in 1988 and declined dramatically thereafter as a result of an international agreement to phase out these ozone-destroying chemicals. (Data from Worldwatch Institute)

examining the relationship between how much economic activity changes in relation to changes in energy use.

The link between economic and environmental approaches over the past two decades can be seen in trends of economic efficiency of energy use. The amount of energy used to produce a dollar of GDP (also known as *energy intensity*) is shown for a few representative countries for the period 1980–2004 in Table 13-1. In this period, the PPP-based GDP of the United States grew at an average rate of 5.0 percent per year (adjusted for inflation), yet energy use grew at only 1.0% per year.

TABLE 13-1	Energy Use per Dollar of GDP (PPP; constant dollars)		
COUNTRY	GDP GROWTH RATE, 1980–2004 (PPP)	ENERGY GROWTH RATE, 1980–2004	PERCENT CHANGE IN ENERGY INTENSITY, 1980–2004
China	11.7	5.2	−61.4
United States	5.0	1.0	−38.5
Canada	4.6	1.4	−27.7
France	4.8	1.2	−17.7
Japan	4.9	1.7	−15.6
Zimbabwe	1.8	1.2	−10.7
India	6.9	5.6	−3.1
Mexico	3.6	2.4	−1.8
South Korea	9.0	7.2	8.7
Nigeria	3.3	3.7	18.4
Turkey	5.3	5.5	31.8
Malaysia	6.7	7.8	44.7
Thailand	7.7	8.3	67.7

Sources: World Resources Institute; Energy Information Administration.

In 2004 a unit of energy use produced about 62 percent more income in the United States than it did in 1980. Most of the wealthy industrial powers experienced a similar pattern. Income grew faster than energy use. In contrast, in many poorer countries, energy use grew faster than income. Thailand, for example, experienced economic growth averaging 7.7 percent per year (in constant U.S. dollars), but energy use grew at an average of 8.3 percent per year. In some poor countries, notably China, economic growth dramatically exceeded growth in energy use. Still, the general trend over that time period is for faster growth of energy use in the poorer world than in the wealthy world.

Energy use is an imperfect measure of pollution, but it is a useful one. In general, the more fossil fuels we use, the more pollution we produce. Certainly this is true for carbon dioxide emissions. For other pollutants, the correspondence is not so close, because pollution depends a lot on how the energy is used. Nevertheless, to the extent that energy use can be equated with pollution output, both energy use and pollution are increasing rapidly in the poor world, especially in those countries with high economic growth rates.

Development, Pollution, and the Quality of Life

Both monetary wealth and environmental conditions contribute to our quality of life. They contribute in different ways, however. Monetary income provides us with the ability to purchase sufficient quantities of good-quality food and electric or gas stoves for cooking, routine medical checkups, and safe water delivered to our homes. It also makes available automobiles, telephones, computers, and large, warm houses and vacations at the shore. Good environmental quality gives us clean air to breathe, fish that are free from

disease-carrying bacteria and toxins, and soil that produces good crops without excessive inputs of fertilizer. It also gives us diversity of wildlife to observe and enjoy, parks for relaxing, and the satisfaction of knowing that future generations will have at least some of the same opportunities to enjoy these things as we do.

Economic activity and environmental quality both contribute to make life materially and observably better in ways that virtually everyone agrees are important, especially health-related factors such as occurrence of disease, infant mortality, or longevity. Wealth helps prevent disease by paying for educational expenses and salaries of medical personnel, as well as the offices, laboratories, medicines, and supplies they need to do their jobs well. A clean environment also helps to prevent diseases such as chronic lung problems that occur in areas with bad air pollution. Wealth helps prevent infant mortality and improves longevity by making possible balanced diets that include sufficient protein and fresh vegetables regardless of season. A clean environment that yields drinking water that is free from chemical and biological contaminants also helps reduce infant mortality and lengthen lifespans.

While everyone needs good health and sufficient food supplies, the value of other things may vary considerably from one individual to another, or one culture to another. Clean water in a lake may be especially important to someone who likes to swim in lakes or who likes to eat fish. But for someone who spends every day in a city and prefers movies and basketball games for entertainment, clean lakes may not be very important. Wealth and environmental quality provide amenities that are valued differently by different cultures. Wealth brings us televisions, on which we can watch comedy or sports or the news. Such entertainment is a central part of some cultures today, but it would be hard to argue objectively that this makes us better people or that it makes our lives materially better than alternative

Water Privatization

As the population increases around the world, and as economic growth stimulates new demands for water, availability problems are increasing. Water supply problems are also worsened by pollution, which can make otherwise useful water unacceptable. Water supply problems—whether issues of quality or quantity or both—do have technological solutions. In most of the world water availability is strongly seasonal: there is a time of year when rivers have high flow and a time when flow is low. But human water use is less variable, so we often need to store water from the wet season in order to make it available in the dry season. We accomplish this by building reservoirs and the pipe or canal systems that deliver water from the reservoirs to where it is used.

Fortunately water, unlike oil or coal, doesn't lose its fundamental properties when it is used. The addition of pollutants that takes place when water goes through industrial or other uses can be reversed. We can increase water supply by treating polluted water to remove unhealthful substances so that it can be used again. In some areas total water use exceeds total renewable water supply (Figure 9-24). This is achieved by reusing water, and normally some water treatment is necessary before water can be reused.

Increasing water supplies by building reservoirs or treatment systems takes money. Typically the construction of water supply or treatment systems is undertaken by government agencies because water is something upon which everyone depends, and governments usually have the necessary financial resources. In addition, in most societies water is something that is commonly owned, as opposed to minerals or manufactured goods that are usually the property of miners or manufacturers before they are sold in a market.

Many of the world's governments, however, do not have sufficient capital to build new water systems. As a result, they are turning to private corporations to make the investments needed. This is especially true in poor countries, but governments in many wealthy regions are also finding it desirable to "outsource" construction and operation of water systems to private companies. These companies are motivated by the prospect of profits earned by selling water or water-related services (such as pipeline operation and maintenance) to water users. In other words, a resource that many consider to be publicly owned, to be provided to all at minimal cost, is being privatized.

Water privatization is going on all around the world. The key players are a handful of large multinational corporations, mainly owned in Europe or North America. They bid for and win contracts with government agencies to construct water systems and operate them, charging water users for the service. In many cases the fees these companies collect are substantial, and are greater than those that water users paid before privatization. It is inevitable that wealthier people will find it easier to pay for water than poor people, and so privatization leads to a situation where rich people's water needs are met while those of the poor are not.

In Bolivia, water privatization has been especially controversial, and the dispute over water contributed to a major political upheaval. In 1999, under pressure from the World Bank, Bolivia granted a 40-year contract to a subsidiary of London-based multinational International Water Ltd. to operate water systems serving a half-million people in Cochabamba. Water prices quickly rose, and widespread protests ensued. At first the protests were relatively peaceful, but they became increasingly violent as the government responded with troops and tear gas. After two months of demonstrations, during which hundreds were injured and six were killed, then Bolivian president Hugo Banzer Suarez finally cancelled the contract. The Cochabamba fiasco awakened many citizens in Bolivia to the privatization controversy, which continues today about the issue of the nationalization of the natural gas industry.

activities such as conversation, participant sports, or reading. A wilderness park can be an exhilarating place to hike, learn, or simply contemplate the beauty of nature, but who is to say that this is a more meaningful experience than attending an opera? The value of many of the goods we produce through extracting resources and polluting the environment depends solely on individual or cultural attitudes.

Balancing development and environmental preservation If there were simple and effective ways to increase both wealth and environmental quality simultaneously, then the process of development would be a straightforward one. Many of the paths to greater monetary wealth, however, actually reduce environmental quality, and improving environmental quality often has significant costs.

In the wealthy countries of North America and Western Europe, industrialization generated a standard of living and level of technology that were high enough so that people could give cleaner air a high priority. These countries could afford to make the needed investments in pollution control. Wealth created the pollution, as consumers bought the things that were made in the smoke-belching factories, but wealth also provided the impetus for eliminating the pollution.

In many poorer countries the situation is quite different. The contrast in wealth between rich and poor countries is enormous. People in poor countries are anxious to raise their standards of living, and rapid industrial development is seen as the quickest way to gain that wealth. Governments thus channel any available funds toward industrial development and actively seek foreign investment in new plants. Environmental protection may be a concern, but it is certainly a secondary one, so environmental regulations that would slow development are few and weak.

The industrial growth that is taking place in poorer countries is heavily concentrated in urban areas, where the physical, political, and economic infrastructure necessary to support it is most available. Seeking employment opportunities, people migrate to these cities from rural areas, thus swelling urban populations. The resulting urban growth is rapid and uncontrolled, and governments find it difficult to provide the basic services that urban dwellers in rich countries expect, such as safe drinking water, sewage collection and treatment, and solid-waste removal.

The result is that some of the worst air and water pollution in the world occurs not in the rich countries, which consume most of the world's resources, but in the poor ones. Many newly industrializing cities, such as Mexico City, Beijing, Cairo, Yangon, and São Paolo, have severe air-pollution problems. The situation for water pollution is no better. Only about 70 percent of the urban population of Asia is served by sewage collection; in Africa the figure is only 53 percent. Similarly, "safe" drinking water reaches only about 80 percent of urban dwellers in Asia and 70 percent in Africa.

In rural areas of poor countries air quality is usually better than in urban areas, but the problem persists. This is because indoor air pollution is perhaps even more a problem in rural areas than in urban ones. Clean energy sources like electricity or gas are rarely available in rural areas, and most people use wood, dung, or other biomass fuels for cooking and heating. These fuels are burned in simple stoves or fireplaces, often poorly ventilated. Therefore, chronic lung diseases are common, especially among women and children.

Can we trade wealth for environmental quality?
It is estimated that at least five million people, mostly children, die each year in Asia, Africa, and Latin America from diarrhea caused by contaminated drinking water. A similar number is estimated to die from chronic respiratory problems caused by poor air quality. These numbers are very crude estimates, but they certainly highlight some key issues regarding wealth, environmental quality, and the quality of life. Why do these deaths occur? Are they caused by poverty or pollution? How can they be prevented?

One way to eliminate these deaths would be to increase wealth. A small amount of money in the hands of a mother in Mumbai would allow her to buy kerosene to boil drinking water for her children, or it might buy simple rehydration therapies that would allow her children to survive the diarrhea. The availability of cheap electricity or gas in Beijing might allow people to cook on electric stoves or heat their homes with gas instead of burning coal, and thus reduce the severe particulate air pollution of that city. If, however, that increase in wealth were achieved through an intensification of coal-fueled industry with few emission controls, as it is in China and India, aren't we just trading one problem for another, with no net gain?

There are no simple answers. Clearly there is a need to increase wealth, especially in the poorer countries, while avoiding the environmental destruction that all too often accompanies that development. As discussed in Chapter 12, such an approach is usually called sustainable development. It has become the stated goal of virtually every governmental, nongovernmental, and international agency involved in promoting economic development.

Sustainable development as a concept makes sense, but making it a reality is quite another matter. Many people involved in promoting sustainability agree that the key is to assign realistic economic values to environmental attributes, and to make sure these values are considered in the development process. This is similar to calculating wealth in terms of gross sustainable product, as was discussed in Chapter 12, but doing so at the project level. For example, when a natural resource such as land covered in forest is converted to another use, say, agriculture, we normally think only of the positive value of the agricultural land created, and perhaps subtract the value of the timber production it replaced. But what of the value of the clean water that came from the forest and that has been replaced by streams polluted with agricultural runoff? What of the value of the biodiversity that may have been lost in the process? Planning projects and measuring their achievements in these terms may help to ensure that the values in the present environment are given due consideration when we plan modifications in the name of economic development.

The central tensions over globalization of economic activity and the liberalization of trade rules under the World Trade Organization (WTO) focus on the equitable distribution of costs and benefits of production. Those who argue that industrial production should not be shifted from wealthy nations to poor ones without adequate environmental protections feel that the benefits of wealth generated by manufacturing jobs are less than the costs endured by those living near the factories. From their perspective, the costs of production are being transferred from rich countries to poor ones while the benefits, in the form of profits and consumer goods, remain in rich countries. Those who argue that free trade is good for poor countries obviously see the benefits of accepting those factories as outweighing their environmental costs.

These tensions were prominent at the World Summit on Sustainable Development, held at Johannesburg, South Africa, in 2002. The summit focused on improving people's lives and conserving natural resources in a world that is experiencing increasing demands for food, water, shelter, sanitation, health services, and economic security. The conference was attended by leaders from around the world; the United States was represented by Secretary of State Colin Powell. When Powell spoke he stated that "the United States is taking action to meet environmental challenges, including global climate change," but a skeptical audience responded with jeers and heckling. Clearly many outside the United States feel that the most powerful and influential nation in the world bears a responsibility to address these issues more effectively.

International Equity in Environmental Management

Measuring the value of environmental quality and natural resources is difficult enough in itself, but it is even more difficult when we recognize that people in different countries value resources differently. Many North Americans and Europeans place a high value on tropical rain forest biodiversity and argue that deforestation should be halted in order to protect that diversity, even if it means curtailing some economic benefits of deforestation. The governments of most tropical nations, however, are much less concerned about biodiversity and more concerned with the economic gains to be made by exploiting the forest. For environmental issues of only regional or national concern, such issues are usually resolved at that level. But when we consider resources of global importance, international negotiations must resolve conflicts derived from differences in valuing resources. In this section we will discuss some examples of major environmental issues that have been the subjects of global-scale negotiations or that are likely to become so in the near future.

Ozone depletion and the Montreal Protocol

As we learned in Chapter 2, several trace gases in the atmosphere are important in absorbing radiation. One of these is ozone, a form of oxygen with three atoms per molecule instead of the usual two. Ozone is relatively abundant in the stratosphere, a layer of the atmosphere located between about 15,000 and 55,000 meters (50,000 and 180,000 feet) above sea level. Stratospheric ozone absorbs significant amounts of incoming solar radiation, especially in the ultraviolet wavelengths. Ultraviolet radiation is harmful to many things at Earth's surface, but our greatest concern about it is that it is a cause of skin cancer.

As early as the 1960s scientists began to raise concerns about the potential impact of air pollution on ozone concentrations in the stratosphere. One of the substances that was particularly worrisome is a class of chemicals called chlorofluorocarbons (CFCs). CFCs are human-made chemicals that are used in a wide variety of applications, including refrigeration and air-conditioning and many industrial processes.

The usefulness of CFCs stems from the fact that they are very stable and do not react with other substances readily. But this very stability allows them to remain in the atmosphere for a long time after they are released—long enough to reach the stratosphere. There they are broken down by the intense solar radiation. When the CFCs break down the chlorine atoms that are released enter reactions that contribute to a loss of ozone in the stratosphere.

In the 1970s there were strong theoretical reasons to believe that CFCs would harm stratospheric ozone, and these were sufficient to cause most nations to ban the use of CFCs as propellants in aerosol spray cans. It was not until the mid-1980s, however, that we accumulated clear evidence that damage to the ozone shield was in fact occurring, especially in high latitudes, and that this damage was attributable to CFCs.

When this became known, international action was obviously necessary. CFCs were being manufactured and used in many, mostly wealthy industrialized, countries. In 1987 an international agreement was reached that has since been ratified by over 125 countries. The agreement, called the Montreal Protocol, called for signatory nations to freeze CFC production at 1989 levels and then cut production 50 percent by 1999. Developing nations were allowed to increase CFC production for 10 more years, and they were not required to phase out CFCs until 2010. As additional evidence confirming damage to the ozone layer accumulated, the agreement was strengthened. Amendments in 1990 and 1992 accelerated the phaseout, resulting in a complete cessation of CFC production in rich countries (which accounted for about two-thirds of all production) in 1996. CFC emissions by the rich countries dropped faster than expected, and by 1997 levels of chlorine in the stratosphere became stabilized and began to decline. Production has risen more rapidly than expected in the poorer countries, however, and there are concerns that this increase may significantly offset the decreases achieved in the rich countries.

Despite these concerns, the Montreal Protocol is generally regarded as a success story in international environmental management. One of the factors in this success was that the principal nations involved had similar stakes in the outcome. (1) The countries for whom skin cancer represents the greatest health problem are generally those in which (a) other health problems, such as those associated with basic sanitation, are less critical, and (b) populations are mainly in high latitudes. It is also noteworthy that skin cancer is most common among fair-skinned people. Thus the wealthy high-latitude nations of Europe and North America are among those with the greatest concern about the negative impacts of ozone destruction. (2) The industrial powers of North America were the major producers and consumers of CFCs. These same wealthy nations were the ones with the greatest chemical research and development capabilities and were thus most likely to be able to develop substitute chemicals quickly. (3) Finally, the agreement established different provisions for poor nations than for rich ones. The two-tiered agreement recognized the greater need of poor countries to continue producing these useful chemicals, and perhaps, these countries' lower level of concern about their environmental impacts. Thus the countries that bore the greatest burden of complying with the terms of the treaty were also the ones that benefited the most from its enactment.

Global warming and CO₂ controls

Global warming and CO_2 controls   Among the international meetings that have been sponsored by the United Nations, two major international conferences stand out: the Conference on the Human Environment held in Stockholm in 1972 and the World Conference on Environment and Development (Earth Summit) held 20 years later, in 1992, in Rio de Janeiro. In both cases the contrasting agendas of rich and poor nations have been major barriers to international agreements and action to protect the global environment, although international friction was less in 1992 than in 1972.

At the 1992 Earth Summit in Rio de Janeiro, an international agreement was signed that committed most countries to undertake serious efforts to reduce global emissions of carbon dioxide and other greenhouse gases. That agreement did not require any country to actually reduce emissions. Rather, it was a promise to talk about reducing emissions, with the expectation that significant action would be agreed to within five years. Nonetheless, it marked the beginning of an ongoing effort to tackle this difficult international environmental issue.

In 1995 negotiators met in Berlin, and they took some small steps toward controlling CO_2. The conference resulted in the Berlin Mandate, which called upon the most industrialized nations to control emissions voluntarily through setting specific emission-reduction targets. It soon became clear that voluntary controls were not going to work for many of the largest emitters. The United States, in particular, is not prepared to shoulder the burden of significantly controlling fossil-fuel use. The European Union, however, committed itself to reducing carbon emissions to a level 15 percent below 1990 emissions by the year 2010. This commitment reflects many Europeans' genuine concern about the consequences of global warming. It was also facilitated by a widespread conversion from coal to natural gas and other cleaner sources of energy that had been under way for some time.

Who would be the winners and who would be the losers if carbon emissions were effectively controlled? We don't know, exactly, and that is part of the problem. The most vocal groups opposing carbon controls are those who benefit most from fossil-fuel use. These include the United States (the world's foremost oil-consuming nation), major oil-producing nations, and the oil industry in general. Arguments in favor of emissions controls come from environmentalists, but also from a coalition of insurance companies and small island nations. These interests are most threatened by sea-level rise. The insurance industry in general has become highly vulnerable to major losses caused by hurricanes and other storms, as discussed in Chapter 2. Such disasters are most likely in nations that have long coastlines and valuable investments such as resorts along those coasts. Hence small island nations like Jamaica, Trinidad, Samoa, and the Seychelles think that preventing global warming is very much in their interest.

Negotiations on reducing CO_2 emissions culminated in a conference at Kyoto, Japan, in December 1997. Under the pressure of the five-year deadline for agreement set in 1992, delegates struggled to hammer out an agreement that would bridge the enormous gap between countries that wanted major reductions in CO_2 output and countries that wanted no restrictions at all.

U.S. representatives went to the conference insisting they would agree only to reduce emissions to 1990 levels, and no lower, by 2010. By the year 2000 U.S. emissions were expected to be 13 percent higher than they were in 1990, so a reduction to 1990 levels by 2010 would be a substantial cut. The United States also demanded meaningful commitments to CO_2 controls on the part of poorer, less-industrialized countries that are expected to produce most of the growth in emissions in the next few decades. A third key U.S. requirement was that countries with high emissions be able to buy credit for reducing emissions by paying for emission reductions in other countries. It would probably be cheaper to pay Brazil to reduce deforestation, for example, than to curtail use of coal in the United States.

For their part, European representatives insisted on a 15 percent reduction in emissions below 1990 levels by 2010. Such a target seems highly ambitious, but Europe had already made significant progress in the early 1990s as a result economic changes there resulting from the collapse of the Soviet Union, combined with efforts to reduce acid rain. The small island nations also demanded substantial cuts in emissions. The less-industrialized bloc of countries refused to commit themselves to any reductions. They insisted that the

Focus ON

Malaria and DDT

In May of 2003 President Bush made a pledge of $15 billion toward fighting infectious diseases, and a major focus of that pledge was AIDS in Africa. While AIDS is indeed the most serious health issue in Africa, malaria remains a serious problem. Malaria kills about one million people per year in Africa, mostly children. Children who survive malaria often suffer from learning impairments or brain damage: The disease is also a leading cause of perinatal death, low birth weight, and maternal anemia.

Although several treatments for malaria are available after an individual has been infected, there is only one way to prevent infection and that is through preventing mosquito bites. And the most effective agent for controlling the mosquitoes is DDT, a pesticide invented in 1937 (its inventor won the Nobel Prize for his work) and used with great success ever since. While DDT has been banned in most developed nations, it is still widely used for mosquito control in many relatively poor tropical nations, where it is considered an essential tool in the battle against malaria.

As discussed in Chapter 4, DDT is a persistent pesticide—one that remains in the environment for a relatively long period of time. DDT is harmless to humans in the low concentrations in which it is normally found, but it is passed up the food chain, reaching lethal concentrations in some carnivores. DDT is one of a group of similar toxic chemicals known as persistent organic pollutants, or POPs. POPs can evaporate in relatively warm conditions, such as are common in the tropics, and be carried on the wind to distant locations where they can be redeposited on Earth's surface. Through this mechanism such chemicals have been redistributed throughout the globe. They are found in even the most remote places, including Arctic regions where they can accumulate because it is too cold for them to evaporate again.

People who live in Arctic regions depend largely on meat for their food, and they are traditionally hunters of seals, whales, and other carnivorous animals that, by their position at the top of the food chain, tend to have high concentrations of POPs. The problem is worsened by the fact that these people have diets that are high in fat, and POPs are most concentrated in animal fat. POPs have been found in mothers' breast milk among the Inuit of northern Canada at concentrations nine times higher than among women in southern Canada. Inuit people are now being warned to reduce their consumption of foods that are staples in their culture, particularly seal.

Health officials thus face a dilemma. The most effective available tool for fighting malaria in the tropics is also one that is poisoning the food upon which arctic people depend. The environmental rules banning DDT that are viewed by North Americans as essential may be wholly unacceptable to many tropical nations. Donor agencies in wealthy nations are reluctant to fund malaria-control programs that include the use of DDT. In 1995, under international pressure, South Africa eliminated DDT from its malaria programs. Mosquitoes quickly became resistant to the substitute chemical, and the number of malaria cases rose dramatically. In 2001, 90 countries signed a treaty to eliminate POPs from the environment, and in that treaty an exemption was made for the use of DDT in poorer tropical nations. South Africa resumed the use of DDT in 2001, and the number of cases quickly dropped to near their 1995 level. It remains to be seen whether the new U.S. initiative to improve health in Africa and elsewhere will be effective at controlling malaria in ways that do not cause DDT to continue to diffuse around the globe.

---

industrialized countries that had caused the problem were primarily responsible for controlling it within their own borders.

With such enormous differences among the countries represented, compromise on many issues was difficult or impossible. The conference finally did reach an agreement, though an incomplete one. The United States agreed to reduce emissions to 7 percent below 1990 levels by 2010. It was also agreed that the treaty would not take effect unless it were ratified by countries whose emissions totaled 55 percent of global emissions. That effectively gave the United States a veto over the treaty, because it contributes 22 percent of total emissions. Without U.S. participation, ratification would be nearly impossible.

Some issues, notably enforcement and the terms under which emission reductions could be traded internationally, were not resolved. The conferees set a deadline of late 1998 for agreement on these issues, but that deadline was ultimately extended to late 2000. Meetings were held in 1998 and 1999, and some issues regarding provisions for trading emissions were worked out at those meetings.

What was to be the final round of negotiations took place in The Hague, Netherlands, in 2000. One of the key remaining issues was whether or not countries would be able to take credit for carbon that was stored, or sequestered, in biomass through activities such as reforestation. The position of the United States, and other industrialized nations including Australia and Japan, was that such carbon storages should count against emissions. This was particularly important to the United States, because forest growth on abandoned farmland is storing a substantial amount of carbon there. The U.S. Senate had made it clear that U.S. ratification of the treaty was highly unlikely even if significant concessions to the U.S. position were made; nonetheless, there was a desire to reach an agreement so that a step toward emissions control, however small, would be made.

In 2000 the scientific evidence of the dangers of global warming became even stronger. The International Panel on Climate Change revised upward its predictions of future warming. A U.S. report documented the increased likelihood of weather-related disasters in a warmed climate, as described in Chapter 2. Then, in November 2000, just weeks before the final conference, a series of massive storms hit Great Britain, causing widespread flooding. The British government warned that this was a sign of things to come, and that more should be done to prepare for future disasters.

At The Hague conference, negotiations focused on the differences between the U.S. position (also supported by Australia, Canada, and Japan) and the European position. Europe insisted that carbon sequestration in forests not be allowed to count against emissions, and that real emissions reductions should be made. The United States, on the other hand, demanded that credits for sequestration in forests be allowed. Britain worked hard to forge a compromise between the United States and Europe, but in the end the parties failed to reach an agreement and negotiations broke down. Finally, in 2001, negotiations resumed. This time agreement was finally reached after Japan won concessions on accounting for carbon sequestration in forests. The result was seen by many as watered-down and ineffective, but it was an important step nonetheless. The United States did not participate in the negotiations in the final round, and is not expected to join the treaty any time soon, if ever. The Bush administration has since acknowledged that global warming is probably occurring and is probably caused by CO_2 emissions, but argues that the solution to the problem lies in adapting to climate change rather than in trying to prevent it.

Perhaps one reason why negotiations over CO_2 have proved so difficult, while those over CFCs were relatively easy, is that with CFCs the nations that would bear the costs of reducing pollution were also those that had much to lose (in the form of increased cancer rates). With CO_2, however, the United States would have to make significant changes in its habits; yet most U.S. citizens see little hazard in global warming.

Europeans and small island nations, on the other hand, perceive a major threat. Europe, as a major emitter of CO_2, is the only bloc of nations that both emits large amounts of CO_2 and perceives a significant benefit from controlling those emissions. Unfortunately the problem is a global one, and Europe acting on its own will not be enough to control global warming.

CONCLUSION: CRITICAL ISSUES FOR THE FUTURE

As we consider the many consequences of global integration and greater cultural and economic interaction between rich and poor nations, the greatest challenges will be those arising from differences in the value systems, both economic and cultural, of the people involved. When Brazilian coffee growers and Chicago commodity traders are linked electronically so they are only seconds away, and if their governments are both members of trade pacts and other international agreements, then they become part of the same political–economic system. That joint membership requires that they share certain values and agree to operate by the same set of rules. The political gulf between the United States and China, on the other hand, is much larger. These two nations are the two largest emitters of carbon dioxide in the world, thus making it more difficult for them to reach agreement on controlling global warming. The enormous economic gaps between rich and poor, both between and increasingly within countries, only multiply the challenges posed by increased global interaction. Will economic liberalization be the key to economic development in the poorer nations of Africa, Asia, and Latin America? Will this economic development be rapid enough to reduce the disparities in national income between these countries and the rich countries? Or will it perpetuate and strengthen the inequities that result from the greater ability of rich countries to capture value added? And will economic development benefit the lower economic strata of the societies it affects, or only those already relatively well off? These questions remain to be answered. Finally, as we grapple with the many social and political tensions created by global change and integration, will we have the opportunity to protect what remains of the natural environment? Or will we instead be so focused on economic development alone that the idea of sustainable development will remain more an idea than a reality?

These questions cannot be answered today. In searching for the answers, however, it is clear that we need to consider a wide range of topics simultaneously: climate, resources, culture, religion, politics, and economics. Such a wide-ranging viewpoint is open to students of any subject, but geography especially encourages such broad-minded analysis.

Chapter Review

SUMMARY

Through the twentieth century, the empires of the European powers and the United States have fragmented. The U.S.S.R.—the Russian Empire in a new name—collapsed only in 1991. New economic blocs are appearing, most notably the European Union and the North American Free Trade Area, but also MERCOSUR in South America and the Association of Southeast Asian Nations. Other international groups seek cultural or political goals.

The United Nations was created after World War II. Today it serves as a forum for international negotiation and coordination of peacekeeping activities, and as a means of organizing social programs, economic development programs, and other international cooperative undertakings. The Law of the Sea Treaty governs activities in the oceans.

Nongovernmental organizations have assumed a leading role on the world stage. Many of these are charitable, but others are guilty of international terrorist acts. Some of these are based in collapsing states, but others may be actively encouraged by hostile states. Under such conditions, the United States and other powers are newly questioning the degree to which any state's territorial sovereignty must be respected absolutely.

Economic growth in the twentieth century has caused dramatic increases in energy use and pollution output, threatening the global environment. International negotiations focus on balancing the need for economic development with concerns for environmental protection. Rich countries focus their attention on pollution control, but poor nations are concerned with their need for immediate economic development. International agreements have been reached on ozone-destroying substances, but efforts to control predicted global warming are likely to be much more difficult.

KEY TERMS

bloc p. 534
common market p. 544
continental shelf p. 548
customs union p. 544
European Union (EU) p. 544
free-trade area p. 544

innocent passage p. 569
international organization
  p. 534
local content requirement p. 553
maquiladora p. 553
Marshall Plan p. 544

nongovernmental organizations
  (NGOs) p. 559
North Atlantic Treaty Organization
  (NATO) p. 543
supranational organization
  p. 534

QUESTIONS FOR REVIEW AND DISCUSSION

1. What small remnants of the British Empire present political problems today?

2. Describe the economic bonds holding together the U.S.S.R. until 1991.

3. What independent countries today were ever ruled by the United States?

4. Name the members of the European Union. Describe the government of the European Union.

5. What activities and results of the European Union prove that it is more than a free-trade area or customs union, but rather is a full common market?

6. Is NAFTA a free-trade area, customs union, or common market?

7. What are the major branches of the United Nations and what are the activities of each?

8. What are territorial waters and what is an EEZ? Give several reasons why these delineations are disputed.

THINKING GEOGRAPHICALLY

1. Investigate and analyze the assignment of seats in the European Parliament. Does it represent population? Economic power?

2. International diplomacy generally observes the principle of nonintervention in sovereign states, and this principle often preserves misgovernment and unnecessary suffering. On what grounds could the principle of nonintervention be overturned?

3. List factors of European cultural unity and disunity. As the European Union expands, how will various factors boost or retard cooperation among the members and economic development?

4. Most Canadian citizens know more about the United States than most U.S. citizens know about Canada. Why?

5. Do you think that each country should have an equal vote in the U.N. General Assembly, or would you recommend a weighted system of voting—by population, wealth, or territory?

6. Investigate a few of the projects of the specialized agencies of the United Nations, such as UNICEF, the FAO, or WHO.

7. French television commentator Christine Ockrent has said, "The only true pan-European culture is American culture." What did she mean by that? Can we say that the only true global culture is American culture?

8. How has physical geography encouraged or discouraged the formation of multinational bonds in Europe, in the CIS states, or in North America? Are there obvious complementarities in products? Do transportation or settlement patterns act as centripetal forces?

9. What might happen to Antarctica if a practical way is found for extracting its resources?

SUGGESTIONS FOR FURTHER LEARNING

Berezin, Mabel, Martin Schain, and Roland Axtmann. *Re-Mapping Europe: Territory, Membership and Identity in a Transnational Age*. Baltimore: Johns Hopkins, 2004.

Blair, Alasdair. *Companion to the European Union*. New York: Routledge, 2006.

Canadian Journal of Regional Science. Vol. 13:2/3. Special double issue devoted to the spatial impact of the U.S.–Canadian free trade agreement. Québec: Université du Québec.

Chaturvedi, Sanjay. *The Polar Regions: A Political Geography*. New York: John Wiley & Sons, 1997.

Churchill, R. R., and A. V. Lowe. *The Law of the Sea*. New York: Juris Publishing, Inc., 3rd ed. 1999.

Combating Terrorism: Axis of Evil, Multilateral Containment or Unilateral Confrontation. Hearings before the Subcommittee on National Security. Washington, D.C.: U.S. Government Printing Office, 2003.

Eilstrup-Sangiovanni, Mette, ed. *Debates on European Integration: A Reader*. London: Palgrave Macmillan, 2006.

Forsythe, David P., Roger Coate, and Thomas G. Weiss. *The United Nations and Changing World Politics*. New York: Westview Publishing, 2000.

Freeland, Chrystia. *The Sale of the Century: Russia's Wild Ride from Communism to Capitalism*. New York: Crown Business, 2000.

Funck, Bernard, ed. *European Integration, Regional Policy, and Growth*. Washington, D.C.: The World Bank, 2003.

Hakim, Peter, ed. *The Future of North American Integration: Beyond NAFTA*. Washington, D.C.: The Brookings Institution, 2002.

Jacoby, Wade. *The Enlargement of the European Union and NATO*. Cambridge, UK: Cambridge University Press, 2004.

Kennedy, Paul. *The Parliament of Man: The Past, Present, and Future of the United Nations*. New York: Random House, 2006.

Kuper, Andrew. *Democracy Beyond Borders: Justice and Representation in Global Institutions*. New York: Oxford University Press, 2006.

Langewiesche, William. *The Outlaw Sea: A World of Freedom, Chaos, and Crime*. New York: North Point Press, 2004.

Macmillan, Margaret. *Paris 1919: Six Months That Changed the World*. New York: Random House, 2001.

Nations of the Earth Report (a three-volume study of the Earth environment country by country). Geneva: United Nations, 1992.

Prescott, John Robert Victor. *Maritime Political Boundaries of the World*. New York: Methuen, 1986.

We the Peoples: The Role of the United Nations in the Twenty-first Century (report of the Secretary General). United Nations Document A/54/2000.

WEB WORK

The United Nations home page will take you to a tremendous variety of economic, political, cultural, and social organizations:

http://www.un.org/

The United Nations High Commissioner for Human Rights is at:

http://www.unhchr.ch/

The United Nations Environment Programme is at:

http://www.unep.org/

Information about the European Union can be found by starting at its home page at:

http://europa.eu.int/

The International Criminal Court is at:

http://www.un.org/law/icc/

Information on the law of the sea and ocean affairs is provided by the Council on Ocean Law at:

http://www.oceanlaw.org/

A number of world organizations have rich websites. Among the good ones are NATO at:

http://www.nato.int/

and the Commonwealth of Nations at:

http://www.thecommonwealth.org/

and the Organization for Economic Cooperation and Development at:

http://www.oecd.org/

The U.S. Department of Commerce has a special page devoted to NAFTA affairs at:

http://www.mac.doc.gov/nafta/

and the home page of the NAFTA Commission on Environmental Cooperation can be found at:

http://www.cec.org/

APPENDIX I
MAP SCALE AND PROJECTIONS
Phillip C. Muehrcke

Unaided, our human senses provide a limited view of our surroundings. To overcome those limitations, humankind has developed powerful vehicles of thought and communication, such as language, mathematics, and graphics. Each of those tools is based on elaborate rules, each has an information bias, and each may distort its message, often in subtle ways. Consequently, to use those aids effectively, we must understand their rules, biases, and distortions. The same is true for the special form of graphics we call maps: We must master the logic behind the mapping process before we can use maps effectively.

A fundamental issue in cartography, the science and art of making maps, is the vast difference between the size and geometry of what is being mapped—the real world, we will call it—and that of the map itself. Scale and projection are the basic cartographic concepts that help us understand that difference and its effects.

MAP SCALE

Our senses are dwarfed by the immensity of our planet; we can sense directly only our local surroundings. Thus, we cannot possibly look at our whole state or country at one time, even though we may be able to see the entire street where we live. Cartography helps us expand what we can see at one time by letting us view the scene from some distant vantage point. The greater the imaginary distance between that position and the object of our observation, the larger the area the map can cover but the smaller the features will appear on the map. That reduction is defined by the *map scale*, the ratio of the distance on the map to the distance on Earth. Map users need to know about map scale for two reasons: (1) so that they can convert measurements on a map into meaningful real-world measures and (2) so that they can know how abstract the cartographic representation is.

Real-World Measures

A map can provide a useful substitute for the real world for many analytical purposes. With the scale of a map, for instance, we can compute the actual size of its features (length, area, and volume). Such calculations are helped by three expressions of a map scale: a word statement, a graphic scale, and a representative fraction.

A *word statement* of a map scale compares X units on the map to Y units on Earth, often abbreviated "X unit to Y units."

For example, the expression "1 inch to 10 miles" means that 1 inch on the map represents 10 miles on Earth (Figure A-1). Because the map is always smaller than the area that has been mapped, the ground unit is always the larger number. Both units are expressed in meaningful terms, such as inches or centimeters and miles or kilometers. Word statements are not intended for precise calculations but give the map user a rough idea of size and distance.

A *graphic scale,* such as a bar graph, is concrete and therefore overcomes the need to visualize inches and miles that is associated with a word statement of scale (see Figure A-1). A graphic scale permits direct visual comparison of feature sizes and the distances between features. No ruler is required; any measuring aid will do. It needs only to be compared with the scaled bar; if the length of 1 toothpick is equal to 2 miles on the ground and the map distance equals the length of 4 toothpicks, then the ground distance is 4 times 2, or 8 miles. Graphic scales are especially convenient in this

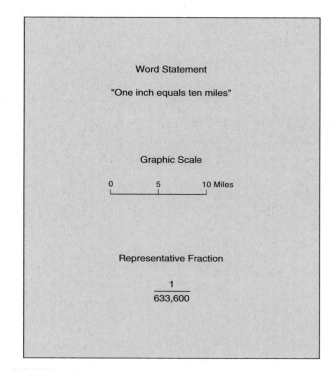

FIGURE A-1 Common expressions of map scale.

age of copying machines, when we are more likely to be working with a copy than with the original map. If a map is reduced or enlarged as it is copied, the graphic scale will change in proportion to the change in the size of the map and thus will remain accurate.

The third form of a map scale is the *representative fraction* (RF). An RF defines the ratio between the distance on the map and the distance on Earth in fractional terms, such as $\frac{1}{633,600}$ (also written 1/633,600 or 1:633,600). The numerator of the fraction always refers to the distance on the map, and the denominator always refers to the distance on Earth. No units of measurement are given, but both numbers must be expressed in the same units. Because map distances are extremely small relative to the size of Earth, it makes sense to use small units, such as inches or centimeters. Thus, the RF 1:633,600 might be read as "1 inch on the map to 633,600 inches on Earth."

Herein lies a problem with the RF. Meaningful map-distance units imply a denominator so large that it is impossible to visualize. Thus, in practice, reading the map scale involves an additional step of converting the denominator to a meaningful ground measure, such as miles or kilometers. The unwieldy 633,600 becomes the more manageable 10 miles when divided by the number of inches in a mile (63,360).

On the plus side, the RF is good for calculations. In particular, the ground distance between points can be easily determined from a map with an RF. One simply multiplies the distance between the points on the map by the denominator of the RF. Thus, a distance of 5 inches on a map with an RF of 1/126,720 would signify a ground distance of 5 × 126,720, which equals 633,600. Because all units are inches and there are 63,360 inches in a mile, the ground distance is 633,600 ÷ 63,360, or 10 miles. Computation of area is equally straightforward with an RF. Computer manipulation and analysis of maps is based on the RF form of map scale.

Guides to Generalization

Scales also help map users visualize the nature of the symbolic relation between the map and the real world. It is convenient here to think of maps as falling into three broad scale categories (Figure A-2). (Do not be confused by the use of the words *large* and *small* in this context; just remember that the larger the denominator, the smaller the scale ratio and the larger the area that is shown on the map.) Scale ratios greater than 1:100,000, such as the 1:24,000 scale of U.S. Geological Survey topographic quadrangles, are large-scale maps. Although those maps can cover only a local area, they can be drawn to rather rigid standards of accuracy. Thus, they are useful for a wide range of applications that require detailed and accurate maps, including zoning, navigation, and construction.

At the other extreme are maps with scale ratios of less than 1:1,000,000, such as maps of the world that are

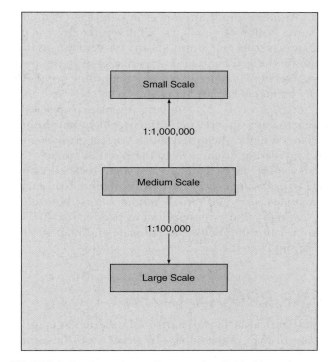

FIGURE A-2 The scale gradient can be divided into three broad categories.

found in atlases. Those are small-scale maps. Because they cover large areas, the symbols on them must be highly abstract. They are therefore best suited to general reference or planning, when detail is not important. Medium- or intermediate-scale maps have scales between 1:100,000 and 1:1,000,000. They are good for regional reference and planning purposes.

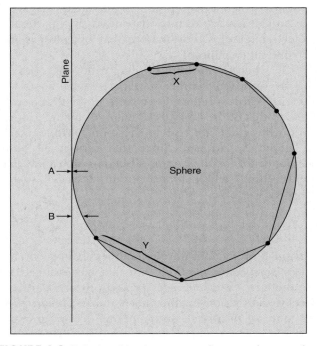

FIGURE A-3 Relationships between surfaces on the round Earth and a flat map.

Another important aspect of map scale is to give us some notion of geometric accuracy; the greater the expanse of the real world shown on a map, the less accurate the geometry of that map is. Figure A-3 shows why. If a curve is represented by straight line segments, short segments (*X*) are more similar to the curve than are long segments (*Y*). Similarly, if a plane is placed in contact with a sphere, the difference between the two surfaces is slight where they touch (*A*) but grows rapidly with increasing distance from the point of contact (*B*). In view of the large diameter and slight local curvature of Earth, distances will be well represented on large-scale maps (those with small denominators) but will be increasingly poorly represented at smaller scales. This close relationship between map scale and map geometry brings us to the topic of map projections.

MAP PROJECTIONS

The spherical surface of Earth is shown on flat maps by means of map projections. The process of "flattening" Earth is essentially a problem in geometry that has captured the attention of the best mathematical minds for centuries. Yet no one has ever found a perfect solution; there is no known way to avoid spatial distortion of one kind or another. Many map projections have been devised, but only a few have become standard. Because a single flat map cannot preserve all aspects of Earth's surface geometry, a mapmaker must be careful to match the projection with the task at hand. To map something that involves distance, for example, a projection should be used in which distance is not distorted. In addition, a map user should be able to recognize which aspects of a map's geometry are accurate and which are distortions caused by a particular projection process. Fortunately, that objective is not too difficult to achieve.

It is helpful to think of the creation of a projection as a two-step process (Figure A-4). First, the immense Earth is reduced to a small globe with a scale equal to that of the desired flat map. All spatial properties on the globe are true to those on Earth. Second, the globe is flattened. Since that cannot be done without distortion, it is accomplished in such a way that the resulting map exhibits certain desirable spatial properties.

Perspective Models

Early map projections were sometimes created with the aid of perspective methods, but that has changed. In the modern electronic age, projections are normally developed by strictly mathematical means and are plotted out or displayed on computer-driven graphics devices. The concept of perspective is still useful in visualizing what map projections do, however. Thus,

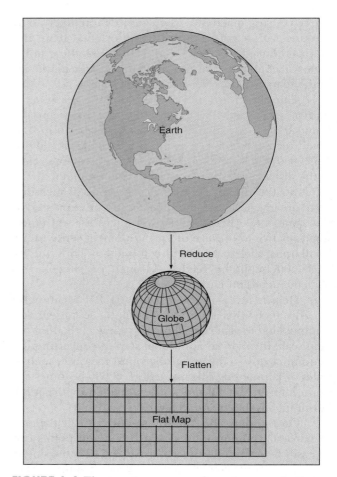

FIGURE A-4 The two-step process of creating a projection.

projection methods are often illustrated by using strategically located light sources to cast shadows on a projection surface from a latitude/longitude net inscribed on a transparent globe.

The success of the perspective approach depends on finding a projection surface that is flat or that can be flattened without distortion. The cone, cylinder, and plane possess those attributes and serve as models for three general classes of map projections: *conic, cylindrical,* and *planar* (or *azimuthal*). Figure A-5 shows those three classes, as well as a fourth, a false cylindrical class

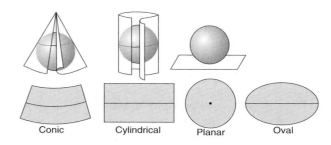

FIGURE A-5 General classes of map projections. (Courtesy of ACSM)

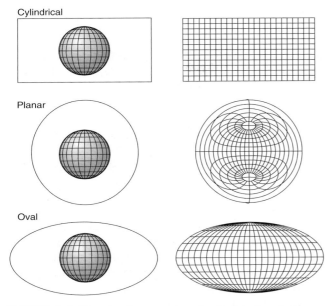

FIGURE A-6 The visual properties of cylindrical and planar projections combined in oval projections. (Courtesy of ACSMJ)

with an oval shape. Although the *oval* class is not of perspective origin, it appears to combine properties of the cylindrical and planar classes (Figure A-6).

The relationship between the projection surface and the model at the point or line of contact is critical because distortion of spatial properties on the projection is symmetrical about, and increases with distance from, that point or line. That condition is illustrated for the cylindrical and planar classes of projections in Figure A-7. If the point or line of contact is changed to some other position on the globe, the distortion pattern will be recentered on the new position but will retain the same symmetrical form. Thus, centering a projection on the area of interest on Earth's surface can minimize the effects of projection distortion.

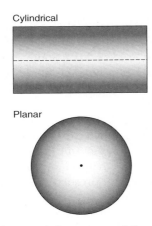

FIGURE A-7 Characteristic patterns of distortion for two projection classes. Here, darker shading implies greater distortion. (Courtesy of ACSM)

And recognizing the general projection shape, associating it with a perspective model, and recalling the characteristic distortion pattern will provide the information necessary to compensate for projection distortion.

Preserved Properties

For a map projection to truthfully depict the geometry of Earth's surface, it would have to preserve the spatial attributes of *distance, direction, area, shape,* and *proximity.* That task can be readily accomplished on a globe, but it is not possible on a flat map. To preserve area, for example, a mapmaker must stretch or shear shapes; thus, area and shape cannot be preserved on the same map. To depict both direction and distance from a point, area must be distorted. Similarly, to preserve area as well as direction from a point, distance has to be distorted. Because Earth's surface is continuous in all directions from every point, discontinuities that violate proximity relationships must occur on all map projections. The trick is to place those discontinuities where they will have the least impact on the spatial relationships in which the map user is interested.

We must be careful when we use spatial terms, because the properties they refer to can be confusing. The geometry of the familiar plane is very different from that of a sphere; yet, when we refer to a flat map, we are in fact making reference to the spherical Earth that was mapped. A shape-preserving projection, for example, is truthful to local shapes—such as the right-angle crossing of latitude and longitude lines—but does not preserve shapes at continental or global levels. A distance-preserving projection can preserve that property from one point on the map in all directions or from a number of points in several directions, but distance cannot be preserved in the general sense that area can be preserved. Direction can also be generally preserved from a single point or in several directions from a number of points but not from all points simultaneously. Thus, a shape-, distance-, or direction-preserving projection is truthful to those properties only in part.

Partial truths are not the only consequence of transforming a sphere into a flat surface. Some projections exploit that transformation by expressing traits that are of considerable value for specific applications. One of those is the famous shape-preserving *Mercator projection* (Figure A-8). That cylindrical projection was derived mathematically in the 1500s so that compass bearing (called rhumb lines) between any two points on Earth would plot as straight lines on the map. That trait let navigators plan, plot, and follow courses between origin and destination, but it was achieved at the expense of extreme areal distortion toward the margins of the projection (see Antarctica at the bottom of

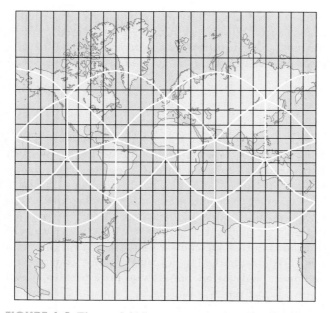

FIGURE A-8 The useful Mercator projection, showing extreme area distortion in the higher latitudes. The white lines are great circles—the shortest route between two points on Earth's surface. (Courtesy of ACSM)

Figure A-8). Although the Mercator projection is admirably suited for its intended purpose, its widespread but inappropriate use for nonnavigational purposes has drawn a great deal of criticism.

The *gnomonic projection* is also useful for navigation. It is a planar projection with the valuable characteristic of showing the shortest (or great circle) route between any two points on Earth as straight lines. Long-distance navigators first plot the great circle course between origin and destination on a gnomonic projection (Figure A-9, top). Next they transfer the straight line to a Mercator projection, where it normally appears as a curve (Figure A-9, bottom). Finally, using

straight-line segments, they construct an approximation of that course on the Mercator projection. Navigating the shortest course between origin and destination then involves following the straight segments of the course and making directional corrections between segments. Like the Mercator projection, the specialized gnomonic projection distorts other spatial properties so severely that it should not be used for any purpose other than navigation or communications.

Projections Used in Textbooks

Although a map projection cannot be free of distortion, it can represent one or several spatial properties of Earth's surface accurately if other properties are sacrificed. The two projections used for world maps throughout this textbook illustrate that point well. Goode's homolosine projection, shown in (Figure A-10), belongs to the oval category and shows area accurately, although it gives the impression that Earth's surface has been torn, peeled, and flattened. The interruptions in Figure A-10 have been placed in the major oceans, giving continuity to the land masses. Ocean areas could be featured instead by placing the interruptions in the continents. Obviously, that type of interrupted projection severely distorts proximity relationships. Consequently, in different locations the properties of distance, direction, and shape are also distorted to varying degrees. The distortion pattern mimics that of cylindrical projections, with the equatorial zone the most faithfully represented (Figure A-11).

An alternative to special-property projections such as the equal-area Goode's homolosine is the compromise projection. In that case no special property is achieved at the expense of others, and distortion is rather evenly distributed among the various properties, instead of being focused on one or several properties. The *Robinson projection*, which is also used in this textbook, falls into that category (Figure A-12). Its oval projection has a global feel, somewhat like that of Goode's homolosine. But the Robinson projection shows the North Pole and the South Pole as lines that are slightly more than half the length of the equator, thus exaggerating distances and areas near the poles. Areas look larger than they really are in the high latitudes (near the poles) and smaller than they really are in the low latitudes (near the equator). In addition, not all latitude and longitude lines intersect at right angles, as they do on Earth, so we know that the Robinson projection does not preserve direction or shape either. However, it has fewer interruptions than the Goode's homolosine does, so it preserves proximity better. Overall, the Robinson projection does a good job of representing spatial relationships, especially in the low to middle latitudes and along the central meridian.

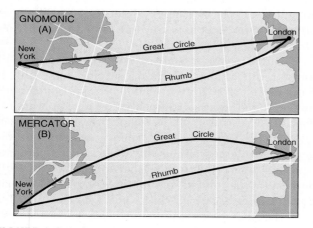

FIGURE A-9 A gnomonic projection (A) and a Mercator projection (B). (A rhumb line describes the course of a ship that crosses all meridians of longitude at the same angle.)

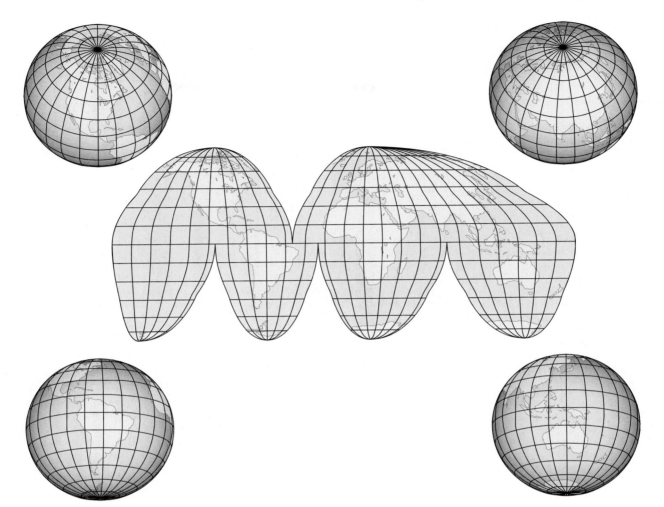

FIGURE A-10   An interrupted Goode's homolosine, an equal-area projection. (Courtesy of ACSM)

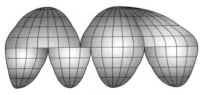

FIGURE A-11   The distortion pattern of the interrupted Goode's homolosine projection, which mimics that of cylindrical projections. As in Figure A-7, the darker shading implies greater distortion. (Courtesy of ACSM)

SCALE AND PROJECTIONS IN MODERN GEOGRAPHY

Computers have drastically changed the way in which maps are made and used. In the preelectronic age, maps were so laborious, time-consuming, and expensive to make that relatively few were created. Frustrated, geographers and other scientists often found themselves trying to use maps for purposes not intended by the map designer. But today anyone with access to computer mapping facilities can create projections in a flash. Thus, projections will be increasingly tailored to specific needs, and more and more scientists will do their own mapping rather than have someone else guess what they want in a map.

Computer mapping creates opportunities that go far beyond the construction of projections, of course. Once maps and related geographical data are entered into computers, many types of analyses can be carried out involving map scales and projections. Distances, areas, and volumes can be computed; searches can be conducted; information from different map-scan be combined; optimal routes can be selected; facilities can be allocated to the most suitable sites; and so on. The term used to describe such processes is *geographical information system*, or GIS (Figure A-13). Within a GIS, projections provide the mechanism for linking data from different sources, and scale provides the basis for size calculations of all sorts. Mastery of both projection and scale becomes the user's responsibility because the map user is also the mapmaker. Now more than ever, effective geography depends on knowledge of the close association between scale and projection.

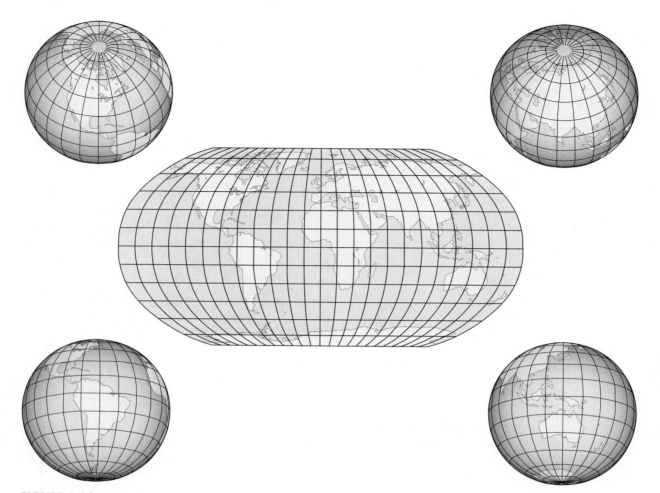

FIGURE A-12 The compromise Robinson projection, which avoids the interruptions of Goode's homolosine but preserves no special properties. (Courtesy of ACSM)

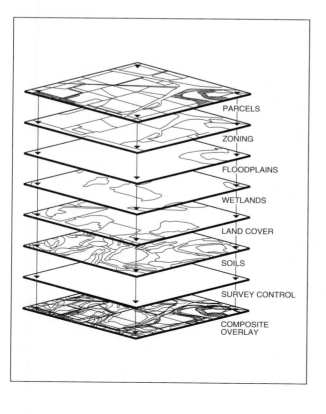

FIGURE A-13 Within a GIS, environmental data attached to a common terrestrial reference system, such as latitude/longitude, can be stacked in layers for spatial comparison and analysis.

APPENDIX II
THE KÖPPEN CLIMATE CLASSIFICATION SYSTEM

Climate classification is a challenging intellectual and scientific enterprise. Climates vary in many different ways, the most significant of which are temperature, precipitation, and seasonal variations in these variables. Each of these dimensions of variation should be portrayed by a useful classification system.

In addition to having multiple criteria that must be considered, none of these variables falls into simple categories, like male and female. Temperature and precipitation vary through wide ranges, with gradual transitions from place to place. Climate classification is a little like naming colors. Most people would agree on what is meant by "blue," and what is meant by "green." Finding agreement on the difference between bluish green and greenish blue (not to mention names like "turquoise" or "teal") is not so simple.

Wladimir Köppen (1846–1940) was a German climatologist who undertook this difficult task. Köppen recognized that there is a strong correlation between climate and vegetation types, and he reasoned that if climate types corresponded to significant vegetation types then the classification would be more useful and meaningful. The correlation between vegetation and climate is also useful in mapping climates in areas where few weather data are available but where vegetation can be mapped. In this book we use a modified version of Köppen's system, summarized in this appendix.

Köppen's approach was to identify five basic climate types and name them A, B, C, D, and E, in rough order of their latitudinal location on Earth (A in the tropics; E at the poles). Four of these are humid climate types and one, B, is an arid type. Each is divided into subtypes, with the second letter providing information about rainfall, and the third indicating temperature, except for polar climates in which only two letters are used and the second letter pertains to temperature. All climates are distinguished on the basis of mean monthly data.

For the humid climates, the second letter of the classification (f, s, m, or w) is used to describe the seasonal distribution of rainfall. In midlatitude climates,

f indicates evenly distributed rainfall, s indicates summer dry, and w indicates winter dry. In tropical (A) climates, the letter m is used to indicate a monsoonal precipitation regime, and the summer (or high-sun) dry type, s, does not exist. The third letter (a, b, c, or d) is used to distinguish warmer and cooler variations of the subtropical and midlatitude (C and D) climates, with a being the warmest and d the coolest.

For dry climates (B), the second letter (W or S) distinguishes desert from semiarid climates. The specific criteria for distinguishing these two from humid climates and from each other are based on mathematical formulas comparing temperature and rainfall. This is because plant water needs are temperature dependent. In a cool climate less rainfall is needed to support vegetation than in a warm climate. This is also the reason for differences in rules for identifying "winter dry" and "summer dry" climates.

Some of the climates distinguished by Köppen do not relate clearly to vegetation differences. In eastern North America, for example, the transition from Dfa to Dfb climates passes through the middle of the broadleaf deciduous forest region; neither does the line distinguish any major agricultural regions. However, the boundary between Dfb and Dfc corresponds relatively closely to the northern limit of commercial agriculture in this area.

In this book we have simplified Köppen's scheme somewhat, especially in Figure 2-32 and the section describing climate types. We chose to identify 11 different major climate types that correspond reasonably well to significant vegetation differences. These types are named humid tropical, seasonally humid tropical, desert, semiarid, humid subtropical, marine west coast, Mediterranean, humid continental, subarctic, tundra, and ice cap and ice sheets. Each of these corresponds to one to four subtypes of a Köppen major type. Our simplifications are designed to retain distinctions that relate clearly to natural vegetation and agricultural regions, while eliminating those that divide major vegetation regions.

LETTER						
FIRST	SECOND	THIRD	DERIVATION	DESCRIPTION	DEFINITION	CLIMATE TYPES AS USED IN THIS BOOK
A			Alphabetical	Tropical humid climates	Average temperature of each month above 18° C (64° F)	
	f		German: *feucht* ("moist")	No dry season	Average rainfall of each month at least 6 cm (2.5 in.)	Humid tropical (Af)
	m		Monsoon	Monsoonal; short dry season compensated by heavy rains in other months	1 to 3 months with average rainfall less than 6 cm (2.5 in.)	Humid tropical (Am)
	w		Winter dry	Dry season in low-sun season	3 to 6 months with average rainfall less than 6 cm (2.5 in.)	Seasonally humid tropical (Aw)
B			Alphabetical	Dry climates; potential evaporation exceeds precipitation	Dry climates are determined by a set of formulas relating temperature and precipitation. Different formulas are used depending on whether precipitation is concentrated in the summer, winter, or neither. If the wettest winter month has three or more times the rainfall of the driest summer month, the climate is considered to have a winter concentration of rainfall. If the wettest summer month has 10 or more times the rainfall of the driest winter month, the climate is considered to have a summer concentration of rainfall. If neither of these is true, rainfall is considered evenly distributed.	
	W		German: *wuste* ("desert")	Arid climates	If precipitation is concentrated in the high-sun season, average annual precipitation (cm) less than mean annual temperature (°C). If precipitation is concentrated in the low-sun season, average annual precipitation (cm) less than mean annual temperature (°C) plus 14. If precipitation is evenly distributed through the year, average annual precipitation (cm) less than mean annual temperature (°C) plus 7.	Desert
	S		Steppe, or semiarid	Semiarid climates	If precipitation is concentrated in the high-sun season, average annual precipitation (cm) less than 2 times mean annual temperature (°C). If precipitation is concentrated in low-sun season, average annual precipitation (cm) less than 2 times mean annual temperature (°C) plus 14. If precipitation is evenly distributed through the year, average annual precipitation (cm) less than 2 times mean annual temperature (°C) plus 7.	Semiarid
		h	German: *heiss* ("hot")	Low-latitude dry climate	Average annual temperature above 18°C (64°F)	Hot desert (BWb) or hot semiarid (steppe) climate (BSk)
		k	German: *kalt* ("cold")	Midlatitude dry climate	Average annual temperature below 18°C (64°F)	Cool desert (BWk) or cool semiarid (steppe) climate (BSk)
C			Alphabetical	Midlatitude humid climates	Average temperature of coldest month between 18°C (64°F) and −3°C (27°F)	

(continued)

LETTER						
FIRST	SECOND	THIRD	DERIVATION	DESCRIPTION	DEFINITION	CLIMATE TYPES AS USED IN THIS BOOK
	s		Summer dry	Dry summer	Driest summer month has less than one-third the precipitation of the wettest winter month	Mediterranean (Csa, Csb)
	w		Winter dry	Dry winter	Driest winter month has less than one-tenth the precipitation of the wettest summer month	Humid subtropical (Cwa)
	f		German: *feucht* "moist"	No dry season	Does not fit either s or w above	Marine west coast (Cfb, Cfc)
		a	Alphabetical	Hot summers	Average temperature of warmest month above 22°C (72°F)	Humid subtropical (Cfa)
		b	Alphabetical	Warm summers	Average temperature of warmest month below 22°C (72°F); at least 4 months with average temperature above 10°C (50°F)	Marine west coast (Cfb, Cfc)
		c	Alphabetical	Cool summers	Average temperature of warmest month below 22°C (72°F); fewer than 4 months with average temperature above 10°C (50°F)	
D			Alphabetical	Humid midlatitude climates with cold winters	4 to 8 months with average temperatures above 10°C (50°F)	Humid continental (Dfa, Dfb, Dwa, Dwb); Subarctic (Dfc, Dfd, Dwc, Dwd)

Second and third letters same as in C climates

FIRST	SECOND	THIRD	DERIVATION	DESCRIPTION	DEFINITION	CLIMATE TYPES AS USED IN THIS BOOK
		d	Alphabetical	Very cold winters	Average temperature of coldest month below −38°C (−36°F)	
E			Alphabetical	Polar climates	No month with average temperature above 10°C (50°F)	
	T		Tundra	Tundra climates	At least one month with average temperature above 0°C (32°F)	Tundra (ET)
	F		Frost	Ice cap climates	No month with average temperature above 0°C (32°F)	Ice cap and ice sheets (EF)

	TOTAL POPULATION (MILLIONS— 2004)	AVERAGE ANNUAL POPULATION GROWTH RATE % (1990–2004)	TOTAL FERTILITY RATE (BIRTHS PER WOMAN— 2006)	LIFE EXPECTANCY AT BIRTH (YEARS— 2006)	INFANT MORTALITY RATE (PER 1,000 LIVE BIRTHS— 2006)	GROSS NATIONAL INCOME PER CAPITA ($—2004)	CARBON DIOXIDE EMISSIONS (PER CAPITA METRIC TONS—2002)	COMMERCIAL ENERGY USE PER CAPITA (KG OF OIL EQUIVALENT— 2003)	FRESHWATER RESOURCES (TOTAL RENEWABLE RESOURCES PER CAPITA CU. M—2004)
Afghanistan	—	—	6.7	43.3	160.2	—	—	—	—
Albania	3.3	−0.4	2.2	77.4	20.8	2,080	0.8	674	8,645
Algeria	32.4	1.8	1.9	73.3	29.9	2,280	2.9	1,036	348
Angola	15.5	2.8	6.3	38.6	185.4	1,030	0.5	606	9,555
Argentina	38.4	1.2	2.2	76.1	14.7	3,720	3.5	1,575	7,193
Armenia	3.0	−1.1	1.3	71.8	22.5	1,120	1.0	660	2,998
Australia	20.1	1.2	1.8	80.5	4.6	26,900	18.1	5,668	24,464
Austria	8.2	0.4	1.4	79.1	4.6	32,300	7.9	4,086	6,729
Azerbaijan	8.3	1.1	2.5	63.9	79	950	3.4	1,493	977
Bangladesh	139.2	2.1	3.1	62.5	60.8	440	0.3	159	754
Belarus	9.8	−0.3	1.4	69.1	13	2,120	6.0	2,613	3,786
Belgium	10.4	0.3	1.6	78.8	4.6	31,030	8.9	5,701	1,152
Benin	8.2	3.3	5.2	53	79.6	530	0.3	292	1,260
Bolivia	9.0	2.1	2.8	65.8	51.8	960	1.2	504	33,692
Bosnia and Herzegovina	3.9	−0.7	1.2	78	9.8	2,040	4.7	1,137	9,080
Botswana	1.8	1.5	2.8	33.7	53.7	4,340	2.3	—	1,357
Brazil	183.9	1.5	1.9	72	28.6	3,090	1.8	1,065	29,460
Bulgaria	7.8	−0.8	1.4	72.3	19.9	2,740	5.3	2,494	2,706
Burkina Faso	12.8	2.9	6.5	48.9	91.3	360	0.1	—	975
Burundi	7.3	1.8	6.5	50.8	63.1	90	0.0	—	1,382
Cambodia	13.8	2.5	3.4	59.3	68.8	320	0.0	—	8,738
Cameroon	16.0	2.3	4.4	51.2	63.5	800	0.2	429	17,022
Canada	32.0	1.0	1.6	80.2	4.7	28,390	16.5	8,240	89,134
Central African Republic	4.0	2.0	4.4	43.5	85.6	310	0.1	—	35,374
Chad	9.4	3.2	6.2	47.5	91.5	290	0.0	—	1,588
Chile	16.1	1.4	2.1	76.8	8.6	4,910	3.6	1,647	54,826
China	1,296.2	0.9	1.7	72.6	23.1	1,290	2.7	1,094	2,170
Hong Kong, China	6.9	1.3	0.9	81.6	3	26,810	5.2	2,428	—
Colombia	44.9	1.8	2.5	72	20.4	2,000	1.3	642	47,022
Congo, Democratic Rep.	55.9	2.8	6.4	51.5	88.6	120	0.0	293	16,114
Congo, Republic	3.9	3.2	6.1	52.8	85.3	770	0.6	273	57,173
Costa Rica	4.3	2.3	2.2	77	9.7	4,670	1.4	880	29,428
Côte d?voire	17.9	2.5	4.5	48.8	89.1	770	0.4	374	4,299
Croatia	4.4	−0.5	1.4	74.7	6.7	6,590	4.8	1,976	8,487
Cuba	11.2	0.5	1.7	77.4	6.2	—	2.1	1,000	3,390

(continued)

	TOTAL POPULATION (MILLIONS— 2004)	AVERAGE ANNUAL POPULATION GROWTH RATE % (1990–2004)	TOTAL FERTILITY RATE (BIRTHS PER WOMAN— 2006)	LIFE EXPECTANCY AT BIRTH (YEARS— 2006)	INFANT MORTALITY RATE (PER 1,000 LIVE BIRTHS— 2006)	GROSS NATIONAL INCOME PER CAPITA ($—2004)	CARBON DIOXIDE EMISSIONS (PER CAPITA METRIC TONS—2002)	COMMERCIAL ENERGY USE PER CAPITA (KG OF OIL EQUIVALENT— 2003)	FRESHWATER RESOURCES (TOTAL RENEWABLE RESOURCES PER CAPITA CU. M—2004)
Czech Republic	10.2	−0.1	1.2	76.2	3.9	9,150	11.2	4,324	1,287
Denmark	5.4	0.4	1.7	77.8	4.5	40,650	8.9	3,853	1,110
Dominican Republic	8.8	1.5	2.8	71.7	28.2	2,080	2.5	923	2,395
Ecuador	13.0	1.7	2.7	76.4	22.9	2,180	2.0	708	33,129
Egypt, Arab Republic	72.6	1.9	2.8	71.3	31.3	1,310	2.1	735	25
El Salvador	6.8	2.0	3.1	71.5	24.4	2,350	1.0	675	2,625
Eritrea	4.2	2.4	5.1	59	46.3	180	0.2	—	662
Estonia	1.3	−1.1	1.4	72	7.7	7,010	11.7	3,631	9,423
Ethiopia	70.0	2.2	5.2	49	93.6	110	0.1	299	1,744
Finland	5.2	0.3	1.7	78.5	3.5	32,790	12.0	7,204	20,466
France	60.4	0.4	1.8	79.7	4.2	30,090	6.2	4,519	2,956
Gabon	1.4	2.5	4.7	54.5	54.5	3,940	2.6	1,256	120,382
Gambia, The	1.5	3.3	5.3	54.1	71.6	290	0.2	—	2,030
Georgia	4.5	−1.4	1.4	76.1	18	1,040	0.7	597	12,866
Germany	82.5	0.3	1.4	78.8	4.1	30,120	10.3	4,205	1,297
Ghana	21.7	2.4	4.0	58.9	55	380	0.4	400	1,399
Greece	11.1	0.6	1.3	79.2	5.4	16,610	8.5	2,709	5,246
Guatemala	12.3	2.3	3.8	69.4	30.9	2,130	0.9	608	8,882
Guinea	9.2	2.8	5.8	49.5	90	460	0.2	—	24,561
Guinea-Bissau	1.5	3.0	4.9	46.9	105.2	160	0.2	—	10,392
Haiti	8.4	1.4	4.9	53.2	71.7	390	0.2	270	1,548
Honduras	7.0	2.6	3.6	69.3	25.8	1,030	0.9	522	13,610
Hungary	10.1	−0.2	1.3	72.7	8.4	8,270	5.6	2,600	594
India	1,079.7	1.7	2.7	64.7	54.6	620	1.2	520	1,167
Indonesia	217.6	1.4	2.4	69.9	34.4	1,140	0.4	753	13,043
Iran, Islamic Republic	67.0	1.5	1.8	70.3	40.3	2,300	5.5	2,055	1,918
Iraq	—	—	4.2	69	48.6	—	—	—	—
Ireland	4.1	1.1	1.9	77.7	5.3	34,280	11.0	3,777	12,045
Israel	6.8	2.7	2.4	79.5	6.9	17,380	10.6	3,086	110
Italy	57.6	0.1	1.3	79.8	5.8	26,120	7.5	3,140	3,170
Jamaica	2.6	0.7	2.4	73.2	16	2,900	4.1	1,543	3,556
Japan	127.8	0.2	1.4	81.2	3.2	37,180	9.4	4,053	3,366
Jordan	5.4	3.9	2.6	78.4	16.8	2,140	3.2	1,027	125
Kazakhstan	15.0	−0.6	1.9	66.9	28.3	2,260	9.9	3,342	5,030
Kenya	33.5	2.5	4.9	48.9	59.3	460	0.2	494	619
Korea, Democratic Rep.	22.4	0.9	2.1	71.7	23.3	—	6.5	896	2,993
Korea, Republic	48.1	0.8	1.3	77	6.2	13,980	9.4	4,291	1,349
Kuwait	2.5	1.0	2.9	77.2	9.7	17,970	25.6	9,566	0
Kyrgyz Republic	5.1	1.0	2.7	68.5	34.5	400	1.0	528	9,121
Lao PDR	5.8	2.4	4.7	55.5	83.3	390	0.2	—	32,878
Latvia	2.3	−1.0	1.3	71.3	9.3	5,460	2.7	1,881	7,238
Lebanon	3.5	1.8	1.9	72.9	23.7	4,980	4.7	1,700	1,356
Lesotho	1.8	0.9	3.3	34.4	87.2	740	—	—	2,909
Liberia	3.2	3.0	6.0	39.6	155.8	110	0.1	—	61,717
Libya	5.7	2.0	3.3	76.7	23.7	4,450	9.1	3,191	105

(continued)

	TOTAL POPULATION (MILLIONS— 2004)	AVERAGE ANNUAL POPULATION GROWTH RATE % (1990–2004)	TOTAL FERTILITY RATE (BIRTHS PER WOMAN— 2006)	LIFE EXPECTANCY AT BIRTH (YEARS— 2006)	INFANT MORTALITY RATE (PER 1,000 LIVE BIRTHS— 2006)	GROSS NATIONAL INCOME PER CAPITA ($—2004)	CARBON DIOXIDE EMISSIONS (PER CAPITA METRIC TONS—2002)	COMMERCIAL ENERGY USE PER CAPITA (KG OF OIL EQUIVALENT— 2003)	FRESHWATER RESOURCES (TOTAL RENEWABLE RESOURCES PER CAPITA CU. M—2004)
Lithuania	3.4	−0.5	1.2	74.2	6.8	5,740	3.6	2,585	4,529
Macedonia, FYR	2.0	0.4	1.6	74	9.8	2,350	5.1	—	2,659
Madagascar	18.1	2.9	5.6	57.3	75.2	300	0.1	—	18,606
Malawi	12.6	2.1	5.9	41.7	94.4	170	0.1	—	1,280
Malaysia	24.9	2.4	3.0	72.5	17.2	4,650	6.3	2,318	23,298
Mali	13.1	2.8	7.4	49	107.6	360	0.0	—	4,572
Mauritania	3.0	2.7	5.9	53.1	69.5	420	1.1	—	134
Mauritius	1.2	1.1	1.9	72.6	14.6	4,640	2.6	—	2,229
Mexico	103.8	1.6	2.4	75.4	20.3	6,770	3.8	1,564	3,940
Moldova	4.2	−0.2	1.8	65.7	38.4	710	1.6	772	237
Mongolia	2.5	1.3	2.2	64.9	52.1	590	3.4	—	13,839
Morocco	29.8	1.6	2.7	70.9	40.2	1,520	1.5	378	972
Mozambique	19.4	2.6	4.6	39.8	129.2	250	0.1	403	5,164
Myanmar	50.0	1.5	2.0	61	61.9	—	0.2	276	17,611
Namibia	2.0	2.6	3.1	43.4	48.1	2,370	1.1	635	3,066
Nepal	26.6	2.4	4.1	60.2	65.3	260	0.2	336	7,454
Netherlands	16.3	0.6	1.7	79	5	31,700	9.3	4,982	676
New Zealand	4.1	1.2	1.8	78.8	5.8	20,310	8.6	4,333	80,522
Nicaragua	5.4	2.2	2.7	70.6	28.1	790	0.8	588	35,293
Niger	13.5	3.3	7.5	43.8	118.2	230	0.1	—	259
Nigeria	128.7	2.5	5.5	47.1	97.1	390	0.4	777	1,717
Norway	4.6	0.6	1.8	79.5	3.7	52,030	13.9	5,100	83,205
Oman	2.5	2.3	5.8	73.4	18.9	7,890	12.1	4,975	389
Pakistan	152.1	2.4	4.0	63.4	70.5	600	0.8	467	345
Panama	3.2	2.0	2.7	75.2	16.4	4,450	2.0	836	46,426
Papua New Guinea	5.8	2.4	3.9	65.3	50	580	0.5	—	138,775
Paraguay	6.0	2.5	3.9	75.1	24.8	1,170	0.7	679	15,622
Peru	27.6	1.7	2.5	69.8	30.9	2,360	1.0	442	58,631
Philippines	81.6	2.1	3.1	70.2	22.8	1,170	0.9	525	5,869
Poland	38.2	0.0	1.2	75	7.2	6,090	7.7	2,452	1,404
Portugal	10.5	0.4	1.5	77.7	5	14,350	6.0	2,469	3,618
Puerto Rico	3.9	0.7	1.7	78.4	9.1	—	3.5	—	1,823
Romania	21.7	−0.5	1.4	71.6	25.5	2,920	4.0	1,794	1,951
Russian Federation	143.8	−0.2	1.3	67.1	15.1	3,410	9.9	4,424	29,981
Rwanda	8.9	1.6	5.4	47.3	89.6	220	0.1	—	1070
Saudi Arabia	24.0	2.7	4.0	75.7	12.8	10,430	15.0	5,607	100
Senegal	11.4	2.5	4.4	59.2	52.9	670	0.4	287	2,266
Sierra Leone	5.3	1.9	6.1	40.2	160.4	200	0.1	—	29,982
Singapore	4.2	2.4	1.1	81.7	2.3	24,220	13.7	5,359	142
Slovak Republic	5.4	0.1	1.3	74.7	7.3	6,480	6.8	3,443	2,341
Slovenia	2.0	0.0	1.2	76.3	4.4	14,810	7.7	3,518	9,349
Somalia	8.0	1.3	6.8	48.5	114.9	—	—	—	753
South Africa	45.5	1.8	2.2	42.7	60.7	3,630	7.6	2,587	984
Spain	42.7	0.7	1.3	79.7	4.4	21,210	7.4	3,240	2,605
Sri Lanka	19.4	0.9	1.8	73.4	14	1,010	0.5	421	2,575
Sudan	35.5	2.2	4.7	58.9	61	530	0.3	477	845
Swaziland	1.1	2.7	3.5	32.6	71.8	1,660	0.9	—	2,357
Sweden	9.0	0.4	1.7	80.5	2.8	35,770	5.8	5,754	19,017

(continued)

	TOTAL POPULATION (MILLIONS— 2004)	AVERAGE ANNUAL POPULATION GROWTH RATE % (1990–2004)	TOTAL FERTILITY RATE (BIRTHS PER WOMAN— 2006)	LIFE EXPECTANCY AT BIRTH (YEARS— 2006)	INFANT MORTALITY RATE (PER 1,000 LIVE BIRTHS— 2006)	GROSS NATIONAL INCOME PER CAPITA ($—2004)	CARBON DIOXIDE EMISSIONS (PER CAPITA METRIC TONS—2002)	COMMERCIAL ENERGY USE PER CAPITA (KG OF OIL EQUIVALENT— 2003)	FRESHWATER RESOURCES (TOTAL RENEWABLE RESOURCES PER CAPITA CU. M—2004)
Switzerland	7.4	0.7	1.4	80.5	4.3	48,230	5.6	3,689	5,467
Syrian Arab Republic	18.6	2.6	3.4	70.3	28.6	1,190	2.8	986	377
Tajikistan	6.4	1.4	4.0	64.9	106.5	280	0.8	501	10,311
Tanzania (mainland only)	37.6	2.6	5.0	45.6	96.5	270	0.1	465	2,232
Thailand	63.7	1.1	1.6	72.2	19.5	2,540	3.7	1,406	3,297
Togo	6.0	3.0	5.0	57.4	60.6	380	0.3	445	1,920
Trinidad and Tobago	1.3	0.5	1.7	66.8	25.1	8,580	31.8	8,553	2,951
Tunisia	9.9	1.4	1.7	75.1	23.8	2,630	2.3	837	422
Turkey	71.7	1.7	1.9	72.6	39.7	3,750	3.0	1,117	3,165
Turkmenistan	4.8	1.9	3.4	61.8	72.6	1,340	9.1	3,662	285
Uganda	27.8	3.2	6.7	52.7	66.2	270	0.1	—	1,402
Ukraine	47.5	−0.6	1.2	70	9.9	1,260	6.4	2,772	1,119
United Arab Emirates	4.3	6.4	2.9	75.4	14.1	—	25.0	9,707	35
United Kingdom	59.9	0.3	1.7	78.5	5.1	33,940	9.2	3,893	2,422
United States	293.7	1.2	2.1	77.8	6.4	41,400	20.2	7,843	9,535
Uruguay	3.4	0.7	1.9	76.3	11.6	3,950	1.2	738	17,154
Uzbekistan	26.2	1.7	2.9	64.6	70	460	4.8	2,023	623
Venezuela, RB	26.1	2.0	2.2	74.5	21.5	4,020	4.3	2,112	27,652
Vietnam	82.2	1.5	1.9	70.8	25.1	550	0.8	544	4,461
West Bank and Gaza	3.5	4.1	4.3	73.3	19.1	1,120	—	—	13
Yemen, Republic	20.3	3.7	6.6	62.1	59.9	570	0.7	289	202
Yugoslavia, Fed. Republic (Serbia & Montenegro, 2006)	8.1	0.1	1.7	75	12.5	2,620	—	1,991	5,401
Zambia	11.5	2.3	5.4	40	86.8	450	0.2	592	6,987
Zimbabwe	12.9	1.4	3.1	39.3	51.7	—	1.0	752	948

Source: The United Nations.

GLOSSARY

A

absolute location   The location of a place as pinpointed in terms of the global geographic grid

acculturation   The process of adopting some aspect of another culture

acid deposition   The deposition of acidic substances on the ground, primarily as a result of sulfur and nitrogen oxide pollution of the atmosphere

actual evapotranspiration (ACTET)   The amount of water evaporated and/or transpired in a given environment

adiabatic cooling   The cooling of air as a result of expansion of rising air; *adiabatic* means "without heat being involved"

advection   The horizontal movements of air or substances by wind or ocean currents

African diaspora   The migration of black peoples out of Africa

agglomeration   The bringing of people and activities together in one place for greater convenience

Agricultural Revolution   The application of science and technology to agriculture, resulting in greatly increased yields and releasing workers for other occupations

alluvial fan   A fan-shaped deposit of sand and gravel formed where a stream emerges from a narrow canyon onto a wider valley floor

alpine glacier   A glacier occupying a valley in a mountainous area. The movement of an alpine glacier is primarily governed by the underlying topography

angle of incidence   The angle at which solar radiation strikes a particular place at a point in time

apartheid   A policy of racial segregation enforced in South Africa 1948–1993

aquaculture   Herding or domesticating aquatic animals and farming aquatic plants

arithmetic density   The number of people per unit of area

artifact   A material object of culture; literally, "a thing made by skill"

asylum   The safety a country grants refugees

atmosphere   A thin layer of gases surrounding Earth to an altitude of less than 480 kilometers (300 miles)

autumnal equinox   September 22 or 23, one of two dates when, at noon, the perpendicular rays of the Sun strike the equator (meaning that the Sun is directly overhead along the equator)

B

basic sector   The part of a city's economy that is producing exports

beach   A deposit of wave-carried sediment along a shoreline, on which waves break

behavioral geography   The study of how we perceive our environment and of how our thoughts and perception influence our behavior

biodiversity   The amount of variety of living things in a given environment

biogeochemical cycle   The environmental recycling process that supplies essential substances such as carbon, nitrogen, and other nutrients to the biosphere

biogeochemical oxygen demand   The amount of dissolved oxygen in a water body that is consumed by decay of organic pollutants added to the water

biomagnification   The tendency for substances that accumulate in body tissues to increase in concentration as they are passed to higher levels in a food chain

biomass   The dry mass of living or formerly living matter in a given environment

biome   A large grouping of ecosystems characterized by particular plant and animal types

biosphere   All living organisms on Earth

biotechnology   New techniques for modifying biological organisms and their physiological processes for applied purposes

bloc   A group of countries that presents a common policy when dealing with other countries

blog   A "Web log;" an individual on-line diary or newsletter on the Internet

boreal forest   An evergreen needleleaf forest characteristic of cold continental climates

brain drain   The emigration of a country's best-educated people and most skilled workers

broadleaf deciduous forest   A forest with broadleaved trees that lose their leaves in the winter; characteristic of humid, midlatitude environments

brownfields   Abandoned polluted industrial sites in central cities, many of which are today being cleaned and redeveloped

C

capital-intensive activity   An activity in which a large amount of capital is invested per worker

capitalist economic system   One in which the state defers to private enterprise and a stock market raises and allocates capital

carbon cycle   The movement of carbon among the atmosphere, hydrosphere, biosphere, and lithosphere as a result of processes such as photosynthesis and respiration, sedimentation, weathering, and fossil-fuel combustion

carbon dioxide   A trace gas found in the atmosphere; a major contributor to the greenhouse effect

carbon monoxide   A pollutant formed by the incomplete combustion of fossil fuels

carnivore   An animal whose primary food supply is other animals

carrying capacity   The ability of agricultural land to support population

cartel   An organization formed to control the market for a particular commodity, usually by restricting supply

cartogram   A maplike image designed to convey the magnitude of something rather than exact spatial locations

cartography   Map making

caste   A group in the rigid social hierarchy of Hinduism

central business district (CBD)   The traditional core of the city where office buildings and retail shops tend to be concentrated

central place theory   A model of the distribution of cities across an isotropic plain

centrifugal forces   Forces that tend to pull states apart

centripetal forces   Forces that bind a state together

chemical weathering The breakdown of rocks or minerals through chemical reactions at Earth's surface

city A concentrated nonagricultural human settlement

climate The totality of weather conditions over a period of several decades or more

climax community The end-point of community succession

cloning The production of identical organisms by asexual reproduction from a single cell of a preexisting organism

cognate A word that clearly looks or sounds like one in another language to which it is related historically

cognitive behavioralism The theory that people react to their environment as they perceive it

cold front The boundary formed when a cold air mass advances against a warmer one

collapsed state A state that has proven incapable of providing its citizens with either economic development or even peace and security

commercial agriculture Raising food to sell

commercial revolution The tremendous expansion of global trade between about 1650 and 1750

common market A customs union within which common laws create similar conditions of production

community succession A process of ecosystem change in which organisms modify their immediate environments in ways that allow other species to establish themselves and dominate

composite cone volcano A volcano formed by a mixture of lava eruptions and more explosive ash eruptions

concentration The distribution of a phenomenon within a given area

condensation Water changing from a gas state (vapor) to a liquid or solid state

conformal map A map that distorts size but preserves shapes

congregation Residential clustering by choice

consolidated metropolitan statistical area (CMSA) Two or more contiguous MSAs

contiguous diffusion Diffusion that occurs from one place directly to a neighboring place

continental glacier A thick glacier hundreds to thousands of kilometers across, large enough to be only partly guided by underlying topography

continental shelf An area of relatively shallow water that surrounds most continents before the continental slope drops more sharply to the deep sea floor

convection Circulation in a fluid caused by temperature-induced density differences, such as the rising of warm air in the atmosphere

convergent plate boundary A boundary between tectonic plates in which the two plates move toward one another, destroying or thickening the crust

core area The historic homeland of a nation or area of greatest settlement

Coriolis effect The tendency of an object moving across Earth's surface to be deflected from its apparent path as a result of Earth's rotation

corporativism Government by representatives of interest groups

creole A pidgin language that has become a mother tongue

crude birthrate The annual number of live births per thousand people

crude death rate The annual number of deaths per thousand people

cultural diffusion The spreading of cultural attributes

cultural ecology The study of the ways societies adapt to environments

cultural geography The study of the geography of human cultures

cultural imperialism The substitution of one set of cultural traditions for another, either by force or by degrading those who fail to acculturate and rewarding those who do

cultural landscape A landscape that reveals the many ways people modify their local environment

cultural mosaic A phrase describing Canada and suggesting that various cultural groups retain individuality there

cultural realm The region throughout which a culture prevails

cultural subnationalism The splitting of a state population among several local primary allegiances

culture A bundle of attributes of shared behavior or belief. These may include virtually anything about the way a people live

culture region The region throughout which a culture prevails

customs union A free-trade area that enforces a common external tariff

cyberspace The extension of reality through global electronic means of communication

cyclone Large lowpressure areas in which winds converge in a counterclock-wise swirl in the Northern Hemisphere (or clockwise in the Southern Hemisphere)

D

delta A deposit of sediment formed where a river enters a lake or ocean

demographic transition model A model that describes the historical experience of population growth in the countries that are today rich

demography The analysis of a population in terms of specific characteristics, such as age or income levels

density The frequency of occurrence of a phenomenon in relation to its geographic area

dependency ratio The ratio of the combined population less than 15 years old and adult population over 65 years old to the population of those between 15 and 64 years of age

desert A vegetation type with sparsely distributed plants, specifically adapted for moisture-gathering and moisture-retention

desert climate A climate with low precipitation and temperatures warm enough to cause potential evapotranspiration to be substantially higher than precipitation

desert pavement The stony surface of a desert soil formed by selective removal of fine particles by surface erosion

desertification The process of a region's soil and vegetation cover becoming more desertlike as a result of human land use, usually by overgrazing or cultivation

dialect A minor variation within a language

diffusion The process of an item or feature spreading through time

diffusionism The theory that aspects of civilization were developed in very few places and then diffused from those places to the rest of the world

diminishing returns A condition that exists when, upon adding equal amounts of one factor of production, such as fertilizer or labor, each successive application yields a smaller increase in production than the application just preceding

discharge The quantity of water flowing past a point on a stream per unit time

dissolved oxygen Oxygen found in dissolved form in water; it is essential for aquatic animals, and depleted by pollution

distance The extent of space between two objects or places; it can be measured absolutely, in terms of miles or kilometers, or in terms of other units, such as time or cost to cross

distance decay   The diminution of the presence or impact of any cultural attribute away from its hearth area

distribution   The position, placement, or arrangement of a phenomena

divergent plate boundary   A boundary between tectonic plates in which the two plates move away from each other, and new crust is created between them

domestication   The process of adapting plants and animals to obtain their intimate association with humankind, to the advantage of humankind

double cropping   Harvesting two crops from each field per year

doubling time   The number of years it would take any country's population to double at present rates of increase

downstream activities   Economic activities that are second, third, or even fourth steps in the transformation of a raw material into goods for ultimate consumers

drainage basin   The geographical area that contributes runoff to a particular stream, defined with respect to a specific location along that stream

drainage density   The total length of streams in a drainage basin divided by the drainage area

dune   An accumulation of windblown sand, shaped by the wind

E

earthquake   A sudden release of energy within Earth, producing a shaking of the crust

ecology   The scientific study of ecosystems

economic development   The process of progressively increasing the value of goods and services that a place is able to produce in order to enjoy or export

economic geography   The study of how various peoples make their living and what they trade

economies of scale   The economic factors determining that as the number of units of a good produced increases, the production cost per unit generally falls

ecosystem   An interrelated collection of plants and animals and the physical environment with which they interact

ecotourism   Travel to see distinctive examples of scenery, unusual natural environments, or wildlife

ejido   A Mexican form of land tenure in which a peasant community collectively owns a piece of land along with the natural resources and houses on it

El Niño   A circulation change in the eastern tropical Pacific Ocean, from westward flow to eastward flow, that occurs every few years

electoral geography   The study of voting districts and voting patterns

electronic highway   Electronic methods of communication, such as e-mail, facsimile machines, and computer modems

emigration   Movement away from a place

enculturation (socialization)   Teaching youngsters a society's values and traditions, its political and social culture

endogenic processes   Forces within Earth that affect its surface, such as plate tectonics, volcanic eruptions, and earthquakes

environmental determinism   The simplistic belief that human events can be explained entirely as the result of the effects of the physical environment

epicenter   The location on Earth's surface immediately above the focus of an earthquake

epidemiological transition   The shift within a country of the principal causes of death from infectious to degenerative diseases

epidemiology   The study of the incidence, distribution, and control of disease

equal-area map   A map projection that preserves size but distorts shape

equator   Earth's imaginary midline everywhere equidistant between the poles

ethnic group   A dubious term suggesting that a particular cultural group is in the minority or "not normal" for a particular place and time

ethnocentrism   A tendency to judge foreign cultures by the standards and practices of one's own; and usually to judge them unfavorably

etymology   The study of the origin and history of words

European Union (EU)   A bloc of European countries enjoying free trade and who are committed to eventual full political union

evapotranspiration   The sum of evaporation and transpiration

evolutionism   The theory that a culture's sources of change were embedded in the culture from the beginning, so the course of development was internally determined

exclusive economic zone (EEZ)   A 200-mile zone within which a coastal state controls both mining and fishing rights from its shores

exogenic processes   Forces originating in the atmosphere that, aided by gravity, shape Earth's surface; erosion by running water, glaciers, wind, and waves are examples

export-led economic growth   A national economic policy of welcoming foreign investment to build factories that will manufacture goods for international markets

external economies   The range of goods and services in a city that can be rented or hired temporarily

externality   An exchange of a good or service that takes place without agreement of the parties in a market; pollution is generally considered an externality

exurbs   The settlements that make up the outermost ring of expanding metropolitan areas

F

failed state   A state that has proven incapable of providing its citizens with either economic development or even peace and security

fault   A fracture in Earth's crust along which displacement of rocks has occurred

federal government   A form of government in which a central government shares power with subunits

felt needs   Things people have become convinced that they need

fertility rate   The number of children born per year per thousand females

First World   Defined during the cold war, all reasonably well-to-do, non-Communist countries

fishery   A concentration of aquatic species suitable for commercial harvesting

floodplain   A low-lying surface adjacent to a stream channel and formed by materials deposited by the stream

focus (of an earthquake)   The location in Earth where motion originates in an earthquake

folk culture   A culture that is handed down and preserves traditions

food chain   The sequential consumption of food in an ecosystem, beginning with green plants, followed by herbivores and carnivores, and ending with decomposers

footloose activities   Activities that can move or relocate freely

foreign direct investment (FDI)   Investment by foreigners in wholly owned enterprises that are operated by the foreigner

formal region   A region defined by essential uniformity in one or more physical or cultural features

fossil fuel   A source of chemical energy stored in formerly living plant and animal tissue. Coal, oil, and natural gas are fossil fuels

free-trade area   An international territory having no internal tariffs, but that give its members the freedom to set their own tariffs on trade with the rest of the world

friction of distance   The effort, time, or cost necessary to move or transport items

front   A boundary between warm air and cold air

frontier   An undeveloped region that may offer potential for settlement

functional region   A region defined by interaction among places, such as trade and communication

fundamentalism   The strictest adherence to traditional religious beliefs

G

Gaia hypothesis   A holistic view that likens Earth to a living organism with the ability to regulate critical functions, such as climate, through interactions between the atmosphere, hydrosphere, biosphere, and lithosphere

gene splicing (or recombinant DNA)   The joining of the genes of two organisms to produce recombinant (recombined) genetic material

genetic engineering   The manipulation of species' genetic material through selective breeding or recombinant DNA

gentrification   The occupation and restoration of select urban residential neighborhoods by successful urban white-collar workers

geographic information systems (GIS)   The use of computer systems to organize, store, analyze, and map information

geography   The study of the interaction of all physical and human phenomena at individual places and of how interactions among places form patterns and organize space

geomorphology   The study of the shape of Earth's surface and the processes that modify it

geopolitics   A pseudoscience studying "the natural and necessary trend towards [national] expansion as a means of selfpreservation;" today often loosely applied to studies of military strategies

gerrymandering   The drawing of voting district lines in ways that include or exclude specific groups of voters, so that one group gains an unfair advantage

glacier   A large mass of flowing, perennial ice

globalization   The organization of any activity treating the entire globe as one place

global positioning system (GPS)   A navigational tool consisting of a fleet of satellites orbiting Earth, broadcasting digital codes, and a portable receiver that can "hear" those codes and determine its location

global warming   A general increase in temperatures over a period of at least several decades believed to be caused primarily by increased levels of carbon dioxide in Earth's atmosphere

grade   A condition in which a stream's ability to transport sediment is balanced by the amount of sediment delivered to it

green revolution   An intensive effort, starting about 1950, to develop new grain varieties and associated agronomic systems and to establish them in developing countries. It focused on certain crops (wheat, rice) and certain techniques (breeding for responses to fertilizer inputs) and was driven largely by private foundations

greenhouse effect   Atmospheric warming that results from the passage of incoming shortwave energy and the capture of outgoing longwave energy

greenhouse gases   Trace substances in the atmosphere that contribute to the greenhouse effect; water vapor, carbon dioxide, ozone, methane, and chlorofluorocarbons are important examples

Greenwich Mean Time (GMT)   The time at the prime meridian or 0° longitude at Greenwich, England

gross domestic product (GDP)   The total value of all goods and services produced within a country

gross national income (GNI)   A country's GDP plus any income that residents receive from foreign investments, minus any money paid out of the country to foreign investors

groundwater   The water beneath Earth's surface at a depth where rocks and/or soils are saturated with water

gyre   A circular ocean current beneath a subtropical high-pressure cell

H

hearth   The place where a distinctive culture originated

herbivore   An animal whose primary food supply is plants

hierarchical diffusion   Diffusion that occurs downward or upward through a organizational hierarchy; when mapped, it shows up as a network of spots

hinterland   The region to which any city provides services and upon which it draws for its needs

historical consciousness   A people's consciousness of past events insofar as that consciousness influences their present behavior

historical geography   The study of the geography of the past and how geographic distributions have changed

historical materialism   The belief that technology has historically increased humankind's control over the environment and improved material welfare, and that this improvement prompts other historical events and movements

horizon   A layer in the soil with distinctive characteristics derived from soil-forming processes

human development index (HDI)   A statistic that combines statistics of life expectancy, school enrollment, literacy, and income to compare the quality of life around the world

human geography   The study of the geography of human groups and activities

humid continental climate   A climate characterized by cold winters and warm summers, with moderate levels of precipitation

humid subtropical climate   A climate with cool winters, hot summers, and moderately high levels of precipitation

humid tropical climate   A climate with high temperatures and high rainfall amounts all the year

hunter-gatherers   People who live on what they can hunt or harvest from Earth

hurricane   An intense tropical cyclone that develops over warm ocean areas in the tropics and subtropics, primarily during the warm season. Hurricanes in the Pacific Ocean are called typhoons; in the Indian Ocean they are called cyclones

hydrocarbon   A chemical substance composed of carbon and hydrogen; hydrocarbons in the atmosphere contribute to the formation of photochemical smog

hydroelectric power   Electricity generated by water passing through turbines at a dam

hydrologic cycle   The movement of water from the atmosphere to Earth's surface, across that surface, and back to the atmosphere

hydrosphere   The water realm of Earth's surface, including the oceans, surface waters on land (lakes, streams, rivers), groundwater in soil and rock, water vapor in the atmosphere, and ice in glaciers

I

ice-cap climate   A climate with very cold temperatures all the year, including summer temperatures that are rarely above freezing

iconography   A state's set of symbols, including a flag and anthem

igneous rocks   Rocks formed by crystallization of a magma

immigration   Movement into a place

import-substitution economic growth   A national economic policy of protecting domestic infant industries

incorporation   The process of defining a city territory and establishing a government

indigenous peoples   As defined by the U.N., "descendants of the original inhabitants of a land who were subjugated by another people coming after them"

indirect rule   The imperialist use of native rulers as intermediaries between the imperialists and the people

Industrial Revolution   The evolution, which first occured in Europe between about 1750 and 1850, from an agricultural and commercial society to an industrial society relying on inanimate power and complex machinery

industrial society   A society with a significant share of its output from the secondary sector

inertia   The force that keeps things stable or fixed in place

infant industries   Newly developing industries that probably cannot compete with imports

infant mortality rate   The number of infants per thousand who die before reaching 1 year of age

infiltration capacity   The maximum amount of water that can soak into a soil per unit time

informal or underground sector   The economic activities that do not appear in official accounts

infrastructure   Fixed assets in place, such as buildings, dams, and roads

innocent passage   The internationally guaranteed right of the ships of one state to pass through the territorial waters of another on their way to a third

insolation   The amount of solar energy intercepted by a particular area of Earth

internal economies   Those goods and services that a large company can provide for itself

International Date Line   Imaginary line on Earth's surface where, by international agreement, travelers traveling eastward subtract one calendar day, and travelers traveling westward add one calendar day. The line generally follows the 180° meridian, but it deviates for political and economic convenience

international organization   An organization that coordinates activities among two or more countries

intertropical convergence zone (ITCZ)   A low-pressure zone between the Tropic of Cancer and the Tropic of Capricorn where surface winds converge

irredenta   Territory that one state claims from another

isogloss   A line around places where speakers use a linguistic feature in the same way

isostatic adjustment   A vertical movement of Earth's crust, caused by the loading or unloading of the buoyant crust

isotropic plain   A theoretical perfectly flat surface with absolutely no variations across it

K

karst   An assemblage of landforms found in areas of intense subsurface chemical weathering that often include features such as caves and underground drainage

kleptocracy   Government by thieves or theft

L

labor-intensive activity   An activity that employs a high ratio of workers to invested capital

laissez-faire capitalism   A capitalist system that minimizes the government's role in the economy

landform   A characteristic shape of the land surface, such as a hill, valley, or floodplain

language   A set of words, plus their pronunciation and methods of combining them, that is used and understood as communication within a group of people

language family   The languages that are related by descent from a common protolanguage

large-scale map   Map that shows a given area in a large space

latent heat   Heat stored in water and water vapor, not detectable by people; *latent* means "hidden"

latent heat exchange   The exchange of energy necessary to change water from one of its states to another—solid, liquid, or gaseous

latitude   The location of a place measured as angular distance north and south of the equator

lava   Magma that reaches Earth's surface and erupts

liberation theology   The belief in putting the problems of overcoming poverty at the heart of Christian theology

life expectancy   The average number of years that a newborn baby within a given population can expect to live

lingua franca   A second language held in common for international discourse

liquidity   The quality of being readily convertible into cash

lithosphere   The solid Earth, composed of rocks and sediments overlying them

Little Ice Age   The period between about 1500 to 1750, when climates on Earth were especially cool

local content requirement   A definition of the percentage of the total value of a good entering one country that must have been added in the second country for that product to qualify as a product of the second country

location   The place where a thing is; it can be defined absolutely or relatively

loess   An accumulation of windblown silt

longitude   The location of a place measured as angular distance east and west from the prime meridian

longshore current   A current in the surf zone along a shoreline, parallel to the shore

longshore transport   Sediment transport by a longshore current

longwave energy   Energy reradiated by Earth in wavelengths of about 5.0 to 30.0 microns

M

magma   Molten rock beneath Earth's surface

Malthusian theory   The pessimistic argument that population increases will always outpace increases in food production, causing cycles of war, famine, and disease; articulated by Thomas Malthus (1766–1834)

mantle   The portion of Earth above the core and below the crust

map A two-dimensional (flat) representation of some portion of Earth's surface

maquiladora A factory in Mexico specializing in assembling items for export to the U.S. market

marine terrace A nearly level surface along a shoreline, elevated above present sea level, formed by coastal erosion at a time when sea level at the location was higher than at present

marine west coast climate A climate with moderately cool winters, moderately warm summers, and moderate to high rainfall all year

market-oriented manufacturing Manufacturing that locates close to the market either because the processing increases the perishability of the product or because the processing adds bulk or weight to the product

Marshall Plan A plan of financial aid for the economic rehabilitation of Europe after World War II; named for U.S. Secretary of State George C. Marshall

mass movement Downslope movement of rock and soil at Earth's surface, driven mainly by the force of gravity acting on those materials

material-oriented manufacturing Manufacturing that locates close to the source of the raw material either because the raw material is heavy or bulky or because it is perishable

meandering The tendency of flowing water to follow a sinuous course with alternating right- and left-hand bends

mechanical weathering The breakdown of rocks into smaller particles caused by application of physical or mechanical forces

Mediterranean climate A climate with warm, dry summers and cool, moist winters

melting pot Name given to the United States in a 1914 novel of that title, suggesting that ethnic and racial differences among immigrants melt together to form one culture

meltwater channel A river channel carved by water from a melting glacier

mental map The ideas that we have about places, regardless of whether our ideas are true or false

meridians Imaginary lines extending from pole to pole and crossing all parallels at right angles

meritocracy A society in which the most capable people can rise to the top on merit alone

metamorphic rocks Rocks formed by modification of other rock types, usually by heat and/or pressure

metropolitan statistical area (MSA) According to the U.S. Census Bureau, "an integrated economic and social unit with a recognized large population nucleus"

microclimate Local climatic conditions, as in cities, that differ from those of surrounding areas

midlatitude cyclone A storm characterized by a center of low pressure in the midlatitudes usually associated with a warm front and a cold front

midlatitude low-pressure zones Regions of low pressure and air converging from the subtropical and polar high-pressure zones

migration chain A network of social and communication linkages that attracts migrants to follow others who have previously migrated

model An idealized, simplified representation of reality

monoculture The specialized production of one crop

monotheism Belief in the existence of only one god

monsoon circulation Seasonal reversal of pressure and wind in Asia, in which winter winds from the Asian interior produce dry winters, and summer winds blowing inland from the Indian and Pacific oceans produce wet summers

moraine An accumulation of rock and sediment deposited by a glacier, usually in or near the melting area

multinational corporation An enterprise that produces and markets goods in several countries

multinational state A state containing several nations

multiplier effect The fact that jobs in a city's basic sector multiply jobs in the nonbasic sector

N

nation A group of people who want to have their own government and rule themselves

national self-determination The idea of the nation-state as defended by U.S. President Woodrow Wilson after World War I

nation-state A state ruling over a territory containing all the people of a nation and no others

natural landscape A landscape without evidence of human activity

natural population increase (decrease) The difference between the number of births and the number of deaths

natural resource Something that is useful and that exists independent of human activity

network hypothesis The theory that central city unemployment is caused by a lack of social networks

nitrogen oxide (NOx) A compound of nitrogen and oxygen; a component of air pollution

nonbasic sector The part of a city's economy serving the needs of the city itself

nongovernmental organizations (NGOs) Organized international interest groups that cross state boundaries and put pressure on the existing governments of states

nonrenewable resource A resource that is either not being produced by nature or is produced at rates much slower than it is used by humans

North Atlantic Treaty Organization (NATO) A military bloc founded in 1949. Its membership and activities have expanded significantly, bringing its very purpose into question

O

official language The language in which legal documents are kept in a country

oligarchy An elite privileged clique

omnivore An animal that feeds on both plants and other animals

opportunity cost A capital return sacrificed by leaving capital invested in one form or activity rather than another

orthography The study of writing, or a system of writing

outwash plain An accumulation of sand and gravel carried by meltwater streams from a glacier, usually deposited immediately beyond the terminal moraine from the glacier

overland flow Water flowing across the soil surface on a hillslope, usually resulting from precipitation falling faster than the ground can absorb it

Overseas Chinese Chinese migrants who often remain linked across international borders by bonds of family, clan, and common home province

oxygen cycle The movement of oxygen between the biosphere and atmosphere through photosynthesis and oxydation or respiration

ozone A gas composed of molecules with three oxygen atoms; it is a highly corrosive gas at ground level, but in the upper atmosphere essential to protecting life on Earth by absorbing ultraviolet radiation

P

parallels   Lines connecting all points of the same latitude

parent material   Mineral matter such as rocks or transported sediments from which soil is formed

particulate   A small solid particle in the air; a component of air pollution

pastoral nomadism   A group's style of life that does not have fixed residences; the group drives flocks from place to place to find grazing lands and water

pathogen   A disease-causing organism

pattern   The arrangement of objects within an area

permafrost   Soil or rock with a temperature below 0°C all year

photochemical smog   A mixture of air pollutants including oxidants such as ozone, formed by interaction of sunlight and pollutants such as nitrogen oxides and hydrocarbons

photosynthesis   A chemical reaction that occurs in green plants in which carbon dioxide and water are converted to carbohydrates and oxygen

photovoltaic cell   A device that converts light to electricity

physical geography   The study of the characteristics of the physical environment

physiological density   The density of population per unit of arable land

plate tectonics theory   The movement of large, continent-sized slabs of Earth's crust relative to one another

Pleistocene Epoch   A period of geologic time consisting of the first part of the Quaternary Period beginning about three million years ago and ending about 12,000 years ago

polar front   A boundary between cold polar air and warm subtropical air that circles the globe in the midlatitudes

polar high-pressure zones   Regions of high pressure and descending air near the North and South poles

political community   A willingness to join together and form a government to solve common problems

political culture   The set of unwritten ways in which written rules are interpreted and actually enforced

political economy   The study of individual countries' organization and regulation of their economies

political geography   The study of the interaction between political processes and the distributions of all other activities and transformations of the landscape

pollution   A human-caused increase in the amount of a substance in the environment

pollution prevention   A strategy for reducing pollution that focuses on reducing the amount of pollutants created, rather than on removing them from waste streams

polyculture   The raising of a variety of crops

polyglot state   A country that grants legal equality to two or more languages

polytheism   The worship of many gods

popular culture   The culture of people who embrace innovation and conform to changing norms

population geography   The study of the distribution of humankind across Earth

population projection   A forecast of the future population, assuming that current trends remain the same or else change in defined ways

population pyramid   A graphic device that shows the shares of a nation's population by age groups

possibilism   The theory that the physical environment itself will neither suggest nor determine what people will attempt, but it may limit what people can profitably achieve

postindustrial society   A society with the bulk of its economic activity in the tertiary sector

potential evapotranspiration (POTET)   The maximum amount of water that could be evaporated from a moist surface and/or transpired by plants if it were available

potential resource   Something that is not useful today but may become so in the foreseeable future

prairie   A vegetation type characterized by dense grass up to 2 meters high, found in midlatitude semiarid climates

preindustrial society   A society with the bulk of its economic activity in the primary sector

primary sector   That part of the economy that extracts resources directly from Earth, including agriculture, fishing, forestry, and mining

primate city   A large city concentrating a national population or national political, intellectual, or economic life

prime meridian   The meridian passing through the Royal Observatory in Greenwich, England, from which longitude is measured

privatize   To give or sell government assets to private individuals or investors

projection   A method of portraying Earth or any portion of it on a flat map

proselytize   To try to convert others to your religious beliefs

protolanguage   The ancestor that is common to any group of several of today's languages

proxemics   The study of how people perceive and use space

psychological value added   An increase in the cost of an item due not to an increase in its actual functionality or usefulness, but in its design, packaging, or "status" advertising

pull factors   Considerations that attract people to new destinations

push factors   Anything that makes a person want to leave a place and seek a better life elsewhere

Q

Quaternary Period   The period of geologic time encompassing approximately the last 3 million years

R

race   A group of people variously defined by relatively minor biological differences within the human species

racism   The false belief in the inherent superiority of one race over another and the linking of human ability, potential, and behavior to racial inheritance

radiation   Energy in the form of electromagnetic waves that radiate in all directions

recombinant DNA (or gene splicing)   The joining of the genes of two organisms to produce recombinant (recombined) genetic material

refugee   As defined by the 1951 Geneva Convention, someone with "a well-founded fear of being persecuted in his country of origin for reasons of race, religion, nationality, membership of a particular social group or political opinion"

region   A territory that exhibits a certain uniformity

regional geography   An inventory analysis of all characteristics of any individual place

relative humidity   The *actual* water content of the air expressed as a percentage of how much water the air *could* hold at a given temperature

relative location   The location of a place relative to other places

relocation diffusion   Diffusion from one widely separated point to another

remote sensing The acquisition of data about Earth's surface from a satellite orbiting the planet or from high-flying aircraft

renewable resource Something that is produced by nature at rates similar to those at which it is consumed by humans

replacement rate A total fertility rate of 2.1, which stabilizes a population

respiration A chemical reaction that occurs in plants and animals in which carbohydrates and oxygen are combined, releasing water, carbon dioxide, and heat

runoff Flow of water from the land, either on the soil surface or in streams

S

sanitary landfill A site at which solid waste is deposited and covered with layers of earth

saturation vapor pressure The maximum water vapor that air can hold

savanna A vegetation type characterized by grasses and scattered trees, characteristic of seasonally dry tropical climates

sawah The Indonesian term for a flooded rice field

scale A quantitative statement of the relative sizes of an object on a map and in reality

scientific revolution in agriculture The continuing application of science to agriculture

sea level The general elevation of the sea surface, averaging out variations caused by waves, storms, and tides

seafloor spreading The creation of new oceanic crust where two tectonic plates are diverging on the seafloor

seasonally humid tropical climate A climate with warm temperatures all the year, a season with high rainfall, and a pronounced dry season

Second World Defined during the cold war, the Soviet bloc dominated by totalitarian Communist governments

secondary sector That part of the economy that transforms raw materials into manufactured goods

sectoral evolution A shift in the concentration of activity from an economy's primary sector to its secondary and tertiary sectors

secularism A lifestyle or policy that deliberately ignores or excludes religious considerations

sediment transport The movement of rock particles by surface erosion

sedimentary rocks Rocks formed through accumulation of many small rock fragments at Earth's surface

segregation Residential clustering as the result of discrimination

seismic waves Vibrations or shock waves originating at the focus of an earthquake and transmitted through Earth

seismograph A device for recording movements of Earth's crust, such as earthquakes

semiarid climate A climate with precipitation slightly less than potential evapotranspiration for most of the year

sensible heat Heat detectable by sense of touch, or with a thermometer

service sector That part of the economy that services the primary and secondary sectors

shamanism A belief in the power of mediums (shamans) who characteristically go into autohypnotic trances, during which they are thought to be in communion with the spirit world

Sharia Islamic teachings that are often incorporated into civil law in Islamic countries

shield The ancient core of a continent

shield volcano A volcano with relatively gentle slopes formed by eruption of relatively fluid lavas

shortwave energy Radiant energy emitted by the Sun in wavelengths about 0.2 to 5.0 microns

SIAL Crust formed of relatively less dense minerals, dominated by silicon and aluminum (for SIlicon-ALuminum)

SIMA A crust formed of relatively dense minerals, dominated by silicon and magnesium (an acronym for SIlicon-MAgnesium)

site The characteristics of the absolute location of a place

situation The characteristics of the relative location of a place

small-scale map A map that shows the land in a very small space

social Darwinism The theory of British sociologist Herbert Spencer that "Nature's law" calls for "the survival of the fittest," even among cultures and entire peoples

socialization (enculturation) Teaching youngsters a society's values and traditions, its political and social culture

soil A dynamic, porous layer of mineral and organic matter at Earth's surface

soil creep The slow downslope movement of soil caused by many individual, near-random particle movements such as those caused by burrowing animals or freeze and thaw

soil fertility The ability of a soil to support plant growth through the storing and supplying water, air, and nutrients

soil order A major category in the U.S. soil classification system

sojourner A migrant who intends to stay in a new location only long enough to save capital to return home to a higher standard of living

solar energy The radiant energy from the Sun

spatial mismatch hypothesis The hypothesis that central city unemployment is caused by the removal of job opportunity to the suburbs and the concentration of the poor in the central city

speech community A group of people who speak together

state An independent political unit that claims exclusive jurisdiction over a defined territory and over all of the people and activities within it

steppe A vegetation type characterized by relatively short, sparse grasses, found in midlatitude semiarid climates

storm surge An area of elevated sea level in the center of a hurricane that may be several meters high, and which does most of the damage when a hurricane comes ashore

stratus clouds Flat layers of clouds formed along a warm front

structural landform A landform whose major characteristics are derived from endogenic processes, or by erosional exposure of rock structures

subarctic climate A high-latitude climate characterized by brief, cool summers and long, cold winters

subsistence agriculture Raising food only for oneself, not to sell

substitutability The degree to which one commodity can be substituted for another in various uses

subtropical high-pressure (STH) zones Regions of high pressure and descending air at about 25° north and south latitudes

succession Community succession

sulfur oxide An air pollutant consisting of compounds of sulfur and oxygen, derived mainly from combustion of coal and oil

summer solstice For places in the Northern Hemisphere, June 20 or 21 is the date when at noon the Sun is directly overhead along the parallel of 23.5° north latitude; for places in the Southern Hemisphere, December 21 or 22 is the date when at noon the Sun is directly overhead at places along the parallel of 23.5° south latitude

superimposed boundaries   The boundaries that imperialists drew to establish colonies over natives' territorial demarcations

supply and demand   The interplay of buyers and sellers of a commodity in the marketplace

supranational organization   An organization that exercises power over countries

surface erosion   The downslope movement of rock and soil at Earth's surface, driven mainly by air, water, or ice moving across the surface

sustainable development   Economic development that increases wealth but avoids environmental destruction

sustained yield   A way of managing a renewable natural resource such that harvest can continue indefinitely

swidden   Slash-and-burn clearing and cultivation

syncretic religions   Religions that combine Christianity with traditional practices

system   An interdependent group of items that interact in a regular way to form a unified whole

systematic geography   The study of universal laws or principles that apply to all places; topics may be as diverse as the geography of soils (pedology), of life forms (biogeography), of politics (political geography), of economic activities (economic geography), and of cities (urban geography)

T

tectonic plate   A large, continent-sized piece of Earth's crust that moves in relation to other pieces

telecommuting   Working at home at a computer terminal connected to an office

temperature inversion   A layer in the atmosphere in which relatively warm air lies above cooler air

terminal moraine   An accumulation of rock and sediment at the toe of a glacier

territoriality   A drive—possibly biological—to lay claim to territory and defend it against members of one's own species

terrorism   Violent acts intended to frighten and to intimidate for political ends a civilian population beyond the immediate victims

tertiary sector   That part of the economy that services the primary and secondary sector

theocracy   A form of government where a church rules directly

Third World   Defined during the cold war, the countries characterized by what is seen as underdevelopment, overpopulation, irrationality, religion, and political chaos

threshold   The minimum number of potential customers that is needed for a product or service to be offered

threshold principle   The belief that a nation must have some minimal population and territory to merit self-determination

topical geography   The study of universal laws or principles that apply to all places; topics may be as diverse as the geography of soils (pedology), of life forms (biogeography), of politics (political geography), of economic activities (economic geography), and of cities (urban geography)

topography   Surface relief

toponymy   The study of place names

tornado   A rapidly rotating column of air usually associated with a thunderstorm, often having winds in excess of 300 kilometers/hour (185 miles/hour)

total fertility rate   The average number of children that would be born to each woman in a given society if, during her childbearing years (15–49), she bore children at the current year's rate for women of that age

toxic substance   A pollutant that can be harmful even at very low concentrations

trade wind   The prevailing wind in subtropical and tropical latitudes that blows toward the Intertropical Convergence Zone, typically from the northeast in the Northern Hemisphere and from the southeast in the Southern Hemisphere

transform plate boundary   A boundary between tectonic plates in which the two plates pass one another in a direction parallel to the plate boundary

transnational corporation   An enterprise that produces and markets goods in several countries

transpiration   The use of water by plants, normally drawing it from the soil via their roots, evaporating it in their leaves and releasing it to the atmosphere

trophic level   A position in the food chain relative to other organisms, such as producer, herbivore, or carnivore

Tropic of Cancer   The parallel of 23.5° north latitude

Tropic of Capricorn   The parallel of 23.5° south latitude

tropical rain forest   Broadleaf evergreen vegetation characteristic of humid tropical environments

tsunami   An extremely long wave created by an underwater earthquake; the wave may travel hundreds of kilometers per hour

tundra   A low, slow-growing vegetation type found in high-latitude and high-altitude conditions in which snow covers the ground most of the year

tundra climate   A climate characterized by long, very cold winters and short, cool summers. Summer temperatures are only a few degrees above freezing

typhoon   The name applied to a hurricane in the Pacific Ocean

U

undocumented, irregular, or illegal immigrants   People who cross borders without completing legal papers

unitary government   A form of government in which the balance of power lies with the central government

untouchable   One of a group of people considered so low that their status is below the formal structure of the Hindu caste system

urban enterprise zones   Areas within which governments create generous conditions for enterprises to encourage the creation of jobs

urban geography   The geographic study of cities

urban heat island   A dome of heat over a city created by urban activities and conditions

urbanization   The process of concentrating people in cities

V

value added by manufacturing   The difference between the value of a raw material and the value of a product manufactured from that raw material

vernacular region   A region defined by widespread popular conception of its existence

vernal (spring) equinox   March 20 or 21, one of two dates when at noon the perpendicular rays of the Sun strike the equator (the Sun is directly overhead along the equator)

virtual reality   A mental state—a theoretical "place"—created by intense involvement either with interactive electronic devices or else with distant people through electronic devices

volcano   A vent in Earth's surface where lava emerges

W

warm front A boundary formed when a warm air mass advances against a cooler one

water budget An accounting of the amounts of precipitation, evapotranspiration, soil moisture storage and runoff at a given place

wavelength The distance between successive waves of radiant energy, or of successive waves on a water body

weather Patterns of atmospheric circulation, temperature, and precipitation over short time periods such as hours to days

weathering The chemical and/or mechanical breakdown of rocks into smaller particles at Earth's surface

Web logs, *See* blogs

winter solstice For places in the Southern Hemisphere, June 20 or 21 is the date when at noon the Sun is directly overhead at places along the parallel of 23.5° north latitude; for places in the Northern Hemisphere, December 21 or 22 is the date when at noon the Sun is directly overhead at places along the parallel of 23.5° south latitude

workers' remittances Money that migrant workers send home from elsewhere

X

xerophyte A plant adapted to living in arid conditions

Z

zero population growth A stabilized total population

Zionism The belief that the Jews should have a homeland of their own

zoning Restricting or prescribing the use to which parcels of land may be put

INDEX

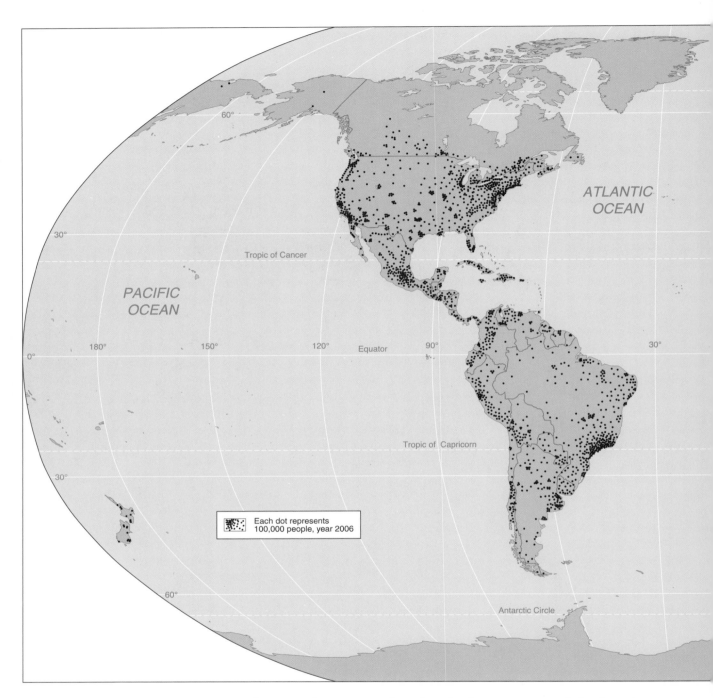

Each dot represents
100,000 people, year 2006

Dot maps like this cannot be exact, because dots cover other dots in dense areas, and the locations of some dots must be generalized. Nevertheless, the map conveys the pattern of concentration.

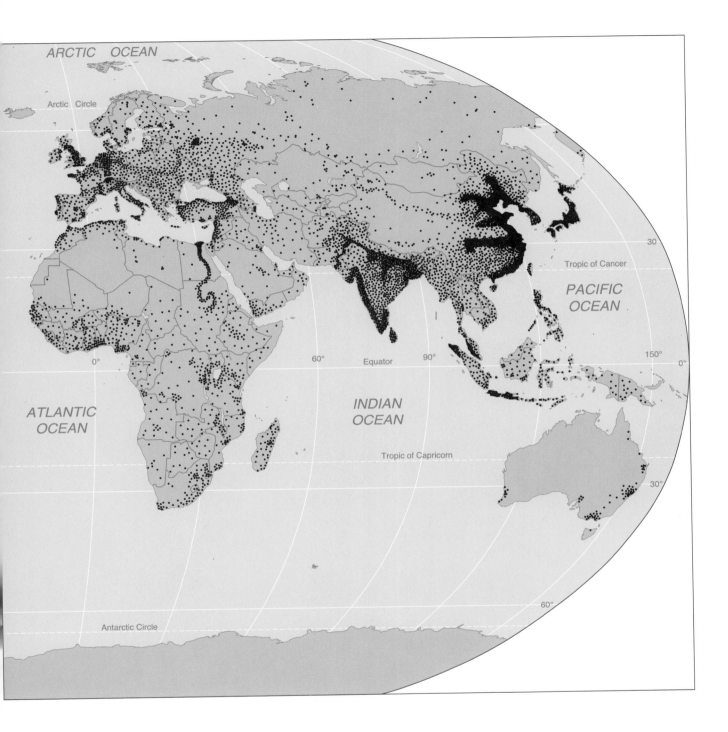